Test File
to accompany

Life
Sixth Edition
The Science of Biology

Test File
to accompany
Life
Sixth Edition
The Science of Biology

Purves, Sadava, Orians, and Heller

Prepared by

Charles Herr
Eastern Washington University

Betty McGuire
Smith College

Asa Oudes
Washington State University

Eric Strate
Eastern Washington University

Sinauer Associates, Inc.

W. H. Freeman and Company

The Cover
Giraffes *(Giraffa camelopardalis)* near Samburu, Kenya.
Photograph © BIOS/Peter Arnold, Inc.

Test File to accompany ***Life: The Science of Biology,*** **Sixth Edition**

Address editorial correspondence to:
Sinauer Associates, Inc., P.O. Box 407
23 Plumtree Road, Sunderland, MA, 01375 U.S.A.
Fax: 413-549-1118
Internet: www.sinauer.com; publish@sinauer.com

Address orders to:
W. H. Freeman and Company
VHPS/W. H. Freeman & Co. Order Department
16365 James Madison Highway,
U. S. Route 15, Gordonsville, VA 22942 U.S.A.
www.whfreeman.com

ISBN 0-7167-4356-6
Printed in U.S.A.

4 3 2 1

Contents

PART FOUR *The Evolution of Diversity*

PART FIVE *The Biology of Flowering Plants*

PART SIX *The Biology of Animals*

PART SEVEN *Ecology and Biogeography*

★ *Indicates a **difficult** question*

■ *Indicates a **conceptual** question*

1 An Evolutionary Framework for Biology

Fill in the Blank

1. **Buffon** noticed the similarities of the limbs of different mammals.
2. All scientific study begins with observations and the formation of testable **hypotheses**.
3. Under present conditions on Earth, cells do not arise from **noncellular** material.
4. The properties exhibited by a group of organisms that are not characteristic of the individual organisms are called **emergent** properties.
5. In contrast to eukaryotic cells, bacteria lack **organelles**.
6. According to **Lamarck's** theory, if humans continue to use their brains, their brains will become larger and more developed.
7. **Darwin** and **Wallace** were the first to suggest that organisms change gradually through the natural selection of variable characteristics.
8. Single-celled organisms that lack a nucleus belong to the two kingdoms called **Archaea** and **Bacteria**.
9. Eukaryotes can be described as **cells** within a cell.
10. As many as **30 million** species inhabit Earth.
11. Multicellular organisms that are photosynthetic belong to the kingdom called **Plantae**.
12. Multicellular organisms that are nonphotosynthetic and that digest their food outside their body's cells and absorb the products belong to the kingdom called **Animalia**.
13. About 2.5 **billion** years ago, prokaryotes acquired the ability to photosynthesize.
14. Currently, scientists agree with the estimate that life first appeared approximately **4 billion** years ago.
15. Both cellular and multicellular forms of life attempt to maintain a consistent internal environment despite changing external conditions. The term for this process is **homeostasis**.
16. There are three domains used to categorize life forms that have evolved separately for about a billion years: **Archaea**, **Bacteria,** and **Eukarya**.
17. Life can be categorized into six kingdoms: plants, animals, fungi, **eubacteria, archaebacteria** and **protists**.
18. The **null** hypothesis states that no difference exists due to the variable under investigation.
19. Fungi and animals are both **heterotrophs**.
20. Amphipods crawl to the surface because **nematodes** influence their behavior.
21. A species of **nematode** infects both sandpipers and amphipods.

Multiple Choice

1. Before Darwin, one scientist who wrote extensively about evolution was
   a. Hooke.
   b. Leeuwenhoek.
   c. Lamarck.
   d. Pasteur.
   e. Virchow.

2. Count de Buffon thought the reason pigs have small functionless toes is because
   a. they are evolving toward having functioning toes but have not yet reached the goal.
   b. they have defective toe-producing information in their DNA.
   c. they evolved from ancestors that had functioning toes.
   d. constant parasitization over generations caused the loss of their toes.
   e. All were proposed as possible reasons.

★3. The evolution of organisms on Earth was a
   a. directional process with definite goals.
   b. directionless process with no goals.
   c. directional process without goals.
   d. directionless process with goals.
   e. pointless process without logic.

4. Darwin noted that all populations have the potential to grow __________, but that in nature most populations __________ over time.
   a. linearly, are stable
   b. exponentially, grow more slowly
   c. linearly, fluctuate unpredictably
   d. exponentially, are stable
   e. linearly, decrease slowly

5. Plants are
   a. eukaryotic, multicellular photosynthesizers.
   b. eukaryotic, unicellular autotrophs.
   c. eukaryotic, multicellular heterotrophs.
   d. prokaryotic, multicellular autotrophs.
   e. prokaryotic, unicellular heterotrophs.

▪6. What distinguishes living organisms from nonliving matter?
   a. Living organisms are characterized by the processes of metabolism and reproduction.
   b. Only living organisms change in response to their environment.
   c. Only living organisms are composed of molecules.
   d. Living organisms do not obey physical and chemical laws.
   e. Only living organisms increase in size.

7. Metabolism is
   a. the consumption of energy.
   b. the release of energy.
   c. all conversions of matter and energy taking place in an organism.
   d. the production of heat by chemical reactions.
   e. the exchange of nutrients and waste products with the environment.

8. All living organisms obtain their energy from
   a. food.
   b. external sources.
   c. sunlight.
   d. heterotrophs.
   e. autotrophs.

9. The smallest complete units of matter are
   a. cells.
   b. lipids.
   c. molecules.
   d. hydrogen.
   e. atoms.

10. Sexual reproduction enhances chances for adaptation by
   a. causing mutation.
   b. recombination.
   c. random distribution.
   d. gene loss.
   e. allowing gene flow.

11. Heterotrophs cannot obtain their energy from
   a. food.
   b. autotrophs.
   c. other heterotrophs.
   d. complex chemical substances.
   e. sunlight.

12. The proper order of objects, from simple to complex, is
   a. atom, molecule, cell, tissue, organ, organism, population, community.
   b. cell, molecule, atom, tissue, organ, organism, population, community.
   c. molecule, cell, organ, atom, tissue, organism, population, community.
   d. atom, molecule, cell, tissue, organ, population, organism, community.
   e. atom, molecule, tissue, organ, cell, population, organism, community.

13. A cell is composed of compounds that include proteins, nucleic acids, lipids, and carbohydrates. A cell is capable of reproduction, but when the compounds that make up a cell are isolated, none of them can reproduce. Thus, cell reproduction is an example of
   a. growth.
   b. a molecule.
   c. an emergent property.
   d. adaptation.
   e. metabolism.

14. Earth is approximately __________ years old.
   a. 4 billion
   b. 4 trillion
   c. 4 million
   d. 6,000
   e. 40 trillion

15. The fundamental unit of life is the
   a. aggregate.
   b. organelle.
   c. organism.
   d. membrane.
   e. cell.

16. A species is
   a. all the organisms that live together in a particular area.
   b. a group of similar organisms that cannot interbreed.
   c. a group of similar organisms capable of interbreeding.
   d. an adult organism and all of its offspring.
   e. a group of similar organisms that live in the same area.

17. In multicellular organisms, cells of specific types are organized to form
   a. molecules.
   b. populations.
   c. organelles.
   d. tissues.
   e. communities.

18. Which of the following is *not* a major stage of the hypothetico-deductive method?
   a. Controlling an environment
   b. Making an observation
   c. Forming a hypothesis
   d. Making a prediction
   e. Testing a prediction

▪19. Which of the following questions *cannot* be answered using the hypothetico-deductive method?
a. Are bees more attracted to red roses than to yellow roses?
b. Are red roses more beautiful than yellow roses?
c. Why are red roses red?
d. Do red roses bloom sooner than yellow roses?
e. Are red roses more susceptible to mildew than yellow roses?

▪20. After observing that fish live in clean water but not in polluted water, you make the statement, "polluted water kills fish." Your statement is an example of
a. a controlled experiment.
b. a field experiment.
c. a laboratory experiment.
d. a hypothesis.
e. comparative analysis.

21. Which of the following is *not* a feature of scientific hypotheses?
a. They are true.
b. They make predictions.
c. They are based on observations.
d. They can be tested by experimentation.
e. They can be tested by observational analysis.

▪22. Based on the large numbers of offspring produced by many organisms, Darwin proposed that mortality was high and only a few individuals survived to reproduce. He called the differential reproductive success of individuals with particular variations
a. evolution.
b. artificial selection.
c. the cell theory.
d. natural selection.
e. inheritance of acquired characteristics.

23. Members of the kingdom Animalia obtain their energy from
a. decomposing organic matter.
b. photosynthesis.
c. other organisms.
d. sunlight.
e. inorganic molecules.

24. A typical cell
a. can be composed of many types of tissues.
b. creates an exact copy of itself when it reproduces.
c. is the smallest entity studied by biologists.
d. is the simplest biological structure capable of independent existence and reproduction.
e. is found only in plants and animals.

▪25. A biologist is most likely to identify emergent properties when studying how
a. big things are made up of smaller things working together.
b. organelles are composed of cells working together.
c. important experiments are performed by scientists working together.
d. paradigms become accepted by scientists working together.
e. organelles are composed of tissues working together.

26. A more common name for prokaryotic cells is
a. plants.
b. animals.
c. bacteria.
d. viruses.
e. fungi.

27. Which kingdom contains eukaryotic, unicellular organisms?
a. Plantae
b. Archaebacteria
c. Animalia
d. Protista
e. Fungi

28. An example of an organ is
a. milk.
b. a protein.
c. the heart.
d. an infant.
e. wood.

29. The key feature of experimentation is to
a. obtain accurate quantitative measurements.
b. prove unambiguously that some hypothesis is correct.
c. avoid comparative analysis.
d. answer as many key questions in one experiment as possible.
e. control factors that might affect a result so the influence of factors that do vary can be seen more clearly.

30. When compared to laboratory experiments, one advantage of field experiments is that they
a. better apply to actual events of nature.
b. more easily control most factors that may affect the outcome of the experiment.
c. require less money and use of technology.
d. achieve results more quickly.
e. demonstrate organisms' reactions to ideal environments.

31. A key point in Darwin's explanation of evolution is that
a. biological structures most likely inherited are those that have become better suited to the environment by their constant use.
b. mutations that occur are those that will help future generations fit into their environments.
c. slight variations among individuals significantly affect the chance that a given individual will survive in its environment and be able to reproduce.
d. genes change in order to help organisms cope with problems encountered within their environments.
e. extinction is nature's way to weed out undeserving organisms.

32. The smallest entities studied by biologists are
a. cells.
b. tissues.
c. organelles.
d. molecules.
e. membranes.

33. It is thought that some prokaryotes invaded other prokaryotes
a. about 4,000 years ago.
b. about 10,000 years ago.
c. more than one million years ago.
d. more than one billion years ago.
e. more than one trillion years ago.

34. All __________ must obtain their energy from the sun.
a. plants
b. autotrophs
c. organisms
d. heterotrophs
e. bacteria

35. Heterotrophs obtain their energy from
a. fungi.
b. water.
c. other organisms.
d. vitamins.
e. heat.

36. The two important developments necessary for the existence of multicellular life forms were
a. differentiation and cell clumping.
b. symmetry and asymmetry.
c. gene regulation and the evolution of the nucleus.
d. sexual reproduction and recombination.
e. isolation and compartmentalization.

37. The term "metamorphosis" in biology means a
a. change over extended periods.
b. physical transformation.
c. change in the structure of rock due to heat and/or pressure.
d. change in the life cycle to a dormant state.
e. All of the above

38. For metamorphosis to occur
a. there must be an increase in pressure and heat.
b. there must be a lack of nutrients.
c. a few genes are required to control the process.
d. many genes are required to control the process.
e. there must be an adequate supply of nutrients.

39. The initial accumulation of oxygen in the atmosphere was the result of photosynthesis from an organism most like modern
a. cyanobacteria.
b. algae.
c. mosses.
d. kelp.
e. eukaryotes.

40. A prerequisite for life to survive on land was the accumulation of
a. oxygen.
b. carbon dioxide.
c. water vapor.
d. ozone.
e. bacteria in the soil.

41. The chemical formula for ozone is
a. O.
b. O_2.
c. H_2O_2.
d. O_3.
e. ultraviolet light.

42. Ozone is important to life on Earth because it
a. is toxic to all forms of life.
b. can be used in place of oxygen.
c. blocks much ultraviolet radiation.
d. provides energy to some basic forms of life.
e. disinfects.

43. When biologists organize species in groups, they attempt to do so based on
a. physical similarities.
b. ecological niches.
c. chronological order.
d. evolutionary relationships.
e. All of the above

44. When attempting to group species, scientists use
a. fossils.
b. physical structures.
c. gene similarities.
d. All of the above
e. None of the above

45. Eukarya include
a. Protista.
b. Plantae.
c. Fungi.
d. Animalia.
e. All of the above

46. Which of the following is *not* a characteristic of most multicellular organisms?
a. Cells change structure and function during development.
b. Cells stick together after division.
c. Cells specialize.
d. Certain cells specialize for purposes of sexual reproduction.
e. Cells can grow without regulation.

47. The best understanding of a biological phenomena comes from
a. field observations.
b. laboratory research.
c. both field observations and laboratory research.
d. applying syllogistic logic.
e. None of the above.

48. Creation science is
a. true science.
b. another scientific theory.
c. not true science.
d. based on absolute fact.
e. proven.

Study Guide Questions

1. Life arose on earth approximately how many years ago?
   a. 3.8 billion
   b. 3.8 million
   c. 4000
   d. 1.5 billion
2. Which of the following molecules are components of all living organisms?
   a. Carbohydrates
   b. Lipids
   c. Nucleic Acids
   d. All of the above
3. Which of the following statement(s) regarding evolutionary biology are not true?
   a. Darwin and Wallace postulated that mortality rates match reproductive rates, that characteristics are heritable, and that those characteristics most suited to the environment are selected for.
   b. Darwin postulated that characteristics could be acquired in a single generation.
   c. Darwin was the first to come up with an evolutionary theory.
   d. b and c
4. Photosynthesis was a major evolutionary milestone because
   a. photosynthetic organisms contributed oxygen to the environment, which led to the evolution of aerobic organisms.
   b. photosynthesis led to conditions that allowed life to arise on land.
   c. photosynthesis is the only metabolic process that can convert light energy to chemical energy.
   d. All of the above
5. Sexual reproduction was evolutionarily important because
   a. previously all "daughter" organisms were clones of the parent organism.
   b. reproduction with variation is a major characteristic of life.
   c. variation is necessary for evolution.
   d. All of the above
6. The evolution of multicellularity led to an increased need for maintaining a constant internal environment. This is referred to as
   a. homeopathy.
   b. homeostasis.
   c. environmental maintenance.
   d. adaptation.
7. Which of the following are necessary for speciation to occur?
   a. Reproductive isolation of two groups of organisms.
   b. A large amount of mutation within a population.
   c. A reduction in the number of individuals in a population.
   d. None of the above
8. Which of the following represents a size hierarchy from smallest to largest?
   a. Atoms, cells, molecules, communities, organs
   b. Atoms, molecules, cells, organs, communities
   c. Molecules, atoms, cells, organs, communities
   d. Atoms, cells, communities, organs, molecules
9. The deepest division of the evolutionary history of life is the
   a. kingdom.
   b. genus.
   c. domain.
   d. species.
10. Organisms are named using first their __________ and then their __________.
    a. species, genus
    b. genus, species
    c. domain, genus
    d. genus, domain
11. In the amphipod example in your book, researchers isolated amphipods and placed them in containers in the laboratory where all conditions between test groups were identical except for the variable of interest. This is an example of a __________ experiment.
    a. repeated
    b. controlled
    c. laboratory
    d. None of the above
12. For a hypothesis to be scientifically valid it must be __________ and it must be possible to __________it.
    a. testable, prove
    b. testable, reject
    c. controlled, prove
    d. controlled, reject

Student Web Site Self-Quiz Questions

1. Observing that the functionally different mammalian forelimbs are all composed of similar bones suggested to Buffon that
   a. natural selection retained the basic forelimb structure but modified it for special purposes.
   b. the limb bones of mammals had been inherited from a common ancestor.
   c. modifications of the limbs had occurred through use and disuse.
   d. each species had been divinely created for different ways of life.
   e. homeostasis was an important force in evolution.
2. Assume that you discover the fossil of a seed fern frond in rocks at two widely separated locations. Which of the following is a logical assumption that can be made?
   a. The rocks at the two locations are of approximately the same age.
   b. The rocks were originally together and were separated by some geological event.
   c. The organism that produced the fossil must have existed for a very long time.
   d. Earth could not have been created in 4004 B.C.
   e. None of the above

3. Which of the following is a *false* statement regarding early cells such as "rock-eating" bacteria?
   a. They maintained control over the entry of molecules into their interior.
   b. They maintained control over the exit of molecules from their interior.
   c. They maintained control over the retention of molecules within their interior.
   d. They maintained control over the chemical reactions taking place within their interior.
   e. They were capable of creating energy.

4. Which of the following statements regarding eukaryotes is *false*?
   a. Early eukaryotes evolved from the first predators.
   b. The first eukaryotes appeared on Earth approximately 1.5 billion years ago.
   c. Eukaryotic cells contain specialized, membrane-enclosed compartments.
   d. All eukaryotic organisms are multicellular.
   e. Eukaryotic cells contain their genetic material inside a nucleus.

5. A species, such as a group of eagles, is defined as any group of organisms
   a. with a common ancestral organism.
   b. that have more than one shared derived feature.
   c. that do not interbreed with other groups of organisms.
   d. with more than 1,000 members.
   e. that is adapted to their environment.

6. If you arranged the units of life in the table below into a hierarchy, from smallest (least inclusive) to largest (most inclusive), which unit would be found three levels below a fish?

Organism	Community	Molecule
Atom	Population	Organ
Tissue	Cell	Biosphere

   a. Organ
   b. Biosphere
   c. Molecule
   d. Cell
   e. Population

7. Of the following units of life, which would have properties determined most directly from a fish?
   a. Organ
   b. Organism
   c. Tissue
   d. Population
   e. Community

8. The division of all living organisms into the three major domains (Bacteria, Archaea, and Eukarya) is based on
   a. direct fossil evidence from each group.
   b. the existence of a common ancestor.
   c. recent advances in our ability to obtain and analyze molecular evidence from existing species.
   d. methods to determine the proportions of radioactive and stable isotopes in rocks.
   e. visible differences between organisms found in each of the three domains.

9. The cactus is in the __________ domain and the __________ group.
   a. Bacteria, Bacteria
   b. Eukarya, Protists
   c. Eukarya, Animalia
   d. Eukarya, Fungi
   e. Eukarya, Plantae

10. You observe that cryptically colored caterpillars are more abundant in a particular area in your garden. Based upon this observation, you begin to wonder why this is the case. You propose that these caterpillars have a higher survival rate due to their ability to blend in with their environment. Your proposal would be considered analogous to which step in the hypothetico-deductive approach?
    a. Making observations
    b. Asking questions
    c. Forming a hypothesis
    d. Making predictions
    e. Testing predictions

TEST FILE

★ *Indicates a* ***difficult*** *question*

■ *Indicates a* ***conceptual*** *question*

2 Small Molecules: Structure and Behavior

Fill in the Blank

1. Every atom except **hydrogen** has one or more neutrons in its nucleus.

2. The sum of the atomic weights in any given molecule is called its **molecular weight**.

3. **Radioactive decay** occurs when one atom, such as ^{14}C, is transformed into another atom, such as ^{14}N, with an accompanying emission of energy.

4. The nutritionist's Calorie, which biologists call a kilocalorie, is the equivalent of **1,000** heat-energy calories.

5. The water strider skates along the surface of water due to a property of liquids called **surface tension**.

6. The chemical properties of an element are determined by the number of **electrons** its atoms contain.

7. The attraction between a slight positive charge on a hydrogen atom and the slight negative charge of a nearby atom is a **hydrogen bond**.

8. A chemical reaction that can proceed in either direction is called a **reversible reaction**.

9. A **molecule** is two or more atoms linked by chemical bonds.

10. A **calorie** is the amount of heat needed to raise the temperature of 1g of pure water from 14.5°C to 15.5°C.

Multiple Choice

1. 1.0 *M* solution of HCl has a pH of
   a. 1.0.
   b. 7.0.
   c. 14.0.
   d. 11.2.
   e. None of the above

2. The part of the atom of greatest biological interest is the
   a. proton.
   b. electron.
   c. neutron.
   d. innermost shell.
   e. None of the above

3. Hydrogen, deuterium, and tritium all possess the same
   a. atomic weight.
   b. atomic number.
   c. mass.
   d. density.
   e. All of the above

4. Which component of an atom does not significantly influence mass?
   a. Proton
   b. Neutron
   c. Electron
   d. Nucleotron
   e. Centron

5. Which of the following describes an unusual property of water?
   a. Three different physical phases
   b. A solid state that is less dense than its liquid state
   c. Heat loss when changing from a liquid to a gas
   d. Limited ionization
   e. All of the above

6. In addition to covalent and ionic bonds, which of the following interactions is of interest to biologists?
   a. van der Waals interactions
   b. Hydrogen bonds
   c. Hydrophobic interactions
   d. a and b
   e. a, b, and c

■7. What is the difference between an atom and an element?
   a. An atom is made of protons, electrons, and sometimes neutrons, and an element is a substance composed of only one kind of atom.
   b. An element is made of protons, electrons, and sometimes neutrons, and an atom is a substance composed of only one kind of element.
   c. an atom does not contain electrons while an element does.
   d. an atom contains protons and electrons while an element contains protons, electrons, and neutrons.
   e. None of the above

8. The number of protons in an atom equals the number of
   a. neutrons.
   b. electrons.
   c. electrons plus neutrons.
   d. neutrons minus electrons.
   e. quarks.

9. Which of the following statements about atoms is *true*?
   a. An electron has a more negative charge than a proton has positive charge, which is why there are usually more protons than electrons in an atom.
   b. Neutrons simply add mass to an atom without ever influencing other properties.
   c. When protons equal electrons, an atom has a neutral charge.
   d. Atoms of an element are radioactive whenever they vary in their number of neutrons.
   e. All the above are sometimes true.

10. Because atoms can have the same number of protons but a different number of neutrons, elements have
   a. isotopes.
   b. more than one atomic mass listed on the periodic table.
   c. more than one atomic number.
   d. None of the above
   e. Two of the above

11. An element has a mass weight of 131.3. The reason the number includes 0.3 is because
   a. of the weight of electrons.
   b. of the average of naturally occurring isotopes.
   c. the neutrons do not have a single unit weight.
   d. All of the above
   e. None of the above

12. The atomic number of an element is the same as the number of
   a. neutrons in each atom.
   b. protons plus electrons in each atom.
   c. protons in each atom.
   d. neutrons plus protons in each atom.
   e. a and c

13. The mass number of an element is the same as the number of
   a. electrons in each atom.
   b. protons in each atom.
   c. neutrons in each atom.
   d. protons plus neutrons in each atom.
   e. electrons plus neutrons in each atom.

14. When ^{14}C decays by releasing a particle, it becomes
   a. significantly lighter.
   b. nitrogen by gaining a proton.
   c. boron by losing a proton.
   d. nitrogen by losing a proton.
   e. ^{13}C by losing a neutron.

15. The best reference for the atomic number and mass number for elements is
   a. a good chemistry text.
   b. the Merck Index.
   c. the periodic table.
   d. a general physics book.
   e. All of the above

*16. All except __________ abide by the octet rule.
   a. sodium
   b. chlorine
   c. carbon
   d. hydrogen
   e. nitrogen

17. In a hydrogen molecule, the two atoms are held together by
   a. hydrogen bonds.
   b. a shared pair of electrons.
   c. van der Waal forces.
   d. ionic attractions.
   e. neutron gravity.

18. Hydrogen bonds
   a. form between two hydrogen atoms.
   b. form between hydrogen and oxygen atoms within a molecule.
   c. form between different molecules.
   d. involve sharing of electrons.
   e. are the strongest bonds because of their length.

19. What determines if a molecule is polar, nonpolar, or ionic?
   a. The number of protons
   b. The bond distances
   c. The differences in the electronegativities of the atoms
   d. The ionic charges
   e. All of the above

20. Which pair has similar chemical properties?
   a. ^{1}H and ^{22}Na
   b. ^{12}C and ^{28}Si
   c. ^{16}O and ^{32}S
   d. ^{12}C and ^{14}C
   e. ^{1}H and ^{2}He

▪21. $^{31}_{15}P$ and $^{32}_{15}P$ have virtually identical chemical and biological properties because they have the same
   a. half-life.
   b. number of neutrons.
   c. atomic weight.
   d. mass number.
   e. **number of electrons.**

22. The number of different elements that exist is closest to
   a. 12.
   b. 24.
   c. 48.
   d. 96.
   e. 192.

▪23. Why is the atomic weight of hydrogen 1.008 and not exactly its mass number, 1.000?
   a. Atomic weight does not take into account the weight of the rare isotopes of an element.
   b. Atomic weight is the average of the mass numbers of a representative sample of the element that includes all its isotopes.

c. The atomic weight includes the weight of the electrons.
d. The atomic weight does not include the weight of the protons.
e. The mass number of an element is always lower than its atomic weight.

24. Which of the following elements is the most chemically reactive?
a. Hydrogen
b. Helium
c. Neon
d. Argon
e. They all have the same chemical reactivity.

25. A single covalent chemical bond represents the sharing of how many electrons?
a. One
b. Two
c. Three
d. Four
e. Six

26. The molecular weight of water is 18.0154. One mole of water weighs exactly __________ grams.
a. 9
b. 18
c. 18.0154
d. 36.031
e. 6.023×10^{23}

27. Ionic bonds are
a. attractions between oppositely charged ions.
b. the result of electron sharing.
c. the strongest of the chemical bonds.
d. caused by partial electrical charges.
e. All of the above

▪28. What is the difference between covalent and ionic bonds?
a. Covalent bonds are the sharing of neutrons, and ionic bonds are the sharing of electrons.
b. Ionic bonds are the sharing of electrons between atoms, and covalent bonds are the electric attraction between two atoms.
c. Covalent bonds are the sharing of protons between atoms, and ionic bonds are the electric attraction between two atoms.
d. Covalent bonds are the sharing of protons between atoms, and ionic bonds are the sharing of electrons between two atoms.
e. Covalent bonds are the sharing of electrons between atoms, and ionic bonds are the transfer of electrons from one atom to the other.

29. Which contains more molecules, a mole of hydrogen or a mole of carbon?
a. A mole of carbon
b. A mole of hydrogen
c. Both contain the same number of molecules.
d. Inadequate information is provided.
e. It depends on the pressure.

30. Sweating is a useful cooling device for humans because
a. water takes up a great deal of heat in changing from its liquid state to its gaseous state.
b. water takes up a great deal of heat in changing from its solid state to its liquid state.
c. water can exist in three states at temperatures common on Earth.
d. water is an outstanding solvent.
e. water ionizes readily.

31. To maintain a normal body temperature when exposed to extreme heat, __________ is used to absorb excess calories of heat.
a. evaporation
b. condensation
c. respiration
d. transpiration
e. perspiration

▪32. What two characteristics make water different from most other compounds?
a. Its solid state is less dense than its liquid state, and it takes up large amounts of heat to change to its gaseous state.
b. Its solid state is more dense than its liquid state, and it takes up only small amounts of heat to change to its gaseous state.
c. Its solid state is less dense than its liquid state, and it takes up only small amounts of heat to change to its gaseous state.
d. Its solid state is more dense than its liquid state, and it takes up large amounts of heat to change to its gaseous state.
e. Its solid state is just as dense than its liquid state, and it takes no heat to change to its gaseous state.

33. Specific heat is defined as
a. the heat released from ice during freezing.
b. the heat needed to raise the temperature of water.
c. one calorie.
d. the calories needed to raise a gram 1°C.
e. 4.184 joules.

▪34. Which characteristic of water contributes to the relatively constant temperatures of the oceans?
a. Water ionizes only slightly.
b. It takes a small amount of heat energy to raise the temperature of water.
c. Water can contain large amounts of salt.
d. Water has the ability to ionize readily.
e. It takes a large amount of heat energy to raise the temperature of water.

35. Ice floats because
a. the crystal takes up more space than the liquid.
b. substances expand when cooled.
c. heat is released and heat makes water expand.
d. hydrogen bonds must break.
e. heat is absorbed.

36. Ice is used in beverages because
   a. it is composed only of water.
   b. it is cold.
   c. more water is needed in the drink.
   d. people like to chew it.
   e. it absorbs a lot of heat when it melts.

▪37. If you place a paper towel in a dish of water, the water will move up the towel by capillary action. What property of water gives rise to capillary action?
   a. Water molecules ionize.
   b. Water is a good solvent.
   c. Water molecules have hydrophobic interactions.
   d. Water can form hydrogen bonds.
   e. Water takes up large amounts of heat when it vaporizes.

38. An alkaline solution contains
   a. more OH^- ions than H^+ ions.
   b. more H^+ ions than OH^- ions.
   c. the same number of OH^- ions as H^+ ions.
   d. no OH^- ions.
   e. None of the above

39. The pH 6.0 contains
   a. 10^6 hydrogen ions.
   b. 6^{10} hydrogen ions.
   c. 6^{10} moles of hydrogen ions.
   d. more OH^- than H^+.
   e. 10^{-6} moles of hydrogen ions.

40. Solutions that contain buffers tend to resist pH changes because buffers
   a. are bases.
   b. change from ionic to non-ionic in solution.
   c. change from non-ionic to ionic in response to changes in pH and release or absorb H^+.
   d. are weak acids or bases.
   e. are ionic, polar molecules that add or absorb H^+ in solutions.

▪41. What is the difference between an acid and a base?
   a. An acid undergoes a reversible reaction, while a base does not.
   b. An acid releases OH^- ions in solution, while a base accepts OH^- ions.
   c. An acid releases H^+ ions in solution, while a base releases OH^- ions.
   d. An acid releases OH^- ions in solution, while a base releases H^+ ions.
   e. An acid releases H^+ ions in solution, while a base accepts H^+ ions.

42. All except which of the following is nonpolar?
   a. O_2
   b. N_2
   c. CH_4
   d. NaCl
   e. H_2

43. Polar molecules
   a. have an overall negative electric charge.
   b. have an equal distribution of electric charge.
   c. have an overall positive electric charge.
   d. have an unequal distribution of electric charge.
   e. are ions.

44. The electron cloud of a water molecule
   a. is equally dense throughout the molecule.
   b. is most dense near the hydrogen atoms.
   c. is most dense near the oxygen atom.
   d. covers only the positive portion of the molecule.
   e. is separated by an angle of 104.5°.

45. Which is *not* a consequence of hydrogen bonding?
   a. The attraction between water molecules
   b. The shape of proteins and DNA
   c. The high solubility of a sugar in water
   d. Capillary action
   e. All of the above

46 What is a van der Waals interaction?
   a. The attraction between the electrons of one molecule and the nucleus of a nearby molecule
   b. The attraction between the electrons of one molecule and the nucleus of the same molecule
   c. The attraction between the electrons of one molecule and the electrons of a nearby molecule
   d. The attraction between nonpolar molecules due to the exclusion of water
   e. The attraction between nonpolar molecules because they are surrounded by water molecules

47. Oil remains as a droplet in water because of
   a. the van der Waals interactions of the nonpolar oil molecules.
   b. the hydrophobic interactions of the nonpolar oil molecules.
   c. the hydrogen bonds formed between the nonpolar molecules of the oil and the water molecules.
   d. the covalent bonds formed between the nonpolar molecules of the oil.
   e. the covalent bonds formed between the nonpolar molecules of the oil and the water molecules.

▪48. A drop of oil in water disperses when detergent is added because
   a. the nonpolar parts of the detergent molecules associate with the oil and the polar parts of the detergent molecules associate with the water.
   b. the polar parts of the detergent molecules associate with the oil and the nonpolar parts of the detergent molecules associate with the water.
   c. the nonpolar parts of the detergent molecules associate with the oil and with the water.
   d. the polar parts of the detergent molecules associate with the oil and with the water.
   e. the detergent lowers the surface tension of the water.

49. When potassium hydroxide (KOH) is added to water, it ionizes, releasing hydroxide ions. This solution is
   a. acidic.
   b. basic.
   c. neutral.
   d. molar.
   e. a buffer.

50. H_2SO_4 can ionize to yield two H^+ ions and one SO_4^{2-} ion. H_2SO_4 is
a. molar.
b. a base.
c. a buffer.
d. a solution.
e. an acid.

51. Acid rain is a serious environmental problem. A sample of rainwater collected in the Adirondack Mountains had an H^+ concentration of 10^{-4} mol/L. The pH of this sample was
a. .0001.
b. –4.
c. 4.
d. 0.
e. 10,000.

*52. Carbonic acid and sodium bicarbonate act as buffers in the blood. When a small amount of acid is added to this buffer, the H^+ ions are used up as they combine with the bicarbonate ions. When this happens, the pH of the blood
a. becomes basic.
b. becomes acidic.
c. doesn't change.
d. is reversible.
e. ionizes.

53. Cholesterol is composed primarily of carbon and hydrogen atoms. What property would you expect cholesterol to have?
a. It is insoluble in water.
b. It is soluble in water.
c. It is a base.
d. It is an acid.
e. It is a buffer.

54. The compound inositol has six hydroxyl groups attached to a six-carbon backbone. Thus inositol can be classified as a(n)
a. amine.
b. acid.
c. ketone.
d. alcohol.
e. buffer.

55. Butane and isobutane have the same chemical formula but different arrangements of atoms. These two compounds are called
a. ionic.
b. alcohols.
c. functional groups.
d. amines.
e. isomers.

56. One dalton is the same as the mass of one
a. electron.
b. carbon atom.
c. proton.
d. gram.
e. charge unit.

57. Oxygen and carbon are defined as different elements because they have atoms with a different
a. number of electrons.
b. number of protons.
c. number of neutrons.
d. mass.
e. charge.

58. The mass of an atom is primarily determined by the
a. number of electrons it contains.
b. number of protons it contains.
c. sum of the number of protons and electrons it contains.
d. sum of the number of protons and neutrons it contains.
e. number of charges it contains.

*59. Two atoms are held together in four covalent bonds because of forces between
a. electrons and protons.
b. electrons and electrons.
c. protons and neutrons.
d. protons and protons.
e. neutrons and neutrons.

60. Two carbon atoms held together in a double covalent bond share __________ electron(s).
a. one
b. two
c. four
d. six
e. eight

61. Particles having a net negative charge are called
a. electronegative.
b. cations.
c. anions.
d. acids.
e. bases.

62. Avogadro's number is the number of
a. grams in a mole.
b. molecules in a mole.
c. moles in gram molecular weight.
d. molecules in a gram.
e. moles in a gram.

63. A mole is
a. 6.02×10^{23} molecules.
b. the molecular weight of a compound.
c. an abbreviation for molecule.
d. 6.02×10^{-23} atoms.
e. None of the above

64. To determine the number of molecules in a teaspoon of sugar you need
a. the density of the sugar.
b. the weight of the sugar.
c. the molecular weight of the sugar.
d. Avogadro's number.
e. the weight and molecular weight of the sugar, and Avogadro's number.

*■65. How would you make 100 ml of an aqueous solution with a 0.25 M concentration of a compound that has a molecular weight of 200 daltons?
a. Add 0.25 grams of the compound to 100 ml of water.
b. Add 250 grams of the compound to 100 ml of water.
c. Take 250 grams of the compound and add water until the volume equals 100 ml.
d. Take 50 grams of the compound and add water until the volume equals 100 ml.
e. Take 5 grams of the compound and add water until the volume equals 100 ml.

66. Of the following amounts of compounds containing ^1H, ^{12}C, and ^{16}O, the one with the greatest number of molecules is 2 grams of
a. CO.
b. CO_2.
c. HCOOH.
d. C_2H_5OH.
e. $C_6H_{12}O_6$.

67. Which of the following molecules is held together primarily by ionic bonds?
a. H_2O
b. $C_6H_{12}O_6$
c. NaCl
d. H_2
e. NH_3

68. The hydrogen bond between two water molecules arises because water is
a. polar.
b. nonpolar.
c. a liquid.
d. a small molecule.
e. hydrophobic.

69. The functional group written as —COOH is called the _________ group.
a. hydroxyl
b. carbonyl
c. amino
d. ketone
e. carboxyl

70. The functional group diagramed =O is a(n)
a. carbonyl.
b. carboxyl.
c. aldehyde.
d. amino.
e. hydroxyl.

71. The more acidic of two solutions has
a. more hydroxyl ions.
b. more hydrogen acceptors.
c. more H+ ions per liter.
d. a higher pH.
e. None of the above

72. Which of the following atoms usually has the greatest number of covalent bonds with other atoms?
a. Carbon
b. Oxygen
c. Sulfur
d. Hydrogen
e. Nitrogen

73. Of the following atomic configurations, the one that has an atomic mass of 14 is the atom with
a. 14 neutrons.
b. 14 electrons.
c. 7 neutrons and 7 electrons.
d. 7 protons and 7 electrons.
e. 6 protons and 8 neutrons.

74. An atom that is neutrally charged contains
a. only neutrons.
b. the same number of neutrons as electrons.
c. the same number of neutrons as protons.
d. the same number of positive particles as negative particles.
e. no charged particles.

75. The four elements most common in organisms are
a. calcium, iron, hydrogen, and oxygen.
b. water, carbon, hydrogen, and oxygen.
c. carbon, oxygen, hydrogen, and nitrogen.
d. nitrogen, carbon, iron, and hydrogen.
e. phosphorus, water, carbon, and oxygen.

76. The notation $[H^+]$ means the
a. number of H^+ ions present in a solution.
b. number of protons in an H^+ ion.
c. charge of an H^+ ion.
d. concentration of H^+ ions in moles per liter.
e. chemical reactivity of H^+ ions.

77. Of the following types of chemical bonds, the strongest is the _________ bond.
a. hydrogen
b. van der Waals
c. ionic
d. acidic
e. covalent

■78. Select the statement that describes a difference between ionic bonds and covalent bonds.
a. An ionic bond is stronger.
b. Electron sharing is more equal in the covalant bond.
c. An ionic bond occurs more often in aqueous solutions.
d. An ionic bond occurs only in acids.
e. A covalent bond occurs in nonpolar molecules.

79. Which of the following statements about the number of bonds formed by atoms is *true*?
a. Oxygen forms one, carbon forms four, and hydrogen forms one.
b. Oxygen forms four, carbon forms four, and hydrogen forms four.
c. Oxygen forms two, carbon forms four, and hydrogen forms none.
d. Oxygen forms two, carbon forms four, and hydrogen forms one.
e. Oxygen forms two, carbon forms two, and hydrogen forms two.

80. Of the following types of molecules, the one always containing nitrogen is
   a. thiol.
   b. sugar.
   c. hydrocarbon.
   d. alcohol.
   e. amino acid.

81. A typical distance between two atoms held together by a covalent bond is
   a. 1 millimeter.
   b. 2 micrometers.
   c. 108 meters.
   d. 0.2 nanometer.
   e. 10^{-12} meter.

82. Surface tension and capillary action occur in water because it
   a. is wet.
   b. is dense.
   c. has hydrogen bonds.
   d. is nonpolar.
   e. has ionic bonds.

83. Which of the following is the correct order for the relative strengths of chemical bonds?
   a. Covalent, ionic, hydrogen, van der Waal forces
   b. Ionic, covalent, hydrogen, van der Waal forces
   c. Van der Waal forces, covalent, ionic, hydrogen
   d. Hydrogen, covalent, van der Waal forces, ionic
   e. Ionic, covalent, van der Waal forces, hydrogen

Study Guide Questions

1. The atomic number of an element refers to the number of
   a. protons and neutrons in an atom.
   b. protons in an atom.
   c. electrons in an atom.
   d. neutrons in an atom.

2. Which of the following statements concerning electrons is *not* correct?
   a. Electrons orbit the nucleus of an atom in defined orbitals.
   b. The outer shell of all atoms must contain eight electrons.
   c. An atom may have more than one valence shell.
   d. Electrons are negatively charged particles.

3. The element with which of the following atomic number would be most stable?
   a. 1
   b. 3
   c. 12
   d. 18

4. How can you differentiate between an element and a molecule?
   a. Molecules may be composed of different types of atoms whereas elements are always composed of only one type of atom.
   b. Molecules are composed of only one type of atom whereas elements are composed of different types of atoms.
   c. Molecules are elements.
   d. Molecules always have larger atomic weights than elements.

5. The strongest chemical bonds occur when
   a. two atoms share electrons in a covalent bond.
   b. two atoms share electrons in an ionic bond.
   c. hydrogen bonds are formed.
   d. van der Waals forces are in effect.

6. You have discovered that a molecule is hydrophilic. What else do you know about this molecule?
   a. It cannot form hydrogen bonds.
   b. It is a polar molecule.
   c. It has a partial positive region and a partial negative region.
   d. b and c

7. The stability of the three-dimensional shape of many large molecules is dependent on
   a. covalent bonds.
   b. ionic bonds.
   c. hydrogen bonds.
   d. van der Waals attractions.

8. The molecular weight of glucose is 180. If you added 180 grams of glucose to a 0.5 liter of water, what would be the molarity of the resulting solution? (See Figure 2.1 for a periodic table.)
   a. 18
   b. 1
   c. 9
   d. 2

9. Which of the following properties makes water an excellent solvent?
   a. The pH of water is 7.
   b. Water is a polar molecule.
   c. Water has a high surface tension.
   d. Water has a high heat capacity.

10. Cola has a pH of 3; blood plasma has a pH of 7. The hydrogen ion concentration of cola is __________ than the hydrogen ion concentration of blood plasma.
   a. 4 times greater
   b. 4 times lesser
   c. 400 times greater
   d. 10,000 times greater

11. If solution A has a pH of 2 and solution B has a pH of 8, which of the following statements is true?
   a. A is basic and B is acidic.
   b. A is acidic and B is basic.
   c. A is a base and B is an acid.
   d. A has a greater [OH-] than B.

12. One mole of glucose ($C_6H_{12}O_6$) weighs
   a. 180 grams.
   b. 42 atomic mass units.
   c. 96 grams.
   d. 342 grams.

13. Which of the following is not an organic molecule?
   a. $C_6H_{12}O_6$
   b. CH_3OH
   c. H_2SO_4
   d. CH_3COOH

End of Chapter Questions

1. The atomic number of an element
   a. equals the number of neutrons in an atom.
   b. equals the number of protons in an atom.
   c. equals the number of protons minus the number of neutrons.
   d. equals the number of neutrons plus the number of protons.
   e. depends on the isotope.

2. The atomic weight (atomic mass) of an element
   a. equals the number of neutrons in an atom.
   b. equals the number of protons in an atom.
   c. equals the number of electrons in an atom.
   d. equals the number of neutrons plus the number of protons.
   e. depends on the relative abundances of its isotopes.

3. Polar molecules
   a. have an overall negative charge.
   b. have an equal distribution of electric charge.
   c. have an overall positive electric charge.
   d. have an unequal distribution of eletric charge
   e. are ions.

4. Covalent bonds differ from ionic bonds in that covalent bonds
   a. are more easily broken in water.
   b. orm ions in solution.
   c. involve one element giving up electrons to another.
   d. link positively charged to negatively charged elements.
   e. result from the sharing of electrons between elements.

5. Hydrophobic interactions
   a. are stronger than hydrogen bonds.
   b. are stronger than covalent bonds.
   c. can hold two ions together.
   d. can hold two nonpolar molecules together.
   e. are responsible for the surface tension of water.

6. Which of the following statements about water is *not* true?
   a. It absorbs a large amount of heat when changing from liquid into vapor.
   b. Its solid form is less dense than its liquid form.
   c. It allows nonpolar substances to dissolve in it.
   d. It is typically the most abundant substance in an active organism.
   e. It takes part in some important chemical reactions.

7. If the pH of solution A is 2 and the pH of solution B is 4, which statement is true?
   a. The OH^- concentrations of A and B are equal.
   b. The H^+ concentration of A is twice that of B.
   c. The H^+ concentration of A is half that of B.
   d. The H^+ concentration of A is 100 times that of B.
   e. The H^+ concentration of A is 1/100 that of B.

8. The hydrogen bond between two water molecules arises because water is
   a. polar.
   b. nonpolar.
   c. a liquid.
   d. a small molecule.
   e. hydrophobic.

9. Which of the following statements about the carboxyl group is *not* true?
   a. It has the chemical formula —COOH.
   b. It is an acidic group.
   c. It can ionize.
   d. It is found in amino acids.
   e. It has an atomic weight of 45.

10. The three most abundant chemical elements in human cells are
   a. calcium, carbon, and oxygen.
   b. carbon, hydrogen, and oxygen.
   c. carbon, hydrogen, and sodium.
   d. carbon, nitrogen, and oxygen.
   e. nitrogen, hydrogen, and oxygen.

Student Web Site Self-Quiz Questions

1. Which of the following lists of characteristics is true for helium?
   a. Mass number = 4; atomic number = 2; net charge = 0
   b. Mass number = 4; atomic number = 4; net charge = 0
   c. Mass number = 4; atomic number = 2; net charge = +2
   d. Mass number = 4.003; atomic number = 2; net charge = 0
   e. Mass number = 4; atomic number = 2; net charge = +1

2. Which of the following statements about the shape of a water molecule is *false*?
   a. The four pairs of electrons repel each other.
   b. Two orbitals have nonbonding electron pairs.
   c. Two orbitals have electrons that are bonded to the hydrogens.
   d. The overall shape of water is a tetrahedron.
   e. The unbonded electron pairs give that end of the molecule a partial positive charge.

3. In comparing solid water (ice) and liquid water, select the *false* statement from the following choices.
   a. The density of liquid water is less than that of solid water.
   b. Solid water has a crystalline structure.
   c. Hydrogen bonds hold water molecules in a rigid state in solid water.
   d. Each water molecule can be hydrogen-bonded to four other water molecules.
   e. Hydrogen bonds continue to break and reform in liquid water.

4. Which of the following properties of water is *not* caused by the hydrogen bonding between two water molecules?
   a. High surface tension
   b. Great cohesive strength
   c. Polar nature
   d. Excellence as a solvent
   e. High heat of evaporation

★ *Indicates a **difficult** question*

■ *Indicates a **conceptual** question*

3 Macromolecules: Their Chemistry and Biology

Fill in the Blank

1. Fluidity and melting point of fatty acids are determined in part by the number of **unsaturated (or carbon double) bonds**.

2. Many monosaccharides like fructose, mannose, and galactose have the same chemical formula as glucose ($C_6H_{12}O_6$), but the atoms are combined differently to yield different structural arrangements. These varying forms of the same chemical formula are called **isomers**.

3. The highly branched polysaccharide that stores glucose in the muscle and the liver of animals is **glycogen**.

4. In proteins, amino acids are linked together by **peptide** bonds.

5. The only amino acid that has no stereoisomer is **glycine.**

6. The amino acid that most limits rotation around the a carbon is **proline**.

7. The bonds between the units in a carbohydrate polymer are called **glycosidic** bonds.

8. The linear arrangement of amino acids in the polypeptide chain is referred to as the **primary** structure of the protein.

9. Starch is a polymer of glucose subunits. The subunits of any polymer are called **monomers**.

10. Fatty acids with more than one carbon-carbon double bond are called **polyunsaturated**.

11. Cholesterol, vitamin D, and testosterone all have a multiple-ring structure and are members of a family of lipids known as **steroids**.

12. Carbohydrates made up of two simple sugars are called **disaccharides**.

13. The covalent bond forces between the sulfur atoms of two cysteine side chains is called a **disulfide bridge**.

14. Disulfide bonds can form between **cysteine** residues in proteins.

15. All amino acids have a hydrogen atom, a carboxyl group, and an amino group attached to a carbon atom. The variability in the 20 different amino acids lies in the structure of their **R group**.

16. A(n) **ester** linkage connects the fatty acid molecule to glycerol.

17. The **glucose** molecules found in humans are the same as those found in tomato plants.

18. The bonds that link sugar monomers together in a starch molecule are **glycosidic** bonds.

19. Cholesterol is classified as a(n) **lipid**.

★20. The reaction $A—H + B—OH \rightarrow A—B + H_2O$ represents a **condensation reaction**.

Multiple Choice

1. The major classes of biologically significant large molecules include which of the following?
   a. Proteins
   b. Nucleic acids
   c. Carbohydrates
   d. Lipids
   e. All of the above

2. Lipids are
   a. insoluble in water.
   b. readily soluble in organic solvents.
   c. characterized by their solubility.
   d. important constituents of biological membranes.
   e. All of the above

3. Which of the following is characteristic of proteins?
   a. Some function as enzymes.
   b. They provide structural units in the form of keratin.
   c. They possess glycosidic linkages between amino acids.
   d. a and b
   e. a, b, and c

4. Molecules with molecular weights greater than 1,000 daltons are usually called
   a. proteins.
   b. polymers.
   c. nucleic acids.
   d. macromolecules.
   e. monomers.

5. Polymerization reactions in which proteins are synthesized from amino acids
   a. require energy.
   b. result in the formation of water.
   c. are condensation reactions.
   d. are dehydration reactions.
   e. All of the above

6. In condensation reactions, the atoms that make up a water molecule are derived from
   a. oxygen.
   b. only one of the reactants.
   c. both of the reactants.
   e. carbohydrates.
   e. enzymes.

7. Which of the following is *not* a macromolecule?
   a. RNA
   b. DNA
   c. An enzyme
   d. A protein
   e. Salt

8. The bonds that form between the units of polymeric macromolecules are
   a. hydrogen.
   b. peptide.
   c. disulfide.
   d. covalent.
   e. ionic.

9. Which of the following is *not* a characteristic of lipids?
   a. They are readily soluble in water.
   b. They are soluble in organic solvents.
   c. They release large amounts of energy when broken down.
   d. They form two layers when mixed with water.
   e. They act as an energy storehouse.

▪10. You have isolated an unidentified liquid from a sample of beans. You add the liquid to a beaker of water and shake vigorously. After a few minutes, the water and the other liquid separate into two layers. To which class of large biological molecules does the unknown liquid most likely belong?
   a. Carbohydrates
   b. Lipids
   c. Proteins
   d. Enzymes
   e. Nucleic acids

*▪11. Lipids form the barriers surrounding various compartments within an organism. Which property of lipids makes them a good barrier?
   a. Many biologically important molecules are not soluble in lipids.
   b. Lipids are polymers.
   c. Lipids store energy.
   d. Triglycerides are lipids.
   e. Lipids release large amounts of energy when broken down.

*▪12. You look at the label on a container of shortening and see "hydrogenated vegetable oil." This means that during processing the number of carbon-carbon double bonds in the oil was decreased. What is the result of decreasing the number of double bonds?
   a. The oil is now a liquid at room temperature.
   b. The oil is now a solid at room temperature.
   c. There are more "kinks" in the fatty acid chains.
   e. The oil is now a derivative carbohydrate.
   e. The fatty acid is now a triglyceride.

13. The portion of a phospholipid that contains the phosphorous group has one or more electric charges. That makes this region of the molecule
   a. hydrophobic.
   b. hydrophilic.
   c. nonpolar.
   d. unsaturated.
   e. saturated.

▪14. Cholesterol is soluble in ether, an organic solvent, but is not soluble in water. Based on this information, what class of biological macromolecules does cholesterol belong to?
   a. Nucleic acids
   b. Carbohydrates
   c. Proteins
   d. Enzymes
   e. Lipids

15. Which of the following is *not* a function in which lipids play an important role?
   a. Vision
   b. Storing energy
   c. Membrane structure
   d. Storing genetic information
   e. Chemical signaling

16. In a biological membrane, the phospholipids are arranged with the fatty acid chains facing the interior of the membrane. As a result, the interior of the membrane is
   a. hydrophobic.
   b. hydrophilic.
   c. charged.
   d. polar.
   e. filled with water.

17. The monomers that make up polymeric carbohydrates like starch are called
   a. nucleotides.
   b. trisaccharides.
   c. monosaccharides.
   d. nucleosides.
   e. fatty acids.

18. The atoms that make up carbohydrates are
   a. C, H, and N.
   b. C and H.
   c. C, H, and P.
   d. C, H, and O.
   e. C, H, O, and N.

19. Glucose and fructose both have the formula $C_6H_{12}O_6$, but the atoms in these two compounds are arranged differently. Glucose and fructose are known as
    a. isomers.
    b. polysaccharides.
    c. oligosaccharides.
    d. pentoses.
    e. steroids.

20. A nucleotide contains a pentose, a phosphate, and a(n)
    a. lipid.
    b. acid.
    c. nitrogen-containing base.
    d. amino acid.
    e. glycerol.

21. A simple sugar with the formula $C_5H_{10}O_5$ can be classified as a
    a. hexose.
    b. polysaccharide.
    c. disaccharide.
    d. pentose.
    e. lipid.

22. Lactose, or milk sugar, is composed of one glucose unit and one galactose unit. It can be classified as a
    a. disaccharide.
    b. hexose.
    c. pentose.
    d. polysaccharide.
    e. simple sugar.

23. Two important polysaccharides made up of glucose monomers are
    a. guanine and cytosine.
    b. RNA and DNA.
    c. sucrose and lactose.
    d. cellulose and starch.
    e. testosterone and cortisone.

24. Polysaccharides that serve as energy storage molecules tend to have __________ linkages.
    a. α-1,4
    b. α-2,3
    c. β-1,4
    d. β-2,3
    e. a and c

25. Cellulose is the most abundant organic compound on Earth. Its main function is
    a. to store genetic information.
    b. as a storage compound for energy in plant cells.
    c. as a storage compound for energy in animal cells.
    d. as a component of biological membranes.
    e. to provide mechanical strength to plant cell walls.

26. In animals, glucose is stored in the compound
    a. cellulose.
    b. amylose.
    c. glycogen.
    d. fructose.
    e. cellobiose.

27. Which of the following monomer/polymer pairs is *not* correct?
    a. Monosaccharide/polysaccharide
    b. Amino acid/protein
    c. Triglyceride/lipid
    d. Nucleotide/DNA
    e. Nucleotide/RNA

28. Amino acids can be classified by the
    a. number of monosaccharides they contain.
    b. number of carbon-carbon double bonds in their fatty acids.
    c. number of peptide bonds they can form.
    d. number of disulfide bridges they can form.
    e. characteristics of their side chains.

29. During the formation of a peptide linkage, which of the following occurs?
    a. A molecule of water is formed.
    b. A disulfide bridge is formed.
    c. A hydrophobic bond is formed.
    d. A hydrophilic bond is formed.
    e. An ionic bond is formed.

30. The side chain of leucine is a hydrocarbon. In a folded protein, where would you expect to find leucine?
    a. In the interior of a cytoplasmic enzyme
    b. On the exterior of a protein embedded in a membrane
    c. On the exterior of a cytoplasmic enzyme
    d. a and b
    e. a and c

31. What is the theoretical number of different proteins that you could make from 50 amino acids?
    a. 50^{20}
    b. 20×50
    c. 20^{50}
    d. 10^{50}
    e. 2^{50}

32. The shape of a folded protein is often determined by
    a. its tertiary structure.
    b. the sequence of its amino acids.
    c. whether the peptide bonds have α or β linkages.
    d. the number of peptide bonds.
    e. the base-pairing rules.

33. The amino acids of the protein keratin are arranged in an a helix. This secondary structure is stabilized by
    a. covalent bonds.
    b. peptide bonds.
    c. glycosidic linkages.
    d. polar bonds.
    e. hydrogen bonds.

34. What is the nucleotide sequence of the complementary strand of this DNA molecule: A A T G C G A?
    a. T T A C G C T
    b. A A T G C G A
    c. G G C A T A G
    d. C C G T T A T
    e. A G C G T A A

35. Which of the following is *not* a difference between DNA and RNA?
a. DNA has thymine, and RNA has uracil.
b. DNA usually has two polynucleotide strands, and RNA usually has one strand.
c. DNA has deoxyribose sugar, and RNA has ribose sugar.
d. DNA is a polymer, and RNA is a monomer.
e. In DNA, A pairs with T, and in RNA, A pairs with U.

36. DNA molecules that carry different genetic information can be distinguished by looking at
a. the number of strands in the helix.
b. how much uracil is present.
c. the sequence of nucleotide bases.
d. differences in the base-pairing rules.
e. the shape of the helix.

37. The "backbone" of nucleic acid molecules is made of
a. nitrogenous bases.
b. alternating sugars and phosphate groups.
c. purines.
d. pyrimidines.
e. nucleosides.

38. According to the base-pairing rules for nucleic acids, purines always pair with
a. deoxyribose sugars.
b. uracil.
c. pyrimidines.
d. adenine.
e. guanine.

*39. What type of amino acid side chains would you expect to find on the surface of a protein embedded in a cell membrane?
a. Cysteine
b. Hydrophobic
c. Hydrophilic
d. Charged
e. Polar, but not charged

40. A molecule with the formula $C_{16}H_{32}O_2$ is a
a. hydrocarbon.
b. carbohydrate.
c. lipid.
d. protein.
e. nucleic acid.

41. A molecule with the formula $C_{16}H_{30}O_{15}$ is a
a. hydrocarbon.
b. carbohydrate.
c. lipid.
d. protein.
e. nucleic acid.

42. Fatty acids are molecules that
a. contain fats.
b. are carboxylic acids.
c. are carbohydrates.
d. contain glycerol.
e. are always saturated.

43. Sucrose is
a. a hexose.
b. a lipid.
c. a disaccharide.
d. a glucose.
e. a simple sugar.

44. DNA and RNA contain
a. pentoses.
b. hexoses.
c. fructoses.
d. maltoses.
e. amyloses.

45. The 20 different common amino acids have different
a. amino groups.
b. R groups.
c. acid groups.
d. peptide linkages.
e. primary structures.

46. The primary structure of a protein is determined by its
a. disulfide bridges.
b. α-helix structure.
c. sequence of amino acids.
d. branching.
e. three-dimensional structure.

47. When a protein becomes nonfunctional as a result of a change in its environment, it is
a. permanent.
b. reversible.
c. denatured.
d. egg white.
e. environmentalized.

*48. A β pleated sheet organization in a polypeptide chain is an example of
a. primary structure.
b. secondary structure.
c. tertiary structure.
d. quaternary structure.
e. coiled structure.

49. A protein can best be defined as
a. a polymer of amino acids.
b. containing one or more polypeptide chains.
c. containing 20 amino acids.
d. containing 20 peptide linkages.
e. containing double helices.

50. The four nitrogenous bases of RNA are abbreviated as
a. A, G, C, and T.
b. A, G, T, and N.
c. G, C, U, and N.
d. A, G, U, and T.
e. A, G, C, and U.

51. Polysaccharides, polypeptides, and polynucleotides have in common that they all
a. contain simple sugars.
b. are formed in condensation reactions.
c. are found in cell membranes.

d. contain nitrogen.
e. have molecular weights less than 30,000 daltons.

52. DNA carries genetic information in its
a. helical form.
b. sequence of bases.
c. tertiary sequence.
d. sequence of amino acids.
e. phosphate groups.

53. A molecule often spoken of as having a head and tail is a(n)
a. phospholipid.
b. oligosaccharide.
c. RNA.
d. steroid.
e. triglyceride.

54. A molecule that has an important role in limiting what gets into and out of cells is
a. glucose.
b. maltose.
c. phospholipid.
d. fat.
e. phosphohexose.

55. Waxes are formed by
a. adding water to fatty acids.
b. removing water from fatty acids.
c. combining fatty acids with alcohol.
d. condensing fatty acids with glycerol.
e. vitamin P.

56. A molecule that has an important role in long-term storage of energy is a(n)
a. steroid.
b. RNA.
c. glycogen.
d. amino acid.
e. hexose.

57. A peptide linkage (peptide bond) holds together two
a. protein molecules.
b. amino acid molecules.
c. sugar molecules.
d. fatty acid molecules.
e. phospholipid molecules.

58. In DNA molecules,
a. purines pair with pyrimidines.
b. A pairs with C.
c. G pairs with A.
d. purines pair with purines.
e. C pairs with T.

59. A type of molecule very often drawn with a single six-sided ring structure is a(n)
a. sucrose.
b. amino acid.
c. glucose.
d. fatty acid.
e. steroid.

60. Maltose and lactose are similar in that they both are
a. simple sugars.
b. amino acids.
c. insoluble in water.
d. disaccharides.
e. hexoses.

61. Starch and glycogen are different in that only one of them
a. is a polymer of glucose.
b. contains ribose.
c. is made in plants.
d. is an energy storage molecule.
e. can be digested by humans.

62. Enzymes are
a. DNA.
b. lipids.
c. carbohydrates.
d. protein.
e. amino acids.

63. The type of bond that holds two amino acids together in a polypeptide chain is a(n)
a. ionic bond.
b. disulfide bridge.
c. hydrogen bond.
d. peptide linkage.
e. dehydration bond.

64. Peptides have ________ and ________ end.
a. a start; a stop
b. a +; a –
c. an N terminus; a C terminus
d. 5′; 3′
e. an A; a Z

65. The ________ structure of a protein relates to how separate polypeptides assemble together.
a. primary
b. secondary
c. tertiary
d. quaternary
e. helical

66. Quaternary structure is found in proteins
a. composed of subunits.
b. of membranes.
c. of the quadruple complex.
d. that change over time.
e. None of the above

67. In DNA, A hydrogen bonds with T and G with C; these are examples of a specific type of reaction called
a. complementary base pairing.
b. a dehydration reaction.
c. a reduction reaction.
d. a hydrophobic reaction.
e. a purine-purine reaction.

68. A fat contains fatty acids and a(n)
a. glycerol.
b. base.
c. amino acid.
d. a phosphate.
e. None of the above

69. There are _________ different types of tripeptides (molecules with three amino acids linked together) that can exist using the 20 common amino acids.
   a. 3
   b. 20
   c. 60
   d. 900
   e. 8,000

70. Chitin is a polymer of
   a. galactosamine.
   b. glucose.
   c. glucosamine.
   d. glycine.
   e. All of the above

71. Two common amino sugars are
   a. glucose and fructose.
   b. glucosamine and galactosamine.
   c. glycine and glutamine.
   d. trehelose and sucrose.
   e. grapeamine and citricamine.

72. Prosthetic groups are molecules that
   a. associate with proteins but are not proteins.
   b. are proteins that associate with lipids.
   c. are lipids that associate with fatty acids.
   d. associate with the surface of cells.
   e. are found free, and rarely associate with other molecules.

73. A type of protein that functions by helping the correct folding of other proteins is called
   a. foldzyme.
   b. renaturing protein.
   c. chaperonin.
   d. hemoglobin.
   denaturing protein.

Study Guide Questions

1. Which of the following statements concerning polymers is *not* true?
   a. Polymers are synthesized from monomers during condensation.
   b. Polymers are synthesized from monomers during dehydration.
   c. Polymers consist of at least two types of monomers.
   d. Both b and c

2. You are a biochemist and have recently discovered a new macromolecule. Studies of the bond types found in this macromolecule reveal many hydrogen bonds and peptide linkages. You most likely have found what type of macromolecule?
   a. Carbohydrate
   b. Lipid
   c. Protein
   d. Nucleic Acid

3. An α helix is an example of which level of protein structure?
   a. Primary
   b. Secondary
   c. Tertiary
   d. Quaternary

4. You have isolated a monomer with the following components: a phosphate group, a sugar, and a nitrogen containing base. Polymers synthesized from this monomer belong to what class of macromolecule?
   a. Carbohydrate
   b. Lipid
   c. Protein
   d. Nucleic Acid

5. Cellulose and starch are composed of the same monomers. Which of the following result in their being structurally and functionally different?
   a. They have different types of glycosidic linkages.
   b. They have different numbers of glucose monomers.
   c. They are held together by different bond types.
   d. None of the above

6. DNA utilizes the bases guanine, cytosine, thymine, and adenine. In RNA, _________ is replaced by _________.
   a. adenine, arginine.
   b. thymine, uracil.
   c. cytosine, uracil.
   d. cytosine, arginine.

7. The pairing of purines with pyrimidines to create a double-stranded DNA molecule is called
   a. complementary base pairing.
   b. phosphodiester bonding.
   c. antiparallel synthesis.
   d. dehydration.

8. Lipids, as a complex class of macromolecules, are similar in that they are synthesized from
   a. glycerol.
   b. fatty acids.
   c. steriod precursors.
   d. cholesterol.

9. Amino acids are linked together into proteins by which of the following bond types?
   a. Covalent bonds
   b. Peptide linkages
   c. Phosphodiester bonds
   d. a and b

10. Which of the following characteristics differentiate carbohydrates from other macromolecule types?
   a. Carbohydrates are constructed of monomers that always have a ring structure.
   b. Carbohydrates never contain nitrogen.
   c. Carbohydrates consist of a carbon bonded to a hydrogen and a hydroxyl group.
   d. None of the above

11. Which of the following statements about carbohydrates is *not* true?
   a. Monomers of carbohydrates have 6 carbon atoms.
   b. Monomers of carbohydrates are linked together during dehydration.
   c. Carbohydrates are energy storage molecules.
   d. None of the above

12. What would you expect to be true of the R groups of amino acids located on the surface of protein molecules found within the interior of biological membranes?
    a. **The R groups would be hydrophobic.**
    b. The R groups would be hydrophilic.
    c. The R groups would be polar.
    d. The R groups would be able to form disulfide.

End of Chapter Questions

1. All lipids are
   a. triglycerides.
   b. polar.
   c. hydrophilic.
   d. polymers.
   e. **more soluble in nonpolar solvents than in water.**

2. Lipids in membranes are composed of
   a. amino acids.
   b. **glycerol attached to fatty acids.**
   c. proteins attached to sugars.
   d. phosphate groups attached to proteins.
   e. polysaccharides.

3. All carbohydrates
   a. are polymers.
   b. are simple sugars.
   c. **consist of one or more simple sugars.**
   d. are found in biological membranes.
   e. are more soluble in nonpolar solvents than in water.

4. Polysaccharides, polypeptides, and polynucleotides all
   a. **are formed from condensation reactions.**
   b. are found in cell membranes.
   c. contain simple sugars.
   d. contain nitrogen.
   e. are not soluble in water.

5. All proteins
   a. are enzymes.
   b. contain the same amino acid sequence.
   c. have a linear, flat shape.
   d. **consist of one or more polypeptide chains.**
   e. are more soluble in nonpolar solvents than in water.

6. The primary structure of a protein is determined by its
   a. disulfide bridges.
   b. α-helix structures.
   c. **order of amino acids**
   d. degree of branching.
   e. three-dimensional nature.

7. The amino acid arginine (see Table 3.2)
   a. is found in all proteins.
   b. cannot form peptide linkages.
   c. is likely to appear in the part of a membrane protein that lies within the phospholipid bilayer.
   d. **is likely to appear in the part of a membrane protein that lies outside the phospholipid bilayer.**
   e. is identical to the amino acid lysine.

8. The quaternary structure of a protein
   a. consists of four subunits—hence the name quaternary.
   b. is unrelated to the function of the protein.
   c. may be either a or b.
   d. depends on covalent bonding among the subunits.
   e. **depends on the primary structures of the subunits.**

9. All nucleic acids
   a. **are polymers of nucleotides.**
   b. are polymers of amino acids.
   c. are double-stranded.
   d. are double-helical.
   e. contain deoxyribose.

10. Which is *not* involved in maintaining the three-dimensional structure of a protein?
    a. Hydrogen bonding between polar R groups
    b. **Covalent bonding between nitrogen-containing R groups**
    c. Covalent bonding between sulfur containing R groups
    d. Hydrophobic interactions between nonpolar R groups
    e. Ionic bonding between charged R groups

Student Web Site Self-Quiz Questions

1. Which of the following statements about the two amino acids methionine and phenylalanine is *true*?
   a. One of the two can participate in forming a disulfide bridge.
   b. Both have a charged side chain.
   c. **Both have side chains that would not be involved in forming hydrogen bonds.**
   d. One of the two has a ringed prosthetic group.
   e. They would not cluster together within a protein.

2. Which of the following statements about proteins is *false*?
   a. The sequence of amino acids is the primary structure of the protein.
   b. **For each amino acid joined to a polypeptide chain, two water molecules are formed.**
   c. A peptide linkage is a covalent bond that joins amino acids together.
   d. The peptide backbone consists of the repeating sequence —N—C—C—.
   e. The total number of different polypeptides containing seven amino acids is equal to 207.

3. Which of the following statements about starch is *false*?
   a. The subunits in starch are all glucose.
   b. **The linkages in starch are all α-1,4 glycosidic linkages.**
   c. Branching limits the hydrogen bonding that can occur between different starch molecules.
   d. The degree of branching varies in different types of starch.
   e. Starch is more similar to glycogen than it is to cellulose.

4. Which of the following components are to nucleic acids what amino acids are to proteins?
   a. Sugar
   b. Sugar-phosphate
   c. Purines and pyrimidines
   d. Nucleoside
   e. Nucleotide

5. Which of the following components are to nucleic acids what side chains are to amino acids?
   a. Sugar
   b. Sugar-phosphate
   c. Purine/pyrimidine
   d. Nucleoside
   e. Nucleotide

6. Select a choice below to make the following statement *false*. Vitamin D _________.
   a. is a water-soluble vitamin
   b. is a steroid
   c. is strongly hydrophobic
   d. deficiency causes a bone-softening disease called rickets
   e. can be produced in the skin

★ *Indicates a **difficult** question*

■ *Indicates a **conceptual** question*

4 The Organization of Cells

Fill in the Blank

1. A measure of the smallest distance that distinguishes two individual objects is the **resolution**.
2. In biology, we call the basic unit of life the **cell**.
3. Photosynthetic membrane systems and mesosomes are internal membrane components of certain organisms termed **prokaryotes**.
4. The light microscope has glass lenses for focusing light (photons) for imaging, whereas the electron microscope has **magnets** for focusing electrons for imaging.
5. Membranous cellular subsystems are termed **organelles**.
6. **Photosynthesis** is the process whereby light energy is converted into chemical bonds.
7. The **ER (or endoplasmic reticulum)** is the organelle or structure with the most lipid membrane in eukaryotic cells.
8. The **Golgi apparatus** is an organelle that serves as a sort of "postal depot" where some of the proteins synthesized on ribosomes and rough ER are processed.
9. RNA carries information for protein synthesis from the DNA in the nucleus to the ribosomes in the cytoplasm. To get from the nucleoplasm to the cytoplasm, RNA must pass through **nuclear pores**.
10. All organisms are composed of cells; all cells come from preexisting cells. These statements are called **the cell theory**.
11. When you cut an orange in half, you **increase** the surface area-to-volume ratio.
12. The DNA in a prokaryotic cell can be found in the **nucleoid** region.
13. The **capsules** of some bacteria help them avoid being detected by the human immune system.
14. The meshwork of intermediate filaments found on the interior surface of the nuclear membrane is called the **nuclear lamina**.
15. Steroids, fatty acids, phospholipids, and carbohydrates are synthesized in the **smooth ER**.
16. The side of the Golgi facing the ER is the ***cis*** face.
17. The substances that enter the Golgi come from the **ER**.
18. Toxic peroxides that are unavoidably formed as side products of important cellular reactions are found and neutralized in **peroxisomes**.
19. The **actin (or microfilament)** is the cytoskeletal component with the smallest diameter.
20. Keratin is classified as an **intermediate** type of filament.

Multiple Choice

★1. The surface area to volume ratio of an object can be decreased by
 a. cutting it into smaller pieces.
 b. flattening it.
 c. stretching it.
 d. making it spherical.
 e. All of the above

2. What must cells do in order to survive?
 a. Obtain and process energy
 b. Convert genetic information into proteins
 c. Keep certain biochemical reactions separate from each other
 d. a and b
 e. a, b, and c

3. Cholesterol is synthesized by
 a. eggs.
 b. prawns.
 c. the SER.
 d. the Golgi.
 e. mitochondria.

4. Examples of cellular "appendages" include
 a. the Golgi apparatus.
 b. cilia.
 c. flagella.
 d. pili.
 e. b, c, and d

5. Roles of biological membranes in eukaryotic cells include which of the functions listed below?
   a. Trafficking of molecules
   b. Serving as staging areas for cellular interaction
   c. Mediating adhesion-recognition reactions between cells
   d. Participating in energy transformations
   e. All of the above

6. The utilization of "food" in the mitochondria, with the associated formation of ATP, is termed
   a. cellular respiration.
   b. metabolic rate.
   c. diffusion.
   d. metabolic processing of fuels.
   e. catabolism.

7. The DNA of mitochondria is located in the
   a. intermembrane space.
   b. matrix.
   c. cristae.
   d. stroma.
   e. granum.

8. The DNA of a chloroplast is located in the
   a. intermembrane space.
   b. matrix.
   c. cristae.
   d. stroma.
   e. granum.

9. Components of chloroplasts include
   a. grana.
   b. thylakoids.
   c. cristae.
   d. a and b
   e. a, b, and c

▪10. The cell is the basic unit of function and reproduction because
   a. subcellular components cannot regenerate whole cells.
   b. cells are totipotent.
   c. single cells can sometimes produce an entire organism.
   d. cells can only come from preexisting cells.
   e. a cell can arise by the fusion of two cells.

▪11. What is the major distinction between prokaryotic and eukaryotic cells?
   a. A prokaryotic cell does not have a nucleus, and a eukaryotic cell does.
   b. A prokaryotic cell does not have DNA, and a eukaryotic cell does.
   c. Prokaryotic cells are smaller than eukaryotic cells.
   d. Prokaryotic cells have not prospered, while eukaryotic cells are evolutionary "successes."
   e. Prokaryotic cells cannot obtain energy from their environment.

12. Which of the following is *not* a characteristic of a prokaryotic cell?
   a. A plasma membrane
   b. A nuclear envelope
   c. A nucleoid
   e. Ribosomes
   e. Enzymes

13. All members of the kingdom Eubacteria
   a. have nuclei.
   b. have chloroplasts.
   c. are multicellular.
   d. are prokaryotes.
   e. have flagella.

14. Ribosomes are made up of
   a. DNA and RNA.
   b. DNA and proteins.
   c. RNA and proteins.
   d. proteins.
   e. DNA.

15. Which of the following are found in prokaryotic cells?
   a. Mitochondria
   b. Chloroplasts
   c. Nuclei
   d. Enzymes
   e. Endomembrane system

16. The infoldings of the plasma membrane of certain prokaryotic organisms can form which of the following structures?
   a. Photosynthetic system
   b. Cell wall
   c. Nuclear membrane
   d. Capsule
   e. Ribosome

17. The DNA of prokaryotic cells is found in the
   a. plasma membrane.
   b. nucleus.
   c. ribosome.
   d. nucleoid region.
   e. mitochondria.

18. Which structure supports the plant cell and determines its shape?
   a. Capsule
   b. Flagellum
   c. Cell wall
   d. Cytosol
   e. Cytoplasm

19. Some bacteria are able to propel themselves through liquid by means of a structure called the
   a. flagellum.
   b. pili.
   c. cytoplasm.
   d. cell wall.
   e. peptidoglycan molecule.

▪20. If you removed the pili from a bacterial cell, which of the following would you expect to happen?
   a. The bacterium could no longer swim.
   b. The bacterium would not adhere to other cells as well.
   c. The bacterium could no longer regulate the movement of molecules into and out of the cell.

d. The bacterium would dry out.
e. The shape of the bacterium would change.

21. The shortest distance that can be resolved with a normal unaided eye is
a. 200 μm.
b. 0.2 μm.
c. 20 μm.
d. 0.2 nm.
e. 20 nm.

22. Ribosomes are not visible under a light microscope, but can be seen with an electron microscope. This is because
a. electron beams have more energy than light beams.
b. electron microscopes focus light with magnets.
c. electron microscopes have more resolving power than light microscopes.
d. electrons have such high energy that they pass through biological samples.
e. living cells can be observed under the electron microscope.

▪23. Using a light microscope it is possible to view cytoplasm streaming around the central vacuole in cells of the green alga *Nitella*. Why would you use a light microscope instead of an electron microscope to study this process?
a. Electron microscopes have less resolving power than light microscopes.
b. Structures inside the cell cannot be seen with the electron microscope.
c. Whole cells cannot be viewed with the electron microscope.
d. The electron microscope cannot be used to observe living cells.
e. The central vacuole is too small to be seen with a scanning electron microscope.

24. Which of the following is a general function of all cellular membranes?
a. They regulate which materials can cross the membrane.
b. They support the cell and determine its shape.
c. They produce energy for the cell.
d. They produce proteins for the cell.
e. They move the cell.

25. Which statement about the nuclear envelope is true?
a. It contains pores for the passage of large molecules.
b. It is composed of two membranes.
c. It contains ribosomes on the inner surface.
d. a and b
e. All of the above

26. What is the purpose of the folds of the inner mitochondrial membrane?
a. They increase the volume of the mitochondrial matrix.
b. They create new membrane-bounded compartments within the mitochondrion.
c. They increase the surface area for the exchange of substances across the membrane.
d. They anchor more of the mitochondrial DNA.
e. The folds have no known purpose.

27. Which type of organelle is found in plants but not in animals?
a. Ribosomes
b. Mitochondria
c. Nuclei
d. Plastids
e. None of the above

28. Where in the cell do you *not* find DNA?
a. Mitochondrial matrix
b. Chloroplast stroma
c. Cell cytosol
d. Cell nucleus
e. You find DNA in all of the above.

29. Which of the following statements about cells is *true*?
a. Animal cells do not produce chloroplasts.
b. Animal cells do not have mitochondria.
c. All plant cells contain chloroplasts.
d. Plant cells do not have plastids.
e. None of the above

▪30. Which of the following is *not* an argument for the endosymbiotic theory?
a. Mitochondria and chloroplasts have double membranes.
b. Mitochondria and chloroplasts cannot be grown in culture free of a host cell.
c. Mitochondria and chloroplasts have DNA and ribosomes.
d. Mitochondrial ribosomes are similar to bacterial ribosomes.
e. All of the above

▪31. What is the difference between "free" and "attached" ribosomes?
a. Free ribosomes are in the cytoplasm, while attached ribosomes are anchored to the endoplasmic reticulum.
b. Free ribosomes produce proteins in the cytosol, while attached ribosomes produce proteins that are inserted into the ER.
c. Free ribosomes produce proteins that are exported from the cell, while attached ribosomes make proteins for mitochondria and chloroplasts.
d. a and c
e. a and b

32. The carotenoid pigments that give ripe tomatoes their red color are contained in organelles called
a. chloroplasts.
b. proplastids.
c. protoplasts.
d. leucoplasts.
e. chromoplasts.

33. Which is a function of a plant cell vacuole?
a. Storage of wastes
b. Support for the cell
c. Excretion of wastes
d. a and b
e. b and c

34. Microtubules are made of
a. actin and function in locomotion.
b. tubulin and are found in cilia.
c. tubulin and are found in microvilli.
d. actin and function to change cell shape.
e. polysaccharides and function in locomotion.

35. The width of a typical animal cell is closest to
a. 1 millimeter.
b. 15 micrometers.
c. 1 micrometer.
d. 10 nanometers.
e. 10^{-10} meter.

36. The smallest structure that can be clearly seen through a light microscope is
a. 1 millimeter.
b. 0.1 millimeter.
c. 10 micrometers.
d. 2 micrometers.
e. 0.2 micrometer.

37. The two major types of cells are
a. human and nonhuman.
b. prokaryotic and eukaryotic.
c. blood and muscle.
d. plant and animal.
e. warm-blooded and cold-blooded.

38. The one type of cell always lacking a cell wall is the
a. bacterial cell.
b. plant cell.
c. animal cell.
d. fungal cell.
e. prokaryotic cell.

39. A structure found only in plant cells is the
a. cilium.
b. nucleus.
c. mitochondrion.
d. glyoxysome.
e. cell membrane.

40. An organelle found in all eukaryotic cells during some portion of their lives is the
a. chloroplast.
b. nucleus.
c. lagellum.
d. vacuole.
e. centriole.

41. Ribosomes are important because they are the structures where
a. chemical energy is stored in making ATP.
b. cell division is controlled.
c. genetic information is used to make proteins.
d. sunlight energy is captured into chemical energy.
e. new organelles are made.

42. Chloroplasts are important because they are the structures where
a. chemical energy is stored by making ATP.
b. cell division is controlled.
c. genetic information is used to make proteins.
d. energy from the sun is converted to chemical energy.
e. new organelles are made.

43. An organelle consisting of a series of flattened sacks stacked somewhat like pancakes is the
a. mitochondrion.
b. chloroplast.
c. Golgi apparatus.
d. rough endoplasmic reticulum.
e. flagellum.

44. An organelle with an internal cross section showing a characteristic "9 + 2" morphology is the
a. mitochondrion.
b. vacuole.
c. Golgi apparatus.
d. flagellum.
e. cytoskeleton.

45. An organelle bounded by two distinct membranes is the
a. nucleus.
b. Golgi apparatus.
c. endoplasmic reticulum.
d. flagellum.
e. lysosome.

46. Chromatin is a series of entangled threads composed of
a. microtubules.
b. DNA and protein.
c. fibrous proteins.
d. cytoskeleton.
e. membranes.

47. Ribosomes are not found in
a. a mitochondrion.
b. a chloroplast.
c. the rough endoplasmic reticulum.
d. a prokaryotic cell.
e. the Golgi apparatus.

48. The overall shape of a cell is determined by its
a. cell membrane.
b. cytoskeleton.
c. nucleus.
d. cytosol.
e. endoplasmic reticulum.

49. Of the following structures of an animal cell, the one with the largest volume is the
a. cilium.
b. mitochondrion.
c. lysosome.
d. nucleus.
e. ribosomes.

50. Of the following structures of a plant cell, the one that most often has the greatest volume is the
    a. glyoxysome.
    b. lysosome.
    c. chromosome.
    d. ribosome.
    e. vacuole.

51. Of the following structures, the one that an animal cell will usually have the greatest number of is the
    a. vacuole.
    b. nucleus.
    c. ribosome.
    d. flagellum.
    e. plastid.

52. Of the following structures, the one that contains both a matrix and cristae is the
    a. plastid.
    b. lysosome.
    c. Golgi apparatus.
    d. mitochondrion.
    e. chromatin.

53. Light energy for conversion to chemical energy is trapped in the
    a. mitochondrion.
    b. chromoplast.
    c. thylakoid.
    d. endoplasmic reticulum.
    e. Golgi apparatus.

54. Proteins that will function outside of the cytosol are made by
    a. the Golgi apparatus.
    b. ribosomes within the mitochondrion.
    c. the smooth endoplasmic reticulum.
    d. ribosomes on the rough endoplasmic reticulum.
    e. ribosomes within the nucleus.

55. Cilia contain
    a. microtubules.
    b. microfilaments.
    c. intermediate filaments.
    d. ribosomes.
    e. plasmodesmata.

56. You would *not* expect to find RNA in which of the following structures?
    a. Nucleus
    b. Mitochondrion
    c. Vacuole
    d. Ribosome
    e. Prokaryotic cell

57. Of the following, the structure involved with the movement of organelles within a cell is/are the
    a. Golgi apparatus.
    b. endoplasmic reticulum.
    c. mitochondrion.
    d. microfilaments.
    e. intermediate filaments.

58. Chloroplasts are a kind of
    a. leucoplast.
    b. endoplasmic reticulum.
    c. chromoplast.
    d. Golgi apparatus.
    e. plastid.

59. Which of the following is *not* a component of the endomembrane system?
    a. Rough endoplasmic reticulum
    b. Smooth endoplasmic reticulum
    c. Golgi apparatus
    d. Lysosomes
    e. Plastids

60. A prokaryotic cell does not have a
    a. nucleus or organelles.
    b. nucleus or DNA.
    c. nucleus or ribosomes.
    d. nucleus or membranes.
    e. cell wall or membranes.

61. The pores found in the nuclear membrane are composed of
    a. one large protein.
    b. eight large protein granules.
    c. keratin.
    d. intermediate filaments.
    e. lipids.

62. Starch molecules are stored inside
    a. chromoplasts.
    b. granularplasts.
    c. chloroplasts.
    d. potatoplasts.
    e. leucoplasts.

63. Some organelles in eukaryotic cells are thought to have
    a. originated from extracellular symbiotic relationships.
    b. their own endoplasmic reticulum.
    c. their own mitochondria.
    d. originated from endosymbiotic relationships.
    e. the ability to live free from the host cell.

64. The membranes of the endoplasmic reticulum are continuous with the membranes of the
    a. nucleus.
    b. Golgi apparatus.
    c. nucleolus.
    d. plasma membrane.
    e. mitochondria.

65. The rough ER is the portion of the ER that
    a. has a bearded appearance.
    b. is older and was once the smooth ER.
    c. has ribosomes attached to it.
    d. is connected to the Golgi apparatus.
    e. is the site of steroid synthesis.

▪66. The difference in the structure of the Golgi of plants, protists, and fungi when compared to that of vertebrates is that the vertebrates' Golgi
  a. forms a large apparatus from a few stacked sacks.
  b. forms small widely distributed sacks.
  c. forms a single large sack.
  d. connects directly to the ER.
  e. lacks integral membrane proteins.

67. Materials that enter and leave the Golgi
  a. are transported by proteins.
  b. are packaged on or in vesicles.
  c. are destined for export from the cell.
  d. require a docking protein.
  e. originated in the nucleus.

68. Proteins from the Golgi are transported to the correct location due to
  a. signals found on the packaged proteins.
  b. the direction all vesicles travel within the cell.
  c. the control provided by the nucleus.
  d. motor proteins.
  e. microtubules.

69. A secondary lysosome is a lysosome that
  a. provides a backup to the primary lysosomes.
  b. is smaller than a primary lysosome.
  c. will become a primary lysosome after it fuses with a phagosome.
  d. is a primary lysosome that has fused with a phagosome.
  e. has exocytosed.

70. Lysosomes are important to eukaryotic cells because they contain
  a. photosynthetic pigments.
  b. starch molecules for energy storage.
  c. their own DNA molecules.
  d. the cell's waste materials.
  e. digestive enzymes.

71. Which of the following cellular components are most important for stabilizing the shape of an animal cell?
  a. Golgi apparatus
  b. Nuclear lamina
  c. Microfilaments
  d. Microtubules
  e. Cell wall

72. The surface area of some eukaryotic cells is greatly increased by
  a. microtubules.
  b. pili.
  c. thylakoid membranes.
  d. myosin.
  e. microvilli.

73. Microvilli are created by projections of
  a. microtubules.
  b. actin.
  c. myosin.
  d. intermediate filaments.
  e. None of the above

74. Hair and intermediate filaments are composed of
  a. microtubules.
  b. microfilaments.
  c. collagen.
  d. hydroxyapatite.
  e. keratin.

75. Microtubules are composed of subunits of
  a. α- and β-tubulin.
  b. d and l actin.
  c. r and s myosin.
  d. kappa tubules.
  e. kappa actinomin.

76. The usefulness of microvilli to cells possessing them is
  a. to aid in their locomotion.
  b. to help concentrate food particles.
  c. for intracellular trafficking of molecules.
  d. to greatly increase their surface area.
  e. for intercellular communications.

77. This cellular component can be found at the base of each cilium.
  a. Centriole
  b. Basal body
  c. Nucleolus
  d. Flagellum
  e. Microvillus

78. The cellular structures that are most like centrioles are
  a. basal bodies.
  b. microbodies.
  c. chromoplasts.
  d. microfilaments.
  e. centromeres.

▪79. What would you expect would happen if you removed a plant cell's wall and placed it into a drop of water?
  a. The cell would begin to grow.
  b. The cell would shrink in size.
  c. The cell would burst.
  d. The cell would first swell and then shrink.
  e. The cell would first shrink and then swell.

Study Guide Questions

1. You have found a mass of cells in the sediment surrounding a thermal vent in the ocean floor. The salinity in the area is quite high. Upon microscopic examination of the cells you find no evidence of membrane-bound organelles. How would you classify this cell?
  a. As a eukaryotic cell
  b. As a prokaryotic cell
  c. As a member of domain Archaea or Bacteria
  d. b and c

2. Centrifugation of a cell results in the rupture of the cell membrane and the contents compacting into a pellet in the bottom of the centrifuge tube. Bathing this pellet with a glucose solution yields metabolic activity including the production of ATP. One of the contents of this pellet is most likely which of the following?

a. Cytosol
b. Mitochondria
c. Lysosomes
d. Golgi bodies

3. Eukaryotic cells are thought to be derived from prokaryotic cells that underwent phagocytosis without digestion of the phagocytized cell. This mutualistic relationship is explained by the
   a. endosymbiotic theory.
   b. cell theory.
   c. evolutionary theory.
   d. parasite theory.

4. Though science fiction has produced stories like "The Blob," we don't see very many large single-celled organisms. Which of the following tends to limit cell size?
   a. Ability to maintain a continuous large membrane
   b. Ability to reproduce a large cell
   c. Surface area-to-volume ratios
   d. All of the above

5. Microscopes are used to resolve images that cannot be seen with the unaided eye. Electron microscopes use __________ to resolve images, whereas light microscopes use __________ to resolve images .
   a. light and lenses, diffraction of electron beams
   b. diffraction of electron beams, light and lenses
   c. lasers, light and lenses
   d. None of the above

6. Match the following cellular functions to the organelle in which they occur:

<u>e.</u> Lysozome	a. DNA synthesis
<u>a.</u> Nucleus	b. Protein folding
<u>d.</u> RER	c. Packaging of materials for export
<u>b.</u> SER	d. Protein synthesis
<u>c.</u> Golgi apparatus	e. Breakdown of phagocytized material
<u>h.</u> Chromoplast	f. Photosynthesis
<u>g.</u> Mitochondrion	g. Cellular respiration
<u>f.</u> Chloroplast	h. Pigment production

7. Which of the following organelles are double membrane-bound?
   a. Nucleus
   b. Chloroplast
   c. Mitochondrion
   d. All of the above

8. Movement of cells is accomplished in both prokaryotes and eukaryotes with which of the following structures?
   a. Cilia
   b. Pili
   c. Dynein
   d. Flagella

9. Which of the following statements regarding mitochondria and chloroplasts is *true*?
   a. Some cells produce chloroplasts.
   b. Mitochondria and chloroplasts may be found in the same cell.
   c. Mitochondria and chloroplasts are not found in the same cell.
   d. Chloroplasts can revert to mitochondria in certain conditions.

10. Which of the following best describes ribosomes?
    a. Ribosomes guide protein synthesis.
    b. Ribosomes are found only in the nucleus or on the RER.
    c. There are no ribosomes in the mitochondria.
    d. All of the above

11. Nuclear DNA exists as a complex of proteins called __________ that condenses into __________ during cellular division.
    a. chromosomes, chromatin
    b. chromatids, chromosomes
    c. chromophors, chromatin
    d. chromatin, chromosomes

12. Rough endoplasmic reticulum and smooth endoplasmic reticulum differ
    a. only by the presence or absence of ribosomes.
    b. both in the presence or absence of ribosomes and in their function.
    c. only in microscopic appearance.
    d. None of the above

End of Chapter Questions

1. Which is present in both prokaryotic and eukaryotic animal cells?
   a. Chloroplast
   b. Mitochondrion
   c. Cell wall
   d. Nucleus
   e. Ribosomes

2. The major factor limiting cell size is the
   a. concentration of water in the cytoplasm.
   b. need for energy.
   c. presence of organelles surrounded by membranes.
   d. ratio of surface area to volume.
   e. thickness of the plasma membrane.

3. Which statement about mitochondria is *not* true?
   a. Their inner membrane folds to form cristae.
   b. They are usually 1 μm or less in diameter.
   c. They are green because of the chlorophyll they contain.
   d. Energy-rich substances from the cytosol are oxidized in them.
   e. Much ATP is synthesized in them.

4. Which organelle is *not* surrounded by a membrane?
   a. Endoplasmic reticulum
   b. Golgi complex
   c. Chloroplast
   d. Microfilament
   e. Nucleus

5. Which statement about the endoplasmic reticulum is *not* true?
   a. It is of two types: rough and smooth.
   b. It is a network of tubes and flattened sacs.
   c. It is found in all living cells.
   d. Some of it is sprinkled with ribosomes.
   e. Parts of it modify proteins.

6. The Golgi apparatus
   a. is found only in animals.
   b. is found in prokaryotes.
   c. is the appendage that moves a cell around in its environment.
   d. is a site of rapid ATP production.
   e. packages and modifies proteins.

7. Which is *not* a component of the cytoskeleton?
   a. Intermediate filaments
   b. Microfilaments
   c. Kinesin
   d. Microfibrils
   e. Microtubules

8. Eukaryotic flagella
   a. are composed of a protein called flagellin.
   b. rotate like propellers.
   c. cause the cell to contract.
   d. have the same internal structure as cilia.
   e. cause the movement of chromosomes.

9. Microfilaments
   a. are composed of polysaccharides.
   b. are composed of actin.
   c. provide the motive force for cilia and flagella.
   d. make up the spindle that aids the movement of chromosomes.
   e. maintain the position of the nucleus in the cell.

10. Which statement about the plant cell wall is *not* true?
    a. Its principal chemical components are polysaccharides.
    b. It lies outside the plasma membrane.
    c. It provides support for the cell.
    d. It completely isolates adjacent cells from one another.
    e. It is semirigid.

Student Web Site Self-Quiz Questions

1. How does the surface area-to-volume ratio of eight 3-mm cubes compare to the surface area-to-volume ratio of one 6-mm cube?
   a. The same
   b. Two times larger
   c. Four times larger
   d. Four times smaller
   e. Eight times larger

2. Which one of the structures listed below is found in eukaryotic cells but not in prokaryotic cells?
   a. Cytosol
   b. Ribosomes
   c. Internal membranes
   d. Internal cytoskeleton
   e. Cell wall

3. In the eukaryotic cell, which one of the following is not a membranous compartment?
   a. Nucleus
   b. Vacuole
   c. Mitochondrion
   d. Lysosome
   e. Ribosome

4. The figure to the right shows a portion of the nuclear envelope. Which of the following statements about the nuclear envelope is *false*?
   a. The space between the inner and outer membranes contains the nuclear lamina.
   b. The inner and outer membranes are continuous.
   c. The nuclear pores connect the nucleoplasm and the cytoplasm.
   d. The inner and outer membranes are standard phospholipid bilayers.
   e. Nucleic acids can move through the nuclear pores.

5. Which of the following statements about the endoplasmic reticulum (ER) is *false*?
   a. Ribosomes are located within the lumen of the rough ER.
   b. Cells that produce a lot of protein for export are packed with ER.
   c. Within the rough ER, many proteins fold and assume their normal tertiary structure.
   d. Carbohydrates are added to proteins to produce glycoproteins in the ER.
   e. Chemical modification of small molecules such as drugs or pesticides occurs within the smooth ER.

TEST FILE

★ *Indicates a **difficult** question*

■ *Indicates a **conceptual** question*

5 Cellular Membranes

Fill in the Blank

1. The **lipid** molecules of membranes act as barriers to the passage of many materials and serve to maintain the membrane's physical integrity.

2. In a complex solution, the **diffusion** of each substance is independent of that of the other substances.

3. Lipids and proteins can move **laterally** but not across biological membranes.

4. **Receptor-mediated endocytosis** is the movement of specified macromolecules into a cell; it involves coated pits, clathrin, and coated vesicles.

5. Some materials move through biological membranes more readily than others. This characteristic of biological membranes is called selective **permeability**.

6. The major lipids in biological membranes are called **phospholipids**.

7. You place a cell into a solution, and the cell shrinks. This solution is **hypertonic** relative to the cell.

8. The cells of the intestinal epithelium are joined to one another by **tight junctions** that prevent substances from passing between the cells of this tissue.

9. The coupled transport system by which glucose and sodium ions enter intestinal epithelial cells is called **symport**.

10. Mammalian embryos have protein complexes that couple cells together, allowing communication by small molecules between cells. These complexes of proteins are called **gap junctions**.

11. Diffusion occurs **down** a concentration gradient.

12. Three things that influence the diffusion rate are the **size of the molecule, temperature,** and the **concentration gradient**.

13. The force that increases inside a plant cell when it is placed in water, which finally prevents further net movement of water molecules into the cell, is called **turgor** pressure.

14. The sodium–potassium pump of cell membranes is a(n) **antiport** active transport system.

15. Biological membranes are organized in the manner described by the **fluid mosaic** model.

16. Membrane proteins with carbohydrates attached are called **glycoproteins**.

17. Membrane lipids with carbohydrates attached are called **glycolipids**.

18. The membranes of the cells of the myelin sheath have less, **more**, or the same amount of protein as the membranes of the mitochondria.

19. The cell adhesion molecule of sponges is a **glycoprotein**.

20. When cell adhesion molecules that connect cells are of the same type, they are called **homotypic**.

21. The type cell adhesion molecule found in sponges is **homotypic**.

22. Of carrier and channel proteins, **channel** proteins were found to be faster; they also do not saturate.

23. Membrane synthesized on the ER moves to other points of the cell as **vesicles**.

24. The process of one cell engulfing another is called **phagocytosis**.

Multiple Choice

1. The chemical makeup, physical organization, and function of a biological membrane depend on which of the following classes of biochemical compounds?
   a. Proteins
   b. Lipids
   c. Fats
   d. Carbohydrates
   e. a, b, and d

2. Integral membrane proteins have
   a. hydrophobic regions within the lipid portion of the bilayer.
   b. hydrophilic regions that protrude in aqueous environments on either side of the membrane.
   c. lateral but not vertical movement within the bilayer.
   d. a and b
   e. a, b, and c

3. Specialized cell junctions include
   a. gap junctions.
   b. tight junctions.
   c. desmosomes.
   d. a, b, and c
   e. a and b

4. Substances move through biological membranes against concentration gradients via
   a. simple osmosis.
   b. active transport.
   c. reverse osmosis.
   d. a and b
   e. None of the above

5. Because the sodium–potassium pump imports K^+ ions while exporting Na^+ ions, it is a coupled transport system termed a(n)
   a. symport.
   b. antiport.
   c. secondary active transporter.
   d. facilitated transport.
   e. diffusion mechanism.

6. Whether a membrane protein can be integral and traverses the membrane depends on the presence of __________ R groups.
   a. primary
   b. secondary
   c. tertiary
   d. quaternary
   e. hydrophobic

▪7. When placed in water, wilted plants lose their limpness because of
   a. active transport of salts from the water into the plant.
   b. active transport of salts into the water from the plant.
   c. osmosis of water into the plant cells.
   d. osmosis of water from the plant cells.
   e. diffusion of water from the plant cells.

*▪8. Houseplants adapted to indoor temperatures might die when accidentally left outdoors in the cold because their
   a. DNA cannot function.
   b. membranes lack adequate fluidity.
   c. photosynthesis is impaired.
   d. chloroplasts malfunction.
   e. membranes need more cholesterol.

9. The compounds in biological membranes that form a barrier to the movement of materials across the membrane are
   a. integral membrane proteins.
   b. carbohydrates.
   c. lipids.
   d. nucleic acids.
   e. peripheral membrane proteins.

10. The interior of the phospholipid bilayer is
   a. hydrophilic.
   b. hydrophobic.
   c. aqueous.
   d. solid.
   e. charged.

11. In biological membranes, the phospholipids are arranged in a
   a. bilayer with the fatty acids pointing toward each other.
   b. bilayer with the fatty acids facing outward.
   c. single layer with the fatty acids facing the interior of the cell.
   d. single layer with the phosphorus-containing region facing the interior of the cell.
   e. bilayer with the phosphorus groups in the interior of the membrane.

12. A protein that forms an ion channel through a membrane is most likely to be
   a. a peripheral protein.
   b. an integral protein.
   c. a phospholipid.
   d. an enzyme.
   e. entirely outside the phospholipid bilayer.

▪13. When a mouse cell and a human cell are fused, the membrane proteins of the two cells become uniformly distributed over the surface of the hybrid cell. This occurs because
   a. many proteins can move around within the bilayer.
   b. all proteins are anchored within the membrane.
   c. proteins are asymmetrically distributed within the membrane.
   d. all proteins in the plasma membrane are peripheral.
   e. different membranes contain different proteins.

14. The hydrophilic regions of a membrane protein are most likely to be found
   a. only in muscle cell membranes.
   b. associated with the fatty acid region of the lipids.
   c. in the interior of the membrane.
   d. exposed on the surface of the membrane.
   e. either on the surface or inserted into the interior of the membrane.

15. Biological membranes are composed of
   a. nucleotides and nucleosides.
   b. enzymes, electron acceptors, and electron donors.
   c. fatty acids.
   d. monosaccharides.
   e. lipids, proteins, and carbohydrates.

▪16. The LDL receptor is an integral protein that crosses the plasma membrane, with portions of the protein extending both outside and into the interior of the cell. The amino acid side chains in the region of the protein that crosses the membrane are most likely to be
   a. charged.
   b. hydrophilic.
   c. hydrophobic.
   d. carbohydrates.
   e. lipids.

17. Which of the following compounds functions as recognition signals between cells?

a. RNA
b. Phospholipids
c. Cholesterol
d. Fatty acids
e. Glycolipids

18. When a membrane is prepared by freeze-fracture and examined under the electron microscope, the exposed interior of the membrane bilayer appears to be covered with bumps. These bumps are
a. integral membrane proteins.
b. ice crystals.
c. platinum.
d. organelles.
e. vesicles.

*19. Structures that contain networks of keratin fibers and hold adjacent cells together are called
a. extracellular matrices.
b. glycoproteins.
c. gap junctions.
d. desmosomes.
e. phospholipid bilayers.

20. The electric signal for contraction passes rapidly from one muscle cell to the next by way of
a. tight junctions.
b. desmosomes.
c. gap junctions.
d. integral membrane proteins.
e. freeze fractures.

21. Tight junctions serve an important function in epithelial cell layers by
a. restricting the extracellular movement of molecules between the adjacent cells.
b. allowing the movement of nerve impulses from one cell to the next.
c. providing cytoplasmic channels between adjacent cells.
d. providing channels between the cytoplasm and the extracellular environment.
e. acting as recognition sites for foreign substances.

▪22. You fill a shallow pan with water and place a drop of red ink in one end of the pan and a drop of green ink in the other end. Which of the following is true at equilibrium?
a. The red ink is uniformly distributed in one half of the pan, and the green ink is uniformly distributed in the other half of the pan.
b. The red and green inks are both uniformly distributed throughout the pan.
c. Each ink is moving down its concentration gradient.
d. The concentration of each ink is higher at one end of the pan than at the other end.
e. No predictions can be made without knowing the molecular weights of the pigment molecules.

23. Which of the following does *not* affect the rate of diffusion of a substance?
a. Temperature
b. Concentration gradient
c. Electrical charge of the diffusing material
d. Presence of other substances in the solution
e. Molecular diameter of the diffusing material

24. For cells where carbon dioxide crosses the plasma membrane by simple diffusion, what determines the rate at which carbon dioxide enters the cell?
a. The concentration of carbon dioxide on each side of the membrane
b. The amount of ATP being produced by the cell
c. The amount of carrier protein in the membrane
d. The amount of energy available
e. The concentration of hydrogen ions on each side of the membrane

25. Plant cells transport sucrose across the vacuole membrane against its concentration gradient by a process known as
a. simple diffusion.
b. active transport.
c. passive transport.
d. facilitated diffusion.
e. cellular respiration.

26. When placed in a hypertonic solution, plant cells
a. shrink.
b. swell.
c. burst.
d. transport water out.
e. concentrate.

*▪27. You place cells in a solution of glucose and measure the rate at which glucose enters the cells. As you increase the concentration of the glucose solution, the rate at which glucose enters the cells increases. However, when the glucose concentration of the solution is increased above 10 *M*, the rate at which glucose enters the cells no longer increases. Which of the following is the most likely mechanism for glucose transport into the cell?
a. Facilitated diffusion via a carrier protein
b. Facilitated diffusion via a channel protein
c. Pinocytosis
d. Secondary active transport
e. Symport

28. Transporting substances across a membrane from an area of lower concentration to an area of higher concentration requires
a. phospholipids.
b. diffusion.
c. gap junctions.
d. facilitated diffusion.
e. energy.

29. In the parietal cells of the stomach, the uptake of chloride ions is coupled to the transport of bicarbonate ions out of the cell. This type of transport system is called
a. a uniport.
b. a symport.
c. an exchange channel.
d. diffusion.
e. an antiport.

30. When a red blood cell is placed in an isotonic solution, which of the following will occur?
a. The cell will shrivel.
b. The cell will swell and burst.
c. The cell will shrivel, and then return to normal.
d. The cell will swell, and then return to normal.
e. Nothing

31. When a plant cell is placed in a hypotonic solution, which of the following occurs?
a. The cell takes up water until the osmotic potential equals the pressure potential of the cell wall.
b. The cell takes up water and eventually bursts.
c. The cell shrinks away from the cell wall.
d. There is no movement of water into or out of the cell.
e. Water moves out of the cell.

▪32. When vesicles from the Golgi apparatus deliver their contents to the exterior of the cell, they add their membranes to the plasma membrane. Why doesn't the plasma membrane increase in size?
a. Some vesicles from the Golgi apparatus fuse with the lysosomes.
b. Membrane vesicles carry proteins from the endoplasmic reticulum to the Golgi apparatus.
c. Membrane is continually being lost from the plasma membrane by endocytosis.
d. New phospholipids are synthesized in the endoplasmic reticulum.
e. The phospholipids become more tightly packed together in the membrane.

33. Receptor-mediated endocytosis is the mechanism for transport of
a. clathrin.
b. all macromolecules.
c. ions.
d. specific macromolecules.
e. integral membrane proteins.

34. During the formation of muscle, the association of individual muscle cells with one another to form a tissue requires specific membrane proteins. These proteins are called
a. coated vesicles.
b. cell adhesion molecules.
c. glycolipids.
d. carrier molecules.
e. transport proteins.

▪35. The neurotransmitter acetylcholine can activate a muscle cell and cause it to contract, even though the acetylcholine molecule never enters the cell. How is this possible?
a. The acetylcholine receptor protein is a peripheral protein.
b. Acetylcholine can bind to all proteins in the plasma membrane.
c. The acetylcholine receptor protein spans the plasma membrane.
d. Acetylcholine is hydrophobic.
e. Acetylcholine enters the cell by receptor-mediated endocytosis.

*36. Insulin is a protein secreted by cells of the pancreas. What is the pathway for the synthesis and secretion of insulin?
a. Rough ER, Golgi apparatus, vesicle, plasma membrane
b. Golgi apparatus, rough ER, lysosome
c. Lysosome, vesicle, plasma membrane
d. Plasma membrane, coated vesicle, lysosome
e. Rough ER, cytoplasm, plasma membrane

37. Carbohydrates associated with cellular membranes are expected to be found associated with
a. proteins inside cells.
b. lipids inside cells.
c. proteins outside cells.
d. proteins of the internal organelles.
e. the nuclear membrane.

38. The site where transmembrane proteins are embedded into membranes is the
a. smooth ER.
b. rough ER.
c. Golgi apparatus.
d. nuclear membrane.
e. lysosomes.

39. The site where carbohydrates are initially added to membrane proteins is the
a. ER.
b. mitochondria.
c. Golgi apparatus.
d. nuclear membrane.
e. lysosomes.

*▪40. You are studying how low-density lipoproteins (LDL) enter cells. When you examine cells that have taken up LDL, you find that the LDL is inside clathrin-coated vesicles. What is the most likely mechanism for the uptake of LDL?
a. Facilitated diffusion
b. Proton antiport
c. Receptor-mediated endocytosis
d. Gap junctions
e. Ion channels

▪41. If you compare the proteins of the plasma membrane and the proteins of the inner mitochondrial membrane, which of the following will be *true*?
a. Both membranes will have only peripheral proteins.
b. Only the mitochondrial membrane will have integral proteins.
c. Only the mitochondrial membrane will have peripheral proteins.
d. All of the proteins from both membranes will be hydrophilic.
e. The proteins from the two membranes will be different.

42. The molecules in a membrane that limit its permeability are the
a. carbohydrates.
b. phospholipids.
c. proteins.

d. negative ions.
e. water.

43. The plasma membrane of animals contains carbohydrates
a. on the side of the membrane facing the cytosol.
b. on the side of the membrane facing away from the cell.
c. on both sides of the membrane.
d. on neither side of the membrane.
e. within the membrane.

44. Integral membrane proteins tend to
a. be of small molecular weight.
b. be very soluble in water.
c. have hydrophobic amino acids on much of their intramembrane surface.
d. be rare.
e. contain much cholesterol.

*45. An important function of certain integral proteins of a eukaryotic cell's plasma membrane is
a. movement of the cell.
b. binding with signals in the cell's environment.
c. usage of genetic information.
d. digestion of food molecules.
e. generation of ATP.

46. Cholesterol molecules act to
a. help hold a membrane together.
b. transport ions across membranes.
c. attach to carbohydrates.
d. disrupt membrane function.
e. alter the fluidity of the membrane.

▪47. The rate of facilitated diffusion of a molecule across a membrane does not continue to increase as the concentration difference of the molecule across the membrane increases. Why?
a. Facilitated diffusion requires the use of ATP.
b. As the concentration difference increases, molecules interfere with one another.
c. The transport protein must be of the carrier type.
d. The transport protein must be of the channel type.
e. The diffusion constant depends on the concentration difference.

48. Active transport is important because it can move molecules
a. from their high concentration to a lower concentration.
b. from their low concentration to a higher concentration.
c. that resist osmosis across the membrane.
d. with less ATP than might otherwise be used to move the molecules.
e. by increasing their diffusion coefficient.

49. Active transport usually moves molecules
a. in the same direction as does diffusion.
b. in the opposite direction as does diffusion.
c. in a direction that tends to bring about equilibrium.
d. toward higher pH.
e. toward higher osmotic potential.

50. Osmosis is a specific form of
a. diffusion.
b. facilitated transport.
c. active transport.
d. secondary active transport.
e. movement of water by carrier proteins.

51. Osmosis moves water from a region of
a. high concentration of dissolved material to a region of low concentration.
b. low concentration of dissolved material to a region of high concentration.
c. hypertonic solution to a region of hypotonic solution.
d. negative osmotic potential to a region of positive osmotic potential.
e. low concentration of water to a region of high concentration of water.

52. Clathrin-coated pits are structures associated with
a. active transport.
b. phagocytosis.
c. flagellar movement.
d. receptor-mediated endocytosis.
e. secretory vesicles.

▪53. Which of the following molecules is probably the most likely to diffuse across a cell membrane?
a. Glucose
b. Na^+
c. A steroid
d. A protein common to blood
e. A peripheral protein

▪54. Cell growth could involve movement of membrane material from
a. the cell membrane to the vesicles.
b. the Golgi apparatus to the cell membrane.
c. the smooth ER to the rough ER.
d. coated pits to the inside of the cell.
e. lysosomes to the cell membrane.

55. An important function of specialized membranes found in certain organelles is to
a. help the organelles move.
b. protect the organelles from increased temperatures.
c. transform energy.
d. use their internal genetic information.
e. destroy cellular waste products.

56. Membrane molecules that help individual cells organize themselves into tissues are known as
a. peripheral proteins.
b. cell adhesion molecules.
c. clathrins.
d. secondary active transport proteins.
e. symports.

57. For each molecule of ATP consumed during active transport of sodium and potassium,
    a. 2 sodium ions are imported and 3 potassium ions are exported.
    b. 2 sodium ions are imported and 1 potassium ion is exported.
    c. 1 potassium ion is imported and 3 sodium ions are exported.
    d. 2 potassium ions are imported and 3 sodium ions are exported.
    e. 3 potassium ions are imported and 2 sodium ions are exported.

58. Materials can be limited from moving through the spaces between cells by
    a. coated pits on the surfaces of the cells.
    b. desmosomes between the cells.
    c. carbohydrates on the surfaces of the cells.
    d. the extracellular matrix around the cells.
    e. tight junctions between the cells.

59. Keratin is a protein found in
    a. plasmodesmata.
    b. desmosomes.
    c. gap junctions.
    d. clathrin pits.
    e. microfilaments.

60. A concentration gradient of glucose across a membrane means
    a. there are more moles of glucose on one side of the membrane than the other.
    b. glucose molecules are more crowded on one side of the membrane than the other.
    c. there is less water on one side of the membrane than the other.
    d. the glucose molecules are chemically more tightly bonded together on one side than the other.
    e. there are more glucose molecules within the membrane than outside of the membrane.

61. Connexons occur in
    a. the cytoskeleton.
    b. tight junctions.
    c. desmosomes.
    d. plasmodesmata.
    e. gap junctions.

62. Transport proteins that simultaneously move two molecules across a membrane in the same direction are called
    a. uniports.
    b. symports.
    c. antiports.
    d. active transporters.
    e. diffusive ports.

63. The only process that could possibly bring glucose molecules into cells that does *not* involve the metabolic energy of ATP is
    a. phagocytosis.
    b. pinocytosis.
    c. active transport.
    d. diffusion.
    e. osmosis.

64. The functional roles for different proteins found in membranes include all except which of the following?
    a. Allowing movement of molecules that would otherwise be excluded by the lipid components of the membrane
    b. Transferring signals from outside the cell to the inside of the cell
    c. Maintaining the shape of the cell
    d. Facilitating the transport of macromolecules across the membrane
    e. Stabilizing the lipid bilayer

65. The tensile strength of connections between adjacent cells in tissues comes from
    a. gap junctions.
    b. tight junctions.
    c. desmosomes.
    d. slip junctions.
    e. weld junctions.

66. Desmosomes include or associate with all except
    a. dense plaque-like regions.
    b. keratin fibers.
    c. external cell adhesion molecules.
    d. internal channel proteins.
    e. All of the above

67. Secondary active transport involves all the following except
    a. the direct use of ATP.
    b. coupling to another transport system.
    c. use of regained energy from an existing gradient.
    d. the requirement for energy.
    e. the ability to concentrate the transported molecule.

*68. How do amino acids get into cells against concentration gradients?
    a. Simple diffusion
    b. Facilitated diffusion
    c. Primary active transport
    d. Secondary active transport
    e. Antiport transport

69. Receptor-mediated endocytosis includes the involvement of
    a. clathrin.
    b. coatomer.
    c. vesiclease.
    d. LDH.
    e. cholesterol.

Study Guide Questions

1. Which of the following statements regarding cellular membranes is *not* true?
    a. The hydrophobic nature of the phospholipid tails limits the migration of polar molecules across the membrane.
    b. Integral proteins and phospholipids move rapidly and fluidly throughout the membrane.

c. **Phospholipids flip back and forth from one side of the bilayer to the other.**
d. Glycolipids and glycoproteins serve as recognition sites on the cell membrane.

2. Which of the following contributes to differences in the two sides of the cell membrane?
a. Differences in peripheral proteins
b. Different domains expressed on the ends of integral proteins
c. Differences in phospholipid types
d. **All of the above**

3. Which of the following cell membrane components serve as recognition signals for interactions between cells?
a. Recognition proteins
b. **Glycolipids or glycoproteins**
c. Phospholipids
d. Integral proteins

4. Which of the following types of junctions are responsible for communication between cells?
a. Tight junctions
b. Desmosomes
c. **Gap junctions**
d. None of the above

5. You are monitoring the diffusion of a colored molecule across a membrane. Which of the following will result in the fastest rate of diffusion?
a. An internal concentration of 5 percent and an external concentration of 60 percent.
b. An internal concentration of 60 percent and an external concentration of 5 percent.
c. An internal concentration of 35 percent and an external concentration of 40 percent.
d. **a and b**

6. If a red blood cell with an internal salt concentration of about 0.85 percent is placed in a saline solution (salt solution) that is 4 percent, which of the following will most likely happen?
a. **The red blood cell will loose water and shrivel.**
b. The red blood cell will gain water and burst.
c. The turgor pressure in the cell will greatly increase.
d. The cell will remain the same.

7. Which statement characterizes a hypotonic solution?
a. The solution has a greater solute concentration than the one it is being compared to.
b. **The solution has a lesser solute concentration than the one it is being compared to.**
c. The solution has an equal solute concentration to the one it is being compared to.
d. None of the above

8. Which of the following statements regarding osmosis is *not* true?
a. Osmosis refers to the movement of water along a concentration gradient.
b. In osmosis, water moves to equalize solute concentrations on either side of the membrane.
c. **If osmosis occurs across a membrane, then diffusion is not occurring.**
d. The movement of water across a membrane can affect the turgor pressure of some cells.

9. Channel proteins allow ions that would not normally pass through the cell membrane to go through the channel. What properties of the proteins are responsible for this?
a. **The channels are often composed of charged or polar R groups.**
b. The channels are often composed of hydrophobic R groups.
c. a and b
d. None of the above

10. Which of the following limits the movement of molecules when carrier-mediated facilitated diffusion is involved?
a. Concentration gradient
b. Availability of carrier molecules
c. Temperature
d. **All of the above**

11. Active transport differs from passive transport in that active transport
a. requires energy.
b. always requires direct input of ATP.
c. moves molecules against a concentration gradient.
d. **a and c**

12. Single-celled animals like amoeba engulf entire cells for food. Which of the following represents the manner in which amoeba "eat"?
a. The amoeba binds only what it has receptors for, the amoeba's cell membrane surrounds the cell to be digested, a vesicle forms, and the vesicle fuses with a lysosome for digestion.
b. **The amoeba's cell membrane surrounds the cell to be digested, a vesicle forms, and the vesicle fuses with a lysosome for digestion.**
c. The cell is taken into the amoeba's vacuole, a vesicle is formed, and the vesicle fuses with a lysosome for digestion.
d. None of the above

13. Sodium and potassium pumps are common in many cells. Which of the following are necessary for the pumps to work?
a. ATP-driven pumping proteins
b. A signal to activate the pumps
c. A concentration gradient to work against
d. **All of the above**

14. Bacterial cells are often found in very hypotonic environments. Which of the following characteristics keeps them from continuing to take on water from their environment?
a. **The presence of a cell wall allows a buildup of turgor pressure that prevents any more water from entering the cell.**
b. The presence of a cell wall allows a buildup of tonic pressure that prevents any more water from entering the cell.
c. The cell expels water as fast as it takes it up.
d. None of the above

15. Which of the following may affect the rate of diffusion?
    a. Temperature
    b. Molecule size
    c. Concentration gradient
    d. All of the above

End of Chapter Questions

1. Membrane phospholipids
   a. are dissolved in the water of the cell.
   b. are totally surrounded by membrane proteins in a "sandwich."
   c. encircle the cell in a double layer.
   d. have their nonpolar regions facing the cell exterior.
   e. have phosphates ionically bound to polar lipids.

2. The phospholipid bilayer
   a. is readily permeable to large, polar molecules.
   b. is entirely hydrophobic.
   c. is entirely hydrophilic.
   d. has different lipids in the two layers.
   e. is made up of polymerized amino acids.

3. Which statement about membrane proteins is *not* true?
   a. They all extend from one side of the membrane to the other.
   b. Some serve as channels for ions to cross the membrane.
   c. Many are free to migrate laterally within the membrane.
   d. Their position in the membrane is determined by their tertiary structure.
   e. Some play roles in photosynthesis.

4. If a plant cell is put in an environment which is hypertonic to the cytoplasm, water will
   a. diffuse into the cell less than out of the cell.
   b. diffuse out of the cell less than into the cell.
   c. diffuse into and out of the cell at the same rates
   d. not diffuse either in or out of the cell.
   e. dissolve into the plasma membrane.

5. Which statement about animal cell junctions is *not* true?
   a. Tight junctions are barriers to the passage of molecules between cells.
   b. Desmosomes allow cells to adhere strongly to one another.
   c. Gap junctions block communication between adjacent cells.
   d. Connexons are made of protein.
   e. The fibers associated with desmosomes are made of protein.

6. Starch, a polymer of glucose, remains inside plant cells because it
   a. easily crosses the chloroplast and plasma membranes.
   b. is a protein and so is a component of the membrane.
   c. is too large to diffuse across the plasma membrane.
   d. is complexed with proteins.
   e. lacks the energy for active transport.

7. Which statement about membrane channels is *not* true?
   a. They are pores in the membrane.
   b. They are proteins.
   c. All ions pass through the same type.
   d. Movement through them is from high concentration to low.
   e. Movement through them is by simple diffusion.

8. Facilitated diffusion and active transport both
   a. require ATP.
   b. require the use of proteins as carriers.
   c. carry solutes in only one direction.
   d. move substances from high to lower concentration
   e. depend on the solubility of the solute in lipid.

9. Primary and secondary active transport both
   a. generate ATP.
   b. are based on passive movement of sodium ions.
   c. include the passive movement of glucose molecules.
   d. use ATP directly.
   e. can move solutes against their concentration gradients.

10. Large molecules such as proteins can move into cells by
    a. diffusion.
    b. endocytosis.
    c. exocytosis.
    d. active transport.
    e. facilitated transport by a gated channel.

Student Web Site Self-Quiz Questions

1. Which of the following statements regarding homotypic cell binding is *false*?
   a. Molecules of the same protein are found on the surfaces of cells that bind together to form most tissues.
   b. Tissue formation generally depends upon homotypic cell binding.
   c. Most molecules involved in cell binding are glycoproteins.
   d. Cell binding requires that only one cell in a binding pair contain the binding protein.
   e. Separated cells from two different species of sponge will re-aggregate only with cells of the same type due to homytypic cell binding.

2. Which of the following is not a function of tight junctions?
   a. Forcing substances to pass through an epithelial cell membrane rather than through the intercellular space
   b. Restricting the movement of materials within the intercellular spaces
   c. Attaching the epithelial cell firmly to the basal lamina
   d. Restricting the movement of proteins and phospholipids within the cell membrane
   e. Dividing the cell into an apical and basolateral region

3. Which of the following is a major function of a desmosome?
   a. Allowing direct passage of electrical signals between cells
   b. Providing a channel between two cells for movement of small molecules
   c. Providing mechanical stability to epithelial tissues
   d. Connecting cells to an extracellular matrix
   e. Forcing materials through a plasma membrane

4. Which of the following statements regarding channel proteins is *false*?
   a. Channel proteins can assist polar molecules to cross the plasma membrane in a process called facilitated diffusion.
   b. The "gated" ion channel can allow many ions to pass when stimulated to open.
   c. Ion channels allow only a specific ion to pass through, restricting the movement of all other molecules.
   d. A channel protein must span the entire width of the plasma membrane.
   e. The channel of most channel proteins is just wide enough to allow the passage of a specific molecule.

5. Select the *false* statement about receptor-mediated endocytosis.
   a. Coated pits can form only where receptor proteins are located.
   b. Coated pits are lined on their cytoplasmic side by the protein clathrin.
   c. Receptor-mediated endocytosis is important in cellular uptake of cholesterol.
   d. Coated vesicles eventually become part of the endoplasmic reticulum.
   e. Coated vesicles contain material that had been extracellular.

★ *Indicates a* ***difficult*** *question*

■ *Indicates a* ***conceptual*** *question*

6 Energy, Enzymes, and Metabolism

Fill in the Blank

1. A **spontaneous** reaction is one that, given enough time, goes largely to completion by itself without the addition of energy.
2. Cells cannot create energy because **energy cannot be created or destroyed**.

*3. Variations of enzymes that allow organisms to adapt to changing environments are termed **isozymes**.

4. Most of earth's energy comes from **the sun**.
5. Although some enzymes consist entirely of one or more polypeptide chains, others possess a tightly bound non-protein portion called a **prosthetic group**.
6. Cells mostly use **ATP** as an immediate source of energy to drive reactions.

*7. A **coupled** reaction, where one reaction is used to drive another, is the major means of carrying out energy-requiring reactions within cells.

8. The second law of thermodynamics states that the **entropy**, or disorder, of the universe is constantly increasing.
9. When a drop of ink is added to a beaker of water, the dye molecules become randomly dispersed throughout the water. This is an example of an increase in **entropy**.
10. For a reaction to be spontaneous, the change in free energy of the reaction, ΔG, must be **negative**.
11. The enzyme phosphoglucoisomerase catalyzes the conversion of glucose 6-phosphate to fructose 6-phosphate. The region on phosphoglucoisomerase where glucose 6-phosphate binds is called the **active site**.
12. The ΔG of a spontaneous reaction is negative, indicating that the reaction releases free energy. Such a reaction is **exergonic**.
13. Enzymes are biological **catalysts**.
14. The zinc ion in the active site of the enzyme thermolysin is called a **prosthetic group**.
15. When an enzyme is heated until its three-dimensional structure is destroyed, the enzyme is said to be **denatured**.
16. Temperature of water above a waterfall is probably **colder** than the temperature where the water falls.
17. **Metabolism** is the term used for all the chemical activity of a living organism.
18. Heat, light, electricity, and motion are all examples of **kinetic** energy.
19. The energy in a system that exists due to position is **potential** energy.
20. Potential energy can be converted to **kinetic** energy, which does work.
21. The building up of molecules in a living system is **anabolism**, while the breaking down is **catabolism**.
22. The primary directional flow of energy in and among earthly life forms is **light** to **chemical** to **heat**.
23. The first law of thermodynamics is that **energy is neither created nor destroyed**.

Multiple Choice

1. Water held back by a dam represents what kind of energy?
   a. Hydroelectric
   b. Irrigation
   c. Potential
   d. Kinetic
   e. At times, all of the above
2. The change in free energy is related to
   a. change in heat.
   b. change in entropy.
   c. change in pressure.
   d. a and b
   e. a, b, and c
3. Of the following choices, which is the greatest percentage of energy that would be available for doing work during an energy transformation?
   a. 110%
   b. 100%
   c. 98%
   d. 60%
   e. 20%

4. Enzymes are sensitive to
   a. temperature.
   b. pH.
   c. irreversible inhibitors such as DIPF.
   d. allosteric effectors.
   e. All of the above

5. End products of biosynthetic pathways often act to block the initial step in that pathway. This phenomenon is called
   a. allosteric inhibition.
   b. denaturation.
   c. branch pathway inhibition.
   d. feedback inhibition.
   e. binary inhibition.

*6. Which of the following identifies a group of enzymes that is important in fine-tuning the metabolic activities of cells?
   a. Isozymes
   b. Alloenzymes
   c. Allosteric enzymes
   d. a and c
   e. a, b, and c

▪7. Competitive and noncompetitive enzyme inhibitors differ with respect to
   a. the precise location on the enzyme to which they bind.
   b. their pH.
   c. their binding affinities.
   d. their energies of activation.
   e. None of the above

▪8. During photosynthesis, plants use light energy to synthesize glucose from carbon dioxide. However, plants do not use up energy during photosynthesis; they merely convert it from light energy to chemical energy. This is an illustration of
   a. increasing entropy.
   b. chemical equilibrium.
   c. the first law of thermodynamics.
   d. the second law of thermodynamics.
   e. a spontaneous reaction.

▪9. The standard free energy change for the hydrolysis of ATP to ADP + P_i is –7.3 kcal/mol. What can you conclude from this information?
   a. The reaction will never reach equilibrium.
   b. The free energy of ADP and phosphate is higher than the free energy of ATP.
   c. The reaction requires energy.
   d. The reaction is endergonic.
   e. The reaction is exergonic.

▪10. The hydrolysis of maltose to glucose is an exergonic reaction. Which of the following statements is true?
   a. The reaction requires the input of free energy.
   b. The free energy of glucose is larger than the free energy of maltose.
   c. The reaction is not spontaneous.
   d. The reaction releases free energy.
   e. At equilibrium, the concentration of maltose is higher than the concentration of glucose.

11. The first law of thermodynamics states that the total energy in the universe is
   a. decreasing.
   b. increasing.
   c. constant.
   d. being converted to free energy.
   e. being converted to matter.

▪12. If the enzyme phosphohexosisomerase is added to a 0.3 *M* solution of fructose 6-phosphate, and the reaction is allowed to proceed to equilibrium, the final concentrations are 0.2 *M* glucose 6-phosphate and 0.1 *M* fructose 6-phosphate. This data gives an equilibrium constant of 2. What is the equilibrium constant if the initial concentration of fructose 6-phosphate is 3 *M*?
   a. 2
   b. 3
   c. 5
   d. 10
   e. 20

▪13. You are studying the effects of temperature on the rate of a particular enzyme-catalyzed reaction. When you increase the temperature from 40°C to 70°C, what effect will this have on the rate of the reaction?
   a. It will increase.
   b. It will decrease.
   c. It will decrease to zero because the enzyme denatures.
   d. It will increase and then decrease.
   e. This cannot be answered without more information.

▪14. If ΔG of a chemical reaction is negative and the change in entropy is positive, what can you conclude about the reaction?
   a. It requires energy.
   b. It is endergonic.
   c. It is exergonic.
   d. It will not reach equilibrium.
   e. It decreases the disorder in the system.

15. Which of the following determines the rate of a reaction?
   a. ΔS
   b. ΔG
   c. ΔH
   d. The activation energy
   e. The overall change in free energy

16. In a chemical reaction, transition-state species have free energies
   a. lower than either the reactants or the products.
   b. higher than either the reactants or the products.
   c. lower than the reactants, but higher than the products.
   d. higher than the reactants, but lower than the products.
   e. lower than the reactants, but the same as the products.

▪17. The hydrolysis of sucrose to glucose and fructose is exergonic. However, if you dissolve sucrose in water and keep the solution overnight at room temperature,

there is no detectable conversion to glucose and fructose. Why?
a. The change in free energy of the reaction is positive.
b. The activation energy of the reaction is high.
c. The change in free energy of the reaction is negative.
d. This is a condensation rection.
e. The free energy of the products is higher than the free energy of the reactants.

18. The enzyme α-amylase increases the rate at which starch is broken down into smaller oligosaccharides. It does this by
a. decreasing the equilibrium constant of the reaction.
b. increasing the change in free energy of the reaction.
c. decreasing the change in free energy of the reaction.
d. increasing the change in entropy of the reaction.
e. lowering the activation energy of the reaction.

19. The enzyme glyceraldehyde 3-phosphate dehydrogenase catalyzes the reaction glyceraldehyde 3-phosphate → 1,3-diphosphoglycerate. The region of the enzyme where glyceraldehyde 3-phosphate binds is called the
a. transition state.
b. groove.
c. catalyst.
d. active site.
e. energy barrier.

*20. The enzyme glucose oxidase binds the six-carbon sugar glucose and catalyzes its conversion to glucono-1,4-actone. Mannose is also a six-carbon sugar, but glucose oxidase cannot bind mannose. The specificity of glucose oxidase is based on the
a. free energy of the transition state.
b. activation energy of the reaction.
c. change in free energy of the reaction.
d. tertiary structure of the enzyme.
e. rate constant of the reaction.

▪21. In the presence of alcohol dehydrogenase, the rate of reduction of acetaldehyde to ethanol increases as you increase the concentration of acetaldehyde. Eventually, the rate of the reaction reaches a maximum, where further increases in the concentration of acetaldehyde have no effect. Why?
a. All of the alcohol dehydrogenase molecules are bound to acetaldehyde molecules.
b. At high concentrations of acetaldehyde, the activation energy of the reaction increases.
c. At high concentrations of acetaldehyde, the activation energy of the reaction decreases.
d. The enzyme is no longer specific for acetaldehyde.
e. At high concentrations of acetaldehyde, the change in free energy of the reaction decreases.

22. When an enzyme catalyzes both an exergonic reaction and an endergonic reaction, the two reactions are said to be
a. substrates.
b. endergonic.
c. kinetic.
d. activated.
e. coupled.

▪23. In glycolysis, the exergonic reaction 1,3-diphosphoglycerate → 3-phosphoglycerate is coupled to the reaction ADP + P_i → ATP. Which of the following is most likely to be true about the reaction ADP + P_i → ATP?
a. The reaction never reaches equilibrium.
b. The reaction is spontaneous.
c. There is a large decrease in free energy.
d. The reaction is endergonic.
e. Temperature will not affect the rate constant of the reaction.

24. X-ray crystallography can be used to obtain which type of information?
a. The rate constant for a reaction
b. ΔG of a reaction
c. ΔS of a reaction
d. The amount of enzyme activation energy
e. The shape of an enzyme

▪25. Trypsin and elastase are both enzymes that catalyze hydrolysis of peptide bonds. But trypsin only cuts next to lysine and elastase only cuts next to alanine. Why?
a. Trypsin is a protein, and elastase is not.
b. ΔG for the two reactions is different.
c. The shape of the active site for the two enzymes is different.
d. One of the reactions is endergonic, and the other is exergonic.
e. Hydrolysis of lysine bonds requires water; hydrolysis of alanine bonds does not.

26. The enzyme catalase has a ferric ion tightly bound to the active site. The ferric ion is called a(n)
a. side chain.
b. enzyme.
c. coupled reaction.
d. prosthetic group.
e. substrate.

*▪27. The addition of the competitive inhibitor mevinolin slows the reaction HMG-CoA → mevalonate, which is catalyzed by the enzyme HMG-CoA reductase. How could you overcome the effects of mevinolin and increase the rate of the reaction?
a. Add more mevalonate.
b. Add more HMG-CoA.
c. Lower the temperature of the reaction.
d. Add a prosthetic group.
e. Lower the rate constant of the reaction.

▪28. How does a noncompetitive inhibitor inhibit binding of a substrate to an enzyme?
a. It binds to the substrate.
b. It binds to the active site.
c. It lowers the activation energy.
d. It increases the ΔG of the reaction.
e. It changes the shape of the active site.

29. Which type of inhibitor can be overcome completely by the addition of more substrate?
    a. Irreversible
    b. Noncompetitive
    c. Competitive
    d. Prosthetic
    e. Isotonic

30. Binding of substrate to the active site of an enzyme is
    a. reversible.
    b. irreversible.
    c. noncompetitive.
    d. coupled.
    e. allosteric.

31–32. Consider the following metabolic pathway. Reactants and products are designated by capital letters; enzymes are designated by numbers.

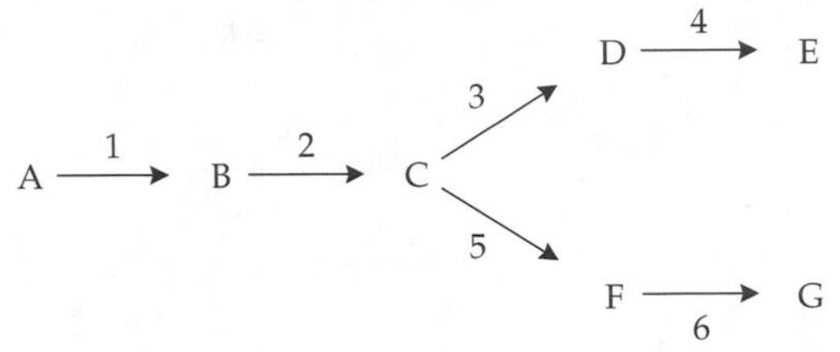

31. Which enzyme is end product G most likely to inhibit?
    a. 1
    b. 2
    c. 3
    d. 5
    e. 6

32. Assume that end product E is a negative feedback regulator of enzyme 1. What happens to a cell when it is grown in the presence of large amounts of E?
    a. The cell can't make G.
    b. The cell can't make A.
    c. The cell makes too much G.
    d. The cell makes too much E.
    e. The cell makes too much D.

33. An allosteric inhibitor
    a. decreases the concentration of inactive enzyme.
    b. decreases the concentration of active enzyme.
    c. increases the concentration of product.
    d. decreases the concentration of substrate.
    e. increases the concentration of enzyme–substrate complex.

34. An RNA molecule that has enzyme activity is called
    a. RNAse.
    b. ribonuclease.
    c. an allosteric enzyme.
    d. a regulatory enzyme.
    e. a ribozyme.

35. Energy present in a system that is unusable to do work relates to the system's
    a. temperature.
    b. entropy.
    c. work.
    d. thermodynamics.
    e. equilibrium.

36. The molecules that are acted on by an enzyme are called
    a. products.
    b. substrates.
    c. carriers.
    d. prosthetics.
    e. effectors.

37. The sum total of all the chemical reactions in a living structure is called its
    a. energetics.
    b. activity.
    c. digestive power.
    d. entropy.
    e. metabolism.

38. The rate of a chemical reaction in a cell is how
    a. often the reaction occurs.
    b. quickly it reaches equilibrium.
    c. much energy must be added to have the reaction occur.
    d. much activation energy is required to have the reaction occur.
    e. easily the reaction is inhibited.

39. The statement "Enzymes are highly specific" means certain
    a. enzymes are found in certain cells.
    b. reactions involving certain substrates are catalyzed by certain enzymes.
    c. enzymes require certain concentrations of substrates.
    d. reactions with certain activation energies are catalyzed by certain enzymes.
    e. concentrations of substrates work with certain enzymes.

40. An active site is
    a. the part of the substrate that binds with an enzyme.
    b. the part of the enzyme that binds with a substrate.
    c. where energy is added to an enzyme catalyst.
    d. where enzymes are found in cells.
    e. None of the above

▪41. Enzymatic reactions can become saturated as substrate concentration increases because
    a. enzymes have the maximum possible number of hydrogen atoms attached to them.
    b. the concentration of substrate cannot increase any higher.
    c. substrates are inhibitors of enzymes.
    d. the activation energy of the reaction cannot be further lowered.
    e. there are a limited number of the enzyme molecules present.

*42. Trypsin is an enzyme whose substrate consists of
    a. polysaccharide chains of bacteria.
    b. sucrose.
    c. another protein.
    d. $FADH_2$.
    e. carboxyl groups.

43. Competitive inhibitors of enzymes work by
 a. **fitting into the active site.**
 b. fitting into a site other than the active site.
 c. altering the shape of the enzyme.
 d. changing the enzyme into an inactive form.
 e. increasing the activation energy of the enzyme-catalyzed reaction.

44. Allosteric inhibitors act by
 a. decreasing the amount of enzyme molecules present.
 b. increasing the amount of enzyme molecules present.
 c. decreasing the amounts of inactive form of the enzyme present.
 d. **decreasing the amounts of active form of the enzyme present.**
 e. increasing the amounts of substrate present.

45. Negative feedback in a sequence of chemical reactions involves a chemical occurring
 a. **late in the sequence inhibiting an earlier reaction.**
 b. early in the sequence inhibiting a later reaction.
 c. early in the sequence activating a later reaction.
 d. late in the sequence activating an earlier reaction.
 e. late in the sequence inhibiting a later reaction.

46. Denatured enzymes are the same as
 a. ribozymes.
 b. abzymes.
 c. isozymes.
 d. **destroyed enzymes.**
 e. coenzymes.

47. The inhibition of enzyme activity by noncompetitive inhibitors can be reduced
 a. by decreasing the concentration of allosteric enzymes.
 b. by decreasing the concentration of substrate.
 c. by increasing the concentration a competitive inhibitor.
 d. by increasing the concentration of substrate.
 e. **only when they become unbound.**

48. The concentration of a substrate present in a reaction at equilibrium depends most strongly on the concentration(s) of
 a. enzyme present.
 b. active form of the enzyme present.
 c. activator present.
 d. other substrates and products present.
 e. **products present.**

49. An allosteric site of an enzyme is where a(n)
 a. competitive inhibitor may bind.
 b. substrate may bind.
 c. prosthetic may group-bind.
 d. **activator may bind.**
 e. coenzyme may bind.

50. The substrate can form a complex with the enzyme and in some cases this complex is stabilized by
 a. hydrogen bonds.
 b. covalent bonds.
 c. ionic attractions.
 d. hydrophobic interactions.
 e. **All of the above**

51. Once initiated, how quickly a spontaneous reaction reaches equilibrium without a catalyst is influenced by
 a. the equilibrium constant.
 b. **a change in free energy.**
 c. a change in entropy.
 d. activation energy.
 e. standard free energy change.

52. Factors that can either activate or inhibit allosteric enzymes are called
 a. proteins.
 b. coenzymes.
 c. sites.
 d. **effectors.**
 e. competitors.

53. The diversity of chemical reactions occurring in a cell depends mostly on certain molecules present in the cell termed
 a. isozymes.
 b. coenzymes.
 c. ribozymes.
 d. abzymes.
 e. **enzymes.**

*54. When organisms move from one environment to another, they sometimes synthesize variations of existing enzymes termed
 a. coenzymes.
 b. abzymes.
 c. **isozymes.**
 d. effectors.
 e. activators.

55. A type of inhibitor of an enzyme that binds within the enzyme's active site is termed
 a. allosteric.
 b. noncompetitive.
 c. **competitive.**
 d. extracompetitive.
 e. None of the above

*56. The catalysis mechanism used by lysozyme is
 a. acid–base catalysis.
 b. covalent catalysis.
 c. metal cofactor redox catalysis.
 d. **induced strain.**
 e. It has yet to be discovered.

57. The fact that an enzyme's active site can sometimes bind inhibitors that are larger than the substrate is called
 a. **induced fit.**
 b. enzyme flex.
 c. lock and key paradox.
 d. substrate-induced active site shaping.
 e. enzyme retrofit.

58. The process that involves an end product acting as an inhibitor of an earlier step in a metabolic pathway is called
  a. feedback activation.
  b. feedback inhibition.
  c. positive feedback.
  d. concerted activation.
  e. competitive inhibition.

59. What can never be created or destroyed?
  a. Entropy
  b. Energy
  c. Free energy
  d. Thermal energy
  e. Potential energy

60. The maximum possible rate of an enzyme reaction influenced by a competitive inhibitor depends on the concentration of _________ present.
  a. inhibitor
  b. substrate
  c. product
  d. enzyme
  e. free energy

*61. Enzymes of the acid-base catalysis type contain
  a. a metal ion bound to a side chain.
  b. a prosthetic group.
  c. a coenzyme.
  d. an acid or base amino acid residue in the active site.
  e. a covalent-activated active site.

*62. The liver enzyme catalase can turn over molecules at a rate of _________ per second.
  a. 400 million
  b. 40 million
  c. 400 thousand
  d. 40 thousand
  e. 400

Study Guide Questions

1. ATP is necessary for the conversion of glucose to glucose 6-phosphate. The energy released by the splitting of ATP into ADP and P_i is what type of energy?
  a. Potential
  b. Kinetic
  c. Entropic
  d. Enthalpic

2. Before ATP is split into ADP and P_i it holds what type of energy?
  a. Potential
  b. Kinetic
  c. Entropic
  d. Enthalpic

3. Which of the following statements concerning energy transformations are true?
  a. Increases in entropy reduce usable energy.
  b. Energy may be created during transformation.
  c. Potential energy increases with each transformation.
  d. Increases in temperature decrease total amount of energy available.

4. A reaction has a ΔG of –20kcal/mole. This reaction is
  a. endergonic and equilibrium is far toward completion.
  b. exergonic and equilibrium is far toward completion.
  c. endergonic and the forward reaction occurs at the same rate as the reverse reaction.
  d. exergonic and the forward reaction occurs at the same rate as the reverse reaction.

5. ATP hydrolysis releases energy to fuel cellular functions. ATP hydrolysis is
  a. endergonic.
  b. exergonic.
  c. chemoautotrophic.
  d. None of the above

6. Enzymes are biological catalysts and function by
  a. increasing free energy in a system.
  b. lowering activation energy of a reaction.
  c. lowering entropy in a system.
  d. increasing temperature near a reaction.

7. Which of the following contribute to the specificity of enzymes.
  a. Each enzyme has a narrow range of temperature and pH optima.
  b. Each enzyme has a specific active site that interacts specifically with a particular substrate.
  c. Substrates themselves may alter the active site slightly for optimum catalysis.
  d. All of the above

8. Coenzymes and cofactors as well as prosthetic groups assist enzyme function by
  a. stabilizing three-dimensional shape and maintaining active sites.
  b. assisting with the binding of enzyme and substrate.
  c. a and b
  d. None of the above

9. Which of the following are unique features of enzymes?
  a. They are not consumed by the enzyme-mediated reaction.
  b. They are not altered by the enzyme-mediated reaction.
  c. They lower activation energy.
  d. All of the above

10. Ascorbic acid, found in citrus fruits, acts as an inhibitor to catecholase, the enzyme responsible for the browning reaction in fruits such as apples, peaches, and pears. One possibility for its function could be that ascorbic acid is very similar in size and shape to catechol, the substrate of the browning reaction. If this is true, then this inhibition is most likely an example of _________ inhibition.
  a. indirect
  b. competitive
  c. direct
  d. noncompetitive

11. Refer to question 10. Suppose further studies indicate that ascorbic acid is not similar to catechol in size and shape but that the pH of the ascorbic acid solution is denaturing catecholase. If this is true, then this inhibition is most likely an example of ________ inhibition.
    a. competitive
    b. irreversible
    c. direct
    d. noncompetitive

12. Metabolism is organized into pathways. The pathway is linked in which of the following manners?
    a. All cellular functions feed into a central pathway.
    b. All steps in the pathway are catalyzed by the same enzyme.
    c. The product of one step in the pathway functions as the substrate in the next step.
    d. Products of the pathway accumulate and are secreted from the cell.

13. Which of the following represents an enzyme-catalyzed reaction?
    a. $E + P \rightarrow E + S$
    b. $E + S \rightarrow E + P$
    c. $E + S \rightarrow P$
    d. $E + S \rightarrow E$

14. Which of the following graphs of enzyme mediated reactions represents an allosteric enzyme?
    a. A graph similar to 6.22*a*
    b. A graph similar to 6.22*b*
    c. A graph similar to 6.26 (change axis labels to those in 6.22)
    d. None of the above

15. Allosteric enzymes differ in that
    a. they have multiple subunits.
    b. changes in one subunit affect the functions of the other subunits.
    c. some subunits are catalytic and others are regulatory.
    d. All of the above

End of Chapter Questions

1. The overall reaction:
   glucose + $O_2 \rightarrow CO_2 + H_2O$ + energy
   a. cannot occur without an enzyme.
   b. is endergonic.
   c. is exergonic.
   d. occurs in a single step in cells.
   e. transforms the glucose to a higher energy state.

2. Which statement about thermodynamics is *not* true?
   a. Free energy is given off in an exergonic reaction.
   b. Free energy can be used to do work.
   c. A spontaneous reaction is exergonic.
   d. Free energy tends always to a minimum.
   e. Entropy tends always to a minimum.

3. In general,
   a. all molecules have the same energy content.
   b. ADP has more energy than ATP.
   c. oxidized compounds have more free energy than reduced compounds.
   d. reduced compounds have more free energy than oxidized compounds.
   e. substances capable of oxidation require ATP hydrolysis.

4. Which statement about enzymes is *not* true?
   a. They act to speed up a biochemical reaction.
   b. They are made up of RNA or protein.
   c. They change the value of ΔG of the reaction.
   d. They are sensitive to heat.
   e. They are sensitive to pH.

5. The active site of an enzyme
   a. never changes shape.
   b. forms no chemical bonds with substrates.
   c. determines, by its structure, the specificity of the enzyme.
   d. looks like a lump projecting from the surface of the enzyme.
   e. changes the value of ΔG of the reaction.

6. Coenzymes differ from enzymes in that coenzymes are
   a. active only outside the cell.
   b. polymers of amino acids.
   c. smaller.
   d. higher in energy state.
   e. specific for a single reaction.

7. The rate of an enzyme-catalyzed reaction
   a. is constant under all conditions.
   b. decreases as substrate concentration increases.
   c. cannot be measured.
   d. depends on pH of the environment.
   e. can be reduced by inhibitors.

8. Which statement about enzyme inhibitors is *not* true?
   a. A competitive inhibitor binds the active site of the enzyme.
   b. An allosteric inhibitor binds a site on the active form of the enzyme.
   c. A noncompetitive inhibitor binds a site other than the active site.
   d. Noncompetitive inhibition cannot be completely overcome by the addition of more substrate.
   e. Competitive inhibition can be completely overcome by the addition of more substrate.

9. Which statement about feedback inhibition of enzymes is *not* true?
   a. It is exerted through allosteric effects.
   b. It is directed at the enzyme that catalyzes the first committed step in a branch of a pathway.
   c. It affects the rate of reaction, not the concentration of enzyme.
   d. It acts very slowly.
   e. It is an example of negative feedback.

10. Which statement about temperature effects is *not* true?
    a. Raising the temperature may reduce the activity of an enzyme.
    b. Raising the temperature may increase the activity of an enzyme.
    c. Raising the temperature may denature an enzyme.
    d. Some enzymes are stable at the boiling point of water.
    e. All enzymes have the same optimal temperature.

Student Web Site Self-Quiz Questions

1. Which one of the following features is not a reason that a cell is considered an open system?
    a. Energy can enter and leave a cell.
    b. Matter can enter and leave a cell.
    c. A cell can use energy from other sources to increase its order and complexity.
    d. All biological processes result in the cell's uptake of free energy.
    e. Biological processes, such as biochemical reactions, have a tendency to increase entropy in the cell.

2. Which of the following statements regarding the first law of thermodynamics is *false*?
    a. The total amount of energy in a closed system remains the same before and after a transformation.
    b. Energy is neither created nor destroyed during the energy transformation shown.
    c. The energy transformations shown results in a reduction in the total energy within the system.
    d. The energy transformation shown can involve the conversion of potential energy into kinetic energy.
    e. The diagram shown represents a closed system.

3. At the end of a chemical reaction involving glucose 1-phosphate and glucose 6-phosphate, there is 95% glucose 6-phosphate and 5% glucose 1-phosphate. In general, the equilibrium concentrations of reactants and products in a chemical reaction is determined by
    a. the activation energy.
    b. enthalpy changes.
    c. ΔG.
    d. whether the reaction is spontaneous or not.
    e. whether the reaction is exergonic or endergonic.

4. Suppose you adjust the pH so that all of the charges on the phosphate groups in the molecule adenosine triphosphate (ATP) are neutralized. What would be the expected effect of this change?
    a. Increases the amount of free energy in the ATP molecule
    b. Decreases the amount of free energy in the ATP molecule
    c. Converts ATP into ADP
    d. Phosphate groups are released from the molecule
    e. There would be no change

5. Complete the following sentence: ADP + $P_i \rightarrow$ ATP is an __________ reaction, ATP $\rightarrow$ ADP + P_i is an __________ reaction, and the conversion of ADP + P_i to ATP __________ energy.
    a. exergonic, endergonic, releases
    b. exergonic, endergonic, requires
    c. endergonic, exergonic, releases
    d. endergonic, exergonic, requires
    e. exergonic, endergonic, does not involve

6. Which of the following statements about enzyme-catalyzed reactions is *false*?
    a. Most enzymes are highly specific for only one or several chemically related substrates.
    b. The enzyme must be chemically unchanged after the reaction is complete.
    c. The free energy of the enzyme-substrate complex (ES) is less than the free energy of transition state species in an uncatalyzed reaction.
    d. The enzyme cannot chemically interact with the substrate, but can change shape (induced fit).
    e. Some enzymes can interact with molecules at two different sites.

TEST FILE

★ *Indicates a **difficult** question*
■ *Indicates a **conceptual** question*

7 Cellular Pathways That Harvest Chemical Energy

Fill in the Blank

1. The breakdown (hydrolysis) of ATP, which yields ADP and an inorganic phosphate ion, is an exergonic reaction yielding approximately **12** kcal of free energy per mole of ATP under biological conditions.

2. Part of the unusually large amount of free energy that results from the hydrolysis of ATP derives from the large number of **negative** charges near each other on neighboring phosphate groups.

3. Oxidation and **reduction** occur together.

4. Thanks to its ability to carry electrons and free energy, **NAD** is the major and universal energy intermediary in cells.

5. Pyruvate is **oxidized** to form acetate.

6. The chemiosmotic formation of ATP during the operation of the respiratory chain is called **oxidative phosphorylation**.

7. The loss of an electron by a ferrous ion (Fe^{2+}) to give a ferric ion (Fe^{3+}) is called **oxidation**.

8. In a redox reaction, the reactant that becomes oxidized is called a **reducing agent**.

9. A chemical reaction resulting in the transfer of electrons or hydrogen atoms is called a **redox** reaction.

10. The pathway for the oxidation of glucose to pyruvate is called **glycolysis**.

11. The conversion of glucose to lactic acid is a form of **fermentation**.

12. Fatty acids must be converted to **acetyl CoA** before they can be used for respiratory ATP production.

13. During alcoholic fermentation, NAD^+ is regenerated by the reduction of acetaldehyde to **ethanol**.

★14. NAD is an abbreviation for **nicotinamide adenine dinucleotide**.

15. An enzyme that transfers phosphorous from ATP to another protein is called a **kinase**.

16. The earliest forms of life probably used **glycolysis** to generate ATP.

Multiple Choice

1. ATP is
   a. a short-term, energy-storage compound.
   b. the cell's principle compound for energy transfers.
   c. synthesized within mitochondria.
   d. the molecule all living cells rely on to do work.
   e. All of the above

2. When a molecule loses hydrogen atoms (not hydrogen ions), it becomes
   a. reduced.
   b. oxidized.
   c. redoxed.
   d. hydrogenated.
   e. hydrolyzed.

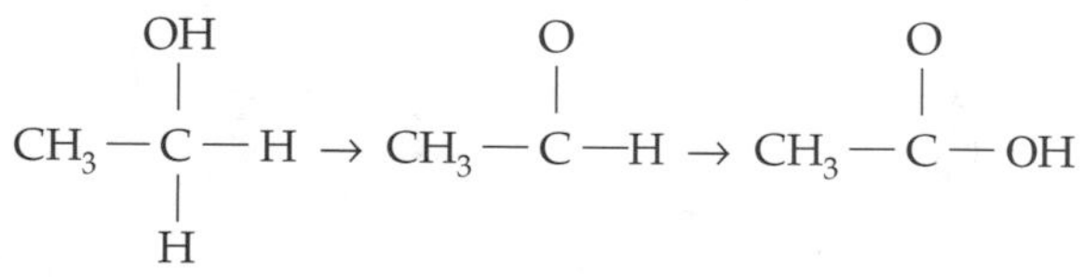

★3. In the diagram shown above, Reaction 1 is __________, and Reaction 2 is __________.
   a. oxidation, oxidation
   b. oxidation, reduction
   c. reduction, reduction
   d. reduction, oxidation
   e. oxiduction, oxiduction

4. The end product of glycolysis is
   a. pyruvate.
   b. the starting point for the citric acid cycle.
   c. the starting point for the fermentation pathway.
   d. a and b
   e. a, b, and c

5. The ΔG of glucose oxidation is __________ kcal/mol.
   a. close to 686
   b. –686
   c. 53
   d. –53
   e. –12

6. In the conversion of succinate to fumarate, hydrogen atoms are transferred to FAD. The conversion of succinate and FAD to fumarate and $FADH_2$ is an example of
   a. hydrolysis.
   b. an allosteric reaction.
   c. a metabolic pathway.
   d. an aerobic reaction.
   e. a redox reaction.

7. The oxidation of malate to oxaloacetate is coupled to the reduction of NAD+ to NADH + H^+. NAD^+ is a(n)
   a. reducing agent.
   b. oxidizing agent.
   c. vitamin.
   d. phosphate ester.
   e. phosphorylating agent.

8. During respiration, NADH donates two electrons to ubiquinone. When this happens, ubiquinone is
   a. reduced.
   b. oxidized.
   c. phosphorylated.
   d. aerobic.
   e. hydrolyzed.

9. Which of the following oxidizes other compounds by gaining free energy and hydrogen atoms and reduces other compounds by giving up free energy and hydrogen atoms?
   a. Vitamins
   b. Adenine
   c. ATP
   d. NAD
   e. Riboflavin

10. Isocitrate dehydrogenase is an enzyme of the citric acid cycle. Where in the cell is this enzyme located?
   a. In the thylakoids
   b. In the cytoplasm
   c. In the chloroplast
   d. In the mitochondrial matrix
   e. In the plasma membrane

11. In the first reaction of glycolysis, glucose receives a phosphate group from ATP. This reaction is
   a. respiration.
   b. a redox reaction.
   c. exergonic.
   d. endergonic.
   e. fermentation.

*▪12. The reduction of pyruvate to lactic acid during fermentation allows glycolysis to continue in the absence of oxygen. Why?
   a. Water is formed during this reaction.
   b. This reaction is a kinase reaction.
   c. This reaction is coupled to the oxidation of NADH to NAD^+.
   d. This reaction is coupled to the formation of ATP.
   e. This reaction is coupled to the reduction of NAD^+ to NADH.

13. During glycolysis, for each mole of glucose oxidized to pyruvate
   a. 6 moles of ATP are produced.
   b. 4 moles of ATP are used, and 2 moles of ATP are produced.
   c. 2 moles of ATP are used, and 4 moles of ATP are produced.
   d. 2 moles of NAD^+ are produced.
   e. no ATP is produced.

*▪14. In steps 6 through 10 of glycolysis, the conversion of one mole of glyceraldehyde 3-phosphate to pyruvate yields 2 moles of ATP. But the oxidation of glucose to pyruvate produces a total of 4 moles of ATP. Where do the remaining 2 moles of ATP come from?
   a. One mole of glucose gives 2 moles of glyceraldehyde 3-phosphate.
   b. Two moles of ATP are used during the conversion of glucose to glyceraldehyde 3-phosphate.
   c. Glycolysis produces 2 moles of NADH.
   d. Fermentation of pyruvate to lactic acid yields 2 moles of ATP.
   e. Fermentation of pyruvate to lactic acid yields 2 moles of NAD^+.

15. For glycolysis to continue, all cells require
   a. a respiratory chain.
   b. oxygen.
   c. mitochondria.
   d. chloroplasts.
   e. NAD^+.

*16. The free energy released during the oxidation of glyceraldehyde 3-phosphate to 1,3 bisphosphoglycerate is
   a. used to oxidize NADH.
   b. lost as heat.
   c. used to synthesize ATP.
   d. used to reduce NAD^+.
   e. stored in lactic acid.

17. The oxidation of pyruvate to carbon dioxide is called
   a. fermentation.
   b. the citric acid cycle.
   c. glycolysis.
   d. oxidative phosphorylation.
   e. the respiratory chain.

*18. During the energy-priming portion of glycolysis, the phosphates from ATP molecules are
   a. added to the first and sixth carbons.
   b. added to the second and fourth carbons.
   c. wasted, as an energy investment.
   d. used to make pyruvate.
   e. used to make lactate.

19. Which of the following is produced during the citric acid cycle?
   a. FAD
   b. Pyruvate
   c. Reduced electron carriers
   d. Lactic acid
   e. Water

▪20. Some of the free energy released by oxidation of pyruvate to acetate is stored in acetyl CoA. How does acetyl CoA store free energy?
a. **Acetyl CoA has a higher free energy than acetate.**
b. Acetyl CoA is an electron carrier.
c. Acetyl CoA is a phosphate donor.
d. Acetate + CoA → acetyl CoA is an exergonic reaction.
e. Reduction of acetyl CoA is coupled to ATP synthesis.

21. The oxidizing agent at the end of the respiratory chain is
a. **O^2.**
b. NAD^+.
c. ATP.
d. FAD.
e. ubiquinone.

22. During the citric acid cycle, energy stored in acetyl CoA is used to
a. create a proton gradient.
b. drive the reaction ADP + P_i – ATP.
c. reduce NAD+ to NADH.
d. **drive the reaction oxaloacetate → citric acid.**
e. reduce FAD to $FADH_2$.

23. During the citric acid cycle, oxidative steps are coupled to
a. oxidative phosphorylation.
b. the oxidation of water.
c. the oxidation of electron carriers.
d. the hydrolysis of ATP.
e. **the reduction of electron carriers.**

*24. Animals breathe in air containing oxygen and breathe out air with less oxygen and more carbon dioxide. The carbon dioxide comes from
a. the carbon from hydrocarbons and the oxygen from the air.
b. **the citric acid cycle.**
c. glycolysis.
d. waste products.
e. All of the above

*▪25. The drug 2,4-dinitrophenol (DNP) destroys the proton gradient across the inner mitochondrial membrane. What would you expect to be the effect of incubating isolated mitochondria in a solution of DNP?
a. Oxygen would no longer be reduced to water.
b. **No ATP would be made during transport of electrons down the respiratory chain.**
c. Mitochondria would show a burst of increased ATP synthesis.
d. Glycolysis would stop.
e. Mitochondria would switch from glycolysis to fermentation.

26. Electron transport within NADH-Q reductase, cytochrome reductase, and cytochrome oxidase can be coupled to proton transport from the mitochondrial matrix to the space between the inner and outer mitochondrial membranes because those protein complexes are located
a. in the mitochondrial matrix.
b. **within the inner mitochondrial membrane.**
c. in the space between the inner and outer mitochondrial membranes.
d. in the cytoplasm.
e. loosely attached to the inner mitochondrial membrane.

27. According to the chemiosmotic theory, the energy for the synthesis of ATP during the flow of electrons down the respiratory chain is provided directly by the
a. hydrolysis of GTP.
b. reduction of NAD^+.
c. **diffusion of protons.**
d. reduction of FAD.
e. hydrolysis of ATP.

28. In the absence of oxygen, cells capable of fermentation
a. accumulate glucose.
b. no longer produce ATP.
c. accumulate pyruvate.
d. oxidize FAD.
e. **oxidize NADH to produce NAD^+.**

29. For bacteria to continue growing rapidly when they are shifted from an environment containing oxygen to an anaerobic environment, they must
a. increase the rate of the citric acid cycle.
b. produce more ATP per mole of glucose during glycolysis.
c. produce ATP during the oxidation of NADH.
d. increase the rate of transport of electrons down the respiratory chain.
e. **increase the rate of the glycolytic reactions.**

30. In alcoholic fermentation, NAD^+ is produced during the
a. oxidation of pyruvate to acetyl CoA.
b. reduction of pyruvate to lactic acid.
c. **reduction of acetaldehyde to ethanol.**
d. hydrolysis of ATP to ADP.
e. oxidation of glucose.

31. During the fermentation of 1 molecule of glucose, the net production of ATP is
a. 1 molecule.
b. **2 molecules.**
c. 3 molecules.
d. 6 molecules.
e. 8 molecules.

32. The portion of aerobic respiration that produces the most ATP per mole of glucose is
a. **oxidative phosphorylation.**
b. the citric acid cycle.
c. glycolysis.
d. lactic acid fermentation.
e. alcoholic fermentation.

*33. More free energy is released during the citric acid cycle than during glycolysis, but only 1 mole of ATP is produced for each mole of acetyl CoA that enters the cycle. What happens to most of the remaining free energy that is produced during the citric acid cycle?
a. It is used to synthesize GTP.
b. It is used to reduce electron carriers.
c. It is lost as heat.
d. It is used to reduce pyruvate.
e. It is converted to kinetic energy.

*34. Animals inhale air-containing oxygen and exhale air with less oxygen and more carbon dioxide. Later, the oxygen from the air will mostly be found in
a. the carbon dioxide that is exhaled.
b. water.
c. organic molecules.
d. ethanol.
e. lactate.

35. When the supply of acetyl CoA being produced exceeds the demands of the citric acid cycle, some of the acetyl CoA is diverted to the synthesis of
a. pyruvate.
b. NAD.
c. proteins.
d. fatty acids.
e. lactic acid.

36. The site of oxygen utilization is the
a. nucleus.
b. chloroplasts.
c. endoplasmic reticulum.
d. mitochondria.
e. cytosol.

37. Before starch can be used for respiratory ATP production, it must be hydrolyzed to
a. pyruvate.
b. fatty acids.
c. amino acids.
d. glucose.
e. oxaloacetate.

38. When yeast cells are switched from aerobic to anaerobic growth conditions, the rate of glycolysis increases. The rate of glycolysis is regulated by the concentration of _________ in the cell.
a. ATP
b. acetyl CoA
c. oxaloacetate
d. FAD
e. protein

39. When acetyl CoA builds up in the cell, it increases the activity of the enzyme that synthesizes oxaloacetate from pyruvate and carbon dioxide. Acetyl CoA is acting as a(n)
a. electron carrier.
b. substrate.
c. allosteric activator.
d. acetate donor.
e. proton pump.

40. In yeast, if the citric acid cycle is shut down because of a lack of oxygen, glycolysis will probably
a. shut down.
b. increase.
c. produce more ATP per mole of glucose.
d. produce more NADH per mole of glucose.
e. produce acetyl CoA for fatty acid synthesis.

41. The function of NAD^+ is to
a. cause the release of energy to adjacent cells when energy is needed in aerobic conditions.
b. hasten the release of energy when the cell has been deprived of oxygen.
c. carry hydrogen atoms and free energy from compounds being oxidized and to give hydrogen atoms and free energy to compounds being reduced.
d. block the release of energy to adjacent cells.
e. None of the above

42. The end result of glycolysis is that
a. 38 ATP molecules are created.
b. 8 NAD molecules get reduced.
c. 2 molecules of pyruvate are formed.
d. 1 molecule of glucose is converted to lactic acid.
e. None of the above

*43. The results of the first five reactions of the glycolytic pathway are
a. adding phosphates, modifying sugars, and forming G3P.
b. oxidative steps, proton pumping, and reactions with oxygen.
c. oxidation of pyruvate and formation of acetyl CoA.
d. removal of hydrogen and protons from glucose.
e. None of the above

*44. For the citric acid cycle to proceed, it is necessary for
a. pyruvate to bind to oxaloacetate.
b. carbon dioxide to bind to oxaloacetate.
c. an acetyl group to bind to oxaloacetate.
d. water to be oxidized.
e. None of the above

45. Which of the following events occurs in the respiratory chain?
a. Carbon dioxide is released.
b. Carbon dioxide is reduced.
c. Cytochromes, FADH, and NADH are oxidized.
d. Only NAD^+ is reduced.
e. None of the above

46. The respiratory chain contains three large enzymes: NADH-Q reductase, cytochrome reductase, and cytochrome oxidase. The function of these enzymes is to
a. allow electrons to be transported.
b. ensure the production of water and oxygen.
c. regulate the passage of water through the chain.
d. oxidize NADH.
e. None of the above

47. Oxygen is used by
a. glycolysis.
b. the citric acid cycle.

c. the electron transport chain.
d. substrate level phosphorylation.
e. ATP sythase.

48. Pyruvate oxidation generates
a. acetate.
b. NADH + H^+ from NAD^+.
c. a change in free energy.
d. a capture of energy.
e. All of the above

*49. Water is a by-product of cellular respiration. The water is produced as a result of
a. combining carbon dioxide with protons.
b. the conversion of pyruvate to acetyl CoA.
c. the degradation of glucose to pyruvate.
d. the reduction of oxygen at the end of the electron transport chain.
e. None of the above

50. The formation of ethanol from pyruvate is an example of
a. an exergonic reaction.
b. providing an extra source of energy from glycolysis.
c. a fermentation process that takes place in the absence of oxygen.
d. cellular respiration.
e. None of the above

51. Regardless of the electron or hydrogen acceptor employed, fermentation always produces
a. AMP.
b. DNA.
c. P_i.
d. NAD^+.
e. None of the above

▪52. Yeast cells tend to create anaerobic conditions because they use oxygen faster than it can be replaced by diffusion through the cell membrane. For this reason, yeast cells
a. exhibit a red pigment.
b. exhibit a green pigment.
c. die.
d. produce ethanol.
e. None of the above

53. In human cells (muscle cells), the fermentation process produces
a. lactic acid.
b. 12 moles of ATP.
c. pyruvic acid.
d. an excessive amount of energy.
e. None of the above

54. If a cell has an abundant supply of ATP, acetyl CoA may be used
a. to enhance fermentation.
b. to enhance oxidative metabolism.
c. for fatty acid synthesis.
d. to convert glucose to glycogen.
e. None of the above

55. In order for glucose to be used as an energy source, it is necessary that
a. glucose be formed from fructose.
b. glucose phosphate be formed from fructose phosphate.
c. glucose be degraded to carbon dioxide.
d. 2 ATP molecules be invested in the system.
e. None of the above

56. Many species derive their energy from fermentation. The function of fermentation is to
a. reduce NAD^+.
b. oxidize carbon dioxide.
c. oxidize NADH + H^+, ensuring a continued supply of ATP.
d. produce acetyl CoA.
e. None of the above

▪57. Proteins must be consumed to provide energy to the brain when glucose and starch stores are depleted because
a. the brain must have glucose.
b. fatty acids cannot get to the brain.
c. amino acids are needed for gluconeogenesis.
d. All of the above
e. None of the above

58. The chemiosmotic generation of ATP is driven by
a. osmotic movement of water into an area of high solute concentration.
b. the addition of protons to ADP and phosphate via enzymes.
c. oxidative phosphorylation.
d. a difference in H^+ concentration of both sides of a membrane.
e. None of the above

59. When a cell needs energy, cellular respiration is regulated by the citric acid cycle enzyme isocitrate dehydrogenase, which is stimulated by
a. H^+.
b. heat.
c. oxygen.
d. ADP.
e. None of the above

60. Substrate-level phosphorylation is transfer of
a. a phosphate to a protein.
b. a phosphate to a substrate.
c. a phosphate to an ADP.
d. an ATP to a protein.
e. a phosphate from ATP to a substrate.

61. The proton-motive force is
a. the force a proton has on the motive.
b. the proton concentration gradient and electric charge difference.
c. a metabolic pathway.
d. ATP synthase.
e. a redox reaction.

62. Most ATP made in our bodies is made
    a. by glycolysis.
    b. in the citric acid cycle.
    c. using ATP synthase.
    d. from photosynthesis.
    e. burning fat.

63. Brown fat is "burned" to raise the body temperature of some small mammals by
    a. thermogenin uncoupling respiration.
    b. increasing the rate of glycolysis.
    c. shivering.
    d. hydrogen ions leaking across the cell's plasma membrane.
    e. cytochrome reductase.

64. In eukaryotic cells, some glycolytic enzymes are found to be associated with the
    a. mitochondrial membrane.
    b. mitochondrial matrix.
    c. nucleus.
    d. cytoskeleton.
    e. None of the above

65. ATP can be used to drive nonspontaneous reactions because
    a. nonspontaneous reactions are exergonic.
    b. the breakdown of ATP to ADP is exergonic.
    c. the breakdown of ATP to ADP is endergonic.
    d. when ATP is broken down to ADP, P_i is released.
    e. ADP possesses more free energy than ATP.

Study Guide Questions

1. Which of the following cellular metabolic processes can occur in the presence *or* the absence of oxygen?
    a. The citric acid cycle
    b. Electron transport
    c. Glycolysis
    d. Fermentation

2. Which of the following statements regarding glycolysis is/are true?
    a. A 6-C sugar is broken down to a 3-C product.
    b. Two ATP molecules are consumed.
    c. A net sum of two ATP molecules is generated.
    d. All of the above

3. During which process is most ATP generated in the cell?
    a. Glycolysis
    b. The citric acid cycle
    c. Electron transport coupled with chemiosmosis
    d. Fermentation

4. The main purpose of the electron transport chain is to
    a. cycle NADH + H^+ back to NAD^+.
    b. use the intermediates from the citric acid cycle.
    c. break down pyruvate.
    d. All of the above

5. Cellular respiration is allosterically controlled. Which of the following act as inhibitors at the various control points?
    a. ATP
    b. NADH + H^+
    c. a and b
    d. None of the above

6. Which of the following best describes the role of the mitochondrial membrane?
    a. The membrane acts as an anchor for the membrane-associated enzymes of cellular respiration.
    b. The membrane allows for the establishment of a proton-motive force.
    c. a and b
    d. None of the above

7. In a redox reaction between G3P and NAD^+ yielding BPG and NADH + H^+, _________ is oxidized and _________ is reduced.
    a. G3P, NAD^+
    b. BPG, NADH + H^+
    c. G3P, NADH + H^+
    d. NAD^+, NADH + H^+

8. Which of the following are true regarding redox reactions?
    a. Oxidizing agents accept electrons.
    b. A molecule that accepts electrons is said to be reduced.
    c. Redox reactions involve electron transfers.
    d. All of the above

9. Cyanide poisoning inhibits aerobic respiration at cytochrome *c* oxidase. Which of the following are results of cyanide poisoning at the cellular level?
    a. Oxygen is not reduced to water.
    b. ATP cannot be synthesized because electron transport is never completed.
    c. Cells (with the exception of brain cells) must switch to anaerobic respiration.
    d. All of the above

10. Not all cells can switch between aerobic and anaerobic respiration. If a cell cannot switch to aerobic respiration in the absence of oxygen, what happens?
    a. The cell concentrates oxygen to continue aerobic respiration.
    b. The cell dies.
    c. The cell uses fermentation.
    d. None of the above

11. The main purpose of cellular respiration is to
    a. convert energy stored in the chemical bonds of glucose to an energy form that the cell can use.
    b. convert potential to kinetic energy.
    c. convert kinetic to potential energy.
    d. create energy in the cell.

12. Which of the following statements concerning the synthesis of ATP is true?
    a. ATP synthesis cannot occur without the presence of ATP synthase.
    b. The proton-motive force is the establishment of a

charge and concentration gradient across the mitochondrial membrane.
c. The proton-motive force is necessary to drive protons back across the membrane through channels established by the ATP synthase channel protein.
d. All of the above

End of Chapter Questions

1. In the presence of oxygen, the energy-rich product(s) of glycolysis in human cells
   a. is carbon dioxide.
   b. is ADP.
   c. are ATP and NADH.
   d. are NAD and FAD.
   e. are lactic acid and ATP.

2. If oxygen is not present, yeast cells break down glucose to
   a. carbon dioxide and water.
   b. carbon dioxide and lactic acid.
   c. carbon dioxide and ethanol.
   d. NADH and ATP.
   e. lactic acid and ATP.

3. NAD^+ is
   a. a type of organelle.
   b. a protein.
   c. an oxidizing agent.
   d. a reducing agent.
   e. formed only under aerobic conditions.

4. Glycolysis
   a. takes place in the mitochondrion.
   b. has many steps, each catalyzed by a different enzyme.
   c. occurs in plants only at night.
   d. is the same thing as fermentation.
   e. occurs only in the presence of oxygen.

5. Fermentation
   a. takes place in the mitochondrion.
   b. takes place in all animal cells.
   c. does not require O_2.
   d. requires lactic acid.
   e. prevents glycolysis.

6. The terminal electron acceptor of electron transport is
   a. CO_2
   b. $C_6H_{12}O_6$.
   c. pyruvate.
   d. H_2O.
   e. O_2 .

7. The citric acid cycle
   a. takes place in the mitochondrion.
   b. produces no ATP.
   c. has no connection with the respiratory chain.
   d. is the same thing as fermentation.
   e. reduces two molecules of NAD^+ for every glucose molecule processed.

8. Which statement about the respiratory chain is true?
   a. It operates only in the absence of oxygen.
   b. It uses O_2 as an oxidizing agent.
   c. It leads to the production of ADP.
   d. It regenerates reducing agents for glycolysis and the citric acid cycle.
   e. It operates simultaneously with fermentation.

9. Which statement about the chemiosmotic mechanism is *not* true?
   a. Protons are pumped across a membrane.
   b. Protons return through the membrane by way of a channel protein.
   c. ATP is required for the protons to return.
   d. Proton pumping is associated with the respiratory chain.
   e. The membrane in question is the inner mitochondrial membrane.

10. In terms of efficiency at converting the energy of glucose to energy in ATP,
    a. aerobic glycolysis is better than the citric acid cycle.
    b. aerobic respiration is about 40% efficient.
    c. anaerobic conditions are much less efficient than aerobic.
    d. eukaryotic cells are more efficient than prokaryotic cells.
    e. the electron transport chain is not necessary for high efficiency.

Student Web Site Self-Quiz Questions

1. Which of the following statements about cellular energy-harvesting pathways is *false*?
   a. Autotrophs can produce their own food but must obtain energy from it by glycolysis and cellular respiration.
   b. Fermentation usually only occurs under anaerobic conditions.
   c. The products of fermentation can be used to feed glycolysis under anaerobic conditions.
   d. The amount of usable energy produced from aerobic metabolism is greater than the energy produced from anaerobic metabolism.
   e. Pyruvate oxidation can only occur under aerobic conditions.

2. In a mitochondrion, which of the following matchings of pathways and cellular locations is *not* correct?
   a. Pyruvate oxidation → cytosol
   b. Glycolysis → cytosol
   c. Citric acid cycle → matrix
   d. Respiratory chain → inner mitochondrial membrane
   e. Fermentation → cytosol

3. Which of the following statements about glycolysis is *false*?
   a. All living cells can perform glycolysis.
   b. The inputs to glycolysis include glucose, ADP, and NAD^+; outputs include ATP, water, NADH + H^+, and pyruvate.
   c. Glycolysis can operate in the presence or absence of oxygen.
   d. Per molecule of glucose, glycolysis produces 4 ATP, 2 $NADH^+$ H+, and 2 pyruvate.
   e. Glycolysis may be followed by fermentation in some cells.

4. Which of the following events does *not* take place during pyruvate oxidation?
   a. A three-carbon compound is oxidized to a two-carbon compound.
   b. The pyruvate dehydrogenase complex is attached to the plasma membrane.
   c. NAD^+ is reduced.
   d. CO_2 is given off.
   e. Coenzyme A (CoA) is added to acetate.

5. Per molecule of glucose, which one of the following is *not* the correct number of product molecules from the citric acid cycle?
   a. 6 CO_2
   b. 2 GTP
   c. 2 $FADH_2$
   d. 6 NADH + H^+
   e. 2 ATP

6. Which one of the following statements about the respiratory chain is *false*?
   a. The integral membrane proteins (NADH-Q reductase, cytochrome reductase, and cytochrome oxidase) are involved in active transport.
   b. Ubiquinone (Q) differs from the other respiratory chain components in that it is a nonprotein.
   c. All hydrogens enter the respiratory chain via NADH-Q reductase.
   d. Electrons enter the respiratory chain via NADH-Q reductase and ultimately end up in water molecules.
   e. ATP synthase uses proton-motive force to generate ATP.

7. Which one of the following would be expected to *increase* the amount of ATP produced through operation of the respiratory chain and oxidative phosphorylation?
   a. Increasing the permeability of the inner mitochondrial membrane to protons (H^+)
   b. Decreasing the pH in the space between the inner and outer mitochondrial membranes
   c. Application of respiratory uncouplers, such as thermogenin
   d. Reducing the supply of NADH + H^+
   e. Increasing the H+ concentration in the mitochondrial matrix

8. Which of the following statements regarding metabolic pathways is *false*?
   a. Many of the steps of glycolysis can run in reverse.
   b. Starch must be hydrolyzed before it can enter glycolysis.
   c. After fats are digested, glycerol enters the glycolytic pathway.
   d. After fats are digested, fatty acids can no longer participate in cellular respiration.
   e. Citric acid intermediates are used in the synthesis of many important cellular components.

9. Which of the following events would not be expected to take place when a cell starts to produce more NADH + H^+ than the respiratory chain can use?
   a. Activation of isocitrate dehydrogenase by NAD^+
   b. Inhibition of phosphofructokinase by citrate
   c. Inhibition of pyruvate dehydrogenase by citrate
   d. Activation of the enzyme converting acetyl CoA into fatty acids
   e. Inhibition of pyruvate dehydrogenase by NADH + H^+

TEST FILE

★ *Indicates a* ***difficult*** *question*

■ *Indicates a* ***conceptual*** *question*

8 Photosynthesis: Energy from the Sun

Fill in the Blank

1. During the light reactions of photosynthesis, the synthesis of **ATP** is coupled to the diffusion of protons.

2. Atmospheric carbon dioxide enters plant leaves through openings called **stomata**.

*3. In the 1800s, the summarized chemical reaction for photosynthesis was incorrect because it left out **water** as a product.

4. In noncyclic photophosphorylation, the electrons for the reduction of chlorophyll in photosystem II come from **water**.

5. When **C_4 plants** are exposed to light and CO_2, four-carbon compounds are the first carbon-containing products.

6. During the process of **photorespiration,** rubisco catalyzes the reaction of RuBP with oxygen.

7. A group of scientists led by **Calvin** conducted experiments demonstrating that RuBP is the CO_2 acceptor in the dark reactions of photosynthesis.

8. When isolated chloroplasts are shifted from a low pH solution to a more alkaline solution, ATP synthesis occurs, even in the absence of light. This experiment was used to support the **chemiosmotic** mechanism of ATP formation in chloroplasts.

9. During cyclic photophosphorylation, the energy of photons is converted to the chemical energy of the product, **ATP**.

*10. In C_3 plants, the Calvin–Benson cycle occurs in the chloroplasts of **mesophyll** cells, but in C_4 plants the cycle occurs in the **bundle sheath** cells.

11. In both photosynthesis and respiration, **ATP** synthesis is coupled to the diffusion of protons across a membrane.

12. The dark reactions take place in the **light**.

*13. NADP is the abbreviation for **nicotinamide adenine dinucleotide phosphate.**

14. The Calvin-Benson cycle is sometimes called the **dark reactions**.

*15. The oxygen found in Earth's atmosphere is generated from the photosystem **II** of noncyclic photophosphorylation.

*16. **Plastoquinone** instead of $NADP^+$ receives the electron from ferredoxin during cyclic photophosphorylation.

17. The most abundant enzyme in the biosphere is **rubisco (or RuBP carboxylase)**.

Multiple Choice

*1. To obtain free energy, heterotrophs require a source of
   a. partially reduced organic compounds.
   b. light energy.
   c. kinetic energy.
   d. carbon dioxide.
   e. water.

■2. How do red and blue light differ from one another?
   a. They differ in intensity.
   b. They have a different number of photons in each quantum.
   c. Their wavelengths are different.
   d. They differ in duration.
   e. Red is radiant, and blue is electromagnetic.

■3. The wavelength of X rays is shorter than the wavelength of infrared rays. Which of the following is true?
   a. X rays have more energy per photon than infrared rays.
   b. X rays have a smaller value for Planck's constant than infrared waves.
   c. X rays have a different absorption spectrum than infrared waves.
   d. X rays and infrared waves have the same frequency.
   e. Infrared waves are in the ground state, and X rays are in the excited state.

4. Brightness of light is a function of
   a. the wavelengths of the light.
   b. the photons striking an area per unit time.
   c. wattage.
   d. the color of an environment.
   e. All of the above

5. A molecule has an absorption spectrum that shows maximum absorption within the wavelengths of visible light. This molecule is
   a. a reducing agent.
   b. a quantum.
   c. a photon.
   d. electromagnetic radiation.
   e. a pigment.

6. When white light strikes a blue pigment, blue light is
   a. reduced.
   b. absorbed.
   c. converted to chemical energy.
   d. reflected or transmitted.
   e. used to synthesize ATP.

7. A graph that plots the rate at which carbon dioxide is converted to glucose versus the wavelength of light illuminating a leaf is called
   a. a Planck equation.
   b. an absorption spectrum.
   c. enzyme kinetics.
   d. an electromagnetic spectrum.
   e. an action spectrum.

8. The photosynthetic pigment chlorophyll *a* absorbs
   a. infrared light.
   b. red and blue light.
   c. X rays.
   d. gamma rays.
   e. white light.

9. Accessory pigments
   a. play no role in photosynthesis.
   b. transfer energy from chlorophyll to the electron transport chain.
   c. absorb only in the red wavelengths.
   d. allow plants to harvest visible light of intermediate wavelengths.
   e. transfer electrons to NADP.

▪10. Why are the absorption spectrum of chlorophyll *a* and the action spectrum of photosynthesis *not* identical?
   a. Accessory pigments contribute energy to drive photosynthesis.
   b. Chlorophyll *a* absorbs both red and blue light.
   c. Chlorophyll *a* reflects green light.
   d. Different wavelengths of light have different energies.
   e. Chlorophyll *a* can be activated by absorbing a photon of light.

▪11. Excited chlorophyll is a better reducing agent than ground-state chlorophyll because
   a. excited chlorophyll can release energy by fluorescence.
   b. one of the electrons is farther from the atomic nucleus.
   c. excited chlorophyll is reduced by NADPH.
   d. excited chlorophyll absorbs light in the green wavelengths.
   e. only excited chlorophyll contains a porphyrin ring.

*12. In cyclic photophosphorylation, chlorophyll is reduced by
   a. NADPH.
   b. a chemiosmotic mechanism.
   c. plastoquinone.
   d. ATP.
   e. hydrogens liberated by the splitting of a water molecule.

13. The energy difference between an electron excited by a photon and its ground state is
   a. lost.
   b. less than the photon.
   c. the same as what was in the photon.
   d. greater than the energy of the photon.
   e. greater when released by the electron.

*14. The precise moment when light energy is captured in chemical energy is when
   a. light shines on chlorophyll.
   b. water is hydrolyzed.
   c. chlorophyll is oxidized.
   d. chlorophyll is reduced.
   e. the CO_2 from air is captured in a sugar.

15. Free energy is released in cyclic photophosphorylation
   a. by the formation of ATP.
   b. during the excitation of chlorophyll.
   c. during the fluorescence of chlorophyll.
   d. during each of the redox reactions of the electron transport chain.
   e. when electrons are transferred from photosystem I to photosystem II.

16. During cyclic photophosphorylation, the energy to produce ATP is provided by
   a. heat.
   b. NADPH.
   c. ground state chlorophyll.
   d. the redox reactions of the electron transport chain.
   e. the Calvin–Benson cycle.

17. In noncyclic photophosphorylation, water is used for the
   a. hydrolysis of ATP.
   b. excitation of chlorophyll.
   c. reduction of chlorophyll.
   d. oxidation of NADPH.
   e. synthesis of chlorophyll.

18. Photophosphorylation provides the Calvin–Benson cycle with
   a. protons and electrons.
   b. CO_2 and glucose.
   c. water and photons.
   d. light and chlorophyll.
   e. ATP and NADPH.

19. In noncyclic photophosphorylation, the chlorophyll in photosystem I is reduced by
   a. water.
   b. an electron from the transport chain of photosystem II.
   c. two photons of light.

d. NADPH.
e. ATP.

20. The enzyme ATP synthase couples the synthesis of ATP to
a. **the diffusion of protons.**
b. the reduction of $NADP^+$.
c. the excitation of chlorophyll.
d. the reduction of chlorophyll.
e. carbon dioxide fixation.

21–22. A suspension of algae is incubated in a flask in the presence of both light and CO_2. When transferred to the dark, you block the reduction of 3-phosphoglycerate to glyceraldehyde 3-phosphate.

*▪21. Why does this reaction stop when the algae are placed in the dark?
a. It requires carbon dioxide.
b. It is an exergonic reaction.
c. **It requires ATP and NADPH + H^+.**
d. It requires oxygen.
e. Chlorophyll is not synthesized in the dark.

*▪22. When the reduction of 3-phosphoglycerate to glyceraldehyde 3-phosphate is blocked, the concentration of ribulose bisphosphate declines. Why?
a. Ribulose bisphosphate is synthesized from glyceraldehyde 3-phosphate.
b. Glyceraldehyde 3-phosphate is converted to glucose.
c. Ribulose bisphosphate is used to synthesize 3-phosphoglycerate.
d. **a and c**
e. a and b

23. The enzyme rubisco is found in
a. **chloroplasts.**
b. mitochondria.
c. the cytoplasm.
d. the nucleus.
e. yeast.

24. During carbon dioxide fixation, carbon dioxide combines with
a. NADPH.
b. 3PG.
c. G3P.
d. water.
e. **ribulose bisphosphate.**

25. __________ mole(s) of carbon dioxide must enter the Calvin–Benson cycle for the synthesis of one mole of glucose.
a. One
b. Two
c. Three
d. **Six**
e. Twelve

▪26. In the experiments conducted to identify the first compound that is formed during CO_2 fixation, why weren't all of the compounds of the Calvin–Benson cycle labeled with ^{14}C?
a. **The cells were incubated with $^{14}CO_2$ for a very short time.**
b. The cells were incubated in the dark.
c. The cells were incubated in the absence of CO_2.
d. The cells were incubated at very low concentrations of CO_2.
e. The cells were incubated in the absence of water.

27. The NADPH required for the reduction of 3PG to G3P comes from
a. the dark reactions.
b. **the light reactions.**
c. the synthesis of ATP.
d. the Calvin–Benson cycle.
e. oxidative phosphorylation.

28. In C_4 plants, the function of the four-carbon compound that is synthesized in the mesophyll cells is to
a. reduce $NADP^+$.
b. combine with CO_2 to produce glucose.
c. **carry CO_2 to the bundle sheath cells.**
d. drive the synthesis of ATP.
e. close the stomata.

*29. In the mesophyll layer of C_4 plants, light energy is used to synthesize
a. O_2.
b. G3P.
c. 3PG from RuBP.
d. CO_2.
e. **oxaloacetate.**

30. In C_4 plants, starch grains are found in the chloroplasts of
a. the thylakoids.
b. mesophyll cells.
c. the intracellular space.
d. the stroma.
e. **bundle sheath cells.**

31. During photorespiration, rubisco uses __________ as a substrate.
a. carbon dioxide
b. **oxygen**
c. glyceraldehyde 3-phosphate
d. 3-phosphoglycerate
e. NADPH

32. Photorespiration starts
a. in mitochondria.
b. **in chloroplasts.**
c. only in C_4 plants.
d. in the microbodies.
e. in the cytoplasm.

*33. In plants, the reactions of glycolysis occur
a. only in C_3 plants.
b. in the mitochondria.
c. in the chloroplasts.
d. only in the presence of light.
e. **in the cytosol.**

34. In both photosynthesis and respiration, protons are pumped across a membrane during
a. electron transport.
b. photolysis.
c. CO_2 fixation.
d. reduction of oxygen.
e. glycolysis.

35. The enzyme PEP carboxylase
a. can trap CO_2 even at relatively low CO_2 concentrations.
b. catalyzes the synthesis of RuBP.
c. catalyzes the synthesis of 3PG.
d. is found in the chloroplasts of bundle sheath cells.
e. couples the synthesis of ATP to the diffusion of protons.

*36. The function of photorespiration is
a. CO_2 fixation.
b. unknown.
c. ATP production.
d. to generate a proton gradient.
e. to synthesize glucose.

37. The NADPH required for CO_2 fixation is formed
a. by the reduction of oxygen.
b. by the hydrolysis of ATP.
c. during the light reactions.
d. only in C_4 plants.
e. in the mitochondria.

38. The O_2 gas produced during photosynthesis is derived from
a. carbon dioxide.
b. glucose.
c. water.
d. carbon monoxide.
e. bicarbonate ions.

▪39. Photosynthesis and respiration have which of the following in common?
a. In eukaryotes, both processes reside in specialized organelles.
b. ATP synthesis in both processes relies on the chemiosmotic mechanism.
c. Both use electron transport.
d. Both require light.
e. a, b, and c

▪40. The expression "We are creatures of the chloroplasts" means that
a. all life possesses chloroplasts.
b. all life depends ultimately on photosynthesis.
c. chloroplasts are models of all organelles.
d. a and c
e. None of the above

41. When a photon interacts with molecules such as those within chloroplasts, the photons may
a. bounce off the molecules, having no effect.
b. pass through the molecules, having no effect.
c. be absorbed by the molecules.
d. a and c
e. a, b, and c

*42. Which of the following scientific tools "cracked" the Calvin–Benson cycle?
a. Isotopes
b. Paper chromatography
c. Crystallography
d. Centrifugation and electron microscopy
e. a and b

43. In noncyclic photophosphorylation, electrons from which source replenish chlorophyll molecules that have given up electrons?
a. Carbon dioxide
b. Water
c. NADPH + H^+
d. Gaseous oxygen
e. None of the above

44. In our past evolutionary history, noncyclic photophosphorylation by cyanobacteria, algae, and plants poured enough gas into the atmosphere to make possible the evolution of cellular respiration. That gas was
a. oxygen.
b. methane.
c. hydrogen.
d. nitrogen.
e. chlorine.

*45. After World War II, the Berkeley group (Calvin et al.) made progress in understanding the dark reactions when these researchers devised a way to grow dense quantities of algae in a flask that looked like a
a. long test tube.
b. condenser.
c. pipette.
d. lollipop.
e. culture dish.

46. Which of the following biological groups is dependent on photosynthesis for its survival?
a. Vertebrates
b. Class Mammalia
c. Fish
d. a and b
e. a, b, and c

47. Photosynthesis is divided into two main phases, the first of which is a series of reactions that requires the absorption of photons. This phase is referred to as the
a. reduction phase.
b. dark reactions phase.
c. carbon fixation phase.
d. light reactions phase, or photophosphorylation.
e. None of the above

48. Based on its electronic structure, a molecule has a range of photon energies. These photon energies are known as the molecule's
a. chloroplasts.
b. light reactions.
c. absorption spectrum.
d. photosystems I and II.
e. None of the above

*49. The energy to hydrolyze water comes from
a. **oxidized chlorophyll.**
b. reduced chlorophyll.
c. the proton gradient.
d. ATP.
e. $NADPH+H^+$.

50. Heterotrophs are dependent on autotrophs for their food supply. Autotrophs can make their own food by
a. feeding on bacteria and converting the nutrients into usable energy.
b. **using light and an inorganic carbon source to make reduced carbon compounds.**
c. synthesizing it from water and carbon dioxide.
d. All of the above
e. None of the above

51. Photosynthesis is the process that uses light energy to extract H atoms from which of the following sources?
a. Glucose
b. Chlorophyll
c. Carbon dioxide
d. **Water**
e. None of the above

52. Which of the following occurs during the dark reactions of photosynthesis?
a. Water is converted into hydrogen and water.
b. **Carbon dioxide is converted into sugars.**
c. Chlorophyll acts as an enzyme only in the dark.
d. Nothing occurs; the plant rests in the dark.
e. None of the above

53. In bright light, the pH of the thylakoid space
a. **can become more acidic.**
b. can become more alkaline.
c. stays the same; the pH of thylakoid spaces never changes.
d. can become neutral.
e. None of the above

▪54. When a photon is absorbed by a molecule, what exactly happens to the photon?
a. It loses its ability to generate any energy.
b. **It raises the molecule from a ground state of low energy to an excited state.**
c. The exact relationship of the photon to the molecule is not clearly understood.
d. It causes a change in the velocity of the wavelengths.
e. None of the above

55. A range of energy that cannot be seen by human eyes, but has slightly more energy per photon than visible light, is
a. adaptive radiation.
b. solar radiation.
c. gamma radiation.
d. **ultraviolet radiation.**
e. None of the above

56. The main photosynthetic pigments in plants are
a. chlorophyll *s* and chlorophyll *a*.
b. chlorophyll *x* and chlorophyll *y*.
c. retinal pigment and accessory pigment.
d. **chlorophyll *a* and chlorophyll *b*.**
e. None of the above

57. If one were to compare long wavelengths to short wavelengths, it would be evident that short-wavelength photons have
a. an insignificant amount of energy.
b. **more energy.**
c. energy not available to plant cells.
d. a ladder of energy.
e. None of the above

58. The chemiosmotic hypothesis states that the energy for the production of ATP comes from
a. the transfer of phosphate from intermediate compounds.
b. the reduction of NADP.
c. **a proton gradient set up across the thylakoid membrane.**
d. the oxidation of carbon dioxide.
e. None of the above

*59. When carbon dioxide is added to RuBP, the first stable product synthesized is
a. pyruvate.
b. glyceraldehyde 3-phosphate.
c. **phosphoglycerate.**
d. ATP.
e. None of the above

60. Photosynthesis takes place in plants only in the light. Respiration takes place
a. in the dark only.
b. in the light only.
c. in all organisms except for plants.
d. **both with and without light.**
e. None of the above

61. The revised, balanced equation for the generation of sugar from sunlight, water, and carbon dioxide is
a. $6\ CO_2 + 6\ H_2O \rightarrow C_6H_{12}O_6 + O_2$.
b. $\mathbf{6\ CO_2 + 12\ H_2O \rightarrow C_6H_{12}O_6 + 6\ O_2 + 6\ H_2O}$.
c. $6\ CO_2 + 6\ H_2O \rightarrow C_6H_{12}O_6 + 6\ O_2$.
d. $12\ CO_2 + 12\ H_2O \rightarrow 2\ C_6H_{12}O_6 + 2\ O_2$.
e. None of the above

*62. After removal of the carbon, the oxygen in the carbon dioxide ends up in
a. the air.
b. the sugar.
c. water molecules.
d. the air and sugar.
e. **the sugar and water molecules.**

63. The net energy outcome of cyclic photophosphorylation is
a. **ATP.**
b. ATP and NADH.
c. ATP and NADPH.
d. sugar.
e. NADPH.

*64. The air contains around __________ carbon dioxide.
a. 36%
b. 3.6%
c. 0.36%
d. 0.036%
e. 0.0036%

65. The concentration of oxygen in the air is
a. 21%.
b. 2.1%.
c. 0.21%.
d. 0.02%.
e. It depends on where in the world you are.

66. In C_4 plants, CO_2 is first fixed into a compound called
a. pyruvate.
b. glucose.
c. oxaloacetate.
d. ribulose bisphosphate.
e. 3-phosphoglycerate.

67. In cacti, CO_2 is stored for use in the Calvin–Benson cycle
a. in the stems, roots, and leaves.
b. during the evening.
c. in glucose molecules.
d. in the stroma.
e. None of the above

*68. Plants classified as CAM store CO_2
a. by making oxaloacetate.
b. by making PEP carboxylase.
c. in malic acid.
d. in crassulacean acid.
e. a and c.

69. The energy a plant acquires from the sun
a. can all be found in the plant.
b. is mostly consumed by the plant.
c. is mostly consumed by heterotrophs.
d. accumulates in the soil.
e. All of the above are sometimes true.

Study Guide Questions

1. The main purpose of photosynthesis is to
a. consume carbon dioxide.
b. produce ATP.
c. convert light energy to chemical energy.
d. produce starch.

2. Which of the following best represent the components that are necessary for photosynthesis to take place?
a. Mitochondria, accessory pigments, visible light, water, and carbon dioxide
b. Chloroplasts, accessory pigments, visible light, water, and carbon dioxide
c. Mitochondria, chlorophyll, visible light, water, and oxygen
d. Chloroplasts, chlorophyll, visible light, water, and carbon dioxide

3. Chlorophyll is suited for the capture of light energy because
a. certain wavelengths of light raise it to an excited state.
b. in its excited state chlorophyll gives off electrons.
c. chlorophyll's structure allows it to attach to thylakoid membranes.
d. All of the above

4. Plants give off oxygen because
a. oxygen results from the incorporation of carbon dioxide into sugars.
b. plants do not respire since they photosynthesize.
c. water is the initial proton donor, leaving oxygen as a photosynthetic by-product.
d. All of the above

5. Cyclic and noncyclic electron flow is used in plants to
a. meet the ATP demands of the Calvin–Benson cycle.
b. avoid producing excess NADPH + H^+.
c. balance ATP and NADPH + H^+ ratios in the chloroplast.
d. All of the above

6. Which of the following statements concerning the light reactions of photosynthesis are true?
a. Photosystem I can operate independently of Photosystem II.
b. Photosystems I and II are activated by different wavelengths of light.
c. Photosystems I and II transfer electrons and create proton gradients across the thylakoid membrane.
d. All of the above

7. ATP is produced during the light reactions via
a. carbon dioxide fixation.
b. chemiosmosis.
c. reduction of water.
d. All of the above

8. Because of the properties of chlorophyll, plants need adequate __________ light to grow properly.
a. green
b. blue and red
c. infrared
d. ultraviolet

9. Which of the statements concerning the Calvin–Benson cycle are *false*.
a. Light energy is not required for the cycle to proceed.
b. Carbon dioxide is assimilated into sugars.
c. RuBP is regenerated.
d. None of the above

10. Which of the following statements concerning rubisco are *true*.
a. Rubisco is an enzyme.
b. Rubisco catalyzes both the beginning steps of photorespiration and the Calvin–Benson cycle.
c. Rubisco is the most abundant protein on earth.
d. All of the above

11. Which of the following begins the Calvin–Benson cycle and is the commitment step that results in the entire pathway being carried out?
    a. 3PG is reduced to G3P using ATP and $NADPH + H^+$.
    b. The regeneration of RuBP.
    c. Carbon dioxide and RuBP join forming 3PG.
    d. As a cycle, it can start at any point.

12. The Calvin–Benson Cycle results in the production of
    a. glucose.
    b. starch.
    c. rubisco.
    d. G3P.

13. Which of the following statements regarding photorespiration are true?
    a. Photorespiration is a metabolically expensive pathway.
    b. Photorespiration is avoided when carbon dioxide is abundant.
    c. Photorespiration results in a loss of usable carbon dioxide.
    d. All of the above

14. The fixation of carbon dioxide by PEP carboxylase functions to
    a. concentrate carbon dioxide for use in photosynthetic cells.
    b. allow plants to close stomata without having photorespiration occur.
    c. allow plants to photosynthesize in the dark.
    d. a and b

15. CAM plants differ from C_4 plants in that
    a. carbon dioxide is stored as malic acid.
    b. photosynthesis can occur at night in these plants.
    c. their stomata close during periods that favor photorespiration.
    d. they use PEP carboxylase to fix carbon dioxide.

16. Which of the following statements are true regarding the relationship between photosynthesis and cellular respiration in plants?
    a. Photosynthesis occurs in specialized photosynthetic cells.
    b. Cellular respiration occurs in specialized respiratory cells.
    c. Cellular respiration and photosynthesis can occur in the same cell.
    d. a and c

End of Chapter Questions

1. The overall reaction $CO_2 + H_2O + NADPH + ATP \rightarrow$ sugar + NADP + ADP + P
    a. describes the light reactions of photosynthesis.
    b. is exergonic.
    c. occurs only at night
    d. requires many enzymes.
    e. occurs in the thylakoid membranes

2. The "light" reactions of photosynthesis
    a. convert light energy into chemical energy.
    b. occur in all plant and animal cells.
    c. occur only at night.
    d. occur in both chloroplasts and mitochondria.
    e. produce ADP from ATP.

3. Which statement about chlorophylls is *not* true?
    a. They absorb light near both ends of the visible spectrum.
    b. They can accept energy from other pigments, such as carotenoids.
    c. Excited chlorophyll can either reduce another substance or fluoresce.
    d. Excited chlorophyll is an oxidizing agent.
    e. They contain magnesium.

4. In cyclic electron flow,
    a. oxygen gas is released.
    b. ATP is formed.
    c. water donates electrons and protons.
    d. $NADPH + H^+$ forms.
    e. CO_2 reacts with RuBP.

5. The chemical source of molecular oxygen (O_2) in photosynthesis is
    a. CO_2.
    b. $C_6H_{12}O_6$.
    c. H_2O.
    d. NADP.
    e. chlorophyll.

6. The NADPH required for CO_2 fixation is formed
    a. by the reduction of oxygen.
    b. by the hydrolysis of ATP.
    c. during the light reactions.
    d. only in C_4 plants.
    e. in the mitochondria.

7. Which statement about the Calvin–Benson cycle is *not* true?
    a. CO_2 reacts with RuBP to form 3PG.
    b. RuBP forms by the metabolism of 3PG.
    c. ATP and $NADPH + H^+$ form when 3PG is reduced.
    d. The concentration of 3PG rises if the light is switched off.
    e. Rubisco catalyzes the reaction of CO_2 and RuBP.

8. In C_4 photosynthesis,
    a. 3PG is the first product of CO_2 fixation.
    b. rubisco catalyzes the first step in the pathway.
    c. four-carbon acids are formed by PEP carboxylase in bundle sheath cells.
    d. photosynthesis continues at lower CO_2 levels than in C_3 plants.
    e. CO_2 released from RuBP is transferred to PEP.

9. Photosynthesis in green plants occurs only during the day. Respiration in plants occurs
    a. only at night.
    b. only when there is enough ATP.
    c. only during the day.
    d. all the time.
    e. in the chloroplasts after photosynthesis.

10. Photorespiration
    a. takes place only in C_4 plants.
    b. includes reactions carried out in peroxisomes.
    c. increases the yield of photosynthesis.
    d. is catalyzed by PEP carboxylase.
    e. is independent of light intensity.

Student Web Site Self-Quiz Questions

1. Which of the following statements about photosynthetic reactants and products is *false*?
   a. Most of the water in a plant comes from the soil.
   b. Carbon atoms in the organic compounds of plants come from atmospheric CO_2.
   c. Oxygen produced during photosynthesis comes from water.
   d. During photosynthesis, light is directly required for both ATP production and carbon fixation.
   e. CO_2 enters and O_2 exits the leaf through pores called stomata.

2. Which one of the following statements about the two pathways of photosynthesis is *false*?
   a. The light reactions are known as photophosphorylation.
   b. The dark reactions are known as the Calvin-Benson cycle.
   c. The products of photophosphorylation are ATP and NADPH + H^+.
   d. The Calvin-Benson cycle takes place in the cytosol; photophosphorylation occurs within chloroplasts.
   e. Both pathways are stopped in the absence of light.

3. Which of the following materials are *not* cycled between the Calvin-Benson cycle and the light reactions?
   a. NADPH + H^+
   b. ADP
   c. CO_2
   d. O_2
   e. Both C and D

4. Which one of the following statements about absorption of photons by pigments (below) is *false*?
   a. The pigment molecule gains an amount of energy equal to that of the absorbed photon.
   b. The color of a pigment is determined by the wavelengths of light that are reflected or transmitted.
   c. An absorption spectrum shows the biological effectiveness of light as a function of wavelength.
   d. During fluorescence, a molecule in an excited state returns to its ground state with the emission of a photon.
   e. Pigments are compounds that absorb light in the region of about 400-700 nm.

5. Which one of the following statements about photosynthetic pigments is *false*?
   a. Magnesium is found in both chlorophyll *a* and *b*.
   b. Accessory pigments in leaves absorb mostly wavelengths intermediate between red and blue.
   c. The antenna system consists of chlorophyll and accessory pigments combined into a functional unit.
   d. The reaction center of the antenna is the molecule that absorbs the shortest light wavelength.
   e. The reaction center acts as a reducing agent and passes its electrons to an oxidizing agent.

6. Which one of the following choices is a *false* statement about the activities associated with photosystems I and II in noncyclic photophosphorylation?
   a. Water is oxidized in photosystem II, but not photosystem I.
   b. Photons are required to activate both photosystems.
   c. The reaction centers (P_{680} and P_{700}) differ in photosystems I and II.
   d. Sustained ATP production requires the activities of both photosystem I and photosystem II.
   e. Production of NADPH + H^+ is associated with photosystem II, but not photosystem I.

7. Which choice completes the following sentence to create a true statement about noncyclic photophosphorylation? Electrons lost from photosystem I are replaced with electrons from __________ electrons lost from photosystem II are replaced with electrons from __________.
   a. photosystem II, water
   b. water, photosystem I
   c. photosystem II, NADPH + H^+
   d. NADPH + H^+, photosystem I
   e. ferrodoxin, water

8. Which one of the following statements correctly describes cyclic photophosphorylation?
   a. Cyclic photophosphorylation has both photosystems I and II.
   b. Cyclic photophosphorylation produces neither ATP nor NADPH + H^+.
   c. Electrons are cycled in cyclic photophosphorylation.
   d. Water is the ultimate source of electrons in cyclic photophosphorylation.
   e. For every two pairs of electrons moved through cyclic photophosphorylation, one O_2 is produced.

9. Two groups of isolated thylakoids are placed in an acidic bathing solution so that H^+ diffuses into the thylakoids. They are then transferred to a basic bathing solution, and one group is placed in the light, while the other group is kept in the dark. Select below the choice that describes what you expect each group of thylakoids to produce.

	In Light	In Dark
a.	ATP only	Nothing
b.	**ATP, O_2**	**ATP only**
c.	ATP, O_2, glucose	ATP, O_2
d.	ATP, O_2	O_2
e.	ATP, O_2, glucose	ATP

10. Which of the following statements about the localization of the photosynthetic reactions within the chloroplast is true?
    a. The Calvin-Benson cycle occurs in the thylakoid interior.
    b. Photosystems I and II are built into the outer chloroplast membrane.
    c. Protons diffuse through ATP synthases from the outside to the inside of the thylakoids.
    d. Carbon fixation occurs in the stroma.
    e. ATP is formed in the interior of the thylakoids.

11. Which of the following statements about the Calvin-Benson cycle is *false*?
    a. The reactions of the Calvin-Benson cycle can occur in the light or the dark.
    b. Carbon fixation produces a compound called 3-phosphoglycerate (3PG).
    c. Ribulose bisphosphate (RuBP) and CO_2 form an unstable six-carbon compound.
    d. The production of carbohydrates from CO_2 yields additional ATP and NADPH.
    e. RuBP carboxylase or rubisco catalyzes the carbon fixation reaction of the Calvin-Benson cycle.

12. Which of the following statements about photorespiration is *false?*
    a. Rubisco can function as both a carboxylase and an oxygenase.
    b. The carboxylase function of rubisco is favored at higher temperatures.
    c. The reaction of RuBP with O_2 forms a two-carbon compound called glycolate.
    d. Glycolate formed in chloroplasts diffuses to peroxisomes, where it is oxidized by O_2.
    e. The photorespiratory pathway appears to reverse photosynthesis because it releases CO_2.

13. Select the choice below that correctly characterizes the conditions of a CAM plant at midnight in terms of its stomata and major active carbon-fixation enzyme.

	Stomata	Enzyme
a.	Open	Rubisco
b.	**Open**	**PEP carboxylase**
c.	Open	Rubisco and PEP carboxylase
d.	Closed	Rubisco
e.	Closed	PEP carboxylase

TEST FILE

★ *Indicates a **difficult** question*
■ *Indicates a **conceptual** question*

9 Chromosomes, the Cell Cycle, and Cell Division

Fill in the Blank

1. When a DNA molecule doubles, a chromosome is then comprised of two joined **chromatids**.

★2. Prokaryotic DNA molecules are packaged by **basic** proteins, which associate with DNA.

3. In general, the division of the cell, called **cytokinesis**, follows immediately upon mitosis.

★4. The bacteria have a short sequence called ***ori*** where DNA synthesis begins.

★5. The bacteria have a short sequence called ***ter*** where DNA synthesis ends.

6. During prophase I of meiosis a unique event occurs that results in the formation of recombinant chromosomes. This event is termed **crossing over**.

7. The structure present during mitosis that is composed of two identical DNA molecules complexed with proteins and joined at the centromere is called a **chromosome**.

8. The stage of the cell cycle during which DNA replicates is called the **S phase**.

9. **Fertilization** is the fusion of two gametes.

10. During prometaphase, microtubules associate with specialized structures in the centromere regions called **kinetochores**.

11. Occasionally, a homologous chromosome pair will fail to separate during anaphase I of meiosis. The resulting cells, one of which lacks a copy of this chromosome while the other contains both members of the homologous pair, are called **aneuploid** cells.

12. During a process known as **translocation**, a piece of one chromosome can break off and become joined to a different chromosome.

13. The orderly distribution of genetic information occurs in prokaryotic cells by a process known as **fission**.

14. A cell with three homologous sets of chromosomes is called a **triploid** cell.

15. The heritable information of the cell is **DNA**.

16. The process that ensures the genetic information is passed on to a cell's daughter cells is **mitosis**.

17. The process that ensures that only one of each pair of chromosomes is included in a gamete is **meiosis**.

18. The main role of nucleosomes in eukaryotic cells is to **package** the DNA.

19. The G2 phase always follows **S**.

20. The G in G1 and G2 is short for **gap**.

21. The chromatin **condenses** during prophase.

22. The milestone event that defines entry into prometaphase is **loss of the nuclear envelop**.

23. In plants a **cell plate** forms at the equatorial region of the cell.

24. The cell plate is derived from the **Golgi apparatus** of the cell.

25. The "invisible thread" that pinches cells into two during cell division is made of **actin** and **myosin**.

26. A zygote usually has **two** copies of each chromosome.

27. A **karyotype** is the number, form, and type of chromosomes found in a cell, usually similar within species and different between species.

28. A **homolog** is one of a pair of chromosomes having the same overall genetic composition and sequence.

29. Nondisjunction causes the production of **aneuploid** cells.

30. Down syndrome can be caused by an extra chromosome **21**.

★31. During meiosis in a tetraploid or a triploid plant, more problems are expected in **triploid** plants.

Multiple Choice

*1. During prokaryotic cell division, two chromosomes separate from each other and distribute into the daughter cells by
a. attachments to microtubules.
b. a mitotic spindle.
c. repellent forces.
d. attachment to separating membrane regions.
e. All of the above

2. Prokaryotic cells take at least __________ minute(s) to divide.
a. 1
b. 2
c. 20
d. 200
e. 1000

3. Bacteria typically have __________ while eukaryotes have __________.
a. one chromosome that is circular, many that are linear
b. several chromosomes that are circular, many that are linear
c. one chromosome that is linear, many that are circular
d. two chromosome that are circular, eight that are linear
e. None of the above

4. Chromosomes contain large amounts of five different interacting proteins, all of which are known as
a. pentanes.
b. hexosamines.
c. histones.
d. protein hormones.
e. histamines.

5. The appropriate decisions to enter the S phase and the M phase of the cell cycle depend on a pair of biochemicals called
a. actin and myosin.
b. Cdk's and cyclin.
c. ligand and receptor.
d. MSH and MSH-receptor.
e. ATP and ATPase.

6. Sex and reproduction
a. are not the same thing.
b. are identical.
c. both involve combining genetic material from two cells.
d. both involve the formation of new individuals.
e. None of the above

7. The molecules that make up a chromosome are
a. DNA and RNA.
b. DNA and proteins.
c. proteins and lipids.
d. nucleotides and nucleosides.
e. proteins and phospholipids.

▪8. During mitosis and meiosis the chromatin compacts. Which of the following processes is made easier by this compaction?
a. The orderly distribution of genetic material to two new nuclei
b. The replication of the DNA
c. Exposing the genetic information on the DNA
d. The unwinding of DNA from around the histones
e. The disappearance of the nuclear membrane

9. The basic structure of chromatin has sometimes been referred to as beads on a string of DNA. These beads are called
a. chromosomes.
b. chromatids.
c. supercoils.
d. interphases.
e. nucleosomes.

10. DNA replication occurs
a. during both mitosis and meiosis.
b. only during mitosis.
c. only during meiosis.
d. during the S phase.
e. during G2.

*11. Around __________ base pairs wrap around each core particle in a __________.
a. 1000, solenoid
b. 20000, chromosome
c. 146, nucleosome
d. 1000, chromosome
e. 20,000, solenoid

▪12. When cyclin binds Cdk,
a the cell transitions from G2 to S.
b. kinase activation occurs.
c. chromosomes condense.
d. the cell quickly enters M phase.
e. the cell begins apoptosis.

13. Mature nerve cells are incapable of cell division. These cells are probably in
a. G1.
b. the S phase.
c. G2.
d. mitosis.
e. meiosis.

14. A cell cycle consists of
a. mitosis and meiosis.
b. G1, the S phase, and G2.
c. prophase, metaphase, anaphase, and telophase.
d. interphase and mitosis.
e. meiosis and fertilization.

*15. Evidence in yeast suggest that the maturation promoting factor of sea urchins was
a. a cyclin.
b. MFP.
c. an S nuclease.
d. a Cdk.
e. a Cdk/cyclin phosphatase.

16. The cells of the intestinal epithelium are continually dividing, replacing dead cells lost from the surface of the intestinal lining. If you examined a population of intestinal epithelial cells under the microscope, most of the cells would
    a. be in meiosis.
    b. be in mitosis.
    c. be in interphase.
    d. have condensed chromatin.
    e. b and d

*17. DNA damage by UV radiation causes the synthesis of
    a. p53.
    b. DNA.
    c. Cdk.
    d. cyclin.
    e. p21.

*18. The uncondensed length of human DNA found in chromosomes is __________, whereas a typical cell's is 10 µm in length.
    a. 5 µm
    b. 2 µm
    c. 2 meters
    d. 20 meters
    e. 2.54 inches

19. The products of mitosis are
    a. one nucleus containing twice as much DNA as the parent nucleus.
    b. two genetically identical cells.
    c. four nuclei containing half as much DNA as the parent nucleus.
    d. four genetically identical nuclei.
    e. two genetically identical nuclei.

20. The mitotic spindle is composed of
    a. chromosomes.
    b. chromatids.
    c. microtubules.
    d. chromatin.
    e. centrosomes.

21. Centrosomes are
    a. constricted regions of phase chromosomes.
    b. regions where microtubules polymerize.
    c. the central region of the same cell.
    d. the region where the membrane constricts during cytokinesis.
    e. part of cilia.

22. When dividing cells are examined under a light microscope, chromosomes first become visible during
    a. interphase.
    b. the S phase.
    c. prophase.
    d. G1.
    e. G2.

23. The structures that line up the chromosomes on the equatorial plate during metaphase are called
    a. asters.
    b. polar and kinetochore microtubules.
    c. centrosomes.
    d. centrioles.
    e. histones.

24. The microtubules of the mitotic spindle attach to a specialized structure in the centromere region of each chromosome, called the
    a. kinetochore.
    b. nucleosome.
    c. equatorial plate.
    d. aster.
    e. centrosome.

25. After the centromeres separate during mitosis, the chromatids, now called __________, move toward opposite poles of the spindle.
    a. centrosomes
    b. kinetochores
    c. half spindles
    d. asters
    e. daughter chromosomes

26. In plant cells, cytokinesis is accomplished by the formation of a(n)
    a. aster.
    b. membrane furrow.
    c. equatorial plate.
    d. cell plate.
    e. spindle.

*27. The distribution of mitochondria between the daughter cells during cytokinesis
    a. is random.
    b. is directed by the mitotic spindle.
    c. is directed by the centrioles.
    d. results in all of the mitochondria remaining in the parent cell.
    e. occurs only during meiosis.

28. Genetically diverse offspring result from
    a. mitosis.
    b. cloning.
    c. sexual reproduction.
    d. cytokinesis.
    e. anaphase.

29. During asexual reproduction, the genetic material of the parent is passed on to the offspring by
    a. homologous pairing.
    b. meiosis and fertilization.
    c. mitosis and cytokinesis.
    d. karyotyping.
    e. chiasmata.

*30. Meiosis can occur
    a. in all organisms.
    b. only when an organism is diploid.
    c. only in multicellular organisms.
    d. only in haploid organisms.
    e. only in single-celled organisms.

31. All zygotes are
    a. multicellular.
    b. diploid.
    c. animals.
    d. clones.
    e. gametes.

32. In all sexually reproducing organisms, the diploid phase of the life cycle begins at
    a. spore formation.
    b. gamete formation.
    c. meiosis.
    d. mitosis.
    e. fertilization.

33. The members of a homologous pair of chromosomes
    a. are identical in size and appearance.
    b. contain identical genetic information.
    c. separate to opposite poles of the cell during mitosis.
    d. are found only in haploid cells.
    e. are present only after the S phase.

34. The diagnosis of Down syndrome is made by examining the individual's
    a. spores.
    b. karyotype.
    c. chromatin.
    d. nucleosomes.
    e. kinetochores.

35. During meiosis, the sister chromatids separate during
    a. anaphase II.
    b. anaphase I.
    c. the S phase.
    d. synapsis.
    e. telophase II.

36. The exchange of genetic material between chromatids on homologous chromosomes occurs during
    a. interphase.
    b. mitosis and meiosis.
    c. prophase I.
    d. anaphase I.
    e. anaphase II.

37. At the end of the first meiotic division, each chromosome consists of
    a. chiasmata.
    b. a homologous chromosome pair.
    c. four copies of each DNA molecule.
    d. two chromatids.
    e. a pair of polar microtubules.

38. The four haploid nuclei found at the end of meiosis differ from one another in their exact genetic composition. Some of this difference is the result of
    a. cytokinesis.
    b. replication of DNA during the S phase.
    c. separation of sister chromatids at anaphase II.
    d. spindle formation.
    e. crossing over during prophase I.

39. Diploid cells of the fruit fly *Drosophila* have 10 chromosomes. How many chromosomes does a *Drosophila* gamete have?
    a. 1
    b. 2
    c. 5
    d. 10
    e. 20

40. During meiosis I in humans, one of the daughter cells receives
    a. only maternal chromosomes.
    b. a mixture of maternal and paternal chromosomes.
    c. the same number of chromosomes as a diploid cell.
    d. a sister chromatid from each chromosome.
    e. one-fourth the amount of DNA in the parent nucleus.

▪41. The fact that most monosomies and trisomies are lethal to human embryos illustrates
    a. the importance of the orderly distribution of genetic material during meiosis.
    b. the exchange of genetic information during crossing over.
    c. the advantage of sexual reproduction to the survival of a population.
    d. that each chromosome contains a single molecule of DNA.
    e. that meiosis results in the formation of haploid gametes.

*42. A triploid nucleus cannot undergo meiosis because
    a. the DNA cannot replicate.
    b. not all of the chromosomes can form homologous pairs.
    c. the sister chromatids cannot separate.
    d. cytokinesis cannot occur.
    e. a cell plate cannot form.

*43. A bacterial cell gives rise to two genetically identical daughter cells by a process known as
    a. nondisjunction.
    b. mitosis.
    c. meiosis.
    d. fission.
    e. fertilization.

44. Which of the following is *not* part of sexual reproduction?
    a. The segregation of homologous chromosomes during gamete formation
    b. The fusion of sister chromatids during fertilization
    c. The fusion of haploid cells from a diploid zygote
    d. The reduction in chromosome number during meiosis
    e. The production of genetically distinct gametes during meiosis

45. Chromatin condenses to form discrete, visible chromosomes
    a. early in G1.
    b. during S.
    c. during telophase.
    d. during prophase.
    e. at the end of cytokinesis.

46. Chromosomes "decondense" into diffuse chromatin
   a. at the end of telophase.
   b. at the beginning of prophase.
   c. at the end of interphase.
   d. at the end of metaphase.
   e. only in dying cells.

47. Microtubules that form the mitotic spindle tend to originate from or terminate in
   a. centromeres and telomeres.
   b. euchromatin.
   c. centrioles and telomeres.
   d. the nuclear envelope.
   e. centrioles and centromeres (kinetochores).

▪48. Asexual reproduction produces genetically identical individuals because
   a. chromosomes do not have to replicate.
   b. it involves chromosome replication without cytokinesis.
   c. no meiosis or fertilization takes place.
   d. the only cell division that occurs is meiosis.
   e. the mitotic spindle prevents nondisjunction.

▪49. One difference between mitosis and meiosis I is that
   a. homologous chromosome pairs synapse during mitosis.
   b. chromosomes do not replicate in the interphase preceding meiosis.
   c. homologous chromosome pairs synapse during meiosis but not mitosis.
   d. spindles composed of microtubules are not required during meiosis.
   e. sister chromatids separate during meiosis but not mitosis.

50. Genetic recombination occurs during
   a. prophase of meiosis I.
   b. interphase preceding meiosis II.
   c. mitotic telophase.
   d. fertilization.
   e. formation of somatic cells.

51. The number of chromosomes is reduced to half during
   a. anaphase of mitosis and meiosis.
   b. meiosis II.
   c. meiosis I.
   d. fertilization.
   e. interphase.

▪52. The total DNA content of each daughter cell is reduced during meiosis because
   a. chromosomes do not replicate during the interphase preceding meiosis I.
   b. chromosomes do not replicate between meiosis I and II.
   c. half of the chromosomes from each gamete are lost during fertilization.
   d. sister chromatids separate during anaphase of meiosis I.
   e. chromosome arms are lost during crossing over.

★▪53. Many chromosome abnormalities (trisomies and monosomies) are not observed in the human population because
   a. they are lethal and cause spontaneous abortion of the embryo early in development.
   b. all trisomies and monosomies are lethal early in childhood.
   c. meiosis distributes chromosomes to daughter cells with great precision.
   d. they are so difficult to count.
   e. the human meiotic spindle is self-correcting.

★54. In a haploid organism, most mitosis occurs
   a. after fertilization and before meiosis.
   b. after meiosis and before fertilization.
   c. between meiosis I and II.
   d. during G1.
   e. in diploid cells.

★55. The event in the cell division process that clearly involves microfilaments rather than microtubules is
   a. chromosome separation during anaphase.
   b. movement of chromosomes to the metaphase plate.
   c. chromosome condensation during prophase.
   d. disappearance of the nuclear envelope during prophase.
   e. cytokinesis in animal cells.

56. Trisomies and monosomies can result from accidents that occur during meiosis called
   a. nondisjunctions.
   b. inversions.
   c. reciprocal translocations.
   d. recombinations.
   e. acrocentricities.

▪57. Chromosome number is reduced during meiosis because the process consists of
   a. two cell divisions without any chromosome replication.
   b. a single cell division without any chromosome replication.
   c. two cell divisions in which half of the chromosomes are destroyed.
   d. two cell divisions and only a single round of chromosome replication.
   e. four cell divisions with no chromosome replication.

58. Which of the following phases of the cell cycle is *not* part of interphase?
   a. M
   b. S
   c. G1
   d. G2
   e. G0

59. During mitotic anaphase, chromatids migrate
   a. from the poles of the cell toward the metaphase plate.
   b. from the metaphase plate toward the poles.
   c. toward the nuclear envelope.
   d. along with their sister chromatids toward one pole.
   e. along with the other member of the homologous pair toward the metaphase plate.

60. Which of the following does *not* occur during mitotic prophase?
a. Disappearance of the nuclear envelope
b. Chromosome condensation
c. Migration of centrioles toward the cell poles
d. Synapsis of homologous chromosomes
e. Formation of the mitotic spindle

*61. Human males have _________ different types of chromosomes.
a. 23
b. 24
c. 46
d. 48
e. 92

62. Which of the following is *not* true of homologous chromosome pairs?
a. They come from only one of the individual's parents.
b. They usually contain slightly different versions of the same genetic information.
c. They segregate from each other during meiosis I.
d. They synapse during meiosis I.
e. Each contains two sister chromatids at the beginning of meiosis I.

63. Which of the following is *not* true of sister chromatids?
a. They arise by replication during S phase.
b. They segregate from each other during each mitotic anaphase.
c. They usually contain identical versions of the same genetic information.
d. They segregate from each other during meiosis I.
e. They are joined during prophase and metaphase at their common centromere.

64. Chromatin consists of
a. DNA and histones.
b. DNA, histones, and many other nonhistone proteins.
c. mostly RNA and DNA.
d. RNA, DNA, and nonhistone proteins.
e. DNA alone.

65. The DNA of an eukaryotic cell is
a. double stranded.
b. single stranded.
c. circular.
d. complex inverted.
e. conservative.

66. Nucleosomes are made of
a. RNA.
b. lipid.
c. histones.
d. chromatin.
e. All of the above

67. A cell that is postreproductive will remain in
a. S.
b. G1.
c. G2.
d. M.
e. prophase.

68. The kinetochore is made of
a. protein.
b. microtubules.
c. DNA.
d. a and b.
e. All of the above

69. The milestone that defines metaphase is when the chromosomes
a. separate.
b. come together.
c. are at opposite poles.
d. line up.
e. cross over.

70. The milestone that defines anaphase is when the chromosomes
a. separate.
b. come together.
c. are at opposite poles.
d. line up.
e. cross over.

71. The milestone that defines telophase is when the chromosomes
a. separate.
b. come together.
c. are at opposite poles.
d. line up.
e. cross over.

72. The major drawback of asexual reproduction is
a. that it takes too little time.
b. the variation generated.
c. it prevents change.
d. it requires cytokinesis.
e. the lack of variation among the progeny.

73. Haploid means
a. the genes are arranged haphazardly.
b. containing only one copy of each chromosome.
c. the process of meiosis.
d. half again the number of chromosomes.
e. All of the above

*74. A reduction step during meiosis is important because
a. it returns the chromosome number to normal before fertilization.
b. there is a mechanism for this.
c. only one copy of each chromosome is necessary.
d. otherwise chromosome copies would double each fertilization.
e. fertilization requires this.

75. Interleukins and erythropoietin are
a. growth factors.
b. Cdk's.
c. cyclins.
d. antitumor agents.
e. intracellular signaling molecules.

76. Half of all human cancers have defective __________ associated with their cells.
   a. p53
   b. p21
   c. Cdk
   d. cyclin
   e. DNA polymerase

77. Each diploid cell of a human female contains
   a. one of each type of chromosome.
   b. two of each type of chromosome.
   c. four of each type of chromosome.
   d. a total of 23 of each type of chromosome.
   e. a total of 46 of each type of chromosome.

78. The importance of synapsis and the formation of chiasmata is that
   a. reciprocal exchange of chromosomal sections occurs.
   b. the DNA on homologous chromosomes mix.
   c. as a result, an increase in the variation of progeny occurs.
   d. overt evidence is provided for maternal and paternal chromosomes mixing.
   e. All of the above

79. A triploid plant has
   a. one extra chromosome.
   b. one extra set of chromosomes.
   c. three chromosomes.
   d. three times the chance of a monoploid.
   e. three attempts at ploidy.

80. Nucleosomes are found
   a. only in interphase cells.
   b. only during mitosis.
   c. only during meiosis.
   d. in all chromatin.
   e. in the cytoplasm.

*81. The energy to move chromosomes during mitosis is provided by
   a. centrioles.
   b. DNA polymerization.
   c. migration of the centrosomes.
   d. formation of the cell plate.
   e. ATP.

82. During bacterial cell division, the two DNA molecules are separated by
   a. centrosomes.
   b. spindle fibers.
   c. nucleosomes.
   d. cell elongation.
   e. aneuploidy.

83. The process of programmed cell death is called
   a. necrosis.
   b. lysis.
   c. apoptosis.
   d. cell displacement.
   e. cellular suicide.

84. An indicator of programmed cell death is
   a. fragmented chromatin.
   b. swelling of the membrane.
   c. cell lysis.
   d. loss of transcription control.
   e. All of the above

Study Guide Questions

1. Which of the following is true of mitosis?
   a. The chromosome number in the resulting cells is halved.
   b. DNA replication is completed prior to the beginning of this phase.
   c. The chromosome number of the resulting cells is the same as that of the parent cell.
   d. b and c

2. Which of the following is true of meiosis?
   a. The chromosome number in the resulting cells is halved.
   b. DNA replication occurs after the beginning of this phase.
   c. The homologs do not pair during prophase I.
   d. The chromosome number of the resulting cells is the same as that of the parent cell.

3. Which of the following is true of kinetochores?
   a. They are localized at the centromere of each chromosome.
   b. They are the sites where microtubules attach to separate the chromosomes.
   c. They are organized so that there is one per sister chromatid in meiosis.
   d. All of the above

4. Which of the following is true of the mitotic spindle?
   a. It is composed of actin and myosin microfilaments.
   b. It is composed of kinetochores at the metaphase plate.
   c. It is composed of microtubules, which help separate the chromosomes to opposite poles of the cell.
   d. It originates only at the centrioles in the centrosomes.

5. Imagine that there is a mutation in the Cdk2 gene such that its gene product is nonfunctional. What kind of effect would this mutation have on a nondividing differentiated cell?
   a. The cell would be unable to replicate its DNA.
   b. The cell would not be able to enter G1.
   c. The cell would be unable to reproduce itself.
   d. There would be no effect, because differentiated cells do not enter the cell cycle.

6. Imagine that there is a mutation in the Cdk2 gene such that its gene product is nonfunctional. What kind of effect would this mutation have on a dividing cell?
   a. The cell would be unable to replicate its DNA.
   b. The cell would be unable to enter G1.
   c. The cell would be unable to reproduce itself.
   d. a and c

7. The cyclin D-Cdk4 complex is only active during G1 of the cell cycle in humans. Why?
   a. The environmental signals to stimulate the cell cycle are active only in G2.
   b. Cyclin D is destroyed at the end of G1 and thus cannot form a catalytically active cyclin D-Cdk4 complex.
   c. Cdk4 is destroyed at the end of G1 and cannot form a catalytically active protein cyclin D-Cdk4 complex.
   d. Cyclin D is only needed to activate DNA synthesis.

8. Which of the following is true of chromatids?
   a. They are replicated chromosomes still joined together at the centromere.
   b. They are identical in mitotic chromosomes.
   c. They are identical in meiotic chromosomes.
   d. a and b

9. Histones are positively charged because
   a. the majority of the ions in the nucleus of the cell are negatively charged.
   b. histones interact with acidic residues of proteins found in the nucleus.
   c. the basic side chains of histone proteins interact with the negatively charged DNA.
   d. histones have a majority of acidic residues in their protein sequence.

10. Chromosome movement during anaphase is the result of
    a. the molecular motors at the kinetochores that move the chromosomes toward the poles.
    b. molecular motors at the centrosome that pull the microtubules toward the poles.
    c. shortening of the microtubules at the centrosome that pull the chromosomes toward the poles.
    d. a and c

11. Programmed cell death (apoptosis)
    a. occurs in cells that have been deprived of essential nutrients.
    b. occurs only in cells that have damaged DNA.
    c. is a natural process during development.
    d. is signaled by the initiation of mitosis.

End of Chapter Questions

1. The appropriate decisions to enter S phase and M phase of the cell cycle depend on
   a. actin and myosin.
   b. Cdk's and cyclins.
   c. ligand and receptor.
   d. ATP and ATPase.
   e. histones and nucleosomes.

2. Nucleosomes
   a. are made of chromosomes.
   b. consist entirely of DNA.
   c. consist of DNA wound around a histone core.
   d. are present only during mitosis.
   e. are present only during prophase.

3. DNA replication occurs
   a. before both mitosis and meiosis.
   b. only before mitosis.
   c. only before meiosis.
   d. during chromosome condensation.
   e. during G2.

4. In human cells, chromosomes are
   a. always condensed during the entire cell cycle.
   b. different in different tissues.
   c. connected by at centromeres, except for the X and Y.
   d. the same size and length.
   e. visible only during mitosis and meiosis.

5. Which statement about cytokinesis is *true*?
   a. In animals, a cell plate forms.
   b. In plants, it is initiated by furrowing of the membrane.
   c. It generally immediately follows mitosis.
   d. In plant cells, actin and myosin play an important part.
   e. It is the division of the nucleus.

6. In sexually reproducing organisms, the diploid phase of the life cycle begins at
   a. spore formation.
   b. gamete formation.
   c. meiosis.
   d. mitosis
   e. fertilization.

7. Durng meiosis, sister chromatids separate during
   a. anaphase I.
   b. anaphase II.
   c. S phase.
   d. synapsis.
   e. telophase II.

8. The number of chromatids in a cell in prophase I of meiosis of a person with Turner syndrome (XO) is
   a. 23.
   b. 45.
   c. 46.
   d. 90.
   e. 92.

9. Which statement about aneuploidy is *not* true?
   a. It results from chromosomal nondisjunction.
   b. It does not happen in humans.
   c. An individual with an extra chromosome is trisomic.
   d. Trisomies are common in human zygotes.
   e. A piece of one chromosome may translocate to another chromosome.

10. Apoptosis
    a. does not occur in normal tissues.
    b. involves a programmed series of events for cell death.
    c. does not involve reclycling of cell contents.
    d. results in the lysis of an intact cell.
    e. results in the release of intact DNA molecules from the cell.

Student Web Site Self-Quiz Questions

1. Which one of the following statements about cell division in prokaryotes is *false*?
   a. The bacterial chromosome is attached to the plasma membrane.
   b. Prokaryotic cells undergo mitosis, but not meiosis.
   c. The initial step in prokaryotic cell division involves replication of the single, usually circular chromosome.
   d. Division of the prokaryotic cell and separation of the sister chromosomes occurs simultaneously.
   e. Cell division in some prokaryotes can occur in as little as 20 minutes.

2. Typically, cells that no longer undergo mitosis have chromosomes with ________ strand(s) of DNA and they remain in the ________ subphase of interphase.
   a. one, G1
   b. two, G1
   c. two, S
   d. one, G2
   e. two, G2

3. Which one of the following statements about protein signals involved in the control of the cell cycle is *false*?
   a. Transition from G1 to S and G2 to M depends on the phosphorylation of certain key proteins by cyclin-dependent kinase, or Cdk.
   b. Cdk's must be bound with a specific cyclin to be active as kinases.
   c. In humans, cyclin D-cdk4 triggers the G1-to-S transition, whereas cyclin B-cdk1 triggers the G2-to-M transition.
   d. External control mechanisms involving growth factors and hormones can activate cells that are arrested or are cycling slowly.
   e. Almost all cancers act by disrupting internal control of the cell cycle, not by external means.

4. At metaphase of mitosis, each chromosome consists of ________ chromatid(s), ________ centromere(s), and ________ kinetochores.
   a. 1, 1, 2
   b. 2, 1, 2
   c. 2, 2, 2
   d. 1, 2, 2
   e. 2, 1, 4

5. Which of the following statements about mitosis is *true*?
   a. Only diploid cells can divide mitotically.
   b. Crossing over can occur during prophase of mitosis.
   c. Cells produced by mitosis are almost always genetically identical.
   d. Each mitotically produced cell has one-quarter the mass of DNA of the cell that produced it.
   e. At metaphase of mitosis, each chromosome has a single kinetochore microtubule attached to it.

6. Which of the following statements about homologous chromosomes is *false*?
   a. Haploid cells have only one homolog from each of the pairs of homologous chromosomes.
   b. Each of the homologs of a pair of homologous chromosomes came from a different parent.
   c. Generally, the number of pairs of homologous chromosomes is related to the complexity of the organism.
   d. Homologs of a pair of homologous chromosomes contain the same types of genetic information.
   e. Maternal and paternal homologs associate during meiosis I.

7. Which of the statements comparing mitosis and meiosis is *false*?
   a. Chiasmata only form during meiosis, not mitosis.
   b. DNA replicates before mitosis, but not before meiosis II.
   c. The total mass of DNA present in all four products of meiosis is the same as the total mass in the two cells produced by mitosis.
   d. Centromeres divide during mitosis metaphase, but not metaphase II of meiosis.
   e. Differences in metaphase of mitosis and metaphase I of meiosis explain why mitosis preserves the ploidy (haploid or diploid) of the cell, while meiosis always produces haploid cells from a diploid cell.

8. Which of the following statements about meiosis is *false*?
   a. Haploid cells cannot divide meiotically.
   b. DNA replication does not occur during interkinesis.
   c. Assortment of homologous chromosomes on the equatorial plate during metaphase I is independent.
   d. Of the four cells produced by meiosis, none are genetically identical to the parent cell, but two are identical to each other.
   e. The mass of DNA in one of the four products of meiosis is one-fourth of that in the original parent cell.

TEST FILE

★ *Indicates a **difficult** question*

■ *Indicates a **conceptual** question*

10 Genetics: Mendel and Beyond

Fill in the Blank

1. A **heritable** trait is one that can be passed from one generation to another.

2. A **gene** is a portion of DNA that resides at a particular locus or site on a chromosome and encodes a particular function.

3. To determine the overall probability of independent events, **multiply** the probabilities of the individual events.

4. One particular allele of a gene may be defined as **wild type** or standard, because it is present in most individuals and gives rise to an expected trait, or phenotype.

5. Geneticists make use of **recombinant** frequencies to map chromosomes, that is, to locate genetic loci on the chromosome.

6. A cross between two parents that differs by a single trait is a **monohybrid** cross.

7. The physical appearance of a character is the **phenotype**, while the genetic constitution is the **genotype**.

8. A cross between two parents that differs by two independent traits is a **dihybrid** cross.

9. For genes that fail to independently assort, phenotypes that appear in combinations that are not present in either parent are **recombinant**.

10. The region of the chromosome occupied by a gene is called a **locus**.

11. When a cross is made and a trait disappears in the F_1 generation, only to reappear in the F_2, the trait is probably **recessive**.

12. A female who is heterozygous for a recessive, sex-linked character is a **carrier**.

13. When the expression of one gene depends on the expression of another gene, the genes demonstrate **epistasis**.

14. When many genes contribute to the phenotype, variation is said to be **continuous**.

15. A **character** is a feature, such as flower color; a **trait** is a particular form of a character, such as a white flower.

Multiple Choice

1. Gregor Mendel presented his genetics project orally in
   a. 1565.
   b. 1665.
   c. 1765.
   d. 1865.
   e. 1965.

2. Mendel concluded that each pea has two units for each character, and each gamete contains one unit. Mendel's "unit" is now referred to as a(n)
   a. gene.
   b. character.
   c. allele.
   d. transcription factor.
   e. None of the above

3. Mendel's research was rediscovered when studies by __________ were published.
   a. de Vries
   b. Correns
   c. Tschermak
   d. All of the above
   e. None of the above

4. A particular genetic cross in which the individual in question is crossed with an individual known to be homozygous for a recessive trait is referred to as a
   a. parental cross.
   b. dihybrid cross.
   c. filial generation mating.
   d. reciprocal cross.
   e. test cross.

5. Incomplete dominance occurs when
   a. chromosomes are deleted.
   b. heterozygotes synthesize a reduced amount of an enzyme, producing an intermediate phenotype.
   c. the genes fail to segregate.
   d. the law of independent assortment is upheld.
   e. one gene is epistatic to the other.

6. Although the law of independent assortment is generally applicable, when two loci are on the same chromosome the phenotypes of the progeny sometimes do not fit the phenotypes predicted. This is due to
   a. translocation.
   b. inversions.
   c. chromatid affinities.
   d. linkage.
   e. reciprocal chromosomal exchanges.

7. When a given trait is the result of multigene action, one of the genes may mask the expression of one or all other genes. This phenomenon is termed
   a. epistasis.
   b. epigenesis.
   c. dominance.
   d. incomplete dominance.
   e. None of the above

8. Which of the following is *not* a characteristic that makes an organism suitable for genetic studies?
   a. A small number of chromosomes
   b. A short generation time
   c. Ease of cultivation
   d. The ability to control crosses
   e. The availability of a variation for traits

9. A key factor that allowed Mendel to interpret his breeding experiments was that
   a. the varieties of peas he started with were "true-breeding."
   b. peas naturally self-pollinate.
   c. peas can reproduce asexually.
   d. pollination could be controlled.
   e. a and d

10. Crossing spherical-seeded pea plants with wrinkled-seeded pea plants resulted in progeny that all had spherical seeds. This indicates that the wrinkled-seed trait is
    a. codominant.
    b. dominant.
    c. recessive.
    d. a and b
    e. a and c

11–13. Imaginary schmoos live in geographically separated groups and rarely interbreed. On one occasion, two from the different groups did mate. A big-footed white schmoo mated with a small-footed brown schmoo. Three offspring resulted: one big-footed brown schmoo and two small-footed brown schmoos.

▪11. Which statement about the inheritance of color in schmoos is most likely to be correct?
    a. Brown is dominant to white.
    b. White is dominant to brown.
    c. White and brown are codominant.
    d. a and c
    e. You cannot reach any conclusions.

*▪12. Which statement about the inheritance of footedness in schmoos is most likely to be correct?
    a. Big is dominant to small.
    b. Small is dominant to big.
    c. Big and small are codominant.
    d. a and c
    e. You cannot reach any conclusions.

▪13. If big feet (*B*) in schmoos is dominant to small feet (*b*), what is the genotype of the big-footed white parent schmoo with respect to the foot gene?
    a. *bb*
    b. *BB*
    c. *Bb*
    d. a and b
    e. a and c

14. When reciprocal crosses produce identical results, the trait is
    a. sex-linked.
    b. not sex-linked.
    c. not autosomally inherited.
    d. a and b
    e. b and c

15. The physical appearance of a character is called
    a. the genotype.
    b. the phenotype.
    c. an allele.
    d. a trait.
    e. a gene.

16. Different forms of a gene are called
    a. traits.
    b. phenotypes.
    c. genotypes.
    d. alleles.
    e. None of the above

17. When genes for two different characters segregate in a cross, what type of cross is it?
    a. Monohybrid
    b. Dihybrid
    c. Trihybrid
    d. F_1
    e. F_2

▪18. In Mendel's experiments, if the allele for tall (*T*) plants was incompletely dominant over the allele for short (*t*) plants, what would be the result of crossing two *Tt* plants?
    a. 1/4 would be tall; 1/2 intermediate height; 1/4 short.
    b. 1/2 would be tall; 1/4 intermediate height; 1/4 short.
    c. 1/4 would be tall; 1/4 intermediate height; 1/2 short.
    d. All the offspring would be tall.
    e. All the offspring would be intermediate.

19. The region of the chromosome occupied by a gene is called a(n)
    a. allele.
    b. region.
    c. locus.
    d. type.
    e. phenotype.

★▪20. In Mendel's experiments, the spherical seed character (*SS*) is completely dominant over the wrinkled seed character (*ss*). If the characters for height were incompletely dominant, such that *TT* are tall, *Tt* are intermediate, and tt are short, what would be the result of crossing a spherical-seeded, short (*SStt*) plant to a wrinkled-seeded, tall (*ssTT*) plant?
a. 1/2 would be smooth-seeded and intermediate height; 1/2 would be smooth-seeded and tall.
b. All the progeny would be smooth-seeded and tall.
c. All the progeny would be smooth-seeded and short.
d. All the progeny would be smooth-seeded and intermediate height.
e. You cannot predict the outcome.

▪21. If Mendel's crosses between spherical-seeded tall plants and wrinkled-seeded short plants had produced many more than 1/16 wrinkled-seeded short plants in the F_2 generation, he might have concluded that
a. the spherical seed and tall traits are linked.
b. the wrinkled seed and short traits are unlinked.
c. all traits in peas assort independently of each other.
d. all traits in peas are linked.
e. He would not have concluded any of the above.

★22. At a certain locus of the human genome, it is found that 200 different alleles exist in the population. Each person would have at the most __________ alleles.
a. 1
b. 2
c. 100
d. 200
e. 400

23. An organism that produces either male gametes or female gametes, but not both, is called
a. monoecious.
b. dioecious.
c. heterozygous.
d. homozygous.
e. parthenogenic.

▪24. Why would you predict that half of the human babies born will be males and half will be females?
a. Because of the segregation of the X and Y chromosomes during male meiosis
b. Because of the segregation of the X chromosomes during female meiosis
c. Because all eggs contain an X chromosome
d. a and b
e. a and c

25. A human male carrying an allele for a trait on the X chromosome is
a. heterozygous.
b. homozygous.
c. hemizygous.
d. monozygous.
e. holozygous.

▪26. Cleft chin is a sex-linked dominant trait. A man with a cleft chin marries a woman with a round chin. What proportion of their female progeny will show the trait?
a. 0%
b. 25%
c. 50%
d. 75%
e. 100%

▪27. What proportion of their male progeny will show the trait?
a. 0%
b. 25%
c. 50%
d. 75%
e. 100%

28. A linkage group corresponds to
a. a group of genes on different chromosomes.
b. the linear order of chromomeres on a chromosome.
c. the length of a chromosome.
d. a group of genes on the same chromosome.
e. None of the above

▪29. It would have been very difficult for Mendel to draw conclusions about the patterns of inheritance if he had used cattle instead of peas. Why?
a. Cattle reproduce asexually.
b. Cattle have small numbers of offspring.
c. Cattle do not have observable phenotypes.
d. Cattle do not have genotypes.
e. Cattle do not have autosomes.

30. Epistasis refers to
a. a group of genes that are close together.
b. the interaction of two genes so that a new phenotype is produced.
c. the expression of two genes in the same individual.
d. the linear order of genes on a chromosome.
e. the expression of one gene masking the expression of another.

31–33. An agouti mouse that is heterozygous at the agouti and albino loci (*AaBb*) is mated to an albino mouse that is heterozygous at the agouti locus (*aaBb*). (Non-albino mice without the dominant agouti allele are black.)

★▪31. What proportion of the progeny do you expect to be albino?
a. 0%
b. 12.5%
c. 37.5%
d. 50%
e. 100%

★▪32. What proportion of the progeny do you expect to be agouti?
a. 0%
b. 12.5%
c. 37.5%
d. 50%
e. 100%

★■33. What proportion of the progeny do you expect to be black? Black is the alternate phenotype to agouti.
a. 0%
b. 12.5%
c. 37.5%
d. 50%
e. 100%

34. The complete phenotype of an organism is dependent on
a. genotype.
b. penetrance.
c. expressivity.
d. polygenes.
e. All of the above

35. When a dihybrid black, straight-winged fly is crossed to a double-recessive brown, curly-winged fly, the frequency at which black curly-winged and brown straight-winged flies are seen in the progeny is called the
a. mutation frequency.
b. mitotic frequency.
c. meiotic frequency.
d. allele frequency.
e. recombinant frequency.

36. Alleles for genes located on mitochondrial DNA are said to be maternally inherited. What is the reason for this pattern of inheritance?
a. The egg and sperm contribute equal numbers of cytoplasmic organelles to the zygote.
b. The egg contributes virtually all of the cytoplasmic organelles to the zygote.
c. Half of the nuclear chromosomes in the zygote come from the father.
d. Half of the nuclear chromosomes in the zygote come from the mother.
e. All of the nuclear chromosomes in the zygote come from the mother.

37. Which of the following methods was *not* used by Mendel in his study of the genetics of the garden pea?
a. Maintenance of true-breeding lines
b. Cross-pollination
c. Microscopy
d. Production of hybrid plants
e. Quantitative analysis of results

38. In Kölreuter's studies, reciprocal crosses
a. always gave identical results.
b. only involved heterozygous individuals.
c. supported the blending hypothesis of inheritance.
d. could only be done with homozygous individuals.
e. consist of an F_1 and an F_2 generation.

39. Which of the following statements about Mendelian genetics is *false*?
a. Alternate forms of genes are called alleles.
b. A locus is a gene's location on its chromosome.
c. Only two alleles can exist for a given gene.
d. A genotype is a description of the alleles that represent an individual's genes.
e. Individuals with the same phenotype can have different genotypes.

40. Segregation of alleles occurs
a. during gamete formation.
b. at fertilization.
c. during mitosis.
d. during the random combination of gametes to produce the F_2 generation.
e. only in monohybrid crosses.

■41. A pea plant with red flowers is test crossed, and one half of the resulting progeny have red flowers, while the other half have white flowers. You know that the genotype of the test-crossed parent was
a. *RR*.
b. *Rr*.
c. *rr*.
d. either *RR* or *Rr*.
e. You cannot tell unless the genotypes of both parents are known.

★■42. Mendel performed a cross between individuals heterozygous for three different traits: yellow versus green seeds (green is dominant), red versus white flowers (red is dominant), and green versus yellow pods (green is dominant). What fraction of the offspring are expected to have green seeds, red flowers, and green pods?
a. 27/64
b. 12/64
c. 9/64
d. 6/64
e. 3/64

■43. Classical albinism results from a recessive allele. Which of the following is the expected ratio for the progeny, when a normally pigmented male with an albino father has children with an albino woman?
a. 3/4 normal; 1/4 albino
b. 3/4 albino; 1/4 normal
c. 1/2 normal; 1/2 albino
d. All normal
e. All albino

■44. In humans, a widow's peak is caused by a dominant allele *W*, and a continuous hairline, by a recessive allele *w*. Short fingers are caused by a dominant allele *S*, and long fingers, by a recessive allele *s*. Suppose a woman with a continuous hairline and short fingers and a man with a widow's peak and long fingers have three children. One child has short fingers and a widow's peak, one has long fingers and a widow's peak, and one has long fingers and a continuous hairline. What are the genotypes of the parents?
a. Female *wwSS*; male *WWss*
b. Female *wwSs*; male *Wwss*
c. Female *wwSs*; male *WWss*
d. Female *WwSs*; male *WwSs*
e. None of the above

▪45. In garden peas, the allele for tall plants is dominant over the allele for short plants. A true-breeding tall plant is crossed with a short plant, and one of their offspring is test crossed. Out of 20 offspring resulting from the test cross, about ________ should be tall.
a. 0
b. 5
c. 10
d. 15
e. 20

★▪46. Which of the following phenomena *cannot* be observed using only dihybrid crosses?
a. Crossing over
b. Segregation of alleles
c. Independent assortment of alleles
d. Recessive lethal alleles
e. None of the above

47. Separation of the alleles of a single gene into different gametes is called
a. synapsis.
b. segregation.
c. independent assortment.
d. heterozygous separation.
e. recombination.

▪48. In mice, short hair is dominant to long hair. If a short-hair individual is crossed with a long-hair individual and both long- and short-hair offspring result, you can conclude that
a. the short-hair individual was homozygous.
b. the short-hair individual was heterozygous.
c. the long-hair individual was homozygous.
d. the long-hair individual was heterozygous.
e. more offspring are required in order to decide the genotypes of the parents.

★▪49. In dogs, erect ears and barking while following a scent are due to dominant alleles; droopy ears and silence while following a scent are due to recessive alleles. A dog homozygous for both traits is mated to a droopy-eared, silent follower. If the two genes are unlinked, the expected F_1 phenotypic ratios should be
a. 9:3:3:1.
b. 1:1.
c. 100% of one phenotype.
d. 1:2:1.
e. None of the above

★▪50. In cocker spaniels, black color (*B*) is dominant over red (*b*), and solid color (*S*) is dominant over spotted (*s*). If the offspring between *BBss* and *bbss* individuals are mated with each other, what fraction of their offspring will be expected to be black and spotted? Assume the genes are unlinked.
a. 1/16
b. 9/16
c. 1/9
d. 3/16
e. 3/4

51. If the same allele has two or more phenotypic effects, it is said to be
a. codominant.
b. a marker.
c. linked.
d. pleiotropic.
e. hemizygous.

52. In the ABO blood type system,
a. A, B, and O are codominant.
b. A, B, and O are incompletely dominant.
c. A and B are codominant.
d. O is incompletely dominant to A and B.
e. A is dominant to B, and B is dominant to O.

▪53. In Netherlands dwarf rabbits, a gene showing intermediate inheritance produces three phenotypes. Rabbits that are homozygous for one allele are small rabbits; individuals homozygous for the other allele are deformed and die; heterozygous individuals are dwarf. If two dwarf rabbits are mated, what proportion of their surviving offspring should be dwarf?
a. 1/4
b. 1/3
c. 1/2
d. 2/3
e. 3/4

★▪54. In tomatoes, tall is dominant to short, and smooth fruits are dominant to hairy fruits. A plant homozygous for both dominant traits is crossed with a plant homozygous for both recessive traits. The F_1 progeny are tested and crossed with the following results: 78 tall, smooth fruits; 82 dwarf, hairy fruits; 22 tall, hairy fruits; and 18 dwarf, smooth fruits. These data indicate that the genes are
a. on different chromosomes.
b. linked, but do not cross over.
c. linked and show 10% recombination.
d. linked and show 20% recombination.
e. linked and show 40% recombination.

▪55. Remembering that white eyes is a recessive, sex-linked trait, if a white-eyed female fruit fly is mated to a red-eyed male, their offspring should be
a. 50% red-eyed, 50% white-eyed for both sexes.
b. all white-eyed for both sexes.
c. all white-eyed males, all red-eyed females.
d. all white-eyed females, all red-eyed males.
e. 50% red-eyed males, 50% white-eyed males, all red-eyed females.

▪56. A dominant allele *K* is necessary for normal hearing. A dominant allele *M* of a different locus results in deafness no matter what other alleles are present. If a *kkMm* individual is crossed with a *Kkmm* individual, what percentage of the offspring will be deaf?
a. 0%
b. 25%
c. 50%
d. 75%
e. None of the above

▪57. The genetic disease blue sclera is determined by an autosomal dominant allele. The eyes of individuals with this allele have bluish sclera. These same individuals may also suffer from fragile bones and deafness. This is an example of
a. incomplete dominance.
b. pleiotropy.
c. epistasis.
d. codominance.
e. linkage.

*▪58. The blue sclera allele has 90% penetrance for producing blue sclera, 60% penetrance for fragile bones, and 40% penetrance for deafness. If these probabilities of penetrance are independent, what is the probability that an individual with the blue sclera allele will be deaf, have blue sclera, and fragile bones?
a. 22%
b. 40%
c. 60%
d. 90%
e. None of the above

▪59. Y-linked genes include a gene that produces hairy pinna (the external ear). A male with hairy pinna should pass this trait
a. usually to his sons, but rarely also to a daughter.
b. only to his sons.
c. only to his daughters.
d. only to his grandsons.
e. to all his children if the mother is a carrier.

60. Which of the following organelles contain DNA?
a. Nucleus
b. Chloroplast
c. Mitochondria
d. Ribosome
e. a, b, and c

61. Two strains of true-breeding plants that have different alleles for a certain character are crossed. Their progeny are called
a. the P generation.
b. the F_1 generation.
c. the F_2 generation.
d. F_1 crosses.
e. F_2 progeny.

62. A mutation at a single locus causes a change in many different characters. This an example of a
a. polygene effect.
b. epigenetic effect.
c. cytoplasmic effect.
d. multiple negativity effect.
e. pleiotropic effect.

*▪63. In guppies, fan tail is dominant to flesh tail, and rainbow color is dominant to pink. F_1 female guppies are crossed to flesh-tailed, pink-colored males, and the following progeny are observed: 401 fan-tailed, pink-colored; 399 flesh-tailed, rainbow-colored; 98 flesh-tailed, pink-colored; and 102 fan-tailed, rainbow-colored guppies. The map distance between these two genes is
a. 80 cM.
b. 25 cM.
c. .8 cM.
d. 20 cM.
e. None of the above

64. How many linkage groups are present in a female human?
a. 46
b. Thousands
c. 23
d. 2
e. 496

▪65. Tall pea plants are crossed to short, and the progeny are medium height. The F_1 plants are crossed together, but the progeny observed among the F_2 have nine different size classes. This character's mode of inheritance is
a. pleiotropic.
b. epistasis.
c. multiallelic
d. polygenic.
e. hypostatic.

66. The approximate total number of genes in the human genome is closest to
a. 60.
b. 600.
c. 6000.
d. 60,000.
e. 600,000.

67. Sex in humans is determined by
a. a gene called *SRY* found on the Y chromosome.
b. a gene called *SRY* found on the X chromosome.
c. a gene found on an autosomal chromosome called SDG.
d. the simple presence of a Y chromosome.
e. a gene called *SDG* found on the Y chromosome.

68. Humans have ________ genes in their mitochondria.
a. 600
b. 6,000
c. 60,000
d. 600,000
e. 37

Study Guide Questions

1. In the beginning of Chapter 10, hemophilia is mentioned as a trait carried by the mother and passed to her sons. What is the pattern of inheritance for this trait?
a. Hemophilia is an allele carried on one of the mother's autosomal chromosomes.
b. Hemophilia is an allele carried on the Y chromosome because more males have this genetic disorder than females.

c. Hemophilia is an allele carried on the X chromosome and can be directly inherited by the son from the father or the mother.
d. Hemophilia is carried on the X chromosome and can only be inherited by the son if the mother is a carrier.

2. Originally, genetic inheritance was thought to be a function of the blending of traits from the two parents. Which exception to Mendel's rules is an example of blending?
a. Polygenic inheritance
b. Incomplete dominance
c. Codominance
d. Pleiotropism

3. True-breeding plants
a. produce the same offspring when crossed for many generations.
b. result from a monohybrid cross.
c. result from a dihybrid cross.
d. result from crossing over during prophase I of meiosis.

4. What is the probability that a cross between a true-breeding pea plant with smooth seeds and a true-breeding pea plant with wrinkled seeds will produce F_1 progeny with smooth seeds?
a. 1/2
b. 1/4
c. 0
d. 1

5. What is the pattern of inheritance for a rare recessive allele?
a. Every affected person has an affected parent.
b. Unaffected parents can produce children who are affected.
c. Unaffected mothers have affected sons and daughters who are carriers.
d. None of the above

6. What is the pattern of inheritance for a rare dominant allele?
a. Every affected person has an affected parent.
b. Unaffected parents can produce children who are affected.
c. Unaffected mothers have affected sons and daughters who are carriers.
d. None of the above

7. What is the pattern of inheritance for a sex-linked allele?
a. Every affected person has an affected parent.
b. Unaffected parents can produce children who are affected.
c. Unaffected mothers have affected sons and daughters who are carriers.
d. None of the above

8. Penetrance and expressivity are related to
a. the increased expression of a particular trait when a hybrid species is formed.
b. quantitative traits that diminish or intensify a particular phenotype.
c. the organism's environment affecting the expression of a particular genotype.
d. the expression of one gene masking the effects of another gene.

9. Sex determination in grasshoppers, humans, and *Drosophila* is similar because
a. females are hemizygous.
b. males have one X chromosome and females have two X chromosomes.
c. all males always have one Y chromosome in all three species.
d. the ratio of autosomes to sex chromosomes is the same in all three organisms.

10. Linked genes are genes that
a. assort independently.
b. segregate equally in the gametes during meiosis.
c. always contribute the same trait to the zygote.
d. are found on the same chromosome.
e. recombine during mitosis.

11. Cytoplasmic inheritance
a. results from polygenic nuclear traits.
b. is the result of gametes contributing equal amounts of cytoplasm to the zygote.
c. is determined by genes on DNA molecules in mitochondria and chloroplasts.
d. follows Mendel's law of segregation.

End of Chapter Questions

1. In pea plants, the gene for yellow seeds, *R* is dominant to the allele for green seeds, *r*. A plant with green seeds must have the genotype
a. *RRr.*
b. *Rrr.*
c. *Rr.*
d. *RR.*
e. *rr.*

2. The phenotype of an individual
a. depends at least in part on the genotype.
b. is either homozygous or heterozygous.
c. determines the genotype.
d. is the genetic constitution of the organism.
e. is either monohybrid or dihybrid.

3. What fraction of offspring of the cross *AaBbDdEEFf* × *AaBBDdEeFf* are homozygous for all the genes, assuming that they are on different chromosomes?
a. 1/4
b. 1/8
c. 1/2
d. 1/32
e. 1/64

4. In humans, brown spotting of teeth is caused by a dominant sex-linked gene. If a man with brown teeth marries a woman with normal teeth,
   a. all of their children have brown teeth.
   b. all of their sons have brown teeth.
   c. all of their daughters have brown teeth.
   d. half of their daughters have brown teeth.
   e. none of their children have brown teeth.

5. Which statement about a test cross is *not* true?
   a. It tests whether an unknown individual is homozygous or heterozygous.
   b. The test individual is crossed with a homozygous recessive individual.
   c. If the test individual is heterozygous, the progeny will have a 1:1 ratio.
   d. If the test individual is homozygous, the progeny will have a 3:1 ratio.
   e. Test cross results are consistent with Mendel's model of inheritance.

6. Linked genes
   a. must be immediately adjacent to one another on a chromosome.
   b. have alleles that assort independently of one another.
   c. never show crossing over.
   d. are on the same chromosome.
   e. always have multiple alleles.

7. In the F_2 generation of a dihybrid cross,
   a. four phenotypes appear in the ratio 9:3:3:1 if the loci are linked.
   b. four phenotypes appear in the ratio 9:3:3:1 if the loci are unlinked.
   c. two phenotypes appear in the ratio 3:1 if the loci are unlinked.
   d. three phenotypes appear in the ratio 1:2:1 if the loci are unlinked.
   e. two phenotypes appear in the ratio 1:1 whether or not the loci are linked.

8. The sex of a human is determined by
   a. ploidy, the male being haploid.
   b. the Y chromosome, the male having a Y.
   c. X and Y chromosomes, the male being XX.
   d. the number of X chromosomes, the male being XO.
   e. Z and W chromosomes, the male being ZZ.

9. In epistasis,
   a. nothing changes from generation to generation.
   b. one gene alters the effect of another.
   c. a portion of a chromosome is deleted.
   d. a portion of a chromosome is inverted.
   e. the behavior of two genes is entirely independent.

10. Incomplete dominance
   a. results in a phenotypic ratio of 1:2:1 in a monohybrid cross.
   b. is far less common than complete dominance.
   c. does not occur in multiple allele series.
   d. may involve genes coding for several enzymes in a pathway.
   e. occurs only in X-linked genes.

Student Web Site Self-Quiz Questions

1. Which of the following is *not* a characteristic of pea plants that caused Mendel to choose them for his studies of inheritance?
   a. Flower morphology that made controlled cross-pollination possible
   b. Prior studies on meiosis in this species by J. G. Kölreuter
   c. Existence of true-breeding strains
   d. Presence of many well-defined, contrasting traits
   e. Ease of cultivation

2. Which of the following statements about Mendel's studies of monohybrid crosses is *false*?
   a. Plants developing from seeds produced by crossing the two parent plants would be the F_1 generation.
   b. In his studies of single characters, reciprocal crosses always gave the same results.
   c. For each of the seven different monohybrid crosses, the trait that disappeared in the F_1 generation reappeared in about 75% of the F_2 generation.
   d. In Mendel's monohybrid crosses, all F_1 plants are heterozygous for the gene being studied.
   e. In Mendel's monohybrid crosses, approximately 50% of the F_2 plants are homozygous for the gene being studied.

3. Which of the following statements about the interpretation of Mendel's monohybrid studies is *false*?
   a. Mendel's monohybrid crosses allowed him to study a single gene in isolation.
   b. A plant's phenotype depends largely on which alleles are present for the gene being studied.
   c. For each gene, the alleles from the two parents blend together to form a new phenotype.
   d. Although F_2 plants showed only two phenotypes, three genotypes were present in this generation.
   e. The alleles for a gene segregate independently of each other during gamete formation.

4. Using a polygenic inheritance model based on four genes for the determination of skin color in humans, what is the expected number of skin pigmentation phenotypic classes, and what would be the expected proportion of individuals with the genotype *AABBCCDD* from matings between parents that are heterozygous for all four genes?
   a. 8, 1/64
   b. 9, 1/64
   c. 9, 1/256
   d. 16, 1/64
   e. 16, 1/256

5. Which of the following statements about sex determination in humans (below) is *false*?
   a. In both humans and *Drosophila*, the genetic constitution of the sperm determines the sex of the zygote.
   b. XO individuals in both humans and *Drosophila* are female and sterile.
   c. XXY individuals are female in *Drosophila*, but male in humans.
   d. In humans, XX individuals with the *SRY* portion of the Y chromosome translocated onto another chromosome are male.
   e. The Y chromosome plays no role in sex determination in *Drosophila*, whereas in humans it does.

6. Sex is determined in different ways in different species. Which of the following statements about sex determination in animals is *false*?
   a. Reciprocal crosses involving genes on the X chromosome do not give identical results.
   b. Sex determination mechanisms are only seen in dioecious species.
   c. Male birds are hemizygous for genes on the sex chromosomes.
   d. Male honeybees have only one set of autosomes.
   e. In both *Drosophila* and humans, males are XY and females are XX.

TEST FILE

★ *Indicates a **difficult** question*

■ *Indicates a **conceptual** question*

11 DNA and Its Role in Heredity

Fill in the Blank

1. The X-ray crystallographs of the English chemist **Rosalind Franklin** were essential for the discovery of the structure of the DNA molecule.

2. Since the DNA molecule runs on and on, without kinks or bulges, nucleotide pair after nucleotide pair, information must lie in the **linear** sequence of the nitrogenous bases.

3. Arthur Kornberg showed that DNA could replicate in the test tube if it contained intact DNA for a template, a mixture of the four precursors—the four nucleoside triphosphates, and **DNA polymerase**.

4. The material that changed R strain pneumococcus into the virulent S strain was originally referred to as the **transforming principle**.

5. The basic units of DNA and RNA molecules are the **nucleotides** (or **nitrogenous bases**).

6. The experiments of Meselson and Stahl established the **semiconservative replication** of DNA.

*7. The nitrogenous bases classified as purines are **adenine** and **guanine**.

*8. The nitrogenous bases classified as pyrimidines are **cytosine** and **thymine**.

*■9. Using Meselson and Stahl's experimental system for studying the mode of replication of DNA, the genetic information from a life form from Mars is analyzed. It is found that after the first round of replication, two distinct bands appear in the cesium chloride gradient. This is consistent with **conservative** replication.

10. The enzyme that replicates the lagging strand is **DNA polymerase III.**

11. The enzyme that replicates the leading strand is called **DNA polymerase III** in prokaryotes.

12. The region of DNA where replication begins is an **origin of replication**.

13. The fragments of RNA and DNA found on the lagging strand of DNA prior to RNA removal and ligation are called **Okazaki fragments**.

*14. The purines take up (**more**/less) space in the center of a DNA molecule.

15. In a sequencing reaction, the shortest sequences are those which end closer to the (**5′**/3′) of the synthesized molecule.

16. The **proofreading** function of the DNA polymerase reduces the number of mistakes by the square of the frequency of the error rate.

Multiple Choice

■1. Before the discovery of DNA, why was the hereditary material thought to be made of proteins and not nucleic acids?
a. Nucleic acids are made up of 20 different bases, while proteins are made up of only 5 amino acids.
b. Protein subunits can combine to form larger proteins.
c. Proteins seemed to be much more diverse chemically.
d. Proteins can be enzymes.
e. None of the above

■2. How can DNA, made up of only 4 different bases, encode the information necessary to specify the workings of an entire organism?
a. DNA molecules are extremely long.
b. DNA molecules are found in the nucleus.
c. DNA is transcribed into RNA and then into proteins with specific functions.
d. DNA is eventually translated into proteins, which are made up of 20 different amino acids.
e. None of the above

3. In Griffith's experiments, what happened when heat-killed S strain pneumococci were injected into a mouse along with live R strain pneumococci?
a. DNA from the live R was taken up by the heat-killed S, converting it to R and killing the mouse.
b. DNA from the heat-killed S was taken up by the live R, converting it to S and killing the mouse.
c. Proteins released from the heat-killed S killed the mouse.
d. RNA from the heat-killed S was translated into proteins that killed the mouse.
e. Nothing

4. Experiments designed to identify the transforming principle were based on
   a. purifying each of the macromolecule types from a cell-free extract.
   b. removing each of the macromolecules from a cell, then testing its type.
   c. selectively destroying the different macromolecules in a cell-free extract.
   d. a and b
   e. a and c

5. The Hershey–Chase experiment determined that
   a. protein and DNA are the hereditary materials of viruses.
   b. protein, not DNA, is the hereditary material of viruses.
   c. viruses do not contain hereditary material.
   d. DNA, not protein, is the hereditary material of viruses.
   e. the blender is useful in the kitchen.

6. The rules formulated by Erwin Chargaff state that
   a. A = T and G = C in any molecule of DNA.
   b. A = C and G = T in any molecule of DNA.
   c. A = G and C = T in any molecule of DNA.
   d. A = U and G = C in any molecule of RNA.
   e. DNA and RNA are made up of the same four nitrogenous bases.

7. Purines include
   a. cytosine, uracil, and thymine.
   b. adenine and cytosine.
   c. adenine and thymine.
   d. cytosine and thymine.
   e. adenine and guanine.

8. The structure of DNA is characterized by a
   a. right- or left-handed double helix and antiparallel strands.
   b. right-handed double helix and antiparallel strands.
   c. right-handed single helix.
   d. right-handed single helix and parallel strands.
   e. All of the above

9. The antiparallel relationship of the two strands of DNA refers to the
   a. strands being the opposite of parallel—they are twisted.
   b. strands providing alternative branching.
   c. strands aligning such that one strand starts with a 3′ carbon, the other with a 5′ carbon.
   d. view looking at one end of the molecule: one strand has an A wherever the other has a T, and one has a G wherever the other has a C.
   e. All of the above

10. The nitrogenous bases (and the two strands of the DNA double helix) are held together by
    a. weak van der Waals forces.
    b. covalent bonds.
    c. hydrogen bonds.
    d. a and b
    e. a and c

11. The base-paired structure of DNA implies that it
    a. can replicate to form identical molecules.
    b. can be used as a template to make RNA.
    c. is the hereditary material.
    d. a and b
    e. a and c

12. Why must RNA be incorporated into the DNA molecule initially during DNA replication?
    a. RNA primase adds bases that act as primers.
    b. RNA primase is able to use DNA as a template.
    c. RNA primase is incorporated into the holoenzyme complex.
    d. DNA polymerase I and III can only add on to an existing strand.
    e. All of the above

13. What are the three major properties of genes that are explained by the structure of DNA?
    a. They contain information, direct the synthesis of proteins, and are contained in the cell nucleus.
    b. They contain nitrogenous bases, direct the synthesis of RNA, and are contained in the cell nucleus.
    c. They replicate exactly, are contained in the cell nucleus, and direct the synthesis of cellular proteins.
    d. They encode the organism's phenotype, are passed on from one generation to the next, and contain nitrogenous bases.
    e. They contain information, replicate exactly, and change to produce a mutation.

14. Semiconservative replication of DNA involves
    a. each of the original strands acting as a template for a new strand.
    b. only one of the original strands acting as a template for a new strand.
    c. the complete separation of the original strands, the synthesis of new strands, and the reassembly of double-stranded molecules.
    d. the use of the original double-stranded molecule as a template, without unwinding.
    e. None of the above

15. The molecules that function to replicate DNA in the cell are
    a. DNA nucleoside triphosphates.
    b. DNA polymerases.
    c. nucleoside polymerases.
    d. DNAses.
    e. ribonucleases.

16. The correct order of events for synthesis of the lagging strand is:
    a. Primase adds RNA primer, DNA polymerase III creates a stretch, DNA polymerase I removes the primer, and ligase seals the gaps.
    b. Primase adds primer, DNA polymerase I removes the primer, DNA polymerase III extends the segment, and ligase seals the gap.
    c. Ligase adds bases to the primase, the primase generates the polymerase I, polymerase III adds to the stretch, helicase winds the DNA.

d. Helicase unwinds the DNA, primase creates a primer, DNA polymerase I elongates the stretch, DNA polymerase III removes the primer, and ligase seals the gaps in the DNA.
e. None of the above

17. During replication, the new DNA strand is synthesized
a. in the 3′ to 5′ direction.
b. in the 5′ to 3′ direction.
c. in both the 3′ to 5′ and 5′ to 3′ directions from the replication fork.
d. from one end to the other, in the 3′ to 5′ or the 5′ to 3′ directions.
e. None of the above

▪18. Why were fragments like those now called Okazaki fragments expected before they were discovered?
a. DNA replicates in the 5′ to 3′ direction.
b. The replication fork moves forward along a double-stranded DNA molecule.
c. DNA replicates in the 3′ to 5′ direction on the lagging strand.
d. RNA primase places short RNA primer sequences along the DNA molecule.
e. DNA Polymerase I can connect short segments.

▪19. DNA replication in eukaryotes differs from replication in bacteria because
a. synthesis of the new DNA strand is from 3′ to 5′ in eukaryotes and from 5′ to 3′ in bacteria.
b. synthesis of the new DNA strand is from 5′ to 3′ in eukaryotes and from 3′ to 5′ in bacteria.
c. there are many replication forks in each eukaryotic chromosome and only one in bacterial DNA.
d. synthesis of the new DNA strand is from 5′ to 3′ in eukaryotes and is random in prokaryotes.
e. Okazaki fragments are produced in eukaryotic DNA replication but not in prokaryotic DNA replication.

20. In Chapter 11, the benefit mentioned for methylation of cytosine to the cells of some eukaryotic species was
a. increased rates of DNA replication.
b. slowing of DNA replication.
c. improved separation of DNA strands.
d. improved proofreading.
e. improved mismatch repair.

*21. In eukaryotes, Okazaki fragments are about ________ long.
a. 50 base pairs
b. 150 base pairs
c. 1,500 base pairs
d. 150,000 base pairs
e. 15,000,000 base pairs

22. Which of the following molecules functions to transfer information from one generation to the next?
a. DNA
b. mRNA
c. tRNA
d. Proteins
e. Lipids

23. Mutations are
a. heritable changes in the sequence of DNA bases that produce an observable phenotype.
b. heritable changes in the sequence of DNA bases.
c. mistakes in the incorporation of amino acids into proteins.
d. heritable changes in the mRNA of an organism.
e. None of the above

24. In order to show that DNA is the "transforming principle," Avery, MacLeod, and McCarty showed that DNA could transform avirulent strains of pneumococcus. This hypothesis was strengthened by their demonstration that
a. enzymes that destroyed proteins also destroyed transforming activity.
b. enzymes that destroyed nucleic acids also destroyed transforming activity.
c. enzymes that destroyed complex carbohydrates also destroyed transforming activity.
d. the transformation activity was destroyed by boiling.
e. other strains of bacteria could also be successfully transformed.

25. During infection of *E. coli* cells by bacteriophage T2,
a. proteins are the only phage components that actually enter the infected cell.
b. both proteins and nucleic acids enter the cell.
c. only protein from the infecting phage can also be detected in progeny phage.
d. only nucleic acids enter the cell.
e. more than one infecting phage particle is required to produce infection.

26. Bacteriophage nucleic acids were labeled by carrying out an infection of *E. coli* cells growing in
a. ^{14}C-labeled CO_2.
b. ^{3}H-labeled water.
c. ^{32}P-labeled phosphate.
d. ^{35}S-labeled sulfate.
e. ^{18}O -labeled water.

27. Information used by Watson and Crick to determine the structure of DNA included
a. electron micrographs of individual DNA molecules.
b. light micrographs of bacteriophage particles.
c. light micrographs of individual bacteria chromosomes.
d. nuclear magnetic resonance analysis of DNA.
e. X-ray crystallography of double-stranded DNA.

28. Double-stranded DNA looks a little like a ladder that has been twisted into a helix, or spiral. The side supports of the ladder are
a. individual nitrogenous bases.
b. alternating bases and sugars.
c. alternating bases and phosphate groups.
d. alternating sugars and phosphates.
e. alternating bases, sugars, and phosphates.

29. The steps of the ladder are
   a. individual nitrogenous bases.
   b. pairs of bases.
   c. alternating bases and phosphate groups.
   d. alternating sugars and bases.
   e. alternating bases, sugars, and phosphates.

30. If a double-stranded DNA molecule contains 30% T, how much G does it contain?
   a. 20%
   b. 30%
   c. 40%
   d. 50%
   e. 60%

★▪31. You have analyzed the DNA isolated from a newly discovered virus and found that its base composition is 32% A, 17% C, 32% G, and 19% T. What would be a reasonable explanation of this observation?
   a. The virus must be extraterrestrial.
   b. In some viruses, double-stranded DNA is made up of base pairs containing two purines or two pyrimidines.
   c. Some of the T was converted to C during the isolation procedure.
   d. The genome of the phage is single-stranded, not double-stranded.
   e. The genome of the phage must be circular, not linear.

▪32. The base composition of DNA that is complementary to the viral DNA described in the previous question would be
   a. 32% A, 17% C, 32% G, 19% T.
   b. 32% T, 17% G, 32% C, 19% A.
   c. 32% C, 17% A, 32% G, 19% T.
   d. 25% A, 25% G, 25% C, 25% T.
   e. 32% A, 32% T, 18% C, 18% G.

★33. In the Meselson-Stahl experiment, the conservative model of DNA replication is ruled out by which of the following observations?
   a. No completely heavy DNA is observed after the first round of replication.
   b. No completely light DNA ever appears, even after several replications.
   c. The product that accumulates after two rounds of replication is completely "heavy."
   d. Completely "heavy" DNA is observed throughout the experiment.
   e. Three different DNA densities are observed after a single round of replication.

34. During DNA replication
   a. one parental strand must be degraded to allow the other strand to be copied.
   b. the parental strands must separate so that both can be copied.
   c. the parental strands come back together after the passage of the replication fork.
   d. origins of replication always give rise to single replication forks.
   e. two replication forks diverge from each origin but one always lags behind the other.

▪35. The enzyme DNA ligase is required continuously during DNA replication because
   a. fragments of the leading strand must be joined together.
   b. fragments of the lagging strand must be joined together.
   c. the parental strands must be joined back together.
   d. 3′-deoxynucleoside triphosphates must be converted to 5′-deoxynucleoside triphosphates.
   e. the complex of proteins that work together at the replication fork must be kept from falling apart.

36. In DNA replication, each newly made strand is
   a. identical in DNA sequence to the strand from which it was copied.
   b. complementary in sequence to the strand from which it was copied.
   c. oriented in the same 3′ to 5′ direction as the strand from which it was copied.
   d. an incomplete copy of one of the parental strands.
   e. a hybrid molecule consisting of both ribo- and deoxyribonucleotides.

▪37. Which feature of the Watson–Crick model of DNA structure explains its ability to function in replication and gene expression?
   a. Each strand contains all the information present in the double helix.
   b. Structural and functional similarities of DNA and RNA.
   c. The double helix is right-handed and not left-handed.
   d. DNA replication does not require enzyme catalysts.
   e. Exposure of the bases in the major groove of the double helix.

38. The Hershey–Chase experiment convinced most scientists that
   a. bacteria can be transformed.
   b. DNA is indeed the carrier of hereditary information.
   c. DNA replication is semiconservative.
   d. the transforming principle requires host factors.
   e. All of the above

39. Which of the following features summarizes the molecular architecture of DNA?
   a. The two strands run in opposite directions.
   b. The molecule twists in the same direction as the threads of most screws.
   c. The molecule is a double-stranded helix.
   d. DNA has a uniform diameter.
   e. All of the above

40. The fidelity of DNA replication is astounding. During DNA synthesis, the error rate is on the order of one wrong nucleotide per
   a. 10,000.
   b. 100,000.
   c. 10^8–10^{12}.
   d. 10^{13}–10^{16}.
   e. one trillion.

41. Chargaff's rule states that
    a. DNA must be replicated before a cell can divide.
    b. viruses enter cells without their protein coat.
    c. only protein from the infecting phage can also be detected in progeny phage.
    d. only nucleic acids enter the cell during infection.
    e. the amount of cytosine equals the amount of guanine.

42. The first scientist(s) to suggest a mode of replication for DNA was (were)
    a. Linus and Pauling.
    b. Hershey and Chase.
    c. Albert Leverman.
    d. Watson and Crick.
    e. Meselson and Stahl.

43. In eukaryotic cells, each chromosome has
    a. one origin of replication.
    b. two origins of replication.
    c. many origins of replication.
    d. only one origin of replication per nucleus.
    e. None of the above

44. When adding the next monomer to a growing DNA strand, the monomer is added to the
    a. 1′ carbon of the deoxyribose.
    b. 2′ carbon of the deoxyribose.
    c. 3′ carbon of the deoxyribose.
    d. 4′ carbon of the deoxyribose.
    e. 5′ carbon of the deoxyribose.

45. The energy necessary for making a DNA molecule comes directly from the
    a. sugar.
    b. ATP.
    c. release of phosphates.
    d. NADPH.
    e. NADH.

46. A deoxyribose nucleotide is a
    a. deoxyribose plus a nitrogenous base.
    b. sugar and a phosphate.
    c. deoxyribose plus a nitrogenous base and a phosphate.
    d. ribose plus a nitrogenous base.
    e. nitrogenous base bonded at the 5′ end to a sugar–phosphate backbone.

47. Synthesis of DNA is
    a. spontaneous.
    b. endergonic.
    c. exergonic.
    d. pseudogonic.
    e. quasigonic.

48. The enzyme that removes the RNA primers is called
    a. DNA ligase.
    b. primase.
    c. reverse transcriptase.
    d. helicase.
    e. DNA polymerase I.

49. The enzyme that restores the phosphodiester linkage between adjacent fragments in the lagging strand during DNA replication is
    a. DNA ligase.
    b. primase.
    c. reverse transcriptase.
    d. helicase.
    e. DNA polymerase I.

50. A deoxyribose nucleoside is a
    a. deoxyribose plus a nitrogenous base.
    b. sugar and a phosphate.
    c. deoxyribose plus a nitrogenous base and a phosphate.
    d. ribose plus a nitrogenous base.
    e. nitrogenous base bonded at the 5′ end to a sugar-phosphate backbone.

51. The building blocks for a new DNA molecule are
    a. deoxyribose nucleoside monophosphates.
    b. deoxyribose nucleoside diphosphates.
    c. deoxyribose nucleoside triphosphates.
    d. deoxyribose nucleotide diphosphates.
    e. deoxyribose nucleotide triphosphates.

52. The force that holds DNA together in a double helix is
    a. the force of the twist.
    b. covalent bonds.
    c. ionic bonds.
    d. ionic interactions.
    e. hydrogen bonds.

53. The enzyme that unwinds the DNA prior to replication is called
    a. DNA polymerase III.
    b. DNA ligase.
    c. single-stranded DNA binding protein.
    d. primase.
    e. helicase.

54. The first repair of mistakes made during DNA replication is made by
    a. the mismatch repair system.
    b. DNA polymerase.
    c. excision repair.
    d. SOS repair.
    e. postreplication repair.

55. The error rate of changing an incorrect base with another incorrect base during proofreading is
    a. 1 in 10 bases.
    b. 1 in 100 bases.
    c. 1 in 1,000 bases.
    d. 1 in 10,000 bases.
    e. 1 in 1,000,000 bases.

56. The difference between DNA and RNA is that
    a. DNA has thymine and RNA has uracil.
    b. DNA has no oxygen bonded to the 2′ carbon; RNA does.
    c. DNA is the genetic material; RNA is not.
    d. DNA is double stranded and RNA cannot hydrogen bond.
    e. DNA is older than RNA.

*■57. If Hershey and Chase had found ^{35}S in both the pellet and the supernatant, the conclusion about the nature of DNA replication would be that
a. protein must be the information molecule.
b. it would be difficult to conclude anything from these results.
c. DNA is the genetic information molecule.
d. phage must have stuck to the bacteria.
e. phosphorus was in the information molecule.

*■58. An alien DNA-like molecule is isolated from the frozen remains of a martian life form that was found beneath the martian polar ice caps. It is found that for every base designated Q, there is 2 times the amount of the base R, and for every base Z there is 2 times the amount of the base designated S. If the molecule contains 12% R, how much Z would you expect?
a. 6%
b. 12%
c. 24%
d. 27.33%
e. 54.66%

*■59. An alien DNA-like molecule was isolated from the frozen remains of a martian life form, which was found beneath the martian polar ice caps. It is established that for every base designated Q, there is 2 times the amount of the base R, and for every base Z, there is 2 times the amount of the base designated S. Select the molecular model that would best fit these data.
a. The molecule is single stranded.
b. The molecule is antiparallel.
c. The molecule is triple stranded.
d. The molecule is helical.
e. The molecule is helical, double stranded, and antiparallel.

*■60. Had Meselson and Stahl observed one intermediate, somewhat-smeared band after growing bacteria for one generation, and then after two generations, again found one somewhat-smeared band, what would have been concluded about the mode of DNA replication?
a. DNA replicates semiconservatively.
b. DNA replicates conservatively.
c. DNA replicates semidiscontinuously.
d. DNA replicates dispersively.
e. None of the above

*61. Pyrophosphate is a
a. building block for DNA synthesis.
b. by-product of DNA synthesis.
c. precursor to DNA synthesis.
d. fire phosphate used in nucleic acid metabolism.
e. All of the above

62. Which one of the following is *not* found in DNA?
a. Carbon
b. Oxygen
c. Nitrogen
d. Hydrogen
e. Sulfur

63. Boiling DNA causes it to become
a. single stranded.
b. monomers.
c. destroyed.
d. smaller the longer it is boiled.
e. All of the above

■64. What was most remarkable about the Griffith experiment?
a. Griffith obtained his results despite the fact that he failed his medical board exam.
b. DNA, not protein, was found to be the genetic molecule.
c. Something from a dead organism could change living cells.
d. Viruses, which were nonliving, could change living cells.
e. Smooth bacteria could survive heating.

65. The Hershey–Chase experiment
a. proved semiconcervative replication is the mode for DNA replication.
b. used ^{32}P to label protein.
c. used ^{35}S to label DNA.
d. helped prove DNA was the genetic molecule.
e. b and c

66. Griffith could distinguish the two strains of pneumococcus due to
a. colony appearance in culture.
b. differences in their lethality in mice.
c. their sizes.
d. surprisingly, their odor.
e. a and b

67. Labeled dideoxynucleotides are used in sequencing reactions to
a. create truncated replication products.
b. create sequences with known ends.
c. add as building blocks for DNA synthesis.
d. a and b
e. a, b, and c

68. The maximum length sequence that can be read at a time using current technology is approximately
a. 50 base pairs.
b. 100 base pairs.
c. 500 base pairs.
d. 1,000 base pairs.
e. 5,000 base pairs.

69. In PCR, it is ________ that creates single-stranded template molecules.
a. heat
b. high salt concentration
c. DNA polymerase
d. exonuclease
e. a primer

70. Ideally, PCR ________ increases the amount of DNA during additional cycles.
a. additively
b. gradually

c. linearly
d. systematically
e. exponentially

Study Guide Questions

1. Griffith's experiments showing the transformation of R strain pneumococcus bacteria to S strain pneumococcus bacteria in the presence of heat-killed S strain bacteria gave evidence that
   a. **an external factor was affecting the R strain bacteria.**
   b. DNA was definitely the transforming factor.
   c. S strain bacteria could be reactivated after heat killing.
   d. All of the above

2. Experiments by Avery, MacLeod, and McCarty supported DNA as the genetic material by showing that
   a. both protein and DNA samples provided the transforming factor.
   b. DNA was not complex enough to be the genetic material.
   c. **only samples with DNA provided transforming activity.**
   d. even though DNA was molecularly simple, it provided adequate variation to act as the genetic material.

3. Hershey and Chase used radioactive ^{35}S and ^{32}P in experiments to provide evidence that DNA was the genetic material. These experiments pointed to DNA because
   a. progeny viruses retained ^{32}P but not ^{35}S.
   b. retention of ^{32}P in progeny viruses indicated that DNA was passed on.
   c. loss of ^{35}S in progeny viruses indicated that proteins were not passed on.
   d. **All of the above**

4. X-ray crystallography provides information about the ____________ of DNA but is limited because of the ____________ of DNA. The technique is based on the pattern of ____________ off the atoms in the molecule.
   a. structure; difficulty of purification; light absorption
   b. **dimensions; molecular weight; diffraction**
   c. molecular weight; shape; diffraction
   d. dimensions; linearity; light absorption

5. Chargaff's rules of base pairing states that
   a. the ratio of purines to pyrimidines is roughly equal in all tested organisms.
   b. the ratio of A to T is roughly equal in all tested organisms.
   c. the ratio of A + T and G + C is roughly equal in all tested organisms.
   d. **a and b**

6. Watson and Crick's model allowed them to visualize
   a. the molecular bonds of DNA.
   b. how the purines and pyrimidines fit together in a double helix.
   c. that the two strands of the DNA double helix were antiparallel.
   d. **All of the above**

7. A fundamental requirement for the function of genetic material is that it must be
   a. conserved among all organisms with very little variation.
   b. passed intact from organism to organism.
   c. **replicable.**
   d. found outside of the nucleus.

8. Evidence indicating that DNA replication was semiconservative came from
   a. DNA staining techniques.
   b. DNA sequencing.
   c. **density gradient studies using "heavy" nucleotides.**
   d. None of the above

9. Current evidence indicates that replication complexes are attached to stationary nuclear components and that DNA is threaded through these complexes. Which of the following best describes the role of the replication complex?
   a. The complex acts as an enzymatic center for DNA replication.
   b. The complex binds specifically to replication origins, then controls the rate at which replication occurs.
   c. The complex is the initiating site of replication forks.
   d. **All of the above**

10. The primary function of DNA polymerase is to
    a. **add nucleotides to the growing daughter strand.**
    b. seal nicks along the sugar–phosphate backbone of the daughter strand.
    c. unwind the parent DNA double helix.
    d. prevent reassociation of the denatured parent DNA strands.

11. The lagging daughter strand of DNA is synthesized in what appears to be the "wrong" direction. This synthesis is accomplished by
    a. ligating (connecting) short Okazaki fragments that are synthesized in short spurts in the "right" direction.
    b. primase.
    c. using multiple primers and DNA polymerase I.
    d. **a and b**

12. RNA primers are necessary in DNA synthesis because
    a. **DNA polymerase can only add to an existing strand of nucleotides.**
    b. DNA polymerase can only add to an existing RNA strand.
    c. DNA primase is the first enzyme in the replication complex.
    d. All of the above

13. Proofreading and repair occur
    a. **at anytime during or after synthesis of DNA.**
    b. only before DNA methylation occurs.
    c. only in the presence of DNA polymerase.
    d. only in the presence of an excision repair mechanism.

14. DNA replication is an ____________ process and ____________ energy.
    a. exergonic; does not require
    b. endothermic; does require
    c. **endergonic; does require**
    d. endodontic; does not require

15. *T. aquaticus* DNA polymerase is not denatured during the heat cycling required to denature DNA. This property allowed advances in what technique?
    a RFLP analysis
    b. PCR
    c. Sequencing
    d. EPA

16. Thirty percent of the bases in a sample of DNA extracted from eukaryotic cells is adenine. What percentage of cytosine is present in this DNA?
    a. 10%
    b. 20%
    c. 30%
    d. 40%

17. Which of the following represents a bond between a purine and a pyrimidine (in that order)?
    a. C–T
    b. G–A
    c. G–C
    d. T–A

18. Which of the following statements about DNA replication is false?
    a. Okazaki fragments are the initiators of continuous DNA synthesis along the leading strand.
    b. Replication forks represent areas of active DNA synthesis on the chromosomes.
    c. Error rates for DNA replication are often less than one in every billion base pairings.
    d. Ligases and polymerases function in the vicinity of replication forks.

19. Which of the following would not be found in a DNA molecule?
    a. Purines
    b. Ribose sugars
    c. Phosphates
    d. Sulfur

20. If a nucleotide lacking a hydroxyl group at the 3′ end is added to a PCR, what would be the outcome?
    a. No additional nucleotides would be added to a growing strand containing that nucleotide.
    b. Strand elongation would proceed as normal.
    c. Nucleotides would only be added at the 5′ end.
    d. *T. aquaticus* DNA polymerase would be denatured.

End of Chapter Questions

1. To show that DNA in cell extracts was responsible for genetic transformation in pneumococcus, important corroborating evidence was that
    a. enzymes that destroyed proteins also destroyed transforming activity.
    b. enzymes that destroyed DNA also destroyed transforming activity.
    c. enzymes that destroyed polysaccharides also destroyed transforming activity.
    d. boiling destroyed transforming activity.
    e. enzymes that destroyed RNA also destroyed transforming activity.

2. In the Hershey–Chase experiment,
    a. DNA from parent bacteriophages appeared in progeny bacteriophages.
    b. most of the phage DNA never entered the bacteria.
    c. more than three-fourths of the phage protein appeared in progeny phages.
    d. DNA was labeled with radioactive sulfur.
    e. DNA formed the coat of the bacteriophages.

3. Which statement about complementary base pairing is *not* true?
    a. It plays a role in DNA replication.
    b. In DNA, T pairs with A.
    c. Purines pair with purines, and pyrimidines pair with pyrimidines.
    d. In DNA, C pairs with G.
    e. The base pairs are of equal length.

4. In semiconservative replication of DNA,
    a. the original double helix remains intact and a new double helix forms.
    b. the strands of the double helix separate and act as templates for new strands.
    c. polymerization is catalyzed by RNA polymerase.
    d. polymerization is catalyzed by a double helical enzyme.
    e. DNA is synthesized from amino acids.

5. In dideoxy-DNA sequencing, what is the role of the dideoxy-nucleoside triphosphates?
    a. Replication in the 3′ to 5′ direction.
    b. Formation of Okazaki fragments.
    c. Inhibition of all replication after the reaction is complete.
    d. Stopping replication after partial completion of the strand.
    e. Primer for replication.

6. Which of the following is *not* required for DNA replication?
    a. A short strand of RNA to act as a primer.
    b. DNA to act as a template.
    c. Deoxyribonucleoside triphosphates.
    d. ATP for energy.
    e. DNA polymerase.

7. The 3′ end of a DNA strand is defined as the place where
    a. the phosphate group is not bound to another nucleotide.
    b. both DNA strands end opposite each other.
    c. DNA polymerase binds to begin replication.
    d. there is a free —OH group at the 3′ carbon of deoxyribose.
    e. three A residues are present.

8. The leading strand and lagging strand in DNA replication differ in that only on the lagging strand
    a. DNA is replicated as short fragments.
    b. RNA primer is present.
    c. replication proceeds in the 5′ to 3′ direction.
    d. DNA ligase is not needed.
    e. replication is slower.

9. The polymerase chain reaction
   a. is a method for sequencing DNA.
   b. is used to transcribe specific genes.
   c. amplifies specific DNA sequences.
   d. does not require DNA replication primers.
   e. uses a DNA polymerase that denatures at 55°C.

10. The following events occur in excision repair of DNA. What is their proper order?
   1 Base-paired DNA is made complementary to the template.
   2 Damaged bases are recognized.
   3 DNA ligase seals the new strand to existing DNA.
   4 Part of a single strand is excised.
   a. 1234
   b. 2134
   c. 2413
   d. 3421
   e. 4231

Student Web Site Self-Quizzes

1. Which of the following statements about the work of Griffith, and then Avery, MacLeod, and McCarty, on *Streptococcus pneumoniae* is *false*?
   a. Only the S strain has a cell wall-like capsule.
   b. The mouse in Griffith's experiments would also have died if injected with living S strain and heat-killed R strain.
   c. The transforming principle is associated with the S strain's capsule.
   d. Transformation of living R strain into S strain could also occur in a test tube without involving a mouse.
   e. The transforming principle carried heritable information.

2. Which of the following was *not* important evidence that helped advance the idea that DNA is the genetic material?
   a. DNA is present in the head of the T2 bacteriophage, but protein is only in its coat.
   b. The amount of DNA staining is constant in the cells of a species, except that gametes have one-half as much as somatic cells.
   c. That heat-killed S strain pneumococci could transform living R strain into virulent S strain.
   d. The demonstration by Avery, MacLeod, and McCarthy that the transforming principle is DNA.
   e. That only ^{32}P ended up in the bacteria in the Hershey–Chase "blender" experiment.

3. Which of the following statements about the Hershey–Chase experiment is *false*?
   a. Sulfur is present in protein, but not in DNA; phosphorus is present in DNA, but not in protein.
   b. The ^{32}P will end up in the supernatant after centrifugation.
   c. The purpose of the blender is to detach viruses from the bacteria.
   d. Progeny generations of T2 bacteriophage contained ^{32}P but no ^{35}S.
   e. A conclusion of the Hershey–Chase experiment was that T2 injects the DNA from its head into the bacterium.

4. You estimate that cytosine represents 30% of the total pyrimidine content of DNA from a mouse cell. Using Chargaff's rules, you determine that the percentage of total nucleotides in the DNA represented by adenine is
   a. unknown.
   b. 70%.
   c. 35%.
   d. 30%.
   e. 15%.

5. Which of the following statements about the structure of DNA is *false*?
   a. The width of the DNA molecule is variable since it can accommodate nucleotides containing varying numbers of nitrogen-based "rings."
   b. Hydrogen bonds determine which nitrogenous bases can pair together.
   c. A total of 10 pairs of nucleotides are included in each complete "turn" of the double helix.
   d. Minor and major grooves alternate along the length of the molecule.
   e. The distance between the nitrogenous base pair "rungs" of the helix is about 0.34 nm.

6. Of the following steps occurring in the formation of a replication fork and the initiation of replication, select the step that would be fifth in sequence.
   a. Attachment of DNA polymerase III
   b. Attachment of single-stranded DNA-binding proteins
   c. Unwinding of DNA by helicase
   d. Formation of RNA primer by RNA primase
   e. Synthesis of leading and lagging strands

7. What happens in the third step in processing Okazaki fragments in replication of the lagging strand?
   a. DNA ligase replaces the Okazaki fragment with DNA.
   b. DNA ligase catalyzes the formation of phosphodiester linkages.
   c. DNA polymerase I replaces the RNA primer with DNA.
   d. DNA polymerase I replaces the Okazaki fragment with DNA.
   e. DNA polymerase III synthesizes Okazaki fragments.

8. Which of the following statements regarding the polymerase chain reaction (PCR) is *false*?
   a. Short regions of DNA are copied many times in a test tube by DNA polymerase.
   b. PCR requires the input of heat to separate template DNA strands prior to synthesis.
   c. PCR requires prior knowledge of the DNA sequence for the segment to be amplified.
   d. Each cycle of PCR essentially doubles the number of DNA molecules in the test tube.
   e. Primers used in PCR can be of any length and sequence, as long as they contain only deoxyribonucleotides.

TEST FILE

★ *Indicates a **difficult** question*

■ *Indicates a **conceptual** question*

12 From DNA to Protein: Genotype to Phenotype

Fill in the Blank

1. Prototrophs ("original eaters") grow on minimal media whereas **auxotrophs** ("increased eaters") require specific additional nutrients.

2. The basic units of DNA and RNA molecules are the **nucleotides**.

3. RNA differs from DNA in base composition because it contains **uracil** instead of thymine.

4. The strand of DNA that is transcribed into RNA is the **template** strand.

5. The part of a protein that determines whether translation will continue in the cytosol or at the endoplasmic reticulum is the **signal** sequence.

6. The excess of codons (64) over amino acids (20) indicates that the genetic code is **redundant**.

7. A mRNA molecule with several ribosomes attached at the same time is called a(n) **polysome**.

8. Small ribosomal subunits are dispersed into smaller components by placing them into a detergent solution. Upon removal of the detergent, the components interact to create new intact subunits by a process called **self-assembly**.

9. A mutation that causes a change in the nitrogenous base sequence of a DNA molecule but no change in the amino acid sequence it codes for is called a(n) **silent** mutation.

10. The fact that some tRNA molecules do not have to pair exactly is called **wobble**.

11. The portion of the tRNA molecule that complementary base base-pairs with the mRNA is called the **anticodon**.

Multiple Choice

1. After irradiating *Neurospora*, Beadle and Tatum collected mutants that require arginine to grow. These mutants
   a. will not grow on minimal media but will grow on minimal media with arginine.
   b. will grow on minimal media and on minimal media with arginine.
   c. will not grow on minimal media and will not grow on minimal media with arginine.
   d. will grow on minimal media but will not grow on minimal media with arginine.
   e. None of the above

■2. Within a group of mutants with the same growth requirement, that is, the same overt phenotype, mapping studies determined that individual mutations were on different chromosomes. This indicates that
   a. the same gene governs all the steps in a particular biological pathway.
   b. different genes can govern different individual steps in the same biological pathway.
   c. different genes govern the same step in a particular biological pathway.
   d. all biological pathways are governed by different genes.
   e. genes do not govern steps in biological pathways.

3. The study of *Neurospora* mutants that grew on various supplemented media led to
   a. determining the steps in biological pathways.
   b. the "one-gene, one-enzyme" theory.
   c. the idea that genes are "on" chromosomes.
   d. a and b
   e. a and c

4. The rates of DNA mutations are ________ in different organisms.
   a. the same
   b. constant
   c. different
   d. dependent on health
   e. dependent on temperature

5. Genes code for
   a. enzymes.
   b. polypeptides.
   c. RNA.
   d. All of the above
   e. None of the above

▪6. Why is it that DNA, made up of only four different bases, can encode the information necessary to specify the workings of an entire organism?
a. DNA molecules are extremely long.
b. DNA molecules form codons of three bases that code for amino acids.
c. The same DNA sequence can be used repeatedly.
d. DNA can be replicated with low error rates.
e. All of the above

▪7. RNA differs from DNA because
a. RNA contains uracil instead of thymine and it is (usually) single-stranded.
b. RNA contains uracil instead of thymine.
c. RNA contains thymine instead of uracil and it is (usually) single-stranded.
d. RNA contains uracil instead of cytosine.
e. None of the above

8. What are the three major properties of genes that are explained by the structure of DNA?
a. They contain information, direct the synthesis of proteins, and are contained in the cell nucleus.
b. They contain nitrogenous bases, direct the synthesis of RNA, and are contained in the cell nucleus.
c. They replicate exactly, are contained in the cell nucleus, and direct the synthesis of cellular proteins.
d. They encode the organism's phenotype, are passed on from one generation to the next, and contain nitrogenous bases.
e. They contain information, replicate exactly, and change to produce a mutation.

9. Which of the following statements about the flow of genetic information is *true*?
a. Proteins encode information that is used to produce other proteins of the same amino acid sequence.
b. RNA encodes information that is translated into DNA, and DNA encodes information that is translated into proteins.
c. Proteins encode information that can be translated into RNA, and RNA encodes information that can be transcribed into DNA.
d. DNA encodes information that is translated into RNA, and RNA encodes information that is translated into proteins.
e. None of the above

10. RNA polymerase uses the _________ DNA template to synthesize a _________ mRNA.
a. 5′ to 3′, 5′ to 3′
b. 3′ to 5′, 3′ to 5′
c. 5′ to 3′, 3′ to 5′
d. 3′ to 5′, 5′ to 3′
e. Examples of all these have been found.

11. Which of the following molecules functions to transfer information from the nucleus to the cytoplasm?
a. DNA
b. mRNA
c. tRNA
d. Proteins
e. Lipids

*12. mRNA is sythesized in the _________ direction which corresponds to the _________ of the protein.
a. 5′ to 3′, N terminus to C terminus
b. 3′ to 5′, C terminus to N terminus
c. 5′ to 3′, C terminus to N terminus
d. 3′ to 5′, N terminus to C terminus
e. Examples of all these have been found.

13. Which of the following molecules functions to transfer information from one generation to the next?
a. DNA
b. mRNA
c. tRNA
d. Proteins
e. Lipids

14. Which of the following molecules functions to transfer information from mRNA to protein?
a. DNA
b. mRNA
c. tRNA
d. Proteins
e. Lipids

15. A sequence of three RNA bases can function as a
a. codon.
b. anticodon.
c. gene.
d. a and b
e. a and c

▪16. The difference between mRNA and tRNA is that
a. tRNA has a more elaborate three-dimensional structure due to extensive base pairing.
b. tRNAs are usually very much smaller than mRNAs.
c. mRNA has a more elaborate three-dimensional structure due to extensive base pairing.
d. a and b
e. None of the above

17. Termination of transcription involves a
a. stop codon.
b. terminator sequence.
c. termiproteator.
d. hairline slip.
e. series of A's.

18. Ribosomes are a collection of
a. small proteins that function in translation.
b. proteins and small RNAs that function in translation.
c. proteins and tRNAs that function in transcription.
d. proteins and mRNAs that function in translation.
e. mRNAs and tRNAs that function in translation.

19. The endoplasmic reticulum functions in protein synthesis
a. as the site where mRNA attaches.
b. as the site where all ribosomes bind.
c. as the site of translation of membrane-bound and exported proteins.
d. to produce tRNAs.
e. to bring together mRNA and tRNA.

▪20. Retroviruses do not follow the "central dogma" of DNA→RNA→protein because they
a. **contain RNA that is used to make DNA.**
b. contain DNA that is used to make more RNA.
c. contain DNA that is used to make tRNA.
d. contain only RNA as the genetic material.
e. do not contain either DNA or RNA as the genetic material.

21. Mutations are
a. heritable changes in the sequence of DNA bases that produce an observable phenotype.
b. **heritable changes in the sequence of DNA bases.**
c. mistakes in the incorporation of amino acids into proteins.
d. heritable changes in the mRNA of an organism.
e. None of the above

22. The type of mutation that stops translation of a protein is a(n)
a. missense mutation.
b. **nonsense mutation.**
c. frame-shift mutation.
d. aberration.
e. None of the above

23. The type of mutation that is an insertion or a deletion of a single base is a(n)
a. missense mutation.
b. nonsense mutation.
c. **frame-shift mutation.**
d. aberration.
e. None of the above

24. The "central dogma" of molecular biology states that
a. information flow between DNA, RNA, and protein is reversible.
b. information flow in the cell is from protein to RNA to DNA.
c. **information flow in the cell is unidirectional from DNA to RNA to protein.**
d. the DNA sequence of a gene can be predicted if we know the amino acid sequence of the protein it encodes.
e. the genetic code is ambiguous but not degenerate.

25. Transcription is the process of
a. synthesizing a DNA molecule from an RNA template.
b. assembling ribonucleoside triphosphates into an RNA molecule without a template.
c. **synthesizing an RNA molecule using a DNA template.**
d. synthesizing a protein using information from a messenger RNA.
e. replicating a single-stranded DNA molecule.

26. A transcription start signal is called a(n)
a. initiation codon.
b. **promoter.**
c. origin.
d. operator.
e. nonsense codon.

27. Initiation of transcription requires
a. a temporary stoppage of DNA replication.
b. **a temporary separation of the strands in the template DNA.**
c. destruction of one of the strands of the template DNA.
d. relaxation of positive supercoils in the DNA template.
e. induction of positive supercoils in the DNA template.

28. The adapters that allow translation of the four-letter nucleic acid language into the 20-letter protein language are called
a. aminoacyl tRNA synthetases.
b. **transfer RNA's.**
c. ribosomal RNA's.
d. messenger RNA's.
e. ribosomes.

29. The number of codons that actually specify amino acids is
a. 20.
b. 23.
c. 45.
d. 60.
e. **61.**

30. The genetic code is best described as
a. **redundant but not ambiguous.**
b. ambiguous but not redundant.
c. both ambiguous and redundant.
d. neither ambiguous nor redundant.
e. nonsense.

31. The three codons in the genetic code that do not specify amino acids are called
a. missense codons.
b. start codons.
c. **stop codons.**
d. promoters.
e. initiator codons.

32. In eukaryotes, ribosomes become associated with endoplasmic reticulum membranes when
a. a signal sequence on the mRNA interacts with a receptor protein on the membrane.
b. a signal sequence on the ribosome interacts with a receptor protein on the membrane.
c. a signal sequence at the amino terminus of the protein being synthesized interacts with a receptor protein on the ribosome.
d. **a signal sequence on the protein being synthesized interacts with a membrane attachment site on the ribosome.**
e. the messenger RNA passes through a pore in the membrane.

33. Viruses that violate the "central dogma" through the use of an enzyme that makes DNA copies of an RNA molecule are called
a. bacteriophage.
b. **retroviruses.**

c. RNA viruses.
d. DNA viruses.
e. enveloped viruses.

34. Sickle-cell disease is caused by a ________ mutation.
a. nonsense
b. missense
c. frame-shift
d. temperature sensitive
e. silent

35. The classic work of Beadle and Tatum, later refined by others, provided evidence for
a. the one-gene, one-enzyme hypothesis.
b. the one-gene, one-polypeptide hypothesis.
c. explaining how information in genes is translated into traits.
d. explaining how some mutations affect organisms.
e. All of the above

36. RNA polymerase is a
a. RNA-directed DNA polymerase.
b. RNA-directed RNA polymerase.
c. DNA-directed RNA polymerase.
d. typical enzyme.
e. form of RNA.

37. The direction of synthesis for a new mRNA molecule is
a. 5′ to 3′ from a 5′ to 3′ template strand.
b. 5′ to 3′ from a 3′ to 5′ template strand.
c. 3′ to 5′ from a 5′ to 3′ template strand.
d. 3′ to 5′ from a 3′ to 5′ template strand.
e. 5′ to 5′ from a 3′ to 3′ template strand.

38. The region of DNA in prokaryotes that RNA polymerase binds to most tightly is the
a. promoter.
b. poly C center.
c. enhancer.
d. operator site.
e. minor groove.

39. Promoters are made of
a. protein.
b. carbohydrate.
c. lipid.
d. nucleic acids.
e. amino acids.

*40. The energy for transcription is derived from
a. ATP.
b. GTP.
c. ATP and GTP.
d. the phosphodiester linkages in the nucleoside triphosphates incorporated.
e. glucose.

▪41. DNA is composed of two strands, and only one strand is typically used as a template for RNA synthesis. The correct strand is chosen because
a. both are tried and the one that works is remembered.
b. only one strand has the start codon.
c. the promoter acts to aim the RNA polymerase.
d. a start factor informs the system.
e. the correct strand is picked half the time.

▪42. There are differences in the amount of transcription that takes place for different genes. One cause for this would be that
a. different promoters have different affinities for RNA polymerase.
b. longer genes take longer to transcribe.
c. the outcome is influenced by random chance.
d. ribosomes tend to attach to transcripts even before transcription is completed.
e. None of the above

43. There are ________ different RNA polymerases in eukaryotes.
a. 1
b. 2
c. 3
d. 4
e. 5

44. In eukaryotes, the RNA polymerase that synthesizes mRNA is
a. I.
b. II.
c. III.
d. IV.
e. V.

45. Proteins are synthesized from the ________, in the ________ direction along the mRNA.
a. N terminus to C terminus, 5′ to 3′
b. C terminus to N terminus, 5′ to 3′
c. C terminus to N terminus, 3′ to 5′
d. N terminus to C terminus, 3′ to 5′
e. N terminus to N terminus, 5′ to 5′

*▪46. Two mutants, 1 and 2, are found that have defects in different genes that code for enzymes, which are part of the same biochemical pathway. Substance X makes it possible for both mutants to grow. Substance Z is only effective for mutant 2. Which mutant has the defective enzyme that is earliest in the biochemical pathway?
a. Mutant 1
b. Mutant 2
c. Z is before X.
d. X is before Z.
e. None of the above

▪47. Imagine that a novel life form is found deep within Earth's crust. Evaluation of its DNA yields no surprises. However, it is found that a codon for this life form is just two bases in length. How many different amino acids could this organism be composed of?
a. 4
b. 8
c. 16
d. 32
e. 64

48. The error rate for RNA polymerase is __________ that for most DNA polymerases.
a. less than
b. equal to
c. greater than
d. greater for frame shifts but less for base substitutions than
e. greater for base substitutions but less for frame shifts than

49. Poly uracil codes for
a. three different amino acids.
b. poly tryptophan.
c. mRNA.
d. a fatty acid.
e. poly phenylalanine.

▪50. It is known that fewer different tRNA molecules exist than the otherwise expected 61. This is possible because
a. the third position of the codon does not have to pair conventionally.
b. the second position of the codon does not have to pair conventionally.
c. the anticodon do not have the conventional bases at the anticodon region.
d. there are fewer amino acids than there are possible codons.
e. the code is degenerating.

51. The enzyme that charges the tRNA molecules with appropriate amino acids is
a. tRNA chargeatase.
b. amino tRNA chargeatase.
c. transcriptase.
d. aminoacyl-tRNA synthetase.
e. None of the above

52. __________ is the addition of sugar residues to the protein after translation.
a. Glycation
b. Glycosylation
c. Phosphorylation
d. Proteolysis
e. Exonuclease digestion

*53. In eukaryotic cells, proteins that contain covalently attached sugar residues are translated
a. in the nucleus.
b. in the cytoplasm.
c. in mitochondria.
d. on the endoplasmic reticulum.
e. on the Golgi apparatus.

54. It is currently believed that the enzyme that catalyses formation of the peptide bond during translation is composed of
a. amino acids.
b. protein.
c. carbohydrate.
d. RNA.
e. DNA.

55. The termination of transcription is signaled by
a. the stop codon.
b. a sequence of nitrogenous bases.
c. a protein bound to a certain region of DNA.
d. rRNA.
e. tRNA.

56. During translation initiation, the first site occupied by a charged tRNA is the
a. A site.
b. B site.
c. large subunit.
d. T site.
e. P site.

57. Some proteins are processed by __________ after translation, which is cleavage of the protein to make a shortened finished protein.
a. glycation
b. glycosylation
c. phosphorylation
d. proteolysis
e. exonuclease digestion

58. During translation elongation, the existing polypeptide chain is transferred to
a. the tRNA occupying the A site.
b. the tRNA occupying the P site.
c. the ribosomal rRNA.
d. a signal recognition particle.
e. None of the above

59. The stop codons code for
a. no amino acid.
b. methionine.
c. glycine.
d. halt enzyme.
e. DNA binding protein.

60. Breaking and rejoining of chromosomes can lead to
a. deletions.
b. duplications.
c. inversions.
d. translocations.
e. All of the above.

*61. Damage to DNA can be caused by __________ absorbed by thymine in DNA and causing interbase covalent bonds, damaging the DNA.
a. x-rays
b. cosmic radiation
c. ultraviolet radiation
d. smoke
e. cigarettes

Study Guide Questions

1. Transcription in prokaryotic cells is
   a. initiated at a promoter using one of three RNA polymerases (RNA polymerase II).
   b. initiated at a start codon with the help of initiation factors and the small subunit of the ribosome.
   c. initiated at a promoter and uses only one strand of DNA, the template strand, to synthesize a complementary RNA strand.
   d. is terminated at stop codons.

2. Which of the following about RNA polymerase is *not* true?
   a. It synthesizes mRNA in a 5′-to-3′ direction reading the DNA strand 3′ to 5′.
   b. It synthesizes mRNA in a 3′-to-5′ direction reading the DNA strand 5′ to 3′.
   c. It binds at the promoter and unwinds the DNA.
   d. It uses the energy of pyrophosphate hydrolysis to synthesize RNA.

3. Translation of messenger RNA into protein occurs
   a. in a 3′ to 5′ direction and from N terminus to C terminus.
   b. in a 5′ to 3′ direction and from N terminus to C terminus.
   c. in a 3′ to 5′ direction and from C terminus to N terminus.
   d. in a 5′ to 3′ direction and from N terminus to C terminus.

4. If a codon were read two bases at a time instead of three bases at a time, how many different possible amino acids could be specified?
   a. 16
   b. 64
   c. 8
   d. 32

5. Translate the following mRNA:
   3′-GAUGGUUUUAAAGUA-5′
   a. NH2 met—lys—phe—leu—stop COOH
   b. NH2 met—lys—phe—trp—stop COOH
   c. NH2 asp—gly—phe—lys—val COOH
   d. NH2 asp—gly—phe—lys —stop COOH

6. What would happen if a mutation occurred in the DNA such that the second codon of a polypeptide, UGG, was changed to a UAG?
   a. Nothing. The ribosome would skip that codon and translation would continue.
   b. Translation would continue, but the reading frame of the ribosome would be shifted.
   c. Translation would stop at the second codon and no functional protein would be made.
   d. Translation would continue, but the second amino acid in the protein would be different.

7. If the synthetic RNA listed below were added to a test tube containing all of the components necessary for protein translation to occur, what would the amino acid sequence be?
   5′-AUAUAUAUAUAU-3′
   a. Polyphenylalanine
   b. Isoleucine-tyrosine-isoleucine-tyrosine
   c. Isoleucine-isoleucine-isoleucine-isoleucine
   d. Tyrosine-tyrosine-tyrosine-tyrosine

8. What part of the tRNA base-pairs with the codon in the mRNA?
   a. The 3′ end where the amino acid is covalently attached
   b. The 5′ end
   c. The anticodon
   d. The promoter

9. Peptidyl transferase is an enzyme that is an
   a. enzyme found in the nucleus of the cell which assists in the transfer of the RNA to the cytoplasm.
   b. enzyme found in the large subunit of the ribosome which catalyzes the formation of the peptide bond in the growing polypeptide.
   c. RNA molecule that is catalytic.
   d. b and c

10. Termination of translation requires
    a. release factor, initiator tRNA, and ribosomes.
    b. initiation factors, the small subunit of the ribosome, and mRNA.
    c. elongation factors and charged tRNAs.
    d. a stop codon positioned at the A site of the ribosome, and peptidyl transferase and release factor.

11. Which mutation is the most deleterious?
    a. A missense mutation in the second codon
    b. A frameshift mutation in the second codon
    c. A nonsense mutation in the last codon
    d. A silent mutation in the second codon

12. If the DNA encoding a nuclear signal sequence were placed in the gene for a cytoplasmic protein, what would happen?
    a. The protein would be localized to the lysosomes.
    b. The protein would be localized to the nucleus.
    c. The protein would be directed to the cytoplasm.
    d. The protein would stay in the endoplasmic reticulum.

13. Auxotrophs are mutant strains that
    a. can grow on minimal media.
    b. require the addition of an essential nutrient to grow on minimal plate.
    c. behave like wild-type strains.
    d. can only grow if arginine is added to the growth media.

14. The central dogma of molecular biology states that _________ is transcribed into _______ which is translated into ________.
    a. genes, polypeptides, gene product
    b. protein, DNA, RNA
    c. DNA, mRNA, tRNA
    d. DNA, RNA, protein

15. A gene product is
    a. an enzyme.
    b. a polypeptide.

c. RNA's.
d. All of the above.

16. The enzyme that catalyzes the synthesis of RNA is
a. DNA polymerase.
b. tRNA synthetase.
c. ribosomal RNA.
d. RNA polymerase.

End of Chapter Questions

1. Which of the following is *not* a difference between RNA and DNA?
a. RNA has uracil, and DNA has thymine.
b. RNA has ribose, and DNA has deoxyribose.
c. RNA has 5 bases, and DNA has 4.
d. RNA is a single polynucleotide strand, and DNA is a double strand.
e. RNA is relatively smaller than human chromosomal DNA.

2. A *Neurospora* mutant cannot grow unless the amino acid, leucine, is added to its growth medium. This strain is
a. dependent on leucine for energy.
b. mutated in genes coding for the synthesis of all 20 amino acids
c. inhibited by leucine.
d. wild-type for the synthesis of all amino acids except for leucine.
e. mutated for the synthesis of all amino acids except for leucine.

3. A region of DNA template strand has the sequence 3′-ATTCGC-5′. What is the sequence of RNA transcribed from this DNA?
a. 3′-AUUCGC-5′
b. 3′-TAAGCG-5′
c. 5′-UAAGCG-3′
d. 5′-AUUCGC-3′
e. 5′-ATTCGC-3′

4. Which of the following is an anticodon found in the tRNA for serine?
a. 5′-CGA-3′
b. 5′-UCU-3′
c. 5′-ACU-3′
d. 5′-AGC-3′
e. 5′-CCU-3′

5. The termination of transcription is signaled by
a. the stop codon.
b. a sequence of bases.
c. a protein bound to DNA.
d. RNA polymease running out of template.
e. lack of deoxyribonucleoside triphosphates.

6. Transcription
a. produces only mRNA.
b. requires ribosomes.
c. requires tRNAs.
d. produces RNA growing from the 5′ end to the 3′ end.
e. takes place only in eukaryotes.

7. Which of the following single base substitutions in the sense strand of DNA would cause a change in phenotype?
a. CCT to CCC
b. GAG to TAG
c. ATG to ATT
d. CTG to CTT
e. CAA to CAG

8. Which statement about RNA is *not* true?
a. Transfer RNA functions in translation.
b. Ribosomal RNA functions in translation.
c. RNA's are produced in transcription.
d. Messenger RNA's are produced on ribosomes.
e. DNA codes for mRNA, tRNA, and rRNA.

9. The genetic code
a. is different for prokaryotes and eukaryotes.
b. has changed a lot during the course of recent evolution.
c. has 64 codons that code for amino acids.
d. is degenerate.
e. is ambiguous.

10. Which of the following mutations does not result in altered structure and function of the protein?
a. Deletion mutation
b. Neutral mutation
c. Frame-shift mutation
d. Nonsense mutation
e. nversion mutation

Student Web Site Self-Quiz Questions

1. Beadle and Tatum worked with cultures of the bread mold, *Neurospora crassa*. Which of the following statements about the work of Beadle and Tatum is *false*?
a. Prototrophic strains of *Neurospora* could grow on minimal medium.
b. Each auxotrophic strain had a specific nutritional requirement not displayed by the prototroph.
c. Addition of essential enzymes to the minimal medium allowed the auxotrophs to satisfy their nutritional requirements.
d. Several different classes of auxotrophs exist for a nutrient such as arginine that is the product of a biosynthetic pathway.
e. Each of the arginine auxotrophs had a defective allele for one gene coding for one enzyme involved in one step of the arginine biosynthetic pathway.

2. RNA and DNA differ in a number of ways. Which one of the following is *not* a difference?
a. Chargaff's equality of purines and pyrimidines
b. The type of sugar present
c. The ability of RNA to base pair within its own sequence
d. The nitrogenous bases present
e. RNA is a polynucleotide whereas DNA is not.

3. Which of the following statements about the initiation of transcription is *false*?
   a. A gene's promoter is a sequence of DNA where RNA polymerase binds.
   b. Promoters are closest to the 5′ end of the gene on the complementary strand.
   c. **All promoters have the exact same DNA sequence.**
   d. The same DNA strand can be the template strand for one gene, but the complementary strand for another.
   e. Promoters differ in their RNA polymerase binding effectiveness.

4. Which of the following is *not* a valid difference between transcription in prokaryotes and eukaryotes?
   a. **Eukaryotes produce only mRNA through transcription; prokaryotes also produce rRNA and tRNA.**
   b. Eukaryotic promoters initiate the transcription of a single gene; prokaryotic promoters may "turn on" the transcription of several, related genes.
   c. Prokaryotes have a single RNA polymerase; eukaryotes have three functionally different RNA polymerases.
   d. In prokaryotes, translation may begin before transcription ends; in eukaryotes, translation always occurs after transcription is complete.
   e. In eukaryotes, there is a spatial separation of translation and transcription; in prokaryotes, there is not.

5. Which of the following statements about transfer RNA (tRNA) is *true*?
   a. Each tRNA anticodon is specific for one codon.
   b. The 3′ end of the tRNA molecule varies depending on the amino acid it will bind.
   c. **Activating enzymes (aminoacyl-tRNA synthetases) link together a specific amino acid with a specific tRNA.**
   d. A DNA master codon of TAC (3′ to 5′) would have an anticodon on a tRNA of CAU.
   e. tRNA binds to mRNA (via its anticodon), but not to rRNA.

6. Which of the following statements regarding mutations is *false*?
   a. Mutation is the primary source of evolutionary change in a population.
   b. **Mutation occurs frequently in a natural population.**
   c. Mutation can be either spontaneous or induced.
   d. Mutations can be beneficial.
   e. Sources of mutation include chemicals and radiation.

★ *Indicates a **difficult** question*

■ *Indicates a **conceptual** question*

13 The Genetics of Viruses and Prokaryotes

Fill in the Blank

1. Bacteria that house bacteriophages that are not lytic are called **lysogenic**.
2. **Plasmids** are nonessential genetic elements that exist as free, independently replicating circular DNA molecules, separate of from the bacterial chromosome.
3. When the synthesis of an enzyme is turned off in response to an external biochemical cue (such as an excess in tryptophan), the enzyme is said to be **repressible**.
4. Bacteriophage DNA that is stably integrated into a bacterial chromosome is called a **prophage**.
5. Bacteriophages that can integrate their DNA into the host bacterial chromosome are called **temperate**.
6. Pieces of DNA that move from place to place in the bacterial chromosome are called **transposable elements (or transposons)**.
7. Genes that produce single mRNA's containing information for more than one protein are **operons**.
8. The region of the gene that binds RNA polymerase is the **promoter**.
9. The site on the operon DNA where a repressor binds is the **operator** sequence.
10. In prokaryotes, **regulatory** genes encode repressor proteins.
11. **Catabolic repression** is a positive control process that relies on increasing the affinity of promoters for RNA polymerase.
12. The first virus that was discovered was the **tobacco mosaic** virus.
13. Antibiotics are ineffective against **viral** infections.
14. A **virion** is the name of an individual viral particle when it is outside its host.
15. A **viroid** is just a single circular RNA molecule with no protein component, but which is infectious.

Multiple Choice

*1. All the following are advantages to working experimentally with bacteria over mice except
   a. bacteria have a smaller amount of DNA.
   b. bacteria are easy to contain.
   c. a large number of individuals can be produced quickly.
   d. bacteria are inexpensive to produce.
   e. bacteria take up very little space.

*2. A typical virus has __________ the DNA of a bacteria, which has __________ the amount of DNA of a human cell.
   a. 1/10, 1/10
   b. 1/2, 1/4
   c. 1/100, 1/100
   d. 1/1,000, 1/1,000
   e. 1/100, 1/1,000

3. The transfer of genes by a bacteriophage vector characterizes which type of gene transfer in bacteria?
   a. *Hfr*
   b. Conjugation
   c. Transduction
   d. Transformation
   e. None of the above

4. The term "lysogeny" refers to a(n)
   a. condition in which the bacteriophage DNA is stably integrated in the bacterial chromosome.
   b. condition in which the bacteriophage DNA is excised from the bacterial chromosome.
   c. condition in which a bacteriophage lyses a bacterium.
   d. mutation induced by a bacteriophage.
   e. exchange of genetic material between a bacteriophage and a bacteria.

5. Viruses were first discovered in 1892 when
   a. a filtrate was found to be infectious.
   b. Albert Virusa patient became ill from smoking infected tobacco.
   c. Stanley found that crystallized viral preparations contained protein and DNA.
   d. they were first observed using a light microscope.
   e. they were observed as plaques for the first time.

6. The genetic information of viruses is
   a. DNA.
   b. RNA.
   c. single-stranded.
   d. double-stranded.
   e. All of the above

7. A strain of virus that always lyses the cells it infects is called
   a. temperate.
   b. virulent.
   c. lytic.
   d. lysogenic.
   e. deadly.

8. The HIV virus that causes AIDS is a(n)
   a. arbovirus.
   b. double-stranded DNA virus.
   c. single-stranded DNA virus.
   d. porcine virus.
   e. retrovirus.

9. The polio virus and the HIV virus differ because only the polio virus can
   a. produce and use reverse transcriptase.
   b. use RNA as its genetic material.
   c. use RNA as genetic information and mRNA, without generating a DNA molecule.
   d. infect both horses and humans.
   e. All of the above

10. Plants have a tough cell wall that viruses cannot pass through. Viruses enter plants
   a. through wounds.
   b. using insect vectors.
   c. by digesting the cell wall.
   d. a and b
   e. b and c

11. Once inside a plant cell, viruses spread by
   a. diffusion.
   b. Brownian motion.
   c. plasmodesmata.
   d. attaching to shared endoplasmic reticulum.
   e. the movement of water each time it rains.

12. A patient is told that he has contracted a disease that is caused by an arbovirus. He most likely
   a. ate something he should not have.
   b. kissed someone he should not have.
   c. should have used insect repellent.
   d. got it because someone who already had it sneezed around him.
   e. will be contagious.

13. Viroids are composed of
   a. DNA and protein.
   b. protein.
   c. RNA.
   d. RNA and protein.
   e. DNA.

14. Viruses replicate __________ than bacteria.
   a. slower than
   b. at the same rate as
   c. faster than
   d. some faster, some slower
   e. Viruses can not replicate.

15. Viruses are composed of
   a. just nucleic acids.
   b. just proteins.
   c. nucleic acids and proteins.
   d. nucleic acids, proteins, and a few organelles.
   e. Some have organelles, but most just have nucleic acids and proteins.

16. A viroid differs from a virus because it
   a. is non-infectious.
   b. is composed of just nucleic acids.
   c. is man-made.
   d. has both protein and lipid components.
   e. All of the above

*17. The scientist who first crystallized tobacco mosaic virus was
   a. Dmitri Ivanovsky.
   b. Martinus Beijerinick.
   c. Jackson Roberston.
   d. Wendell Stanley.
   e. Rosalyn Franklin.

18. A virion with a lipid and protein membrane is likely to infect
   a. animal cells.
   b. plant cells.
   c. bacteria.
   d. fungi.
   e. All of the above

19. Lysis of the host cell is caused by
   a. the cells simply bursting due to the amount of viral particles.
   b. the cells opening up in an attempt to dump out the viruses.
   c. a product of a viral gene attacking the cell wall.
   d. a yet unknown mechanism.
   e. None of the above

20. Beginning with a single bacterium, how many cells would be present after four hours of growth if they can double every 20 minutes?
   a. 12
   b. 24
   c. 64
   d. 4,096
   e. 34,217,728

21. The term "auxotroph" refers to
   a. a mutant form of a bacteria that requires nutrient(s) not required by the wild-type bacteria.
   b. a mutant form of a bacteria that requires no nutrients.
   c. a mutant form of a bacteria that can synthesize a nutrient that the wild-type bacteria cannot.
   d. a mutant form of a bacteria that can metabolize a nutrient that the wild-type bacteria cannot.
   e. a bacteria that can metabolize sugars.

▪22. When a *met⁻bio⁻*- strain of bacteria is mixed with a *thr⁻leu⁻*- strain, wild-type bacteria result at a rate of one in every 10^7 cells. Why can't this be explained by mutation?
   a. The wild type would have to be a double mutation, which is extremely rare.
   b. The wild type would have to be a mutation in four genes, which is extremely rare.
   c. The wild type would have to be a deletion, which is extremely rare.
   d. Mutations can only occur in response to a mutagen.
   e. Bacterial genes do not mutate.

23. The met^+leu^+ bacteria that result from a cross between met^+leu^- and met^-leu^+ bacteria are called
   a. Hfr strains.
   b. auxotrophs.
   c. parentals.
   d. recombinants.
   e. None of the above

24. The function of the pili is
   a. to store the F plasmid.
   b. to make the initial contact between an F^+ and an F^-- cell that precedes conjugation.
   c. to uptake DNA during transformation.
   d. to transfer the DNA between mating partners during conjugation.
   e. The pili have no function.

25. The transfer of genes by uptake of DNA from the medium characterizes which type of gene transfer in bacteria?
   a. Hfr
   b. Sexduction
   c. Transduction
   d. Transformation
   e. None of the above

26. Which of the following is *not* a mechanism of bacterial genetic recombination?
   a. Transformation
   b. Conjugation
   c. Transduction
   d. Catabolite repression
   e. None of the above

27. In transduction,
   a. only a particular part of the bacterial chromosome can be transferred.
   b. a part of the bacterial chromosome might be transferred.
   c. only the F plasmid can be transferred.
   d. only the part of the bacterial chromosome near the F plasmid can be transferred.
   e. None of the above

28. An R factor is a(n)
   a. plasmid that carries genes for antibiotic resistance.
   b. episome that carries genes for antibiotic resistance.
   c. region of the bacterial chromosome that carries genes for antibiotic resistance.
   d. small portion of the F plasmid.
   e. measure of how well the bacterium is insulated from the cold.

29. For a plasmid to survive within the cytoplasm of a cell as it continually divides, it must
   a. be integrated in the bacterial chromosome.
   b. have an R factor.
   c. have an F plasmid.
   d. have an Hfr.
   e. have an origin of replication.

▪30. What is the difference between a plasmid and a transposable element?
   a. Plasmids exist in many copies in the cell, while a transposable element exists in a single copy.
   b. A plasmid replicates independently of the cell; a transposable element does not.
   c. A plasmid has an origin of replication; a transposable element does not.
   d. A plasmid exists independently in the cell cytoplasm, while a transposable element is integrated into another larger DNA molecule.
   e. None of the above

31. A population of genetically identical bacteria that arose from a single cell is known as a
   a. plaque.
   b. lysogen.
   c. clone.
   d. Hfr.
   e. K/4.

32. Which of the following is *not* true of the F plasmid?
   a. The F plasmid can replicate independent of the bacterial chromosome.
   b. The F plasmid can be transferred from a donor to a recipient bacterium during mating.
   c. The F plasmid is involved in conjugation.
   d. The F plasmid contains genes that encode the pili.
   e. The F plasmid contains genes that encode antibiotic resistance.

33. Which of the following statements is *true* for both F plasmids and bacteriophages?
   a. They have a protein coat enclosing the DNA.
   b. They can lyse bacteria.
   c. They can participate in sexductionF-duction.
   d. They can incorporate into the bacterial chromosome.
   e. None of the above

34. Which of the following is *not* true of conjugation between an F^- bacteria and an F^+ bacteria?
   a. A conjugation tube is formed between the two bacteria.
   b. The F^+ cell transfers a copy of the F plasmid to the F^-- cell.
   c. The F^- cell receives a copy of the F plasmid.
   d. The F^- cell becomes F^+.
   e. The F^+ cell becomes F^-.

*35. R factors, or resistance factors, carried by plasmids or bacteria may be more widespread now than in the past because
   a. the presence of antibiotics has stimulated the evolution of R factors.
   b. we are better at detecting them now than in the past.
   c. current bacterial culture conditions favor the selection of R factors.
   d. a and c
   e. None of the above

36. When *E. coli* are grown in a medium with little lactose,
   a. all of the enzymes of the lactose operon are present in very small quantities.
   b. all of the enzymes of the lactose operon are present in large quantities.
   c. none of the enzymes of the lactose operon are present.
   d. β-galactosidase and permease are present in small quantities, but transacetylase is present in large quantities.
   e. the mRNA of the lactose operon is not present at all.

37. A promoter is the region of
   a. a plasmid that binds the enzymes for replication.
   b. the mRNA that binds to a ribosome.
   c. DNA that binds RNA polymerase.
   d. the mRNA that binds tRNA's.
   e. None of the above

38. The frequency of transcription of a particular bacterial gene is controlled by the
   a. DNA sequence of the particular promoter.
   b. availability of the promoter to RNA polymerase.
   c. number of ribosomes that are available in the cell.
   d. a and b
   e. a, b, and c

39. The three basic parts of an operon are the
   a. promoter, the operator, and the structural gene(s).
   b. promoter, the structural gene(s), and the termination codons.
   c. promoter, the mRNA, and the termination codons.
   d. structural gene(s), the mRNA, and the tRNA's.
   e. None of the above

40. An inducer
   a. combines with a repressor and prevents it from binding the promoter.
   b. combines with a repressor and prevents it from binding the operator.
   c. binds to the promoter and prevents the repressor from binding to the operator.
   d. binds to the operator and prevents the repressor from binding at this site.
   e. binds to the termination codons and allows protein synthesis to continue.

41. The RNA transcribed from an operon is
   a. transcribed by the ribosomes to make more repressor.
   b. translated by the ribosomes to make more inducer.
   c. translated by the ribosomes to make the enzymes.
   d. translated by the ribosomes to make more RNA polymerase.
   e. usually not translated by the ribosomes at all.

42. The genes that encode repressor proteins are
   a. repressor genes.
   b. operons.
   c. inducer genes.
   d. regulatory genes.
   e. None of the above

43. When an operon is turned "off" in response to molecules present in the environment of the cell, it is
   a. repressible.
   b. suppressible.
   c. impressible.
   d. inducible.
   e. degraded.

44. In a repressible operon, the repressor molecule
   a. must first be activated by a corepressor.
   b. can repress the transcription of the operon on its own.
   c. is a molecule made from the operon.
   d. binds to the mRNA.
   e. must first be made negative to control the operon.

45. Catabolite repression refers to the
   a. increased transcription seen from many operons when glucose is present in the medium.
   b. shutdown of transcription from many operons when glucose is present in the medium.
   c. increased activity of inducers caused by glucose in the medium.
   d. a and b
   e. a and c

46. To be activated, the CRP must first bind
   a. the repressor molecule.
   b. the repressor protein.
   c. the activator protein.
   d. the corepressor molecule.
   e. cAMP.

*47. When the operator is unbound, the binding of the CRP–cAMP complex __________ the binding of RNA polymerase at the promoter.
   a. increases by two-fold
   b. decreases by tenfold
   c. increases by 50-fold
   d. blocks
   e. All of the above are sometimes true, depending on the concentration of lactose.

48. When the concentration of glucose in the medium falls, the concentration of __________ rises.
   a. CRP
   b. cAMP
   c. repressors
   d. inducers
   e. None of the above

49. The function of the promoter is to signal the RNA polymerase as to
    a. where to start transcribing the DNA.
    b. which strand of the DNA to read.
    c. where to stop transcribing the DNA.
    d. All of the above
    e. a and b

50. The mechanism by which the inducer causes the repressor to detach from the operator is an example of
    a. catabolite repression.
    b. transcription.
    c. transposition.
    d. allosteric modification.
    e. recombination.

51. __________ acts as a corepressor to block transcription of the tryptophan operon.
    a. cAMP
    b. Lactose
    c. Tryptophan
    d. Methionine
    e. CRP

52. The CRP-cAMP complex binds __________ of the operon.
    a. close to the RNA polymerase binding site
    b. close to the operator
    c. inside one of the structural genes
    d. at the termination point
    e. None of the above

*▪53. It is found that a certain enzyme is synthesized whenever the solution in which the cells are growing in lacks substance X. This is most likely __________ gene regulation.
    a. inducible
    b. positive
    c. negative
    d. repressible
    e. positive-negative

*▪54. It is found that a certain enzyme is synthesized whenever the solution in which the cells are growing in contains substance X. This is most likely __________ gene regulation.
    a. inducible
    b. positive
    c. negative
    d. repressible
    e. positive-negative

55. Cells control the amount of enzymes by
    a. blocking transcription.
    b. hydrolyzing the RNA.
    c. preventing translation.
    d. hydrolyzing the protein.
    e. All of the above

56. What effect does glucose concentration have on the amount of *lac* operon transcript?
    a. It increases cAMP.
    b. It decreases cAMP.
    c. It increases the rate of transcription.
    d. It decreases the rate of transcription.
    e. a and c

57. The *trp* operon
    a. codes for proteins needed for typtophan synthesis.
    b. codes for the proteins to metabolize tryptophan.
    c. is activated by the presence of tryptophan.
    d. is inducible.
    e. All of the above

*58. The effects of Cro and cI on lambda are
    a. competitive.
    b. cooperative.
    c. coordinate.
    d. inverted.
    e. additive.

*59. The total minimum number of genes necessary for a life form grown in laboratory conditions is
    a. 48.
    b. 337.
    c. 470.
    d. 3,876.
    e. 10,082.

Study Guide Questions

1. Viruses consist of
    a. a protein core and a nucleic acid capsid.
    b. a cell wall surrounding nucleic acid.
    c. RNA and DNA enclosed in a membrane.
    d. a nucleic acid core surrounded by a protein capsid and in some cases a membrane.

2. Bacterial cells that are resistant to viruses
    a. lack a cell surface receptor that the virus must bind in order to infect the cell.
    b. harbor a prophage in their chromosome, making the bacterial cell immune to further viral infection.
    c. have enzymes (restriction enzymes) that cut up the viral nucleic acid so that the viral replication cannot occur in the host cell.
    d. All of the above

3. Bacterial viruses
    a. infect the cell, replicate their genomes, and lyse the cell.
    b. infect the cell, replicate their genomes, transcribe and translate their genes, and lyse the cell.
    c. infect the cell, replicate their genomes, transcribe and translate their genes, package those genomes into viral capsids, and lyse the cell.
    d. infect the cell, translate their RNA, and lyse the cells.

4. Animal viruses that integrate their DNA into the host chromosome
    a. are RNA viruses.
    b. are prophages.
    c. copy their RNA genome into DNA using reverse transcriptase.
    d. a and c

5. During conjugation,
   a. DNA from one bacterial cell is transferred to another bacterial cell using a bacteriophage.
   b. mutants that are auxotrophic for one nutrient can be converted to prototrophs when mixed with mutants that are auxotrophic for another nutrient.
   c. a pilus is synthesized and DNA is transferred from one bacterium across the conjugation tube to the recipient bacterium.
   d. a and b

6. Plasmid DNA contains genes that
   a. confer drug resistance to the host cell.
   b. regulate conjugation.
   c. can confer resistance to heavy metals.
   d. All of the above

7. An operon
   a. is regulated by a repressor binding at the promoter.
   b. has structural genes that are all transcribed from same promoter.
   c. has several promoters, but all of the structural genes are related biochemically.
   d. is a set of structural genes all under the same translational regulation.

8. If the gene encoding the *lac* repressor is mutated so that it can no longer bind the operator, will transcription of that operon occur?
   a. Yes, but only when lactose is present.
   b. No, because RNA polymerase is need to transcribe the genes.
   c. Yes, because the operator will not be bound by repressor and RNA polymerase can transcribe the *lac* operon.
   d. No, because cAMP levels are low when the repressor is nonfunctional.

9. If the gene encoding the *trp* repressor is mutated such that it can no longer bind tryptophan, will transcription of the *trp* operon occur?
   a. Yes, because the *trp* repressor can only bind the *trp* operon and block transcriptional initiation when it is bound to tryptophan.
   b. No, because this mutation does not affect the part of the repressor that can bind the operator.
   c. No, because the *trp* operon is repressed only when tryptophan levels are high.
   d. Yes, because the *trp* operon can allosterically regulate the enzymes needed to synthesize the amino acid tryptophan.

10. Transcriptional regulation in prokaryotes can occur by
   a. a repressor binding an operator and preventing transcription.
   b. an activator binding upstream from a promoter and positively affecting transcription.
   c. different promoter sequences binding RNA polymerase more tightly, resulting in more effective transcriptional initiation.
   d. All of the above

End of Chapter Questions

1. In the lysogenic cycle of a bacteriophage,
   a. a repressor, cI, blocks the lytic cycle.
   b. bacteriophage carries DNA between bacterial cells.
   c. both early and late phage genes are transcribed.
   d. the viral genome is made into RNA, which stays in the host cell
   e. many new viruses are made immediately, regardless of host health.

2. Which of the following is not representative of a type of virus reproduction?
   a. DNA virus by a lytic cycle
   b. DNA virus by a lysogenic cycle
   c. RNA virus by a double-stranded RNA intermediate
   d. RNA virus by reverse transcription of a cDNA
   e. RNA virus by acting as a tRNA

3. Which method is *not* used to make bacterial cells that have new genes?
   a. Conjugation
   b. Cloning
   c. Transformation
   d. Transduction
   e. Transposition

4. An operon is
   a. a molecule that can turn genes on and off.
   b. an inducer bound to a repressor.
   c. regulatory sequences controlling protein-coding genes.
   d. any long sequence of DNA with related genes.
   e. a group of linked genes.

5. Which statement about transformation and transduction is *true*?
   a. DNA is transferred by a virus between bacteria.
   b. Both occur in the lab but not in nature.
   c. Small fragments of DNA are moved from one cell to another.
   d. Recombination between the incoming DNA and host cell DNA is not needed.
   e. A conjugation tube is used to transfer DNA between cells.

6. Plasmids
   a. are circular protein molecules.
   b. are required by bacteria.
   c. are tiny bacteria.
   d. may confer resistance to antibiotics.
   e. are a form of transposable element.

7. The "minimal genome" can be estimated for a prokaryote
   a. by counting up the total number of genes.
   b. by comparative genomics.
   c. as about 5000 genes
   d. by transposon mutagenesis one gene at a time.
   e. does not include any genes coding for tRNA

8. In the lac operon,
   a. the repressor binds to the operator when *lac* is present.
   b. a corepressor unites with the repressor.
   c. transcription is inhibited when the repressor binds to the operator.
   d. lactose binding alters the shapoe of the operator.
   e. the control mechanism is positive.

9. Which is *not* true of the promoter region?
   a. It is made of DNA.
   b. It is where RNA polymerase binds in prokaryotes to begin transcription.
   c. All RNA polymerases bind equally well to all promoters.
   d. It is near to or overlaps the operator region.
   e. It is the site for *lac* repressor binding.

10. The CRP–cAMP system
   a. produces many catabolites.
   b. requires ribosomes.
   c. operates by an operator-repressor mechanism.
   d. is an example of positive control of transcription.
   e. relies on operators.

Student Web Site Self-Quiz Questions

1. Which of the following statements about the genetics of viruses is *false*?
   a. A single human cell has about 100,000 times as much DNA as a typical bacterial virus.
   b. Viruses usually have only a single copy of each gene in their genome.
   c. A virion consists of a central core of both DNA and RNA surrounded by a capsid.
   d. All viruses are acellular.
   e. All viruses are obligate intracellular parasites.

2. Which of the following statements about bacteriophages is *false*?
   a. Bacteriophages can reproduce via a lytic or lysogenic cycle, but not both.
   b. Bacteriophages are DNA viruses with capsids of the binal type.
   c. Temperate phage are not actively lytic.
   d. Lysogenic bacteria can be induced to activate its prophage.
   e. A prophage is replicated with the bacterial chromosome and passed to the two new daughter cells.

3. Which of the following statements regarding the reproductive cycle of viruses is *false*?
   a. Some viruses can be surrounded by membrane derived from the host cell.
   b. Naked virions enter cells via endocytosis.
   c. Membrane-enclosed viruses (enveloped viruses) cannot be imported into the host cell via endocytosis.
   d. Membrane-enclosed viruses may be taken up into the host cell when the membranes of the virus and host cell fuse.
   e. Virion assembly occurs only in the cytoplasm of the host cell.

4. Colonies result
   a. if a dense suspension of bacteria is spread on solid nutrient medium.
   b. if a dense suspension of bacteria is spread on solid medium lacking nutrients.
   c. if a small number of bacteria is spread on solid nutrient medium.
   d. if a dense suspension of bacteria is added to liquid nutrient medium.
   e. if a small number of bacteria is added to liquid nutrient medium.

5. When two bacteria are connected by a **pilus,** this is an example of ____.
   a. transduction
   b. transformation
   c. conjugation
   d. transposition
   e. mutation

6. Following conjugation between bacteria, which of the following events does *not* occur?
   a. The circular DNA must first be fragmented before transfer.
   b. Communications between conjugating bacteria is transient.
   c. Recombination may occur between donor and recipient DNA.
   d. The DNA fragments from the donor cell may line up with homologous genes in the recipient cell.
   e. Homologous genes in the transferred DNA can attach themselves and spontaneously integrate into the host cell's DNA.

7. Which of the following statements regarding the process of transformation is *true*?
   a. Transformation can occur between living and non-living cells.
   b. Transformation only occurs in living cells.
   c. Transformation requires the assistance of viruses.
   d. Transformation occurs only with the entire genome of the donor cell.
   e. Transformation requires intimate contact between bacterial cells.

8. Which of the following is *not* a true statement regarding the transduction process?
   a. Phage DNA must first be incorporated into the host bacterial DNA.
   b. During the lytic cycle, genomic DNA is packaged into capsids.
   c. Bacterial DNA may also be packaged along with phage DNA into bacteriophage particles.
   d. Phage particles can contain either phage DNA or bacterial DNA, but not both.
   e. Bacterial DNA can be inserted into new host chromosomes during subsequent infections.

9. Which of the following statements about plasmids is *false*?
   a. The fertility factor in *E. coli* is a plasmid.
   b. Plasmids act like viruses in that they take over the bacterial cells biochemical machinery.
   c. All plasmids are chromosomes.
   d. An episome is a type of plasmid that can exist independently.
   e. Transposable elements move genes between plasmids and chromosomes.

TEST FILE

★ *Indicates a **difficult** question*

■ *Indicates a **conceptual** question*

14 The Eukaryotic Genome and Its Expression

Fill in the Blank

1. DNA in eukaryotes is wrapped around special proteins to form a structure that looks like beads on a string, which are called **nucleosomes**.

2. Segments of a gene that are transcribed but do not encode part of the protein product are called **introns**.

3. DNA sequences called **transposable elements** can replicate and insert themselves into different parts of the genome.

4. Members of a gene family that do not make a functional gene product are called **pseudogenes**.

5. The RNA of snRNP's are complementary to **consensus** sequences located at the intron–exon boundaries.

■6. When you stain a preparation of epithelial cells from a female rat, you observe one highly condensed chromosome in interphase nuclei. This is the inactive X chromosome, or **Barr body**.

7. During the development of amphibian oocytes, the number of genes encoding the rRNA increases. This increase is called **gene amplification**.

8. Those DNA sequences that are not represented in the mRNA are referred to as intervening sequences, or **introns**.

9. Most **pseudogenes** are probably duplicate genes that changed so much during evolution that they no longer function.

10. During RNA splicing, different snRNP's constitute a "splicing machine" called a **spliceosome**.

11. There are at least three ways in which **transcription** may be controlled: (1) genes may be inactivated, (2) specific genes may be amplified, and/or (3) specific genes may be selectively transcribed.

12. In both prokaryotes and eukaryotes, the **promoter** is a stretch of DNA to which RNA polymerase binds to initiate transcription.

Multiple Choice

1. Among eukaryotes __________ between complexity and genome size
   a. there is an inverse relationship
   b. there is a direct relationship
   c. there fails to always be an inverse relationship
   d. there fails to always be a direct relationship
   e. there is never a relationship

2. In eukaryotic cells, transcription and translation
   a. are separated. Translation occurs in the nucleus and transcription in the cytoplasm.
   b. occur together in the cytosol.
   c. occur together in the nucleus.
   d. are separated. Translation occurs in the cytoplasm and transcription in the nucleus.
   e. are separated, except for proteins that bind to the DNA and ribosomes, which are translated in the nucleus.

★3. A lily, which has 18 times more DNA than a human, codes for __________ proteins.
   a. more
   b. the same number of
   c. fewer
   d. 337
   e. 6,330

★4. The number of base pairs in a human chromosome ranges from
   a. 20×10^6 to 100×10^6.
   b. 2×10^6 to 10×10^6.
   c. 20×10^4 to 100×10^4.
   d. 2×10^4 to 10×10^4.
   e. 20×10^2 to 100×10^2.

5. The DNA from a human cell, stretched out, would be __________ in length.
   a. 2 mm.
   b. 2 m
   c. 2 km
   d. 2 μm
   e. None of the above

6. Human cells have approximately __________ the DNA found in prokaryotic cells.
   a. 10 times
   b. 100 times
   c. 1,000 times
   d. 10,000 times
   e. 100,000 times

7. Eukaryotic DNA is organized into chromosomes that must have
   a. DNA sequences that make up telomeres and centromeres.
   b. proteins that are centromeres and DNA that form telomeres.
   c. a 5′ G cap.
   d. ubiquitin bound.
   e. an inactivation center.

8. The phrase "beads on a string" describes the structural appearance of
   a. condensed chromosomes.
   b. lampbrush chromosomes.
   c. nucleosomes.
   d. the solenoid structure of DNA.
   e. the 30 nm DNA fibers.

9. Cancer cells are often found to have abnormally high amounts of
   a. nucleosomes.
   b. spliceosomes.
   c. centromereosomes.
   d. topoisomerase.
   e. telomerase.

10. Telomerase is important to eukaryotic cells because
   a. telomeres tend to get shortened with each cell division.
   b. telomeres tend to get longer with each cell division.
   c. telomerase digests telomeres to proper length.
   d. the leading strand of DNA causes the telomeres to shorten.
   e. it aids in making artificial chromosomes.

11. Each human chromosome must have
   a. a centromeric sequence.
   b. telomeric sequences.
   c. an origin of replication.
   d. All of the above
   e. None of the above

12. RNA synthesis in eukaryotes occurs
   a. solely in mitochondria.
   b. primarily in the Golgi apparatus.
   c. primarily in the endoplasmic reticulum.
   d. primarily in the nucleus.
   e. solely in the cytoplasm.

13. The three RNA polymerases of eukaryotes catalyze __________ synthesis.
   a. rRNA
   b. mRNA
   c. tRNA
   d. All of the above
   e. None of the above

14. RNA polymerase II by itself cannot bind to the chromosome and initiate transcription. It can bind and act only after the assembly of regulatory proteins called __________ factors.
   a. translation
   b. posttranslation
   c. initiation
   d. transcription
   e. None of the above

15. RNA translation in eukaryotes occurs
   a. solely in mitochondria.
   b. primarily in the Golgi apparatus.
   c. primarily in the endoplasmic reticulum.
   d. primarily in the nucleus.
   e. solely in the cytoplasm.

16. The RNA polymerase that produces mRNA molecules in eukaryotic cells is
   a. RNA polymerase I.
   b. RNA polymerase II.
   c. RNA polymerase III.
   d. primase.
   e. reverse transcriptase.

17. In bacteria, transcription
   a. is separate from translation.
   b. occurs in the nucleus.
   c. is not separated from translation.
   d. is the process of synthesizing proteins.
   e. is continuous for all genes.

18. The number of genes in *Saccharomyces cerevisiae* has __________ genes and 12,068,000 base pairs.
   a. 62
   b. 620
   c. 6,200
   d. 620,00
   e. 620,000

19. *Caenorhabditis elegans* is a
   a. yeast.
   b. virus.
   c. fruit fly.
   d. bacterium.
   e. roundworm.

20. *C. elegans* has approximately __________ genes.
   a. 19,000
   b. 6,000
   c. 4,200
   d. 3,200
   e. 370

*21. *Drosophila melanogaster*, the fruit fly, has __________ genes than a simple roundworm.
   a. about the same number of
   b. fewer
   c. many more
   d. a few more
   e. This is currently not known.

22. The genome size of fruit flies is __________ roundworms.
   a. about the same as
   b. smaller than
   c. much larger than
   d. a bit larger than
   e. This is currently not known.

23. __________ is the study involved with assigning functions to DNA sequences.
   a. Biology
   b. Genetics
   c. Cytogenetics
   d. Genomics
   e. All of the above

*24. *Arabidopsis thaliana* is an organism with 130 million base pairs of DNA that
   a. are being sequenced to provide information on plants.
   b. grows under high salt conditions.
   c. is useful for production of paper.
   d. has a large genome for a plant.
   e. All of the above

25. Transposable elements can
   a. alter transcription of a gene.
   b. cause a mutation.
   c. cause gene duplication.
   d. All of the above
   e. None of the above

26. SINEs and LINEs are both
   a. highly repetitive DNA sequences.
   b. transposable elements.
   e. control sequences.
   d. retrotransposons.
   e. DNA transposons.

27. DNA transposons
   a. work by using reverse transcriptase.
   b. are transcribed but not translated.
   c. are also called retrotransposons.
   d. move to new locations from old locations.
   e. are circular DNA molecules.

▪28. Transposable elements are found in both prokaryotes and eukaryotes. One of the major differences between the copying of transposable elements in some eukaryotes and prokaryotes is that in eukaryotes the elements
   a. must be denatured before they can be copied.
   b. all require an RNA intermediate.
   c. are not always copied.
   d. do not need an enzyme.
   e. None of the above

29. The importance of transposons are to
   a. provide mobile promoters.
   b. alter gene activities.
   c. provide for genetic change when required.
   d. inactivate unnecessary or undesirable genes.
   e. It is not known.

▪30. The mRNA will form hybrids only with the template strand of DNA because
   a. DNA will not reanneal at high temperatures.
   b. the salt concentration will affect DNA reannealing.
   c. DNA will not reanneal at low temperatures.
   d. RNA–DNA hybridization follows the base-pairing rules.
   e. denatured DNA will not reanneal after it is diluted.

31. DNA sequences found in introns provide
   a. amino acid sequence information.
   b. regulatory information.
   c. no known useful information.
   d. structure for the gene.
   e. alternative DNA splicing possibilities.

32. A gene family is a set of genes that over time has changed slightly, extensively, or not at all. Which of the following statements is *true* for gene families?
   a. They must always be on the same chromosome.
   b. They must always code for the exact same proteins.
   c. One copy must retain the original function.
   d. They usually differ in their exons because these are coding segments.
   e. None of the above

33. The different members of the β-globin gene family
   a. are expressed differently in different tissues.
   b. are expressed differently at different times of development.
   c. are expressed the same in different tissues.
   d. are expressed the same at different times of development.
   e. have no known function.

34. Which of the following is a posttranscriptional modification of mRNA found in eukaryotes?
   a. A 5′ cap
   b. A 3′ cap
   c. A poly T tail
   d. A polyadenylation at the 5′ end
   e. None of the above

35. Which of the following is *not* part of RNA processing in eukaryotes?
   a. Splicing of exons
   b. Reverse transcription
   c. Addition of a 5′ cap
   d. Addition of a poly A tail
   e. Intron removal

36. Consensus sequences (short segments of DNA) appear in the transcribed regions of various genes. These sequences appear to be involved in
   a. directing the polymerases to the appropriate place on the DNA for transcription to begin.
   b. the splicing of introns out of the DNA.
   c. allowing the transcription to stop at the appropriate spot.
   d. catalyzing the synthesis of a protein.
   e. None of the above

37. The pre-mRNA gets a tail added,
a. which is coded for by DNA.
b. which is composed of poly T.
c. which is 100 to 300 bases in length.
d. to its 5′ end.
e. All of the above

38. The expression of some genes can be regulated in part by the pattern of RNA splicing. This is an example of
a. DNA methylation.
b. transcriptional regulation.
c. catalytic RNA's.
d. posttranscriptional control.
e. the endosymbiotic theory.

39. Which of the following best describes the function of the addition of a methylated guanosine cap to the 5′ end of primary mRNA?
a. It contains all of the coding and noncoding sequences of the DNA template.
b. It provides the mRNA molecule with a poly A tail.
c. It facilitates the binding of mRNA to ribosomes.
d. It forms hydrogen bonds.
e. It helps transfer amino acids to the ribosomes.

40. Energy for RNA splicing comes from
a. the nucleotides in the RNA.
b. ATP.
c. directly from photosynthesis.
d. always from glucose.
e. the splicing enzyme.

41. RNA processing in eukaryotes involves
a. the addition of a G cap.
b. polyadenylation.
c. removal of introns.
d. splicing of exons.
e. All of the above

42. Exons are
a. spliced out of the original transcript.
b. spliced together from the original transcript.
c. spliced to introns to form the final transcript.
d. much larger than introns.
e. larger than the original coding region.

43. The binding of snRNP's to consensus sequences is necessary for
a. gene duplication.
b. addition of a poly A tail.
c. capping an hnRNA.
d. RNA splicing.
e. transcription.

44. Three processes that must be completed before transcripts can be translated in eukaryotes are
a. binding of snRNP's, adding a poly A tail, and splicing introns.
b. binding of snRNP's, transporting, and synthesizing ribose.
c. capping, transporting, and synthesizing ribose.
d. binding of snRNP's, capping, and splicing.
e. splicing, capping, and adding a poly A tail.

45. Snurps (snRNP's) are
a. exon–intron boundary regions.
b. small nuclear ribonucleoprotein particles.
c. protein fragments removed from the snRNA molecules.
d. signal ribosomal nuclear proteins.
e. glucose conjugated trapezoids.

46. The theoretical basis for DNA–DNA or DNA–RNA hybridization studies is
a. base complementarity between nucleic acid strands.
b. enzyme action.
c. DNA and RNA looping.
d. a and c
e. None of the above

47. When eukaryotic DNA is hybridized with mRNA, the hybrid molecules contain loops of double-stranded DNA. These regions of DNA are called
a. retroviruses.
b. introns.
c. exons.
d. transcripts.
e. puffs.

48. The regions of DNA in a eukaryotic gene that encode a polypeptide product are called
a. enhancers.
b. mRNAs.
c. hnRNAs.
d. exons.
e. leader sequences.

49. Exons are
a. translated.
b. found in most prokaryotic genes.
c. removed during RNA processing.
d. a and b
e. a and c

50. The globin __________ includes a number of genes that encode similar proteins.
a. pseudogene
b. intron
c. gene family
d. retrovirus
e. proto-oncogene

51. Most gene families probably arose by
a. RNA processing.
b. RNA splicing.
c. DNA methylation.
d. transcriptional regulation.
e. gene duplication.

52. The modified G cap on eukaryotic mRNAs is found
a. at the 5′ end.
b. at the 3′ end.
c. in the consensus sequence.
d. in the poly A tail.
e. in snRNA.

53. Poly A tails
    a. are added after transcription.
    b. are encoded by a sequence of thymines in the DNA.
    c. are found in all mRNAs.
    d. have no function.
    e. are removed during RNA processing.

54. Which of the following is an enhancer?
    a. Protein
    b. RNA
    c. DNA
    d. Carbohydrate
    e. Enzyme

55. Which of the following is a transcription factor?
    a. Protein
    b. RNA
    c. DNA
    d. Carbohydrate
    e. Enzyme

56. In eukaryotic cells, a repressor
    a. is made of DNA.
    b. binds to the enhancer region to block transcription.
    c. is located both upstream and downstream from the promoter.
    d. binds to the operator to block RNA polymerase.
    e. binds to a silencer to reduce transcription rates.

*57. In eukayrotes, the TATA box is __________ base pairs.
    a. –10
    b. –25
    c. –40
    d. +10
    e. +40

58. The region of a gene that binds RNA polymerase to initiate transcription is called
    a. an exon.
    b. the consensus sequence.
    c. heterochromatin.
    d. the cap.
    e. the promoter.

59. When an enhancer is bound, it
    a. increases the stability of a specific mRNA.
    b. stimulates transcription of a specific gene.
    c. stimulates transcription of all genes.
    d. stimulates splicing of a specific mRNA.
    e. stimulates splicing of all mRNAs.

60. Which of the following is *not* a feature of TATA boxes?
    a. They bind a specific transcription factor.
    b. They are found in the region of the promoter.
    c. They are part of the intron consensus sequence.
    d. They help specify the starting point for transcription.
    e. They contain thymine–adenine base pairs.

61. Transcription of eukaryotic genes requires
    a. binding of RNA polymerase to the promoter.
    b. binding of several transcription factors.
    c. capping of mRNA.
    d. a and b
    e. a, b, and c

62. A DNA sequence, which can be distant to the gene, stimulates transcription when bound by a protein. This sequence is a(n)
    a. TATA box.
    b. operon.
    c. enhancer.
    d. promoter.
    e. consensus sequence.

63. The TATA box is a(n)
    a. sequence common to the promoter region of many genes.
    b. square-shaped sequence.
    c. enhancer consensus sequence.
    d. activator sequence necessary for proper translation.
    e. None of the above

64. Remote DNA sequences that bind activator proteins and stimulate specific promoters, thus facilitating the transcription of specific genes, are called
    a. enhancers.
    b. stimulators.
    c. inducers.
    d. allosteric modulators.
    e. derepressor elements.

65. What holds the zipper of the leucine zipper motif proteins are
    a. ionic interactions.
    b. hydrophobic interactions.
    c. hydrogen bonds.
    d. covalent bonds.
    e. van der Waals forces.

66. DNA binding motifs described in the chapter include all except
    a. helix-straight-helix.
    b. helix-turn-helix.
    c. helix-loop-helix.
    d. leucine zipper.
    e. zinc finger.

67. The DNA binding proteins that are associated with activation of genes by steroids have the __________ motif.
    a. helix-straight-helix
    b. helix-turn-helix
    c. helix-loop-helix
    d. leucine zipper
    e. zinc finger

68. Steroids interact with DNA binding proteins with the __________ motif.
    a. zinc finger
    b. leucine zipper
    c. helix-turn-helix
    d. helix-loop-helix
    e. helix-helix

69. Switching of mating type in yeast involves
    a. selection.
    b. mutation.
    c. differential gene expression.
    d. aging.
    e. transposition.

70. A cell is found that contains three Barr bodies. This cell has
a. three Y chromosomes.
b. three X chromosomes.
c. four Y chromosomes
d. four X chromosomes.
e. three nucleoli.

71. DNA methylation in eukaryotic chromosomes involves adding a methyl to the
a. 5′ position of G.
b. 5′ position of C.
c. DNA binding proteins.
d. RNA molecules.
e. ribose.

72. One of the genes that is known to be transcribed from the inactive X chromosome is
a. *XIST*.
b. *ZIST*.
c. inactivation controller protein.
d. lithozist.
e. methyl-X.

73. Mary Lyon, Liane Russell, and Ernest Beutler discovered
a. the basis of hormone action.
b. X chromosome inactivation.
c. Barr bodies.
d. melanosomes.
e. heterochromatin.

74. The Barr body is evidence for
a. X chromosome inactivation.
b. cell death.
c. ion pumps.
d. posttranslational control of eukaryote gene expression.
e. None of the above

▪75. You stain a preparation of normal rat epithelial cells and examine them under the microscope. Each interphase nucleus contains a single Barr body. What can you conclude about these cells?
a. The cells are in meiotic prophase.
b. All of the chromatin in these cells is inactive.
c. The DNA in these cells has replicated.
d. The cells are not transcribing any genes.
e. The cells came from a female rat.

76. The interphase cells of normal female mammals have a stainable nuclear body called a Barr body. This body is
a. an inactive X chromosome.
b. made up of fat droplets.
c. fragments of mRNA.
d. extra chromosomal pieces.
e. None of the above

77. Heterochromatin
a. contains poly A tails.
b. is usually not transcribed.
c. does not contain any DNA.
d. is not replicated during the S phase.
e. is found only in prokaryotes.

78. You are studying the expression of the globin gene. In neurons, this gene contains many methylated cytosines. What might this mean about the expression of the globin gene in neurons?
a. It is expressed only in males.
b. It is expressed only in females.
c. It is not expressed.
d. It is regulated by posttranscriptional control.
e. It is regulated by posttranslational control.

79. A chemical modification that adds methyl groups to cytosine residues in some genes acts to
a. enhance transcription.
b. amplify the gene.
c. inactivate the gene.
d. stabilize the mRNA.
e. None of the above

80. DNA methylation
a. is a mechanism of gene inactivation.
b. adds methyl groups to cytosine residues in certain genes.
c. inhibits transcription.
d. a, b, and c
e. a and b

81. In fish and frog eggs, __________ go(es) from being 0.2% to 68% of the total genomic DNA.
a. tRNA
b. rRNA
c. rRNA gene clusters
d. tRNA gene clusters
e. mRNA gene clusters

82. The cells in *Drosophila* that make eggshell (chorion) proteins increase the number of genes encoding these proteins. This is an example of
a. gene amplification.
b. DNA methylation.
c. posttranscriptional control.
d. RNA splicing.
e. a gene family.

*83. Ubiquitin complexes proteins and delivers them to
a. the extracellular space.
b. mitochondria.
c. the molecular chamber of doom.
d. lysosomes.
e. the Golgi apparatus.

▪84. Transferrin is a protein that transports iron into cells. The iron concentration in a cell regulates the amount of transferrin protein being made, but has no effect on the levels of transferrin mRNA. This is an example of
a. translational control.
b. an enhancer.
c. transcriptional regulation.
d. RNA processing.
e. a promoter.

▪85. Some metabolic pathways are regulated in part by changing the rate of degradation of key enzymes. This is an example of
a. an operon.
b. transcriptional control.
c. liquid hybridization.
d. feedback inhibition.
e. posttranslational control.

▪86. Expression of some eukaryotic genes can be regulated by translational control. What is an advantage of translational control?
a. It provides a means for rapid change in protein concentrations.
b. It prevents synthesis of excess RNA.
c. It directs proteins to their proper subcellular location.
d. It occurs only in zygotes.
e. It degrades proteins that are no longer needed.

87. The regulation of gene expression via differential degradation of proteins is an example of _________ control.
a. transcriptional
b. translational
c. posttranslational
d. nuclear
e. None of the above

88. In eukaryotic cells, promoters are
a. trans cribed.
b. transcribed and translated.
c. neither transcribed nor translated.
d. transcribed and then removed.
e. sequences of RNA that are spliced out.

89. An interesting feature of the globin genes is
a. different ones are expressed during different stages of prenatal development.
b. only one copy is functional.
c. the lengths of the mRNA's are very different.
d. they are the result of differential posttranscriptional splicing.
e. the transcripts are longer than the coding region.

90. Potential control points for regulating the amount of protein synthesized in eukaryotic cells include all of the following *except*
a. transcription regulation.
b. DNA amplification.
c. transcript processing.
d. breakdown of the synthesized protein.
e. stabilization of the mRNA.

91. Although eukaryotic genes are solitary and not transcribed together using the operon system found in prokaryotic cells, coordinated gene expression occurs due to
a. a single transcript containing the coding region for more than one protein.
b. response elements that are similar among the genes that are expressed in a coordinated fashion.
c. sigma and rho factors that switch banks of genes on and off.
d. estrogen.
e. TATA boxes.

92. Transcription factors are
a. RNA sequences that bind to RNA polymerase.
b. DNA sequences that up regulate transcription.
c. protein that bind to DNA near the promoter sequence.
d. polysaccharides that bind to the transcripts.
e. factors that bind to enhancers.

93. Coordinated gene expression in eukaryotic cells can occur because
a. one event often leads to another similar event.
b. more than one protein is coded for by a transcript.
c. of similarities in promoters for different genes.
d. of the universal nature of the genetic code.
e. all promoters are different.

94. Which of the following is a promoter?
a. Protein
b. RNA
c. DNA
d. Carbohydrate
e. Enzyme

*95. The heat shock response is an example of
a. a survival tactic.
b. a survival strategy.
c. a way to increase body temperature.
d. coordinated gene expression.
e. None of the above

*▪96. An interesting characteristic involving the control of tubulin production is that
a. a translational repressor protein binds to the tubulin mRNA, which prevents ribosomes from attaching.
b. alternative splicing of pre-mRNA results in the production of several different proteins.
c. tubulin can recognize and bind to tubulin mRNA, causing an acceleration of its breakdown.
d. ubiquitin forms a complex with the tubulin, which causes its breakdown.
e. the 5′ guanosine cap is added to the mRNA but is not modified so the mRNA is not expressed. When needed, it gets modified.

*▪97. The interesting characteristic involving tropomyosin production is that
a. a translational repressor protein binds to the tropomyosin mRNA, which prevents ribosomes from attaching.
b. alternative splicing of pre-mRNA results in the production of several different proteins.
c. tropomyosin can recognize and bind to tropomyosin RNA causing an acceleration of its breakdown.
d. ubiquitin forms a complex with the tropomyosin, which causes its breakdown.
e. the 5′ guanosine cap is added to the mRNA but is not modified so the mRNA is not expressed. When needed, it gets modified.

*98. The interesting characteristic involving the control of protein synthesis in tobacco hornworms is that
a. a translational repressor protein binds to the mRNA, which prevents ribosomes from attaching.
b. alternative splicing of pre-mRNA results in the production of several different proteins.
c. proteins can recognize and bind to mRNA causing an acceleration of its breakdown.
d. ubiquitin forms a complex with the mRNA, which causes its breakdown.
e. the 5′ guanosine cap is added to the mRNA but is not modified so the mRNA is not expressed. When needed, it is modified.

*▪99. The interesting characteristic involving the control of ferritin synthesis is that
a. a translational repressor protein binds to the ferritin mRNA, which prevents ribosomes from attaching.
b. alternative splicing of pre-mRNA results in the production of several different proteins.
c. ferritin can recognize and bind to ferritin RNA causing an acceleration of its breakdown.
d. ubiquitin forms a complex with the ferritin, which causes its breakdown.
e. the 5′ guanosine cap is added to the mRNA but is not modified so the mRNA is not expressed. When needed, it is modified.

Study Guide Questions

1. Eukaryotic chromosomes
a. are circular and contain origins and terminator sequences.
b. are linear and have origins and telomeres.
c. contain coding and noncoding sequences.
d. b and c

2. Model eukaryotic organisms have helped biologists understand gene(s)
a. involved in development.
b. families.
c. encoding proteins that are essential for all cells.
d. All of the above

3. Moderately repetitive DNA includes
a. only coding sequences.
b. only noncoding sequences.
c. coding and noncoding sequences.
d. satellites, minisatellites, and microsatellites.

4. The ends of eukaryotic chromosomes
a. are single-stranded and must be bound by transcription complexes.
b. are highly repetitive sequences that hold replicated chromosomes together.
c. shorten every time a cell divides.
d. require telomerase to move to other sites in the chromosome.

5. Transposable genetic elements
a. always affect the cell adversely, because when they move, they inactivate genes.
b. are retroviruses.
c. provide a mechanism for moving genetic material from organelle genomes to the nuclear genome.
d. always replicate their DNA when they more.

6. Introns are DNA sequences that
a. code for functional domains in proteins.
b. are removed from pre-mRNA by spliceosomes.
c. allow one gene to make different gene products depending on which introns are removed during splicing.
d. b and c

7. Globin genes are
a. only transcribed in bone marrow cells.
b. translationally regulated by excess heme.
c. part of a family of globin genes, with different genes expressed at different times in development.
d. All of the above

8. Pre-messenger RNAs must be processed in the nucleus in order to
a. increase their stability in the cytoplasm.
b. allow transcription to begin.
c. permit coding sequences to be joined to adjacent noncoding sequences.
d. facilitate ribosome recognition in preparation for DNA synthesis

9. Coordinated regulation of genes in eukaryotic cells
a. is the result of positioning the same regulatory sequence in front of each gene.
b. results from all of those genes being under the control of one promoter.
c. occurs because related genes all have the same operons.
d. occurs because enhancers cause DNA to bend.

10. The transcription complex includes ________ and ________.
a. transcription factors, promoters
b. regulator proteins, regulators
c. repressor proteins, silencers
d. a and b

11. DNA binding proteins
a. have distinct three-dimensional structures that allow them to bind to the DNA.
b. are transcription factors.
c. inhibit the loss of ends from the chromosome and help DNA condense in the nucleus.
d. All of the above

12. Chromatin structure must be altered in order for gene expression to occur because
a. condensed chromatin is replicated but not transcribed.
b. condensed chromatin makes most DNA sequences inaccessible to the transcription complex.
c. decondensed chromatin has more nucleosomes per DNA molecule.
d. heterochromatin is actively transcribed and euchromatin is not transcribed.

13. When DNA sequences are moved to new sites on a chromosome
    a. genes can be transcribed.
    b. genes can be inactivated.
    c. new genes can be created.
    d. All of the above

14. rRNA gene copies are amplified
    a. to ensure that gene expression in somatic cells will continue.
    b. to make more copies of DNA origins for replication.
    c. to ensure that there is enough translational machinery for the rapid development of the embryo.
    d. to help heterochromatin become more accessible to the transcription complex.

15. Posttranscriptional regulation includes
    a. binding of repressor on silencer regions.
    b. transport of messenger RNA into the cytoplasm.
    c. decreasing messenger RNA stability in the cytoplasm.
    d. b and c

16. Genes can be inactivated by
    a. inaccurate removal of introns.
    b. transposable genetic elements.
    c. movement of genes to heterochromatic regions of the chromosome.
    d. All of the above

End of Chapter Questions

1. Eukaryotic protein-coding genes differ from their prokaryotic counterparts in that only eukaryotic genes
    a. are double-stranded.
    b. are present in only a single copy.
    c. contain introns.
    d. have a promoter.
    e. transcribe mRNA.

2. Comparison of the genomes of yeast and bacteria show that only yeast has many genes for
    a. energy metabolism.
    b. cell wall synthesis.
    c. intracellular protein targeting.
    d. proteins that bind to DNA.
    e. RNA polymerases.

3. The genomes of the fruit fly and nematode worm are similar to that of yeast, except that the former have many genes for
    a. intercellular signaling.
    b. syntheis of complex polysaccharides.
    c. cell cycle regulation.
    d. intracellular protein targeting.
    e. transposable elements.

4. Which of the following does *not* occur after mRNA is transcribed?
    a. Binding of RNA polymerase II to the promoter
    b. Capping of the 5′ and
    c. Addition of a poly A tail to the 3′ end
    d. Splicing out of the introns
    e. Transport to the ribosome

5. Which statement about RNA splicing is *not* true?
    a. It removes introns.
    b. It is performed by small nuclear ribonucleoprotein particles (snRNPs).
    c. It always removes the same introns.
    d. It is directed by consensus sequences.
    e. It shortens the RNA molecule.

6. Telomeres are
    a. present in equal length in all cells.
    b. removed by telomerase.
    c. essential for the stability of chromosomes.
    d. located at the ends and middle of eukaryotic chromosomes.
    e. caused by errors in DNA replication.

7. Transcription factors in eukaryotes
    a. bind to DNA, and then RNA polymerase can bind.
    b. always repress transcription.
    c. are indetical at every gene.
    d. are always bound to DNA.
    e. have only helix-loop-helix domains foir DNA binding and recognition.

8. Heterochromatin
    a. contains more DNA than does euchromatin.
    b. is transcriptionally inactive.
    c. is responsible for all negative transcriptional control.
    d. clumps the X chromosome in human males.
    e. occurs only during mitosis.

9. Translational control
    a. is not observed in eukaryotes.
    b. is a slower form of regulation than transcriptional control.
    c. can be achieved by only one mechanism.
    d. requires that mRNA be uncapped.
    e. ensures that heme synthesis equals globin synthesis.

10. Which of the following are *not* used in transcriptional regulation in eukaryotes?
    a. Enhancers
    b. Silencers
    c. Transcription factors
    d. RNA polymerase subunits
    e. Promoters

Student Web Site Self-Quiz Questions

1. Which of the following is not a characteristic of the chromosomes of most eukaryotic cells?
    a. Telomeres
    b. The presence of a single, large chromosome in each nucleus
    c. Centromeres
    d. Recognition sequences
    e. Nuclear organization

2. The genomes of prokaryotes like *E. coli* differ in many features from the genomes of eukaryotes like the fruit fly. Select from the following the choice that is not a valid difference.
   a. Base pair size of the genome
   b. Percentage of DNA not coding for protein
   c. Presence of promoters associated with coding DNA
   d. Presence of enhancers and silencers associated with coding DNA
   e. Degree of protein binding to genomic DNA

3. Which of the following statements about gene families is *false*?
   a. Most members of a gene family differ only slightly in terms of their mature mRNA transcripts.
   b. Usually gene families are clustered together on the same chromosome.
   c. At least one member of the gene family must be functional.
   d. Nonfunctional gene family members are called pseudogenes.
   e. The figure below shows that the α- and β-globin gene families have some members that are pseudogenes.

4. Which of the following is *not* a way in which pre-mRNA and mature RNA differ?
   a. Presence of introns
   b. Presence of G cap
   c. Presence of poly A tail
   d. Presence of promoter and terminator sequences
   e. Presence of consensus sequences

5. Which of the following statements about intron removal is *false*?
   a. Consensus sequences define the boundaries between introns and exons.
   b. Intron removal occurs immediately after transcription is complete.
   c. ATP energy must be expended to cut the RNA during splicing.
   d. The closed loop, or lariat, includes the consensus 5′ sequence, the consensus 3′ sequence, and the enclosed exon.
   e. The snRNPs involved in creating the spliceosome are composed of both protein and RNA.

6. Which of the following general statements about the regulation of eukaryotic gene expression is *false*?
   a. Examination of RNA sequences within the nucleus shows that posttranscriptional control is more commonly employed than transcriptional control.
   b. "Housekeeping" genes are always transcribed in all cell types of a multicellular organism.
   c. Eukaryotic genes are not organized into operons.
   d. Unlike prokaryotes, eukaryotes employ more than one type of RNA polymerase.
   e. Eukaryotic promoters are more numerous and variable than the promoters found in prokaryotes.

7. Which of the following statements about X chromosome inactivation is *false*?
   a. The probability that a given X chromosome will become inactive is 50%.
   b. Inactivation of the X chromosome seems to involve methylation of cytosine on DNA.
   c. The Barr body is a clump of euchromatin representing the inactivated X chromosome.
   d. An abnormal male with the genotype XXY would have one Barr body.
   e. The inactive X chromosome has a gene that is transcriptionally active, called *XIST*.

TEST FILE

★ *Indicates a **difficult** question*

■ *Indicates a **conceptual** question*

15 Cell Signaling and Communication

Fill in the Blank

1. Nitric oxide and **acetylcholine** are needed to relax the smooth muscle cells of the blood vessels.

★2. **Nitric oxide (NO)** could be eliminated as a needed signal to relax smooth muscle cells if a membrane permeable cGMP was used.

3. Cells of many multicellular animals communicate directly by coupling their cytoplasms using **gap junctions.**

4. Signals that act on the same cells as those that generated them are called **autocrine**.

5. Signals that diffuse to other types of cells are called **paracrine**.

6. The molecule that binds to the receptor is called a **ligand**.

7. Phosphatidylinositol is a **lipid**.

8. In the signal pathway that includes phospholipase C, **inositol triphosphate (IP$_3$)** opens calcium channels.

Multiple Choice

■1. Why might a certain signaling molecule affect one cell type in an organism and not another?
   a. Each different cell type is capable of interpreting just one kind of signal to prevent confusion.
   b. Each different cell type is capable of interpreting signals necessary for its functions.
   c. All cells can interpret the different signaling molecules only in different ways.
   d. Not all the cells are exposed to all the different signals.
   e. It is not the cells that must interpret the signals, but the signals that must interpret the cells.

2. Cells receive which of the following signals?
   a. Light
   b. Sound
   c. Odorants
   d. Hormones
   e. All of the above

3. Two ways signals get to target cells in multicellular organisms are via
   a. circulation and diffusion.
   b. conduction and diffusion.
   c. circulatory and lymphatic.
   d. chaperon trafficking and transmembrane transport.
   e. cytoskeletal trafficking and perfusion.

4. A signal pathway is
   a. the path a signal takes to find its target cell.
   b. a group of signals along a concentration gradient.
   c. the coordinates created from concentration gradients that tell cells where they are located in multicellular organisms.
   d. the series of events that occur in response to a signal being detected.
   e. a nerve propagation.

★5. The envZ protein of *E. coli* changes shape in response to
   a. ion concentration.
   b. ligand binding.
   c. light.
   d. sound.
   e. O_2.

★6. The envZ protein is a(n)
   a. channel protein.
   b. kinase.
   c. phosphorylase.
   d. environmental gene.
   e. All of the above

★7. The ompR protein is a(n)
   a. DNA binding protein.
   b. channel protein.
   c. channel-blocking protein.
   d. osmotic pressure-detecting protein.
   e. kinase.

8. In general, all cell signaling causes
   a. altered gene expression.
   b. influx of ions.
   c. protein kinase activity.
   d. G protein activation.
   e. a change in receptor conformation.

9. The major categories of signal receptors are
   a. inside and outside.
   b. enzyme and ion channel.
   c. transmembrane and cytoplasmic.
   d. protein kinase and cAMP.
   e. sensory and molecular.

10. Steroids bind
    a. to the outer face of transmembrane proteins.
    b. to cytoplasmic receptors.
    c. within the lipid bilayer.
    d. around the nuclear membrane.
    e. directly to DNA.

11. Steroids typically affect
    a. gene transcription.
    b. ion channels.
    c. enzyme activity.
    d. biochemical pathways.
    e. aggression.

12. Typically, large polar signals directly interact with
    a. cytosolic receptors.
    b. transmembrane receptors.
    c. G proteins.
    d. adenylyl cyclase.
    e. calmodulin.

13. Insulin is a(n)
    a. transmembrane protein kinase.
    b. intracellular signaling molecule.
    c. form of sugar.
    d. extracellular ligand.
    e. derivative of glucose.

14. Some signal receptors are
    a. ion channels.
    b. protein kinases.
    c. G protein-linked.
    d. DNA binding proteins.
    e. All of the above

▪15. Different cell types that respond to a certain signal do not all respond the same way because
    a. different types of cells might have different signal pathways for the same signal.
    b. different cells have different metabolic needs.
    c. not all cells respond to all signals.
    d. the cell might have already received the signal previously.
    e. None of the above

16. Protein kinase is
    a. an enzyme that makes cAMP.
    b. the enzyme that makes cGMP.
    c. the substrate molecule for kinase.
    d. an enzyme that phosphorylates.
    e. None of the above

17. One of the substrates for protein kinase is
    a. cAMP.
    b. cGMP.
    c. G proteins.
    d. ATP.
    e. GTP.

18. In eukaryotic cells, a substrate for phosphorylation is
    a. cAMP.
    b. cGMP.
    c. specific tyrosines in target proteins.
    d. specific glycines in target proteins.
    e. All of the above

19. Insulin alters cellular metabolism by
    a. binding with two receptor subunits on the outer cell surface.
    b. alerting cells about the availability of glucose.
    c. kinase activity.
    d. G protein activation.
    e. opening glucose channels in the membrane.

20. The insulin receptor is a
    a. G protein.
    b. cAMP molecule.
    c. kinase.
    d. phosphodiesterase.
    e. phosphatase.

21. The receptor-associated proteins called G proteins
    a. bind GTP.
    b. are bound to GDP.
    c. interact with membrane-associated internal proteins to influence their function.
    d. a and b
    e. All of the above

22. The Ras protein is
    a. a G protein.
    b. a protein that activates a kinase.
    c. part of a pathway called a phosphorylation cascade.
    d. the cause of a certain cancer when it is defective.
    e. All of the above

▪23. In order for a G protein to play its part at moving events forward in a signal pathway,
    a. GDP must be released and a GTP must occupy the nucleotide-binding site.
    b. GTP must be released and a GDP must occupy the nucleotide-binding site.
    c. cGMP must occupy the otherwise empty nucleotide-binding site.
    d. cGMP must leave the otherwise occupied nucleotide-binding site.
    e. the G protein must interact with a receptor protein.

▪24. If a G protein was unable to release its bound nucleotide but was able to hydrolyze it, signal transduction would
    a. not move beyond this point.
    b. be continuous beyond this point.
    c. be unaffected.
    d. be constantly switching on and off.
    e. be unpredictable.

▪25. If a G protein was able to release its bound nucleotide but was unable to hydrolyze it, signal transduction would
    a. not move beyond this point.
    b. be continuous beyond this point.
    c. be unaffected.

d. be constantly switching on and off.
e. be unpredictable.

*26. In heart muscles, the G protein that associates with the epinephrine receptor
a. inhibits adenylate cyclase.
b. activates adenylate cyclase.
c. causes a rise in cGMP.
d. causes a decline in cGMP.
e. None of the above

*27. In smooth muscle cells of the blood vessels, the G protein that associates with the epinephrine receptor
a. inhibits adenylate cyclase.
b. activates adenylate cyclase.
c. causes a rise in cGMP.
d. causes a decline in cGMP.
e. None of the above

28. Cytoplasmic receptors are used for
a. epinephrine.
b. cAMP.
c. steroids.
d. insulin.
e. All of the above

29. Cytoplasmic receptors are or are closely associated with
a. ligands.
b. G proteins.
c. transmembrane receptors.
d. the endoplasmic reticulum.
e. DNA binding proteins.

30. When a cell with a steroid receptor is exposed to that steroid, the one thing certain to happen is that the
a. receptor will change shape upon binding the steroid.
b. cell will produce much more of a muscle cell protein, actin.
c. cell will be masculinized.
d. cell will be feminized.
e. cell will get switched to an "on" state.

31. Transducers
a. change signals from one form to another.
b. alter gene expression.
c. are simple switches.
d. pass a certain signal forward.
e. All of the above

*32. The Ras protein is part of a
a. steroid receptor.
b. receptor-linked enzyme system.
c. cytoplasmic signal receptor system.
d. protein kinase cascade.
e. random, accelerated sorting system.

*33. The point in the pathway where Ras is located is
a. early.
b. middle.
c. late.
d. all along it.
e. None of the above

*34. The Ras protein is a
a. G protein.
b. protein kinase.
c. cell surface receptor.
d. a, b, and c are all correct.
e. None of the above

*35. The reason the defective Ras protein found in the many cases of bladder cancer causes uncontrolled cell division is that
a. GDP fails to be released.
b. kinase activity is defective.
c. the cGMP site remains occupied.
d. GTP fails to be hydrolyzed.
e. All of the above

36. The molecule cAMP is called
a. coupled adenine monophosphate.
b. a signal molecule.
c. a second messenger.
d. adenylyl adenylate cyclase.
e. cyclase adenylyl-phosphate.

▪37. Imagine that it is found that the membrane is necessary for a signal to elicit its effect, but it need be present only after signal and cytoplasm have been incubated together. If the experimenter removes the signal from the cytoplasm before adding the membrane, the effect can still be detected. Which of the following conclusions is consistent with these observations?
a. The signal interacts with a cytoplasmic receptor.
b. The signal pathway involves an interaction with a membrane-associated component.
c. The signal is unnecessary, but the membrane is necessary.
d. The signal pathway starts at the membrane and then progresses into the cytosol.
e. a and b

▪38. There are analogs to cAMP, which can pass through a membrane unimpeded. If you added this form of cAMP to liver cells, what would you expect to happen?
a. The same events that would occur if you added epinephrine.
b. The cell would open its Na^+ channels.
c. The receptor-associated G protein would activate.
d. The receptor-associated G protein would inactivate because it is what normally causes the elevated cAMP.
e. Adenylyl cyclase would activate.

39. Adenylate cyclase
a. is a cyclic nucleotide.
b. produces a cyclic nucleotide.
c. produces a G protein.
d. is a protein kinase.
e. is a second messenger molecule.

▪40. There are analogs to cAMP, which can pass through a membrane unimpeded. If you added this form of cAMP to odorant receptor nerve cells, what would you expect to happen?
a. The same events that would occur if you added epinephrine.
b. The cell would open its Na^+ channels.
c. The receptor associated G protein would activate.
d. The receptor associated G protein would inactivate because it is what normally causes the elevated cAMP.
e. The cAMP is odorless, so nothing would happen.

41. The concentration of cAMP in a cell is increased by a(n)
a. phosphodiesterase.
b. cGMP.
c. protein kinase.
d. ion channel.
e. G protein.

42. Phospholipase C is activated by
a. cAMP.
b. an elevated Ca^{2+} concentration.
c. a specific G protein.
d. cGMP.
e. GTP.

43. Activated phospholipase C
a. is a second messenger molecule.
b. opens Ca^{2+} channels.
c. hydrolyzes phosphatidylinositol.
d. activates G proteins.
e. cleaves cAMP's phosphodiester linkage.

44. The second messenger IP_3
a. opens Ca^{2+} channels.
b. activates protein kinase C.
c. hydrolyzes phosphatidyl inositol.
d. activates G proteins.
e. is released from the cell as a paracrine-signaling molecule.

45. Diacylglycerol
a. opens Ca^{2+} channels.
b. activates protein kinase C.
c. hydrolyzes phosphatidylinositol.
d. activates G proteins.
e. is released from the cell as a paracrine-signaling molecule.

46. To be activated, protein kinase C must
a. interact with IP_3.
b. bind Ca^{2+}.
c. bind DAG.
d. Both a and b
e. Both b and c

*47. The point of signal divergence in the phospholipase C signal pathway is when
a. the G protein activates the enzyme.
b. the receptor binds the ligand.
c. the enzyme hydrolyzes the lipid.
d. the Ca^{2+} and DAG bind protein kinase C.
e. calcium binds to it.

48. The point of signal convergence in the phospholipase C signal pathway is when
a. the G protein activates the enzyme.
b. the receptor binds the ligand.
c. activation of the enzyme hydrolyzes the lipid.
d. the Ca^{2+} and DAG bind protein kinase C.
e. calcium binds to it.

49. An example of Ca^{2+} acting as a second messenger is
a. G protein activation.
b. IP_3 opening calcium channels.
c. activation of the enzyme that hydrolyzes the lipid.
d. the Ca^{2+} binding to protein kinase C.
e. **b and d**

50. Calmodulin is activated when it
a. binds a cAMP molecule.
b. interacts with IP_3.
c. binds three K^+.
d. binds four Ca^{2+}.
e. binds GTP.

Study Guide Questions

1. Which of the following is not required for signal transduction to take place?
a. Ligand binding to receptor
b. Conformational change to the receptor protein
c. Conformational change to the signal
d. Alteration of cellular activity

2. Signals produced in the organism but not in the target cell are referred to as ____________ signals.
a. paracrine
b. autocrine
c. environmental
d. intrinsic

3. When looking at the osmotic change in *E. coli* as an example of how signals are transduced, the ultimate goal of the pathway is
a. to change the permeability of the membrane.
b. to change what DNA is transcribed.
c. phosphorylation of OmpR.
d. binding of the ligand to EnvZ.

4. G protein-linked receptors, protein kinases, and ion channels all have which of the following in common?
a. Ligand binding
b. Conformational change once the ligand is bound
c. Amplification of the signal
d. All of the above

5. Signal binding to a receptor differs from enzyme–substrate binding in what way(s)?
a. The signal is not altered in any way during the process.
b. The process is reversible.
c. The signal is consumed in the binding.
d. None of the above

6. Secondary messengers function to
   a. amplify the signal.
   b. bind to the active site of the receptor.
   c. result in multiple effects from a single signal.
   d. a and c
7. In the beginning section of this chapter, Carol is relying on caffeine to keep her awake. The caffeine works because it acts as ______________ to the adenosine receptors in Carol's brain, and by stimulating ______________ in her heart and liver that bypass hormonal regulation.
   a. an alloster, a pathway
   b. an inhibitor, cascades
   c. an inhibitor, ligands
   d. a signal, inhibitors
8. Cytoplasmic receptors are available only to
   a. small signals that can permeate the plasma membrane.
   b. secondary messengers like cAMP.
   c. hydrophilic molecules.
   d. All of the above
9. Protein kinase cascades are significant because
   a. amplification can occur at each step in the path.
   b. information at the plasma membrane is communicated to the nucleus.
   c. the multiple steps allow for specificity of the process.
   d. All of the above
10. G proteins provide an example of the many ways in which signal transduction is regulated. Which of the following are mechanisms by which regulation can occur?
    a. The amount of signal present can be regulated.
    b. Enzymes convert active forms of proteins to inactive forms.
    c. Signals are denatured.
    d. None of the above
11. Which of the following statements regarding receptors is *true*?
    a. Only one type of receptor is utilized throughout a cascade.
    b. Cascades may involve many types of receptors including ion channel receptors, protein kinase receptors, and G protein-linked receptors.
    c. Receptors amplify signals in a one-to-one ratio.
    d. All of the above
12. Gap junctions and plasmodesmata differ in that
    a. gap junctions are connected by protein tubules called connexons and plasmodesmata are connected by extensions of the plant plasma membrane.
    b. plasmodesmata allow much larger molecules to pass through them.
    c. gap junctions have no real physical connection but are the space between adjacent cell membranes.
    d. one is of animal origin and the other is of plant origin; otherwise they are physically the same.

End of Chapter Questions

1. What is the correct order for these events in the interaction of a cell with a signal?
   1 Alteration of cell function
   2 Signal binds to receptor
   3 Signal released from source
   4 Signal transduction
   a. 1234
   b. 2314
   c. 3214
   d. 3241
   e. 2431
2. The major difference between a cell that responds to a signal and one that does not is the presence of a
   a. DNA sequence that responds to the signal.
   b. nearby circulatory vessel.
   c. receptor.
   d. second messenger.
   e. transduction pathway.
3. Which of the following is *not* a consequence of binding of a signal to a receptor?
   a. Change in conformation of the receptor
   b. Activation of receptor enzyme activity
   c. Diffusion of the receptor in the plasma membrane
   d. Breakdown of the receptor to amino acids
   e. Release of the signal from the receptor
4. Which of the following is *not* a common type of receptor?
   a. Ion channel
   b. Protein kinase
   c. G protein-linked
   d. Transcription factor
   e. Adenylate cyclase
5. A nonpolar molecule such as a steroid hormone usually binds to a(n)
   a. cytoplasmic receptor.
   b. protein kinase.
   c. ion channel.
   d. phospholipid.
   e. second messenger.
6. What is the correct order of events in the activation of a G protein–coupled receptor?
   1 Effector becomes activated
   2 G protein is activated
   3 G protein subunit-GTP diffuses to effector
   4 GDP replaces GTP on G protein
   5 GTP replaced GDP on G protein
   6 Signal binds receptor
   a. 526341
   b. 621354
   c. 625314
   d. 643125
   e. 523614

7. Which of the following is *true* of the protein kinase cascade?
   a. The signal is amplified.
   b. A second messenger is formed.
   c. Target proteins are phosphorylated.
   d. The cascade ends up in the nucleus where gene transcription is altered.
   e. The cascade begins outside the plasma membrane.

8. Which of the following is *not* a second messenger for signal transduction?
   a. Calcium ions
   b. Nitric oxide gas
   c. ATP
   d. Cyclic AMP
   e. Diacylglycerol

9. A protein kinase is an enzyme that
   a. becomes active in all signal transduction events.
   b. adds phosphate groups to certain proteins.
   c. cannot be part of an actual receptor.
   d. only activates target proteins.
   e. is not affected by second messengers.

10. Plasmodesmata and gap junctions
   a. allow small molecules and ions to pass rapidly between cells.
   b. are both membrane lined channels.
   c. are channels about 1 mm in diameter.
   d. are present on few copies per cell.
   e. are involved in cell recognition in signaling.

Student Web Site Self-Quiz Questions

1. Which of the following statements regarding chemical signaling systems is *false*?
   a. Autocrine signals have an effect on the cells that secrete them.
   b. Paracrine signals have an effect on nearby cells.
   c. Most signals in larger organisms affect only neighboring cells.
   d. Most signals received by cells in a larger organism are chemical in nature.
   e. The cells of our body can respond to a diverse array of chemical signals.

2. Which of the following statements regarding cell signaling is *false*?
   a. Cortisol activates the cell by stimulating the release of a chaperone protein from its receptor.
   b. A single signal will produce the same response in all cell types.
   c. Steroid hormones, such as estrogen, generally bind to receptors found in the cytoplasm.
   d. Polar signals, such as insulin, interact with receptors found embedded in the cell's membrane.
   e. Signals are usually specific for a particular receptor type.

3. In the phosphorylase activation experiments conducted by Sutherland, Krebs, and Fischer, investigators incubated liver cell membranes with epinephrine. Following incubation, the investigators found that they could remove the membrane components and mix what was left with liver cell cytoplasm, resulting in activation of phosphorylase enzyme. What critical component was present in their membrane-depleted mixture that caused the activation of the phosphorylase enzyme?
   a. Glycogen
   b. Glucose
   c. cAMP
   d. Adenylyl cyclase
   e. Epinephrine

4. Which of the following statements regarding the IP_3 and DAG second messenger systems is *false*?
   a. Protein kinase C requires DAG and Ca^{2+} to be activated.
   b. Activation of phospholipase C requires the activity of activated G protein.
   c. Phosphatidyl inositol (PTI) is produced by the activity of phospholipase C and stimulates the release of Ca^{2+} from the smooth endoplasmic reticulum.
   d. IP_3 is produced from PTI as an activity of phospholipase C.
   e. Activated protein kinase C elicits a cellular response by phosphorylating target enzymes in the cell.

5. Many signaling pathways in cells make use of the calcium ion. Cells often store large amounts of calcium intracellularly—for example, in the smooth endoplasmic reticulum. Cells stores Ca^{2+} intracellularly because
   a. calcium is much more effective as a signaling molecule than any other compound.
   b. the cell cannot easily produce calcium.
   c. calcium-binding calmodulin molecules present in the cytoplasm quickly remove Ca^{2+} from the cytoplasm, turning the signal off.
   d. excessive calcium concentrations outside the cell are lethal to the cell, and therefore the cell must store large amounts of intracellular Ca^{2+} to counteract the external calcium levels.
   e. calcium is the only second messenger molecule available to eukaryotic cells.

6. Which of the following plays a direct role in the activation of signaling proteins?
   a. Membrane-bound Ca^{2+} ion pumps
   b. Calmodulin
   c. cAMP phosphodiesterase
   d. Protein phosphatases
   e. GTPases

7. In the study conducted by Robert Furchgott on blood vessel constriction, a unique second messenger system was unveiled. This pathway revealed why nitroglycerin is an effective drug for treating angina (insufficient blood flow in the heart). Which of the following statements describes why this drug is effective?
   a. Nitroglycerine directly stimulates the opening of Ca^{2+} channels.
   b. Nitroglycerin binds to guanylyl cyclase enzymes, stimulating the production of cGMP.
   c. Nitroglycerin blocks nitric oxide synthase activity, resulting in activation of guanylyl cyclase.
   d. Nitroglycerin promotes the diffusion of nitric oxide across the membrane of smooth muscle cells.
   e. Nitroglycerin releases nitric oxide, which then activates the signaling pathway without nitric oxide synthase activity.

8. The signal transduction pathway involved in the sense of smell does *not* involve which of the following molecules?
   a. Sodium channels
   b. Receptor proteins
   c. G proteins
   d. cAMP
   e. Nitric oxide

9. Which of the following interactions is *not* an example of an alteration of enzyme activity caused by a signal transduction pathway?
   a. Ca^{2+} binding to troponin molecules in muscle cells, causing their release
   b. Activation of adenylyl cyclase by G protein
   c. Phosphorylase kinase activating protein kinase A in liver cells
   d. cAMP binding to channel proteins, resulting in their opening
   e. Auto-phosphorylation of growth factor receptors, allowing for interaction with adaptor proteins

10. One of the most important functions of plasmodesmata in plant cells is
    a. to allow rapid diffusion of small signaling molecules, such as hormones, between cells.
    b. to provide a mechanism to transport endoplasmic reticulum from cell to cell.
    c. to prevent viral infection between adjacent plant cells.
    d. to allow the transport of large molecules between adjacent plant cells.
    e. to allow the exchange of chloroplast organelles between adjacent plant cells.

TEST FILE

★ *Indicates a **difficult** question*
■ *Indicates a **conceptual** question*

16 Development: Differential Gene Expression

Fill in the Blank

1. Although we often stress the embryo in discussing animal development, development is a process that continues through all stages of life, ceasing only with **death**.
2. Presumed mechanisms of determination include determination by cytoplasm segregation and determination by **induction**, in which certain tissues induce the determinations of other tissues.
3. One important lesson to be learned from the two sequential steps of embryonic induction in the vulval cells of *Caenorhabditis elegans* is that much of development is controlled by switches that allow a cell to proceed down **alternative** tracks.
4. Related to mutant homeotic mutations is an important sequence of 180 base pairs of DNA called the **homeobox**, which has been found in both animals and plants and which encodes a portion of some proteins called a homeodomain.
5. Bicoid protein (a morphogen) is a **transcription factor**.
6. The process by which underlying mesoderm in chick embryos causes the differentiation of ectodermal cells into feathers is called **induction**.
7. When the developmental fate of a cell does not change, even when the cell's surroundings are altered, the cell is said to be **determined**.
8. A sea urchin egg is said to have **polarity** because the distribution of cytoplasm components is different at one end of the egg than at the other end.
9. The number and polarity of segments formed during the development of insect larvae is determined by three classes of **segmentation** genes.
10. The **homeobox** is a small sequence of DNA that is found in some of the segmentation genes of *Drosophila*, as well as in some genes of all animals with segmented body plans.
11. One factor that regulates pattern formation is **positional** information, signals that indicate where one group of cells lies in relation to other cells.
12. The process of cells becoming functionally distinct is called cellular **differentiation**.
13. *MyoD1* and homeotic genes are examples of **selector** genes.
14. Substances that are produced in one place and diffuse to another and that cause pattern formation are called **morphogens**.
15. The **anchor** cell controls the fate of six other cells, which are involved in the formation of the vulva in *Caenorhabditis elegans*.

Multiple Choice

1. Three major processes that reveal a great deal about animal development include
   a. **determination, differentiation, and pattern formation.**
   b. blastulation, gastrulation, and physiology.
   c. immunoregulation, neurulation, and histogenesis.
   d. oncogenesis, differentiation, and histogenesis.
   e. None of the above

2. The earliest stage of development is called __________ stage.
   a. fetal
   b. **embryonic**
   c. germinal
   d gametic
   e. All of the above

3. Development occurs
   a. only during growth of the organism.
   b. **throughout the life of the organism.**
   c. only in nondividing cells.
   d. in ectoderm and endoderm, but not in mesoderm.
   e. only in animals.

4. During cleavage, the number of cells in a developing frog embryo increases. The cytoplasm in these new cells
   a. **comes from the egg cytoplasm.**
   b. is synthesized by the blastomeres.
   c. does not contain any yolk.
   d. is the vegetal pole.
   e. undergoes mitosis.

5. Proteins in the egg cytoplasm that play a role in directing embryonic development are called
   a. instars.
   b. imaginal discs.
   c. mesenchyme proteins.
   d. polar proteins.
   e. cytoplasmic determinants.

6. During initial development plant cells tend to __________ and animal cells tend to __________.
   a. grow, grow
   b. divide, elongate
   c. elongate, divide
   d. differentiate, determine
   e. determine, differentiate

7. The human body has approximately __________ functionally distinct kinds of cells.
   a. 12
   b. 24
   c. 100
   d. 200
   e. 1,000

▪8. The use of nuclear transplantation allowed developmental biologists to address the fundamental question of
   a. why a frog embryo develops into a frog and not some other organism.
   b. whether or not cell differentiation is due to loss of DNA from cells as they divide.
   c. whether or not humankind can create many identical people.
   d. the effect of mitochondria on growth rates.
   e. All of the above

9. The process of cells organizing to create the form of the multicellular organism is called
   a. morphogenesis.
   b. differentiation.
   c. determination.
   d. restriction.
   e. metamorphosis.

10. Humans typically have __________ cells of __________ different types.
    a. 10^{14}; 200
    b. 10^{6}; 200
    c. 10^{9}; 300
    d. 200; 10^{14}
    e. 200; 10^{6}

11. Which of the following is *not* true of animal development?
    a. Genes regulate development.
    b. Development occurs by progressive loss of DNA.
    c. Blastomeres are early embryonic cells.
    d. The sea urchin blastopore forms the archenteron.
    e. Cells actively migrate during development.

12. Cells must become __________ before they __________.
    a. aged, divide
    b. large, divide
    c. elongated, divide
    d. determined, differentiate
    e. developed, shape

*13. Cells of different types
    a. express different genes.
    b. express the same genes.
    d. express some of the same genes.
    d. have different DNA sequences.
    e. None of the above

14. Differentiation is caused by
    a. loss of DNA.
    b. determination.
    c. morphogenesis.
    d. differential gene expression.
    e. nuclear transplantation.

15. Experiments by Briggs and King and by Gurdon provided graphic evidence that
    a. neurulation is fixed.
    b. differentiation is not irreversible.
    c. amphibian embryos can cease development for long periods of time.
    d. a, b, and c
    e. a and b

16. Pattern formation is necessary for
    a. morphogenesis.
    b. differentiation.
    c. determination.
    d. restriction.
    e. metamorphosis.

17. As cells become specialized, they must first
    a. be differentiated.
    b. be determined.
    c. lose broad developmental potential.
    d. gain a fate.
    e. become functionally distinct.

18. Once cells become fixed in a final functional and physical state, they are
    a. determined.
    b. committed.
    c. differentiated.
    d. totipotent.
    e. morphogized.

19. The fertilized egg is
    a. fully competent.
    b. totipotent.
    c. pluripotent.
    d. determined.
    e. always diploid.

▪20. If cells from the neural tube of a frog embryo are transplanted onto the ventral surface of a second embryo, the transplanted tissue will still continue to develop into tissues of the nervous system. The transplanted cells are
a. differentiated.
b. totipotent.
c. discontinuous.
d. determined.
e. endodermal.

21. Dolly is a transgenic sheep that produces _________ in her milk.
a. human growth hormone
b. IGF
c. insulin
d. interferon
e. α-1-antitrypsin

22. Transplantation has been done
a. with nuclei from cells of an adult ewe's udder.
b. with very small amounts of cytoplasm as the graft.
c. in mammals.
d. in frogs.
e. All of the above

23. The Dolly experiment addressed the fundamental question of
a. whether nuclei of mammals irreversibly differentiate.
b. what the effect of cytozymes is on cellular differentiation.
c. whether mankind can create many identical people.
d. what the effect of mitochondria is on growth rates.
e. All of the above

24. The current theory on why the nucleus of an udder cell "dedifferentiated" to generate the famous lamb Dolly is that the
a. cells were cultured.
b. cells were starved and stalled at G1.
c. cells were starved and stalled at G2.
d. cells were fed and dividing.
e. egg failed to have its genetic material properly removed.

25. Nuclear differentiation is characterized as
a. irreversible in the frog nuclear transplant experiments.
b. unimportant in the study of regeneration.
c. a global or universal event in multicellular organisms.
d. due to a major loss of DNA from chromosomes during cleavage.
e. limited only to embryonic development through the gastrula stage.

26. Embryonic induction
a. cannot explain the formation of the vertebrate eye.
b. was first described by Briggs and King.
c. initiates a sequence of differential gene expression.
d. does not occur in the adult.
e. is an example of a tissue inducing itself.

27. The nuclear transplantation experiment conducted by Gurdon's group demonstrated that differentiation
a. is due to changes in gene expression.
b. in animals is irreversible.
c. in plants is reversible.
d. is due to loss of genetic material.
e. is due to loss of nuclei.

*28. The DNA binding motif for *MyoD1* is
a. helix-loop-helix.
b. zinc finger.
c. leucine zipper.
d. helix-turn-helix.
e. hydrogen bonding.

29. When the protein from the *MyoD1* gene is injected into a fat cell, the cell
a. becomes a muscle cell.
b. becomes a muscle cell only until the protein breaks down.
c. becomes a hybrid fat–muscle cell called a factual cell.
d. stays a fat cell because differentiation has already occurred.
e. becomes a selector cell.

*30. Cutting sea urchin eggs into two parts, upper and lower halves results in
a. two normal larvae.
b. one abnormal larva.
c. two dwarfed larvae.
d. no larvae.
e. two abnormal larvae.

31. Development of the vertebrate limb requires
a. cellular differentiation.
b. interactions between embryonic germ layers.
c. coding for positional information.
d. retinoic acid.
e. All of the above

32. Which of the following statements about eye development is *false*?
a. The optic vesicle forms before the lens placode.
b. The lens placode is lateral to the optic vesicle.
c. The lens placode undergoes invagination.
d. The optic cup connects with the optic nerve.
e. The eye develops because of cytoplasmic determinants.

*▪33. If a permeable barrier, one that allows free movement of molecules, is placed between optic vesicles and the surface cells in the future eye region, what will occur?
a. No lenses will form.
b. Lenses will form.
c. Lens placodes but no lenses will form.
d. No lens placodes but lenses will form.
e. No lens placodes or lenses will form.

34. Induction of ectoderm to form a lens placode requires
a. competent ectoderm.
b. an underlying optic vesicle.
c. a signal from the optic vesicle.
d. All of the above
e. None of the above

35. When ectodermal cells are placed next to notochordal mesoderm, they form a neural tube. This is an example of
   a. **induction.**
   b. gastrulation.
   c. division.
   d. totipotency.
   e. cytoplasmic determination.

36. The developmental fate of mesodermal cells in the chick wing bud is determined, at least in part, by the distance of those cells from the apical ectodermal ridge. This is an example of
   a. homeotic mutations.
   b. **positional information.**
   c. cytoplasm determinants.
   d. totipotency.
   e. irreversible differentiation.

37. *Caenorhabditis elegans*
   a. has a fixed number of cells in the adult form.
   b. has no mutations to help in its analysis.
   c. lacks a zygote.
   d. is a nematode roundworm of significant research value.
   e. **a and d**

38. The anchor cell influences the differentiation and morphogenesis of several surrounding cells using the mechanism of
   a. cytoplasm determinants.
   b. random events generator.
   c. **induction.**
   d. P granules.
   e. None of the above

39. Which of the following statements about *Caenorhabditis elegans* is *false*?
   a. It is a roundworm.
   b. Genetic analysis of it has been productive.
   c. It has a fixed number of cells in the nervous system.
   d. **It can reproduce asexually.**
   e. Its life cycle lasts less than one week.

40. In *Caenorhabditis elegans*, if the anchor cell is destroyed,
   a. **no vulva will form.**
   b. no anchorettes will form.
   c. secondary vulval precursors will become primary vulval cells.
   d. another cell will differentiate into an anchor cell.
   e. signals from the roundworm's surface will direct readjustments.

▪41. The roundworm *Caenorhabditis elegans* has been used for the detailed analysis of animal development. Which of the following is *not* a characteristic that makes this organism useful for such studies?
   a. The adult contains fewer than 1,000 cells.
   b. Development from the zygote to the adult takes only a few days.
   c. The body is relatively transparent.
   d. **Symmetry in the adult body is the result of symmetrical cell divisions.**
   e. Mutations in genes that control development have been identified.

42. Large-cell lymphoma is caused by a mutation in __________. This gene is analogous to *ced-9* in *Caenorhabditis elegans*.
   a. *ced-3*
   b. *ced-4*
   c. ***bcl-2***
   d. *lin-s*
   e. *MyoD1*

43. The genes that regulate the differentiation of whorls in *Arabidopsis thaliana* are called
   a. **organ identity genes.**
   b. dimer genes.
   c. whorl locator genes.
   d. sepals, petals, stamens, and carpels.
   e. central axis control genes.

★▪44. In a type 3 whorl, genes of classes B and C are expressed. As a result of this expression, it would be expected that __________ would be found in those whorl cells.
   a. AA dimers
   b. AB dimers
   c. BB dimers
   d. **All of the above**
   e. None of the above

45. If in *Drosophila*, the segment that normally gives rise to the halteres instead develops into a second pair of wings, the *Drosophila* probably has a __________ mutation.
   a. chronogene
   b. gap gene
   c. **homeotic**
   d. segmentation gene
   e. cytoplasm

46. The __________ is eight or more genes that control the development of the abdomen and posterior thorax of *Drosophila* sp.
   a. ***bithorax* mutation**
   b. segment polarity
   c. antennapedia complex
   d. pair rule
   e. imaginal disc

47. Body segmentation in *Drosophila* sp. is controlled by which sequence of gene action?
   a. Segment polarity, pair rule, gap
   b. Gap rule, pair true, segment polarity
   c. **Gap, pair rule, segment polarity**
   d. *Bithorax*, pair rule, segment polarity
   e. None of the above

48. The homeobox gene complex
   a. codes for a transcriptional regulator.
   b. is found in segmented organisms.
   c. is present in some plants.
   d. is found in some genes that show differential expression in development.
   e. **All of the above**

49. Homeotic mutations
   a. do not create developmental abnormalities.
   b. affect the number of body segments.
   c. alter eye color in *Drosophila* sp.
   d. are easily studied in the adult.
   e. alter the developmental path of the imaginal disc.

50. The __________ genes determine the number and polarity of the segments in an insect larva.
   a. homeobox
   b. determinant
   c. segmentation
   d. positional
   e. involution

51. Homeobox DNA is found
   a. in all organisms except humans.
   b. only in tomatoes and sea urchins.
   c. only in organisms with segmented body plans.
   d. in both plants and animals.
   e. only in *Drosophila*.

52. The proteins made from genes containing a homeobox
   a. bind to DNA.
   b. are found only in the cytoplasm.
   c. regulate the transcription of other genes.
   d. a and c
   e. b and c

53. The mutant *Drosophila* called *Antennapedia*
   a. has legs growing in place of antennae.
   b. grows wings in place of eyes.
   c. is a homeotic mutant.
   d. a and c
   e. b and c

54. The fly larva is very different in appearance from the adult. The major changes that occur during the development from larva to adult are called
   a. gastrulation.
   b. involution.
   c. complete metamorphosis.
   d. discontinuous growth.
   e. neurulation.

*55. A syncytium in *Drosophila* is expected
   a. early during development.
   b. near the head region.
   c. in larvae.
   d. in the unfertilized egg.
   e. during the time when segmentation genes are expressed.

56. In a "global" sense, mutations in developmental biology
   a. can be explained by changes in DNA.
   b. can alter body segmentation.
   c. help explain congenital malformations in humans.
   d. can be experimentally studied with molecular biology techniques.
   e. All of the above

57. Programmed cell death is crucial to normal development. The scientific term for this kind of cell death is
   a. apoptosis.
   b. program X.
   c. terminal differentiation.
   d. death by default.
   e. sonic hedgehog.

58. The genes *ced-3*, *ced-4*, and *ced-9* are all involved with regulating
   a. muscle cell differentiation.
   b. positional information.
   c. morphogenesis.
   d. egg polarization.
   e. apoptosis.

59. If the gene *ced-9* becomes nonfunctional, the result is
   a. webbed feet and hands.
   b. death.
   c. a fetus with no muscle cells.
   d. loss of symmetry.
   e. loss of nerve function.

*60. The signals for dorsal-ventral coordinate establishment are
   a. the same in both arthropods and vertebrates.
   b. different in arthropods and vertebrates.
   c. the opposite in arthropods and vertebrates.
   d. *Sog* in vertebrates and *chordin* in arthropods.
   e. *Pax6* in both vertebrates and arthropods.

Study Guide Questions

1. Cell differentiation is a result of
   a. the loss of particular genes from the nucleus of the differentiated cell.
   b. the differential expression of genes that are responsive to environmental signals.
   c. early embryonic cells remaining totipotent in the mature organism.
   d. mutations in genes that control the synthesis of DNA.

2. A totipotent cell is a cell
   a. whose developmental fate has been decided.
   b. that has differentiated into a specialized tissue.
   c. that is expressing morphogens to form a particular structure.
   d. whose developmental potential is extremely broad.

3. Apoptosis is programmed cell death and
   a. only occurs in developing embryos.
   b. only occurs in the mature organism.
   c. is an important process in both development and in mature organisms.
   d. can cause a type of cancer, follicular large-cell lymphoma.

4. Once a cell is committed to a particular developmental fate, it cannot be redirected down a different pathway unless that cell is
   a. fused to an enucleated egg cell.
   b. exposed to environmental factors that activate genes that control a different developmental pathway.
   c. transplanted to a different tissue.
   d. All of the above

5. Mammals can be cloned using somatic cells fused to enucleated eggs if the
   a. donor cell is in the S phase of the cell cycle.
   b. donor cell is in the G1 phase of the cell cycle.
   c. mammalian embryo can be propagated in culture.
   d. somatic cell is a red blood cell.

6. Identify the function of the protein MyoD1.
   a. It is a transcription factor that controls the expression of genes that form muscle.
   b. It controls segment identity in mice.
   c. It is a transcription factor that regulates its own transcription.
   d. a and c

7. Tissue induction occurs because
   a. nuclear genes are lost in some tissues.
   b. signals are secreted in one part of the developing embryo and are transported via the circulatory system to another site in the embryo to activate tissue differentiation.
   c. contact from one tissue allows the environmental signal from those cells to be delivered to the other tissue that must be induced.
   d. when two tissues come into contact with one another, transcription factors are released.

8. Apoptosis
   a. can cause abnormal development of the nervous system when inhibited in developing *C. elegans*.
   b. can cause webbing to be retained between the fingers of developing human embryos.
   c. is a natural part of development.
   d. a and c

9. It may be possible to genetically engineer plants to produce more seeds and fruits by
   a. fertilizing them more in the spring.
   b. manipulating their gap genes.
   c. altering the homeotic genes that control whorl development.
   d. inducing a mutation that eliminates *leafy* gene function.

10. Morphogens
    a. include vitamins and growth factors.
    b. are expressed as positional signals in developing embryos.
    c. direct differentiated cells to form organs.
    d. All of the above

11. Maternal genes
    a. set up positional axes in the egg prior to fertilization.
    b. are mRNA's and proteins that are placed in the egg cell.
    c. are masked by the paternal genes.
    d. a and b

12. Homeotic genes
    a. encode highly conserved protein domains that are important in development.
    b. are found in greatly diverse organisms.
    c. can produce the wrong structure in the wrong place when mutated, and have been highly conserved over evolutionary time.
    d. All of the above

End of Chapter Questions

1. Which statement about determination is true?
   a. Differentiation precedes determination.
   b. All cells are determined after two cell divisions in most organisms.
   c. A determined cell will keep its determination no matter where it is placed in an embryo.
   d. A cell changes its appearance when it becomes determined.
   e. A differentiated cell has the same pattern of transcription as a determined cell.

2. The cloning experiments on sheep and frogs showed that
   a. somatic cell nuclei of an organism are totipotent.
   b. nuclei of embryonic cells can be totipotent.
   c. nuclei of differentiated cells have different genes than zygote nuclei have.
   d. differentiation is fully reversible in all cells of a frog.
   e. differentiation involves permanent changes in the genome.

3. If cells from the neural tube of a frog embryo are transplanted onto the ventral surface of a second embryo, the transplanted tissue will develop into tissues of the nervous system. This experiment shows that the transplanted cells are
   a. differentiated.
   b. totipotent
   c. discontinuous.
   d. determined.
   e. endodermal.

4. A major difference between early human and fruit fly embryology is that only in the latter
   a. does cytokinesis not occur.
   b. are polarity genes not expressed.
   c. is the fertilized egg totipotent.
   d. do nuclei become determined before differentiation.
   e. is *sonic hedgehog* expressed.

5. Differentiation is caused by
   a. loss of DNA.
   b. determination
   c. morphogenesis.
   d. differential gene expression.
   e. nuclear transplantation.

6. In fruit flies, the following genes are used to determine segment polarity: (*k*) gap genes; (*l*) homeotic genes; (*m*) maternal effect genes; (*n*) pair rule genes. In what order are these genes expressed during development?
   a. *klmn*
   b. *lknm*
   c. *mknl*
   d. *nkml*
   e. *nmkl*

7. Which statement about embryonic induction is *not* true?
   a. One tissue induces an adjacent tissue to develop in a certain way.
   b. It triggers a sequence of gene expression in target cells.
   c. It may be either instructive or permissive.
   d. A tissue may induce itself.
   e. The chemical identification of specific inducers has been difficult.

8. In the evolution of arthropods and vertebrates,
   a. homologous genes code for the same dorsal and ventral organs.
   b. genes coding for dorsal in one code for ventral in the other, and vice versa.
   c. the genes coding for eye formation are not homologous.
   d. homeotic genes that determine segmentation in arthropods do not occur in vertebrates.
   e. segmentation is the same as in earthworms.

9. Homeotic mutations
   a. are often so severe that they can be studied only in larvae.
   b. cause subtle changes in the forms of larvae or adults.
   c. occur only in prokaryotes.
   d. do not affect the animal's DNA.
   e. do not occur in plants.

10. The proteins made from genes containing the homeobox
   a. bind to DNA.
   b. are found only in animals.
   c. regulate transcription of single genes.
   d. act in very different ways in mice and fruit flies.
   e. regulate translation of proteins.

Student Web Site Self-Quiz Questions

1. Of the following processes, which is more important in the early development of a typical plant than in that of an animal?
   a. Cell division
   b. Organized expansion
   c. Differentiation
   d. Morphogenesis
   e. Pattern formation

2. Which of the following statements about the process of animal cloning is *false*?
   a. Cloning has been performed successfully in sheep, mice and monkeys.
   b. The nucleus of differentiated cells must be obtained.
   c. Nuclei in the G2 phase of cell division must be obtained.
   d. Cloning requires the use of enucleated eggs from the same species.
   e. Cloning can be performed with virtually any differentiated tissue that has nucleated eggs.

3. The process of growing embryonic stem cells in the lab has great potential for clinical medicine. Which of the following statements about stem cells is *false*?
   a. Stem cells are undifferentiated.
   b. Stem cells have limited differentiation ability.
   c. Environment determines the differentiation potential of stem cells
   d. Stem cells from embryos have the greatest totipotency.
   e. Embryonic stem cells have a short life span when cultured in the lab.

4. In the sequence of events involved in formation of the vertebrate eye, how many different instances of induction occur?
   a. 0
   b. 1
   c. 2
   d. 3
   e. 4

5. Which of the following statements about the roundworm *Caenorhabditis elegans* or the role of inducers in its development is *false*?
   a. The fates of all 959 somatic cells of the worm have been established.
   b. At each successive division, the fate of the cells of *C. elegans* becomes more determined.
   c. Differentiation requires the presence of a single inducer.
   d. Differentiation of early embryonic cells is controlled by an anchor cell.
   e. Cells farthest away from the anchor cell are determined to be epidermal cells.

6. At about the second month of human development, the webbing between the digits disappears. This change is caused by ________.
   a. induction.
   b. morphogenesis.
   c. apoptosis.
   d. organ identity genes.
   e. selector genes.

7. Which of the following statements regarding maternal effect genes in *Drosophila* development is *false*?
   a. Morphogens in the cell are products of maternal effect genes.
   b. Expression of maternal effect genes in different cells determines differentiation.
   c. Maternal effect genes are transcribed in nurse cells.
   d. Mutants of the maternal effect *bicoid* gene produce larvae with two head regions.
   e. Maternal effect genes determine anterior-posterior regions in the embryo regardless of paternal genotype.

8. Which of the following statements about homeobox-containing genes is *false*?
   a. The *antennapedia* gene complex consists of homeobox-containing genes.
   b. Organ identity genes of plants do not contain the homeobox.
   c. The linear sequence of homeobox-containing genes on a chromosome reflects the time sequence of their activity in the course of development.
   d. The homeodomain is part of some transcription factors.
   e. The homeobox codes for the homeodomain.

TEST FILE

★ *Indicates a **difficult** question*
■ *Indicates a **conceptual** question*

17 Recombinant DNA and Biotechnology

Fill in the Blank

1. A **transgenic** animal has recombinant DNA integrated into its own genetic material.

2. Enzymes that cleave double-stranded DNA at specific sites are **restriction endonucleases**.

3. A short, single-stranded region at the end of a DNA fragment is a **sticky end**.

4. An enzyme that can covalently link two DNA fragments is **DNA ligase**.

5. A **cloning vector** is a virus or plasmid that can replicate its DNA and foreign DNA within a cell without being degraded.

6. A circular cloning vector that replicates autonomously within a cell is a **plasmid**.

7. A cloning vector that replicates and destroys a bacterial cell is a **bacteriophage**.

8. **Complementary DNA** is obtained by reverse transcription of RNA.

9. A collection of DNA molecules that represents a population of RNAs is a **cDNA library**.

★■10. DNA of the bacterial host is not cleaved by its own restriction endonucleases because of the activity of specific **methylases**, which add methyl groups to certain bases within the recognition sites.

11. Restriction endonucleases recognize sites by a **specific** sequence of bases.

12. An organism's DNA is isolated, cut by restriction endonucleases, and inserted into vectors where it is cloned. The result is a collection of clones called a **genomic library**.

13. In order to synthesize cDNA from mRNA, an important retroviral enzyme called **reverse transcriptase** is essential.

14. VNTRs are a **variable number of tandem repeats**.

Multiple Choice

1. A fragment of DNA with "sticky" ends
   a. can form hydrogen bonds with another fragment with complementary "sticky" ends.
   b. will be readily degraded in the test tube.
   c. is the starting point for RNA polymerase.
   d. is the recognition sequence for restriction endonucleases.
   e. None of the above

2. The enzyme that can join pieces of DNA together is
   a. RNA polymerase.
   b. DNA polymerase.
   c. DNA ligase.
   d. β-galactosidase.
   e. None of the above

★3. Restriction endonucleases cleave DNA at specific sequences by hydrolyzing
   a. two phosphodiester linkages on opposite strands.
   b. at the 1′ carbons to cleave the nitrogenous bases.
   c. at the 2′ carbons to cleave hydroxyl groups.
   d. two phosphodiester linkages on the same strand.
   e. four phosphodiester linkages, two on each strand.

4. Restriction endonucleases are used naturally in a bacterial cell to
   a. defend against foreign DNA.
   b. cleave large sections of bacterial DNA.
   c. digest extra copies of plasmid DNA.
   d. replicate the bacterial chromosomal DNA.
   e. produce RNA from DNA.

★■5. Why is the DNA of the host cell not cleaved by the restriction endonucleases it produces?
   a. The restriction endonucleases are contained within lysosomes.
   b. The restriction endonucleases can only cleave RNA.
   c. The restriction endonucleases are made on the rough endoplasmic reticulum and exported out of the cell.
   d. The bacterial DNA is altered by methylation and is not a substrate for restriction endonucleases.
   e. None of the above

6. Endonucleases were first identified in
   a. plant cells.
   b. eukaryotic cells.
   c. prokaryotic cells.
   d. blue-green algae.
   e. None of the above

7. Which of the following processes makes use of the nucleic acid base-pairing rules?
   a. DNA replication
   b. Transcription
   c. Translation
   d. Sequencing of genes
   e. All of the above

8. Which of the following sequences is a DNA palindrome?
   a. GCTATCG
   b. AAAAAA
   c. GCATGC
   d. ACGTAC
   e. All of the above

9. In order to join a fragment of human DNA to bacterial or yeast DNA, both the human DNA and the bacterial (or yeast) DNA must be first treated with the same
   a. DNA ligase.
   b. DNA polymerase.
   c. restriction endonuclease.
   d. DNA gyrase.
   e. None of the above

▪10. *Eco*RI makes staggered cuts when it cleaves DNA, creating single-stranded tails called "sticky ends." These ends will form a complementary base pair. Which of the following conditions must exist for this to happen?
   a. Specific helicases
   b. High enough temperatures
   c. Methyl groups at each end
   d. Low enough temperatures
   e. None of the above

11. Genetic engineering is important to botanists. The goal of this technology in plants is to
   a. alleviate the fears of the public regarding genetic engineering.
   b. isolate beneficial genes from one species and introduce them into another plant species for a superior crop.
   c. create new species of plants to replace the extinct ones.
   d. cause plant disease.
   e. None of the above

12. The specific purpose for cloning DNA fragments out of plants or animals is to
   a. splice other DNA fragments from dissimilar genomes together.
   b. locate a specific gene and prepare DNA probes.
   c. collect appropriate vectors for genetic engineering.
   d. prepare synthetic oligonucleotides.
   e. None of the above

13. When a DNA solution is added to a gel, the DNA will migrate
   a. toward the positive pole.
   b. toward the negative pole.
   c. perpendicular to the positive and negative poles.
   d. only due to diffusion.
   e. away from the positive pole.

14. Electrophoresis separates DNA fragments of different sizes, but this technique does not indicate which of the fragments contains the DNA piece of interest. This problem is solved by
   a. measuring the sizes of the bands on the gel.
   b. removing the bands from the gel and hybridizing with a known strand of DNA complementary to the gene of interest.
   c. knowing the isoelectric points of the piece in question.
   d. identifying the molecular weights of the fragments in question.
   e. None of the above

15. When DNA migrates in a gel, the longer DNA
   a. generally fails to move.
   b. migrates quickest because it has a greater charge.
   c. moves slowest.
   d. is found closest to the positive pole.
   e. length affects migration less than base composition.

*▪16. If the bacteriophage T7 DNA does not contain any *Eco*RI sites, how might an *E. coli* cell protect itself from T7 infection?
   a. The cell sequesters the phage DNA in a vesicle.
   b. The cell makes a number of restriction endonucleases, each with a different recognition site.
   c. The cell destroys its ribosomes and protein synthetic machinery.
   d. The cell destroys all of the cellular DNA replication enzymes.
   e. The cell cannot possibly protect itself.

17. Which of the following is *not* a usual characteristic of a cloning vector?
   a. It contains at least one restriction endonuclease recognition sequence.
   b. It must integrate into the host chromosome.
   c. It must be a relatively small piece of DNA.
   d. It must be able to replicate within the host cell.
   e. It carries a gene for antibiotic resistance.

▪18. Transfection of plant cells is more difficult than transfection of prokaryotic cells because
   a. plant cells are larger than prokaryotic cells.
   b. plant cells contain more DNA than prokaryotic cells.
   c. DNA must get through plant cell walls.
   d. a and b
   e. a and c

19. At present, which of the following is *not* a way to insert DNA into cells?
   a. Microinjection
   b. Enclosure in an artificial membrane

c. Projectiles coated with DNA
d. Electroporation
e. None of the above

20. A genomic library
a. is a collection of foreign DNA molecules inserted into cloning vectors.
b. ideally represents the entire genome of an organism.
c. is a collection of RNA molecules inserted into cloning vectors.
d. a and b
e. b and c

21. The expression of a cloned gene in a bacterium is easily regulated if it is
a. inserted anywhere in the bacterial chromosome.
b. incorporated into a phage genome.
c. spliced into a self-replicating plasmid.
d. spliced onto the *lac* operon promoter.
e. left free in the bacterial cell.

22. Which of the following is *not* usually a source of DNA for cloning?
a. Pieces of genomic, chromosomal DNA
b. DNA made by the reverse transcription of mRNA
c. DNA made by the reverse transcription of tRNA
d. DNA synthesized in the laboratory
e. All of the above

23. DNA migrates in an electric field because it
a. is positively charged.
b. is not charged.
c. is negatively charged.
d. combines with the gel molecules.
e. is hydrophobic.

*24. The disease called crown gall is caused by
a. the insertion of a transposable element carried on the Ti plasmid.
b. the transcription of the Ti plasmid in the plant cells.
c. the transfer of bacterial genomes into the plant cell genome.
d. the rampant multiplication of *A. tumefaciens* bacteria within the plant.
e. None of the above

*25. If a cloned DNA molecule is inserted into either of the antibiotic sites in the pBR322 (*amp*^{r} or *tet*^{r}), the antibiotic gene then becomes
a. transferred to a gene upstream of the site on the chromosome.
b. immediately transcribed into mRNA.
c. inactivated.
d. complemented with a corresponding sequence.
e. None of the above

26. Which of the following could be used for detecting genetic disorders?
a. cDNA probes that hybridize with DNA regions that are missing or have mutant nucleotide sequences near or within the DNA sequence responsible for the genetic disorder
b. PCR amplification of the region responsible for the genetic disorder
c. Restriction mapping of sequences that have different restriction sites and are associated with the disease
d. a and c
e. All of the above

27. Complementary DNA (cDNA) is made using
a. synthetic oligonucleotides.
b. amino acid sequences from structural genes.
c. DNA as the template.
d. mRNA as a template.
e. None of the above

28. When DNA is introduced into eukaryotic cells and is integrated, the cell is
a. transformed.
b. transduced
c. conjugated.
d. expressed.
e. transfected.

29. The production of double-stranded cDNA utilizes
a. hybridization between the poly A tails of mRNA and oligo dT.
b. reverse transcriptase.
c. DNA polymerase.
d. a and b
e. a, b, and c

30. In the production of double-stranded cDNA, mRNA is used as
a. a template.
b. a tail.
c. one strand of the completed cDNA.
d. a and b
e. a and c

31. An organism that is modified by introduction of a DNA sequence from another organism is called
a. transformed.
b. transgenic.
c. transfected.
d. tranduced.
e. a multi-species.

32. A YAC is a(n)
a. animal of interest because of its unusual mutations.
b. yeast artificial chromosome.
c. yellow activation center.
d. Y chromosome.
e. None of the above

33. Human artificial chromosomes have
a. a centromere.
b. telomeres.
c. origins of replication.
d. more than 10,000 base pairs.
e. All of the above

■34. If a plasmid has two antibiotic resistance genes, tetracyclin and ampicillin, and DNA is inserted into the ampicillin gene, the bacteria that are transformed should be first placed onto a plate
a. without either antibiotic.
b. with both antibiotics.
c. with just tetracyclin.
d. with just ampicillin.
e. on complete medium.

■35. Sticky ends are "sticky" because they are
a. single-stranded.
b. from one to four bases long.
c. complementary to other sticky ends.
d. poly A tails.
e. part RNA and part DNA.

■36. A second screening step is often necessary to find bacteria that have taken up plasmids with foreign DNA inserts because
a. there are many cells without any plasmids at all.
b. only the plasmids with foreign DNA are taken up into cells.
c. very often a plasmid without foreign DNA is taken up by a cell.
d. a and b
e. a and c

*■37. From the list below, choose a reasonable sequence of steps for cloning a piece of foreign DNA into a plasmid vector, introducing the plasmid into bacteria, and verifying that the plasmid and insert are present.
1. Transform competent cells
2. Select for the lack of antibiotic resistance gene #1 function
3. Select for the plasmid antibiotic resistance gene #2 function
4. Digest vector and foreign DNA with *Eco*RI, which inactivates antibiotic resistance gene #1
5. Ligate the digested DNA together
a. 4, 5, 1, 3, 2
b. 4, 5, 1, 2, 3
c. 1, 3, 4, 2, 5
d. 3, 2, 1, 4, 5
e. None of the above

38. A knockout experiment involves
a. homologous recombination.
b. transfected embryonic cells.
c. introduction of cells into a developing embryo.
d. use of a genetic marker.
e. All of the above

39. DNA chips have
a. lamda DNA libraries attached.
b. DNA sequences up to 20 nucleotides long attached.
c. cDNA attached.
d. radionucleotides attached.
e. All of the above

40. Antisense RNA prevents the expression of specific genes by
a. binding to the complementary RNA.
b. preventing the tRNA from assorting.
c. causing a mutation that interrupts the gene.
d. a and b
e. b and c

41. A cell or an organism that contains foreign DNA inserted into its own genetic material is termed
a. transgenic.
b. polygenic.
c. engineered.
d. foreign.
e. xenophobic.

42. Methods to insert foreign DNA into host cells include which of the following?
a. Bacterial transformation
b. Transfection
c. Electroporation
d. Microprojectiles
e. All of the above

43. Principal sources of genes or DNA fragments used in recombinant DNA work include
a. genomic libraries.
b. cDNA samples.
c. artificially prepared oligonucleotides.
d. a, b, and c
e. a and b

*44. An important vector for manipulation of plant genes comes from the bacterium *A. tumefaciens* and is called
a. an *Eco*RI plasmid.
b. a raze bacteriophage.
c. a pangene-site vector.
d. a Ti plasmid.
e. None of the above

45. The advantage of tissue plasminogen activator (TPA) over streptokinase is that streptokinase
a. is too effective.
b. might trigger an immune response.
c. costs much more than TPA.
d. a and b
e. b and c

46. The advantage of a viral vector over a plasmid vector is
a. viral vectors can often carry much larger DNA inserts.
b. plasmid vectors are unreliable.
c. viral vectors lack the necessary origins of replication.
d. viral vectors require less medium.
e. All of the above

47. A plasmid is isolated, digested with a restriction enzyme, and run on a gel using electrophoresis. The fragments closest to the well where the DNA was loaded are __________ than the fragments farthest away.
a. shorter
b. longer
c. duller

d. less original
e. We can only say that they are different.

48. The actual process of splicing DNA fragments into plasmids takes place in
a. the bacteria.
b. the phage.
c. a test tube.
d. All of the above
e. None of the above

49. Antisense RNA is
a. RNA that investigators find confusing.
b. RNA that makes opposite sense.
c. RNA that is complementary to a certain mRNA.
d. DNA.
e. the noncoding strand of the DNA molecule.

50. The first human drug made using recombinant DNA technology was
a. glyphosatase.
b. tissue plasminogen activator.
c. insulin.
d. human growth hormone.
e. erythropoietin.

▪51. A single hair is found at the scene of a crime. What technology would you use to determine if the hair could have come from the suspect?
a. PCR
b. DNA sequencing
c. Fragment cloning
d. Probing
e. Antisense RNA

Study Guide Questions

1. Cloning a gene involves
a. restriction endonucleases and ligase.
b. plasmids and bacteriophage lambda.
c. yeast artificial chromosomes and complementary base pairing.
d. All of the above

2. Restriction endonucleases
a. are enzymes that process pre -RNA's.
b. are enzymes that degrade DNA.
c. protect bacterial cells from viral infections.
d. All of the above

3. DNA fragments are separated using gel electrophoresis
a. because DNA is pulled through the gel toward the negative end of the field.
b. because larger DNA fragments move faster through the gel than smaller DNA fragments.
c. to identify and isolate DNA fragments.
d. to synthesize DNA for cloning.

4. Complementary base pairing is important for
a. ligation reactions with blunt-end DNA molecules.
b. hybridization between DNA and transcription factors.
c. restriction endonucleases to cut cell walls.
d. synthesizing cDNA molecules from mRNA templates.

5. In order for a prokaryotic vector to be propagated in a host bacterial cell, the vector needs
a. telomeres.
b. centromeres.
c. drug resistant genes.
d. an origin of replication.

6. In order for a eukaryotic vector to be propagated in a host eukaryotic cell, the vector needs
a. telomeres.
b. centromeres.
c. an origin of replication.
d. All of the above

7. Reporter genes
a. include genes for drug resistance.
b. include genes for bioluminescence.
c. include genes for DNA origins.
d. a and b

8. Vectors include
a. bacterial and plant plasmids.
b. viruses.
c. artificial chromosomes.
d. All of the above

9. Cloning means that
a. all of the cells are derived from one cell and are genetically identical.
b. a gene from one organism has been inserted into a vector and successfully introduced into a host cell.
c. all of the cells in a particular organism are identical.
d. All of the above

10. Recombinant DNA can be transferred into host cell by
a. growing the host cell in growth medium containing ampicillin.
b. coating the DNA with carbohydrates so that the cells will engulf the DNA.
c. treating cells with calcium ions or electrical pulses to increase cell permeability.
d. injecting proteins into host cells to make them more permeable.

11. A cDNA clone contains
a. mostly cytosine.
b. a copy of the DNA identical to the nuclear gene.
c. a copy of noncoding DNA.
d. a DNA copy of mRNA.

12. Gene expression can be inhibited by
a. antisense RNA.
b. knockout genes.
c. DNA chips.
d. a and b

13. Expression vectors are different from other vectors because they
a. contain drug resistance markers.
b. contain telomeres.
c. contain regulatory regions that permit the cloned DNA to produce a gene product.
d. contain DNA origins.

14. DNA fingerprinting works because
    a. genes containing the same alleles make it simple to compare different individuals.
    b. PCR allows amplification of proteins from single cells.
    c. there are multiple alleles for some DNA sequences, making it possible to obtain unique patterns for each individual.
    d. DNA in the skin cells is very diverse.

End of Chapter Questions

1. Restriction endonucleases
    a. play no role in bacteria.
    b. cleave DNA at highly specific recognition sequences.
    c. are inserted into bacteria by bacteriophages.
    d. are made only by eukaryotic cells.
    e. add methyl groups to specific DNA sequences.

2. Sticky ends are
    a. double-stranded ends of DNA fragments.
    b. identical for all restriction enzymes.
    c. removed by restriction enzymes.
    d. complementary to other sticky ends generated by the same restriction enzyme.
    e. hundreds of bases long.

3. In gel electrophoresis,
    a. DNA fragments are separated on the basis of size.
    b. DNA does not have an electric charge.
    c. DNA fragments cannot be removed from the gel.
    d. the electric field separates positively charged DNA fragments from negatively charged DNA fragments.
    e. the DNA fragments are naturally fluorescent.

4. Possession of which feature is *not* desirable in a vector for gene cloning?
    a. An origin of DNA replication
    b. Genetic markers for the presence of the vector
    c. Multiple recognition sites for the restriction enzyme to be used
    d. One recognition site each for many different restriction enzymes
    e. Genes other than the target for cloning

5. A labeled fragment of DNA that is complementary to a target regions of DNA and which is used for hybridization to the target is called a
    a. ligase.
    b. restriction fragment.
    c. probe.
    d. vector
    e. VNTR.

6. Complementary DNA (cDNA)
    a. is produced from ribonucleoside triphosphates.
    b. is produced by reverse transcription.
    c. is the "other strand" of single-stranded DNA.
    d. requires no template for its synthesis.
    e. cannot be placed into a vector, since it has the opposite base sequence of the vector DNA.

7. In a genomic library of frog DNA in *E. coli* bacteria,
    a. all bacterial cells have the same sequences of frog DNA.
    b. all bacterial cells have different sequences of DNA.
    c. each bacterial cell has a random fragment of frog DNA.
    d. each bacterial cell has many fragments of frog DNA.
    e. the frog DNA is transcribed into mRNA in the bacterial cells.

8. An expression vector requires all of the following, except
    a. genes for ribosomal RNA.
    b. a selectable genetic marker.
    c. a promoter of transcription.
    d. an origin of DNA replication.
    e. restriction enzyme recognition sites.

9. "Pharming" is a term that describes
    a. animals used in transgenic research.
    b. plants making genetically altered foods.
    c. the synthesis of recombinant drugs by bacteria.
    d. the large scale production of cloned animals.
    e. the synthesis of a protein drug in the milk of a transgenic animal.

10. Which of the following is *not* an argument made against the use of genetically altered plants?
    a. It is not nice to fool with Mother Nature.
    b. The long-term effects of the plants on the environment are not known.
    c. The long-tem effects of genetically altered foods are not known.
    d. Genetically altered foods benefit only producers and not consumers.
    e. The genes inserted to make transgenic crops are unknown.

Student Web Site Self-Quiz Questions

1. On average, you would expect the sticky-ended recognition sequence for EcoRI to occur in a genome about every 4,000 base pairs. The enzyme *Hpa*II has the recognition sequence CCGG. This would be expected to occur every _________ base pairs.
    a. 4
    b. 16
    c. 64
    d. 256
    e. 1024

2. Which of the following is *not* an important characteristic of a plasmid cloning vector?
    a. It should include a single origin of replication.
    b. It should have a selectable marker such as antibiotic resistance.
    c. It should have several recognition sequences for the restriction endonuclease used to isolate the gene to be inserted.
    d. It should be a replicon.
    e. It should have the ability to amplify the gene being carried by producing many copies within the host cell.

3. Which of the following is *not* a reason that the bacterium *Agrobacterium tumefaciens* is much used in recombinant DNA studies?
   a. Plants cannot be infected with the bacterium unless new DNA is spliced into the *Ti* plasmid.
   b. Insertion of the recombinant DNA into the plant cell is accomplished by infection.
   c. The Ti plasmid is a cloning vector.
   d. The Ti plasmid's T DNA is a transposon that can be integrated into the chromosomes.
   e. The Ti plasmid cannot produce tumors if foreign DNA has been spliced into the T DNA.

4. You are working with a plasmid that has two antibiotic resistance genes for tetracycline (tet^r) and ampicillin (amp^r) with a *BamHI* restriction site within the tet^r gene. You restrict a mixture of foreign DNA and your plasmid with *BamHI,* use ligase to create recombinant molecules, and transform *E. coli* sensitive to both antibiotics with the mixture. You then spread the *E. coli* onto plates with four types of media: (1) without any antibiotic, (2) with just tetracycline, (3) with just ampicillin, or (4) with both. The bacteria with the recombinant plasmid would grow on
   a. plate 1 only.
   b. plates 1 and 2 only.
   c. plates 1 and 3 only.
   d. plates 2 and 3 only.
   e. all four plates.

5. Which of the following is *not* a valid statement regarding DNA chip technology?
   a. DNA chip technology allows scientists to determine mRNA expression patterns
   b. DNA chips contain a single type of DNA, allowing for the hybridization to a single gene at a time.
   c. When cellular mRNA is analyzed, cDNA must first be made prior to hybridization with the DNA chip.
   d. The cDNA molecules used in an experiment must first be labeled with a fluorescent marker (dye) prior to hybridization with the DNA chip.
   e. DNA chips may be used for detecting genetic variants.

6. When DNA is transcribed into mRNA, usually the mRNA remains single-stranded, but in some cases an RNA strand can be made that is complementary to the mRNA. This is called __________, and its main function is to __________.
   a. antisense RNA, block gene expression
   b. antisense RNA, amplify mRNA
   c. antisense RNA, enhance translation
   d. reverse transcription, make a cDNA library
   e. reverse transcription, enhance translation

7. Select below the feature that is *not* normally required in an expression vector designed to be inserted into *E. coli.*
   a. *Ori*
   b. Transcription terminator sequence
   c. Transcription factor binding sites
   d. Ribosome binding sequence
   e. Restriction endonuclease recognition site

8. Which of the following statements about the application of biotechnology to produce tissue plasminogen activator (TPA) is *false*?
   a. Antibodies specific to human TPA were used to isolate the mRNA encoding TPA.
   b. The TPA expression vector had to be designed for use in a prokaryotic cell.
   c. The TPA gene was the foreign DNA inserted into the expression vector.
   d. TPA is an enzyme that converts plasminogen into plasmin.
   e. Reverse transcriptase was used to make cDNA from TPA mRNA.

9. Which of the following statements regarding the use of biotechnology to produce artificial insulin with an expression vector is *false*?
   a. Insulin was originally obtained from pancreatic extracts of cows and pigs.
   b. Animal insulin is difficult to obtain.
   c. Animal insulin is identical to human insulin.
   d. Recombinant DNA allowed scientists to insert a human insulin gene into a bacterial expression vector.
   e. Recombinant insulin is actually obtained from *E. coli* bacterial cells.

10. Which of the following is *not* a key reason for the cloning of transgenic animals?
    a. Transgenic animals are capable of expressing human alpha-1-antitrypsin.
    b. Alpha-1 antitrypsin is produced in large quantities in milk when placed under the control of the lactoglobulin promoter.
    c. Alpha-1 antitrypsin can easily be purified from milk.
    d. Cloning transgenic animals was necessary since the creation of transgenic animals is an inexact science.
    e. The cloning process simply involves the transfer of the expression vector containing the human gene directly into the egg cell of a female sheep.

TEST FILE

★ *Indicates a **difficult** question*

■ *Indicates a **conceptual** question*

18 Molecular Biology and Medicine

Fill in the Blank

1. In addition to defective alleles, a second major source of genetic ill health is **chromosomal** aberrations.
2. Sex-linked recessive diseases affect **men** more than **women**.

*3. Sex-linked dominant diseases would affect **women** more than **men**.

4. The name of an X-linked recessive disorder resulting in muscular deterioration is **Duchenne's muscular dystrophy**.
5. **Hemophilia** is an X-linked recessive trait affecting the clotting of blood.
6. **Gene therapy** is the process whereby an abnormal gene is replaced by a normal one.
7. Success was short-lived for a treatment of a patient with defective adenosine deaminase genes who had genes transferred to her white blood cells. It might have been more effective to transfer the genes to bone marrow **stem** cells.
8. In testing a fetus for harmful alleles, DNA obtained from amniocentesis can be amplified by an artificial cycling process called **PCR**.
9. The most thoroughly studied example of a disease that arises via somatic mutation is **cancer**.
10. An estimated half of spontaneous abortions that occur during the first trimester of pregnancy are attributed to **chromosome abnormalities**.

Multiple Choice

1. What two diseases are both caused by defects in phenylalanine metabolism?
   a. Alkaptonuria and cystic fibrosis
   b. Sickle-cell anemia and cystic fibrosis
   c. Alkaptonuria and sickle-cell anemia
   d. Fragile-X syndrome and prion disease
   e. Phenylketonuria and alkaptonuria

2. The disease sickle-cell anemia is caused by a __________ mutation.
   a. frameshift
   b. base substitution
   c. deletion
   d. insertion
   e. tandem repeat

3. The most common form of inherited mental retardation is
   a. Tay-Sachs disease.
   b. cystic fibrosis.
   c. Down syndrome.
   d. fragile-X syndrome.
   e. None of the above

4. Triplet repeats
   a. occur in certain genetic disorders.
   b. are not limited to disease genes.
   c. expand readily in human genes but not in nonhuman genes.
   d. a and b
   e. a, b, and c

5. In addition to inherited genetic diseases, __________ can also undergo mutations that can result in genetic diseases that are not inherited.
   a. somatic cells
   b. germ cells
   c. white blood cells
   d. connective tissue cells
   e. All of the above

*6. Metabolic diseases such as PKU and alkaptonuria are caused by a(n)
   a. nonfunctional or missing enzyme.
   b. abnormal number of chromosomes.
   c. mutation in the mitochondrial DNA.
   d. virus.
   e. abnormal structural protein.

7. Although the exact cause of the mental retardation associated with PKU is not known, what is known is that
   a. it occurs when the X chromosome breaks into pieces.
   b. high levels of phenylalanine are involved.
   c. it is currently untreatable.
   d. All of the above
   e. None of the above

8. The first genetic disease for which an amino acid abnormality was tracked down was
   a. PKU.
   b. alkaptonuria.
   c. fragile-X syndrome.
   d. sickle-cell anemia.
   e. hemophilia.

9. Sickle-cell anemia is the result of a
   a. deletion.
   b. nonsense mutation.
   c. frameshift mutation.
   d. base-pair substitution.
   e. chromosomal deletion.

*10. The probability of carrying a defective allele for a certain disease depends on
   a. environment.
   b. ancestry.
   c. genetic predisposition.
   d. unknown factors.
   e. None of the above

11. The frequency of cystic fibrosis in the human population is
   a. 1 in 2.5.
   b. 1 in 25.
   c. 1 in 250.
   d. 1 in 2,500.
   e. 1 in 25,000.

12. Tissues affected by cystic fibrosis include the
   a. lungs.
   b. liver.
   c. pancreas.
   d. gut.
   e. All of the above

13. Patients with cystic fibrosis often die in their
   a. 10s and 20s.
   b. 20s and 30s.
   c. 30s and 40s.
   d. 40s and 50s.
   e. None of the above

14. The gene that is affected by cystic fibrosis normally encodes a protein that
   a. stimulates mitochondria.
   b. regulates gene expression.
   c. synthesizes mucus.
   d. encodes a chloride channel in the cell membrane.
   e. None of the above

15. Victims of phenylketonuria cannot metabolize
   a. alanine.
   b. phenylalanine.
   c. glutamic acid.
   d. tryptophan.
   e. tyrosine.

16. The principle consequence of phenylketonuria is
   a. muscle atrophy.
   b. kidney failure.
   c. mental retardation.
   d. skeletal problems.
   e. None of the above

17. Following diagnosis of phenylketonuria, infants are restricted to a diet low in
   a. alanine.
   b. phenylalanine.
   c. glutamic acid.
   d. tryptophan.
   e. tyrosine.

*18. Among the African-American population, the frequency of sickle-cell anemia is about
   a. 1 per 2,000.
   b. 1 per 600.
   c. 1 per 60.
   d. 1 per 20.
   e. None of the above

19. Individuals homozygous for the sickle-cell trait produce abnormal
   a. keratin.
   b. hemoglobin.
   c. myosin.
   d. tyrosinase.
   e. None of the above

*20. The gene associated with Duchenne's muscular dystrophy
   a. codes for a protein called dystrophin.
   b. is recessive.
   c. encodes components of plasma membranes of skeletal muscle cells.
   d. is X-linked.
   e. All of the above

*21. Familial hypercholesterolemia is caused by an allele that is classified as
   a. a deletion.
   b. a hot spot.
   c. dominant.
   d. recessive.
   e. X-linked.

*22. The reason most serious genetic diseases are rare is that
   a. each person is unlikely to carry any genetic disease.
   b. each person is unlikely to carry the same mutant allele as the person with whom he or she mates.
   c. genetic diseases are usually dominant.
   d. genetic diseases are usually not serious.
   e. None of the above

23. If a variant for a protein can be found at least 2 percent of the time, the protein is
   a. mutant.
   b. polymorphic.
   c. an RFLP.
   d. a marker.
   e. All of the above

24. Familial hypercholesterolemia is a genetic disease
   a. that causes elevated cholesterol levels in the blood.
   b. in which a liver cell membrane receptor is defective.
   c. that leads to higher likelihood of strokes.

d. All of the above
e. None of the above

25. Transmissible spongiform encephalopathies are the causative agents for
a. Scrapies
b. "Mad cow disease"
c. Kuru
d. All of the above
e. None of the above

26. The infectious prion is a
a. virus.
b. viron.
c. yeast.
d. bacteria.
e. protein.

27. Estimates are that _________ of all people have diseases that are genetically influenced.
a. 0.06%
b. 1%
c. 10%
d. 6%
e. 60%

*28. Normal brain cells have the membrane protein _________, and abnormal TSE-affected brains have _________.
a. PrP^{c}; PrP^{sc}
b. PrP^{sc}; PrP^{c}
c. PrP^{r}; PcP^{c}
d. PcP^{c}; PrP^{r}
e. $PscP^{sc}$; PcP^{c}

29. Genomic imprinting makes
a. male and female genetic contributions balance.
b. expression different for DNA contributed from males than that from females.
c. differential gene expression occur.
d. methylation of DNA.
e. cells differentiate.

*▪30. The advantage of genetic screening for Tay-Sachs disease is
a. it makes treatment possible.
b. it provides a means of reducing the negative effects by controlling diet.
c. it helps prevent the disease by preventing matings between heterozygotes.
d. it allows early treatment with gene therapy.
e. All of the above

31. "Hot spots" are locations where
a. there are cytosine residues.
b. there are 5-methylcytosine residues.
c. there are adenine residues.
d. there is guanidine.
e. all bases are equally prone to mutation.

32. Fragile-X associated mental retardation occurs when the number of CGG repeats
a. declines to 6–54 copies.
b. increases to 52–200 copies.
c. increases to 200–1,300 copies.
d. All of the above
e. b and c

▪33. Which of the following is a step that scientists take to deal with genetic disease?
a. Characterize symptoms of the disease.
b. Develop epidemiological data.
c. Define the pattern of inheritance.
d. Move from the Mendelian to the molecular level of analysis.
e. All of the above

▪34. Differences in RFLP banding patterns indicate that
a. the two different DNA's being tested possess different base pairs.
b. mRNA is not transcribed.
c. the genes map to different chromosomes.
d. a and c
e. None of the above

35. A RFLP
a. is a restriction fragment length polymorphism.
b. is inherited in a Mendelian fashion.
c. can be used as a genetic marker.
d. can be useful to help define a discrete gene.
e. All of the above

36. When a patient with defective adenosine deaminase (ADA) was treated, which of the following steps was performed for gene therapy?
a. Leukocytes were obtained from the patient.
b. Leukocytes were transferred to culture dishes.
c. Leukocytes were transfected with normal ADA genes.
d. The transfected cells were returned to the patient.
e. All of the above

37. Human gene therapy requires
a. gene isolation.
b. introduction of DNA into target cells.
c. inclusion of a promoter sequence.
d. a and b
e. a, b, and c

▪38. Which of the following is an ethical issue that arises from screening fetuses for genetic diseases?
a. The controversial option of abortion may be recommended when screening detects a defective allele in the fetus.
b. The question of privacy of screening results, and the debate over who has rightful access to those results
c. Should we humans be "playing God" with our genetic makeup in the first place?
d. How will we determine which diseases "merit" gene therapy, and which genetic conditions warrant screening?
e. All of the above

39. An example of a genetic disease that causes a defect not in an enzyme, but in a structural protein, is
 a. sickle-cell disease.
 b. hemophilia.
 c. Duchenne's muscular dystrophy.
 d. PKU.
 e. None of the above

40. Cancer is caused by
 a. mutagens.
 b. carcinogens.
 c. viruses.
 d. substances found in natural foods.
 e. All of the above

▪41. How do all forms of cancer differ from other diseases?
 a. Control of cell division is lost in cancerous cells: they divide rapidly and continuously.
 b. Cancers arise in several different tissues.
 c. Cancer cells metastasize.
 d. a, b, and c
 e. a and c

42. Noncancerous tumors that remain in place are termed
 a. malignant.
 b. benign.
 c. harmless.
 d. innocuous.
 e. None of the above

43. The spread of tumor cells through the body is termed
 a. cell focusing.
 b. metastasis.
 c. benign neglect.
 d. orientation.
 e. None of the above

44. Metastasis proceeds via the cancer cells'
 a. extension into surrounding tissues.
 b. differentiation into normal cells.
 c. entrance into the bloodstream or the lymphatic system.
 d. a, b, and c
 e. a and c

45. Carcinomas are defined as cancers that arise in
 a. muscle.
 b. bone.
 c. the liver.
 d. the lung, breast, colon, or liver.
 e. None of the above

46. Sarcomas are cancers of
 a. the brain.
 b. the skin.
 c. bone, blood vessels, or muscle tissue.
 d. the lung, breast, colon, or liver.
 e. None of the above

47. Lymphomas and leukemias affect the cells
 a. that give rise to the white and red blood cells.
 b. of the brain.
 c. of blood vessels.
 d. of connective tissue.
 e. of muscle and connective tissue.

48. The first cancer-causing virus to be identified was
 a. Epstein-Barr virus.
 b. T cell leukemia virus.
 c. Rous sarcoma virus in chickens.
 d. porcine sarc virus.
 e. benzene 1 virus.

49. Which of the following factors is thought to be involved in the development of cancer cells?
 a. Viruses
 b. Chemicals
 c. Ultraviolet radiation and X rays
 d. Excessive exposure to sunlight
 e. All of the above

50. The development of cancer cells results from changes in genes required for
 a. collagen synthesis.
 b. normal growth.
 c. the production of keratin sulfate.
 d. the production of myosin and actin.
 e. None of the above

51. Eighty-five percent of all human cancers fall into which of the following categories?
 a. Carcinomas
 b. Lymphomas
 c. Leukemias
 d. Sarcomas
 e. b and c

52. Agents that can cause mutations in the DNA of host cells include
 a. chemical carcinogens.
 b. radiation.
 c. tumor-induced viruses.
 d. enzyme kinetics.
 e. a, b, and c

53. Cell division is regulated in part by a group of proteins that circulate in the blood and trigger the normal division of cells. These proteins are called
 a. follicle-stimulating hormones.
 b. erythropoietins.
 c. anabolic steroids.
 d. growth factors.
 e. None of the above

54. Cancers originate in the activities of normal but potentially cancer-producing alleles called
 a. homeoboxes.
 b. hormones.
 c. retroalleles.
 d. a and c
 e. proto-oncogenes.

55. An oncogene arises by a mutation from a
 a. normal gene that controls some aspect of the normal development of the cell.
 b. proto-oncogene.
 c. class of cancer genes.
 d. a and b
 e. None of the above

56. A mutation that causes an amino acid change in one of the hemoglobin subunits
   a. always causes a disease when an individual is homozygous for the mutant allele.
   b. always causes a disease even when an individual is heterozygous.
   c. sometimes causes a disease when the individual is homozygous.
   d. is always a dominant mutation.
   e. None of the above

57. The basic concept of the "two-hit" hypothesis is that to get cancer
   a. both copies of a tumor suppressor gene must mutate.
   b. one tumor suppressor and one proto-oncogene must mutate.
   c. both copies of a proto-oncogene must mutate.
   d. All of the above
   e. None of the above

58. The tumor suppressor gene, *p53*, codes for a product that
   a. kills cancerous cells.
   b. causes apoptosis.
   c. stops cell division during G1.
   d. triggers an immune response.
   e. All of the above

59. Today, sufferers of hemophilia A are treated with
   a. screened human-derived blood products.
   b. porcine blood products.
   c. a yeast clotting factor.
   d. a product generated from recombinant DNA technology.
   e. None of the above

60. The entire human genome
   a. will be completed around 2003.
   b. was completed in 1986.
   c. will require another two decades to be completed.
   d. is 10% complete.
   e. will be completed within one decade.

*61. The method used to sequence the entire 180 million base-pair sequence of the fruit fly involved the __________ method.
   a. hierarchical sequencing
   b. Maxwell-Gilbert
   c. random-plasmid-select
   d. shotgun sequencing
   e. All of the above

Study Guide Questions

1. Fragile-X syndrome is
   a. due to a single base change in the DNA.
   b. caused by changing a valine to a glutamic acid.
   c. caused by triplet expansion.
   d. caused by a chromosomal translocation.

2. A nonfunctional membrane protein is responsible for
   a. hemophilia.
   b. sickle-cell anemia.
   c. cystic fibrosis.
   d. PKU.

3. Triplet repeat expansion
   a. occurs during DNA replication due to slippage of DNA polymerase.
   b. is caused by errors in DNA synthesis with reverse transcriptase.
   c. differences can be identified using RFLP mapping.
   d. a and c

4. In sickle-cell anemia the
   a. sixth amino acid is changed from a valine to a glutamic acid.
   b. sixth amino acid is changed to a stop codon.
   c. hemoglobin concentration builds up in the red blood cells.
   d. structure of β-globin is altered and the hemoglobin protein forms aggregates in the red blood cells.

5. Most cancers are
   a. due to the inheritance of mutant genes.
   b. caused by viruses.
   c. caused by mutations in genes whose products help blood clot.
   d. caused by mutations that are spontaneous or induced.

6. Metabolic disorders
   a. refer to genes that have alterations in their centromeres.
   b. are caused by prions.
   c. are due to abnormal membrane proteins that transport chloride ions.
   d. are the result of mutations in enzymes that are required to synthesize particular compounds in the cell (such as an amino acid).

7. Prions
   a. cause scrapie in sheep.
   b. are caused by abnormally folded proteins interfering with the normal brain cell function.
   c. cause "mad cow disease."
   d. All of the above

8. Human genetic disease can be
   a. due to an autosomal recessive trait.
   b. passed from mothers to their sons (if X-linked) 50 percent of the time and also can be caused by translocations.
   c. due to a mutant allele that is dominant to wild type.
   d. All of the above

9. The mutations that result in human disease include
   a. triplet expansions that occur during DNA synthesis.
   b. point mutations that do not change the amino acid sequence of the gene.
   c. prion-like diseases such as kuru.
   d. All of the above

10. Genetic screening has been used to
    a. identify embryos carrying a mutant allele.
    b. identify parents who are carriers for PKU and Tay-Sachs.
    c. identify fathers who are carriers for X-linked diseases.
    d. a and b

11. Genes that when mutated cause cancer are
    a. only those genes involved in the regulation of the cell cycle.
    b. transcription factors only.
    c. only those genes inactivated by viruses.
    d. None of the above

12. In order for gene therapy to be most effective, genes should be inserted in
    a. white blood cells.
    b. red blood cells.
    c. stem cells.
    d. All of the above

13. The human genome allows scientists to
    a. understand regulatory sequences that are important for gene expression.
    b. locate genes that cause disease.
    c. understand evolutionary relationships by comparing human genes to genes in other organisms.
    d. All of the above

End of Chapter Questions

1. Phenylketonuria is an example of a genetic disease in which
    a. a single enzyme is not functional.
    b. inheritance is sex-linked.
    c. two parents without the disease cannot have a child with the disease.
    d. mental retardation always occurs, regardless of treatment.
    e. a transport protein does not work properly.

2. Mutations of the gene for β-globin
    a. are usually lethal.
    b. occur only at amino acid position 6.
    c. number in the hundreds.
    d. always result in sickling of red blood cells.
    e. can always be detected by gel electrophoresis.

3. Multifactorial diseases
    a. are less common than single-gene diseases.
    b. involve the interaction of many genes with the environment.
    c. affect less than 1 percent of humans.
    d. involve the interactions of several mRNAs.
    e. are exempified by sickle-cell anemia.

4. The most common form of inherited mental retardation is
    a. Tay-Sachs disease.
    b. cystic fibrosis.
    c. Down syndrome.
    d. fragile-X syndrome.
    e. phenylketonuria.

5. The reason that most serious genetic diseases are rare is that
    a. each person is unlikely to be a carrier of harmful recessive alleles.
    b. genetic diseases are usually dominant.
    c. genetic diseases are usually sex linked and therefore uncommon in females.
    d. a married couple probably do not carry the same recessive alleles.
    e. mutation rates in humans are very low.

6. Mutational "hot spots" in human DNA
    a. always occur in genes that are transcribed.
    b. are common at cytosines that have been modified to 5-methylcytosine.
    c. involve long stretches on nucleotides.
    d. occur where there are triplet repeats, such as CTG.
    e. are very rare in genes that code for proteins.

7. Newborn genetic screening for PKU
    a. is very expensive.
    b. detects phenylketones in urine.
    c. has not led to the prevention of mental retardation resulting from this disorder.
    d. must be done during the first day of an infant's life.
    e. uses bacterial growth to detect excess phenylalanine in blood.

8. Genetic diagnosis by DNA testing
    a. detects only mutant and not normal alleles.
    b. can be done only on eggs or sperm.
    c. involves hybridization to rRNA.
    d. utilizes restriction enzymes and a polymorphic site.
    e. cannot be done with PCR.

9. Most human cancers
    a. are caused by viruses.
    b. are in blood cells or their precursors.
    c. involve mutations of somatic cells.
    d. spread through solid tissues rather than by the blood or lymphatic system.
    e. are inherited.

10. Oncogenes and tumor suppressor genes
    a. are not mutated in most cancers.
    b. can be compared to the "gas pedal" and "brakes" of cell cycling.
    c. act only at the plasma membrane.
    d. have no roles in normal cells.
    e. code only for transcription factors.

Student Web Site Self-Quiz Questions

1. Which of the following statements about PKU is *false*?
    a. The disease results from a block in the conversion of phenylalanine into phenylpyruvic acid.
    b. PKU is inherited as an antosomal recessive.
    c. This disease illustrates the one gene–one enzyme principle.
    d. A child with the disease must have both parents as carriers for the disease.

e. The exact cause of mental retardation in PKU is not known.

2. Which of the following statements about familial hypercholesterolemia is *false*?
   a. Familial hypercholesterolemia is inherited as an autosomal dominant.
   b. Heterozygotes for this gene have only about 50 percent of the normal number of functional receptors for low-density lipoproteins.
   c. Sufferers from this disease have high cholesterol levels in their liver cells.
   d. Familial hypercholesterolemia is due to a point mutation.
   e. Familial hypercholesterolemia is not a multifactorial disease.

3. Which of the following statements about triplet repeats is *false*?
   a. Triplet repeats are found in many genes not associated with disease.
   b. Expanding triplet repeats are associated with several diseases besides fragile-X syndrome.
   c. The number of repeats in the premutated allele is intermediate.
   d. The mechanism for expansion of triplet repeats usually involves a transposon.
   e. The sons of daughters whose father had a premutated allele for fragile-X syndrome are likely to be mentally retarded.

4. Which of the following statements about PKU and the Guthrie test is *false*?
   a. Newborn babies who are homozygous recessive for the PKU allele have normal levels of phenylalanine hydroxylase activity and only develop symptoms about two weeks later.
   b. Screening for PKU in newborns is required by law in the United States and Canada.
   c. The Guthrie test is most effective if not administered until several days after birth.
   d. A positive result in the test shows more bacterial growth around the blood spot than a negative result.
   e. The Guthrie test indirectly assays the phenylalanine levels in the infant's blood.

5. Which of the following is *not* a valid difference between cancer cells and normal cells?
   a. Cancer cells do not respond to hormones and growth factors that stimulate mitosis in normal cells
   b. Cancer cells do not express the gene for telomerase.
   c. Cancer cells can undergo metastasis.
   d. Cancer cells are often variable in size and shape.
   e. Many cancer cells dedifferentiate.

6. Mutation of a proto-oncogene into an oncogene can stimulate cell division and cause cancer. Which of the following proto-oncogene types causes inactivation of a series of proteins?
   a. Those coding for growth factors
   b. Those coding for growth factor receptors
   c. Those controlling apoptosis
   d. Those coding for signal transduction proteins
   e. Those coding for transcription factors

7. Which of the following steps involved in the physical mapping of DNA would be **second** in sequence?
   a. Restriction endonucleases with 8-12 base pair recognition sequences are used to digest DNA.
   b. A specific chromosome band of interest is isolated.
   c. Fragments are cloned into YAC vectors.
   d. Restriction endonucleases with 3-5 base pair recognition sequences are used to digest DNA.
   e. STS's are identified on fragments.

8. Complete genomic sequencing of several organisms, including humans, is underway. Which of the following is *not* one of the anticipated products of the Human Genome Project and sequencing efforts?
   a. Cancer cell mRNA profiles
   b. Identification of sequence polymorphisms related to diseases
   c. Obtaining the DNA fingerprints of the entire population
   d. Mapping of disease-causing genes
   e. Facilitating the isolation of human genes

★ *Indicates a **difficult** question*

■ *Indicates a **conceptual** question*

19 Natural Defenses against Disease

Fill in the Blank

1. The final large lymph duct connects to a major **vein** near the **heart**.
2. The **humoral** immune response is a specific defense system against pathogens that is carried out by antibodies in the blood.
3. The blood fluid that carries leukocytes but not red blood cells is **lymph**.
4. **T cell receptors** are specific molecules on the surface of T cells that react to antigenic determinants.
5. Immunological **tolerance** is the term that describes the acceptance of foreign tissue.
6. The fusion of lymphocytes and myeloma cells in culture produces **hybridomas**, which make monoclonal antibodies.
7. **Cytokines** are soluble signal proteins released by T cells.
8. When the immune recognition of self fails, a(n) **autoimmune** disease results.
9. HIV-I, the retrovirus that causes AIDS, uses the enzyme **reverse transcriptase** to make a DNA copy of the viral genome.
10. The **constant** regions of the heavy chains of the antibody molecule determine whether the antibody remains part of the plasma membrane of the cell or is secreted into the bloodstream.
11. Of the phagocytes, neutrophils and macrophages, **macrophages** live longer.
12. The concept that antigenic determinants stimulate clones of B cells that were already making specific antibodies against those antigens is called the **clonal selection** theory.
13. The ability of the human body to remember a specific antigen explains why **immunization** has eliminated diseases such as smallpox, diphtheria, and polio.
14. Highly specified protein molecules called **immunoglobulins** carry out the humoral immune response against invaders in the fluids.
15. T cells are educated, and those that have receptors for "self" are eliminated in the **thymus**.
16. Two broad groups of cells—the **B cells**, which sometime differentiate to produce antibodies, and the **T cells**, of which some types are involved with elimination of virus-infected cells—are the important cells of the immune system.
17. Tears, nasal drips, and saliva possess an enzyme called **lysozyme** that degrades the cell walls of many bacteria.
18. Types of defenses that provide general protection against a wide variety of pathogens are classified as **innate** defenses.

Multiple Choice

1. Which of the following is *not* one of the first lines of defense against invading pathogens?
   a. Skin
   b. Mucus secretion
   c. Lysozyme in tears
   d. T cell receptors
   e. Low pH of the stomach
2. Blood is a fluid tissue with a noncellular fluid called
   a. lymph.
   b. leukocyte.
   c. plasma.
   d. lymphocyte.
   e. immunoglobulin.
3. Nonspecific responses include all of the following components except
   a. macrophages.
   b. antibodies.
   c. eosinophils.
   d. neutrophils.
   e. interferons.

4. Which of the following is *not* an adaptation for preventing a pathogen from penetrating the body surface?
   a. Presence of a normal flora
   b. Sneeze reflex
   c. Mucus-covered body surfaces
   d. Low pH
   e. Immunological tolerance

5. Allergies involve
   a. IgE.
   b. mast cells.
   c. basophils.
   d. histamine release.
   e. All of the above

*6. One of the following is *not* a characteristic of interferons. Select the exception.
   a. Bind to receptors on cell surfaces
   b. All are glycoproteins.
   c. Prevent viral replication
   d. Found only in mammals
   e. Confer a generalized resistance to viral diseases

7. _________ is the generalized bodily response to infections and is accompanied by redness, swelling, heat (increased temperature), and pain.
   a. Shock
   b. Inflammation
   c. DNA repair
   d. AIDS
   e. None of the above

8. Which of the following activities is *not* normally involved in controlling pathogens from infecting the mucous membranes of vertebrate animals?
   a. Beating cilia
   b. Lysozyme production
   c. Secretion of HCl
   d. Production of bile salts in the small intestine
   e. Secretion of interferon

9. The bacteria *E. coli* live in our large intestines and do not normally cause disease. These microorganisms are called
   a. pathogens.
   b. antibodies.
   c. normal flora.
   d. phagocytes.
   e. the complement system.

*10. Each milliliter of blood normally contains about _________ red blood cells and _________ white blood cells.
   a. 5 billion, 7 million
   b. 5 million, 7 billion
   c. 5 thousand, 7 thousand
   d. 5 million, 7 thousand
   e. 5 million, 7 million

11. Lysozyme is an enzyme that
   a. digests proteins.
   b. causes viral infected cells to lyse.
   c. attacks cell walls.
   d. produced lysozim.
   e. is produced by bacteria.

12. Ingested pathogens must first get past
   a. the bile salts.
   b. B cells.
   c. stomach acids.
   d. T cells.
   e. All of the above

13. Immunoglobulins are composed of _________, which are composed of _________ chain(s).
   a. octomers, two heavy and two light
   b. tetramers, one large and one small
   c. dimers, one heavy and one light
   d. tetramers, two heavy and two light
   e. dimers, one large and one small

14. The binding of antigen to IgM on a B cell's membrane triggers
   a. histamine release.
   b. antibody production.
   c receptor-mediated endocytosis.
   d. T_H activation
   e. class switching.

15. Phagocytes kill pathogenic bacteria by
   a. endocytosis.
   b. production of antibodies.
   c. complement fixation.
   d. T cell stimulation.
   e. inflammation.

16. Which of the following is a nonspecific defense mechanism used to protect animals against pathogenic microorganisms?
   a. Sealing off the damaged tissue
   b. Production of phytoalexins
   c. Production of antibodies
   d. Humoral immune response
   e. Inflammation

17. An individual with influenza is unlikely to develop a second viral infection. This is because the infected cells are producing a glycoprotein called
   a. phytoalexin.
   b. interferon.
   c. immunoglobulin.
   d. antigen.
   e. IgG.

18. The humoral immune system primarily acts against
   a. intracellular viruses.
   b. circulating bacteria.
   c. tissue transplants.
   d. a and b
   e. a and c

19. B cells will react
   a. nonspecifically with any foreign matter they encounter.
   b. only with tissue transplants.
   c. with all of the antigenic determinants on a specific antigen.

d. with only one specific antigenic determinant on an antigen.
e. only with specific antibody molecules.

20. When an individual is first exposed to the smallpox virus, there is a delay of several days before significant numbers of specific antibody molecules and T cells are produced. However, a second exposure to the virus causes a large and rapid production of antibodies and T cells. This is an example of
a. antigenic determinants.
b. phytoalexins.
c. phagocytosis.
d. interferon production.
e. immunological memory.

21. When an animal encounters an antigen for the second time, it is capable of producing a massive and rapid immune response to the antigen. The cells responsible for this rapid response are called __________ cells.
a. memory
b. effector
c. humoral
d. immunization
e. antigenic

22. According to the clonal selection theory,
a. antibodies are not produced until the animal encounters a specific antigen.
b. antigens determine the three-dimensional structure of antibodies.
c. all B cells have identical genotypes.
d. an antigen stimulates the proliferation of a specific group of B cells.
e. B cells give rise to specific T cells.

23. One explanation for the absence of antiself lymphocytes in the bloodstream is
a. the presence of memory cells.
b. immunological memory.
c. clonal deletion.
d. interferon production.
e. destruction by natural killer cells.

24. Immunological tolerance occurs
a. after an exposure to an antigen early in development.
b. when an antigen has no antigenic determinants.
c. during the clonal growth of B cells.
d. as a result of class switching.
e. when interleukins activate T cells.

25. Failure to distinguish "self" from "nonself" can result in
a. clonal deletion.
b. the production of suppressor T cells.
c. the development of AIDS.
d. an autoimmune disease.
e. a deficiency in complement proteins.

26. When a T cell is activated by an antigen, it
a. secretes antibodies.
b. proliferates.
c. dies.
d. becomes a hybridoma.
e. becomes a plasma cell.

27. Which of the following is *not* a characteristic of plasma cells?
a. They arise from B cells.
b. They are effector cells.
c. They secrete antibodies.
d. They survive in the animal for many years.
e. They have large amounts of endoplasmic reticulum.

28. Hay fever is an allergic response to pollen. What type of antibody molecule is being produced?
a. IgG
b. IgM
c. IgD
d. IgA
e. IgE

29. Monoclonal antibodies
a. recognize a single antigenic determinant.
b. are produced in the spleen.
c. are produced by animals injected with a single antigen.
d. are memory cells.
e. are a complex mixture of different antibody classes.

30. The region of the antibody that binds to the antigen is the
a. constant region of the heavy chain.
b. constant region of the light chain.
c. variable region.
d. a and b
e. b and c

▪31. Since the joining and random deletion mechanisms only account for a part of antibody diversity, what else is involved in producing vast numbers of unique immunoglobulins?
a. Mutation
b. Inversion
c. Cell fusion
d. Cell surface proteins
e. None of the above

32. The genes for immunoglobulin heavy chains are __________ the genes for the light chains.
a. located on the same chromosomes as
b. spliced together with
c. expressed when the cells are exposed to antigens, as are
d. located on different chromosomes from
e. rearranged during development, as are

▪33. A plasma cell is producing IgM molecules that recognize an antigenic determinant on an influenza virus. After several days, the cell begins to produce IgG molecules that recognize the same antigenic determinant. This is called
a. activation.
b. RNA splicing.
c. gene mutation.
d. class switching.
e. an autoimmune disease.

34. T cell receptors recognize and bind
   a. T_C cells.
   b. B cells.
   c. processed antigens.
   d. T_H cells.
   e. IgM antibodies.

35. The class I MHC proteins are __________ in mice.
   a. secreted by B cells
   b. only found on T cells
   c. on the surface of all nucleated cell types
   d. only produced early in development
   e. the same in all individuals of the same strain

36. The retrovirus HIV specifically destroys T_H cells and thus disrupts
   a. the humoral immune response.
   b. the cellular immune response.
   c. both the humoral and cellular immune responses.
   d. the inflammatory response.
   e. the complement cascade.

37. Which of the following cells is *not* normally involved in the functioning of the immune system?
   a. Phagocytes
   b. Red blood cells
   c. Lymphocytes
   d. B cells
   e. T cells

38. Which of the following is *not* associated with a non-specific defense mechanism?
   a. Inflammation
   b. Mucous membranes
   c. Cytokines
   d. Phagocytes
   e. Interferons

39. Select the following feature that is *not* shared by B cells and T cells.
   a. A kind of lymphocyte
   b. Found in the lymph
   c. Antibody secreting
   d. Give rise to both effector and memory cells
   e. Originate in bone marrow

40. Select the following feature that is characteristic of both the humoral and cellular immune responses.
   a. Leads to elimination of infected cells
   b. Cell–cell communication via interleukin
   c. Involved in immunological memory
   d. Activation of the complement system
   e. Activation of phagocytosis by macrophages

41. Which of the following cellular immune components is the functional equivalent of an immunoglobulin?
   a. T cell receptor
   b. An antigenic determinant
   c. An antibody
   d. Processed antigen bound to class II MHC proteins
   e. A complement cascade

42. A foreign cell can
   a. act as just one antigen.
   b. have only one antigenic determinant.
   c. elicit and bind only one antibody.
   d. lead to the production of more than one type of immunoglobulin.
   e. lead to the selection of only a single B cell.

*43. A fundamental postulate of the clonal selection theory is that
   a. the antigen specifies the structure of the antibody directed against it.
   b. an antigen can only lead to the selection of a single line of B cells.
   c. a B cell makes only one specific immunoglobulin.
   d. the production of diverse antibodies was not genetically based.
   e. a mechanism must exist to prevent the production of antiself lymphocytes.

*44. Which of the following statements about the clonal selection theory is *not* true?
   a. An enormous variety of B cells exist in an animal.
   b. All B cells have the same genotype, although the expression of that genotype always differs.
   c. Differences in B cell specificities exist prior to an encounter with antigen.
   d. Exposure to an antigen leads to the production of both memory and effector cells.
   e. An antigen activates a preexisting lymphocyte with receptors for it.

45. Which of the following is *not* specifically involved with attempts to explain the body's recognition of self?
   a. Monoclonal antibodies
   b. The clonal deletion theory
   c. Presence of self-identifying cell surface proteins
   d. Immunological tolerance
   e. Nonidentical twin cattle with blood of mixed types

46. Which of the following features is *not* characteristic of the structure of an immunoglobulin such as IgG?
   a. A protein molecule with a quaternary structure
   b. Subunits held together by hydrogen bonding
   c. A tetramer with two heavy chains and two light chains
   d. Variable and constant regions in each subunit
   e. Two antigen binding sites

*47. Which of the following features is *not* characteristic of an immunoglobulin?
   a. The variable regions determine the type of antigen that will bind.
   b. The constant regions determine the class of the immunoglobulin.
   c. Disulfide bonds occur within and between the polypeptide chains.
   d. The antigen binding sites are formed by the variable portions of the light chains only.
   e. The two halves of an immunoglobulin are identical.

48. Which of the following features is characteristic of the immunoglobulin class IgG?
   a. Found in blood immediately after first exposure to the antigen
   b. Involved in inflammation and allergic reactions
   c. Form multi-immunoglobulin complexes
   d. Produced in greatest amount after the second exposure to the antigen
   e. Found in saliva, tears, milk, and gastric secretions

49. Which of the following features is *not* characteristic of the complement system?
   a. Consists of 20 different proteins
   b. Involved in a cascade of reactions
   c. Forms a lytic complex
   d. Able to lyse foreign cells in the absence of antibody
   e. Interacts with phagocytes to promote endocytosis

50. Which of the following features is *not* characteristic of T cell receptors?
   a. Consist of two polypeptide chains
   b. Are glycoproteins
   c. Are able to bind to free antigen
   d. Are able to bind MHC proteins
   e. Have constant and variable regions

51. Which of the following is *not* a normal activity of helper T cells?
   a. Release of lytic signals when bound to processed antigen on the surface of a virus-infected cell
   b. Binding to class II MHC and processed antigen on the surface of macrophages
   c. Binding to class II MHC and processed antigen on the surface of B cells
   d. Release of helping signals
   e. Proliferation and differentiation into memory and effector cells

52. With respect to AIDS, which of the following statements is *true*?
   a. Chances of getting the disease are far greater if one of the partners already has a sexually transmitted disease.
   b. AIDS stands for "autoimmune deficiency syndrome."
   c. The HIV virus of AIDS does not kill people directly.
   d. a, b, and c
   e. a and c

53. To ensure survival, pathogenic organisms must
   a. enter a host.
   b. multiply in the host.
   c. prepare to infect the next host.
   d. a, b, and c
   e. a and b

54. Interferons
   a. bind to cell receptors.
   b. are glycoproteins.
   c. inhibit the ability of viruses to replicate.
   d. are the subject of intense research because of their potential medical applications.
   e. All of the above

55. The immune system has two general types of responses against invaders:
   a. the humoral immune response and the cellular immune response.
   b. the humoral immune response and the antihumoral immune response.
   c. complementation and clonal deletion.
   d. the cellular immune response and the antihumoral immune response.
   e. None of the above

56. Antibodies are produced by
   a. fibroblasts.
   b. mesenchyme cells.
   c. adrenal cortex cells.
   d. cells of the anterior pituitary called antibodycytes.
   e. plasma cells.

57. A person can produce __________ of distinct antibodies directed against antigenic determinants that it has never encountered.
   a. dozens
   b. hundreds
   c. thousands
   d. millions
   e. trillions

58. There appear to be two mechanisms of self-tolerance in the immune system. They are
   a. clonal selection and clonal deletion.
   b. clonal deletion and clonal anergy.
   c. clonal proliferation and suppressor T cell action.
   d. suppressor T cell action and clonal selection.
   e. None of the above

59. In order to synthesize antibodies, a plasma cell must
   a. be activated by the binding of specific antigens to the antibodies carried on the B cell surface.
   b. interact with a helper T cell.
   c. develop an extensive endoplasmic reticulum and Golgi complex.
   d. a and b
   e. a, b, and c

60. The immunoglobulin class IgA is found
   a. circulating in the blood.
   b. in mucus secretions.
   c. near the surface of the skin associated with mast cells
   d. shortly after first exposure to an antigen.
   e. All of the above

61. All of the following are necessary for a humoral response *except*
   a. T_C cells.
   b. T_H cells.
   c. B cells.
   d. macrophages.
   e. antigenic determinant.

62. Which of the following is *not* a characteristic of an inflammatory reaction?
    a. Release of histamine
    b. Invasion of the region by phagocytes
    c. Binding of IgG
    d. Dilation of capillaries
    e. Escape of blood plasma

63. To make a hybridoma, a plasma cell is fused to a _________ cell.
    a. carcinoma
    b. B
    c. macrophage
    d. myeloma
    e. liver

64. To produce monoclonal antibodies, hybridoma cells are
    a. transferred to animals.
    b. grown in culture.
    c. placed in deep freeze.
    d. a and b
    e. b and c

65. An important organ that helps to protect against autoimmune disease is the
    a. kidney.
    b. spleen.
    c. thymus.
    d. adrenal gland.
    e. tonsils.

66. The core of HIV contains
    a. a protease.
    b. reverse transcriptase.
    c. integrase
    d. two identical RNA molecules.
    e. All of the above

*67. Viral membrane proteins of HIV are synthesized
    a. in the core particle.
    b. in the infected cell's nucleus.
    c. on the surface of the ER.
    d. in the cytosol.
    e. in the Golgi apparatus.

68. Cells that get infected by HIV generally have
    a. CD4 membrane proteins.
    b. gp120.
    c. gp41.
    d. antigen bound T cell receptors.
    e. All of the above

*69. HAART is
    a. a hypersensitive AIDS patient.
    b. highly active antiretroviral therapy.
    c. HIV and AIDS advanced retroviral treatment.
    d. highly advanced AIDS RNA treatment.
    e. hereditary AIDS acquired and retransmitted.

Study Guide Questions

1. Phagocytes
    a. are T and B cells.
    b. present antigen on MHC II complexes.
    c. digest nonself materials.
    d. b and c

2. B cells
    a. secrete antibodies.
    b. present antigen on MHC II.
    c. ingest antigens.
    d. All of the above

3. Nonspecific responses in the immune system include
    a. macrophages.
    b. natural killer cells.
    c. complement.
    d. All of the above.

4. When the receptor of a T_H cell binds to a pathogen being presented on a macrophage it
    a. activates itself.
    b. secretes cytokines.
    c. activates B cells.
    d. All of the above

5. Part of the normal immune response includes
    a. the production of B memory cells.
    b. the production of memory macrophages.
    c. antibody secretion by eosinophils.
    d. the production of B cells that attack the individual's own cells.

6. Antibody molecules are
    a. produced by B cells and have a variable and a constant region.
    b. secreted by B cells, once a signal (a cytokine) is received from a T cell.
    c. produced by T cells and have a variable and a constant region.
    d. a and b

7. Cytotoxic T cells
    a. release cytokines that activate B cells.
    b. attack pathogens by binding to cell surface antigens on those pathogens.
    c. destroy pathogens by lysing them.
    d. destroy host cells that are infected with virus.

8. The humoral response
    a. is due to the secretion of antibodies.
    b. occurs when T cells bind antigen-presenting cells.
    c. is due to T cells secreting their receptors.
    d. occurs when natural killer cells engulf cancer cells.

9. Autoimmunity is
    a. active when organ transplantation is successful.
    b. caused by the failure of anergy.
    c. b and d
    d. the response of the immune cells attacking the body's own tissues.

10. Patients with HIV are susceptible to a variety of infections because
    a. the virus produces cell surface receptors that bind to pathogens, making it easier for those pathogens to be infective.
    b. synthesizing a DNA copy of the viral genome makes a person feel sick.
    c. HIV attacks and destroys the T helper cells which are central to mounting an effective immune response, making those individuals more susceptible to other infections.
    d. HIV destroys B cells so that antibodies cannot be made in response to invading pathogens.

11. DNA rearrangements in the B cell
    a. are responsible for generating single B cells that can express many different antibodies.
    b. lead to mutations in T cells resulting in the elimination of essential T cell genes.
    c. occur only in B memory cells.
    d. are responsible for generating many different antibodies, with each B cell expressing only one set of identical antibodies.

12. Major histocompatibility proteins function in the immune system by
    a. presenting antigen to T cells.
    b. presenting self antigens to T helper cells so those T cells will continue to tolerate self throughout the lifetime of the individual.
    c. generating antibodies to different pathogens.
    d. a and b

End of Chapter Questions

1. Phagocytes kill harmful bacteria by
    a. endocytosis.
    b. producing antibodies.
    c. complement fixation.
    d. stimulation of T cells.
    e. inflammation.

2. Which statement about immunoglobulins is *true*?
    a. They help antibodies do their job.
    b. They recognize and bind antigenic determinants.
    c. They encode cell surface proteins.
    d. They are the chief participants in nonspecific defense mechanisms.
    e. They are a specialized class of white blood cells.

3. Which statement about an antigenic determinant is *not* true?
    a. It is a specific chemical grouping.
    b. It may be part of many different molecules.
    c. It is the part of an antigen to which an antibody binds.
    d. It may be part of a cell.
    e. A single protein has only one on its surface.

4. T cell receptors recognize and bind
    a. T_C cells.
    b. B cells.
    c. processed antigens.
    d. T_H cells.
    e. IgM antibodies.

5. According to the clonal selection theory,
    a. an antibody changes its shape according to the antigen it meets.
    b. an individual animal contains only one type of B cell.
    c. the animal contains many types of B cells, each producing one kind of antibody.
    d. each B cell produces many types of antibodies.
    e. many clones of antiself lymphocytes appear in the bloodstream.

6. Immunological tolerance
    a. depends on exposure to antigen.
    b. develops late in life and is usually life-threatening.
    c. disappears at birth.
    d. results from the activities of the complement system.
    e. results from DNA splicing.

7. Which of the following is *not* a characteristic of immunoglobulins such as IgG?
    a. A protein with a quaternary structure
    b. Subunits held together by hydrogen bonding
    c. Two heavy chains and two light chains
    d. Variable and constant regions in each chain
    e. Two antigen binding sites

8. Which of the following play(s) no role in the antibody response?
    a. Helper T cells
    b. Interleukins
    c. Macrophages
    d. Reverse transcriptase
    e. Products of class II MHC gene loci

9. The major histocompatibility complex
    a. codes for proteins that present antigens on the cell surface.
    b. plays no role in T cell immunity.
    c. plays no role in antibody responses.
    d. plays no role in skin graft rejection.
    e. is encoded by a single locus with multiple alleles.

10. Which of the following plays no role in HIV reproduction?
    a. Integrase
    b. Reverse transcriptase
    c. gp120
    d. Interleukin-1
    e. Protease

Student Web Site Self-Quiz Questions

1. Which one of the following matchings of leukocyte types and functions is incorrect?
    a. B lymphocytes—secrete antibodies
    b. T lymphocytes—secrete T cell receptors
    c. Basophils—produce histamine
    d. Macrophages—engulf and digest bacteria
    e. Natural killer cells—lyse virus-infected or cancerous body cells

2. The clonal selection theory accounts for all of the following *except*
   a. the versatility of the humoral immune response.
   b. the versatility of the cellular immune response.
   c. the basis of antibody diversity.
   d. immunological memory.
   e. recognition of self.

3. If lymphoid cells from strain **A** mice are injected into treated newborn strain **B** mice, while control mice are not injected, which of the following results would be expected?
   a. Control mice would accept skin grafts from strain A mice.
   b. Control mice would reject skin grafts from strain B mice.
   c. The experimental mice would accept skin grafts from only strain B mice.
   d. The experimental mice would accept skin grafts from only strain A mice.
   e. The experimental mice would accept skin grafts from both strain A and strain B mice.

4. Which of the following statements about the production of monoclonal antibodies is *false*?
   a. The cell labeled "a" is a hybridoma.
   b. A mouse injected with a pure antigen will produce a single line of B lymphocytes.
   c. Some myeloma cells grow well in cell culture, but do not produce antibodies.
   d. Each hybridoma produces a single type of antibody.
   e. Monoclonal antibodies can be used for passive immunization.

5. Which of the following statements contrasting T cell receptors and immunoglobulins is *false*?
   a. Unlike immunoglobulins, T cell receptors can only recognize processed antigen.
   b. The genetic system underlying T cell receptor diversity differs from the system specifying immunoglobulins.
   c. T cell receptors are more specific than immunoglobulins; they have dual specificity for antigen and MHC protein.
   d. Both T cell receptors and immunoglobulins consist of multiple polypeptide chains held together with disulfide bonds.
   e. Like immunoglobulins, T cell receptors have constant regions and variable regions.

6. Which one of the following is *not* a way in which cytotoxic T cells (T_C) and helper T cells (T_H) differ?
   a. T_H cells can bind to class II MHC molecules on other cells; T_C cells cannot.
   b. T_C cells can bind to class I MHC molecules on other cells; T_H cells cannot.
   c. T_H cells have class II MHC plasma membrane glycoproteins; T_C cells do not.
   d. The net result of activation of T_H cells is proliferation of B cells; the net result of activation of T_C cells is lysis of infected cells.
   e. T_C cells may participate in the "rejection" of transplants; T_H cells do not.

TEST FILE

★ *Indicates a* ***difficult*** *question*

■ *Indicates a* ***conceptual*** *question*

20 The History of Life on Earth

Fill in the Blank

1. A massive extinction occurred at the end of the **Permian** period. It may have been caused over a long period of time (ten million years) by the coalescing of the continents into the supercontinent, Pangaea.
2. The period within the Mesozoic era in which frogs, salamanders, and lizards first appeared, and in which one lineage of dinosaurs gave rise to birds, is termed the **Jurassic** period.
3. Earth's crust consists of solid **plates** approximately 40 kilometers thick that float on a liquid mantle.
4. Many patterns in the fossil record suggest that there are frequently long periods during which rates of morphological evolutionary change are extremely slow. These periods are called **stasis**.

★5. There have been **six** mass extinctions in the history of life, and **three** periods of rapid diversification of organisms.

6. There is a general tendency for organisms to **increase** in size through the fossil record.
7. **Microevolution** is a change in a species that happens during its lifetime.
8. Changes that involve the appearance of new species or lineages are known as **macroevolution**.
9. Some ancient insects have been perfectly preserved in **amber** formed by tree sap.

Multiple Choice

1. Which of the following is *not* true concerning the history of the evolution of life?
   a. Organisms have increased in size.
   b. Organisms have increased in complexity.
   c. The number of species has increased.
   d. Life evolved early in the history of Earth.
   e. There are more varieties of animal body plans now than there were 600 million years ago.

2. The movements of continents were important causes of extinctions during the history of life on Earth. Which of the following results of continental movements was important in these extinctions?
   a. Changes in sea levels
   b. Separation of biotas
   c. Mixing of biotas
   d. Changes in climates
   e. All of the above

3. Over much of its history, the climate of Earth was
   a. about the same as it is today.
   b. considerably cooler than it is today.
   c. considerably warmer than it is today.
   d. unknown; we have no information about climates in the past.
   e. much more variable annually.

4. Conditions for the preservation of organisms (fossilization) are best in environments
   a. lacking oxygen.
   b. with high levels of oxygen.
   c. that are warm and moist.
   d. that are cold and dry.
   e. with constant temperature.

★5. The fossil record of horses depicts gradual changes over time in North America; however, new forms of horses appear suddenly in Asia. Which of the following hypotheses about horse evolution is accepted by most scientists?
   a. Horses evolved quickly in North America and periodically migrated into Asia.
   b. Horses evolved slowly in North America and quickly in Asia.
   c. Horses evolved quickly in both North America and Asia.
   d. Horses evolved slowly in North America and periodically migrated into Asia.
   e. Horses evolved quickly in Asia and migrated to North America, where they evolved slowly in the absence of predators.

■6. If a new form of organism appears in the fossil record in a particular area, we could conclude that
   a. it evolved there rapidly.
   b. it migrated there from another location.
   c. there is a gap in the fossil record and it evolved there slowly.
   d. All of the above
   e. None of the above

*7. Radioisotopes are often used to determine the time of death of fossilized remains. Tritium has a half-life of 12.3 years. That means that 24.6 years after an organism dies it will have what fraction of the original radioactive tritium?
a. All
b. One-half
c. None
d. One-quarter
e. One-eighth

8. The evolution of hard skeletons occurred in many animal phyla during the Cambrian period. What major ecological factor is thought to have been responsible for this evolution?
a. Competition with other species
b. Predation pressure
c. Protection from desiccation
d. Hard skeletons permitted movement.
e. Hard skeletons provided protection from parasitism.

9. The coal beds we now mine for energy are the remains of trees of the
a. Precambrian period (600 million years ago).
b. Cambrian period (600–500 million years ago).
c. Ordovician period (500–440 million years ago).
d. Silurian period (440–400 million years ago).
e. Carboniferous period (about 300 million years ago).

10–13. The Mesozoic era (about 245–66 million years ago [mya]) had three periods: the Triassic, the Jurassic, and the Cretaceous. Match the following events with the correct time frame from the list below. Each choice may be used once, more than once, or not at all.
a. Mesozoic era (about 245–66 mya)
b. Triassic period (about 245–195 mya)
c. Jurassic period (about 195–138 mya)
d. Cretaceous period (about 138–66 mya)

10. The first mammals evolved from reptiles. **(b)**

11. Individual continents acquired distinctive terrestrial floras and faunas. **(a)**

12. Frogs, salamanders, and lizards first appeared. **(c)**

13. The first birds evolved from reptiles. **(c)**

14. The Cenozoic era (66 mya–present) is often called the age of
a. bacteria, because they cause so many diseases.
b. fishes, because they are an important food resource.
c. mammals, because of the extensive radiation of this group.
d. plants, because they convert radiant energy from the sun into chemical energy.
e. viruses, because viruses such as AIDS have recently evolved.

*15. The last glaciers retreated from temperate latitudes about _________ years ago. As a result, many temperate ecological communities have occupied their current locations for no more than _________ years.
a. 10,000,000, a million
b. 1,000,000, a hundred thousand
c. 100,000, tens of thousands of
d. 10,000, a few thousand
e. 1,000, a few hundred

16. Evolutionists have long puzzled over why no new phyla have evolved since the Cambrian, about 500 million years ago. The most commonly accepted theory is
a. that the Cambrian radiation took place in a world that contained few species, and the ecological setting was favorable for the evolution of many new body plans and different ways of life.
b. that competition resulted in many different species, with different ecological requirements, that were able to coexist.
c. that heavy predation pressures selected for differences among prey; therefore predators could not be as efficient at capturing many diverse species as they could one numerous species.
d. that there are only a certain number of body plans that are functional and all of those types evolved during the Cambrian.
e. that there has been relatively little genetic variation present since the Cambrian, so natural selection has been restricted to creating new species, genera, and families instead of phyla.

17. It appears from the fossil record that there has been a general tendency of increase in size during the course of evolution. Which of the following groups runs counter to that trend; that is, which has not increased in size over time?
a. Echinoderms
b. Foraminiferans
c. Insects
d. Vertebrates
e. Plants

18. One theory about the cause of the mass extinction at the end of the Cretaceous period is that a large meteorite or asteroid collided with Earth. The proposed ecological effect of this collision is
a. that the impact changed the distance between Earth and the sun enough so that Earth became significantly cooler.
b. that the collision threw enough dust into the atmosphere to darken the skies and lower temperatures worldwide.
c. that the impact was like a giant earthquake and caused instantaneous death to most inhabitants of Earth.
d. that the impact, which occurred in what is now the Caribbean Sea, caused giant tidal waves in the oceans. When the waves impacted the land, many organisms died.

19. Which of the following is *not* considered to be a plausible hypothesis concerning the causes of mass extinctions?
a. Extraterrestrial causes, such as meteorite or asteroid collisions

b. Glaciations
c. Massive volcanic activity
d. Competition among organisms
e. All of the above

*20. How many species of fossil organisms have been identified in the fossil record?
a. 300
b. 3,000
c. 30,000
d. 300,000
e. 3,000,000

*21. There are thought to be __________ mass extinctions in the history of life on Earth and __________ times when many new evolutionary lineages originated.
a. 6, 0
b. 3, 3
c. 6, 6
d. 3, 6
e. 6, 3

▪22. The major difference between the earliest explosion in evolutionary lineages, the Cambrian, about 500 million years ago, and the Triassic, which was more recent, is that
a. during the Cambrian, new phyla originated.
b. during the Triassic, new phyla originated.
c. the Cambrian explosions were probably caused by an asteroid colliding with Earth.
d. there were more species, and thus more competition, when the Cambrian explosion occurred.

23. During the Mesozoic era (245–66 million years ago), Pangaea separated into individual continents. As a result
a. many species became extinct.
b. many new phyla evolved.
c. individual continents acquired distinctive terrestrial floras and faunas.
d. flight evolved to allow organisms to migrate among continents.
e. All of the above

24. Which of the following continents is oldest?
a. Laurasia
b. Pangaea
c. Antarctica
d. Australia
e. None of the above

25. During the majority of Earth's history, the climate has been
a. about the same as it is now.
b. uniform and considerably colder than it is now.
c. considerably colder than it is now, but with numerous, short warm periods.
d. considerably warmer than it is now.
e. considerably warmer than it is now, but with numerous, short cold periods.

26. What can be said about the nature of a typical environment that will become the site for a rich fossil bed and the origin of organisms that will become fossils there?
a. The environment is anaerobic, and most organisms originate elsewhere.
b. The environment is anaerobic, and most organisms originate locally.
c. The environment is aerobic, and most organisms originate elsewhere.
d. The environment is aerobic, and most organisms originate locally.
e. The environment is anaerobic, and about equal numbers of organisms originate elsewhere or locally.

*▪27. The half-life of a particular radioactive isotope of element X is 100 years. If you know that 400 years ago a fossil contained 10 milligrams of this isotope, how much would you expect to find in that fossil today?
a. About 5 milligrams
b. About 2.5 milligrams
c. About 0.6 milligrams
d. About 0.3 milligrams
e. About 0.15 milligrams

28. Modern biota evolved during the __________ era.
a. Proterozoic
b. Cenozoic
c. Mesozoic
d. Paleozoic
e. a and b

29. Select the time division during which the world biota became provincialized due to continental drift.
a. Precambrian
b. Paleozoic
c. Mesozoic
d. Cenozoic
e. b and c

30. Select the time division during which the major diversification of angiosperms, birds, and mammals occurred.
a. Precambrian
b. Paleozoic
c. Mesozoic
d. Cenozoic
e. c and d

31. Select the period during which the first insects and amphibians appeared on land.
a. Cambrian
b. Devonian
c. Permian
d. Triassic
e. Tertiary

32. What types of organisms would you *not* expect to see during a walk in a Permian forest?
a. Dragonflies
b. Amphibians
c. Club mosses and horsetails
d. Flowering plants
e. Gymnosperms

33. During this period, the "age of reptiles" began.
   a. Permian
   b. Triassic
   c. Jurassic
   d. Cretaceous
   e. Tertiary

34. During the Cretaceous period of the Mesozoic era,
   a. the continents Laurasia and Gondwana first formed.
   b. horses first appeared.
   c. birds and mammals diverged from a common reptilian stock.
   d. a second major extinction eliminated most land vertebrates.
   e. bony fishes dominated the seas.

35. During the Tertiary period of the Cenozoic era,
   a. the dinosaurs become extinct.
   b. the "age of fishes" began.
   c. many major "ice ages" occurred.
   d. the genus *Homo* first appeared.
   e. the main radiation of birds took place.

36. Which of the following statements about the Cambrian, Paleozoic, or modern faunas is *true*?
   a. The diversity of body plans is greatest in the modern fauna.
   b. Most of the phyla that were present in the Paleozoic fauna are now extinct.
   c. The number of families has steadily increased in the modern fauna.
   d. Most of the species that were part of the Cambrian fauna are alive today.
   e. None of the above

37. Which of the following statements about the diversity of the modern fauna is *true*?
   a. As the diversity of insect pests declined, the diversity of flowering plants increased.
   b. The evolution of more complex ecological communities was important in creating the changes in diversity in the modern fauna.
   c. Provinciality due to continental drift was an unimportant factor after the Cambrian era.
   d. As diversity increased, organisms became more specialized and less interdependent.
   e. The numbers of species during the Cenozoic has been rather stable.

38. Over a wide variety of lineages
   a. size tends to increase with time.
   b. size tends to decrease with time.
   c. size tends to remain constant with time.
   d. size tends to increase with time, but only in the insects.
   e. a decrease in size is usually followed by an adaptive radiation of many smaller forms.

39. Of the six mass extinctions on Earth, the extinction at the end of the _________ eliminated the most phyla.
   a. Cambrian
   b. Devonian
   c. Permian
   d. Triassic
   e. Cretaceous

40. Spine reduction in sticklebacks is usually related to
   a. genetic drift.
   b. a reduction in predation in fresh water.
   c. a dominant allele.
   d. movement into a more stable environment.
   e. differences in the type of predators encountered in freshwater and marine habitats.

▪41. Which of the following statements about evolution would be supported by most biologists?
   a. Evolution is understandable, but not predictable.
   b. Evolution results in greater complexity.
   c. Given specific initial conditions, evolution is predetermined.
   d. As evolution progresses, extinction rates decrease.
   e. Mutation is more important than environment in directing the course of evolution.

42. Evolution of species on Earth
   a. has stopped.
   b. only occurred in the distant past.
   c. occurred after the Cambrian explosion.
   d. has occurred throughout Earth's history and is still underway.
   e. None of the above

*43. Earth's crust consists of solid tectonic plates approximately _________ km thick that "float" on a liquid mantle.
   a. 4
   b. 40
   c. 400
   d. 4,000
   e. None of the above

44. The 1979 Alvarez hypothesis of a collision of a large meteorite with Earth attempts to explain
   a. the theory of Gondwana.
   b. Alfred Wegner's concept of continental drift.
   c. the mass extinction that occurred at the end of the Cretaceous period.
   d. contemporary atmospheric conditions on Earth.
   e. None of the above

45. More than _________ of the species that have ever lived on Earth since the beginning of time are now extinct.
   a. 10%
   b. 50%
   c. 75%
   d. 90%
   e. 99%

46. Three times during the history of life, many new evolutionary lineages originated. Select the correct choice below that represents these periods.
   a. Cambrian explosion, Paleozoic explosion, Triassic explosion
   b. Jurassic revolution, Cambrian explosion, Hadean period

c. Cambrian explosion, Archean extinction, Jurassic explosion
d. Paleozoic explosion, Mesozoic explosion, Cenozoic explosion
e. Archean extinction, Jurassic explosion, Triassic explosion

47. The most probable cause of the mass extinction at the end of the Permian period was
a. a volcano.
b. a giant meteorite hitting Earth.
c. major predation among species.
d. aggregation of the continents into Pangaea.
e. a large dust cloud that cooled Earth's temperature.

48. Luis Alvarez and colleagues proposed that a meteorite was the cause of the mass extinction 66 million years ago. Their hypothesis was based on
a. finding high concentrations of iridium in the rock layer separating the Cretaceous and Tertiary periods.
b. fossils found in the rock layer separating the Cretaceous and Tertiary periods.
c. the finding of a crater site.
d. finding high concentrations of argon 39 in the rock layer separating the Cretaceous and Tertiary periods.
e. finding evidence of unusual volcanic activity during that period.

49. Most kingdoms evolved prior to the Cambrian period (about 600 million years ago). Which kingdom evolved after the Cambrian; that is, which was the last kingdom to evolve?
a. Prokaryotes
b. Protists
c. Fungi
d. Plants
e. Animals

50. The appearance of multicellular organisms coincided with increased __________ levels in Earth's atmosphere.
a. sulfur
b. hydrogen
c. nitrogen
d. carbon
e. oxygen

*51. The Ediacaran fauna consisted of mostly
a. prokaryotic organisms.
b. soft-bodied invertebrates similar to present-day forms.
c. soft-bodied invertebrates unlike present-day forms.
d. trilobites and other hard-shelled forms.
e. primitive horses.

52. Which of the following is believed to be a cause of the mass extinction at the end of the Devonian period?
a. Two continents colliding
b. An asteroid colliding with Earth
c. A large volcano
d. The breaking apart of Pangaea and subsequent rise in the ocean
e. a and b

53. Which of the following did *not* occur during the Cenozoic era?
a. Australia and Antarctica were still attached.
b. Flowering plants dominated world forests.
c. The climate got hotter and wetter.
d. There was an extensive radiation of mammals.
e. *Homo sapiens* arrived in North and South America.

Study Guide Questions

1. The half-life of an isotope is best defined as the
a. time a fixed fraction of isotope material will take to change from one form to another.
b. age over which the isotope is useful for dating rocks.
c. ratio of one isotope species to another in a sample of organic matter.
d. None of the above

2. Mountain ranges are ultimately the result of
a. plates in Earth's crust that move against one another on top of a liquid mantle.
b. climate changes and the movement of glacial ice sheets.
c. leftover debris from ancient collisions with an asteroid or meteor
d. the breakup of Laurasia and Gondwana.

3. Which of the following would likely be the best setting for fossilization to occur?
a. The surf zone along a sand beach
b. A shallow, cool swamp with good deposition rates of mud sediments
c. The bottom of a hot, dry cave with no running water
d. A fast-running mountain stream

4. Despite being incomplete as a whole, the fossil record is rather detailed for
a. soft-bodied insects.
b. cnidarians and sponges.
c. most terrestrial animals.
d. hard-shelled mollusks.

5. When biologists say that Earth's biota became provincialized, they mean that
a. mass extinctions took place.
b. continental drift did not occur.
c. distinctive assemblages of plants and animals arose on different continents.
d. reductions in species diversity took place.

6. All but one of the following occurred during the Ordovician period. Which statement is *not* correct?
a. Marine filter feeders flourished.
b. The number of classes and orders increased.
c. Modern mammals appeared.
d. Many groups became extinct at the end of the period.

7. Most of human evolution has occurred during the
   a. Paleozoic era.
   b. Devonian period.
   c. Quaternary period.
   d. Carboniferous period.

8. For terrestrial animals and plants, the most recent mass extinction event that occurred prior to the evolution of humans took place approximately
   a. 10 mya.
   b. 65 mya.
   c. 220 mya.
   d. 400 mya.

9. One of the main factors that distinguishes the Cambrian explosion from all others is that
   a. evolutionarily, it was the most recent explosion.
   b. many novel lineages (particularly phyla) appeared at this time, but not during other explosions.
   c. it was the time when the dinosaurs became extinct.
   d. there was a dramatic drop in species diversity, especially among marine organisms.

10. During which of the following geologic times did the most new kinds of body plans appear?
    a. Paleozoic
    b. Triassic
    c. Jurassic
    d. Cambrian

11. The event that precipitated the sudden disappearance of the dinosaurs some 65 mya may have been the result of
    a. Earth's collision with a large meteorite.
    b. slow climate changes due to planetary cooling.
    c. competition from better adapted organisms.
    d. the rise of birds and mammals.

12. Which of the following statements about evolution is *not* true?
    a. Predators have evolved more efficient methods of capturing prey over time.
    b. Species have become extinct throughout the history of life.
    c. The fossil record of whales provides a good example of gradual evolutionary change within a lineage of organisms.
    d. There is no evidence in the fossil record for periods of rapid evolutionary change.

13. Which of the following groups of organisms has undergone an overall decrease in body size during evolution?
    a. Mammals
    b. Mollusks
    c. Plants
    d. None of the above

14. Which of the following statements about patterns or processes in the evolution of life is *not* true?
    a. ^{14}C can be used to date the age of dinosaur bones.
    b. The supercontinent Pangaea formed during the Permian period.
    c. All of the major groups of animals that have species living today appeared in the Cambrian period.
    d. Rates of evolutionary change can be rapid during times of dramatic change in physical environments.

15. Which of the following pairs of organisms were *not* present on Earth in their living forms at the same time?
    a. Tree ferns and bony fish
    b. Amphibians and birds
    c. Gymnosperms and insects
    d. Humans and dinosaurs

End of Chapter Questions

1. The number of species of fossil organisms that has been described is about
   a. 50,000.
   b. 100,000.
   c. 200,000.
   d. 300,000.
   e. 500,000.

2. In undisturbed strata of sedimentary rocks, the relative ages of the strata are:
   a. The oldest rocks lie at the top
   b. The oldest rocks lie at the bottom
   c. The oldest rocks are in the middle
   d. The oldest rocks are distributed among the strata of younger rocks
   e. None of the above

3. Radioactive carbon can be used to date the ages of fossil organisms because
   a. all organisms contain many carbon compounds.
   b. radioactive carbon has a regular rate of decay to nonradioactive carbon.
   c. the ratio of radioactive to nonradioactive carbon in living organisms is always the same as that in the atmosphere.
   d. the production of new radioactive carbon in the atmosphere just balances the natural radioactive decay of ^{14}C.
   e. All of the above

4. An important unidirectional change in Earth during its history was
   a. a steady increase in volcanic activity.
   b. a gradual strengthening of the processes that cause continents to move.
   c. a steady increase in the oxygen content of the atmosphere.
   d. that it has gradually become warmer.
   e. a steady increase in Earth's radioactive furnace.

5. The total of all species of organisms living in a region is known as its
   a. biota.
   b. flora.
   c. fauna.
   d. flora and fauna.
   e. diversity.

6. The coal beds we now mine for energy are the remains of
   a. trees that grew in swamps during the Carboniferous period.
   b. trees that grew in swamps during the Devonian period.
   c. trees that grew in swamps during the Permian period.
   d. herbaceous plants that grew in swamps during the Carboniferous period.
   e. None of the above

7. The cause of mass extinctions at the end of the Ordovician was
   a. the collision of Earth with a large meteorite.
   b. massive volcanism.
   c. massive glaciation in Gondwana.
   d. the uniting of all continents to form Pangaea.
   e. changes in Earth's orbit.

8. The cause of the mass extinction at the end of the Mesozoic era probably was
   a. continental drift.
   b. the collision of Earth with a large meteorite.
   c. changes in Earth's orbit.
   d. massive glaciation.
   e. changes in the salt concentration of the oceans.

9. The times during the history of life when many new evolutionary lineages appeared were the
   a. Precambrian, Cambrian, and Triassic.
   b. Precambrian, Cambrian, and Tertiary.
   c. Cambrian, Paleozoic, and Triassic.
   d. Cambrian, Triassic, and Devonian.
   e. Paleozoic, Triassic, and Tertiary.

10. Many scientists believe that the collision of Earth with a large meteorite was a major contributor to the mass extinction at the boundary between the Cretaceous and Tertiary periods because
    a. there is an iridium-rich layer at the boundary of rocks from these two periods.
    b. a crater that may be the site of the collision has been found off the Yucatán Peninsula.
    c. the mass extinction at the end of the Cretaceous may have been very sudden.
    d. new methods have allowed scientists to date the iridium layer precisely.
    e. All of the above

11. We know that organisms can evolve rapidly because
    a. the fossil record reveals periods of rapid evolutionary change.
    b. theoretical models of evolutionary change show that rapid change can be produced by natural selection.
    c. rapid evolutionary changes have been produced under artificial selection.
    d. rapid evolutionary changes have been measured in natural populations of organisms during the past century.
    e. All of the above

12. At which of the following times was there no mass extinction?
    a. The end of the Cretaceous period
    b. The end of the Devonian period
    c. The end of the Permian period
    d. The end of the Triassic period
    e. The end of the Silurian period

Student Web Site Self-Quiz Questions

1. Radioisotopes are used to date rocks and the fossils that they contain. If a radioisotope has a half-life of 2.0 million years and you determine that a rock has only 1/128 of the amount of the radioisotope it had when the rock was formed, then the rock was formed _________ mya.
   a. 7
   b. 14
   c. 64
   d. 128
   e. 256

2. The table below shows the hierarchy of names for Earth's geological history.

Era	Period
Cenozoic	Quaternary
	Tertiary
Mesozoic	Cretaceous
	Jurassic
	Triassic
Paleozoic	Permian
	Carboniferous
	Devonian
	Silurian
	Ordovician
	Cambrian

   Of the following geological time units, which is the most inclusive? And, on what are the boundaries between these divisions based?
   a. Eras, changes in rock types
   b. Eras, changes in fossil types
   c. Periods, changes in rock types
   d. Periods, changes in fossil types
   e. Periods, changes in DNA sequences

3. Which of the following changes in Earth's history is a unidirectional change?
   a. Iridium in rocks
   b. Sea level
   c. Volcanism
   d. Atmospheric oxygen levels
   e. Mean temperature

4. Which of the following statements about the fossil record is *false*?
   a. The number of known fossil species is less than 2% of the number of named living species.
   b. Species last, on average, less than 10 million years.

c. Marine animals with shells are over-represented in the fossil record.
d. The fossil record contains many series of fossils that demonstrate gradual change in a lineage over time.
e. An incomplete fossil record can provide misinformation about the rate of evolution of a lineage.

5. The breakup of Pangaea during the Mesozoic era resulted in
   a. a drop in sea level.
   b. a provincialized biota.
   c. development of an icehouse climate.
   d. a proliferation of flying forms in many lineages.
   e. a long period of stasis.

6. Which of the following statements about evolutionary changes in populations of the three-spined stickleback (*Gasterosteus aculeatus*) is *false*?
   a. Stasis is most common during periods when environmental change is low.
   b. The three-spined stickleback inhabits both marine and freshwater habitats.
   c. Spine reduction in the stickleback is caused by variation in predation in different habitats.
   d. Spine reduction in the stickleback evolved many times in different populations.
   e. Molecular data show that all sticklebacks with reduced spines are closely related.

7. Which of the following has *not* been an evident trend in the evolution of Earth's major faunas?
   a. Overall increase in body size
   b. Evolutionary change is greatest for lineages found in harsh, stable environments
   c. Overall increase in complexity
   d. Overall increase in competition
   e. Predators become more efficient

8. Which of the following statements about the patterns of evolutionary change is *false*?
   a. Most evolutionary innovations are novel and not adaptations of preexisting structures
   b. The feather is an example of an evolutionary innovation.
   c. The figure below shows how differential growth rates can result in evolutionary change.
   d. In the history of Earth, there have only been three major new faunas.
   e. All major groups of present day organisms appeared in the Cambrian fauna.

9. Which of the following statements about the thickening of snail shells over time is *false*?
   a. The efficiency of predation contributed to the thickening of the snail shells.
   b. The development of defense systems is a response to the environment.
   c. The development of thicker shells was due to the extinction of predators.
   d. The development of thicker shells only provided partial protection from predation.
   e. The increase in fossilized shells showing repairs following attacks is evidence that predation increased during the Cretaceous period.

TEST FILE

★ *Indicates a **difficult** question*
■ *Indicates a **conceptual** question*

21 The Mechanisms of Evolution

Fill in the Blank

1. Evolution is the accumulation of **heritable** changes within populations over time.

2. The frequencies of the variants at each locus in a population are called **allele frequencies**.

3. The genetic expression of traits, for example, "homozygous recessive," is called the organism's **genotype**.

4. The physical expression of a trait, for example, height or eye color, describes an organism's **phenotype**.

5. Short-term changes in a population's gene pool are often called **microevolution**.

6. The **gene pool** is the sum total of genetic information present in a population at any given moment. It includes every allele at every locus in every organism in a population.

7. The relative reproductive contribution of an individual to subsequent generations is termed the individual's **fitness**.

8. When individuals with intermediate values of traits have the highest fitness, **stabilizing selection** is said to be operating.

9. A population that is not changing, that has constant genotype and allele frequencies from generation to generation, is said to be in **equilibrium**.

10. **Gene flow** involves movement of individuals to a new location, followed by breeding.

11. The differential contribution of offspring resulting from different heritable traits is called **natural selection**.

12. The idea of natural selection is most closely associated with **Charles Darwin**, who proposed it in his book, *The Origin of Species*, in 1859.

13. A change in a population's allele frequencies that results from chance as a result of small population size is called **genetic drift**.

14. Genetic drift can be brought about by either a severe reduction in population size, known as a population **bottleneck**, or a small number of individuals establishing a new population, which results in a **founder effect**.

15. Populations of a European clover, *Trifolium repens*, produce cyanide, which increases their resistance to herbivores such as mice and slugs; however, it also increases their susceptibility to frost. Populations from northeast Europe produce relatively little cyanide compared with populations in southwest Europe. This geographic change in phenotype is known as a **cline**.

16. A **mutation** is a change in DNA sequence.

Multiple Choice

■1. Assume that a population is in Hardy–Weinberg equilibrium for a trait controlled by one locus and two alleles. If the frequency of the recessive allele is 0.90, what is the frequency of the dominant allele?
a. 0.10
b. 0.19
c. 0.81
d. None of the above
e. Not enough information to tell

■2. There is a gene that causes people to have crumbly earwax. This gene is expressed as a complete dominant: individuals that are homozygous dominants (*CC*) or heterozygous (*Cc*) have crumbly earwax. Homozygous recessives (*cc*) have gooey earwax. On Paradise Island there are 100 people, of which 75 have crumbly earwax. Assuming Hardy–Weinberg conditions, what is the frequency of the *c* allele on Paradise Island?
a. 0.25
b. 0.50
c. 0.87
d. None of the above
e. Not enough information to make a determination

■3. In a population at Hardy–Weinberg equilibrium, the frequency of heterozygotes is 0.64. What is the frequency of the homozygous dominants?
a. 0.08
b. 0.64
c. 0.80
d. None of the above
e. Not enough information to make a determination

▪4. In a population at Hardy–Weinberg equilibrium, the frequency of the *a* allele is 0.60. What is the frequency of individuals heterozygous for the *A* gene?
a. 0.16
b. 0.24
c. 0.48
d. None of the above
e. Not enough information to make a determination

▪5. One in 10,000 babies in the United States is born with phenylketonuria (PKU), a metabolic disorder caused by a recessive allele. What proportion of that human population is likely to be a carrier of the PKU allele?
a. Approximately 0.02
b. Approximately 0.20
c. Approximately 0.98
d. None of the above
e. Not enough information to make a determination

6. There are five conditions that must be met for a population to be in Hardy–Weinberg equilibrium. Which of the following is *not* one of those conditions?
a. Nonrandom mating
b. Large population size
c. No migration
d. No natural selection
e. No mutations

7. Which of the following situations would demonstrate a population bottleneck?
a. The population of El Paso, Texas, moves to Patagonia.
b. Eight male and eight female elephant seals survive the wreck of the Exxon *Valdez*.
c. A million male orangutans
d. Six male orangutans collected from a natural population in Sumatra and moved to the San Diego Zoo
e. None of the above

8. _________ is the effect produced when a bee carries pollen from one population to another.
a. Gene flow
b. Population bottleneck
c. Founder event
d. Genetic equilibrium
e. Assortative mating

9. In a large population, mutation pressure in the absence of selection
a. probably has little effect on the gene pool of a population.
b. produces major evolutionary changes.
c. is what makes us different from the dinosaurs.
d. is usually beneficial.
e. never occurs.

*10. Which of the following is *not* true of mutation?
a. It creates the raw material that makes evolution possible.
b. Most mutations are harmful or neutral.
c. Mutation rates are very low for most loci.
d. Mutations are a likely cause of deviations from Hardy–Weinberg proportions in a population.
e. Mutations probably have little effect on the gene pool of a large population.

▪11. Over the long run, mutations are important to evolution because
a. they are the original source of genetic variation.
b. once an allele is lost through mutation, another mutation to that same allele cannot occur.
c. most mutation rates are one in a thousand.
d. whether good or bad, mutations increase the fitness of an individual.
e. mutations are usually beneficial to the progeny.

12. _________ selection occurs when the extremes of a population contribute relatively few offspring to the next generation than average members of the population.
a. Corrective
b. Directional
c. **Stabilizing**
d. Disruptive
e. Natural

13. _________ selection occurs when one extreme of a population contributes more offspring to the next generation.
a. Corrective
b. Directional
c. Stabilizing
d. Disruptive
e. Natural

14. Which of the following agents of evolution adapts populations to their environment?
a. Nonrandom mating
b. Natural selection
c. Migration
d. Genetic drift
e. Mutation

15. The raw material for evolutionary change is
a. phenotypic variation.
b. genetic variation.
c. geographical variation.
d. environmentally induced variation.
e. behavioral variation.

16. A population evolves when
a. environmentally induced variation is constant between generations.
b. individuals having different genotypes survive or reproduce at different rates.
c. the environment changes on a seasonal basis.
d. members reproduce by cloning.
e. juvenile and adult stages require different environments.

*17. Selection acts on __________ however, evolution depends on __________.
a. **phenotypic variation, genetic variation**
b. genetic variation, phenotypic variation
c. genetic variation, environmentally induced variation
d. environmentally induced variation, phenotypic variation
e. environmentally induced variation, genetic variation

*18. Limpets growing high in the intertidal zone, where they experience heavy wave action, are more conical than individuals of the same species growing in the subtidal zone, where they are protected from waves. Individuals transplanted from the high intertidal zone to the subtidal zone add new growth, which produces a flatter, subtidal shape. This experiment suggests that the difference in
a. **phenotypes is environmentally induced.**
b. genotypes is environmentally induced.
c. phenotypes is genetically based.
d. genotypes is due to natural selection.
e. Not enough information to make a determination

*19. Which of the following is *not* an example of environmentally induced variation?
a. Water fleas grown in cool or calm water develop round heads. If they are moved to warm or turbulent water, they develop pointed "helmets" on their heads.
b. Plants collected along an altitudinal cline vary in size when grown in their native habitat, yet when grown in a greenhouse they are all the same size.
c. Limpets growing high in the intertidal zone, where they experience heavy wave action, are more conical than individuals of the same species growing in the subtidal zone, where they are protected from waves. Individuals transplanted from the high intertidal zone to the subtidal zone add new growth, which produces a flatter, subtidal shape.
d. Leaves on the same tree or shrub often differ in shape and size. In oaks, leaves closer to the top receive more wind and sunlight and are more deeply lobed than leaves lower down.
e. **Beetles that feed on various host plants are different colors, yet when their offspring are grown on different plants than the parents, they are the same color as the parents.**

20–24. Suppose you have a population of flour beetles with 1,000 individuals. Normally the beetles are a red color; however, this population is polymorphic for a mutant autosomal body color, black, designated by *b/b*. Red is dominant to black, so *B/B* and *B/b* genotypes are red. Assume the population is in Hardy-Weinberg equilibrium, with $f(B) = p = .5$ and $f(b) = q = .5$.

▪20. What are the expected frequencies of the red and black phenotypes?
a. .5 red, .5 black
b. **.75 red, .25 black**
c. .25 red, .75 black
d. None of the above
e. Not enough information to make a determination

▪21. What would be the expected frequencies of the homozygous dominant, heterozygous, and homozygous recessive after 100 generations if the population is under the conditions of Hardy–Weinberg?
a. .75, .20, .05
b. **.25, .5, .25**
c. All red because it is the natural color
d. All black because all red alleles would mutate to black
e. .5, .2, .3

▪22. What would be the expected red and black allele frequencies if Hardy–Weinberg conditions were met except that 1,000 black individuals migrated into the population?
a. .75, .25
b. **.25, .75**
c. .5, .5
d. They would not change because the population would still be in Hardy–Weinberg equilibrium.
e. None of the above

▪23. What would be the allele frequencies if a population bottleneck occurred and only four individuals survived, one female red heterozygote and three black males?
a. .875, .125
b. **.125, .875**
c. .25, .75
d. .75, .25
e. .5, .5

▪24. If the population in Question 23 randomly mated, what would be the allele frequencies of their offspring?
a. .875, .125
b. **.125, .875**
c. .25, .75
d. .75, .25
e. .5, .5

25. A gene pool is all the alleles
a. of an individual's genotype.
b. **present in a specific population at a given moment.**
c. that occur in a species throughout its evolutionary existence.
d. that contribute to the next generation of a population.
e. of a biome.

26. An important feature of sexual reproduction is that it
a. is a reproductive process unique to animals.
b. **produces variation through genetic recombination.**
c. uses mitosis in producing gametes.
d. is more efficient than asexual reproduction.
e. is always associated with selective mating.

27. In a West African finch species, birds with large or small bills survive better than birds with intermediate-sized bills. The type of natural selection operating on these bird populations is __________ selection.
a. directional
b. **disruptive**
c. stabilizing
d. nonrandom
e. deme

28–29. Hardy-Weinberg frequencies for the three genotypes *AA*, *Aa*, and *aa*, at all values of *p* and *q*, are shown below.

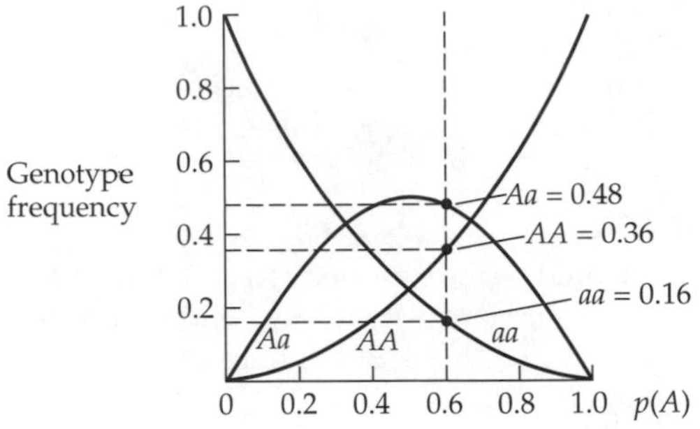

28. As the frequency of the homozygous recessive genotype increases, the frequency of the
   a. heterozygous genotype decreases continuously.
   b. homozygous dominant increases.
   c. heterozygous genotype increases continuously.
   d. dominant allele approaches 1.
   e. recessive allele approaches 1.

29. Which of the following is supported by the graph above?
   a. When *aa* = 0.45, *Aa* = 0.45.
   b. When *aa* = 0.16, *q* = 0.60.
   c. When *p* = 0.33, *q* = 0.77.
   d. When *Aa* = 0.49, *p* = 0.70.
   e. When *AA* = 0.50, *aa* = 0.

30. In a population of organisms, the frequency of the heterozygous genotype is always
   a. greater than the frequency of the homozygous recessive genotype.
   b. equal to one minus the sum of the homozygous genotypes.
   c. equal to two times the frequencies of the homozygous genotypes.
   d. greater than either homozygous genotype.
   e. equal to the frequency of the homozygous dominant genotype minus the frequency of the homozygous recessive genotype.

*31. Cheetahs are a very homogeneous species. The lack of genetic variability among cheetahs may be attributed to
   a. gene flow.
   b. sexual selection.
   c. population bottleneck.
   d. high mutation rate.
   e. high immigration rates.

32. Genetic equilibrium in a population refers to
   a. equal numbers of dominant and recessive alleles.
   b. equal numbers of females and males.
   c. unchanging allele frequencies in successive generations.
   d. lack of mutations that affect the observed phenotypes.
   e. proportional numbers of each genotype.

*33. An example of evolutionary change is
   a. a change from more conical to less conical shape as a limpet moves from turbulent water in the upper intertidal zone to less turbulent water in the subtidal zone.
   b. an increase in edge per unit of surface area in sun-grown leaves versus leaves grown in the shade.
   c. the development of a pointed "helmet" on the head of *Daphnia* when it moves from cooler, calm water to warmer, turbulent water.
   d. the development of "brussels sprouts" when *Brassica oleracea* was selected for lateral buds.
   e. a tadpole metamorphosing into an adult frog.

34. Genetic drift as an evolutionary factor is
   a. a greater force in a population with small numbers than in a population with large numbers.
   b. a greater force in a population with much genetic variation than in a population with little genetic variation.
   c. a force because it is responsible for the selection of mutations.
   d. a force because it involves the movements of alleles between populations of a single species.
   e. a greater force the larger the size of the population.

35. In the Hardy–Weinberg equation, the homozygous dominant individuals in a population are represented by
   a. p^2.
   b. $2pq$.
   c. q^2.
   d. p.
   e. q.

36–40. The two graphs below represent phenotypic distribution of a character in a population of organisms over time. The first curve represents the distribution at an initial sampling time, while the second represents a sampling of a later generation.

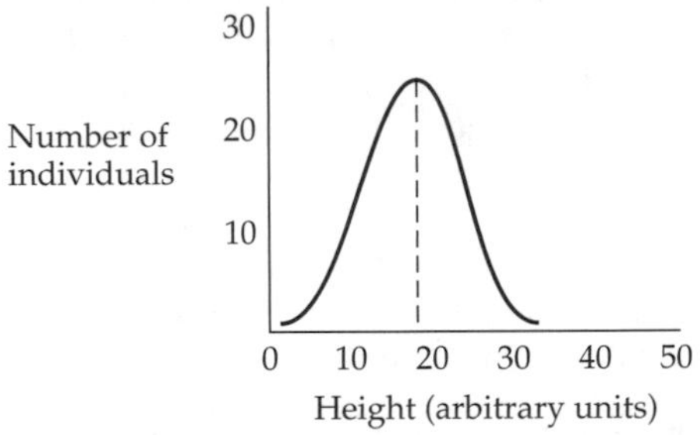

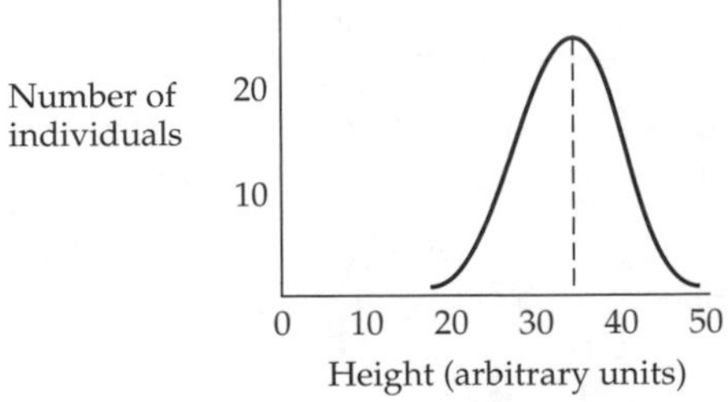

36. The average size of an individual at the initial sampling is __________ units.
   a. 5

b. 10
c. 15
d. 20
e. 25

37. At the initial sampling, there are as many individuals 15 units tall as there are __________ units tall.
a. 5
b. 10
c. 15
d. 20
e. 25

*38. The process illustrated by these graphs is called
a. stabilizing selection.
b. directional selection.
c. disruptive selection.
d. convergent evolution.
e. divergent evolution.

39. The graphs suggest that
a. taller individuals migrated away from the original population.
b. shorter individuals migrated into the original population.
c. taller individuals were more successful at reproducing than short individuals.
d. in the initial population, there were more tall individuals than short ones.
e. the individuals in the population continued to grow between sample times.

40. A mutation occurs in one of your lung cells. Which of the following is *true*?
a. You have evolved to be better adapted to your environment.
b. You will soon die because most mutations are lethal.
c. You will be sterile and no longer be able to have children.
d. The human species will have evolved because this mutation will be passed on to your children.
e. This mutation does not affect human evolution because it will not be passed on to your offspring.

41. Natural selection acts directly on
a. the genotype to produce new mutations.
b. the phenotype to produce new mutations.
c. the genotype to favor existing mutations.
d. the phenotype to favor traits due to existing mutations.
e. the genotype to inhibit new mutations.

42–45. Biologists use the term "fitness" when speaking of evolution. Below are descriptions of four male cats. Answer the following questions based on this information.

Name	Tabby	Chessy	Tony	Tiger
Size	12 lbs	10 lbs	6 lbs	8 lbs
Number of kittens fathered	19	25	20	20
Kittens surviving to adulthood	15	14	14	19
Age at death	13 yrs	16 yrs	12 yrs	9 yrs

Comments: Tabby is the largest and strongest cat Chessy has mated with the most females. Tony was lost on a family vacation but adapted to street life and lived two more years. Tiger died from an infection after a cat fight.

*42. Who contributed the most genes to the next generation?
a. Tabby
b. Chessy
c. Tony
d. Tiger
e. Not enough information to to make a determination

*43. Who contributed the most genes to the gene pool?
a. Tabby
b. Chessy
c. Tony
d. Tiger
e. Not enough information to to make a determination

*44. Which cat contributed the most to the ongoing evolution of the species?
a. Tabby
b. Chessy
c. Tony
d. Tiger
e. Not enough information to to make a determination

*45. The "fittest" cat is
a. Tabby.
b. Chessy.
c. Tony.
d. Tiger.
e. Not enough information to to make a determination

46–48. In a population of 200 individuals, 72 are homozygous recessive for the character of eye color (*cc*). One hundred individuals from this population die due to a fatal disease. Thirty-six of the survivors are homozygous recessive. Answer the following questions.

▪46. In the original population, the frequency of the dominant allele is
a. 0.16.
b. 0.36.
c. 0.40.
d. 0.48.
e. 0.60.

▪47. In the new population, the frequency of the dominant allele is
a. 0.16.
b. 0.36.
c. 0.40.
d. 0.48.
e. 0.60.

▪48. How many heterozygous individuals are expected in the new population?
a. 16
b. 36
c. 40
d. 48
e. 60

49. Which of the following is *not* an effect sexual recombination has on alleles?
    a. Greater evolutionary potential
    b. Change in the frequency of specific alleles
    c. Greater genotypic variety
    d. Creates new combinations of genetic material
    e. Greater phenotypic variety

*50. An oak leaf near the top of a tree has a different shape than a leaf close to the trunk and near the base of the same tree. The leaves are genetically identical; the difference in leaf shape is a result of
    a. polymorphisms.
    b. evolution.
    c. phenotypic plasticity.
    d. natural selection.
    e. frequency-dependent selection.

51. An allele that does *not* affect the fitness of an organism is called a _________ allele.
    a. neutral
    b. directional
    c. distributed
    d. positive
    e. natural

Study Guide Questions

1. Evolution occurs at the level of
    a. the individual genotype.
    b. the individual phenotype.
    c. environmentally based phenotypic variation.
    d. the population.

2. What does natural selection act upon?
    a. The gene pool of the species
    b. The genotype
    c. The phenotype
    d. Multiple gene inheritance systems

3. Suppose a particular species of flowering plant that lives only one year can produce red, white, or pink blossoms, depending on its genotype. Biologists studying a population of this species count 300 red-flowering, 500 white-flowering, and 800 pink-flowering plants in a population. When the population is censused the following year, 600 red-flowering, 900 white-flowering, and 1000 pink-flowering plants are observed. Which color has the highest fitness?
    a. Red
    b. White
    c. Pink

4. Which choice would form a statement about Mendelian populations that is *not* true? A Mendelian population must
    a. consist of members of the same species.
    b. have members that are capable of interbreeding.
    c. show genetic variation.
    d. have a gene pool.

5. In comparing several populations of the same species, the population with the greatest genetic variation would have the
    a. greatest number of genes.
    b. greatest number of alleles per gene.
    c. greatest number of population members.
    d. largest gene pool.

6. The ability to taste the chemical PTC (phenylthiocarbamide) is determined in humans by a dominant allele *T*, with tasters having the genotypes *Tt* or *TT* and nontasters having *tt*. If you discover that 36 percent of the members of a population cannot taste PTC, then according to the Hardy–Weinberg rule, the frequency of the *T* allele should be
    a. 0.4.
    b. 0.6.
    c. 0.64.
    d. 0.8.

7. A gene in humans has two alleles, *M* and *N*, that code for different surface proteins on red blood cells. If you know that the frequency of allele *M* is 0.2, according to the Hardy–Weinberg rule, the frequency of the genotype *MN* in the population should be
    a. 0.16.
    b. 0.32.
    c. 0.64.
    d. 0.8.

8. If the frequency of allele *b* in a gene pool is 0.2, according to the Hardy–Weinberg rule, the expected frequency of the genotype *bbb* in a triploid ($3n$) plant species would be
    a. 0.008.
    b. 0.04.
    c. 0.08.
    d. 0.2.

9. A small, isolated population would most likely be subject over time to
    a. assortative mating.
    b. a founder effect.
    c. genetic drift.
    d. gene flow.

10. Allele frequencies for a gene locus are *least* likely to be changed by
    a. assortative mating.
    b. the founder effect.
    c. mutation.
    d. gene flow.

11. Select all of the following evolutionary agents that produce nonrandom changes in the genetic structure of a population.
    a. Self-fertilization
    b. Population bottlenecks
    c. Mutation
    d. Natural selection

12. The following graph shows the range of variation among population members for a trait determined by multiple genes.

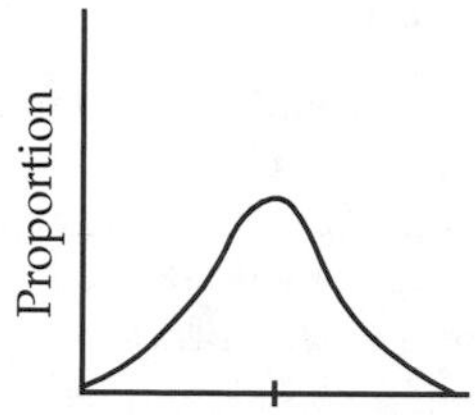

If this population is subject to *stabilizing selection* for several generations, which of the distributions (a–d) is most likely to result?

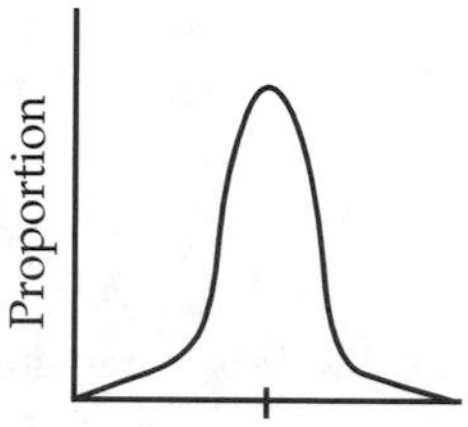

a.

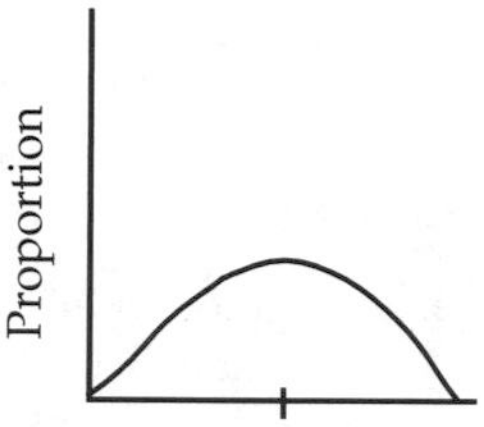

b.

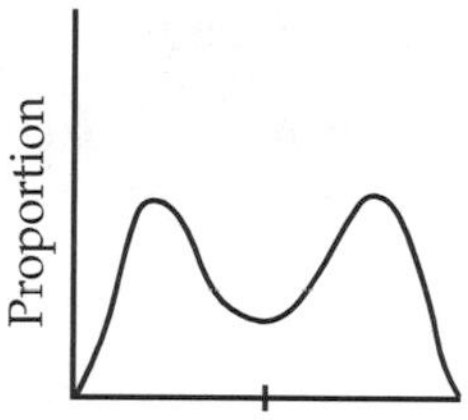

c.

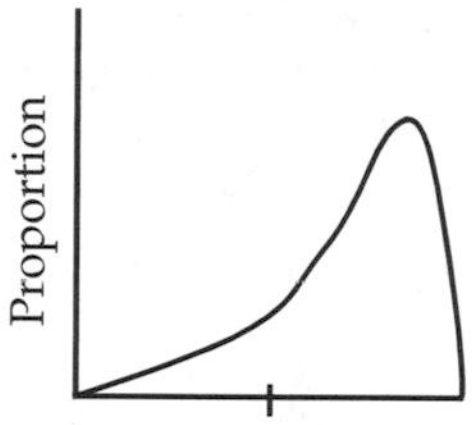

d.

13. In areas of Africa where malaria is prevalent, many human populations exist in which the phenotypes of normal sickle-cell trait (the result of the heterozygous genotype), and severe sickle-cell anemia are constant, despite the fact that sickle-cell anemia frequently causes death at an early age. This is an example of
    a. the founder effect.
    b. a stable polymorphism.
    c. mutation.
    d. assortative mating.

End of Chapter Questions

1. The two major components of Darwin's theory of evolution are
    a. evolution is a fact and mutations are the major agent of evolutionary adaptation.
    b. evolution is a fact and natural selection is the major agent of evolutionary adaptation.
    c. species cannot change into other species but natural selection can modify them.
    d. species cannot change into other species but mutations can modify them.
    e. evolution is a hypothesis and genetic drift may be the major evolutionary agent.

2. To ground his theory, Charles Darwin
    a. developed a comprehensive theory of inheritance
    b. described several evolutionary changes and identified the agents that caused them
    c. used patterns of domestication to show how his theory differed from those patterns
    d. assembled a broad base of supporting information from many fields
    e. developed a mathematical model of evolutionary change

3. The phenotype of an organism is
    a. the type specimen of its species in a museum.
    b. its genetic constitution, which governs its traits.
    c. the chronological expression of its genes.
    d. the physical expression of its genotype.
    e. the form it achieves as an adult.

4. The appropriate unit for defining and measuring genetic variation is the
    a. cell.
    b. individual.
    c. population.
    d. community.
    e. ecosystem.

5. Which statement about allele frequencies is *not* true?
    a. The sum of any set of allele frequencies is always 1.
    b. If there are two alleles at a locus and we know the frequency of one of them, we can obtain the frequency of the other by subtraction.
    c. If an allele is missing from a population, its frequency is 0.
    d. If two populations have the same gene pool for a locus, they will have the same proportion of homozygotes at that locus.
    e. If there is only one allele at a locus, its frequency is 1.

6. In a population at Hardy–Weinberg equilibrium in which the frequency of *A* alleles (p) is 0.3, the expected frequency of *Aa* individuals is
    a. 0.21.
    b. 0.42.
    c. 0.63.
    d. 0.18.
    e. 0.36.

7. Natural selection that preserves existing allele frequencies is called
    a. unidirectional selection.
    b. bidirectional selection.
    c. prevalent selection.
    d. stabilizing selection.
    e. preserving selection.

8. The fitness of a genotype is determined by the
    a. average rates of survival and reproduction of individuals with that genotype.
    b. individuals that have the highest rates of both survival and reproduction.
    c. individuals that have the highest rates of survival.
    d. individuals that have the highest rates of reproduction.
    e. average reproductive rate of individuals with that genotype.

9. Laboratory selection experiments with fruit flies have demonstrated that
   a. bristle number is not genetically controlled.
   b. bristle number is not genetically controlled, but changes in bristle number are caused by the environment in which the fly is raised.
   c. bristle number is genetically controlled, but there is little variation on which natural selection can act.
   d. bristle number is genetically controlled, but selection cannot result in flies having more bristles than any individual in the original population had.
   e. bristle number is genetically controlled, and selection can result in flies having more bristles than any individual in the original population had.

10. Disruptive selection maintains bill size variability in the West African seedcracker because
   a. bills of intermediate shapes are difficult to form.
   b. the two major food sources of the finches differ markedly in size and hardness.
   c. males use their large bills in displays.
   d. migrants introduce different bill sizes into the population each year.
   e. older birds need larger bills than younger birds.

11. A population is said to be polymorphic for a locus if it has at least
   a. three different alleles at that locus.
   b. two different alleles at that locus.
   c. two genotypes for that locus.
   d. three genotypes for that locus.
   e. two genotypes for that locus, the rarest of which is more common than expected by mutation alone.

12. Natural selection may be constrained in its action because
   a. the necessary genetic variation may be lacking.
   b. evolutionary theory does not allow a population to temporarily get worse.
   c. major evolutionary innovations are rare.
   d. insufficient phenotypic plasticity may be present.
   e. All of the above

13. Adaptation is studied by
   a. altering the form of an organism and observing the consequences.
   b. testing predictions by comparing traits in many species.
   c. developing theoretical models.
   d. selecting for traits in the laboratory.
   e. All of the above

Student Web Site Self-Quiz Questions

1. Which of the following statements regarding Darwin and his theory of evolution is *false*?
   a. Darwin noticed that South American species of birds more closely resembled species found in other temperate regions than European species of birds.
   b. Darwin proposed that species adapt over time.
   c. Darwin proposed that natural selection produces changes in species over time.
   d. Darwin was the original and sole developer of the theory of natural selection.
   e. It took Darwin 14 years before he published his theory of natural selection.

2. Which of the following statements regarding genetic variability in a population is *true*?
   a. Genetic variation in populations has never been demonstrated experimentally.
   b. The number of offspring produced by an organism has a great effect on allele frequencies in a population.
   c. Changes in allele frequencies are caused by changes in the number of offspring produced in the population.
   d. The agricultural development of strains of plants and animals carrying desirable traits requires natural genetic variability in the original population.
   e. Reproductive success is determined by evolution.

3. Coloration in the peppered moth is determined by a single gene with two alleles showing complete dominance. Dark moths are homozygous dominant or heterozygous for the gene, light moths are homozygous recessive. In a population of moths, you determine that 64% of the moths are dark. According to the Hardy–Weinberg rule, the expected frequency of the dominant allele is __________.
   a. 0.36
   b. 0.4
   c. 0.6
   d. 0.64
   e. 0.8

4. In a population of moths where coloration is determined by a single gene showing complete dominance, consider that the frequency of the *A* allele is 0.7. According to the Hardy–Weinberg rule, the expected frequency of the heterozygous *Aa* genotype would be __________.
   a. 0.21
   b. 0.3
   c. 0.42
   d. 0.5
   e. undetermined

5. Consider a population of birds where feather color is determined by a single gene with 2 alleles, *R* and *r*. Using the Hardy–Weinberg rule, select a characteristic from the following choices that differs for the two populations (1 and 2) shown below.

	Population 1	Population 2
RR	90	45
Rr	40	130
rr	70	25

   a. *p*
   b. *q*
   c. Gene pool
   d. Genetic structure
   e. Deme size

6. Which of the following is *not* an essential assumption required to maintain Hardy–Weinberg equilibrium in a population?
   a. Random mating
   b. No migration
   c. Natural selection
   d. Large population size
   e. No mutation

7. The California populations of the northern elephant seal are descendants from a very small population of seals that was over-hunted in the 1890s. Genetic variability in this population would be expected to be __________ due to __________.
   a. slight, a population bottleneck
   b. slight, the founder effect
   c. great, disruptive selection
   d. great, a population bottleneck
   e. great, assortive mating

8. The several hundred species of picture-winged fruit flies of the Hawaiian Islands are genetically almost identical, yet they all differ markedly from their ancestral population in Asia. This is an example of
   a. stabilizing selection.
   b. directional selection.
   c. disruptive selection.
   d. founder effect.
   e. gene flow.

9. Studies have shown that species of pigeons select a mating partner to match their own characteristics. This is an example of __________ and it can result in __________ homozygous genotypes than predicted by the Hardy–Weinberg rule.
   a. selfing, more
   b. selfing, fewer
   c. assortive mating, more
   d. assortive mating, fewer
   e. genetic drift, fewer

10. Which of the following statements regarding sexual reproduction and genetic variation in populations is *true*?
    a. Sexual reproduction increases the frequency of certain alleles
    b. Through sexual reproduction, populations maintain the exact same combination of alleles from generation to generation.
    c. Sexual reproduction is not as effective at producing new genotypes as asexual reproduction.
    d. Sexual reproduction in populations increases evolutionary potential.
    e. During sexual reproduction, offspring are almost always identical to the parental organisms.

11. Many species show genetic variation within a population. The angle of the mouth in certain scale-eating fish is an example. Such genetic variation within populations of a species is called __________. When the fitness of a phenotype depends upon the frequency of the other phenotype(s) present in a population, this is __________.
    a. polymorphism, neutral variation
    b. kin selection, neutral selection
    c. polymorphism, frequency dependent selection
    d. kin selection, clinal variation
    e. polymorphism, genetic drift

12. When, of two leaves on an oak tree, one is grown in the sun and one in the shade, the difference in the two leaves' appearance is due to __________.
    a. different genotypes
    b. incomplete dominance
    c. Hardy–Weinberg equilibrium
    d. phenotypic plasticity
    e. differential survival rates

TEST FILE

★ *Indicates a **difficult** question*

■ *Indicates a **conceptual** question*

22 Species and Their Formation

Fill in the Blank

1. The process by which one evolutionary unit splits into two such units that thereafter evolve as distinct lineages is known as **speciation**.

2. **Geographic speciation** occurs when one species evolves into two daughter species after a physical barrier to movement develops within its range.

3. Two populations that are **reproductively isolated** are said to belong to different species.

4. The offspring of genetically dissimilar parents, such as parents from two geographically isolated populations, are called **hybrids**.

5. **Isolating mechanisms** are factors that reduce the possibility of interbreeding between geographically isolated populations that somehow come into contact.

6. **Prezygotic** isolating mechanisms reduce the probability that hybrids will be formed by preventing the formation of a zygote.

7. Once fertilization between isolated populations has occurred, **postzygotic** isolating mechanisms may prevent the hybrids from surviving.

8. A common means of sympatric speciation is **polyploidy**, in which the number of chromosomes is multiplied.

9. The fossil record and current distributions of organisms reveal that some species or groups have given rise to a large number of daughter species, a phenomenon called **evolutionary radiation**.

10. Reproductive isolation between adjacent populations in the absence of a geographic barrier is known as **parapatric speciation**.

11. A factor promoting cohesion of species is **gene flow**, the movement of individuals so that they reproduce in a population other than their original population.

12. Two hybrid species of sunflowers, *Tragopogon mirus* and *T. miscellus*, have four sets of chromosomes; this is called **tetraploidy**.

13. Subdivision of a gene pool when members of the daughter species are not geographically separated is called **sympatric speciation**.

14. Some blackbirds have evolved powerful bill-opening muscles for moving objects to expose food. This activity is called **gaping**.

Multiple Choice

1. Evolutionary biologists believe that all the species present today, along with all the species that lived in the past, are descended from
   a. a single ancestral species.
   b. thousands of different origins.
   c. one ancestral species per kingdom.
   d. one ancestral species per phylum.
   e. two or three ancestral species per kingdom.

2. Ernst Mayr's definition of species states that species are groups of
   a. actually interbreeding natural populations that are reproductively isolated from other such groups.
   b. potentially interbreeding natural populations that are reproductively isolated from other such groups.
   c. actually or potentially interbreeding natural populations that are reproductively isolated from other such groups.
   d. actually or potentially interbreeding natural populations that are geographically isolated from other such groups.
   e. potentially interbreeding natural populations that are geographically isolated from other such groups.

3. Ernst Mayr's definition of species does *not* include
   a. groups that actually or potentially interbreed.
   b. evolutionary units evolving separately from other units.
   c. gene exchange.
   d. members in a single geographic location.
   e. groups that are reproductively separated from other groups.

▪4. The phrase "natural population" is important to the definition of species because
   a. if two populations interbreed only in captivity, they are members of the same species.
   b. if two populations co-occur but do not interbreed in nature, they are separate species.
   c. if two populations do not co-occur, they must be different species because they cannot interbreed.
   d. if two populations can potentially interbreed, they are different species.
   e. if two populations interbreed for the first time, their offspring form a new species.

5–9. Match the following descriptions of speciation with the appropriate terms from the list below. Terms may be used once, more than once, or not at all.
   a. Geographic speciation
   b. Parapatric speciation
   c. Sympatric speciation
   d. Polyploidy

5. Two species of Japanese ladybird beetles occur in the same area. *Epilachnea niponica* feeds on thistles, and *Epilachnea yasutomii* feeds on other plants that grow with thistles. Adults of both species feed and mate on their own host plant. The two species hybridize in the laboratory, but not in nature. Although we cannot be sure which process actually produced speciation in this case, it is likely the two separate species are a result of **(c)**.

6. In Wales there are soils heavily contaminated with lead very close to normal rich pasture land. The pasture grass *Agrostis tenuis* is common to both sites; however, there is a sharp gradient in lead tolerance among individuals separated by less than 20 meters. Nearly complete reproductive isolation exists between plants on contaminated and normal soil. This may eventually lead to **(b)**.

7. If the number of chromosomes of a species of lizard were to double (perhaps through an error in meiosis), resulting in female offspring that could produce young from unfertilized eggs, we would say speciation had occurred through **(d)**.

8. The red tubular-flowered gilias of western North America are a group of species living in the Mojave Desert. Originally considered to be a single species, the group now contains five species: three diploids and two tetraploids (having four sets of chromosomes). These five species are similar in appearance and are sterile in all interspecific mating combinations. **(d)**

9. Different populations of platyfish live in the rivers of eastern Mexico. The subpopulations living in different streams have been diverging since an ancestral platyfish colonized all the streams. Some subpopulations have diverged so much that they cannot interbreed and produce viable offspring with individuals from other subpopulations. Thus, these populations of platyfish exist at various stages in the process of **(a)**.

10. Which of the following factors was probably *not* important in the formation of the 14 species of Darwin's finches on the Galápagos Islands?
   a. Geographical isolation from the mainland
   b. Polyploidy
   c. Different habitats on the different islands
   d. Geographical isolation among the Galápagos Islands
   e. Different food supplies on the different islands

11. Studies of island populations suggest that the smallest area within which a single species has split into two daughter species varies among types of organisms. Which of the following types of organisms requires the largest area to speciate?
   a. Mammals
   b. Birds
   c. Reptiles
   d. Amphibians
   e. Snails

12. Which of the following would *not* result in reproductive isolation between two populations that are reunited following geographic isolation?
   a. The two populations produce successful hybrids.
   b. The two populations have different breeding seasons.
   c. There are physiological differences between the two populations so that they cannot produce viable offspring.
   d. Members of one population do not find members of the other population attractive as mating partners.
   e. The two populations have different courtship behaviors.

13. Which of the following is a postzygotic isolation mechanism?
   a. Temporal isolation
   b. Behavioral isolation
   c. Reduced viability of hybrids
   d. Geographic variation in mating pheromones
   e. Differences in courtship behavior

14. Which of the following is a prezygotic isolation mechanism?
   a. Abnormal meiosis following fertilization
   b. Infertile hybrids
   c. Reduced viability of hybrids
   d. Abnormal mitosis following fertilization
   e. Geographic variation in mating pheromones

15. The development of reproductive isolation among members of a continuous population in the absence of a geographic barrier is called
   a. polyploidy.
   b. hybrid zonation.
   c. geographic speciation.
   d. anagenesis.
   e. parapatric speciation.

16. _________ involves the subdivision of a gene pool even though members of the daughter species overlap in their range during the speciation process. Often this is accomplished by a multiplication of the number of chromosomes.
a. Polyploidy
b. Hybrid zonation
c. Sympatric speciation
d. Geographic speciation
e. Parapatric speciation

17. Which of the following is *not* true about polyploidy?
a. Polyploidy is more common in animals than in plants.
b. Many species of flowering plants arose by polyploidy.
c. Animals that speciate by polyploidy are often parthenogenetic.
d. Polyploidy can create new species quickly if the polyploid individuals can self-fertilize.
e. Polyploid siblings are capable of reproducing with one another.

18. Which mode of speciation is thought to be most important in speciation of large animals?
a. Polyploidy
b. Hybrid zonation
c. Sympatric speciation
d. Geographic speciation
e. Parapatric speciation

*19. Which of the following is *not* true of the genetics of speciation?
a. Sympatric species need not have diverged from each other much genetically.
b. Speciation in *Drosophila* has not involved major reorganization of the genome.
c. Species that show a lot of morphological variation may be relatively similar genetically.
d. Gene flow is important in the creation of new species.
e. Differences between closely related species occur due to the same mechanisms that operate within species.

▪20. Rapid speciation is thought to occur among animals with complex behavior patterns because
a. the timing of reproduction is behaviorally mediated.
b. behavioral differences usually reflect physiological differences.
c. those organisms have difficulty identifying members of their own species and often mate with members of other species.
d. those organisms make sophisticated discriminations among potential mates.
e. those organisms find mates more quickly and efficiently.

21. The deer mouse, *Peromyscus maniculatus*, is the most widely distributed small mammal in North America. It varies greatly geographically, especially in coat color, tail length, and foot length. Over which area would you expect deer mice to be relatively uniform?
a. In mountainous areas, where environmental conditions change dramatically from place to place
b. On islands, where populations are isolated
c. Over large areas with little topographical or vegetational change
d. Between forests and deserts, with significant vegetational change
e. Between forests and prairies, with significant vegetational change

22. Antelope in Africa went through a burst of speciation 2.5–2.9 million years ago. What was the likely trigger of the speciation event?
a. Continental drift
b. Genetic drift
c. A change in the gene pool
d. A change in climate
e. Reproductive incompatibility

*▪23. Which of the following is a reason why evolutionary radiation is likely to occur on islands?
a. Islands may have fewer plant and animal groups and thus present more ecological opportunities.
b. There are many more species on islands.
c. Many organisms disperse easily over oceans.
d. Islands have more diverse habitats than do continents.
e. Islands tend to have more favorable habitats than continents, due to the moderating effects of the oceans.

24. The Hawaiian Islands are useful for studying evolutionary radiation because they are very isolated. Which of the following is *not* true about the biota of these islands?
a. More than 90 percent of the plant species are endemic.
b. Several groups of flowering plants are more diverse on these islands than their mainland counterparts.
c. There are no endemic amphibians on these islands.
d. There are no endemic reptiles on these islands.
e. There are no endemic mammals on these islands.

25. The hypothesized sequence of events in geographic speciation is
a. geographic barrier, reproductive isolation, genetic divergence.
b. geographic barrier, genetic divergence, reproductive isolation.
c. genetic divergence, geographic barrier, reproductive isolation.
d. genetic divergence, reproductive isolation, geographic barrier.
e. reproductive isolation, genetic divergence, geographic barrier.

26. Mules are the offspring of parents of two different species, a horse and a donkey. These hybrids exhibit
a. polyploidy.
b. shortened lifespan.
c. behavioral isolation.
d. reduced viability.
e. sterility.

27. The modern polar bear species evolved from ancestral bear populations in southern Alaska that became separated by glaciers from bear populations in the rest of North America. This type of event is called
   a. **allopatric speciation.**
   b. temporal isolation.
   c. parapatric speciation.
   d. mechanical isolation.
   e. sympatric speciation.

*28. A situation where two very similar animal species have overlapping distributions is most likely due to
   a. parapatric speciation.
   b. **allopatric speciation followed by range expansion.**
   c. sympatric speciation due to polyploidy.
   d. convergent evolution of unrelated species.
   e. mechanical isolation between the two species.

*29. Which of the following is *not* true of a hybrid zone?
   a. It occurs where two different populations come into contact.
   b. It may shift in location due to environmental changes.
   c. Its habitat may differ from that favored by the parent populations.
   d. It may disappear if isolating mechanisms develop.
   e. **It may occur among animals, but does not occur among plants.**

*▪30. American and European sycamores have been isolated from one another for at least 20 million years, but are morphologically very similar and can form fertile hybrids. According to Ernst Mayr's definition of species, these sycamores should belong to
   a. different species because they are geographically isolated.
   b. different species because they lack the opportunity to interbreed in nature.
   c. different species because they are evolving separately.
   d. the same species because they are morphologically similar.
   e. **the same species because they are capable of forming fertile offspring.**

*31. Ginkgo trees occur in Asia and North America. Despite the geographic separation by the Pacific Ocean, biologists consider them the same species. What aspect of Ernst Mayr's definition of species accounts for this?
   a. They are reproductively isolated.
   b. **They are potentially capable of exchanging genes.**
   c. They are exchanging genes across the ocean.
   d. They have different evolutionary ancestry.
   e. They formed a large hybrid zone.

▪32. Why are biological species not always equivalent to taxonomic species?
   a. **Taxonomic species are based on appearance, not reproductive behavior.**
   b. Taxonomic species are based on reproductive behavior, not appearance.
   c. Biological species are based on appearance, not reproductive behavior.
   d. The question has a false premise; biological species are always equivalent to taxonomic species.
   e. Biological species are based on genetic information; taxonomic species are based on ecological information.

33. Which mode of speciation involves a geographic barrier that prevents exchange of genes?
   a. **Allopatric**
   b. Parapatric
   c. Sympatric
   d. Polyploidy
   e. Behavioral isolation

*34. Six platyfish with different tail spotting patterns live in eastern Mexico. Five of them can interbreed and produce fertile offspring. One of them cannot interbreed with the other five. What can you conclude about these platyfish?
   a. They all belong to the same biological species.
   b. They are each a different biological species.
   c. **There are two biological species in the example above.**
   d. There is not enough data given to make a conclusion.
   e. There are six biological species in the example above.

35. What are the progeny of genetically dissimilar parents called?
   a. Infertile
   b. Fertile
   c. **Hybrids**
   d. Mutants
   e. Hermaphrodites

*▪36. If two genetically differentiated populations reestablish contact and the resulting hybrid progeny are successful and reproduce with other members of the two populations, what will probably happen?
   a. Speciation will occur.
   b. Genetic differences will increase between the populations.
   c. Reproductive isolation will occur.
   d. **The two populations will amalgamate; no new species will form.**
   e. Two new species will form.

▪37. If two genetically differentiated populations reestablish contact and the resulting hybrid progeny are *not* as successful in reproduction as the members of the two parental populations, what will probably happen?
   a. **Reproductive isolation will occur.**
   b. A stable hybrid zone will form.
   c. The two populations will amalgamate; no new species will form.
   d. The populations will become genetically identical.
   e. Two new species will form.

38. What form of reproductive isolation occurs between a horse and donkey when they mate and produce a mule?
   a. Prezygotic isolation
   b. Abnormal zygote formation
   c. **Hybrid infertility**

d. Hybrid vigor
e. Hybrid fertility

39. What form of speciation tends to occur when a population exists across an important environmental discontinuity?
a. Hybridization
b. Parapatric
c. Sympatric
d. Dyspatric
e. Extinction

40. Which of the following modes of speciation can occur most quickly?
a. Allopatric
b. Parapatric
c. Sympatric
d. Dyspatric
e. Adaptive radiation

41. Because there are no physical barriers separating the species, many biologists believe that speciation of host-specific insects occurs by
a. sympatric speciation.
b. parapatric speciation.
c. allopatric speciation.
d. hybridization.
e. prezygotic isolation.

42. What is the most important factor promoting the cohesion of a species?
a. Mutation
b. Natural selection
c. Gene flow
d. Genetic drift
e. Reproduction

43. Behavioral complexity and short generation time will tend to __________ the process of speciation.
a. accelerate
b. decelerate
c. have no effect on
d. obscure
e. threaten

44. Some groups or species have given rise to large numbers of daughter species. This phenomenon is known as
a. sympatric speciation.
b. hybridization.
c. evolutionary radiation.
d. gene flow.
e. evolution.

*45. Which of the following statements is *not* true of the genetics of speciation?
a. Speciation does not necessarily involve major reorganization of the genome.
b. Small genetic differences can yield substantial morphological and physiological differences between species.
c. The kinds of genetic differences among species are similar to those found within species.
d. The genetic processes of speciation are different from those of intraspecific genetic differentiation.
e. Genetic changes within a short time period may result in speciation.

46. What group of organisms has the highest prevalence of sympatric speciation?
a. Mammals
b. Insects
c. Plants
d. Fungi
e. Bacteria

47. When individuals from one population disperse to a new geographically isolated location and mate, eventually resulting in the formation of a new species, a __________ event has occurred.
a. foundation
b. founder
c. radiative
d. gene flow
e. hybridization

48. A species that is found only in a certain area of the planet and nowhere else is called
a. endangered.
b. endemic.
c. extinct.
d. emetic.
e. exotic.

49. Which of the following is *not* true of ecology and the rate of speciation?
a. Speciation rates are correlated with the diets of mammals.
b. Speciation rates are not related to birth rates.
c. Speciation rates differ markedly in the fossil record.
d. Speciation rates are consistent across all organisms yet studied.
e. Speciation rates are higher among animal species with more complex behavior patterns.

▪50. Islands are often called "natural laboratories for evolutionary studies." Why is this so?
a. Islands are isolated from other land masses.
b. Islands are geologically very young.
c. Islands have very low speciation rates.
d. Islands are all ecologically similar.
e. Islands always have small numbers of species on them.

51. The adaptive radiation of Hawaiian silverswords has demonstrated what fact about speciation?
a. Major morphological changes can be produced by small genetic changes.
b. Major morphological changes require large-scale genetic changes.
c. Many different silversword ancestors colonized Hawaii.
d. The Hawaiian silverswords are not a different species.
e. Plant speciation always requires polyploidy.

*52. Allele frequencies of malate dehydrogenase-1 show significant differences among city block populations of the snail *Helix aspersa*. What is the most likely factor maintaining these frequency differences?
a. Natural selection—the city blocks are very different ecologically.
b. Genetic drift—founder effects of city block colonization.
c. Lack of gene flow among city blocks—cars run over snails in the roads.
d. Different mutation rates among city blocks.
e. Parapatric effects due to different amounts of pollutants on different city blocks.

*53. Of the following reproductive isolating mechanisms, which one is the most efficient at preventing waste of reproductive effort?
a. Hybrid inviability
b. Hybrid sterility
c. Gamete incompatibility
d. Prezygotic isolation
e. Hybrid cross-reproduction

54. Which mode of speciation is most prevalent among larger animals?
a. Geographic
b. Parapatric
c. Sympatric
d. Polyploidy
e. Genetic drift

55. Which of the following is *not* true of speciation?
a. A new species can be formed in one breeding season.
b. Animals with complex behavior and mating patterns speciate more rapidly.
c. Members of the same species can live on separate continents.
d. Speciation is more rapid on island archipelagos such as Hawaii and the Galápagos.
e. Sympatric speciation is the most common type among animals.

56. The family Asteraceae contains several genera found on the Hawaiian islands. There is very little genetic variation between them, and they are thought to have evolved from a single ancestral species. What is responsible for the large morphological differences?
a. Each island has a unique environment that the plants have adapted to.
b. Predation has eliminated all but the remaining genera on each island.
c. Birds traveling between islands carried different types of seeds that started the plant growth.
d. Polyploidy during reproduction created the variation in features.
e. c and d

57. Populations, such as blue and snow geese, that occasionally interbreed and form hybrid zones
a. cannot be different species.
b. often result in evolutionary radiation.
c. can maintain species differences.
d. have little geographic variation.
e. will produce polyploid offspring.

*▪58. Why is it so difficult to obtain evidence supporting sympatric speciation?
a. Sympatric speciation always takes a very long time to occur.
b. Sympatric speciation always involves polyploidy.
c. It is hard to distinguish true sympatric speciation from allopatric speciation that occurred in the recent past.
d. It is impossible to show genetic differences in sympatric speciation.
e. Sympatric speciation is a rare event.

59. Fertile species produced from asexual reproduction of tetraploid individuals are considered
a. diploid.
b. polyploid.
c. paraploid.
d. allopolyploid.
e. mutants.

Study Guide Questions

1. Select all the correct answers that complete the following sentence: It is difficult to apply the biological species concept to groups of organisms that
a. are asexual.
b. produce hybrids only in captivity.
c. show little morphological diversity.
d. exist only in the fossil record.

2. Which of the following statements about geographic speciation is *not* true?
a. Geographic speciation can sometimes involve small populations.
b. Geographic speciation only occurs in species that are widely distributed.
c. Geographic speciation always involves a physical barrier that interrupts gene flow.
d. Geographic speciation can sometimes involve chance events.

3. The activities of a mining company result in deposition of a new soil type within the range of a widespread plant species. Which of the following phenomena is likely to occur as a result?
a. Geographic speciation
b. Sympatric speciation
c. Parapatric speciation
d. Allopatric speciation

4. Plant species *A* ($2n$=20) and *B* ($2n$=14) hybridize to produce species C, an allopolyploid. How many chromosomes would be present in the cells of species C?
a. 17
b. 28
c. 34
d. 40

5. Which type of speciation is most common among flowering plants?
   a. Geographic
   b. Sympatric
   c. Parapatric
   d. Allopatric

6. Which of the following would *not* be considered an example of a prezygotic reproductive isolating mechanism?
   a. One bird species forages in the tops of trees for flying insects while another forages on the ground for worms and grubs.
   b. The males of one species of moth cannot detect and respond to the sex attractant chemicals produced by the females of another species.
   c. Sperm of one species of sea urchin are unable to penetrate the egg plasma membrane of another species.
   d. Mosquitos of one species are active in foraging and searching for mates at dusk, whereas those of another species are active at dawn.

7. Which of the following factors would *not* be expected to increase the rate of speciation in a group of organisms?
   a. A species range consisting of fragmented populations
   b. A diet consisting of food items whose abundance varies widely
   c. High birth rates
   d. Increased behavioral complexity

8. Which of the following is *not* a suggested reason for the evolutionary radiation of silverswords on the Hawaiian archipelago?
   a. Water is an effective barrier for many organisms.
   b. Because islands are small compared with mainland areas, you would expect more species to develop there.
   c. Frequently, competition is reduced on islands.
   d. More ecological opportunities exist on islands.

9. Studies of species of Hawaiian *Drosophila* show that
   a. sympatric speciation via polyploidy is common in insects.
   b. few genes need be involved in establishing reproductive isolation.
   c. species with nonoverlapping ranges can be very similar.
   d. a geographic barrier is not always necessary for the establishment of reproductive isolation.

10. Which of the following statements about speciation is *not* true?
    a. A small founding population can be involved in speciation.
    b. Speciation always involves interruption of gene flow between different groups of organisms.
    c. The rate of speciation can vary for different groups of organisms.
    d. Speciation cannot occur in a single generation.

11. Which of the following observations constitutes conclusive evidence that two overlapping populations that had been geographically separated have *not* diverged into distinct species?
    a. Matings between members of the two populations produce viable hybrids.
    b. A stable hybrid zone exists where their ranges overlap.
    c. Interbreeding is common between members of the two populations.
    d. None of the above.

End of Chapter Questions

1. A species is a group of
   a. actually interbreeding natural populations that are reproductively isolated from other such groups.
   b. potentially interbreeding natural populations that are reproductively isolated from other such groups.
   c. actually or potentially interbreeding natural populations that are reproductively isolated from other such groups.
   d. actually or potentially interbreeding natural populations that are reproductively connected to other such groups.
   e. actually interbreeding natural populations that are reproductively connected to other such groups.

2. Anagenesis is
   a. a continuous change in a single lineage of organisms.
   b. the formation of two species by the splitting of one evolutionary lineage.
   c. the formation of a new species by the coming together of two evolutionary lineages.
   d. the reduction of two lineages by the extinction of one of them.
   e. the formation of new species by the reclassification of a group.

3. Allopatric speciation may happen when
   a. continents drift apart and separate previously connected lineages.
   b. a mountain range separates formerly connected populations.
   c. different environments on two sides of a barrier cause populations to diverge.
   d. the range of a species is separated by loss of intermediate habitat.
   e. All of the above

4. Finches speciated on the Galápagos Islands because
   a. the Galápagos Islands are a long way from the mainland.
   b. the Galápagos Islands are arid.
   c. the Galápagos Islands are small.
   d. the islands in the Galápagos Archipelago are sufficiently isolated from one another that there is little migration among them.
   e. the islands in the Galápagos Archipelago are close enough to one another that there is considerable migration among them.

5. Which of the following is *not* a potential prezygotic reproductive barrier?
   a. Temporal segregation of breeding seasons
   b. Differences in chemicals that attract individuals
   c. Sterility of hybrids
   d. Spatial segregation of mating sites
   e. Inviability of sperm in female reproductive tracts

6. A common means of sympatric speciation is
   a. polyploidy.
   b. hybrid sterility.
   c. temporal segregation of breeding seasons.
   d. spatial segregation of mating sites.
   e. imposition of a geographic barrier.

7. Sympatric species are often similar in appearance because
   a. appearances are often of little evolutionary significance.
   b. genetic changes accompanying speciation are often small.
   c. genetic changes accompanying speciation are usually large.
   d. speciation usually requires major reorganization of the genome.
   e. the traits that differ among species are not the same as the traits that differ among individuals within species.

8. Which statement about speciation is *not* true?
   a. It always takes thousands of years.
   b. It often takes thousands of years but may happen within a single generation.
   c. Among animals it usually requires a physical barrier.
   d. Among plants it often happens as a result of polyploidy.
   e. It has produced the millions of species living today.

9. Evolutionary radiations
   a. often happen on continents but rarely on island archipelagos.
   b. characterize birds and plants but not other taxonomic groups.
   c. have happened on all continents, as well as on islands.
   d. require major reorganizations of the genome.
   e. never happen in species-rich environments.

10. Speciation is often rapid within lineages in which species have complex behavior because
    a. individuals of such species make very fine discriminations among potential mating partners.
    b. such species have short generation times.
    c. such species have high reproductive rates.
    d. such species have complex relationships with their environments.
    e. None of the above

11. Speciation is an important component of evolution because it
    a. generates the variation upon which natural selection acts.
    b. generates the variation upon which genetic drift and mutations act.
    c. enabled Charles Darwin to perceive the mechanisms of evolution.
    d. generates the high extinction rates that drive evolutionary change.
    e. has resulted in a world with millions of species, each adapted for a particular way of life.

Student Web Site Self-Quiz Questions

1. The identity of some species, such as the red-winged blackbird, is not controversial. Which of the following phrases is *not* an essential part of the species concept?
   a. Morphological similarity
   b. Reproductive isolation
   c. Independent evolutionary unit
   d. Actually or potentially interbreeding
   e. Natural population

2. Which of the following statements about allopatric speciation is *false*?
   a. Allopatric speciation requires that the daughter populations be fairly small when they are first separated.
   b. Allopatric speciation is also called geographic speciation.
   c. Continental drift resulted in allopatric speciation.
   d. The founder effect may result in daughter populations having a more limited set of alleles than the parent population.
   e. Allopatric speciation is the most common form of speciation.

3. Several species of the Galapagos finches arose by the mode of speciation known as
   a. anagenesis.
   b. cladogenesis.
   c. sympatric speciation.
   d. allopatric speciation.
   e. parapatric speciation.

4. Several new tetraploid species of salsifies have arisen in western North America from diploid species by allopolyploidy. This mode of speciation is known as
   a. anagenesis.
   b. cladogenesis.
   c. sympatric speciation.
   d. allopatric speciation.
   e. parapatric speciation.

5. Interbreeding between several species of fruit flies seldom occurs, because they feed on different types of fruit found in different locations. This is an example of __________ isolation.
   a. spatial
   b. temporal
   c. behavioral
   d. mechanical
   e. gametic

6. Several species of sea urchins with overlapping reproductive periods may inhabit the same tide pool. Although they practice external fertilization, they seldom interbreed, because eggs produce chemicals that are only attractive to sperm of their own species. This is an example of a _________ reproductive barrier involving _________.
   a. prezygotic, spatial isolation
   b. prezygotic, temporal isolation
   c. prezygotic, gametic isolation
   d. postzygotic, hybrid infertility
   e. postzygotic, gametic isolation

7. Although warblers have species-specific courtship behavior, hybrids are occasionally produced. Which of the following outcomes would *not* be expected to lead to the evolution of a more effective prezygotic barrier?
   a. The development of a hybrid zone.
   b. Hybrid zygotes fail to develop to sexual maturity.
   c. Hybrids are sterile.
   d. Hybrids are less viable.
   e. Hybrids produce fewer offspring.

8. Many male songbirds have very distinctive songs that are used to attract females. These songs are an example of a _________ isolating mechanism.
   a. spatial
   b. temporal
   c. behavioral
   d. mechanical
   e. gametic

9. Speciation rates are influenced by various factors. For example, the speciation rates among some African antelopes are influenced by their diet and have varied greatly over the last several million years. Which of the following conditions would *not* be expected to increase the speciation rate in a group of organisms?
   a. A species range consisting of fragmented populations
   b. A diet consisting of food items whose abundance varies widely
   c. Long generation times
   d. Increased behavioral complexity
   e. A high degree of mate selection

10. Evolutionary radiations have been common on island chains such as the Hawaiian Archipelago. Which of the following statements about island-based evolutionary radiations is *false*?
   a. Islands usually lack plant and animal groups found on the mainland.
   b. Gene flow among the islands in an archipelago is restricted.
   c. Many species found on islands are endemic—found nowhere else.
   d. Because islands are small, reduced ecological opportunities cause island species to be more specialized than their relatives on the mainland.
   e. Frequently, competition is reduced on islands.

11. Which of the following statements about the evolutionary radiation of Hawaiian silverswords and tarweeds is *false*?
   a. The group shares a recent common ancestor from the Pacific coast of North America.
   b. This group occupies nearly all available habitats on the islands.
   c. Chloroplast DNA analysis shows that this group has differentiated a great deal genetically.
   d. This group is much more diverse in size and shape than mainland relatives.
   e. Although silverswords and tarweeds are non-woody plants, some Hawaiian species have evolved into shrubs and trees.

12. Gaping is an adaptation seen in the American blackbird family. Which of the following statements about the evolutionary significance of this type of adaptations is *false*?
   a. Evolution of novel adaptations can lead to evolutionary radiations.
   b. Gaping allows the blackbirds to use the environment in new and varied ways.
   c. There are no morphological differences between species that can gape and those that cannot.
   d. Gaping is mainly a food procurement adaptation.
   e. Gaping has allowed the blackbirds to occupy a variety of habitats.

TEST FILE

★ *Indicates a **difficult** question*

■ *Indicates a **conceptual** question*

23 Reconstructing and Using Phylogenies

Fill in the Blank

1. **Taxonomy** is the theory and practice of classifying organisms.

2. **Systematics** is the scientific study of the diversity of organisms.

3. **Carolus Linnaeus**, a Swedish biologist in the 1700s, developed the classification system used today.

4. A two-name classification system, referred to as **binomial nomenclature**, is used today throughout biology.

5. A trait, such as the modern horse's single toe, that differs from the trait in an organism's ancestors in the lineage is called a **derived trait**.

6. Any two traits derived from a common ancestral form are **homologous traits**.

7. Traits such as fins in aquatic mammals and fins in fishes exhibit **homoplasy**; that is, they evolved in different lineages, although they are not found in their most recent common ancestor.

8. Traits that evolve by **convergent evolution** were formerly very different, yet now resemble one another because they have undergone selection to perform similar functions.

9. **Cladistic** classification shows evolutionary relationships and expresses them in treelike diagrams.

10. The entire portion of a phylogeny that is descended from a common ancestor is called a **clade**.

11. The term "Rosaceae" is an example of the taxonomic category of **family**.

12. In the Linnaean system, classes are divided into **orders**, which are then divided into families.

13. When designing a cladogram, the operating rule that it is wiser to postulate a minimal number of changes in traits is called **parsimony**.

Multiple Choice

1. In modern systematics, each family name is based on
   a. the name of the order to which it belongs.
   b. a characteristic common to all members.
   c. the name of a member genus.
   d. the name of the largest member species.
   e. the Latin name for the organisms.

2. In the Linnaean system, the suffix "-aceae" refers to a(n)
   a. genus of plants.
   b. genus of animals.
   c. family of plants.
   d. family of animals.
   e. order of plants.

■3. North America and Great Britain both have birds called robins. These birds have brown backs and red breasts, but are incapable of interbreeding, are different sizes, and have different diets and habitats. Based on this information, one would expect these birds to belong to
   a. the same species.
   b. the same genus, but different species.
   c. the same family, but different genera.
   d. different species, but more information is needed for further classification.
   e. different genera, but more information is needed for further classification.

4. Within a family, the number of species placed into each genus is determined by
   a. the evolutionary uniqueness among organisms.
   b. an even distribution of the total number of species.
   c. an even distribution of the species based on their sizes.
   d. grouping species by their geographic ranges.
   e. clustering species based on their habitats.

5. The wing of a bat and the wing of a bird are an example of
   a. divergent evolution.
   b. vertical evolution.
   c. convergent evolution.
   d. a derived trait.
   e. evolutionary reversal.

6–9. Use the cladogram below to answer the following questions.

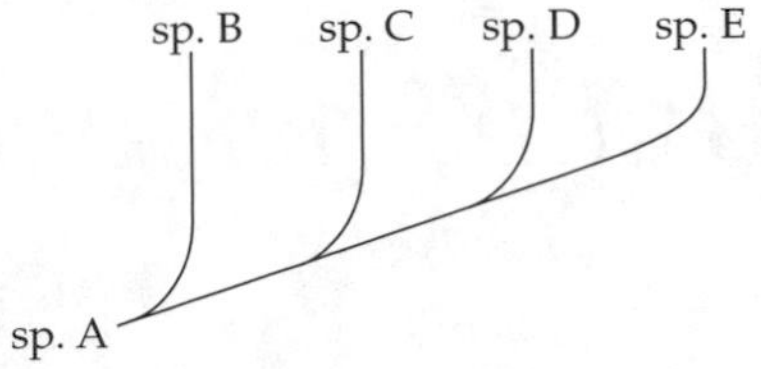

6. Assuming the cladogram includes modern species, which of these species is (are) alive today?
   a. A, B, C, D, E
   b. B, C, D, E
   c. C, D, E
   d. D, E
   e. E

*7. Species D and E share __________ homologous and __________ homoplastic traits.
   a. many, many
   b. many, few
   c. few, many
   d. few, few
   e. no, no

*8. Which species share(s) the most recent common ancestor with species E?
   a. Species A
   b. Species B
   c. Species C
   d. Species D
   e. All species share the same common ancestry.

9. The position of species B in the cladogram indicates that species B
   a. became extinct before species C, D, or E.
   b. has fewer derived traits than species C, D, or E.
   c. has a shorter evolutionary history than species C, D, or E.
   d. shares more common ancestors with species C than with species D or E.
   e. is less fit for its environment than species C, D, or E.

10. When compared to other taxa, a taxon that shares general homologous traits but lacks special homologous traits is called a(n)
   a. clade.
   b. genus.
   c. homoplasy.
   d. outgroup.
   e. population.

*▪11. When comparing humans and chimps with dogs and cats, a general homologous trait would be
   a. hands specialized for grasping.
   b. presence of body hair.
   c. standing on two legs.
   d. lack of a tail.
   e. poor sense of smell.

*▪12. When comparing humans and chimps with dogs and cats, a special homologous trait would be
   a. bony skeleton.
   b. presence of body hair.
   c. hands specialized for grasping.
   d. mouth containing teeth.
   e. presence of circulatory system.

13. Which two statements are often assumed to be *true* in the construction of cladograms?
   1. Derived traits appear only once in a lineage.
   2. Branching points are determined by the number of homologous traits.
   3. Derived traits are never lost.
   4. Descendant species often lose ancestral traits.
   a. 1 and 2
   b. 1 and 3
   c. 1 and 4
   d. 2 and 3
   e. 3 and 4

14–17. Use the table below to answer the following questions.

Ancestral and Derived Traits Among Five Species

	Trait				
Species	**1**	**2**	**3**	**4**	**5**
A	1	1	0	0	1
B	0	0	1	0	1
C	1	1	0	0	1
D	0	1	0	0	1

The ancestral form of each trait is coded 0, and the derived trait is coded 1.

*14. In constructing a cladogram, the oldest divergence would separate
   a. A from B, C, and D.
   b. B from A, C, and D.
   c. C from A, B, and D.
   d. D from A, B, and C.
   e. A and D from B and C.

*15. The two species that diverged most recently from one another are
   a. A and B.
   b. A and C.
   c. A and D.
   d. B and C.
   e. B and D.

*16. Which trait is most recently derived?
   a. 1
   b. 2
   c. 3
   d. 4
   e. 5

*17. Which trait is most ancestral?
   a. 1
   b. 2
   c. 3
   d. 4
   e. 5

*18. What is the most appropriate use of knowledge about behavioral traits in the reconstruction of phylogenies?
a. This knowledge is never relevant.
b. This knowledge is more important than other traits among closely related species.
c. This knowledge supports knowledge of other traits among closely related species.
d. This knowledge supports knowledge of other traits among distantly related species.
e. This knowledge is always relevant.

19. Systematists usually employ parsimony in reconstructing a phylogeny. Parsimony involves arranging the organisms such that
a. the minimal number of changes in traits is postulated in determining the lineage.
b. the maximum number of traits is used in establishing the lineage.
c. molecular information is given priority over other traits in determining the lineage.
d. larval traits are included in establishing the lineage.
e. ancestral traits are given priority over derived traits in determining the lineage.

20. __________ is the study of biological diversity and its evolution.
a. Phylogeny
b. Systematics
c. Taxonomy
d. Classification
e. Nomenclature

21. __________ is the science of biological classification.
a. Phylogeny
b. Systematics
c. Taxonomy
d. Classification
e. Nomenclature

22. Classification systems have many uses. Which of the following is *not* a goal of biological classification?
a. To depict convergent evolution
b. To clarify relationships among organisms
c. To help us remember organisms and their traits
d. To clearly identify organisms being studied
e. To provide predictive powers

23. Taxonomic systems used by biologists are hierarchical; that is,
a. taxonomic groups reflect shared characters, not evolutionary relationships.
b. each higher taxonomic group contains all the groups below it.
c. taxonomic groups reflect common habitats.
d. a hierarchy of traits is used to establish classifications.
e. phylogenetic relationships do not help us understand evolution.

24. Classification systems serve four important roles. Which of the following is *not* one of those roles?
a. To help us remember characteristics of a large number of different things
b. To help us identify shared traits, such as hair, mammary glands, and constant high body temperature in mammals
c. To reveal the harmony of nature
d. To provide stable, unique, unequivocal names for organisms
e. To help reconstruct evolutionary pathways

25. The biological classification system used today is based on the work of
a. Charles Darwin.
b. Barbara McClintock.
c. Gregor Mendel.
d. Lynn Margulis.
e. Carolus Linnaeus.

26. The biological classification system used today is referred to as
a. dichotomous taxonomy.
b. dichotomous nomenclature.
c. dichotomous keys.
d. binomial taxonomy.
e. binomial nomenclature.

27. The hierarchy of categories in the classification system used today is
a. division or phylum, kingdom, order, family, class, genus, and species.
b. division or phylum, kingdom, order, class, family, genus, and species.
c. division or phylum, kingdom, class, order, family, genus, and species.
d. kingdom, division or phylum, order, class, family, genus, and species.
e. kingdom, division or phylum, class, order, family, genus, and species.

28. In referring to an organism in writing, such as in a newspaper, textbook, or lab report, which of these rules should be followed?
1. Underline or italicize genus
2. Underline or italicize species
3. First letter of species should be uppercase
4. First letter of genus should be uppercase
a. 1, 2, 4
b. 1, 2, 3
c. 2, 3, 4
d. 1, 3, 4
e. 1, 2, 3, 4

29. If you saw the name of a family of organisms and it ended with "-idae," such as Formicidae, you would know that it was a family of
a. bacteria.
b. fungi.
c. plants.
d. animals.
e. protists.

30. A consensus tree is produced by
    a. merging multiple phylogenetic trees.
    b. aligning two phylogenetic trees.
    c. subtracting two phylogenetic trees.
    d. using only molecular data.
    e. using only morphological data.

31. If a research team was employing the maximum likelihood method to construct a phylogeny, what information would they be basing it on?
    a. Morphological
    b. Environmental
    c. Molecular
    d. Parsimony
    e. Fossil

32. If the relationship of an adult sea squirt is compared to vertebrate animals using morphological data, it would seem that sea squirts and vertebrates are very much unrelated; what other information could be useful to verify this conclusion?
    a. Fossil
    b. Developmental
    c. Environmental
    d. Asystematic
    e. None

33. Molecular traits most often used in the construction of phylogenies are
    a. protein and nucleic acid structure.
    b. lipid and nucleic acid structure.
    c. carbohydrate and lipid structure.
    d. protein and lipid structure.
    e. carbohydrate and nucleic acid structure.

34. In a cladistic classification, each taxon
    a. includes several lineages from a single common ancestor.
    b. is a single lineage and includes all—and only—the descendants of a single ancestor.
    c. shares common morphologies that are homoplasic.
    d. shares common morphologies that are derived characters.
    e. includes several lineages that share common morphologies.

35. The excellent fossil record of horses shows that modern horses, which have one toe on each foot, evolved from ancestors that had multiple toes. A trait like the modern horse's single toe that differs from the ancestral trait in the lineage is called a(n) ________ trait.
    a. derived
    b. ancestral
    c. morphological
    d. biochemical
    e. fundamental

36. Any two structures derived from a common ancestral trait are said to be
    a. analogous.
    b. morphological traits.
    c. biochemical traits.
    d. homologous.
    e. homoplasic.

*37. Homoplasy can result from convergent evolution. Under convergent evolution,
    a. structures that were formerly very different come to resemble one another because they have undergone selection to perform similar functions.
    b. the same character evolves in different lineages, often from a common basis.
    c. a structure evolves in a lineage that is not found in their common ancestor.
    d. the same character evolves in different lineages, from the same ancestral trait, due to similar selection pressures.
    e. the ancestral trait evolves to different characters in different lineages.

38. Genetic similarities among some vertebrates have been estimated by DNA sequencing. According to these data, humans are most closely related to
    a. gibbons.
    b. chimpanzees.
    c. baboons.
    d. galagos.
    e. rhesus monkeys.

39. If the abbreviation "sp." is used after a genus name, it indicates
    a. there are too many species of that genus to name.
    b. there are no species in that genus.
    c. the species is unknown.
    d. the species was named incorrectly.
    e. the species name is not agreed upon.

▪40. In groups such as birds, ________ make(s) it difficult to resolve phylogenies using only morphological data because the use of different traits produces different phylogenies.
    a. reverse evolution
    b. parallel evolution
    c. convergent evolution
    d. genetic drift
    e. founder events

41. Which of the following statements about common names is *not* true?
    a. The same species can have two different common names in different areas.
    b. Two different species can have the same common name.
    c. According to the Linnaean classification system, common names for two species within the same genus must be different.
    d. Common names do not exist for some species.
    e. Common names can be both ambiguous and redundant.

42. A taxon is
    a. an archaic concept not used in modern classification systems.
    b. any group of organisms treated as a unit.
    c. a single species.
    d. a group of organisms that are reproductively isolated from other such groups.
    e. the smallest grouping in the Linnaean classification system.

43. Which of the following choices is the conventionally correct representation for the name of a sea star common to the rocky intertidal zone?
   a. Pisaster Ochraceous
   b. *Pisaster Ochraceous*
   c. *Pisaster ochraceous*
   d. *pisaster ochraceous*
   e. pisaster ochraceous

44. Which of the following choices is the conventionally correct way to refer to several species of fruit flies?
   a. *Drosophila* spp.
   b. *Drosophila* sp.
   c. *Drosophila's*
   d. Drosophila spp.
   e. Drosophila sp.

45. When writing a scientific paper, which of the following choices shows the conventionally correct way to refer to the binomial name of the English bluebell after the name has been cited earlier?
   a. *Edymion nonscriptus*
   b. Edymion n.
   c. *E. nonscriptus*
   d. E. n.
   e. E. nonscriptus

46. In the Linnaean classification system, which one of the following taxa usually ends in "-idae" when used with animals?
   a. Genus
   b. Order
   c. Division
   d. Class
   e. Family

47. Which of the following statements about classification is *not* true?
   a. Members of a family are less similar than members of an included genus.
   b. An order has more members than the number of members in an included genus.
   c. Families have more members than phyla.
   d. Members of a family share a common ancestor in the more distant past than members of an included genus.
   e. The number of species in a taxon depends on their relative degree of similarity.

48. Homoplasy can result from all but one of the following. Select the exception.
   a. Convergent evolution
   b. Parallel evolution
   c. Reverse evolution
   d. Descent from a common ancestor
   e. Similar selection pressures

49. A taxon consisting of members that do *not* share the same common ancestor is
   a. monophyletic.
   b. polyphyletic.
   c. paraphyletic.
   d. unrelated.
   e. a clade.

50. In horses, presence of multiple toes is the __________ trait, while one toe is the __________ trait.
   a. ancestral, derived
   b. derived, ancestral
   c. ancestral, homologous
   d. homologous, ancestral
   e. ancestral, monophyletic

51. A monophyletic group containing all the descendents of a particular ancestor and no other organisms is also know as a
   a. phylogeny.
   b. clade.
   c. composite.
   d. breeding group.
   e. species.

52. Which of the following techniques for studying the biochemical traits of organisms could used to understand the evolution of cytochrome *c* (a protein)?
   a. Immunological distance determination
   b. Amino acid sequencing
   c. DNA hybridization
   d. Nucleic acid base sequencing
   e. cpDNA and mtDNA comparisons

53. Most taxonomists today believe that classification systems should be
   a. phylogeneitc.
   b. monophyletic.
   c. paraphyletic.
   d. polyphyletic.
   e. changed.

▪54. In systematics and phylogeny, the fossil record is especially important because
   a. most groups are well represented.
   b. it provides the absolute timing of evolutionary events.
   c. random mutations make most biochemical methods unreliable.
   d. DNA can be extracted from the fossils and analyzed.
   e. it is the only type of data useful in reconstructing the past.

55. In reconstructing phylogenies, an outgroup is used to
   a. distinguish general homology from special homology.
   b. exclude a taxon from the phylogenic group.
   c. distinguish homoplasy from convergent traits.
   d. gain knowledge of reverse evolution.
   e. b and c

▪56. Larval stages can help determine relationships among organisms, but care must be taken in that
   a. not all organisms have a larval stage.
   b. the larval form is too morphologically different from the adult form to be useful.
   c. the larval form may closely resemble an organism that the adult stage does not.
   d. larval forms all have a notochord and all adult forms do not.
   e. All of the above

57. Reconstructed phylogenies can be useful for all of the following *except*
    a. evolution of human language.
    b. the migration of human populations.
    c. determining if divergent traits occurred from reverse evolution.
    d. predicting future trends in evolution.
    e. evaluating how a protein has changed between species.

Study Guide Questions

1. Which of the following would not be expected to result in homoplasy?
   a. Convergent evolution
   b. The independent evolution of similar structures in different lineages
   c. Selection for traits that perform similar functions
   d. The inheritance of ancestral traits

2. A derived trait is one that
   a. differs from its ancestral form.
   b. is homologous with another trait found in a related species.
   c. is the product of an evolutionary reversal.
   d. has the same function, but not the same evolutionary origin, as a trait found in another species.

3. Which of the following statements about reconstructing phylogenies is *false*?
   a. Traits found in the outgroup as well as in the focal group are likely to be ancestral traits.
   b. Shared traits are generally assumed to be homoplastic until they can be proven to be homologous.
   c. Phylogenies do not include ancestors of modern groups, or date the splits between lineages.
   d. Nodes (branching points) in phylogenetic trees have only two branches because during speciation a lineage normally splits into only two daughter species.

4. The *most* important attribute of a biological classification scheme is that it
   a. avoids the ambiguity created by using common names.
   b. reflects the evolutionary relationships among organisms.
   c. helps us remember organisms and their traits.
   d. improves our ability to make predictions about the morphology and behavior of organisms.

5. Suppose you are writing a scientific paper about a unicellular green alga called *Chlamydomonas reinhardi*. What would be the proper way to refer to this species after you had used the full binomial earlier in the same paragraph?
   a. *Chlamydomonas reinhardi*
   b. *Chlamydomonas sp*
   c. *Chlamydomonas sp.*
   d. C. reinhardi

6. Organisms in a higher taxon are __________ similar, usually have diverged from a common ancestor __________ recently, and include __________ species than organisms in a lower, included taxon.
   a. less, less, fewer
   b. less, more, fewer
   c. less, less, more
   d. more, more, fewer

7. Which of the following incomplete lists of taxonomic categories ranks them properly from most inclusive to least inclusive?
   a. Phylum, order, family, genus
   b. Class, phylum, order, species
   c. Order, class, family, genus
   d. Family, order, class, kingdom

End of Chapter Questions

1. Which of the following is *not* a major role of a classification system?
   a. To aid memory
   b. To improve predictive powers
   c. To help explain relationships among things
   d. To provide relatively stable names for things
   e. To design identification keys

2. Any group of organisms treated as a unit in a classification system is a
   a. species.
   b. genus.
   c. taxon.
   d. clade.
   e. phylogen.

3. A genus is a
   a. group of closely related species.
   b. group of genera.
   c. group of similar genotypes.
   d. taxonomic unit larger than a family.
   e. taxonomic unit smaller than a species.

4. A trait that differs from its ancestral form is called a(n)
   a. altered trait.
   b. homoplastic trait.
   c. parallel trait.
   d. derived trait.
   e. homologous trait.

5. Outgroups are used in cladistic analyses to
   a. distinguish homoplasies from homologies.
   b. distinguish homoplasies from convergence.
   c. distinguish between general and special homologies.
   d. determine relationships between closely related taxa.
   e. distinguish between general and special homoplasies.

6. The parsimony principle is typically used when reconstructing phylogenies because
   a. evolution is nearly always parsimonious.
   b. it is better to provisionally adopt the simplest hypothesis capable of explaining the known facts.
   c. it is easier to handle parsimonious data with computers.

d. parsimony works well for all kinds of traits, both morphological and molecular.
e. parsimony was used before computers were available and it continues to be used even though other methods are now better.

7. Which of the following is *not* a way of identifying ancestral traits?
a. Determining which traits are found among fossil ancestors
b. Using an outgroup in which the trait is also found
c. Using more than one outgroup that has the trait
d. Determining how many species in the lineage share the trait today
e. Experimentally creating a known lineage

8. Traits that evolve very slowly are useful for determining relationships at the level of
a. phyla.
b. genera.
c. order.
d. family.
e. species.

9. Homologous traits are
a. similar in function.
b. similar in structure.
c. similar in structure but derived from different ancestral structures.
d. derived from a common ancestor whether or not they have the same function today.
e. derived from different ancestral structures and have dissimilar structures.

10. The genes that are most extensively used to determine evolutionary relationships among plants are
a. nuclear genes.
b. chloroplast genes.
c. mitochondrial genes.
d. genes in flowers.
e. genes in roots.

11. Which of the following is *not* a way in which phylogenies are used?
a. To establish evolutionary relationships
b. To determine how rapidly traits evolve
c. To determine historical patterns of movement of organisms
d. To help identify unknown organisms
e. To infer evolutionary trends

Student Web Site Self-Quiz Questions

1. The presence of a notochord in the embryonic stages of all vertebrates is an example of a(n)
a. ancestral trait.
b. general homologous trait.
c. special homologous trait.
d. derived trait.
e. More than one of the above.

2. Use of uric acid as the nitrogenous waste product in both birds and insects is an example of
a. homoplasy.
b. divergence.
c. an ancestral trait.
d. special homology.
e. general homology.

3. The group of vascular plants called the cacti have evolved modified leaves in the form of spines. Select a choice from those below to make the following sentence *true*.
In constructing a phylogeny of vascular plants, possession of spines would be __________.
a. an ancestral trait
b. a general homologous trait
c. a special homologous trait
d. homoplasy
e. divergence

4. Based on the following table showing the ancestral and derived states of five traits (1 through 5) of five species (A through E), which species has been selected as the outgroup in this comparison? **Note**: the ancestral state of each trait is indicated by 0 and the derived state is indicated by 1.

Ancestral and Derived Traits Among Five Species

	Trait				
Species	**1**	**2**	**3**	**4**	**5**
A	1	0	1	1	1
B	0	0	1	1	1
C	0	0	0	0	0
D	1	1	1	0	0
E	0	0	1	1	0

a. Species A
b. Species B
c. Species C
d. Species D
e. Species E

5. Select the statement about fossils that is *true*.
a. By definition, fossils are the preserved remains of an organism that lived in the past.
b. Absence of fossils of an organism in an area is strong evidence that the organism never lived in that area.
c. Systematists rely on fossils of "missing links" to establish the ancestors of a group.
d. The fossil record can be useful in helping to identify the ancestral form of traits.
e. Fossils establish the dates when a group first appears and when it becomes extinct.

6. The distribution of lungfish across the globe follows a distinct pattern. The discovery of fossils from a common ancestor in all continents except Antarctica and South America suggests that
a. each species of lungfish evolved independently.
b. the lungfish all belong to the same species.

c. the divergence of the forms of lungfish shown occurred before the breakup of Gondwana.
d. the divergence of the forms of lungfish shown occurred after the breakup of Gondwana.
e. the Australian variety of lungfish is misclassified.

7. Suppose that you discover an unknown species in the same genus as the Blackburnian warbler (*Dendroica fusca*). Which of the following choices shows the correct designation for this new, unknown species?
a. *Dendroica*
b. *Dendroica* sp.
c. *Dendroica* spp.
d. *Dendroica fusca*
e. *Dendroica* sp. *fusca*

8. According to the traditional classification of vertebrates, the class Reptilia is paraphyletic because it does not include all descendents of a common ancestor. Which group has been left out?
a. Domain Eukarya
b. Kingdom Animalia
c. Class Aves
d. Class Mammalia
e. Domain Bacteria

TEST FILE

★ *Indicates a* ***difficult*** *question*

■ *Indicates a* ***conceptual*** *question*

24 Molecular and Genomic Evolution

Fill in the Blank

1. The mechanisms of morphological change were not understood until discoveries were made in **biochemistry**.
2. Molecular evolution studies the patterns of **evolutionary change** in molecules and the process by which this is achieved.
3. A **substitution** mutation is the partial or complete replacement of a nucleotide base or longer sequence by another throughout an entire population or species.
4. Most of the variability in molecules measured by evolutionists does not affect their functioning. This is the hypothesis of **neutral evolution**.
5. Genes that are responsible for development of specific body segments are called **homeobox** genes.
6. A molecule similar to hemoglobin, **myoglobin**, has a greater affinity for oxygen.
7. Lungfishes and lilies have more **total** DNA than humans but less of it is **coding** DNA.
8. If a protein has consistent changes over time, then its **molecular clock** ticks at a constant rate.
9. DNA that has changed little during evolution probably has the same **function** in all species.
10. A visual representation of the evolutionary relationship between species is called a **gene tree**.
11. Morphological traits are not useful for comparisons between bacteria and humans because they lack **comparable structures**.
12. From **mitochondrial** DNA evidence, it was concluded that Neanderthals contributed few or no genes to the human gene pool.
13. In studying lysozyme, it was found that five amino acid substitutions on its surface affect its biological **activity**.
14. Evidence suggests that the three families of globin genes arose as gene **duplications**.
15. Langurs and cows share a very similar **lysozyme** molecule.
16. To infer times of the most ancient splits in lineages, scientists must use **molecules** that are found in all organisms and that evolve slowly.
17. Scientists can clone only small segments of ancient DNA because it **degrades** with time.
18. A **gene family** is a group of homologous genes with related functions produced by successive rounds of duplication and mutation.

Multiple Choice

1. Molecular evolution
   a. studies the patterns of evolutionary change in molecules.
   b. studies the processes of molecular changes.
   c. compares the molecules of living organisms.
   d. can determine the linkages of various species based on molecular changes.
   e. All of the above

■2. Why is the protein cytochrome *c* important in the study of molecular evolution?
   a. It is found in all eukaryotes.
   b. Its mutation rate is very low.
   c. Its mutation rate fluctuates to a great extent.
   d. It is found in all animals.
   e. All of its mutations are adaptive changes.

3. Using the polymerase chain reaction, DNA has been amplified from fossils up to
   a. 135,000 years old.
   b. 13.5 million years old.
   c. 135 million years old.
   d. 135 billion years old.
   e. DNA cannot be extracted from fossils.

4. Which of the following statements is *not* true of homeobox genes?
   a. All of these genes probably have a common evolutionary origin.
   b. They arose only once in animal development.
   c. They occur in the same order along the chromosome in various organisms.
   d. They are responsible for development.
   e. They vary greatly from species to species.

5. Which of the following is *not* an example of a gene that arose from an ancient gene duplication?
   a. rRNA
   b. tRNA
   c. Hemoglobin
   d. Pyruvate kinase
   e. Lactate dehydrogenase

6. Which of the following is *not* true of lysozyme?
   a. It is produced in tears and saliva.
   b. It can digest bacterial cell walls.
   c. It can digest plant matter.
   d. Langur lysozyme must function at a low pH.
   e. It is found in the whites of bird eggs.

▪7. What changes did scientists discover when they studied the amino acid sequence of lysozyme in langurs and cows?
   a. The changes were on the active site of the molecule.
   b. Arginine was changed to lysine, making it more resistant to trypsin.
   c. Lysine was changed to arginine, making it less resistant to trypsin.
   d. The changes are all neutral and do not affect the function of the molecule.
   e. The lysozyme in langurs and cows can break down plant matter.

▪8. Cows and langurs have nearly identical lysozyme. Why?
   a. Lysozyme is nearly identical in all animals, not just cows and langurs.
   b. They both ferment leafy food in their stomachs.
   c. They both share a common ancestor.
   d. The bacteria in their foregut caused mutations in their lysozyme.
   e. None of the above

9. Select the group that shares the most similar lysozyme molecule.
   a. Horse, cow, pig
   b. Langur, baboon, human
   c. Deer, antelope, horse
   d. Neotropical cuckoo, sparrow, cow
   e. Neotropical cuckoo, cow, langur

10. Select the correct order, from smallest to largest, of total DNA in these organisms.
    a. Bacterium, lungfish, human
    b. Fruit fly, bacterium, human
    c. Lungfish, human, fruit fly
    d. Bacterium, human, lungfish
    e. Yeast, bacterium, fruit fly

11. Which of the following of how the genome size differs between lungfish and humans is *true*?
    a. Lungfish have less total DNA than humans but more coding DNA.
    b. Lungfish have more total DNA than humans but less coding DNA.
    c. Lungfish have more total DNA than humans because they have more genes.
    d. Lungfish have less total DNA than humans because they are simpler organisms.
    e. Lungfish have the same amount of total DNA but less coding DNA.

12. Most changes in nucleic acids or amino acids
    a. change the function of the molecule.
    b. occur at constant locations.
    c. can be predicted.
    d. do not change the function of the molecule.
    e. are influenced by natural selection.

13. A molecular clock is useful for
    a. estimating lineage divergence during evolution.
    b. figuring the rates of change of a molecule.
    c. predicting where a molecule will mutate.
    d. timing when a new gene duplication will arise.
    e. timing the aging in an organism.

14. The consistency of molecular clocks is checked by
    a. finding the rate of mutation of a molecule.
    b. using well-dated fossil records to compare.
    c. artificially creating mutations of molecules in the lab and timing them.
    d. comparing similar molecules.
    e. comparing against cytochrome *c*.

15. Molecules that evolve slowly are used to study
    a. very similar species.
    b. relationships of all organisms.
    c. extinct organisms.
    d. duplicate genes.
    e. organisms that diverged long ago.

16. Molecules that have a rapid mutation rate are used to study
    a. very similar species.
    b. relationships of all organisms.
    c. extinct organisms.
    d. duplicate genes.
    e. organisms that diverged long ago.

▪17. Ribosomal RNA makes a good molecule to study evolution because it
    a. mutates frequently.
    b. mutates very slowly.
    c. is found only in a few organisms.
    d. greatly varies from organism to organism.
    e. does not have functional constraints.

18. The times of lineage splits between the major branches of life—Bacteria, Archaea, and Eukarya—have been estimated using
    a. lysozyme.
    b. cytochrome *c*.
    c. small-subunit rRNA.
    d. hemoglobin.
    e. genome size.

19. The DNA studies done on the flightless moas and kiwis of New Zealand showed that
    a. the two birds have a very recent divergence from each other.

b. kiwis evolved on New Zealand and then migrated to Australia.
c. Moas originated on Australia.
d. Rheas and moas are more similar than kiwis and moas.
e. Kiwis and moas are not each other's closest relative.

★■20. Which of the following statements about why ancient dinosaurs cannot be recreated (as in Jurassic Park) is *false*?
a. Their fossilized DNA is degraded.
b. It is unknown how to regulate expression of DNA during development.
c. We don't know how to piece together the fragmented DNA.
d. Their DNA is different from modern DNA; we can't make proteins from it.
e. Only small amounts of DNA can be amplified.

21. The "out of Africa" hypothesis suggests that
a. *Homo sapiens* diverged from chimpanzees.
b. *Homo sapiens* in Europe and Asia arose from a single origin in Africa.
c. *Homo sapiens* arose from Europe, Asia, and Africa simultaneously.
d. *Homo erectus* arose in Africa, but later *Homo sapiens* arose in Europe.
e. *Homo sapiens* arose in Africa, died off, and then simultaneously arose in Europe, Asia, and Africa.

22. What finding led to the belief in the "out of Africa" hypothesis?
a. Fossil records
b. Y chromosomal DNA sequencing of modern humans
c. Mitochondrial DNA sequencing of modern humans
d. Cytochrome *c* sequencing of modern humans
e. Total DNA comparisons between fossils and modern humans

23. The polymerase chain reaction allows biologists to
a. amplify DNA.
b. sequence DNA.
c. observe DNA.
d. amplify RNA.
e. sequence RNA.

24. If there are about 100 substitutions in the myoglobin molecule per 500 million years, then two organisms differing by 75 amino acids split about
a. 500 million years ago.
b. 125 million years ago.
c. 380 million years ago.
d. 485 million years ago.
e. 65 million years ago.

25. Gene duplication may involve
a. part of a gene.
b. a whole gene.
c. whole chromosomes.
d. a, b, and c
e. None of the above

26. Which of the following is *not* true of gene duplication?
a. All duplications produce functional genes.
b. It can result in evolution of novel functions in proteins.
c. It created much of the diversity in our genome.
d. rRNA and tRNA arose from duplications.
e. It can result in increased complexity.

27. To achieve more accurate estimates of lineage divisions,
a. molecular data is combined with morphological data.
b. molecular data is combined with fossil data.
c. the data from two or more molecules is combined.
d. a and b
e. a, b, and c

28–31. The following questions are based on the hypothetical similarity matrix below:

Human	Horse	Cow	Langur	Baboon	
	9	11	1	0	**Human**
1		1	12	13	**Horse**
1	18		14	12	**Cow**
14	0	0		1	**Langur**
16	0	0	17		**Baboon**

28. What do the numbers on the upper right stand for?
a. Similarities in amino acid sequence
b. Differences in amino acid sequence
c. Number of total amino acids
d. Number of deletions in amino acids
e. Number of additions in amino acids

29. What do the numbers on the bottom left stand for?
a. Similarities in amino acid sequence
b. Differences in amino acid sequence
c. Number of total amino acids
d. Number of deletions in amino acids
e. Number of additions in amino acids

★30. Which of the following organisms are the most similar?
a. Human and horse
b. Horse and cow
c. Human and baboon
d. Langur and cow
e. Human and langur

★31. Which of the following organisms are the most different?
a. Human and horse
b. Horse and cow
c. Human and baboon
d. Langur and cow
e. Human and langur

32. Homeobox genes are responsible for
 a. gene duplications.
 b. gene therapy.
 c. mutations.
 d. turning genes on and off.
 e. a, b, and c

33. Humans have three families of globin genes. The three arose from
 a. spontaneous mutations.
 b. gene duplication.
 c. ancestral DNA.
 d. differential splicing of DNA.
 e. transposable elements.

34. Select the statement about mitochondrial DNA that is *false*.
 a. It is maternally inherited.
 b. It was used to give validation to the "out of Africa" hypothesis.
 c. It mutates very slowly.
 d. It is useful for studying recent evolution of closely related species.
 e. None of the above

▪35. Ribosomal RNA has evolved very slowly because even minor changes in its base sequence will
 a. activate ribosomes.
 b. never occur.
 c. duplicate genes.
 d. inactivate ribosomes.
 e. be advantageous.

36. Which of the following is *not* true of the genomes of organisms?
 a. Eukaryotes have more coding DNA than do prokaryotes.
 b. Mutation rates of DNA can vary within a species.
 c. Some genes are accidentally duplicated.
 d. We know how much genetic information is needed to program development of a structure.
 e. Not all DNA is transcribed into RNA.

37. The hypothesis of neutral evolution asserts that
 a. the rate of mutations in molecules is influenced by natural selection.
 b. the variability in structures of molecules does not affect their functioning.
 c. closely related species have more similar molecular structures than distantly related species.
 d. organisms evolved through neutral changes in their molecules.
 e. mutations neither add nor subtract amino acids from molecules.

38. The earliest organisms known to have both α- and β-globin genes lived about
 a. 500 million years ago.
 b. 1 billion years ago.
 c. 100 million years ago.
 d. 50 million years ago.
 e. 1 million years ago.

39. Homeobox elements have been identified in
 a. flies only.
 b. animals.
 c. plants.
 d. bacteria.
 e. b and c

40. Sometimes molecular data can mislead scientists to conclude that two species are more closely related than other evidence suggests. A reason for this is
 a. convergent evolution.
 b. divergent evolution.
 c. neutral evolution.
 d. gene duplication.
 e. None of the above

41. Which of the following is *not* true of the parallel origins theory of human evolution?
 a. *Homo sapiens* arose simultaneously in Europe, Africa, and Asia.
 b. Ancient human fossils have been found in Europe, Indonesia, and Africa.
 c. Mitochondrial DNA evidence suggests that this theory is correct.
 d. It requires 1 million years of divergence from the last common ancestor.
 e. *Homo sapiens* arose from *Homo erectus*.

42. Ancient lineage data supports the division of living organisms into three major branches:
 a. Bacteria, Archaea, and Eukarya.
 b. Bacteria, Prokaryotes, and Eukarya.
 c. Bacteria, Eukarya, and Plants.
 d. Bacteria, Humans, and Plants.
 e. Archaea, Eukarya, and Animals.

43. X ray crystallography allows investigators to determine
 a. the sequence of amino acids in proteins.
 b. the sequence of nucleic acids in DNA.
 c. a molecule's three-dimensional shape.
 d. the age of fossils.
 e. the composition of ancient bones.

44. The invariant positions 14, 17, 18, and 80 in cytochrome *c* interact with
 a. the positions on the molecule that change rapidly.
 b. the other invariant positions.
 c. the heme group.
 d. hemoglobin.
 e. myoglobin.

45. If there were mutations in the invariant positions of a cytochrome *c* molecule, what would be the result?
 a. The molecule would have a new function.
 b. The molecule would no longer be functional.
 c. The molecule would still bind the iron-containing group.
 d. The changes would not affect the function of the molecule.
 e. The molecule would bind to another metal-containing group.

46. Genes that are nonfunctional and probably resulted from gene duplication are called
   a. invariant genes.
   b. homeobox genes.
   c. homologous genes.
   d. pseudogenes.
   e. exons.

47. Which of the following is *not* true of hemoglobin?
   a. It is a simpler molecule than myoglobin.
   b. It binds four molecules of oxygen.
   c. It is a more refined molecule than myoglobin.
   d. It has a lower affinity for oxygen than myoglobin.
   e. It is a tetramer of two α and two β chains.

48. By studying the similarities of molecules between species, scientists can
   a. determine common ancestry.
   b. determine the position of branches in a cladogram.
   c. determine a similar function in widely divergent species.
   d. a and b
   e. a, b, and c

49. Which of the following must be true in order to use molecular clocks to date events?
   a. There must be relatively few adaptive changes in the molecule.
   b. Scientists must know what proportion of molecular clocks tick at a constant rate.
   c. The molecular clock of the molecule being studied must tick very slowly.
   d. a and b
   e. a, b, and c

50. Ancient lineage divisions can be estimated by
   a. molecular data.
   b. morphological data.
   c. fossil data.
   d. a and c
   e. a, b, and c

Study Guide Questions

1. Which of the following statements about mutations is *false*?
   a. A silent substitution results in no change in the amino acid sequence of a protein.
   b. Since most substitution mutations are selectively neutral, most evolutionary changes in macromolecules are the result of random genetic drift.
   c. A substitution mutation in the third codon position is more likely to be neutral than a substitution at the first or second codon position.
   d. Nonsynonymous substitutions are rarely, if ever, selectively neutral.

2. Which of the following statements about gene families is *false*?
   a. Gene families evolve via gene duplication.
   b. Pseudogenes are quickly removed from gene families by deletion.
   c. Members of a gene family can include several functional genes.
   d. Examples of gene families include the homeotic gene complex in *Drosophila* and the alpha hemoglobin family in humans.

3. Which of the following is *not* a valid conclusion based on studies of the structure of proteins, such as cytochrome *c*, in different species?
   a. The rate of accumulation of neutral substitutions is based on the mutation rate.
   b. Many nucleotide substitutions result in neutral changes in the amino acid sequence of the protein.
   c. Fewer differences were observed in the amino acid sequences of a protein if the organismal sources of the proteins were closely related.
   d. Functionally important regions of a protein can be discovered by identifying the regions with the most amino acid substitutions.

4. Which of the following statements about molecular clocks is *false*?
   a. The molecular clock of cytochrome *c* ticks more slowly than that of rRNA.
   b. The tick rate of a clock for a molecule with great functional constraints would be very slow.
   c. The tick rate of a molecular clock that mostly reflects neutral changes is determined by the mutation rate.
   d. The tick rate of a molecular clock that mostly reflects neutral changes is directly related to the slope of the line showing the relationship between the time of divergence of different taxa and amino acid sequence differences for the protein.

5. Select the correct statement about the enzyme lysozyme.
   a. A small group of closely related mammals have evolved a special form of lysozyme that functions in digestion.
   b. The lysozymes found in the foregut fermenters resulted from convergent evolution.
   c. Lysozyme could not have evolved a secondary function if it had been an enzyme with a vital primary function.
   d. A higher mutation rate in the foregut fermenters allows their lysozymes to evolve rapidly.

6. Which of the following sequences of organisms ranks them from least to most, in terms of the expected total amount of coding DNA in their genomes?
   a. Bacterium, single-celled eukaryote, roundworm, insect, bird
   b. Bacterium, roundworm, bird, insect, single-celled eukaryote
   c. Single-celled eukaryote, bacterium, roundworm, insect, bird
   d. Roundworm, single-celled eukaryote, insect, bird, bacterium

7. Which of the following molecules would be best to use in a study of the domains of the phylogenetic tree of life?
   a. Lysozyme
   b. Mitochondrial DNA
   c. rRNA
   d. Myoglobin

End of Chapter Questions

1. Questions about the process of molecular evolution address
   a. whether molecules evolve.
   b. how molecules change with time.
   c. how to reconstruct phylogenies of extinct organisms.
   d. the evolutionary relationships among molecules.
   e. the origin of organismal diversity.

2. Molecular evolution differs from phenotypic evolution in which of the following ways?
   a. It requires changes in molecules if it is to happen.
   b. Random genetic drift and mutations exert important influences on rates and directions of molecular evolution but not on rates and directions of phenotypic evolution.
   c. Molecular evolution is not influenced by natural selection.
   d. Rates of molecular evolution are much slower than rates of phenotypic evolution because mutation rates typically are low.
   e. There are no important differences between molecular and phenotypic evolution.

3. Choosing the appropriate molecule for phylogenetic reconstruction does *not* require a consideration of the
   a. question being answered.
   b. rate of evolution of the molecule.
   c. phylogenetic distribution of the molecule.
   d. function of the molecule.
   e. completeness of the fossil record.

4. Ribosomal RNA sequences are useful for addressing the evolutionary relationships of highly divergent molecules because they
   a. evolve at a rapid rate.
   b. have undergone convergent evolution in many lineages.
   c. are molecules that all organisms have.
   d. consist of mainly neutral characters.
   e. are difficult to align.

5. Mitochondrial DNA sequences are useful in studying the recent evolution of closely related species because they
   a. accumulate mutations very rapidly.
   b. are paternally inherited.
   c. evolve only in a neutral fashion.
   d. are highly constrained in function.
   e. are easy to sequence.

6. Questions about the pattern of molecular evolution focus on the
   a. evolutionary relationships among molecules.
   b. molecular clock.
   c. rate of mutation for neutral characters.
   d. importance of gene duplication in evolution.
   e. neutral theory of evolution.

7. Molecules are used to reconstruct phylogenies, even if a fossil record is available, because
   a. the more characters the better.
   b. they are more accurate characters than are fossils.
   c. they undergo less homoplasy than do fossil characters.
   d. they are less subjective characters than are fossils.
   e. they give us the "right" phylogeny.

8. Neutral characters
   a. are not evolving under the influence of positive selection.
   b. have a neutral pH.
   c. are not useful in reconstructing phylogenies.
   d. are subject to strong functional constraints.
   e. are not likely to evolve.

9. Jurassic Park is unlikely ever to be a reality because
   a. we cannot obtain DNA from ancient organisms.
   b. we have no parks big enough for such large dinosaurs.
   c. although we can obtain DNA from dinosaur fossils, it is highly fragmented.
   d. the genetic code used by dinosaurs is different from all other organisms.
   e. we wouldn't know what to feed the dinosaurs.

10. The concept of a molecular clock implies that
    a. many proteins show a constancy in rate of change with time.
    b. organisms evolve at a constant rate.
    c. one can date evolutionary events with molecules alone.
    d. all molecules change at the same rate in evolution.
    e. we can predict how fast all genes will evolve.

11. The lysozyme story suggests that
    a. molecules cannot change function in evolution.
    b. selection does not act at the molecular level.
    c. molecules can help us understand the process of organismal evolution.
    d. all organisms are capable of fermenting bacteria.
    e. lysozyme has a very accurate molecular clock.

Student Web Site Self-Quiz Questions

1. Fossils of Neanderthal remains aged 30,000–100,000 years old show mtDNA sequences dissimilar to modern human mtDNA. Which of the following statements regarding these results is *true*?
   a. Neanderthal and human DNA are identical.
   b. Neanderthal and humans co-existed in ancient Europe.
   c. Humans are direct descendants of Neanderthals.

d. Neanderthals did not contribute to the human gene pool.
e. Neanderthals contributed few genes to the human gene pool.

2. Many changes in gene sequences do not alter the proteins encoded by the genes. This is known as silent substitution. Which of the following statements regarding silent substitution is *true*?
a. Silent substitutions accumulate more slowly than nonsynonymous substitutions.
b. Silent substitutions in protein sequences are influenced by natural selection.
c. Silent substitutions in protein sequence are not influenced by natural selection.
d. Nonfunctional changes to nucleotide sequences cannot be detected.
e. Molecular evolution closely mirrors phenotypic evolution.

3. The enzyme cytochrome *c* is found in cells of all eukaryotes, from yeast to chimpanzees and humans. Which of the following statements about the neutral evolution hypothesis is *not* supported by studies of cytochrome *c* and like molecules?
a. The functionally important regions of molecules can be discovered by identifying the nucleotide positions in the coding gene that differ from species to species.
b. Most nucleotide substitutions are neutral.
c. The rate of accumulation of neutral substitutions is not influenced by natural selection.
d. Invariant amino acid positions in cytochrome *c* from a different species determine the molecule's conformation.
e. Amino acids with side chains that interact with the heme group in cytochrome *c* are invariant.

4. Which of the following statements about gene families and the human globin family of genes is *false*?
a. Humans have three families of globin genes.
b. Gene families can arise by abnormal events during crossing over.
c. Gene families can arise as a result of the action of transposable elements.
d. The human globin family includes some pseudogenes.
e. Gene family members usually arise from different genes that have independently converged on some similar function.

5. Which of the following statements about homeotic genes in fruit flies is *false*?
a. All homeotic genes code for a polypeptide 60 amino acids in length called the homeobox.
b. All homeotic genes produce similar effects on development.
c. All homeotic genes specify body segmentation.
d. Genes with homeobox elements have been identified in a wide variety of eukaryotic organisms.
e. The homeotic and *Hox* genes occur in the same order along the chromosome.

6. Which of the following statements about gene duplication, pseudogenes, and the evolution of new molecular function is *false*?
a. Gene duplication is the first step in the evolution of new molecular function.
b. Pseudogenes consist of nonfunctional DNA.
c. Pseudogenes are created when the gene duplication mechanism is faulty.
d. Pseudogenes can be eliminated by deletion.
e. Natural selection acts on the effects of random point mutation on gene copies within a gene family.

7. Ruminants, such as cattle, leaf-eating monkeys called langurs, and a bird called the hoatzin are all foregut fermenters that make use of the protein lysozyme. Which of the following statements about foregut fermentation is *false*?
a. Lysozyme is found in almost all animals, but it has a new function in foregut fermenters.
b. The lysozymes of langurs and cows have five amino acid substitutions, suggesting that they share a recent common ancestor.
c. The presence of protein-digesting enzymes in the foregut led to understandable changes in the amino acid composition of the lysozymes of foregut fermenters.
d. In addition to cows and langurs, hoatzins also utilize lysozyme in their foreguts.
e. In the foregut fermenters, the function of lysozyme has switched from defense against bacteria to digestion of food.

8. Which of the following organisms has the greatest ratio of coding DNA to total DNA?
a. *E. coli*
b. Yeast
c. Roundworm
d. Lungfish
e. Human

9. Ribosomal RNA in combination with various proteins forms the ribosomal subunits. In which of the following situations would it be *most* appropriate to use rRNA as a phylogenetic tool?
a. To determine the possible occurrence of convergence among different hot spring (thermophilic) bacteria
b. To study a molecular clock that ticks very rapidly
c. To study a molecule with few functional constrains
d. To conduct phylogenetic studies at the level of domains
e. To test the "out of Africa" hypothesis

10. Which of the following statements about flightless bird phylogeny is *false*?
a. DNA from extinct organisms was used in helping to develop a phylogeny for flightless birds.
b. Both geography and DNA argue for the relatedness of the moa and the kiwi.
c. All six flightless birds share a common ancestor.
d. The moa was hunted to extinction about 1,000 years ago.
e. Continental drift was a factor in establishing the current distribution of flightless birds.

11. Which of the following statements about the insights that molecular evolution has provided on human origins is *false*?
    a. Studies suggest a common ancestry of all human mtDNA about 200,000 years ago.
    b. mtDNA is useful for studies of human origins because it accumulates mutations rapidly.
    c. The mtDNA sequence divergence was calibrated using mammals with a good fossil record.
    d. Although the mtDNA studies support the "Out of Africa" hypothesis, studies of nuclear genes support the "parallel" origins hypothesis.
    e. Because mtDNA shows maternal inheritance, it is not subject to genetic recombination.

TEST FILE

★ *Indicates a* ***difficult*** *question*

■ *Indicates a* ***conceptual*** *question*

25 The Origin of Life on Earth

Fill in the Blank

1. **Louis Pasteur** performed an experiment in 1862 that finally convinced most people that spontaneous generation does not occur.

2. **Stanley Miller** was one of the first scientists to perform experiments simulating conditions similar to those believed to have prevailed on Earth when life first evolved.

3. The formation of living organisms from nonliving matter, often called **spontaneous generation**, was accepted by most people as an obvious fact of nature for more than 2,000 years.

4. **Coacervates** are drops that spontaneously form in solutions and have several properties relevant to the origin of life. They can contain enzyme molecules, absorb substrates, catalyze reactions, and let the products diffuse back out into the solution.

5. The universe is estimated to be between **10 and 20 billion** years old.

6. Life on Earth is thought to have arisen **1 billion** years after the planet was formed.

7. Oxygen-producing photosynthesis first evolved among prokaryotes called **cyanobacteria**.

8. Molecules that absorb visible light or near-UV light are called **chromophores**.

9. Before life existed on Earth, free oxygen was not available for chemical reactions in the atmosphere; this condition is known as a(n) **reducing** environment.

10. **Ribozymes** were believed to be the first catalysts for synthesizing proteins in early life forms.

11. The larger, eukaryotic organisms were able to flourish only after **oxygen** levels increased.

12. Present day Earth has a(n) **oxidizing** atmosphere.

13. **Replicators** are self-producing molecules.

14. **Protobionts** are aggregates of artificially produced prebiotic molecules that can maintain internal chemical environments that are different than their surroundings.

15. **RNA** molecules have been shown to be able to catalyze the formation of nucleotide polymers from a solution of purine and pyrimidine molecules.

Multiple Choice

1. According to the "big bang" and related theories, our part of the universe developed in the following order:
   a. galaxy, sun, Earth.
   b. sun, Earth, galaxy.
   c. Earth, galaxy, sun.
   d. sun, galaxy, Earth.
   e. galaxy, Earth, sun.

2. Earth's crust is thinnest under the __________ and has a density that is __________ Earth's mantle.
   a. continents, less than
   b. continents, more than
   c. continents, the same as
   d. oceans, less than
   e. oceans, more than

3. Earth's greatest concentration of iron and nickel is found in its
   a. soils.
   b. crust.
   c. mantle.
   d. outer liquid core.
   e. central core.

4. Earth's early atmosphere contained all of the following gases *except*
   a. ammonia.
   b. carbon dioxide.
   c. methane.
   d. nitrogen.
   e. oxygen.

5. The first seas formed from the
   a. upwelling of liquid water from vents in Earth's crust.
   b. condensation of water vapor from the atmosphere.
   c. melting of polar ice caps.
   d. formation of water from hydrogen and oxygen gases.
   e. Earth's assimilation of asteroids composed of ice.

6. In the presence of oxygen, most anaerobic bacteria
   a. are unaffected.
   b. switch to aerobic respiration.
   c. are poisoned.
   d. convert the oxygen to carbon dioxide.
   e. are stimulated.

7. A microsphere is a type of
   a. enzyme.
   b. catalyst.
   c. life.
   d. protobiont.
   e. chromophore.

8. Which of the following types of molecules could be considered a prebiotic replicator?
   a. DNA
   b. RNA
   c. Protein
   d. Carbohydrate
   e. Lipid

9. Stanley Miller showed that, given conditions similar to those of early Earth,
   a. inorganic molecules would react to form organic molecules.
   b. RNA would self-replicate.
   c. organic molecules would form primitive cells.
   d. an oxygen atmosphere would develop.
   e. DNA would be synthesized.

10. Polymerization of molecules is *not* associated with
    a. solutions of hydrogen cyanide.
    b. drying bodies of water-containing organic molecules.
    c. solutions of formaldehyde.
    d. monomers attached to phosphates.
    e. solutions of glucose.

11. A ribozyme is a(n)
    a. small organelle found in all cells.
    b. enzyme that synthesizes ribose sugar.
    c. RNA molecule that catalyzes reactions.
    d. type of proto-cell containing both sugars and nucleic acids.
    e. phosphorylated ribose solution capable of polymerization.

12. In which order did the following evolutionary events most likely occur?
    1. RNA catalyzes the synthesis of lipids and proteins.
    2. Proteins catalyze chemical reactions.
    3. RNA catalyzes its own replication.
    a. 1, 2, 3
    b. 2, 3, 1
    c. 3, 2, 1
    d. 2, 1, 3
    e. 3, 1, 2

13. Which of the following is *not* a characteristic of coacervates?
    a. They have lipid coats.
    b. Their interiors are high in protein.
    c. They can catalyze reactions.
    d. They contain DNA.
    e. They can absorb substrate molecules.

14. The theory that DNA evolved after RNA is supported by the fact that
    a. DNA is stable only in hydrophobic environments.
    b. not all modern cells have DNA, but all have RNA.
    c. DNA is synthesized by RNA in modern cells.
    d. DNA is more similar among different groups of organisms than RNA is.
    e. DNA is found in cells' nuclei, and early cells lacked nuclei.

15. The _________ principle acknowledges that prebiotic processes should leave some traces in contemporary biochemistry.
    a. uncertainty
    b. equilibrium
    c. continuity
    d. no-free-lunch
    e. signature

16. Cyanobacteria liberate oxygen molecules by splitting molecules of
    a. carbon dioxide.
    b. glucose.
    c. sulfate.
    d. nitrate.
    e. water.

17. The "Milky Way" refers to our
    a. solar system.
    b. galaxy.
    c. universe.
    d. nebula.
    e. constellation.

18. Our solar system, including the sun, Earth, and our sister planets, took form about
    a. 10–20 billion years ago.
    b. 5 billion years ago.
    c. 3.5 billion years ago.
    d. 5 million years ago.
    e. 3.5 million years ago.

19. Lightning and other energy sources converted early atmospheric gases into
    a. proteins.
    b. DNA.
    c. simple organic molecules.
    d. RNA.
    e. chlorophyll *a*.

▪20. Which of the following types of information is *not* used as a basis of plausible theories of the origins of life?
    a. The preserved remains of early organisms
    b. The study of a single common ancestor of all present-day organisms
    c. Laboratory experiments investigating chemical reactions under conditions similar to those thought to have existed on early Earth
    d. The observation that all organisms share the same genetic machinery

e. The observation that all organisms have similar basic cellular metabolic pathways

21. The __________ principle states that all living organisms require some from of energy for growth.
a. no-free-lunch
b. signature
c. continuity
d. parallel
e. None of the above

22. Coacervate drops that contain enzymes
a. do not preferentially concentrate substances.
b. have lipids in their interiors.
c. have polysaccharide boundaries with membrane-like structures.
d. absorb substrates, catalyze reactions, and release products back into the solution.
e. are usually unstable.

23. Oparin believed that coacervate drops were possible precursors to cells; however, since they lack __________ outer membranes, they differed from the probable precursors of life.
a. lipid
b. carbohydrate.
c. protein
d. nucleic acid
e. none of the above

*▪24. A problem central to understanding the evolution of life is to determine how the process of transcription of DNA into RNA and the subsequent translation of RNA into proteins evolved. That process provides
a. stability of proteins.
b. a hereditary basis for metabolic pathways.
c. versatility of nucleic acids.
d. a mechanism of nutrient cycling.
e. the ability to use RNA as the hereditary molecule.

25. The oldest remains of living things discovered so far are
a. 3.5 billion years old.
b. 350 million years old.
c. 35 million years old.
d. 3.5 million years old.
e. 5 million years old.

26. Of the following, the first bacteria to increase the oxygen content of the early atmosphere were the
a. archaebacteria.
b. cyanobacteria.
c. green sulfur bacteria.
d. purple sulfur bacteria.
e. purple nonsulfur bacteria.

27. During the first billion years of life on Earth, living things included all of the following *except*
a. anaerobic photosynthesizers.
b. heterotrophic bacteria.
c. sulfate reducers.
d. eukaryotes.
e. archaebacteria.

28. Louis Pasteur designed an experiment to prove that
a. bacterial organisms cannot be killed by heat.
b. life does not arise spontaneously from nonliving matter.
c. Earth was really much older than people of the time thought.
d. the half-life of uranium-238 is 10 billion years.
e. maggots grow in meat.

29. Coacervates are
a. the first catalysts for synthesizing proteins.
b. a mixture of protein and polysaccharides in an aqueous drop believed to be precursors to cells.
c. the fossils of what is believed to be the earliest life form.
d. molecules that are precursors to DNA.
e. early organisms that obtain energy without using oxygen.

30. Two organisms responsible for creating the atmospheric environment conducive to life today are
a. cyanobacteria and anaerobic bacteria.
b. anaerobic bacteria and archaebacteria.
c. filamentous bacteria and archaebacteria.
d. cyanobacteria and coacervates.
e. coacervates and blue-green algae.

31. Which of the following is *not* a benefit of reproduction by sexual recombination?
a. Better repair of damaged chromosomes
b. Better adaptability to different environments
c. Faster reproduction
d. Greater variation in offspring
e. Longer evolutionary life

▪32. Life forming from nonlife, and polymerization of biological molecules, does not happen today because
a. there is too much oxygen in the atmosphere.
b. cyanide and formaldehyde are present in lower concentrations today.
c. those biological molecules are being consumed by existing life.
d. the temperature of Earth is much lower today, so those reactions cannot take place.
e. All of the above

33. The key change necessary for aerobic life to evolve was
a. the ability of organisms to use water as a source of hydrogen.
b. Earth's reducing environment becoming an oxidizing environment.
c. the ability of organisms to break down carbon dioxide.
d. the ability of organisms to break down methane.
e. the extinction of anaerobic bacteria.

34. Fossils of cyanobacteria are called __________ and are still forming in salty areas of Earth today.
a. stromatolites
b. stalactites
c. carotenoids
d. coal
e. carbon deposits

35. Which is the correct order, from surface inward, of the layers of Earth?
a. Core, mantle, crust
b. Mantle, core, crust
c. Mantle, crust, core
d. Crust, core, mantle
e. Crust, mantle, core

36. Before life evolved, a new atmosphere on Earth was created by the release of __________ by the crust and mantle.
a. carbon dioxide
b. nitrogen
c. heavy gases
d. a and b
e. a, b, and c

37. Polymers that came to predominate in the prebiotic soup
a. formed faster than others.
b. were more stable than others.
c. were larger molecules than others.
d. a and b
e. a, b, c

*▪38. The synthesis of sugars was important to early life because they
a. are components of nucleotides.
b. form cell walls.
c. form alcohols and carboxylic acids.
d. are components of lipids.
e. catalyzed the first chemical reactions.

39. In RNA polymerization experiments, when __________ is added to the reaction, larger sequences of RNA form.
a. ribonuclease A
b. DNA
c. zinc
d. nickel
e. tRNA

40. Which of the following is *not* true of proteins as enzymes?
a. They are better catalysts than RNA.
b. They are capable of more diverse specifications.
c. They took over most enzymatic functions from RNA.
d. They catalyze the actions of ribozymes.
e. They are translated by RNA.

41. Aerobic metabolism __________ than anaerobic metabolism.
a. is faster
b. is more efficient
c. uses more energy
d. a and b
e. a and c

Study Guide Questions

1. Which of the following is *not* evidence of the conservatism of biochemical evolution?
a. The storage and release of energy in the living world is associated with the formation and hydrolysis of phosphate bonds.
b. All proteins are made from the same group of amino acids.
c. The first genetic code was based on DNA.
d. Ribosomes are the sites of protein synthesis in all cells.

2. Which of the following gases was probably *least* abundant in Earth's early atmosphere?
a. CO_2
b. NH_3
c. H_2
d. O_2

3. Experiments to synthesize organic compounds under simulated early Earth conditions have demonstrated that
a. organic compounds can be formed in atmospheres containing a variety of gases, provided oxygen is absent.
b. a variety of organic compounds can be formed in these experiments, including amino acids found in most organisms.
c. organic compounds cannot be formed in experiments that simulate an aquatic environment.
d. a and b

4. Concerning the earliest catalysts, scientists studying the origin of life believe that
a. the first catalysts were nonproteins and lacked the specificity shown by enzymes.
b. DNA had no catalytic function, either before or after the evolution of enzymes.
c. RNA had important catalytic functions before the evolution of enzymes.
d. **All of the above**

5. Which of the following statements about coacervates is *false*?
a. Coacervates are clumps of protein and polysaccharide in an aqueous environment.
b. If artificial coacervates contain nucleic acids, they can accurately reproduce themselves.
c. Oparin was able to create coacervates with the ability to absorb substrates and catalyze reactions.
d. Because Oparin's artificially produced coacervates lacked lipid outer membranes, they differed from the probable precursors of the first cells.

6. Which of the following was *not* presented in the text as evidence that catalysis by RNA preceded catalysis by enzymes?
a. RNA is only stable within membrane-enclosed cells, where the water concentration is lower than the surrounding environment.
b. RNA can excise its own introns and splice together its exons.
c. Ribosomes contain a catalytic RNA that functions in protein synthesis.
d. RNA can catalyze the polymerization of RNA.

7. The earliest organisms were probably
   a. able to use the small quantity of oxygen available in the atmosphere for their metabolism but also able to survive without oxygen.
   b. unable to live in the absence of oxygen.
   c. unable to live in the presence of oxygen.
   d. like present-day cyanobacteria.

8. As a result of the evolution of photosynthesis using water as an electron donor,
   a. Earth developed an oxidizing atmosphere.
   b. organisms using aerobic metabolism became predominant.
   c. the atmospheric concentration of CO_2 increased.
   d. a and b

9. Which of the following represents a correct (though not necessarily complete) sequence of events in the origin of life, from earliest to latest?
   a. Organic monomers accumulate; protobionts evolve into cells; DNA becomes the genetic material; oxygen accumulates in the atmosphere.
   b. Monomers are polymerized; ribozymes appear; oxygen accumulates in the atmosphere; protobionts evolve into cells.
   c. Organic monomers accumulate; cyanobacteria appear; oxygen accumulates in the atmosphere; DNA becomes the genetic material.
   d. Monomers are polymerized; DNA becomes the genetic material; ribozymes appear; protobionts evolve into cells.

10. Any planet that supports life will almost certainly have
   a. liquid water on its surface.
   b. a constant supply of radiant energy from a star.
   c. an atmosphere containing oxygen.
   d. a and b

End of Chapter Questions

1. The atmosphere of early Earth consisted largely of
   a. water vapor.
   b. hydrogen.
   c. carbon dioxide.
   d. helium.
   e. nitrogen.

2. The principle of continuity states that
   a. all postulated states in life's evolution should be derivable from preexisting states.
   b. most postulated states in life's evolution should be derivable from preexisting states.
   c. because of historical continuity, prebiotic processes should be preserved in contemporary biochemistry.
   d. all organisms must oxidize some material and obtain energy from that oxidation.
   e. studies of life's origin are inevitably highly speculative.

3. To determine which molecules might have formed spontaneously on early Earth, Stanley Miller used an apparatus with an atmosphere containing
   a. oxygen, hydrogen, and nitrogen.
   b. oxygen, hydrogen, ammonia, and water vapor.
   c. oxygen, hydrogen, and methane.
   d. hydrogen, oxygen, and carbon dioxide.
   e. hydrogen, ammonia, methane, and water vapor.

4. Most biologists think that RNA was the first genetic material because
   a. amino acids were produced in Stanley Miller's apparatus.
   b. DNA is the universal genetic material of eukaryotes.
   c. the existence of ribozymes suggests that early cells could have used RNA to catalyze chemical reactions and transfer information.
   d. RNA is simpler than DNA.
   e. DNA is not stable in hydrophobic environments.

5. Biologists believe that the current DNA → RNA → protein system is the result of a long period of evolution because
   a. the transcription of DNA to mRNA and translation of mRNA into proteins consists of many steps.
   b. DNA replication is complicated but relatively error-free.
   c. the current system is very complex and precise.
   d. evidence indicates that RNA preceded DNA as the genetic material.
   e. All of the above

6. The answer to which of the following questions enabled research on the origin of catalysis to proceed rapidly?
   a. How could complex life have evolved on such a young Earth?
   b. How could the precise duplication of DNA have evolved?
   c. How could catalysis have evolved, given that RNA needs proteins for its synthesis and proteins need RNA for their synthesis?
   d. How could eukaryotes evolve from prokaryotes?
   e. How did the first cells form?

7. The key process in the formation of nucleic acid bases is the
   a. polymerization of hydrogen cyanide.
   b. polymerization of formaldehyde.
   c. spontaneous formation of monomers.
   d. spontaneous formation of proteins.
   e. polymerization of proteins.

8. The metabolism of living prokaryotes provides important insights into the chemical processes used by early organisms because
   a. many prokaryotes live in environments similar to those in which life first evolved.
   b. prokaryotes are simpler to study and hence are better known than are eukaryotes.
   c. many prokaryotes are obligate aerobes.
   d. many prokaryotes use oxygen as their oxidizing agent.
   e. fermentation evolved before aerobic respiration.

9. Pasteur's experiments and similar ones that followed convinced most people that spontaneous generation of life did not happen because
   a. Pasteur was extremely meticulous.
   b. Pasteur used very fine mesh screens to cover his flasks.
   c. Pasteur did not boil his flasks for a long time.
   d. Pasteur's swan-necked flasks ruled out the objection that spoiled air could have contaminated his experiments.
   e. by the time Pasteur performed his experiments, many people no longer believed in spontaneous generation.

10. Many scientists think that complex multicellular life may be rare in the universe because
   a. if complex life existed elsewhere, we would already have discovered it.
   b. only on Earth do conditions suitable for the origin of life exist.
   c. it is better to assume that something is rare if we have no direct evidence of it.
   d. liquid water does not exist on the surface of many planets.
   e. the conditions necessary for multicellular life to evolve and survive are much more stringent than the conditions under which unicellular life could evolve.

Student Web Site Self-Quiz Questions

1. When do scientists believe life first appeared on Earth?
   a. 2 million years ago
   b. 4 million years ago
   c. 4 billion years ago
   d. 3.5 billion years ago
   e. 3.5 million years ago

2. Which of the following gases was *not* present in Earth's atmosphere 3.8 billion years ago?
   a. Methane (CH_4)
   b. Ozone (O_3)
   c. Carbon dioxide (CO_2)
   d. Ammonia (NH_3)
   e. Water vapor (H_2O)

3. Which of the following is *not* one of the guiding scientific principles that apply to the origin of life?
   a. Living organisms require energy for growth.
   b. Current biochemical processes contain traces of ancient processes.
   c. All living forms have descended from the first forms of life.
   d. Large, sudden changes in the Earth's atmosphere contributed to the development of life.
   e. Living forms are based on aqueous solutions.

4. Which of the following is *not* one of the essential characteristics of life?
   a. Life is based on carbon, hydrogen, nitrogen, oxygen, phosphorous, and sulfur.
   b. Energy flow is based upon the ATP molecule.
   c. DNA or RNA is the hereditary material
   d. While proteins are made from the same group of amino acids, RNA and DNA can interchange nucleotide subunits.
   e. All living things are based upon the unit of the cell.

5. Many events help to shape the characteristics of Earth's environments, including volcanic eruptions. Which of the following is *not* a fundamental difference between the early atmosphere of Earth and the present-day atmosphere?
   a. The early atmosphere was reducing.
   b. The present day atmosphere is oxidizing.
   c. Oxygen was present primarily in water in the early atmosphere.
   d. Oxygen molecules are more abundant in the present atmosphere.
   e. Oxygen atoms were not present in the early atmosphere.

6. The chemical reactions that are believed to have occurred during the formation of "life" molecules in the early atmosphere of Earth can be carried out in the lab. Which of the following is *not* a condition required for the development of life?
   a. Replicators must be present.
   b. Small molecules must be present to generate energy.
   c. Replicators must be subject to mutation during replication.
   d. Prebiotic molecules must be formed before any other molecules.
   e. Replicators require partial isolation from the general environment.

7. Which of the following statements about ribozymes is *false*?
   a. Ribozymes are capable of self-replication, but cannot catalyze other types of reactions.
   b. Manfred Eigen found that RNA's could replicate themselves in solution without the aid of proteins.
   c. In *Tetrahymena thermophila*, an intron was found that carries out its own excision and splicing.
   d. A tRNA-processing enzyme containing RNA was found in which the RNA portion provided the catalysis.
   e. "Evolution" of RNA in the test tube has produced ribozymes with reaction rates seven million times faster than the uncatalyzed rate.

8. Select the *false* statement from the following descriptions of the importance of sugars to the prebiotic synthesis of organic compounds.
   a. The first polymers of nucleic acids were based on deoxyribose.
   b. Sugars are produced by the polymerization of formaldehyde.
   c. Five- and six-carbon sugars break down into alcohol and carboxylic acids.
   d. Three- and four-carbon sugars are very stable.
   e. Organic molecules, including sugars, accumulated in the oceans and were concentrated in drying pools.

9. Which of the following statements about the evolution of the central dogma is *false*?
   a The central dogma shows that nucleic acids are required in the synthesis of protein catalysts (enzymes).
   b. Present-day replication, transcription, and translation require protein catalysts.
   c. RNA preceded DNA as the hereditary molecule.
   d. DNA evolved as the hereditary molecule with the appearance of the eukaryotic cell.
   e. RNA catalysis preceded enzyme catalysis.

10. Fossil stromatolites were formed by organisms most like present-day
    a. anaerobic photosynthetic prokaryotes.
    b. cyanobacteria.
    c. green sulfur bacteria.
    d. purple sulfur bacteria.
    e. purple nonsulfur bacteria.

11. Which of the following statements regarding stromatolite-forming cyanobacteria is *false*?
    a. Cyanobacteria appeared following the release of oxygen in the atmosphere.
    b. Cyanobacteria developed the ability to split water.
    c. Cyanobacteria paved the way for the development of the respiratory chain of biochemical reactions.
    d. Cyanobacteria developed a system for reducing carbon dioxide.
    e. Cyanobacteria were the first organisms to be able to use water as a hydrogen source.

12. In the experiments done by Louis Pasteur on spontaneous generation, what was the purpose of the "swan" necks of the flasks he used, and what did the flasks contain?
    a. Keep out oxygen, water
    b. Keep out oxygen, nutrient medium
    c. Keep out oxygen, bacteria culture
    d. Keep out microbes, nutrient medium
    e. Keep out microbes, bacteria culture

13. The Earth has a variety of habitats that are capable of supporting life. Which of the following statements is *not* a requirement for sustaining multicellular life on a planet?
    a. A constant energy source
    b. A planet with associated nearby planets
    c. A planet with a near circular orbit
    d. A planet that is third from a star in a solar system
    e. A planet with an orbiting moon

TEST FILE

★ *Indicates a* ***difficult*** *question*

■ *Indicates a* ***conceptual*** *question*

26 Bacteria and Archaea: The Prokaryotic Domains

Fill in the Blank

1. The most abundant organisms on Earth are bacteria that fall into the kingdoms Archaebacteria and **Bacteria**.
2. Bacteria unable to survive for extended periods in the absence of oxygen are termed **obligate aerobes**.
3. Photoautotrophs use light as their source of energy and **carbon dioxide** as their source of carbon.
4. With respect to the four nutritional categories of bacteria, most bacteria are **chemoheterotrophs**, as are all animals, fungi, and most protists.
5. **Hans Gram** developed a staining process for bacteria in 1884 that is still the single most common tool in the study of bacteria.
6. **Invasiveness** is the ability of a bacterial pathogen to enter into and multiply within the body of the host.
7. **Toxigenicity** is the ability of a bacterial pathogen to produce chemical substances injurious to the tissues of the host.
8. Archaea are closer, evolutionarily, to **eukarya** than to bacteria.
9. *Treponema pallidum*, the causative agent of syphilis, belongs to the group of bacteria known as the **spirochetes**.
10. *Agrobacterium tumefaciens* is an important bacteria, useful for inserting genes into **plant** hosts.
11. One of the smallest bacteria, **chlamydia**, has a complex intracellular reproductive cycle.

*12. The important bacteria *Escherichia coli* has a **rod** shape and a **negative** Gram stain.

13. **Actinomycetes** have a branched system of filaments and were once thought to be fungi.
14. Many Gram-negative bacteria have **endotoxins** comprised of lipopolysaccharides that can cause fever and vomiting.
15. Some Gram-positive bacteria have potent **exotoxins** made of proteins that can be fatal.
16. Gram-positive bacteria stain **violet**, and Gram-negative stain **pink-red**.

Multiple Choice

1. The largest bacteria is
   a. 0.2 µm.
   b. 2 µm.
   c. 7.5 µm.
   d. 75 µm.
   e. 750 µm.

2. Which of the following is *not* a basic unit of a prokaryotic cell?
   a. DNA
   b. RNA
   c. Enzymes for transcription and translation
   d. A system for generating ATP
   e. An immune system

3. Which of the following would be found in both eukaryotes and prokaryotes?
   a. Organelles
   b. A system for generating ATP
   c. A nucleus
   d. Chromatin
   e. Histones

4. Which of the following is *not* a characteristic of prokaryotic cells?
   a. Mesosome
   b. Photosynthetic membrane system
   c. Peptidoglycan
   d. Circular DNA
   e. Organelles

5. Among the modes of locomotion for bacteria are
   a. flagella, gas vesicles, and rolling.
   b. flagella, cilia, and axial filaments.
   c. axial filaments, rolling, and pseudopods.
   d. cilia, pseudopods, and axial filaments.
   e. pseudopods, flagella, and cilia.

6. Some cyanobacteria possess heterocysts, which are
   a. the method of locomotion for the bacteria.
   b. a cell type specialized for nitrogen fixation.
   c. a reproductive state.
   d. an endospore.
   e. None of the above

7. Bacteria participate in
   a. digestion in animals.
   b. processing nitrogen and sulfur in soils.
   c. decomposition in all ecosystems.
   d. many industrial and commercial processes.
   e. All of the above

8. The earliest fossils of bacteria date back at least __________ years.
   a. 35,000
   b. 350,000
   c. 3.5 million
   d. 3.5 billion
   e. 3.5 trillion

9. Which of the following is a broad nutritional category of bacteria that is recognized by biologists?
   a. Physioautotrophs
   b. Heteroautotrophs
   c. Chemoautotrophs
   d. a and b
   e. a and c

▪10. Bacteria differ from one another
   a. structurally.
   b. metabolically.
   c. reproductively.
   d. a and c
   e. a, b, and c

11. The highest categorization of life is
   a. bacteria, archaea and eukarya.
   b. bacteria, fungi, plants, and animals.
   c. bacteria, protists, fungi, planta, and animalia.
   d. planta and animalia.
   e. None of the above

12. Bacteria reproduce
   a. only asexually.
   b. only sexually.
   c. asexually and sexually by transformation, conjugation, and/or transduction.
   d. following mitosis.
   e. a and d

*13. Prokaryotic flagella
   a. are structurally related to those of eukaryotes
   b. are similar to cilia.
   c. are structurally unrelated to those of eukaryotes.
   d. operate the same way as those of spermatozoa.
   e. a, b, and d

*14. The space between the outer membrane and the cell wall of bacteria is called the space.
   a. periplasmic
   b. negative
   c. perimembrane
   d. enzymatic
   e. resistance

15. The chlorophyll of cyanobacteria is
   a. like that of plants.
   b. distinct from chlorophyll of plants.
   c. bacteriochlorophyll.
   d. bacteriorhodopsin.
   e. None of the above

16. The range of time between cell divisions for different bacteria in a vegetative state is
   a. 10 minutes to 100 years.
   b. 1 minute to 60 minutes.
   c. 10 minutes to 60 minutes.
   d. 20 minutes to 120 minutes.
   e. 1 minute to 1000 years.

*▪17. What was learned from sequencing rRNA's was
   a. the clear relationships of different prokaryotic species.
   b. unclear because of lateral gene transfer.
   c. that archaea are similar to bacteria.
   d. that all DNA is relatively the same.
   e. that half of the sequences of archaea were previously unknown.

18. Pathogenic bacteria often face difficulty establishing themselves in a host because of the host's immune system. The bacterium that causes diphtheria, *Corynebacterium diphtheriae,* is able to multiply only within the throat of the host. We say such a bacterium has
   a. low invasiveness.
   b. high invasiveness.
   c. low toxigenicity.
   d. high toxigenicity.
   e. moderate toxigenicity.

19. One factor that determines the consequences of a bacterial infection for the host is the ability of the bacterium to produce chemical substances injurious to the tissues of the host. The anthrax-causing bacterium, *Bacillus anthracis,* produces few toxins; however, it can multiply readily and ultimately invades the entire bloodstream. We say such a bacterium has
   a. low invasiveness and high toxigenicity.
   b. high invasiveness and high toxigenicity.
   c. low invasiveness and low toxigenicity.
   d. high invasiveness and low toxigenicity.
   e. moderate invasiveness and moderate toxigenicity.

20. Bacteria are known for the many roles they play in biological communities. Which of the following roles is the rarest for this group of organisms?
   a. Pathogens
   b. Digestive aids
   c. Nitrogen and sulfur processing in soils
   d. Decomposers
   e. Uses in industry and agriculture

21. Which one of the following is *not* a motivation for classification of biological organisms?
   a. Convenience
   b. Showing evolutionary affinity
   c. Facilitating identification

d. Displaying diversity
e. They are all motivations

*22. An ecosystem based on chemoautotrophs exists 2,500 meters below the ocean surface near the Galápagos Islands. These archaea
a. use light as energy and use carbon dioxide for carbon.
b. use light as energy and get organic compounds from other organisms.
c. oxidize inorganic substances for energy and use carbon dioxide for carbon.
d. get both energy and carbon from organic compounds.
e. oxidize organic compounds for energy and use carbon dioxide for carbon.

23. Which of the following is *not* a property of the Archaebacteria?
a. They include some species that live in environments with extreme salinity and low oxygen.
b. They have peptidoglycan in their cell walls.
c. Their rRNA is as different from that of other prokaryotes as from that of eukaryotes.
d. They include some species that are obligate anaerobes, which produce all the methane in the atmosphere.
e. They include some species that love heat and acid and may die of "cold" at 55°C (131°F).

24. Which of the following statements about bacteria is *true*?
a. Gram-positive bacteria have a lot of peptidoglycan in their cell walls and stain blue to purple.
b. Gram-positive bacteria have relatively little peptidoglycan in their cell walls and stain pink to red.
c. Gram-positive bacteria weigh more than a gram.
d. Gram-positive bacteria weigh more than a milligram.
e. Gram-negative bacteria have no peptidoglycan in their cell walls.

25–30. Match the following descriptions of groups of bacteria with their names from the list below.
a. Cyanobacteria
b. Spirochetes
c. Chlamydia
d. Gram-negative rods
e. Actinomycetes
f. Mycoplasmas

25. These Gram-positive bacteria which form mycelium, were once classified as fungi. They include the bacteria that cause tuberculosis and produce streptomycin. Most antibiotics come from bacteria in this group. **(e)**

26. These bacteria photosynthesize using chlorophyll *a*, are a homogeneous grouping with similar rRNA sequences, and contain photosynthetic lamellae or thylakoids. **(a)**

27. These bacteria use axial filaments to move and include the bacterium that causes syphilis. **(b)**

28. These bacteria are not a natural group with close evolutionary ties; there is much diversity among the members. **(d)**

29. These bacteria are small intracellular parasites that have a unique, complex reproductive cycle and include strains that cause eye infections and venereal disease. **(c)**

30. These lack cell walls and are among the smallest cellular creatures. They are mostly plant and animal parasites. **(f)**

31. A disease-causing bacterium kills many cells within the host, but spreads poorly from one individual to another. It can be said to have
a. low invasiveness.
b. high invasiveness.
c. high toxigenicity.
d. low invasiveness and high toxigenicity.
e. high invasiveness and high toxigenicity.

32. Which of the following is *not* a characteristic of prokaryotic cells?
a. Lack any internal membrane systems
b. Some have cell walls composed of peptidoglycan.
c. DNA not organized into chromatin
d. A circular chromosome
e. Cell division by binary fission

33. The purple sulfur bacteria use H_2S as an electron donor and release pure sulfur as a waste product. They are examples of
a. photoautotrophs.
b. photoheterotrophs.
c. chemoautotrophs.
d. chemoheterotrophs.
e. deep-sea, volcanic vent bacteria.

34. Which of the following nutritional categories of bacteria can exist independently of other organisms?
a. Photoautotrophs
b. Photoheterotrophs
c. Photochemotrophs
d. Chemoheterotrophs
e. More than one of the above

35. The majority of bacteria are
a. photoautotrophs.
b. photoheterotrophs.
c. chemoautotrophs.
d. chemoheterotrophs.
e. disease-causing.

36. The Gram method is useful for classifying all bacteria that
a. have two plasma membranes.
b. have cell walls.
c. have cell walls with at least some peptidoglycan.
d. are prokaryotic.
e. form endospores.

*37. Which of the following is *not* characteristic of endospores?
a. Parent cells can produce more than one.
b. Can survive harsh environmental conditions
c. Contain some cytoplasm and replicated nucleic acid
d. Enclosed within a tough cell wall
e. A resting structure, not a reproductive structure

38. Which of the following is *not* characteristic of Archaebacteria?
a. Live in harsh environments
b. Cell walls without peptidoglycan
c. Unlike most other bacteria
d. Similarities in base sequences of ribosomal RNA's
e. A recently evolved group

39. Which of the following areas or conditions would be favored by thermoacidophiles?
a. The stomachs of many herbivores
b. Hot, alkaline springs
c. Anaerobic conditions
d. Hot, sulfur springs
e. Deep-sea volcanic vents

*40. Which of the following is *not* characteristic of the methanogens?
a. Methane is their preferred carbon source.
b. They are associated with mammalian flatulence.
c. They prefer anaerobic conditions.
d. Some are thermophilic.
e. Some live in volcanic vents on the ocean floor.

41. If you were looking for a new heat-tolerant DNA polymerase enzyme, you would investigate
a. thermoacidophiles.
b. methanogens.
c. strict halophiles.
d. a and b
e. All of the above

42. Which one of the following bacterial groups includes the greatest number of species?
a. Archaebacteria
b. Proteobacteria
c. Gram-positive
d. Mycoplasmas
e. Cyanobacteria

43. Gliding bacteria
a. were once classified as fungi.
b. form erect fruiting structures.
c. include spirochetes.
d. are known to glide using cilia.
e. contain less DNA than any other organism.

44. Spirochetes
a. are all free-living.
b. are the only spiral-shaped bacteria.
c. can form chains of cells.
d. all possess structures called axial filaments.
e. are all parasites of other bacteria.

45. Gram-negative rods
a. are all closely related.
b. include many human pathogens, but also *Escherichia coli.*
c. can only reproduce within the cells of other organisms.
d. all possess structures called axial filaments.
e. include the important genera *Bacillus* and *Clostridium.*

46. Chlamydias
a. can form heterocysts.
b. form a branched, filamentous mycelium.
c. contain less DNA than any other organism.
d. can only reproduce within the cells of other organisms.
e. were once classified as fungi.

47. Cyanobacteria
a. are a diverse, unrelated group of bacteria.
b. use chlorophyll *a* and liberate oxygen during photosynthesis.
c. are the only group of photosynthetic bacteria.
d. can reproduce sexually.
e. are obligate aerobes.

48. Gram-positive cocci
a. include nitrogen-fixing bacteria, such as *Rhizobium.*
b. include the important genus *Staphylococcus.*
c. can produce extremely potent toxins.
d. form chains of cells by mitosis.
e. b and c

49. Actinomycetes
a. are photoheterotrophs.
b. are the source of many important antibiotics.
c. are Gram-negative bacteria.
d. were once classified as protists.
e. are the only bacteria to divide by mitosis.

*50. Mycoplasmas
a. contain less DNA than any other organism.
b. have peptidoglycan cell walls.
c. form a branched, filamentous mycelium.
d. possess elaborate internal membrane systems.
e. can be controlled by penicillin.

51. Which of the following is *not* one of Koch's postulates?
a. The microorganism is always found in the diseased individual.
b. The microorganism taken from the diseased host can be grown in pure culture.
c. A sample of the pure culture of the microorganism produces the disease when injected into an uninfected host.
d. A host infected by injection from the cultured microorganism yields a new culture of the microorganism identical to the original culture.
e. The microorganism can be transmitted by insect vectors.

52. An extremely important set of rules for the determination of bacterial disease transmission is
a. Ehrlich's optimal law.
b. Koch's postulates.
c. Occam's razor.
d. Zeno's paradox.
e. Darwin's law of evolution based on natural selection.

53. The pathogenic bacteria *Clostridium* and *Bacillus*
a. are Gram-positive cocci.
b. are Gram-negative cocci.
c. have endospores.

d. have endotoxins.
e. a and c

*54. Lateral gene transfer is when
a. scientists make transgenic organisms.
b. bacteria within a species exchange genes.
c. bacteria acquire DNA from a different species.
d. genes are passed to daughter cells.
e. None of the above

*55. Which of the following is distinctive to archaea?
a. Many of their genes
b. Two-hydrophilic-head lipid molecules
c. A lack of peptidoglycan
d. Lipids with glycerol-ester linkages
e. All of the above

Study Guide Questions

1. Which of the following characteristics are unique to prokaryotes?
a. Lack of membrane bound organelles
b Presence of cell walls
c. Plasma membranes
d. All of the above

2. A tooth scraping yields large numbers of corkscrew shaped bacteria. These bacteria are referred to as
a. bacilli.
b. cocci.
c. spirilli.
d. helici.

3. Which of the following statements concerning prokaryotes are *true*?
a. Because prokaryotes do not contain organelles they cannot photosynthesize or carry out cellular respiration.
b. Prokaryotes have no chromosomes and therefore lack DNA.
c. Prokaryote flagella are similar in structure to eukaryote flagella.
d. None of the above

4. Gram-negative cells stain pink because
a. they have specialized lipids in their cell walls.
b. their peptidoglycan layer is thin.
c. their peptidoglycan layer is thick.
d. they are receptive to antibiotics.

5. Which of the following statements concerning bacteria are *not* true?
a. Most bacteria are pathogenic.
b. Bacteria live in diverse environments.
c. Bacteria are derived from Archaea.
d. a and b

6. Determining evolutionary lineages of bacteria is difficult because
a. bacteria evolve rapidly due to short life spans.
b. the evolution of bacteria is greatly influenced by mutation due to their single chromosome.
c. Much of the data appears conflicting.
d. All of the above

7. You would find several pathogenic bacteria in which of the following groups of bacteria?
a. Proteobacteria
b. Cyanobacteria
c. Chlamydias
d. None of the above

8. Parasitic spirilli are found in which of the following bacteria groups?
a. Proteobacteria
b. Cyanobacteria
c. Spirochetes
d. None of the above

9. Autoclaves must pass a "spore test" in many states to be considered sterile. The spore producing bacteria they use for this test is most likely a member of which of the following groups of bacteria?
a. Firmicutes
b. Proteobacteria
c. Cyanobacteria
d. Chlamydia

10. Archaeans often live in harsh environments. An archaean that lives in extremely salty conditions is referred to as a
a. thermophile.
b. halophile.
c. salinophile.
d. None of the above

11. Methane gas is thought to be a threat to ozone. The primary source of methane emission is from grazing cattle. This is because cows harbor a methane producing archaean from which of the following groups?
a. Crenarcheaota
b. Euryarcheaota
c. Anarchaeota
d. Proteobacteria

12. One of the major diagnostic characteristics that distinguishes archaea from bacteria is lack of
a. peptidoglycan cell walls in archaea.
b. peptidoglycan cell walls in bacteria.
c. ribosomes in archaea.
d. chemoautotrophic bacteria.

13. A bacterium that requires a carbon source other than carbon dioxide, yet can convert light energy to chemical energy is called a
a. photoautotroph.
b. photoheterotroph.
c. chemoautotroph.
d. chemoheterotroph.

14. A bacterium that cannot live in the presence of oxygen is called a(n)
a. obligate aerobe.
b. facultative aerobe.
c. obligate anaerobe.
d. facultative anaerobe.

End of Chapter Questions

1. Most prokaryotes
   a. are agents of disease.
   b. lack ribosomes.
   c. evolved from the most ancient protists.
   d. lack a cell wall.
   e. are chemoheterotrophs.

2. The division of the living world into three domains
   a. is strictly arbitrary.
   b. was inspired by key visible differences between archaea and bacteria.
   c. emphasizes the greater importance of eukaryotes.
   d. was proposed by the early microscopists.
   e. is strongly supported by data on rRNA sequences.

3. Which statement about the archaean genome is *true*?
   a. It is much more similar to the bacterial genome than to eukaryotic genomes.
   b. More than half of its genes are genes that are never observed in bacteria or eukaryotes.
   c. It does not undergo mutation as readily as other domains.
   d. It is housed in the nucleus.
   e. No archaean genome has yet been sequenced.

4. Which statement about nitrogen metabolism is *not* true?
   a. Certain prokaryotes reduce atmospheric N^2 to ammonia.
   b. Nitrifiers are soil bacteria.
   c. Denitrifiers are strict anaerobes.
   d. Nitrifiers obtain energy by oxidizing ammonia and nitrite.
   e. Without the denitrifiers, terrestrial organisms would lack a nitrogen supply.

5. All photosynthetic bacteria
   a. use chlorophyll a as their photosynthetic pigment.
   b. use bacteriochlorophyll as their photosynthetic pigment.
   c. release oxygen gas.
   d. produce particles of sulfur.
   e. require a carbon source.

6. Gram-negative bacteria
   a. appear blue to purple following Gram staining.
   b. contain a periplasmic space.
   c. are all either rods or cocci.
   d. contain no peptidoglycan in their walls.
   e. are all photosynthetic.

7. Endospores
   a. are produced by viruses.
   b. are reproductive structures.
   c. are very delicate and easily killed.
   d. are dormant forms of bacteria.
   e. lack cell walls.

8. Actinomycetes
   a. are important producers of antibiotics.
   b. belong to the kingdom Fungi.
   c. are never pathogenic to humans.
   d. are gram-negative.
   e. are the smallest known bacteria.

9. Which statement about mycoplasmas is *not* true?
   a. They lack cell walls.
   b. They are the smallest known cellular organisms.
   c. They contain the same amount of DNA as do other prokaryotes.
   d. Some have a rigid material outside the plasma membrane for structure.
   e. Some are pathogens.

10. Archaea
    a. have cytoskeletons.
    b. have distinctive lipids in their plasma membranes.
    c. survive only at moderate temperatures and near neutrality.
    d. all produce methane.
    e. have substantial amounts of peptidoglycan in their cell walls.

11. Which of the following organism : description pairs is *incorrect*?
    a. Crenarchaeota : prefer acidic environments
    b. Euryarchaeota : produce methane as a metabolic product
    c. Methanogens : contain ester-linked, long-chain fatty acids
    d. Halophiles : prefer salty environments
    e. Thermophiles : undergo aerobic metabolism

Student Web Site Self-Quiz Questions

1. Placing both the bacteria and archaea into a single kingdom within a five-kingdom classification would result in a kingdom that
   a. is monophyletic.
   b. is polyphyletic.
   c. is paraphyletic.
   d. includes all the descendants of their common ancestor.
   e. does not include all prokaryotes.

2. The mode of locomotion of a bacterium in the genus *Ectothiorhodospira* is by
   a. flagella.
   b. cilia.
   c. gliding.
   d. rolling.
   e. gas vesicles.

3. Which of the following does *not* correctly compare the cyanobacteria with the photosynthetic eukaryotes?
   a. Both cyanobacteria and the photosynthetic eukaryotes are photoautotrophs.
   b. Both cyanobacteria and the photosynthetic eukaryotes produce oxygen as a by-product of noncyclic photophosphorylation.
   c. Both cyanobacteria and the photosynthetic eukaryotes use carbon dioxide as their source of carbon.
   d. The cyanobacteria used bacteriochlorophyll as their photosynthetic pigment, whereas the photosynthetic eukaryotes use chlorophyll *a*.
   e. Unlike any photosynthetic eukaryotes, the cyanobacteria are nitrogen fixers.

4. The archaea that live near volcanic vents in the ocean floor are examples of
   a. photoautotrophs.
   b. photoheterotrophs.
   c. chemoautotrophs.
   d. chemoheterotrophs.
   e. nitrifiers.

5. The heterocyst of the cyanobacterium *Nostoc* sp. is specialized for nitrogen fixation. Which of the following statements about the nitrogen cycle is *false*?
   a. The nitrogen fixers convert atmospheric nitrogen gas into ammonia (NH_3).
   b. The denitrifiers return nitrogen to the atmosphere as nitrogen gas (NH_2).
   c. The nitrifiers convert ammonia to nitrate or nitrite.
   d. The nitrifiers are chemoautotrophs.
   e. Plants get most of their nitrogen from nitrogen fixation that occurs in their chloroplasts.

6. Which of the following is *not* one of Koch's postulates used to identify pathogens?
   a. It must be possible to grow a pure culture of the isolated microorganism.
   b. Cultured samples must be able to produce disease symptoms in a newly infected host.
   c. Isolated microorganisms must respond positively to gram-staining procedures.
   d. Newly infected hosts must yield the same suspect microorganism in pure culture.
   e. The suspected microorganism must always be found in diseased hosts.

7. Which of the following groups of terms is *most* applicable to the species *Clostridium botulinum,* which does not readily proliferate within humans but which secretes a deadly toxin?
   a. High invasiveness, high toxigenicity, endotoxin
   b. Low invasiveness, high toxigenicity, endotoxin
   c. High invasiveness, low toxigenicity, exotoxin
   d. Low invasiveness, low toxigenicity, exotoxin
   e. Low invasiveness, high toxigenicity, exotoxin

8. Observable phenotypic characteristics are not nearly as useful as molecular traits for understanding the evolutionary relationships of prokaryotes. Which of the following statements about the importance of ribosomal RNA's (rRNA's) to the classification of organisms is *false*?
   a. rRNA is useful for evolutionary studies because rRNA is ancient.
   b. rRNA is useful for evolutionary studies because all organisms have rRNA's.
   c. rRNA is useful for evolutionary studies because rRNA plays the same role in translation in all organisms.
   d. Signature sequences in rRNA's are useful for distinguishing relationships at the domain level, but not at lower levels, such as phylum.
   e. rRNA is useful for evolutionary studies because rRNA has evolved slowly enough that sequence similarities between groups of organisms are easily found.

9. Recently, scientists have arranged life forms into three domains: Bacteria, Archaea, and Eukarya. Which of the following statements regarding this classification scheme is *false*?
   a. Mutations are an important source of variation.
   b. Lateral gene transfer has played a role in the evolution of the prokaryotic group.
   c. Prokaryotes can acquire new alleles through the process of conjugation.
   d. Mutations in prokaryotes have an immediate impact on the cell.
   e. Lateral gene transfer in prokaryotes occurs only through the process of transduction.

10. Which of the following statements about the proteobacteria is *false*?
    a. The proteobacteria can be classified into five major groups.
    b. The common ancestor of all proteobacteria was a photoautotroph.
    c. All groups of proteobacteria have retained the photoautotrophic nutritional mode.
    d. Some groups of proteobacteria include members that are photoautotrophs, chemoautotrophs, or chemoheterotrophs.
    e. The phylum Proteobacteria shows the closest relationship to the eukaryotic mitochondria.

11. Crown gall disease is caused by *Agrobacterium tumefaciens*. This organism belongs to the group of bacteria known as
    a. mycoplasmas.
    b. actinomycetes.
    c. gram-positive bacteria.
    d. proteobacteria.
    e. thermophiles.

12. Some bacteria have an elaborately branched system of filaments. This feature is characteristic of the group of bacteria know as
    a. mycoplasmas.
    b. actinomycetes.
    c. gram-positive bacteria.
    d. gram-negative rods.
    e. cyanobacteria.

13. Which of the following statements about the domain Archaea (whose members love heat and acid) is *false*?
    a. All archaea lack peptidoglycan in their cell walls.
    b. Unlike the bacteria and eukarya, all archaea have membranes in which branched long-chain hydrocarbons are attached to glycerol at one or both ends, by ester linkages.
    c. The environments where the archaea live tend to be extreme.
    d. Some archaea can reduce CO_2 to produce methane (CH_4).
    e. Some halophilic archaea use a chemiosmotic mechanism to produce ATP that is dependent on the pigment retinal.

★ *Indicates a **difficult** question*
■ *Indicates a **conceptual** question*

27 Protists and the Dawn of the Eukarya

Fill in the Blank

1. All eukaryotes that are not plants, fungi, or animals are defined as **protists**.

2. Animal-like protists are also called **protozoans**.

3. The concept of different organisms living together one inside the other is called **endosymbiosis**.

4. The process whereby a diploid generation that produces spores alternates with a haploid generation that produces gametes is called **alternation of generations**.

5. Members of the protozoan phylum **Foraminifera** secrete shells of calcium carbonate. These protozoans are valuable in the geological dating of sedimentary rocks and in oil prospecting.

6. **Contractile vacuoles** keep some freshwater protists from exploding by taking on too much water.

7. Giardia is a unicellular eukaryote that lacks **mitochondria**.

8. Ciliates, such as paramecia, often contain two kinds of nuclei, a single **macronucleus** with DNA that is translated and transcribed, and several **micronuclei**, which are typical eukaryotic nuclei.

9. Paramecia have an elaborate sexual behavior called **conjugation**, in which sexual recombination but not sexual reproduction occurs.

10. The form taken by acellular slime molds under adverse environmental conditions is called a **sclerotium**. This structure rapidly becomes a **plasmodium** again upon restoration of favorable conditions.

11. The red tides that can kill tons of fish are caused by toxic species of **dinoflagellates**.

12. Some chlorophytes exhibit a variation of the heteromorphic life cycle called the **haplontic** life cycle.

13. Many **diatoms** deposit silicon in their cell walls; architectural magnificence on a microscopic scale is a hallmark of this group, and their taxonomy is based entirely on their cell wall patterns.

14. Giant kelp belongs to the phylum Phaeophyta, whose members are commonly called **brown algae**.

15. **Forimiferans** secrete shells made of calcium carbonate.

16. The causative agent of African sleeping sickness is ***Trypanosoma***, a parasitic kinetoplastid.

17. *Paramecium* uses a precise form of locomotion; by beating its **cilia**, it can move forward or backward.

Multiple Choice

1. Which of the following were essential steps in the origin of eukaryotic cells?
   a. The origin of a flexible cell surface
   b. The origin of a cytoskeleton
   c. The origin of a nuclear envelope
   d. The endosymbiotic acquisition of certain organelles
   e. All of the above

2. Which of the following is *not* true about conjugation?
   a. Micronuclei disintegrate.
   b. Meiosis takes place.
   c. It is a reproductive process.
   d. Mitosis takes place.
   e. None of the above

3. The overall size that unicellular protists can achieve is limited by their
   a. energy-producing potential.
   b. metabolism.
   c. mitochondria.
   d. surface area–to–volume ratio.
   e. None of the above

4. Which of the following is a sexual reproductive process common to organisms in the kingdom Protista?
   a. Simple splitting of the cell
   b. Multiple fission
   c. Budding and spore formation
   d. Union of male and female gametes
   e. All of the above

5. Perhaps the best-known __________ are the malarial parasites of the genus *Plasmodium*.
   a. turbellarians
   b. mastigophorans
   c. dinoflagellates
   d. apicomplexans
   e. poriferans

6. The chemical signal responsible for the aggregation of myxamoebas is
   a. **3′,5′-cyclic adenosine monophosphate.**
   b. alginic acid.
   c. phycoerythrin.
   d. actin.
   e. None of the above

7. Which of the following modes of nutrition is used by protists to fuel their metabolism?
   a. Autotrophic
   b. Absorptive heterotrophic
   c. Ingestive heterotrophic
   d. None of the above
   e. **All of the above**

8. Protists are found in which of the following habitats?
   a. Marine
   b. Freshwater aquatic
   c. Within the body fluids of other organisms
   d. Damp soil
   e. **All of the above**

*9. Contractile vacuoles help some protists cope with a(n) _________ environment. These protists have a more negative osmotic potential than their freshwater environments and constantly take on water by osmosis.
   a. **hypoosmotic**
   b. hyperosmotic
   c. isoosmotic
   d. acidic
   e. basic

10. The _________ are beautiful marine protists that secrete a glassy endoskeleton.
    a. algae
    b. protozoa
    c. **radiolarians**
    d. flagellates
    e. dinoflagellates

*11. Some algae protists demonstrate the phenomenon of alternation of generations, in which a multicellular diploid, spore-producing organism gives rise to a multicellular haploid, gamete-producing organism. Which of the following statements about alternation of generations is *not* true?
    a. The haploid and diploid organisms may or may not differ morphologically.
    b. The haploid and diploid organisms differ genetically.
    c. **Only the haploid organism may also reproduce asexually**.
    d. Haploid gametes can produce new organisms only by fusing with other gametes.
    e. Diploid sporophytes may undergo meiosis to produce haploid spores.

12. Paramecia contain two types of nuclei: a large macronucleus and as many as 80 micronuclei. The micronuclei are typical eukaryotic nuclei, essential for genetic recombination. The macronucleus
    a. is important in sexual recombination (conjugation).
    b. contains several micronuclei.
    c. **contains many copies of the genetic information**.
    d. contains DNA that is not transcribed.
    e. contains DNA that is transcribed but not translated.

13. Paramecia have an elaborate sexual behavior in which they line up against each other and fuse. This is followed by an extensive reorganization and exchange of nuclear material. This process is called
    a. isogamous reproduction.
    b. alternation of generations.
    c. sexual reproduction.
    d. **conjugation.**
    e. None of the above

14. Which of the following is *not* a characteristic of the brown algae?
    a. Leaflike growths called thalli.
    b. **They store the products of photosynthesis as floridean starch**.
    c. A specialized holdfast that aids in attachment to a surface.
    d. The presence of the carotenoid fucoxanthin in their chloroplasts.
    e. All of the above

15. _________ have created vast limestone deposits throughout the world.
    a. Radiolarians
    b. Dinoflagellates
    c. **Foraminiferans**
    d. Heliozoans
    e. None of the above

16. All of the following diseases *except* _________ are caused by the trypanosomes.
    a. African sleeping sickness
    b. **malaria**
    c. leishmaniasis
    d. Chagas' disease
    e. East Coast fever

17. The photosynthetic stramenopiles obtained their chloroplasts, which are surrounded by three membranes, through
    a. primary endosymbiosis.
    b. **secondary endosymbiosis, retaining the chloroplast from a red alga.**
    c. secondary endosymbiosis, retaining the chloroplast from a chlorophyte.
    d. tertiary endosymbiosis.
    e. None of the above

18. The phylum _________ includes the water molds and their terrestrial relatives, such as the downy mildews.
    a. Pyrrophyta
    b. Chrysophyta
    c. Rhodophyta
    d. Phaeophyta
    e. **Oomycota**

19. The feature that distinguishes acellular from cellular slime molds is

a. that cellular slime molds are motile; acellular ones are not.
b. the number of nuclei contained within one plasma membrane.
c. that acellular slime molds ingest food by endocytosis; cellular ones do not.
d. that cellular slime molds ingest food by endocytosis; acellular ones do not.
e. that acellular slime molds prefer cool, moist habitats; cellular slime molds prefer dry, hot conditions.

20. The first true eukaryotic cell possessed which of the following?
a. A cytoskeleton
b. An endoplasmic reticulum
c. A nuclear envelope
d. A mitochondrion
e. a and c

21. This phylum consists of the red algae, almost all of which are multicellular. The accessory pigment phycoerythrin gives this group its characteristic reddish color; however, many species also contain phycocyanin. The color of individuals in this group can vary depending on environmentally variable light conditions.
a. Pyrrophyta
b. Chrysophyta
c. Phaeophyta
d. Rhodophyta
e. Chlorophyta

22. This phylum is commonly known as the green algae. In this group are the only protists that contain the full complement of photosynthetic pigments characteristic of plants.
a. Pyrrophyta
b. Chrysophyta
c. Phaeophyta
d. Rhodophyta
e. Chlorophyta

23. Multicellular brown algae make up this group. The brown algae are composed of either branched filaments or leaflike growths called thalli. Giant kelps, which may be over 60 meters long, and *Sargassum*, which forms dense mats of vegetation in the Sargasso Sea in the mid-Atlantic, are famous members of this group.
a. Pyrrophyta
b. Chrysophyta
c. Phaeophyta
d. Rhodophyta
e. Chlorophyta

24. When some autotrophic *Euglena* are placed in the dark, they
a. stop producing their photosynthetic pigment.
b. produce excess photosynthetic pigment.
c. begin feeding on organic material floating in the water around them.
d. die.
e. a and c

25. When organic material is digested in a food vacuole, the pH in the vacuole
a. decreases to aid in digestion, then increases as digestion is completed.
b. increases to aid in digestion, then decreases as digestion is completed.
c. stays the same during and after digestion.
d. stays the same during digestion, then increases as digestion is completed.
e. None of the above

*26. The plant kingdom is thought to have evolved from the Chlorophyta, the green algae. Which of the following statements is *not* a basis for this belief?
a. Plants and green algae both contain chlorophyll *a*.
b. The photosynthetic storage products of both plants and green algae include oils.
c. Chlorophyll *b*, found in plants, is found in no other group of algae.
d. The principal photosynthetic storage product of both groups is starch.
e. The carotenoids found in green algae are characteristic of those found in plants.

27. The animal kingdom is thought to have evolved from the
a. protozoa.
b. Eubacteria.
c. algae.
d. funguslike protists.
e. None of the above

28. Algal life cycles show extreme variation. Only one phylum, the __________, does *not* have flagellated motile cells in at least one stage of the life cycle.
a. Chlorophyta
b. Rhodophyta
c. Phaeophyta
d. Chrysophyta
e. Pyrrophyta

*29. Chromatic adaptation, found in red algae and several other groups of algae, is the
a. ability to become heterotrophic when light levels are low.
b. ability to bioluminesce when disturbed.
c. ability to change the wavelength of light to that useful to their photosynthetic pigments.
d. capacity to change the relative amounts of their various photosynthetic pigments depending on the light conditions where they are growing.
e. None of the above

*30. Radiolarians contain photosynthetic algae as endosymbionts. The algae provide __________ for the radiolarians, and the radiolarians provide __________ for the algae. Both the algae and the radiolarians are protists.
a. food, protection
b. protection, food
c. food, essential nutrients
d. essential nutrients, protection
e. essential nutrients, food

31. Protists are believed to be ecologically and evolutionary important for many reasons. Which of the following is *not* true of this group?
   a. Multicellular kingdoms evolved from protists.
   b. Photosynthetic protists play a major role in the energy balance of the living world.
   c. None of the protists are parasites.
   d. Saprobic protists are among the important decomposers and thus play a major role in the nutrient cycles of the living world.
   e. Many protists have highly differentiated bodies even though they consist of but a single cell.

32. Which of the following terms does *not* apply to notable members of the kingdom Protista?
   a. Autotroph
   b. Absorptive heterotroph
   c. Ingestive heterotroph
   d. Aquatic
   e. Dikaryotic

33. Select the group of organisms *not* assigned to the kingdom Protista.
   a. Protozoa
   b. Algae
   c. Slime molds
   d. Sponges
   e. Giant kelp

*34. You place two different species of protists in a solution with an unknown osmotic potential. Protist A has a contractile vacuole firing rate of 5 per minute; protist B has a contractile vacuole firing rate of 12 per minute. Select a reasonable conclusion based on these observations.
   a. Protist A has a more negative osmotic potential than protist B.
   b. Protist B has a more negative osmotic potential than protist A.
   c. The solution has a more negative osmotic potential than protist A.
   d. The solution has a more negative osmotic potential than protist B.
   e. No conclusions can be made unless we know the osmotic potential of the solution.

35. In the alternation of generations, the gametophyte generation is
   a. haploid and produces spores.
   b. haploid and produces gametes.
   c. diploid and produces spores.
   d. diploid and produces gametes.
   e. can be either haploid or diploid, but always produces spores.

36. What do the protists that are responsible for sleeping sickness and malaria both have in common?
   a. They are from the same phylum.
   b. They both have insect vectors for transmission to humans.
   c. They cause the same symptoms.
   d. They both have gametocyte life stages.
   e. All of the above

37. *Plasmodium*, the organism that causes malaria, is in the group
   a. Zoomastigophora.
   b. Sarcodina.
   c. Apicomplexa.
   d. Choanoflagellida.
   e. None of the above

*38. Which characteristic does *not* apply to some species of amoebas?
   a. Autotrophic
   b. Free-living
   c. Predators
   d. Parasitic
   e. Shelled

*39. In the disease malaria caused by the protist *Plasmodium*, the gametocytes
   a. develop into merozoites.
   b. inhabit the salivary glands of *Anopheles* mosquitoes.
   c. are the infective stage obtained from the insect vector.
   d. **are found inside red blood cells.**
   e. give rise to zygotes within the mammalian circulatory system.

40. Which of the following statements concerned with the micronucleus and macronucleus is *false*?
   a. The macronucleus is involved in genetic recombination.
   b. The micronucleus is a typical eukaryotic nucleus.
   c. Multiple copies of macronuclear genes are common.
   d. Transcription and translation mostly involves genes found in the macronucleus.
   e. The micronucleus and macronucleus are unique to members of the phylum Ciliophora.

41. Of the following structures select the one that is *not* associated with movement or feeding in *Paramecium*.
   a. Food vacuole
   b. Oral groove
   c. Cilia
   d. Pseudopodia
   e. None of the above

42. Which of the following processes is *not* part of conjugation in *Paramecium*?
   a. Meiosis
   b. Mitosis
   c. Cytokinesis
   d. Fusion of haploid nuclei
   e. Breakdown of some micronuclei

43. Select the following event that is *not* normally associated with the life cycle of an acellular slime mold.
   a. Development of spores into myxamoebas
   b. Formation of a sclerotium when conditions are adverse
   c. Active feeding by the plasmodium as it engulfs food particles
   d. Meiosis to form sporangiophores
   e. Fusion of swarm cells to form a diploid zygote

44. Which of the following features is *not* characteristic of the plasmodium of an acellular slime mold?
   a. Coenocyte
   b. Composed of haploid nuclei
   c. Shows cytoplasmic streaming
   d. Formed by mitosis without cytokinesis
   e. Can become a sclerotium

45. In the cellular slime molds, cAMP causes
   a. aggregation of swarm cells to form a plasmodium.
   b. formation of sporangia.
   c. release of myxamoebas from fruiting bodies.
   d. the onset of cytoplasmic streaming.
   e. aggregation of myxamoebas to form a pseudoplasmodium.

*46. The algal protist groups differ from each other in terms of the
   a. principal photosynthetic storage product.
   b. unicellular/multicellular body plan.
   c. possession of flagella.
   d. principal photosynthetic pigments.
   e. All of the above

47. Which phylum includes the dinoflagellates?
   a. Pyrrophyta
   b. Chrysophyta
   c. Rhodophyta
   d. Phaeophyta
   e. Chlorophyta

48. Members of which group are thought to comprise the closest relatives of the animals?
   a. Euglenozoa
   b. Choanoflagellida
   c. Stramenopila
   d. Alveolata
   e. Chlorophyta

*49. Select the following statement that does *not* apply to diatoms.
   a. Diatom cell walls may be impregnated with silicon.
   b. During mitosis, the top and bottom of the cell become the tops of the two new cells.
   c. Prior to meiosis, the cell sheds its cell wall and increases in size.
   d. Zygotes (auxospores) are formed by gametes that lack cell walls.
   e. Diatoms can show bilateral or radial symmetry.

50. The characteristic color of the red algae is due to the presence of
   a. phycoerythrin.
   b. fucoxanthin.
   c. β-carotene.
   d. chrysolaminarin.
   e. chlorophyll *b*.

*51. Select the algal phylum that has some members showing chromatic adaptation and others that produce mucilaginous polysaccharides important in the production of agar.
   a. Pyrrophyta
   b. Chrysophyta
   c. Rhodophyta
   d. Phaeophyta
   e. Chlorophyta

52. Which of the following statements about the life cycle of the green alga *Ulva* is *not* true?
   a. *Ulva* has an isomorphic life cycle.
   b. The sporophyte and gametophyte can only be differentiated microscopically.
   c. All species of *Ulva* are isogamous.
   d. Gametophytes produce sperm only or eggs only, never both.
   e. The diploid sporophyte produces flagellated spores.

*53. An ecologically and evolutionarily important algae phylum, which is a common endosymbiont of coral, is
   a. Pyrrophyta.
   b. Rhodophyta.
   c. Phaeophyta.
   d. Chrysophyta.
   e. Chlorophyta.

*54. Which of the following criteria is *not* used in separating protists from animals?
   a. Unicellular versus multicellular
   b. Having an ingestive metabolism
   c. Having an embryonic stage
   d. Having an extracellular matrix of actin and collagen
   e. Having membrane-enclosed organelles

Study Guide Questions

1. Which of the following modes of reproduction can be found in at least some protists?
   a. Binary fission
   b. Sexual reproduction
   c. Spore formation
   d. All of the above

2. During the evolution of eukaryotes from prokaryotes, which of the following did *not* occur?
   a. Infolding of the flexible cell membrane
   b. Loss of the cell wall
   c. A switch from aerobic to anaerobic metabolism
   d. Endosymbiosis of once free-living prokaryotes

3. Which of the following statements about protists is *false*?
   a. Apicomplexans are the only protist group without parasitic representatives.
   b. Foraminiferans and radiolarians are shelled protists.
   c. Ciliates have great control over the direction their cilia beat.
   d. Although they appear structurally simple, amoebas are not primitive organisms.

4. Ciliates are characterized by protective structures called
   a. trichonympha.
   b. trichocysts.
   c. trichomes.
   d. throclea.

5. A major difference between the vegetative states of cellular and acellular slime molds is that
   a. acellular slime molds have haploid nuclei while cellular slime molds have diploid nuclei.
   b. acellular slime molds produce fruiting bodies and cellular slime molds do not.
   c. acellular slime molds use cAMP as a chemical cue for aggregation and cellular slime molds do not.
   d. acellular slime molds exist as a coenocytic mass and cellular slime molds exist as individual myxamoebas.

6. Isogamy is best characterized as
   a. indistinguishable male and female gametes.
   b. female gametes being significantly larger than male gametes.
   c. male gametes being significantly larger than female gametes.
   d. a diplontic mode of reproduction.

7. Red tides often cause massive fish kills and human illness in those eating shellfish. Which group of protists is responsible for red tides?
   a. Chlorophyta
   b. Rhodophyta
   c. Phaeophyta
   d. Chrysophyta

8. Holdfasts and alginic acid are characteristic of which group of protists?
   a. Chlorophyta
   b. Rhodophyta
   c. Phaeophyta
   d. Chrysophyta

9. A water mold is discovered growing on the surface of a freshwater green algae. Which two groups of protists are involved?
   a. Myxomycota and Rhodophyta
   b. Oomycota and Chlorophyta.
   c. Foraminifera and Chrysophyta.
   d. Apicomplexa and Ciliophora.

10. Which of the following statements concerning colonial organisms is *false*?
    a. Colonial organisms have division of labor among cells.
    b. Colonial organisms are multicellular.
    c. Colonial organisms are aggregates of single cells.
    d. None of the above

11. Which of the following statements regarding conjugation in *Paramecium* is *false*?
    a. Conjugation results in genetic recombination.
    b. Conjugation results in clones.
    c. Conjugation results in offspring.
    d. b and c

12. Which of the following regarding protists in general is *false*?
    a. Protists are always parasitic.
    b. Protists are single-celled.
    c. Protists are all heterotrophic.
    d. All of the above

End of Chapter Questions

1. Flagellates
   a. appear in several protist phyla.
   b. are all algae.
   c. all have pseudopods.
   d. are all colonial.
   e. are never pathogenic.

2. Which statement about amoebas is *not* true?
   a. Most amoebas are predatorial, parasitic or scavengers.
   b. They use amoeboid movement.
   c. They include both naked and shelled forms.
   d. They possess pseudopods.
   e. They appeared only once in evolutionary history.

3. The Apicomplexa
   a. possess flagella.
   b. possess chloroplasts.
   c. are all parasitic.
   d. are algae.
   e. include the trypanosomes that cause sleeping sickness.

4. The Ciliophora
   a. move by means of flagella.
   b. use amoeboid movement.
   c. include Plasmodium, the agent of malaria.
   d. possess both a macronucleus and micronuclei.
   e. are autotrophic.

5. The Myxomycota
   a. are also called acellular slime molds.
   b. lack fruiting bodies.
   c. consist of large numbers of myxamoebas.
   d. consist at times of a mass called a pseudoplasmodium.
   e. possess flagella.

6. The Cellular Slime Molds
   a. form an irregular mass of cells called a sclerotum.
   b. lack fruiting bodies.
   c. form a plasmodium that is a coenocyte.
   d. use cAMP as a "messenger" to signal aggregation.
   e. possess flagella.

7. Which statement about algae is *not* true?
   a. They differ from plants in lacking life cycles with alteration of generations.
   b. They are photosynthetic autotrophs.
   c. They contain only chlorophyll photosynthetic pigments.
   d. They include both unicellular and multicellular forms.
   e. Their life cycles show extreme variation.

8. Which statement about the Phaeophyta is *not* true?
   a. They are all multicellular.
   b. Their brown color is due to the presence of different types of chlorophyll.
   c. They are almost exclusively marine.
   d. Are a primary source of of ice cream emulsifiers.
   e. Some have extensive tissue differentiation.

9. The Rhodophyta
   a. are mostly unicellular.
   b. are mostly marine.
   c. owe their red color to a special photosynthetic pigment.
   d. have flagella on their gametes.
   e. are all heterotrophic.

10. Which statement about the Chlorophyta is *not* true?
   a. They use the same photosynthetic pigments as do plants.
   b. Some are unicellular.
   c. Some are multicellular.
   d. All are microscopic in size.
   e. They display a great diversity of life cycles.

Student Web Site Self-Quiz Questions

1. The development of a flexible cell surface was an important event in the evolution of the eukaryotic cell. The origin of which of the following eukaryotic cell structures was *not* directly related to this event?
   a. Peroxisome
   b. Nucleus
   c. Endoplasmic reticulum
   d. Lysosome
   e. Food vacuole

2. You count the rate of contraction of a protist's contractile vacuole in solution A. After transferring it to solution B, you note that the rate decreases. You conclude that solution A has an osmotic potential __________ than solution B and that both solutions are __________ to the protist.
   a. less negative, hypoosmotic
   b. less negative, hyperosmotic
   c. more negative, hypoosmotic
   d. more negative, hyperosmotic
   e. less negative, isosmotic

3. The flagellate body form has evolved many times in distantly related groups. Which of the following protist groups (body forms) is found in several different phyla?
   a. Ciliates
   b. Amoebas
   c. Euglenoids
   d. Kinetoplastids
   e. Apicomplexans

4. *Euglena* sp. is a well-known euglenoid. Which of the following statements about the euglenoids is *true*?
   a. Euglenoids are flagellates.
   b. Some euglenoids are colonial.
   c. All euglenoids belong to the same phylum.
   d. All euglenoids have chloroplasts.
   e. *Euglena* placed in continuous darkness lose their photosynthetic pigments and die.

5. Which of the following features is *not* characteristic of the dinoflagellates?
   a. Many dinoflagellates are endosymbionts.
   b. Some dinoflagellates show an alternation of generations.
   c. Dinoflagellates have two flagella.
   d. Certain dinoflagellates are responsible for red tides.
   e. Many dinoflagellates are bioluminescent.

6. Which of the following statements about malaria is *false*?
   a. The protist causing malaria is an apicomplexan.
   b. *Plasmodium falciparum* can only inhabit humans and female mosquitos of the *Anopheles* genus.
   c. Sexual reproduction of the malaria protist occurs within the liver of humans.
   d. Merozoites enter human red blood cells.
   e. *Plasmodium* is an intracellular parasite in humans and an extracellular parasite in the mosquito.

7. Which of the following statements about conjugation in paramecia is *true*?
   a. The most important difference between micronuclei and macronuclei is size.
   b. Some asexual clones of paramecia have permanently lost the ability to conjugate.
   c. In paramecia, reproduction and sex occur at different times.
   d. During conjugation, one cell acts as a donor of micronuclei, and the other cell is the recipient.
   e. Macronuclei remain unchanged during conjugation.

8. Which of the following characteristics is *not* a feature of the brown algae?
   a. Multicellularity
   b. Almost exclusively marine
   c. Large size
   d. Attached forms have holdfasts
   e. Most contain phycoerythrin, an accessory pigment

9. Which of the following statements does *not* characterize the red algae?
   a. Red algae can vary their ratio of photosynthetic pigments depending on the light conditions.
   b. The products of photosynthesis are stored as floridean starch.
   c. No motile flagellated cells exist in any life stage.
   d. The chloroplasts of the red algae may have been derived from endosymbiotic brown algae.
   e. Agar is derived from a metabolic product of red algae.

10. The life cycle of the green alga, *Ulothrix,* can best be described as __________ and __________.
   a. isomorphic, haplontic
   b. isomorphic, diplontic
   c. heteromorphic, haplontic
   d. heteromorphic, diplontic
   e. oogamy, haplontic

11. Which of the following is *not* a major difference between the acellular and cellular slime molds?
    a. The vegetative unit in the acellular slime molds is the plasmodium; in the cellular slime molds it is a myxamoeba.
    b. The acellular slime molds have a coenocytic stage; cellular slime molds do not.
    c. There are two commonly recognized phyla of cellular slime molds, but only one phylum of acellular slime molds.
    d. 3′,5′-cyclic adenosine monophosphate (cAMP) plays an important role in aggregation in the cellular slime molds, but not in the acellular slime molds.
    e. Acellular slime molds produce fruiting structures; cellular slime molds do not.

TEST FILE

★ *Indicates a **difficult** question*

■ *Indicates a **conceptual** question*

28 Plants without Seeds: From Sea to Land

Fill in the Blank

1. In plants (and some algae like *Ulva*), the sporophyte and gametophyte exhibit different levels of ploidy. Sporophytes are **diploid**, while gametophytes are **haploid**.

2. Broadly defined, a **plant** is a multicellular, photosynthetic eukaryote that develops from an embryo protected by tissues of the parent.

3. The vascular system can be said to have been launched by a single evolutionary event. Sometime during the Paleozoic era, the sporophyte generation of a now long-extinct green algae produced a new cell type, the **tracheid**.

4. In the strictest sense, a **leaf** is a flattened photosynthetic structure emerging laterally from a main axis or stem and possessing true vascular tissue.

5. In 1903 a French botanist, E. A. O. Lignier, proposed a currently accepted hypothesis about the origin of **roots**. His hypothesis states that they were derived from dichotomous branches, some of which entered the soil, anchoring the plant and absorbing water and nutrients.

6. **Ferns** are plants with large leaves (fronds) and no seeds, which require water as a medium for the transfer of male gametes. The sporophytes and gametophytes grow independently of each other.

7. The tissues that conduct the products of photosynthesis from sites of production to sites of utilization or storage are called the **phloem**.

8. The **sporophyte** generation extends from the zygote through the adult, multicellular, diploid plant.

9. Plants that have an internal transport system are called **tracheophytes**. The older term was vascular plants.

10. The vascular system in plants consists of specialized tissues. The **xylem** transports water and minerals from soil to aerial parts. The **phloem** transports products of photosynthesis from sites of production to storage sites.

11. **Rhizoids** are water-absorbing elements that are found on the lower surfaces of the simplest liverwort gametophytes.

12. In the nontracheophytes, the conspicuous green structure that can be seen by the naked eye is the **gametophyte**.

13. The sporophytes of the mosses and tracheophytes grow by **apical** cell division, where a region at the growing tip provides an organized pattern of cell division, elongation, and differentiation.

14. Some liverworts, such as *Marchantia*, reproduce sexually and vegetatively. **Gemmae** are lens-shaped clumps of cells that are the means by which vegetative reproduction takes place.

Multiple Choice

1. It is widely agreed that the plant kingdom arose from
   a. Eumycota.
   b. Chrysophyta.
   c. Phaeophyta.
   d. Rhodophyta.
   e. Chlorophyta.

★2. Characteristics of green algae make them attractive candidates for the origin of plants. Which of the following describes such a characteristic?
   a. Photosynthetic pigments in plastids
   b. Active stomata
   c. Starch as a major storage compound
   d. Cellulose in cell walls
   e. a, c, and d

3. Nonvascular plants have never evolved to the size of vascular plants. The most probable explanation is that they could not solve the problems posed by
   a. photosynthesis.
   b. respiration.
   c. nutrient and water transportation within the plants themselves.
   d. nutrient and water absorption.
   e. All of the above

4. In some of the liverworts, there are special elongated cells called elaters that possess a helical thickening of the cell wall. As elaters lose water, they shrink longitudinally and compress the helical thickening like a spring. When the stress reaches a critical point, the compressed "spring" snaps back to its resting position, liberating hundreds of __________ in all directions.
   a. moisture particles
   b. spores
   c. sperm
   d. ova
   e. rhizoids

5. Two important evolutionary consequences of plants having tissues composed of tracheids are
   a. a plant vascular system and structural support.
   b. structural support and increased growth.
   c. enhanced photosynthesis and structural support.
   d. enhanced photosynthesis and a plant vascular system.
   e. None of the above

6–8. Match the phyla from the list below with the following descriptions.
   a. Hepatophyta (liverworts)
   b. Anthocerophyta (hornworts)
   c. Bryophyta (mosses)

6. Sporophytes contain elaters; some species reproduce sexually and vegetatively. **(a)**

7. The gametophyte is a branched, filamentous structure called a protonema; many types contain hydroid cells. **(c)**

8. Exhibits indefinite growth; has a single large chloroplast in each cell; has internal cavities filled with mucilage, often populated by cyanobacteria that fix atmospheric nitrogen gas into a nutrient form usable by the plant. **(b)**

*▪9. Which of the following is *not* consistent with Lignier's hypothesis on the origin of roots?
   a. Ancestors of the first vascular plants branched dichotomously.
   b. Roots evolved from a symbiotic relationship with simple, avascular plants.
   c. Roots were originally stems that just went underground.
   d. The different selective pressures between branches in the air and branches in the ground led to the differences in root and stem structure that we see today.
   e. All of the above

10. Although Earth is estimated to be 5 billion years old, and although life first appeared about 4 billion years ago, plants did not appear until about __________ years ago.
   a. 3 billion
   b. 300 million
   c. 30 million
   d. 3 million
   e. 300 thousand

11. Plants can be broadly defined as multicellular photosynthetic eukaryotes. There are several additional features of this group. Which of the following is *not* a characteristic of plants?
   a. They develop from embryos protected by tissues of the parent plant.
   b. Their cell walls contain cellulose.
   c. Their chloroplasts contain chlorophylls *a* and *b*.
   d. Their storage carbohydrate is starch.
   e. Their respiration is anaerobic.

12. A universal feature of the life cycle of plants is
   a. morphologically identical haploid and diploid stages.
   b. genetically identical haploid and diploid stages.
   c. alteration of generations between heteromorphic haploid gametophytes and diploid sporophytes.
   d. All of the above
   e. None of the above

13. It was on land that plants first appeared and evolved. Two challenges of occupying the land were
   a. development of photosynthetic pigments that are not dependent on an aqueous environment and transport of water and minerals to aerial parts.
   b. development of starch for carbohydrate storage and transport of water and minerals to aerial parts.
   c. physical support and transport of water and minerals to aerial parts.
   d. development of photosynthetic pigments that are not dependent on an aqueous environment and development of starch for carbohydrate storage.
   e. development of photosynthetic pigments that are not dependent on an aqueous environment and physical support.

▪14. Vascular plants are thought to be the result of a single evolutionary event: the evolution of a wholly new cell type, the tracheid. This cell type
   a. provides a mechanism for the storage of a new type of carbohydrate, starch.
   b. is the first cell type to contain chloroplasts.
   c. permits fertilization in the absence of water, thus permitting plants to invade dry habitats.
   d. forms the seed.
   e. is the principal water-conducting element of the xylem in all vascular plants except the angiosperms.

15. Several important ecological changes occurred that helped to make the invasion of land by plants permanent. Which of the following is *not* one of those changes?
   a. Evolution of a water-impermeable cuticle
   b. Evolution of a carbohydrate energy storage molecule
   c. Evolution of protective layers for the gamete-bearing structures
   d. Initial absence of herbivores
   e. All of the above

16. The major difference between a simple leaf and a complex leaf is that in a complex leaf,
    a. the vascular system of the leaf creates a major alteration in the architecture of the stem vascular system.
    b. the leaf can be over one centimeter long.
    c. the leaf possesses true vascular tissue.
    d. the chloroplasts are the only photosynthetic structures.
    e. the vascular strand departs from the vascular system of the stem so that there is scarcely any perturbation in the stem's vascular system.

17. Unlike the broad, loose definition of plants, the narrower definition states that plants
    a. are multicellular.
    b. have photosynthetic pigments.
    c. are eukaryotic.
    d. develop from embryos.
    e. have a vascular system.

18. Several evolutionary adaptations to land are shared by all plants. These shared adaptations do *not* include
    a. waxy protective coverings.
    b. support against gravity.
    c. means of taking up water from the soil.
    d. protective structures for the new sporophyte.
    e. water transport by xylem.

19. You find a green, "leafy" bryophyte growing on your neighbor's front lawn. It is most likely a
    a. liverwort.
    b. hornwort.
    c. moss.
    d. whisk fern.
    e. fern.

20. Fossil vascular plants can be recognized by the presence of
    a. tracheids.
    b. antheridia.
    c. archegonia.
    d. protonemata.
    e. vessels.

21. One important consequence associated with the evolution of xylem was
    a. sugar transport.
    b. sperm transport.
    c. prevention of water evaporation.
    d. rigid structural support.
    e. None of the above

22. One can identify sporangia as part of the fossil *Rhynia* since meiosis produces
    a. two cells.
    b. four cells.
    c. two cells, but only one is functional.
    d. four cells, but only one is functional.
    e. two cells, but they remain attached.

23. In heterosporous plants,
    a. microgametophytes produce eggs.
    b. microgametophytes produce sperm.
    c. microgametophytes produce eggs and sperm.
    d. megagametophytes produce sperm.
    e. megagametophytes produce eggs and sperm.

*24. Heterospory probably affords selective advantages since it
    a. it evolved a number of times.
    b. is only found in the most advanced plants.
    c. is simpler than homospory.
    d. is found in the most primitive plants.
    e. evolved along with swimming sperm.

25. Which of the following plants have simple leaves?
    a. Mosses
    b. Horsetails
    c. Club mosses
    d. Ferns
    e. b and c

26. Ferns are in which phylum?
    a. Lycophyta
    b. Anthocerophyta
    c. Hepatophyta
    d. Pterophyta
    e. Bryophyta

▪27. The bryophytes are dependent on water for reproduction because
    a. sperm are passively transported to eggs by water.
    b. gametogenesis only occurs when the plants are moist.
    c. eggs and sperm are released into water and then unite.
    d. sperm must swim through water to reach and fertilize eggs.
    e. None of the above

28. All plants produce __________ by mitosis and __________ by meiosis.
    a. spores, gametes
    b. gametes, gametes
    c. gametes, spores
    d. spores, spores
    e. spores, gametes and spores

29. The most ancient surviving plant lineage are the
    a. liverworts.
    b. hornworts.
    c. mosses.
    d. ferns.
    e. None of the above

30. Which of the following characteristics and structures are *not* utilized by the nontracheophytes to obtain water and minerals in the absence of a vascular system?
    a. Growing in dense masses through which water can move by capillary action
    b. Leaflike structures that catch and hold water that splashes onto them
    c. Being small enough that minerals can be distributed evenly by diffusion
    d. An extensive root system
    e. None of the above

31. The first tracheophytes belonged to the phylum
    a. Rhyniophyta.
    b. Lycophyta.
    c. Sphenophyta.
    d. Pterophyta.
    e. None of the above

32. Which living tracheophyte phylum diverged earlier than all other living tracheophyte phyla?
    a. Sphenophyta
    b. Pterophyta
    c. Lycophyta
    d. Psilophyta
    e. Coniferophyta

33. The first true ferns appeared during which period?
    a. Permian
    b. Carboniferous
    c. Silurian
    d. Devonian
    e. Triassic

34. Which of the following is *not* true for the phylum Anthocerophyta (the hornworts)?
    a. They do not possess stomata.
    b. Their cells contain only a single, large chloroplast.
    c. Their sporophytes are capable of indefinite growth.
    d. They have internal, mucilage-filled cavities populated with cyanobacteria.
    e. a and c

35. The large and obvious plant within the tracheophytes is the __________; in the nontracheophytes it is the __________.
    a. gametophyte, sporophyte
    b. sporophyte, gametophyte
    c. gametophyte, gametophyte
    d. sporophyte, sporophyte
    e. None of the above

36. The evolutionary origin of simple leaves is thought by some biologists to be
    a. sterile sporangia.
    b. rhizoids.
    c. specialized xylem.
    d. photosynthetic tissue developed between complex branching patterns.
    e. None of the above

*37. Which of the following is *not* consistent with the fossilized tracheophytes found by Kidston and Lang?
    a. The presence of roots
    b. The presence of xylem
    c. Spores that were in groups of two
    d. Dichotomous branching
    e. a and c

38. In a heterosporous life cycle, the microspore develops into the __________ gametophyte, while the megaspore develops into the __________ gametophyte.
    a. female, male
    b. male, female
    c. diploid, haploid
    d. haploid, diploid
    e. None of the above

39–42. Match the phyla from the list below with the descriptions.
    a. Lycophyta (club mosses)
    b. Sphenophyta (horsetails)
    c. Psilotophyta (whisk ferns)
    d. Pterophyta (ferns)

39. Exhibit basal growth; sometimes called the scouring rushes because the silica deposits in their cell walls made them useful for cleaning; simple leaves that form distinct whorls around the stem. **(b)**

40. Originally thought to be evolutionarily ancient descendants of anatomically simple ancestors; gametophytes live below the surface of the ground and lack chlorophyll; now considered to be highly specialized and to have evolved fairly recently. **(c)**

41. Diverged earlier than all other living tracheophytes; they bear simple leaves arranged spirally on a stem; they exhibit apical growth; sporangia are contained within conelike structures called strobili. **(a)**

42. Typically have large leaves with branching vascular strands; some can reach heights of up to 20 meters; sporangia are usually clustered in groups called sori. **(d)**

*43. Stomata appear to have first arisen with which phylum?
    a. Anthocerophyta
    b. Bryophyta
    c. Hepatophyta
    d. Lycophyta
    e. Psilotophyta

44. Which of the following characteristics is *not* shared by the club mosses and the horsetails?
    a. Large, independent sporophytes
    b. Specialized vascular tissue
    c. Apical growth
    d. Simple leaves
    e. b and c

45. The cyanobacteria present in the internal, mucilage-filled cavities of the hornworts serve to
    a. provide additional structural support for the plant.
    b. produce additional carbohydrates for the plant.
    c. convert atmospheric nitrogen gas into a nutrient form usable by the plant.
    d. convert atmospheric carbon gas into a nutrient form usable by the plant.
    e. c and d

46. An abundant type of coal, called Cannel coal, is formed almost entirely from the fossilized spores of a species in which phylum?
    a. Lycophyta
    b. Sphenophyta
    c. Pterophyta
    d. Psilotophyta
    e. Anthocerophyta

47. The moss gametophyte that develops after spore germination is a branched, filamentous structure called a
    a. antheridium.
    b. capsule.
    c. protonema.
    d. hydroid.
    e. None of the above

48. The sporophyte of tracheophyte is __________; the sporophyte of a nontracheophyte is __________.
    a. dependent; independent
    b. independent; independent
    c. dependent; dependent
    d. independent; dependent
    e. None of the above

49. Which of the following observations led to the conclusion that the fossilized plants (*Rhynia*) found by Kidston and Lang were sporophytes?
    a. Thin sections of sporangia revealed that spores were in groups of four.
    b. The presences of rhizomes
    c. Dichotomous branching
    d. The presence of xylem
    e. a and d

Study Guide Questions

1. Plants may be differentiated from green algae by which of the following characteristics?
   a. Chlorophyll type
   b. Presence of multicellular gametangia
   c. Presence of roots
   d. Swimming sperm

2. In order to colonize land, plants needed to acquire which of the following characteristics?
   a. A mechanism for moving water throughout the plant
   b. A mechanism to prevent desiccation of tissues
   c. An ability to screen ultraviolet radiation
   d. b and c

3. The main differentiating factor between nontracheophytes and tracheophytes is
   a. lack of gametophytes.
   b. spore production.
   c. the presence of tracheids.
   d. All of the above

4. In alternation of generations, the sporophyte generation is __________ and the gametophyte generation is __________.
   a. haploid, diploid
   b. diploid, haploid
   c. haploid, haploid
   d. diploid, diploid

5. Which of the following characteristics can help differentiate between Hepatophytes and Byrophytes?
   a. Presence of hydroids
   b. Gametophyte dominance
   c. Sporophyte dominance
   d. Swimming sperm

6. During alternation of generations, meiosis takes place in the __________ to produce haploid __________.
   a. gametophyte, gametes
   b. sporophyte, gametes
   c. sporophyte, spores
   d. gametophyte, spores

7. Assexual reproduction in liverworts is accomplished by
   a. gametophytes.
   b. spores.
   c. gemmae.
   d. tracheids.

8. Hornworts have continually growing sporophytes that rely on the gametophyte for sustenance. Which of the following limit the size of the sporophyte?
   a. Water transport ability
   b. The gametophyte cannot produce enough nutrients
   c. Cells divide a finite number of times
   d. All of the above

9. You are walking along a roadside and find a plant with the following characteristics: very thin waxy cuticle, stomata, simple leaves in whorls around a central stem, independent sporophyte and gametophyte, sporangia in strobili. This plant is most likely a member of which of the following phyla?
   a. Bryophyta
   b. Sphenophyta
   c. Pterophyta
   d. Lycophyta

10. Match the phyla with their diagnostic characteristics.

_f._ Bryophyta	a. no true stomata
_a._ Hepatophyta	b. terminal sporangia
_d._ Anthocerophyta	c. complex branching leaves
_e._ Lycophyta	d. unique chloroplasts
_g._ Sphenophyta	e. first strobili of sterile leaves and sporangia
_b._ Psilotophyta	f. presence of hydroids
_c._ Pterophyta	g. basal growth patterns

End of Chapter Questions

1. Plants differ from algae in that only plants
   a. are photosynthetic.
   b. are multicellular.
   c. possess chlorophyll.
   d. have multicellular embryos protected by the parent.
   e. are eukaryotic.

2. Which statement about the alternation of generations in plants is *not* true?
   a. The plant life cycle is characterized by diploid and haploid forms.
   b. Meiosis occurs in sporangia.
   c. Gametes are always produced by meiosis.
   d. The zygote is the first cell of the sporophyte generation.
   e. The gametophyte and sporophyte differ genetically.

3. Which statement is *not* evidence for the origin of plants from the green algae?
   a. Similarity in rRNA and DNA sequences.
   b. Similarity in chloroplast structure.
   c. Both organisms utilize plasmodesmata.
   d. Both plants and green algae have protected sporophytes.
   e. Both plants and green algae can survive on land as well as in aquatic environments.

4. The nontracheophytes
   a. lack a sporophyte generation.
   b. grow in dense masses, allowing capillary movement of water.
   c. possess xylem and phloem.
   d. possess leaf-like structures.
   e. possess true roots.

5. The rhyniophytes
   a. possessed xylem and phloem.
   b. possessed true roots.
   c. were the original gametophytes.
   d. possessed leaves.
   e. lacked branching stems.

6. Club mosses and horsetails
   a. have larger gametophytes than sporophytes.
   b. possess small leaves.
   c. are represented today primarily by trees.
   d. have never been a dominant part of the vegetation.
   e. have a life cycle which is independent of aqueous environments.

7. Which statement about ferns is *not* true?
   a. The sporophyte is larger than the gametophyte.
   b. Most are heterosporous.
   c. The young sporophyte can grow independently of the gametophyte.
   d. The frond is a large leaf.
   e. The gametophytes produce archegonia and antheridia.

8. Which of the following statements regarding Liverworts is *not* true?
   a. The gametophyte is composed of green, leaf-like layers.
   b. The sporophyte is larger than the gametophyte.
   c. Some liverworts use a spring-like "elaters" to catapult spores.
   d. Some species of liverworts can reproduce both sexually and vegetatively.
   e. Gametophytes bear both antheridia and archegonia.

9. Which of the following features is *not* an evolutionary innovation developed by the tracheophytes?
   a. development of tracheid cells.
   b. structural rigidity.
   c. branching, nutritionally independent sporophytes.
   d. ability to transport water and nutrients to distant parts of the plant.
   e. water absorbing rhizoid filaments for uptake of water.

10. Which of the following statements regarding the difference between homosporous and hetersporous plants is *not* true?
    a. homosporous plants produce only one type of spore.
    b. homosporous plants produce gametophytes containing both archegonium and anteridium.
    c. hetersporous plants produce gametophytes that are either male or female, but not both.
    d. megagametophytes bear both female and male reproductive organs.
    e. microspores develop into gametophytes capable of producing only sperm cells.

Student Web Site Self-Quiz Questions

1. Plants are multicellular, photosynthetic eukaryotes. Which one of the following characteristics is definitive of this group?
   a. Starch is the major storage carbohydrate
   b. Cellulose cell walls
   c. Embryo protected by tissues of the parent
   d. Chloroplasts containing chlorophylls *a* and *b*
   e. Life cycle with alternation of generations

2. All plant life cycles feature alternation of generations. Which of the following statements regarding alternation of generations is *false*?
   a. The sporophyte is the multicellular diploid plant.
   b. The gametophyte is a multicellular haploid plant.
   c. Spores mature to form a diploid plant.
   d. The sporophyte generation extends from the zygote to the adult diploid plant.
   e. The gametophyte generation extends from the spore through the haploid adult plant.

3. Most evidence indicates that the closest living relatives of the plants are a group of green algae called the charophytes, which includes the stoneworts. Which of the following characteristics is *not* shared by plants and at least some green algae?
   a. Chloroplast structure
   b. Vascular system with xylem and phloem
   c. Peroxisome contents
   d. Plasmodesmata
   e. Similar rRNA sequences

4. Which of the following is *not* an adaptation to land by plants?
   a. Photosynthesis
   b. Cuticle
   c. Embryos
   d. Thick spore walls
   e. Gametangia

5. The nontracheophytes (such as mosses)
   a. lack a sporophyte generation.
   b. grow in dense masses, allowing capillary movement of water.
   c. possess xylem and phloem.
   d. have gametophytes that are nutritionally independent.
   e. possess true roots.

6. Which of the following statements about the life cycles of various nontracheophytes is *false*?
   a. In the bryophytes, the sporophyte remains attached to the gametophyte.
   b. The first cell of the sporophyte generation is the zygote.
   c. The liverwort's gemmae are part of the gametophyte generation.
   d. The moss capsule is part of the sporophyte generation.
   e. The moss protonema develops from a spore through mitosis and cytokinesis.

7. Which one of the following features applies to the common nonseed tracheophyte, the horsetail?
   a. Does not have a gametophyte stage
   b. Once dominant plants in Earth's biota
   c. Complex leaves
   d. Sporangia in conelike strobili
   e. Produces "fiddleheads"

8. The genus *Rhynia* was among the earliest tracheophytes. Which of the following list of features was *not* found in *Rhynia*?
   a. Xylem and phloem
   b. Horizontal stems (rhizomes)
   c. Meiotically produced spores
   d. Homospory
   e. True leaves

9. Which statement about ferns is *false*?
   a. The sporophyte is larger than the gametophyte.
   b. Most are heterosporous.
   c. The young sporophyte can grow independently of the gametophyte.
   d. The frond is a large leaf.
   e. The gametophytes produce archegonia and antheridia.

TEST FILE

★ *Indicates a* ***difficult*** *question*

■ *Indicates a* ***conceptual*** *question*

29 The Evolution of Seed Plants

Fill in the Blank

1. Cycadophyta, Ginkgophyta, Gnetophyta, and Coniferophyta are all phyla of the **gymnosperms**, of which there are only about 750 living species.

2. Most gymnosperms are wind-pollinated, and most angiosperms are **animal**-pollinated.

3. One group of seed plants, called the **gymnosperms**, include pines and their relatives and have active secondary growth.

4. The phylum **Angiospermae** is also known as the flowering plants. These plants produce seeds, have double fertilization, and vessel elements.

5. The reproductive organ of angiosperms is the **flower**.

6. The ovary of a flowering plant may develop into a **fruit**. This structure may consist only of the mature ovary and its seeds, or it may include other parts of the flower.

7. Also known as seed leaves, **cotyledons** can digest endosperm or become photosynthetic, or both.

8. In a flower, the male organs are contained in the **stamen**, and the female organs are contained in the **pistil**.

9. **Petals** are modified leaves on a flower that can be showy to attract animals.

10. In the seed plants, male gametophytes are called **pollen grains**.

11. A seed may contain tissue from **three** different generations.

12. The sterile sporophyte structures that surround the megasporangium are called the **integument**.

13. Flowers with both megasporangia and microsporangia are said to be **perfect**.

14. The majority of angiosperm species are included in two large lineages: the **monocots** and the **eudicots**.

15. The **endosperm** is triploid tissue that nourishes the embryonic sporophyte during its early development.

16. A stamen is composed of a **filament** bearing an **anther** that contains pollen-producing microsporangia.

17. The structures that bear the megasporangia in the seed plants are called **carpels**.

18. Flowers have specialized sterile leaves. The inner specialized leaves are called **petals** and the outer, **sepals**.

Multiple Choice

1. The dominant vegetation type 200 million years ago, when dinosaurs inhabited Earth, were
   a. early whisk ferns.
   b. horsetail-tree fern forests.
   c. lycopod-fern forests.
   d. gymnosperm forests.
   e. angiosperm forests.

*2. There has been a change in dominant vegetation type since plants first invaded the terrestrial environment about 400–500 million years ago. What was the order in which the following vegetation types were dominant, from earliest to modern day?
   I. Gymnosperm forests
   II. Horsetail-lycopod-fern forests
   III. Whisk fern forests
   IV. Angiosperm forests
   a. I, II, III, IV
   b. II, III, I, IV
   c. III, II, I, IV
   d. IV, III, I, II
   e. I, III, II, IV

3. A seed of a flowering plant or gymnosperm may contain tissues from __________ generation(s).
   a. 1
   b. 2
   c. 3
   d. 4
   e. 5

4. One reason for the enormous evolutionary success of seed plants is their possession of
   a. complex leaves that can photosynthesize at a faster rate so that they can outcompete non-seed-producing plants.
   b. seeds with food reserves for the young sporophyte.
   c. seeds with a resting stage that can remain viable for many years, germinating when conditions are favorable for growth of the sporophyte.
   d. b and c
   e. All of the above

5. The most abundant gymnosperm phylum today contains the cone-bearing plants such as pines. These plants belong to the phylum
   a. Cycadophyta.
   b. Ginkgophyta.
   c. Gnetophyta.
   d. Coniferophyta.
   e. None of the above

6. A universal feature of plant life cycles is
   a. the seed.
   b. the archegonium.
   c. the elater.
   d. the tracheid.
   e. alternation of generations.

*7. One generation of a plant life cycle is called the gametophyte generation. The gametophyte is a multicellular haploid plant that produces haploid gametes. In some plants, the gametophyte is free-living and photosynthetic. Which group does *not* have a free-living gametophyte generation?
   a. Ferns
   b. Gymnosperms
   c. Angiosperms
   d. b and c
   e. None of the above

8. All plant life cycles have alternation of generations, alternating between the gametophyte generation and the __________ generation. This generation extends from the zygote through the adult, multicellular, diploid plant.
   a. heteromorphic
   b. sporophyte
   c. vascular
   d. archegonium
   e. antheridium

9. Coniferous gymnosperms, such as pines, depend primarily on __________ for pollination; thus, the plants produce large quantities of pollen that disperse over large areas during the spring.
   a. insects
   b. birds
   c. water
   d. wind
   e. mammals

10. An evolutionary trend that runs throughout the plant kingdom is that the sporophyte generation __________ and is more independent of the gametophyte, and the gametophyte generation __________ and is more dependent upon the sporophyte.
   a. becomes smaller, becomes smaller
   b. becomes larger, becomes smaller
   c. becomes smaller, becomes larger
   d. becomes larger, becomes larger
   e. does not change in size, becomes larger

11. Angiosperms differ from other plants in that two male gametes, contained within a single male gametophyte, participate in fertilization events. One sperm nucleus combines with the egg to produce a diploid zygote. The other sperm nucleus combines with two other haploid nuclei of the female gametophyte. This process is called
   a. biparental inheritance.
   b. multiple paternity.
   c. double fertilization.
   d. biparental fertilization.
   e. multiple fertilization.

12. The reproductive organ of angiosperms is the
   a. sporangium.
   b. flower.
   c. cone.
   d. archegonium.
   e. sporophyte.

13. Plant species that produce truly male and female plants are
   a. eudicots.
   b. heterozygous.
   c. perfect.
   d. monoecious.
   e. dioecious.

14. The ovary of a flowering plant, together with its seeds, develops into a fruit after fertilization. This structure may consist only of the mature ovary and its seeds, or it may include other parts of the flower or structures closely related to it. A fruit that develops from several carpels of a single flower, such as a raspberry, is a(n) __________ fruit.
   a. aggregate
   b. simple
   c. multiple
   d. accessory
   e. perfect

15. The two major groups of angiosperms are called monocots and eudicots. These plants differ in the number of
   a. sperm involved in fertilization.
   b. sexes per plant; monocots have one sex per plant, eudicots have both.
   c. sexes per plant; eudicots have male and female plants, monocots have both sexes in one plant.
   d. seed leaves.
   e. None of the above

16. Which of the following might you find in an archegonium?
    a. Sperm
    b. An egg
    c. An egg and a sperm
    d. An embryo
    e. An egg, a sperm, and an embryo

17. From an evolutionary standpoint, pollen is a
    a. microsporophyll.
    b. megasporophyll.
    c. microgametophyte.
    d. megagametophyte.
    e. microspore.

18. The prominent components of Earth's modern land flora in most areas are
    a. angiosperms.
    b. gymnosperms.
    c. ferns.
    d. bryophytes.
    e. club mosses.

19–21. Refer to the diagram of a pine seed below to answer the following questions.

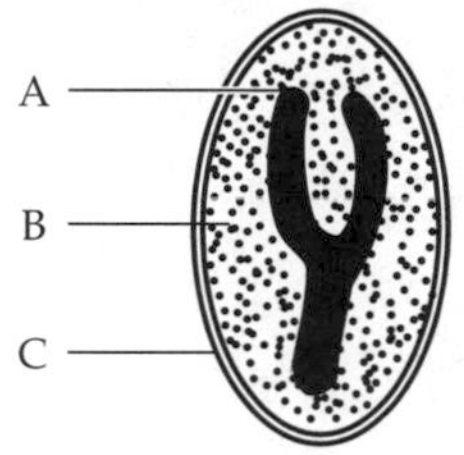

19. Which structure represents the embryo?
    a. A
    b. B
    c. C
    d. A and C
    e. B and C

20. Which structure(s) is(are) diploid?
    a. A
    b. B
    c. C
    d. A and C
    e. B and C

21. The integument portion of the ovule developed into part(s) _________.
    a. A
    b. B
    c. C
    d. A and C
    e. B and C

22. In angiosperms, double fertilization results in the development of
    a. two embryos.
    b. two embryos, but only one survives.
    c. the embryo and the endosperm.
    d. the embryo and the seed coat.
    e. the embryo and the megagametophyte.

23. The seeds in angiosperms are
    a. on the upper surface of the sporophylls.
    b. on the lower surfaces of the sporophylls.
    c. buried within the sporophylls.
    d. enclosed in the ovule.
    e. None of the above

24. Flowers that produce both megasporangia and microsporangia are called
    a. perfect and monoecious.
    b. perfect and dioecious.
    c. imperfect and monoecious.
    d. imperfect and dioecious.
    e. imperfect and monoecious or dioecious.

25. Flowers with insect pollination have showy
    a. petals.
    b. sepals.
    c. tepals.
    d. petals or tepals.
    e. sepals or tepals.

26. In angiosperms, pollen is transferred from the _________ to the _________.
    a. anther, style
    b. filament,ovary
    c. anther, stigma
    d. filament, ovary
    e. anther, ovule

*27. The diploid zygote in angiosperms develops into the
    a. embryonic axis.
    b. embryonic axis and cotyledons.
    c. embryonic axis and endosperm.
    d. embryonic axis, cotyledons, and endosperm.
    e. embryonic axis, cotyledons, endosperm, and seed coat.

28. The pistil consists of
    a. anthers, filaments, and stamen.
    b. ovary, archegonium, and embryo.
    c. stigma, style, and ovary.
    d. sepals, petals, and corolla.
    e. embryo, endosperm, and cotyledons.

29. Seed plants are all
    a. heterosporous.
    b. dioecious.
    c. monoecious.
    d. eudicots.
    e. rely on animals for fertilization.

▪30. One difference between gymnosperms and angiosperms is that gymnosperms
    a. do not form seeds.
    b. do not form flowers.
    c. do not have tracheid cells.
    d. rely on animals for fertilization, while angiosperms do not.
    e. None of the above

31. The pollen tube is used to
   a. eject pollen from the microsporangium.
   b. direct pollen to the megasporangium.
   c. digest the sporophyte tissue toward the female gametophyte.
   d. produce pollen.
   e. attract animals to the plant to spread the pollen.

32–36. Match the phyla below with the following descriptions.
   a. Cycadophyta
   b. Ginkgophyta
   c. Coniferophyta
   d. Gnetophyta
   e. Angiospermae

32. Seeds are in cones; have needlelike or scalelike leaves; include plants such as pines and redwoods. **(c)**

33. Represented today by a single species; common during the Mesozoic era; have fan shaped leaves. **(b)**

34. The palmlike plants of the tropics; the least changed group of the present-day gymnosperms. **(a)**

35. Consists of three very different genera that share characteristics with the angiosperms; have vessels in vascular tissue; opposite, single leaves. **(d)**

36. Seeds in fruit; endosperm; much-reduced gametophytes; flowers. **(e)**

37. Which of the following characteristics is *not* unique to the angiosperms?
   a. The production of triploid endosperm.
   b. They produce fruit.
   c. They have xylem.
   d. They show a reduced gametophyte generation.
   e. **b and d**

38. Bird-pollinated flowers
   a. are often red and odorless.
   b. have characteristic odors.
   c. have conspicuous markings that are evident only in the ultraviolet region of the spectrum.
   d. are always grouped in an inflorescence.
   e. None of the above

*39. The following events in the angiosperm life cycle occur in which order?
   I. A diploid zygote divides repeatedly until an embryonic stage is reached at which growth is temporarily halted.
   II. A pollen grain reaches the appropriate surface of a sporophyte.
   III. Fertilization occurs.
   IV. A slender pollen tube is produced that elongates and digests its way through the sporophytic tissue toward the female gametophyte.
   a. I, II, III, IV
   b. II, IV, III, I
   c. II, I, IV, III
   d. IV, III, I, II
   e. III, I, IV, II

40. You are given two flowers of the same species. One flower has only a pistil, while the other has only a stamen. Based on your observations, you conclude that these flowers are
   a. perfect.
   b. imperfect.
   c. monoecious.
   d. dioecious.
   e. b and d

*41. The angiosperms are sister to which gymnosperm phylum?
   a. Cycadophyta
   b. Coniferophyta
   C. Gnetophyta
   d. Ginkgophyta
   e. The exact gymnosperm phylum is still unknown.

*42. The earliest fossil evidence of gymnosperms is found in rocks from which period?
   a. Devonian
   b. Permian
   c. Silurian
   d. Carboniferous
   e. None of the above

43. Which of the following phyla are *not* classified as gymnosperms?
   a. Cycadophyta
   b. Coniferophyta
   c. Anthocerophyta
   d. Gnetophyta
   e. Ginkgophyta

44. A structure composed of one carpel or two or more fused carpels is called a(n)
   a. stamen.
   b. anther.
   c. pistil.
   d. receptacle.
   e. filament.

45. The corolla and the calyx often play roles in
   a. attracting animal and insect pollinators to the flower.
   b. protecting the immature flower in a bud.
   c. photosynthesis.
   d. spore dispersion.
   e. a and b

46. Which of the following is true of most plant–pollinator interactions?
   a. They are highly specific.
   b. They are not highly specific.
   c. Flowers may have markings or odors to attract certain pollinators.
   d. Pollinators include bees, birds, and bats.
   e. b, c, and d

47. The angiosperm carpel serves to
   a. protect the ovules and seeds.
   b. attract pollinators.
   c. produce sugars via photosynthesis.
   d. prevent self-pollination.
   e. a and d

Study Guide Questions

1. You are enjoying a stroll in the botanical gardens. You notice a plant with a beautiful flower. Upon closer inspection you find that the flower has a pistol but no anthers. You look at several more flowers on the same plant and they are also missing anthers. This plant is ____________ and its flowers are ____________.
   a. monoecious, perfect
   b. monoecious, imperfect
   c. dioecious, perfect
   d. dioecious, imperfect

2. Your sweetheart sent you an incredible rose with many petals on it. When you look closely at the rose, it appears that the petals have sterile anthers attached to them. Which of the following best explains this phenomenon?
   a. Both anthers and petals are derived from leaf structures; therefore, the genes of the anther could be altered to form a petal.
   b. Anthers are derived from petals.
   c. The anthers and the petals probably fused.
   d. None of the above

3. Why are gymnosperms referred to as "naked seed plants?"
   a. They lack ovules.
   b. They lack ovaries.
   c. They don't protect their embryos.
   d. They don't have seed coats.

4. Which of the following functions make seeds useful to plants?
   a. Seeds provide a mechanism for dispersal.
   b. Seeds protect the embryo.
   c. An embryo may remain dormant until optimum growth conditions are available.
   d. All of the above

5. Many angiosperms and animals have coevolved. What roles do animals play in the life cycle of plants?
   a. They act as pollinators.
   b. They assist in dispersal of seeds.
   c. They insure fertilization.
   d. All of the above

6. The product(s) of fertilization in gymnosperms is (are) ____________ and in angiosperms is (are) ____________.
   a. endosperm and embryo, embryo
   b. embryo, endosperm and embryo
   c. embryo, embryo
   d. embryo, endosperm

7. In gymnosperms there is(are) ____________ functional sperm nuclei and in angiosperms there is(are) ____________ functional sperm nuclei.
   a. one, two
   b. two, one
   c. four, one
   d. one, one

8. The diagnostic characteristic of an angiosperm is the presence of
   a. multiple carpels.
   b. woody growth.
   c. the flower.
   d. All of the above

9. Which of the following statements regarding gymnosperms is *true*?
   a. All gymnosperms produce cones.
   b. Gymnosperms are heterosporous.
   c. Gymnosperm seeds have no protection.
   d. Gymnosperms are all woody.

10. Match the following flower structures with their function.

<u>g.</u> Ovule	a. Leaf modification that protects the ovary
<u>j.</u> Anther	b. Colorful display that assists in attracting pollinators
<u>c.</u> Stigma	c. Secretes sticky material to help pollen adhere
<u>h.</u> Style	d. Site of ovule development
<u>a.</u> Carpel	e. Holds the pistol, forms some fruits
<u>d.</u> Ovary	f. Surrounds petals
<u>b.</u> Petal	g. Houses the egg
<u>f.</u> Sepal	h. Nourishes the elongating fertilization tube
<u>e.</u> Receptacle	i. Found in plants in which the petals and sepals cannot be differentiated
<u>i.</u> Tepal	j. Site of microsporogenesis

11. What is the significance of the fruit?
    a. It aids in dispersion of seeds.
    b. It protects seeds until they are mature.
    c. It attracts pollinators.
    d. a and b

12. Vascular tissue in angiosperms is highly developed. The purpose of this vascular tissue is to move
    a. water.
    b. food.
    c. nutrients.
    d. All of the above

End of Chapter Questions

1. Which of the following is *not* a characteristic of conifer reproduction?
   a. cones composed of a cluster of scales/leaves are used for reproduction.
   b. there are (2) types of cones which produce either megaspores or microspores.
   c. pollen grains require water for transport to the female gametophyte.
   d. each pollen grain contains a total of (2) sperm cells.
   e. the megasporangium is enclosed within a structure called integument.

2. Which of the following is *not* a distinguishing characteristic of angiosperms?
   a. They produce diploid endosperm.
   b. They have double-fertilization.
   c. They all have flowers.
   d. They all produce fruit.
   e. Their phloem contains companion cells.

3. Double-fertilization:
   a. requires only a single male gamete.
   b. results in the formation of triploid endosperm.
   c. occurs only in a few species of angiosperms.
   d. requires two egg cells.
   e. results in a tissue with diploid nuclei.

4. Which of the following structures is *not* considered a type of modified leaf?
   a. petals
   b. sepals
   c. stamens
   d. carpels
   e. fibers

5. Which of the following is believed to explain the presence of pistils with long styles and anthers with long filaments?
   a. long anthers increase the likelihood of contact with insects
   b. long styles increase the rate of double-fertilization
   c. long anthers produce more pollen than shorter ones
   d. longer pistils can hold more pollen than shorter ones
   e. longer pistils allow insects to easily feed on the flower

6. Which of the following is *not* a difference between monocot and eudicot plants?
   a. only monocots have a single embryonic cotyledon
   b. only eudicots can store energy in cotyledons
   c. only monocots have triploid endosperm
   d. monocots include species of grasses and palm trees
   e. eudicots include species of trees and shrubs

7. Which of the following did *not* result from co-evolution of angiosperms and animals/insects?
   a. Most angiosperms can be pollinated by either insects or wind.
   b. When insects visit flowers to obtain nectar, they inadvertently pick up pollen grains.
   c. Pollen can be carried from plant to plant by animals/insects that visit flowering plants.
   d. Animals/insects have helped shape the evolution of plants.
   e. Angiosperms have helped shape the evolution of some species of animals.

8. The gymnosperms
   a. dominate all land masses today.
   b. have never dominated land masses.
   c. have active secondary growth.
   d. all have vessel elements.
   e. lack sporangia.

9. Which statement about flowers is *not* true?
   a. Pollen is produced in the anthers.
   b. Pollen is received on the stigma.
   c. An inflorescence is a cluster of flowers.
   d. A species having female and male flowers on the same plant is dioecious.
   e. A flower with both mega- and microsporangia is said to be perfect.

10. Which statement about fruits is *not* true?
    a. They develop from ovaries.
    b. They may include other parts of the flower.
    c. A multiple fruit develops from several carpels of a single flower.
    d. They are produced only by angiosperms.
    e. A cherry is a simple fruit.

Student Web Site Self-Quiz Questions

1. Angiosperms and gymnosperms reproduce using seeds. Which of the following statements about seeds is *false*?
   a. The seed develops on the parent sporophyte.
   b. The seed contains an embryonic stage of a young sporophyte.
   c. The seed may contain tissues from three generations.
   d. The seeds of angiosperms and gymnosperms represent a resting stage, viable for many years.
   e. Endosperm is the female gametophyte.

2. Which of the following is *not* a characteristic of gymnosperm reproduction?
   a. Cones composed of a cluster of scales are used in the process.
   b. Megasporangium is enclosed within integument.
   c. Different cones produce megaspores and microspores.
   d. Water is required to assist transfer of pollen grains to gametophyte.
   e. Two sperm cells are present in each pollen grain.

3. The angiosperms or flowering plants are a diverse group. Which of the following features is *not* mostly restricted to angiosperms?
   a. Flowers and fruits
   b. Both wind and animal pollination
   c. Double fertilization
   d. True seeds
   e. Companion cells in the phloem

4. Double fertilization is a key characteristic of the angiosperms. Which of the following statements regarding double fertilization is *true*?
   a. Double fertilization results in a tissue containing diploid nuclei.
   b. Double fertilization requires two egg cells.
   c. Double fertilization occurs in only a few key species of angiosperms.
   d. Double fertilization results in the formation of a triploid endosperm.
   e. Double fertilization requires a single male gamete.

5. Angiosperms are characterized by possessing specialized cells within their xylem and phloem. Which of the following statements regarding these specialized cells is *false*?
   a. Vessel elements are cells that specialize in the transport of water.
   b. Modified versions of vessel elements are found in some gnetophytes and a few ferns
   c. Fiber cells present in the phloem assist in supporting the plant body.
   d. Vessel elements are primarily located in the xylem.
   e. Companion cells are located in the phloem of angiosperms.

6. Which of the following statements regarding flowers is *false*?
   a. A flower bearing both megasporangia and microsporangia is considered "perfect."
   b. A cluster of flowers is called an "influorescence."
   c. A species having female and male flowers on the same plant is called "dioecious."
   d. The stigma of the flower receives pollen.
   e. The anthers of the flower produce pollen.

7. Which of the following statements explains the presence of anthers with long filaments and pistils with long styles?
   a. Longer pistils allow insects to feed on the flower.
   b. Longer pistils can hold more pollen than shorter ones.
   c. Longer anthers increase contact with insects.
   d. Longer pistils produce more pollen than shorter ones.
   e. Longer styles increase double fertilization rate.

8. Which of the following is *not* a product of the coevolution of angiosperms and animals/insects?
   a. Angiosperms have helped to shape the evolution of some animal species.
   b. Animals and insects both have helped to shape the evolution of some plant species.
   c. Most angiosperms can be pollinated by either insects or wind.
   d. Pollen is inadvertently distributed by animals and insects.
   e. Nectar produced by some plants is the primary attractant for animals and insects.

9. Which of the following is *not* a difference between monocots and eudicots?
   a. Only monocots have a single embryonic cotyledon.
   b. Only eudicots can store energy in cotyledons.
   c. Only monocots have triploid endosperm.
   d. Monocots include species of grasses and palm trees.
   e. Eudicots include species of trees and shrubs.

TEST FILE

★ *Indicates a **difficult** question*

■ *Indicates a **conceptual** question*

30 Fungi: Recyclers, Killers, and Plant Partners

Fill in the Blank

1. Organisms living in mutually beneficial symbiosis with other organisms are called **mutualists**.
2. Sexual reproduction in the fungi is accomplished when two different mating types **fuse**.
3. The cell walls of all fungi consist of the polysaccharide **chitin**, which is also found in the exoskeletons of arthropods.
4. The body of a multicellular fungus is called a **mycelium**.
5. The kingdom Fungi is defined as **heterotrophic** organisms with absorptive nutrition.
6. The cells of a body of a multicellular fungus are organized into rapidly growing individual filaments called **hyphae**.
7. Individual filaments that anchor saprobic fungi to their substrate are called **rhizoids**.
8. There are two types of parasitic fungi: **facultative**, which can grow parasitically but can also grow by themselves, and **obligate**, those that can only grow on their specific hosts.
9. Sexual reproduction in fungi occurs between different **mating types**, which are genetically different, yet can function as either a male or a female.
10. The kingdom of Fungi consists of **four** phyla.
11. Fungi often have a life stage that is **dikaryotic**, which is when a hypha has two different haploid nuclei.
12. Hyphae of conjugating fungi (zygomycetes) are attracted to one another due to the release of chemicals called **pheromones**.
13. Lichens can **reproduce** by producing a thallus or a soredia.
14. The most ancient phylum of fungi, and one that was until recently classified as a protist, is the **Chytridiomycota**.

Multiple Choice

★1. You find an organism that appears to be a fungus and has chitin in its cell wall phagocytizing smaller organisms. The organism is most likely a(n)
   a. plant.
   b. animal.
   c. protist.
   d. bacteria.
   e. fungus.

2. The names of fungal classes are based on important and characteristic structures associated with
   a. reproduction.
   b. nutrition.
   c. ecology.
   d. vegetative growth.
   e. cell division.

3. __________ are organisms that live on dead matter.
   a. Parasites
   b. Saprobes
   c. Anaerobes
   d. Aerobes
   e. Autotrophs

4. The cell walls of all fungi consist of the polysaccharide
   a. chitin.
   b. cellulose.
   c. starch.
   d. silica.
   e. pectin.

5. The body of a multicellular fungus is called a
   a. dikaryon.
   b. hyphae.
   c. rhizoids.
   d. mycelium.
   e. None of the above

6. Individual filaments that anchor saprobic fungi to their substrate are called
   a. dikaryon.
   b. hyphae.
   c. rhizoids.
   d. mycelium.
   e. None of the above

7. The cells of the body of a multicellular fungus are organized into rapidly growing individual filaments called
 a. dikaryon.
 b. hyphae.
 c. rhizoids.
 d. mycelium.
 e. None of the above

8. Which of the following statements is sufficient by itself to identify an unknown organism as belonging to the kingdom Fungi?
 a. It is multicellular and nonphotosynthetic.
 b. It has cell walls and reproduces by spores.
 c. It has filamentous growth and obtains its food by absorption.
 d. It has prokaryotic cells, and cell walls made of chitin.
 e. It is unicellular and eukaryotic.

9. One adaptation that fungi have for absorptive nutrition, in which nutrients are absorbed across the cell surfaces, is
 a. lack of a cell wall.
 b. a low surface area–to–volume ratio.
 c. a high surface area–to–volume ratio.
 d. tolerance of low temperatures.
 e. tolerance of high temperatures.

10. A major role of saprobic fungi in terrestrial ecosystems is to
 a. trap atmospheric carbon dioxide.
 b. return carbon and other elements to the environment for further cycling.
 c. parasitize animals.
 d. parasitize plants.
 e. parasitize protists.

▪11. Fungi have a larger surface area–to–volume ratio than most other multicellular organisms because
 a. every cell along a hypha may be in contact with the environment.
 b. an individual mycelium can grow very large.
 c. hyphae grow together to form a mycelium.
 d. must fungi are microscopic organisms.
 e. chitinous cell walls are more permeable than cellulose cell walls.

12. Which of the following is *not* an economically useful aspect of fungi?
 a. Some species are used commercially to flavor foods.
 b. Some species are edible.
 c. Some species produce alcohol via fermentation.
 d. Some species produce oxygen via fermentation.
 e. Some species produce antibiotics.

13. Many fungi are __________, associating with photosynthetic organisms to form mycorrhizae or lichens.
 a. mutualistic
 b. parasitic
 c. saprobic
 d. photosynthetic
 e. predatory

14. Fungi can be parasitic on
 a. animals.
 b. plants.
 c. protists.
 d. other fungi.
 e. All of the above

15–20. Match the terms from the list below with following descriptions of fungal interactions. Each term may be used once, more than once, or not at all.
 a. Saprobic
 b. Competitive
 c. Predation
 d. Parasitic
 e. Mutualistic

15. Fungus decaying a fallen tree. **(a)**

16. Black stem rust draws nutrition from wheat. The rust damages the wheat plant. **(d)**

17. Fungi grow in association with the roots of soybeans, providing the plants with more minerals. **(e)**

18. A constricting ring formed by *Arthrobotrys* traps a nematode. Fungal hyphae invade and digest the nematode. **(c)**

19. Seed germination in most orchid species depends on the presence of a specific fungus species. The fungus derives nutrients from the seed and seedling. **(e)**

20. Some leaf-cutting ants farm fungi, feeding the fungi and later harvesting and eating them. The ants may even "weed" the fungal gardens by removing other fungal species. **(e)**

▪21. Fungi are more common than bacteria in a jar of jelly in your refrigerator because fungi have a
 a. lower tolerance for highly hypotonic environments.
 b. lower tolerance for highly hypertonic environments.
 c. higher tolerance for highly hypotonic environments.
 d. higher tolerance for highly hypertonic environments.
 e. None of the above

22. Predatory fungi may trap prey with
 a. a constricting ring that traps the prey.
 b. sticky substances secreted by hyphae.
 c. mycorrhizae.
 d. a and b
 e. All of the above

23. The fusion of two different mating types forms a dikaryon or heterokaryon. The term "heterokaryon" emphasizes the fact that
 a. it is haploid.
 b. two nuclei fused in its formation.
 c. their are two different nuclei in a common hypha.
 d. the two nuclei are different.
 e. None of the above

24. What distinguishes the phylum Chytridiomycota from all the other fungi?
 a. It reproduces only asexually.
 b. Its haploid gametes have flagella.

c. It is the only one that is parasitic.
d. It contains a fruiting body.
e. It contains a thallus.

25. Fungi that reproduce only asexually are found in
a. chytridiomycetes.
b. zygomycetes.
c. ascomycetes.
d. basidiomycetes.
e. deuteromycetes.

26. Conidia are
a. Spore produced within sporangia.
b. meiotic products.
c. naked, asexually produced spores.
d. encased diploid spores.
e. a type of basidium.

27. Motile gametes are found in
a. Zygomycota.
b. Ascomycota.
c. Basidiomycota.
d. Deuteromycota.
e. Chytridiomycota.

28. Baker's yeast are
a. hemiascomycetes.
b. euascomycetes.
c. allomyces.
d. basidiomycetes.
e. deuteromycetes.

29. Common morels are classified as
a. basidiomycetes.
b. ascomycetes.
c. zygomycetes.
d. mycorrhizae.
e. lichens.

*30. Which of the following fungi share the same phylum?
a. Cup fungi and bracket fungi
b. Truffles and morels
c. Amanita and powdery mildew
d. Black bread mold and pink bread mold
e. Dutch elm disease and smut fungi

31. Which of the following is *not* true about fungal reproduction?
a. Gamete cells are produced.
b. Only gamete nuclei are produced.
c. There is no true diploid tissue.
d. Gametes are not motile, and thus water is not required for reproduction.
e. All of the above

32. Mycorrhizae are __________ associations of a fungus with __________.
a. mutualistic, an alga or a bacterium
b. mutualistic, plant roots
c. parasitic, an alga or a bacterium
d. parasitic, plant roots
e. saprobic, an alga or a bacterium

33. What is the function of a fruiting structure of a fungus?
a. It distracts predators away from the essential underground parts.
b. It is an important organ for gas exchange with the atmosphere.
c. It is an organ of reproduction.
d. It provides hallucinogens for rodents and mammals.
e. It serves as a landing pad for fungal pollinators.

34. Which of the following plant diseases or parasitic fungi is a basidiomycete?
a. Dutch elm disease
b. Chestnut blight
c. Powdery mildew
d. Green fruit mold
e. Bracket fungus

35. Sexual reproduction may be found in all of the following *except*
a. zygomycetes.
b. ascomycetes.
c. basidiomycetes.
d. deuteromycetes.
e. lichens.

36. Plants with active mycorrhizae
a. benefit nutritionally from this arrangement.
b. display enhanced absorption of water and minerals (especially phosphorus).
c. are heavily parasitized and die.
d. a and b
e. None of the above

37–41. Match the classes of Eumycota from the list below with the following descriptions.
a. Zygomycetes
b. Ascomycetes, subgroup euascomycetes
c. Ascomycetes, subgroup hemiascomycetes
d. Basidiomycetes
e. Deuteromycetes

37. No fleshy fruiting body; no hyphal cross-walls; reproduce sexually by conjugation. **(a)**

38. Perforated cross-walls; no specialized fruiting structures; includes baker's and brewer's yeast. **(c)**

39. Common name is club fungi; complete cross-walls; includes puffballs, mushrooms, wheat rust, smut fungi, mycorrhizae. **(d)**

40. Perforated cross-walls produced in a specialized fruiting structure (called the perithecium); includes molds, parasites such as Dutch elm disease, and epicurean delights such as morels and truffles. **(b)**

41. No known sexual stages, it is likely the sexual stages were lost in evolution. **(e)**

42. Virtually all oak trees and pine trees depend on mycorrhizal fungi to absorb nutrients. The relationship between the tree and the fungus is an example of
a. saprobism.
b. parasitism.
c. mutualism.
d. heterotropism.
e. commensalism.

*43. Rusts and smuts are pathogens of cereal grains classified in the
a. zygomycetes.
b. ascomycetes.
c. basidiomycetes.
d. deuteromycetes.
e. lichens.

44. An unknown fungus was examined microscopically and found to lack cross-walls in its hyphae. It must belong to the
a. zygomycetes.
b. ascomycetes.
c. basidiomycetes.
d. deuteromycetes.
e. lichens.

45. Cells of ascomycetes typically have
a. cellulose cell walls.
b. perforated cross-walls.
c. no sexual stages.
d. diploid nuclei.
e. mycorrhizae.

46. Dikaryotic cells
a. have two identical nuclei per cell.
b. contain pairs of homologous chromosomes.
c. divide only by meiosis.
d. contain two genetically different nuclei per cell.
e. contain diploid nuclei.

47. A sexually produced spore that buds from the surface of a basidium is a
a. zygospore.
b. ascospore.
c. conidiospore.
d. basidiospore.
e. uredospore.

48. Ascomycetes produce conidia
a. asexually.
b. sexually.
c. in response to harsh environmental conditions.
d. Conidia are produced by basidiomycetes, not ascomycetes.
e. Conidia are produced by zygomycetes, not ascomycetes.

49. The gills of a mushroom (basidiomycetes) are specialized for
a. respiration.
b. food production.
c. defense.
d. reproduction.
e. water storage.

50. The algal partner in a lichen symbiosis is responsible primarily for
a. respiration.
b. food production.
c. defense.
d. reproduction.
e. water storage.

51. Lichens obtain nutrients by
a. photosynthesis.
b. engulfing other organisms.
c. absorbing nutrients from the environment.
d. decaying organic material.
e. parasitizing flowering plants.

52. In a lichen, which part of the fungus is involved directly in the symbiosis?
a. Fruiting body
b. Mycelium
c. Spores
d. Spore cases
e. Blue-green bacterium

53. Which chemical should one expect to find as part of the fungal partner in a lichen?
a. Chitin
b. Chlorophyll
c. Reverse transcriptase
d. Silica
e. Cellulose

54. Lichens acquire energy from
a. algae.
b. cyanobacteria.
c. the sun.
d. minerals in the air and precipitation.
e. minerals on rocks.

55. Lichens are __________ associations of a fungus with __________.
a. mutualistic, an alga or a bacterium
b. saprobic, an alga or a bacterium
c. parasitic, an alga or a bacterium
d. mutualistic, plant roots
e. parasitic, plant roots

56. Soredia are characteristic of some
a. zygomycetes.
b. ascomycetes.
c. basidiomycetes.
d. deuteromycetes.
e. lichens.

Study Guide Questions

1. The phyla of fungi are distinguished based on
a. methods of sexual reproduction.
b. methods of asexual reproduction.
c. presence or absence of cross-walls separating the cell-like compartments.
d. All of the above

2. Which of the following would you *not* expect to find in the life cycle of a typical fungus?
a. Haploid cells
b. Diploid cells
c. Spores
d. Chloroplasts

3. All fungi are absorptive heterotrophs. Which of the following is an adaptation shown by many fungi that greatly aids this mode of nutrient procurement?

a. Dikaryosis
b. Large surface-to-volume ratio
c. Conjugation
d. Complex life cycle

4. Assume that two normal hyphal cells of different fungal mating types unite. After a period of time the cell walls between these cells will dissolve, producing a
a. mycelium.
b. fruiting body.
c. zygote.
d. dikaryotic cell, which is also heterokaryotic.

5. The dikaryotic hyphae are
a. $2n$
b. n
c. $n + n$
d. n/n

6. Which of the following statements about sexual reproduction in fungi is *false*?
a. Motile gametes are present in all fungal species.
b. Water is not required for fertilization to occur in fungi.
c. There is no true diploid tissue in the life cycle of most sexually reproducing fungi.
d. Sexual reproduction often begins with contact between opposite mating types.

7. A lichen is
a. an imperfect fungi.
b. the fruiting structure of the Basidiomycete
c. a symbiotic association of fungus and algae.
d. a symbiotic association of fungus and plants.

8. Which of the following is not a life form of fungi?
a. Saprobe
b. Parasite
c. Mutualist
d. Herbivore

End of Chapter Questions

1. Which statement about fungi is *not* true?
a. A hyphal fungus has a body called a mycelium.
b. Hyphae are composed of individual mycelia.
c. Many fungi tolerate highly hyperosmotic environments.
d. Many fungi tolerate low temperatures.
e. Some fungi are anchored to their substrate by rhizoids.

2. The absorptive nutrition of fungi is aided by
a. dikaryon formation.
b. spore formation.
c. the fact that they are all parasites.
d. their large surface area–to–volume ratio.
e. their possession of chloroplasts.

3. Which statement about fungal nutrition is not *true*?
a. Some fungi are active predators.
b. Some fungi form mutualistic associations with other organisms.
c. All fungi require mineral nutrients.
d. Fungi can make some of the compounds that are vitamins for animals.
e. Facultative parasites can grow only on their specific hosts.

4. Which statement about heterokaryosis is *not* true?
a. The cytoplasm of two cells fuses before their nuclei fuse.
b. The two haploid nuclei are genetically different.
c. The two nuclei are of the same mating type.
d. The heterokaryotic stage ends when the two nuclei fuse
e. Not all fungi have a heterokaryotic stage.

5. Reproductive structures consisting of a photosynthetic cell surrounded by fungal hyphae are called
a. ascospores.
b. basidiospores.
c. conidia.
d. soredia.
e. gametes.

6. The zygomycetes
a. have hyphae without regularly occurring septa (cross-walls).
b. produce motile gametes.
c. form fleshy fruiting bodies.
d. are haploid throughout their life cycle.
e. are only capable of asexual reproduction.

7. Which statement about ascomycetes is *not* true?
a. They include the yeasts.
b. They form reproductive structures called asci.
c. Their hyphae are segmented by septa (cross-walls).
d. Many of their species have a dikaryotic stage.
e. All have fruiting structures called ascocarps.

8. The basidiomycetes
a. often produce fleshy fruiting structures.
b. have hyphae without septa (cross-walls).
c. have no sexual stage.
d. never produce large fruiting structures.
e. form diploid basidiospores.

9. The imperfect fungi
a. have distinctive sexual stages.
b. are all parasitic.
c. are also collectively called deuteromycetes.
d. include the ascomycetes.
e. are never components of lichens.

10. Which statement about lichens is *not* true?
a. They can reproduce by fragmentation of their vegetative body.
b. They are often the first colonists in a new area.
c. They render their environment more basic (alkaline).
d. They contribute to soil formation.
e. They may contain less than 10 percent water by weight.

Student Web Site Self-Quiz Questions

1. The body plan of most fungi is a mass of filamentous cells. Which of the following is *not* a feature that applies to all members of this kingdom?
   a. Absorptive heterotrophs
   b. Presence of characterizing sexual structures
   c. Parasitic, saprobic, or mutualistic nutritional mode
   d. Cells with some chitin
   e. Spore forming

2. Which of the following statements about nutrition in the fungi is *false*?
   a. A facultative parasite can grow parasitically or by itself on defined media.
   b. No known fungus gets its nitrogen directly from nitrogen gas.
   c. The filamentous body plan of a fungus is adaptive to an absorptive life style because it provides a small surface area-to-volume ratio.
   d. Some fungi are active predators on protists or animals.
   e. Mycorrhizae are mutualistic associations between fungi and the roots of higher plants.

3. Which of the following statements about reproduction in the fungi is *false*?
   a. Spores can be produced sexually or asexually.
   b. In some fungi, sexual reproduction begins with formation of a dikaryon.
   c. Spores can be produced in structures called sporangia or "naked" at the tips of hyphae (such naked spores are called "conidia").
   d. There are no male or female organisms in fungi—just two or more mating types.
   e. Although the dikaryon has genes from both parental cells, the genes from only one parent may be expressed at any time.

4. Which of the following statements about the chytrids is *false*?
   a. As with other fungi, the chytrids do not have any motile cells.
   b. Some chytrids have an alternation of multicellular haploid and diploid forms.
   c. Most chytrids are aquatic.
   d. Chytrids were once considered to be protists.
   e. The chytrids are the most ancient fungal phylum.

5. The euascomycetes include such forms as cup fungi. Which of the following features does *not* distinguish the euascomycetes from the hemiascomycetes?
   a. Production of conidia
   b. Mostly multicellular forms
   c. Formation of a dikaryon stage
   d. Production of an ascocarp
   e. Asci

6. The ascomycetes include all of the following fungi *except*
   a. yeasts.
   b. bracket fungi.
   c. those producing fruiting structures called morels.
   d. the chestnut blight and Dutch elm disease fungi.
   e. Neurospora.

7. Which one of the following types of fungi is *not* in the phylum Basidiomycetes?
   a. Smuts
   b. Puffballs
   c. Mushroom-producing fungi
   d. Powdery mildews
   e. Bracket fungi

8. The basidiomycetes are a diverse group of fungi with a characteristic reproductive structure called a basidium. The basidium is equivalent to the __________ in the ascomycetes.
   a. sporangiophore
   b. conidia
   c. ascus
   d. zygosporangium
   e. ascocarp

9. Which of the following statements about mycorrhizae is *false*?
   a. In ectomycorrhizae, the association between the fungus and plant is less intimate than in endomycorrhizae.
   b. The mycorrhizal association of fungus and plant may have had importance in the evolution of land plants.
   c. The mycorrhizal association is a mutualistic symbiosis.
   d. Only advanced modern plants (such as the angiosperms) possess mycorrhizae.
   e. Mycorrhizae only involve the roots of the plant.

10. Lichens are composite organisms made up of a fungus and a photosynthetic microorganism. Which combination of organismal types is *least* commonly found in a lichen?
    a. Zygomycete, cyanobacterium
    b. Ascomycete, cyanobacterium
    c. Ascomycete, green alga
    d. Basidiomycete, cyanobacterium
    e. Basidiomycete, green alga

11. Which of the following statements about lichens is *false*?
    a. Lichens are sensitive to air pollution because they have no way to excrete toxic substances.
    b. The association between a fungus and a photosynthetic microorganism is mutualism.
    c. Lichens are restricted to aquatic environments.
    d. Soredia include fungal hyphae, but not ascospores or basidiospores.
    e. The flora of Antarctica is dominated by lichens.

TEST FILE

★ *Indicates a **difficult** question*

■ *Indicates a **conceptual** question*

31 Animal Origins and Lophotrochozoans

Fill in the Blank

1. **Bilateral** symmetry is strongly correlated with the development of sense organs and central nervous tissues at the anterior end of an animal, a process known as cephalization.

2. Animals probably arose from protists whose cells remained together after division, forming a multicellular colony. Among those alive today, the animals that are most similar to the probable ancestral colonial protists are the **sponges**.

3. Cnidaria possess **radial** symmetry, having a cylindrical form with one main axis around which body parts are arranged.

4. Platyhelminthes, nematodes, rotifers, annelids, and arthropods possess a body plan that shows **bilateral** symmetry.

5. **Protostomes** are animals in which cleavage of the fertilized egg is determinate; that is, if the egg is allowed to undergo cell divisions, and the cells are then separated, each cell will develop into only a partial embryo.

6. Animals that lack an internal body cavity, such as Porifera, Cnidaria, and Platyhelminthes, are called **acoelomates**.

7. A distinguishing feature of mollusks is the **mantle**, a sheet of specialized tissue that covers the internal organs like a body wall. This structure secretes the shell.

8. Biologists call the ancestors of the bilaterally symmetrical animals the **urbilateria**.

9. The **ventral** side of the bilaterally symmetrical animal is the surface with the mouth.

10. The oligochaetes are **hermaphrodites**; each organism has both male and female sex organs.

11. Cephalopods and gastropods are lineages of the phylum **Mollusca**.

12. The embryos of **diploblastic** animals have two cell layers: an outer **ectoderm** and an inner **endoderm**.

13. The embryos of **triploblastic** animals have three cell layers: an outer **ectoderm**, an inner **endoderm**, and a middle layer called the **mesoderm**.

14. The fluid-filled body cavities of early animals functioned as **hydrostatic** skeletons.

15. **Spherical** symmetry is the simplest type of symmetry, in which body parts radiate outward from a central point. It is widespread among the protists.

16. Animals that are attached to a substrate and do not move about are said to be **sessile**.

17. The generalized cnidarian life cycle has two stages: the **polyp** and the **medusa**.

18. Evidence suggests that the protostomes split into two major lineages that have been evolving separately since ancient times. These two lineages are the **lophotrochozoans** and the **ecdysozoans**.

19. Many lophotrochozoans have a type of free-living larva called a **trochophore**.

20. Members of the class **Hirudinea** are used medically to reduce fluid pressure and prevent blood clotting.

Multiple Choice

1. A phylum that includes animals with subdivided coeloms is
   a. Ctenophora.
   b. Porifera.
   c. Annelida.
   d. Platyhelminthes.
   e. Cnidaria.

■2. Animals can be divided into two major groups based on major evolutionary lineages that separated in the Cambrian. Those two lineages differ fundamentally in their
   a. modes of reproduction.
   b. early embryological development.
   c. mode of obtaining and storing energy.
   d. environmental conditions needed to live.
   e. metabolism.

3–6. Refer to the diagram below of a cross section of a sponge to answer the following questions.

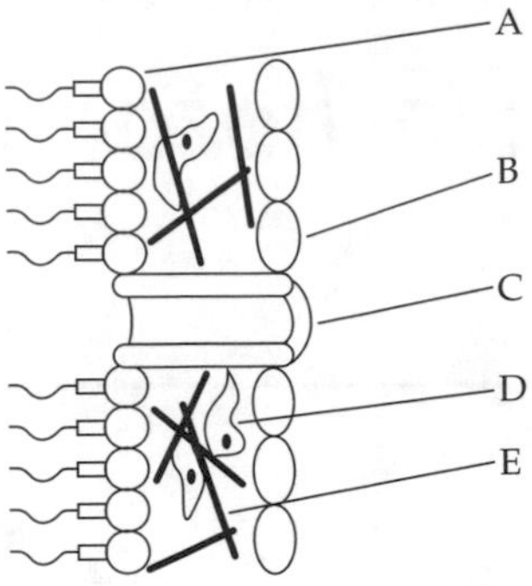

3. Which of the above structures is a choanocyte? **(a)**

4. Which of the above structures is an osculum? **(c)**

5. Which of the above structures is an epidermal cell? **(b)**

6. Which of the above structures is not a cell? **(e)**

7. Which of the following statements concerning nematocysts is *true*?
   a. They are used in the capture of prey by cnidarians.
   b. They are excretory organs in Platyhelminthes.
   c. They are reproductive cells in cnidarians.
   d. They are ciliated cells in Porifera.
   e. They are excretory cells in Porifera.

8. Which class within the Cnidaria is important geologically because the growth of its members can result in the formation of islands and atolls in tropical oceans?
   a. Hydrozoa
   b. Scyphozoa
   c. Anthozoa
   d. Turbellaria
   e. Ctenophora

▪9. Animals in the phylum Platyhelminthes are often parasitic. Which of the following is an adaptation found in Platyhelminthes to cope with a parasitic lifestyle?
   a. A flat shape
   b. A highly branched gastrovascular cavity
   c. An oxygen transport system
   d. Absorption of nutrients through the body surface
   e. b and c

▪*10. The phylum Platyhelminthes differs from the phylum Cnidaria in that
   a. platyhelminthes are radially symmetrical; cnidarians are bilaterally symmetrical.
   b. platyhelminthes have more complex internal organs than cnidarians do.
   c. platyhelminthes are diploblastic; cnidarians are triploblastic.
   d. platyhelminthes have two openings to the gastrovascular cavity; cnidarians have only a single opening to the gastrovascular cavity.
   e. platyhelminthes have an excretory system; cnidarians do not.

▪11. An interesting feature of many members of the class Trematoda is their complex life cycle. Understanding the life cycles of trematode parasites is of great importance to humans because
   a. it is often possible to identify a portion of the life cycle where the parasite is vulnerable and can be controlled.
   b. although these parasites are of little importance in modern society, they must be studied before they are completely eliminated.
   c. it is always important to know the biology of the diseases of humans.
   d. it helps us to understand the evolutionary relationships between the different parasites.
   e. it emphasizes the interrelationship between humans and their environment.

*12. In which phylum did bilateral symmetry first appear?
   a. Porifera
   b. Cnidaria
   c. Platyhelminthes
   d. Nematoda
   e. Rotifera

*13. Which phyla contain acoelomate animals?
   a. Porifera and Rotifera
   b. Porifera and Cnidaria
   c. Porifera, Cnidaria, and Platyhelminthes
   d. Porifera, Cnidaria, Platyhelminthes, and Nematoda
   e. Porifera, Cnidaria, Platyhelminthes, Nematoda, and Rotifera

*14. Which phyla contain coelomate animals?
   a. Arthropods and Rotifera
   b. Arthropods and Annelida
   c. Arthropods, Annelida, and Rotifera
   d. Arthropods, Annelida, Rotifera, and Nematoda
   e. Arthropods, Annelida, Rotifera, Nematoda, and Platyhelminthes

*15. Which phylum or phyla contain animals with no digestive tract?
   a. Porifera
   b. Porifera and Cnidaria
   c. Porifera, Cnidaria, and Platyhelminthes
   d. Porifera, Cnidaria, Platyhelminthes, and Ctenophora
   e. Porifera, Cnidaria, Platyhelminthes, Ctenophora, and Rotifera

*16. Which phylum or phyla contain animals with a complete digestive tract?
   a. Nematoda
   b. Annelida
   c. Annelida and Rotifera
   d. Annelida, Rotifera, and Ctenophora
   e. Annelida, Rotifera, Nematoda, and Platyhelminthes

▪17. The main difference between a coelom and a pseudocoelom is that
   a. a coelom is lined with tissue of mesodermal origin, whereas a pseudocoelom is not.

b. a coelom is a body cavity that is outside of the gastrovascular cavity, whereas a pseudocoelom is an outgrowth of the gastrovascular cavity.
c. a coelom is the developmental end point of the blastocoel, whereas in the pseudocoelomate animals the blastocoel is obliterated.
d. a coelom is lined with tissue of endodermal origin, whereas a pseudocoelom is lined with tissue of ectodermal origin.

18. The main body parts common to all mollusks are the
a. foot, the radula, and the mantle.
b. foot, the mantle, and the shell.
c. visceral mass, the radula, and the mantle.
d. visceral mass, the mantle, and the shell.
e. foot, the visceral mass, and the mantle.

19. Which of the following statements concerning the mantle cavity of mollusks is *false*?
a. It is no longer present in octopuses.
b. It has been modified into internal support in slugs and squids.
c. It is used as a filtering device by bivalves.
d. It is used as the basis of jet propulsion by cephalopods.
e. It secretes the shell that provides external protection in most molluscan groups.

20. Cephalization is most commonly associated with
a. spherical symmetry.
b. radial symmetry.
c. biradial symmetry.
d. bilateral symmetry.
e. sessile animals.

*21. Deuterostomes and protostomes differ in a number of categories. From the following list, select one characteristic in which they do *not* differ.
a. Cleavage type
b. Ability to form a blastopore
c. Origin of the mouth
d. Derivation of mesoderm
e. Manner in which the coelom is formed

22. Which of the following features is *not* characteristic of a true coelom?
a. It is located between the endoderm and ectoderm
b. It is derived from the mesoderm
c. It must arise as an outpocketing of the gut tube
d. It is a body cavity lined with peritoneum
e. It is found in both protostomes and deuterostomes

*23. Which of the following events was probably *not* important in the evolution of animals?
a. Increase in atmospheric oxygen concentration
b. Evolution of division of labor among the cells of colonial protists
c. Evolution of noncyclic photophosphorylation
d. Success of cyanobacteria
e. Abundant zooplankton

24. Which of the following is *not* characteristic of the body plan of members of the phylum Porifera?
a. Gastrovascular cavity
b. No distinct tissue layers or organs
c. No separation between the different cell layers
d. Organization around water tubes
e. Asymmetry

25. Which of the following structures is *not* associated with a typical sponge?
a. Choanocytes
b. Porocytes
c. Spicules
d. Mesoglea
e. Amoebocytes

26. Which of the following characteristics is *not* associated with members of the phylum Cnidaria?
a. Alternation between polyp and medusa
b. Three distinct body layers
c. Nematocysts
d. Gastrovascular cavity
e. Planula larva

27. Which of the following is associated with both the medusa and polyp stage in a typical hydrozoan cnidarian?
a. Asexual reproduction
b. Sessile
c. Gastrovascular cavity
d. Produces planula larva
e. Thick mesoglea

28. Select the following characteristic that is a feature unique to the cnidarians in the class Anthozoa.
a. Most species are colonial.
b. They are commonly known as jellyfish.
c. Only the polyp form exists.
d. Gametes are produced directly by the medusa
e. They possess nematocysts

*29. Which of the following features is *not* considered to be a major factor in the overall success of the cnidarians?
a. Unidirectional movement of food through the gut
b. Symbiosis with dinoflagellates
c. Low metabolic rate
d. Nematocysts
e. Ability to subdue large prey

30. In free-living flatworms, which of the following functions occurs by simple diffusion?
a. Respiration
b. Absorption of nutrients
c. Excretion
d. Distribution of nutrients within the body
e. All of the above

31. Which of the following is *not* associated with a parasitic lifestyle as seen in tapeworms and flukes?
a. Complex digestive systems
b. Flattened body
c. Extensive reproductive organs
d. Life cycles with multiple hosts
e. Reduced gastrovascular cavity

32. Which of the following pseudocoel-containing phyla has members that use cilia for both locomotion and feeding?
   a. Nemertea
   b. Platyhelminthes
   c. Nematoda
   d. Rotifera
   e. Annelida

33. In the molluscan body plan, select the body part that secretes the shell.
   a. Mantle
   b. Foot
   c. Visceral mass
   d. Radula
   e. Spicules

34. Which phylum or phyla do *not* contain wormlike organisms?
   a. Platyhelminthes
   b. Ctenophora
   c. Annelida
   d. Porifera
   e. b and d

35. Which phylum has a circulatory system, a complete digestive tract, and a reduction of the coelom?
   a. Rotifera
   b. Annelida
   c. Mollusca
   d. Ctenophora
   e. Platyhelminthes

36. You discover a dioecious marine animal that has parapodia. Select the phylum to which this animal most likely belongs.
   a. Ctenophora
   b. Annelida
   c. Arthropoda
   d. Mollusca
   e. Platyhelminthes

37. Clues to the evolutionary relationships of animals can be found in
   a. the fossil record.
   b. developmental patterns.
   c. anatomy and physiology.
   d. nucleotide sequence patterns.
   e. All of the above

38. The terms "acoelomate," "pseudocoelomate," and "coelomate" are used to describe
   a. cephalization.
   b. origin of the blastopore.
   c. ectoderm, mesoderm and endoderm.
   d. the origin of the body cavity of animals.
   e. the vertebrate body plan.

▪39. Corals, members of the class Anthozoa, flourish in nutrient-poor, clear, tropical waters because
   a. symbiotic dinoflagellates that reside in that environment provide corals with carbohydrates.
   b. coral cnidoblasts and nematocysts are highly efficient.
   c. their digestive cells contain enzymes similar to those found in mammals.
   d. a and c
   e. a, b, and c

40. Which of the following is *not* true of members of the phylum Platyhelminthes?
   a. Some are parasitic, such as those of the classes Cestoda and Trematoda.
   b. They possess a mouth but no anus.
   c. They are diploblastic.
   d. They possess some cephalization.
   e. Some possess complex life cycles.

41. The body cavity of coelomate animals is called peritoneum and arises from
   a. endoderm tissue.
   b. ectoderm tissue.
   c. mesoderm tissue.
   d. the pseudocoel.
   e. mesenchyme.

*42. Porifera differ from most other animals in that they
   a. have no distinct tissues and no cavities between layers.
   b. have no distinct tissues and are sessile.
   c. are sessile and are pseudocoelomate.
   d. are acoelomates and are triploblastic.
   e. are triploblastic and have no distinct tissues.

43. The Portuguese man-of-war is an example of a deadly
   a. poriferan.
   b. cnidarian.
   c. ctenophore.
   d. nematode.
   e. mollusk.

44. Which phylum or phyla are *not* members of the lophotrochozoan lineage?
   a. Porifera
   b. Annelida
   c. Mollusca.
   d. Ctenophora.
   e. a and d

45. Which of the following is *not* true regarding hydrozoans?
   a. Polyps are often colonial.
   b. All species have a polyp-dominated life cycle.
   c. Colonial polyps can be interconnected and share gastrovascular cavities.
   d. A single planula can give rise to an entire colony of polyps.
   e. c and d

46. Corals are members of the class
   a. Anthozoa.
   b. Cnidaria.
   c. Porifera.
   d. Ctenophora.
   e. Annelida.

47. Which of the following is *not* associated with the phylum Rotifera?
   a. Pseudocoelom
   b. Complete gut
   c. Conspicuous feeding organs
   d. Radial symmetry
   e. Movement by beating cilia

*48. Which of the following classes does *not* belong to the phylum Annelida?
   a. Scyphozoa
   b. Polychaeta
   c. Oligochaeta
   d. Hirudinea
   e. a and d

49. Which of the following is *not* true concerning oligochaetes?
   a. They are hermaphroditic
   b. They have separate nerve centers called ganglia
   c. They have a segmented coelom
   d. They have serially repeated organs
   e. None of the above

50. Which of the following organisms are hermaphroditic?
   a. Leeches
   b. Earthworms
   c. Snails
   d. Squids
   e. a and b

51. Which of the following traits is *not* shared by all animals?
   a. Special types of cell–cell junctions
   b. A common set of extracellular matrix molecules
   c. A complete gut
   d. Similarities in their 5S and 18S ribosomal RNAs
   e. c and d

52. Members of the class Anthozoa (the corals) are able to survive in nutrient-poor tropical waters. They rely on a symbiotic relationship with photosynthetic _________ to accomplish this.
   a. green algae
   b. red algae
   c. brown algae
   d. dinoflagellates
   e. None of the above

53. Fossilized traces of urbilaterian movements from late Precambrian times suggest they may have possessed which of the following characteristics?
   a. Circulatory systems
   b. Bony skeletons
   c. Antagonistic muscles
   d. a and c
   e. a, b, and c

54. Which of the following shared, derived traits are *not* among those that unite the protostomes?
   a. Dorsal nervous system
   b. An anterior brain
   c. Free floating larva
   d. Paired or fused longitudinal nerve cords
   e. a and d

55. Many lineages of lophotrochozoans have a type of free-living larva known as a
   a. polyp.
   b. planula.
   c. trochophore.
   d. nauplius.
   e. None of the above

*56. The largest and most remarkable vestimentiferans live near deep-ocean hydrothermal vents. Their tissues harbor endosymbiotic prokaryotes that fix carbon by oxidizing
   a. CH_4.
   b. H_2S.
   c. $C_6H_{12}O_6$.
   d. H_2O.
   e. a and b

57. Mollusks have a rasping feeding structure known as the
   a. proboscis.
   b. rhynchocoel.
   c. radula.
   d. corona.
   e. mastax.

58. The cephalopods have a modified exit siphon and mantle that allow them to
   a. ingest prey.
   b. alter their buoyancy.
   c. move rapidly through the water.
   d. attach to a substrate.
   e. b and c

59. Which phylum or phyla are *not* members of the lophophorate lineage?
   a. Nemertea
   b. Bryozoa
   c. Phoronida
   d. Pterobranchia
   e. a and c

*60. A fertilized egg that has divided a few times is separated into individual cells. These cells proceed to develop into partial embryos. This egg must be from a _________ species.
   a. protostomate
   b. deuterostomate
   c. hermaphroditic
   d. A conclusion cannot be drawn from this observation.
   e. None of the above

Study Guide Questions

1. Which of the following is *not* a derived trait that is shared by all animals?
   a. Similarities in the 5S and 18S ribosomal RNA of animals
   b. The extracellular matrix molecule collagen
   c. Tight junctions, desmosomes, and gap junctions
   d. Bilateral symmetry

2. Which of the following traits do the Ctenophora share with the Lophotrochozoans?
   a. Complete gut
   b. Three body layers
   c. Indeterminate cleavage
   d. Internal skeleton

3. An animal is known to be a deuterostome. Which of the following does *not* apply to this animal?
   a. Three distinct layers of tissue were present during development.
   b. If a coelom is present, it formed from the embryonic gut.
   c. Its early embryonic cleavage pattern was radial.
   d. It is diploblatic.

4. The primary reason animals are more diverse than plants is the considerable variation in animal
   a. multicellularity.
   b. embryology.
   c. methods of food acquisition.
   d. symmetry.

5. Which of the following statements about the body cavity of animals is *true*?
   a. The body cavity of coelomates develops from the embryonic ectoderm.
   b. The acoelomates' body cavity is filled with liquid.
   c. The pseudocoel of the pseudocoelomates have a peritoneum.
   d. The acoelomates do not have an enclosed body cavity.

6. Sponges have a very simple body plan. Which of the following statements about sponge structure or function is *false*?
   a. Choanocytes are flagellated cells that play a role in feeding.
   b. Large species are found in areas of heavy wave action, where food is most abundant.
   c. Individual sponges are both male and female.
   d. Water enters a sponge through pores and exits via one or more oscula.

7. Which of the following traits is *not* shared by both sea anemones and jellyfish?
   a. A medusa as the dominant stage in the life cycle
   b. Possession of a gastrovascular cavity
   c. Sexual reproduction
   d. Nematocysts on the tentacles

8. Which of the following traits is *not* shared by the Ctenophora and the Cnidaria?
   a. Both are diploblastic.
   b. Both have radial symmetry.
   c. Both have complete guts.
   d. Both have feeding tentacles.

9. The rhyncocoel is a body plan feature found in which of the following groups?
   a. Trematoda only
   b. Rotifera and Nematoda
   c. Rotifera only
   d. Nemertea only

10. Which of the following phylum is *not* a lophophore?
    a. Bryozoa
    b. Phoronida
    c. Annelida
    d. Brachiopoda

11. Which of the following statements is *not* true of the Rotifera?
    a. They have a complete gut with an anterior mouth and posterior anus.
    b. They are coelomates.
    c. The corona is a ciliated organ used in acquiring food.
    d. They use a hydrostatic skeleton.

12. A meal containing snails, clams, and octopus represents which of the following classes of mollusk?
    a. Bivalvia only
    b. Gastropoda and Cephalopoda
    c. Bivalvia and Gastropoda
    d. Bivalvia, Gastropoda, and Cephalopoda

13. The combination of a true coelom and repeating body segmentation allows the annelids (unlike the anatomically "simpler" worms) to do which of the following?
    a. Attain complex body shapes and thus locomote more precisely
    b. Move through loose marine sediments
    c. Be hermaphroditic
    d. Inject paralytic poisons into their prey

14. Which class of Cnidarian contains species that live symbiotically with dinoflagellates?
    a. Hydrozoa
    b. Ctenophora
    c. Scyphozoa
    d. Anthozoa

End of Chapter Questions

1. The body plan of an animal is
   a. its general structure.
   b. the functional interrelationship of its parts.
   c. its general form and the functional interrelationship of its parts.
   d. its general form and its evolutionary history.
   e. the functional interrelationship of its parts and its evolutionary history.

2. A bilaterally symmetrical animal can be divided into mirror images by
   a. any cut through the midline of its body.
   b. any cut from its anterior to its posterior end.
   c. any cut from its dorsal to its ventral surface.
   d. only a cut through the midline of its body from its anterior to its posterior end.
   e. only a cut through the midline of its body from its dorsal to its ventral surface.

3. Among protostomes, cleavage of the fertilized egg is
   a. delayed while the egg continues to mature.
   b. determinate; cells separated after a few divisions develop into only partial embryos.
   c. indeterminate; cells separated after a few divisions develop into complete embryos.

d. triploblastic.
e. diploblastic.

4. The sponge body plan is characterized by
   a. a mouth and digestive cavity but no muscles or nerves.
   b. muscles and nerves but no mouth or digestive cavity.
   c. a mouth, digestive cavity, and spicules.
   d. muscles and spicules but no digestive cavity or nerves.
   e. no mouth, digestive cavity, muscles, or nerves.

5. The phyla of diploblastic animals are
   a. Porifera and Cnidaria.
   b. Cnidaria and Ctenophora.
   c. Cnidaria and Platyhelminthes.
   d. Ctenophora and Platyhelminthes.
   e. Porifera and Ctenophora.

6. Cnidarians are abundant, perhaps because of their ability to
   a. live in both salt and fresh water.
   b. move rapidly in the water column.
   c. capture and consume large numbers of small prey.
   d. capture large prey and their low metabolic rate.
   e. capture large prey and to move rapidly.

7. Many parasites evolved complex life cycles because
   a. they are too simple to disperse readily.
   b. they are poor at recognizing new hosts.
   c. of host defenses.
   d. having an intermediate host usually increases the probability of transfer to a new individual of the primary host.
   e. their ancestors had complex life cycles and they retained them.

8. The phyla whose members have lophophores are
   a. Phoronida, Brachiopoda, Gastrotricha, and Pterobranchia.
   b. Phoronida, Brachiopoda, Bryozoa, and Pterobranchia.
   c. Brachiopoda, Bryozoa, Gastrotricha, and Pterobranchia.
   d. Phoronida, Rotifera, Bryozoa, and Gastrotricha.
   e. Rotifera, Bryozoa, Brachiopoda, and Pterobranchia.

9. Which of the following is *not* part of the molluscan body plan?
   a. Mantle
   b. Foot
   c. Radula
   d. Visceral mass
   e. Jointed skeleton

10. Many lineages of protostomes evolved feeding structures designed to extract small prey from the water because
    a. during much of protostome evolution, the only food available was dissolved organic matter and very small organisms.
    b. during much of protostome evolution, small animals were more abundant than large animals.
    c. large animals were available as food but they were difficult to capture.
    d. to be successful in competition for space, protostomes had to feed on small prey.
    e. water flowed naturally over their feeding structures, so early protostomes did not have to work to get food.

Student Web Site Self-Quiz Questions

1. With the possible exception of the sponges, animals appear to have evolved from a common ancestor. Which of the following statements about the early evolution of animals is false?
   a. Studies of 5S and 18S ribosomal RNA's suggest a polyphyletic origin for the animal kingdom.
   b. The diversity of size and shape evolved as animals become predatorial.
   c. The evolution of complex behaviors may have occurred early in the origin of animals.
   d. Early on, natural selection favored larger and more motile animals.
   e. The division of labor among colonial cells was a major first step in the evolution of the animals.

2. Which one of the following terms would *not* be used to describe the roundworm?
   a. Protostome
   b. Triploblastic
   c. Pseudocoelomate
   d. Indeterminate cleavage
   e. Pseudocoel

3. The earthworm is an example of a coelomate. Which of the following terms does *not* apply to this animal?
   a. Triploblastic
   b. Spiral cleavage
   c. Pseudocoelomate
   d. Segmented
   e. Peritoneum

4. Which of the following structural parts of a sponge is responsible for creating water currents and capturing food?
   a. Choanocytes
   b. Amoeba-like cells
   c. Pores
   d. Spicules
   e. Oscula

5. Cnidarians typically have two body forms, one asexual and the other sexual. The sexual form is the __________ and it differs most from the asexual form in the __________ body layer.
   a. medusa, endoderm
   b. medusa, mesoglea
   c. polyp, endoderm
   d. polyp, mesoglea
   e. planula, endoderm

6. Ctenophores were once classified as cnidarians. Which of the following is *not* a way in which ctenophores differ from cnidarians?
   a. Ctenophores have low metabolic rates.
   b. Ctenophores have complete digestive tracts.
   c. Locomotion in ctenophores is via cilia.
   d. Ctenophores use sticky tentacles to capture prey.
   e. Ctenophores completely lack nematocysts.

7. Which of the following statements about flatworms is *false?*
   a. Flatworms that can move, do so by ciliary action.
   b. Flatworms have a complete tubular digestive system.
   c. Flatworms depend on diffusion to accomplish the functions of a specific circulatory or respiratory system.
   d. Most of the species of living flatworms are parasitic.
   e. Parasitic flatworms such as tapeworms frequently lack a digestive system.

8. Which one of the following structures is *not* present in rotifers?
   a. Corona
   b. Mastax
   c. Pseudocoel
   d. Segments
   e. Complete tubular digestive system.

9. Pterobranchs are in the phylum _________.
   a. Annelida
   b. Platyhelminthes
   c. Nematoda
   d. Pterobranchia
   e. Brachiopoda

10. In the phylum Mollusca, which of the following structures partly encloses an external cavity in which the gills lie?
    a. Radula
    b. Foot
    c. Mantle
    d. Coelom
    e. Visceral mass

11. One group of mollusks undergoes torsion during development, and includes the only members of the phylum that are terrestrial. This group is the _________.
    a. monoplacophorans
    b. chitons
    c. bivalves
    d. gastropods
    e. cephalopods

★ *Indicates a **difficult** question*

■ *Indicates a **conceptual** question*

32 Ecdysozoans: The Molting Animals

Fill in the Blank

1. In Precambrian times, evolution resulted in animals with a chitinous external body covering. This structure is referred to as an **exoskeleton,** and is a characteristic of arthropods.
2. Growth in arthropods is accomplished by **molting**, a periodic shedding of the exoskeleton followed by the rapid hardening of a new and larger exoskeleton that has formed under the old one.
3. A class of arthropods that has few members living in the oceans is the **insects**; however, in fresh water and on land they are a dominant group.
4. The strong, flexible, waterproof polysaccharide body covering characteristic of Precambrian ecdysozoan lineages is called **chitin**.
5. The immature stages of insects between molts are called **instars**.
6. Jointed appendages first appeared in a now-extinct line of arthropods called the **trilobites**.
7. The dominant uniramians belong in the subphylum **Insecta**; currently there are about 1.5 million described species.
8. An insect that exhibits gradual changes between its instars is said to undergo **incomplete metamorphosis**.
9. A caterpillar changing into a butterfly is a example of complete metamorphosis. During this process, a worm-like larva transforms itself during a specialized phase as a **pupa**.
10. Some worm-like ecdysozoans have a relatively thin and flexible exoskeleton called a **cuticle**.
11. *Trichinella* is a parasitic roundworm of the phylum **Nematoda**.
12. A rigid exoskeleton prevents wormlike movement. In order to move, animals with stiff exoskeletons require **appendages** that can be manipulated by muscles.
13. Some familiar marine arthropods include crabs, lobsters, and shrimps. They belong to the phylum **Crustacea**.
14. Insects exchange gases by means of air sacs and tubular channels called **tracheae**.
15. The insects have excretory organs called **Malpighian tubules**.

*16. Spiders are important terrestrial predators belonging to the phylum **Chelicerata**.

17. Many species of crustaceans have a fold of exoskeleton that extends dorsally and laterally back from the head to cover and protect some body segments. This structure is called the **carapace**.

Multiple Choice

*1. In nematodes, or roundworms, the main form of locomotion is
   a. cilia that beat rhythmically to move the animal forward.
   b. antagonistic muscles that work against each other and the pseudocoelom to change the shape of the animal, moving it forward.
   c. contraction of longitudinal muscles.
   d. a series of hairs that project backward and engage the substrate; back-and-forth movements propel the animal forward.
   e. the expelling of water through special ducts, moving the animal forward.

2. Which of the following statements concerning nematodes is *not* true?
   a. Nematode parasites infect many members of the animal kingdom, including many domestic animals.
   b. Nematode parasites infect many members of the plant kingdom, including many crop plants.
   c. Relatively few people in developed countries, such as the United States, have ever been infected by nematodes.
   d. Free-living nematodes are often extremely abundant.
   e. Much of what is known about roundworms has been stimulated by the desire to control parasitic species.

3. Which of the following is *not* an advantage of an exoskeleton?
   a. It is a highly efficient means of anchoring muscles, providing more efficient movement.
   b. It provides protection from predators.
   c. It dominates the development of the animal after it emerges from the egg, since it must be shed for the animal to grow.
   d. It provides protection from water loss.
   e. It provides support for walking on dry land.

4. The class Arachnida includes scorpions, mites, spiders, and a host of other less well-known orders. The bodies of one member of this class are divided into two major regions, the anterior of which bears two pairs of appendages modified to form mouthparts, and four pairs of walking legs. To which phylum does it belong?
   a. Chelicerata
   b. Crustacea
   c. Uniramia
   d. Nematoda
   e. Kinorhyncha

5. Which of the following statements concerning the tracheal system in insects is *not* true?
   a. Tracheae are air sacs and tubular channels.
   b. Tracheae penetrate virtually every part of an insect's body.
   c. Tracheae work by providing oxygen to the blood, which carries it to the other tissues of the insect.
   d. Tracheae provide oxygen to the tissues of the insects.
   e. Tracheae extend from external openings inward to tissues throughout the body.

6. *Trichinella spiralis*, the causative agent of the disease trichinosis, is a member of the phylum
   a. Nemertea.
   b. Platyhelminthes.
   c. Nematoda.
   d. Rotifera.
   e. Annelida.

7. You discover an animal with bilateral symmetry, a pseudocoelom, a tubular digestive system, and a thick multilayer cuticle. Select the phylum to which this animal most likely belongs.
   a. Nemertea
   b. Platyhelminthes
   c. Nematoda
   d. Rotifera
   e. Annelida

8. Which one of the following groups of arthropods includes the dragonflies?
   a. Onychophora
   b. Trilobita
   c. Chelicerata
   d. Crustacea
   e. Uniramia

9. Which of the following characteristics of insects is also found in other arthropods?
   a. Malpighian tubules
   b. Tracheae
   c. Wings
   d. Molting
   e. Incomplete metamorphosis

*■10. The evolution of an exoskeleton affected many aspects of arthropod evolution. From the following list, select the aspect that was *least* affected.
   a. Division of labor among the body parts
   b. Mode of locomotion
   c. Pattern of growth
   d. Type of gas exchange system
   e. Subdivision of the coelom

11. Which of the following is *not* a consequence of a sessile lifestyle?
   a. Decreased predation
   b. Increased competition for space
   c. Production of toxins
   d. Increased motility of the larval stages
   e. Coloniality

12. From the following list of characteristics select one that has *not* been a major theme in protostome evolution.
   a. Predation as a selective pressure
   b. Subdivision of the body cavity
   c. Switch from cilia to muscles for movement
   d. Mechanisms for ingesting large prey
   e. Development of hard, external body parts

13. You discover an animal that has a true coelom and an exoskeleton that includes the modified polysaccharide chitin. Select the phylum to which this animal most likely belongs.
   a. Nematoda
   b. Annelida
   c. Arthropoda
   d. Mollusca
   e. Platyhelminthes
   e. None of the above

*14. Members of which arthropod group have bodies divided into two major regions, with an anterior region bearing two pairs of appendages modified to form mouthparts, and four pairs of walking legs?
   a. Nematoda
   b. Crustacea
   c. Uniramia
   d. Chelicerata
   e. c and d

15. Insects belong to the phylum
   a. Nematoda.
   b. Crustacea.
   c. Uniramia.
   d. Chelicerata
   e. None of the above

*16. A typical crustacean larva is called a
a. trochophore.
b. nauplius.
c. planula.
d. pupa.
e. None of the above

17. Which of the following is *not* a characteristic of insects?
a. Some are parasitic species
b. Three basic body parts
c. Complex respiratory system including lungs
d. Excretory organs
e. c and d.

18. Most of protostome evolution took place in
a. oceans.
b. terrestrial environments.
c. freshwater environments.
d. a and c
e. None of the above

*19. The centipedes and millipedes belong to which group?
a. Insecta
b. Arachnida
c. Myriapoda
d. Merostomata
e. None of the above

20. A consequence of the fact that flowing water brings food with it was the repeated evolution of _________ lifestyles during protostome evolution.
a. predatory
b. parasitic
c. scavenging
d. sessile
e. None of the above

21. Individuals of which group are so numerous they are thought to be the most abundant of all animals?
a. Copepoda
b. Merostomata
c. Arachnida
d. Cirripedia
e. Apterygota

*22. Which of the following is *not* one of the major recognized lineages of winged insects?
a. Winged insects that cannot fold their wings back against the body
b. Winged insects that can fold their wings back against the body and that undergo incomplete metamorphosis
c. Winged insects that do not undergo metamorphosis
d. Winged insects that can fold their wings and that undergo complete metamorphosis
e. None of the above

23. Members of which of the following phyla do *not* have a circulatory system?
a. Nematoda
b. Uniramia
c. Crustacea
d. Nematomorpha
e. a and d

24. Spider webs are known to serve several purposes. Which of the following is *not* a characteristic or purpose of spider webs?
a. A home
b. Produced by modified abdominal appendages connected to internal secretory glands
c. Composed primarily of carbohydrates
d. A snare for catching prey
e. b and c

*25. You are examining a small animal found in marine sands under a microscope. It is about 0.3 mm in length and appears to have fleshy, unjointed legs. To which phylum does this organism most likely belong?
a. Tardigrada
b. Onychophora
c. Chelicerata
d. Crustacea
e. Uniramia

26. Members of which of the following phyla have a pseudocoel?
a. Chaetognatha and Nematomorpha
b. Nematomorpha and Nematoda
c. Nematomorpha, Nematoda, and Chaetognatha
d. Nematoda, Chelicerata, and Chaetognatha
e. Nematomorpha, Nematoda, and Uniramia

27. Members of which of the following phyla have a complete gut?
a. Chaetognatha and Nematomorpha
b. Chaetognatha, Nematomorpha, and Nematoda
c. Chaetognatha, Crustacea, Uniramia, and Nematoda
d. Chaetognatha, Crustacea, Uniramia, and Chelicerata
e. Chaetognatha, Crustacea, Uniramia, and Nematomorpha

*28. The wingless insects belong to which group?
a. Apterygota
b. Pterygota
c. Myriapoda
d. Arachnida
e. Merostomata

29. Organisms of which phylum are probably most similar to ancestral arthropods?
a. Chelicerata
b. Uniramia
c. Crustacea
d. Onychophora
e. None of the above

*30. Butterflies belong to which order of winged insects?
a. Lepidoptera
b. Coleoptera
c. Diptera
d. Trichoptera
e. Isoptera

▪31. A firm exoskeleton has protective and supportive advantages, but poses a problem for insects. What is this problem?
a. The animal must consume large amounts of food to support the growth of the exoskeleton.
b. The exoskeleton prevents the animal from moving rapidly.
c. The exoskeleton cannot grow as the animal body inside it grows.
d. The exoskeleton attracts predators.
e. None of the above

*32. Which insect order includes the grasshoppers?
a. Plecoptera
b. Homoptera
c. Orthoptera
d. Neuroptera
e. Odonata

▪33. Why are animals with thin cuticles generally restricted to moist habitats?
a. The thin cuticle allows water to be lost across the body surface.
b. There are fewer predators in these habitats.
c. These animals rely on flowing water to bring them food.
d. a and c
e. None of the above

Study Guide Questions

1. Lobsters, millipeds, and butterflies all share which of the following traits?
a. Parapodia
b. Setae
c. Gas exchange across the skin
d. Jointed appendages

2. Which of the following phyla or subphyla have a specialized internal gas exchange system?
a. Priapulida
b. Chaetognatha
c. Insecta
d. Nematomorpha

3. Which of the following insect order undergoes complete metamorphosis and can fold their wings back against the body?
a. Coleoptera (beetles)
b. Ephemeroptera (mayflies)
c. Isoptera (termites)
d. Odonata (dragonflies)

4. The bulk of arthropods belong to the phylum Uniramia, and of these, the majority of species belong to the
a. Myriapoda.
b. Insecta.
c. Cirripedia.
d. Tardigrada.

5. The typical crustacean larva is called a
a. nauplius.
b. pupa.
c. instar.
d. cyst.

6. Which class of the phylum Chelicerata is considered a living fossil?
a. Pycnogonida (sea spiders)
b. Arachnida (mites, ticks, and spiders)
c. Merostomata (horseshoe crabs)
d. None of the above

7. Which phylum or subphylum of the arthropods is extinct?
a. Chelicerata
b. Crustacea
c. Onychophora
d. Trilobita

8. Which characteristic is *not* part of the arthropod body plan?
a. Segmentation
b. Jointed appendages
c. Soft exoskeleton
d. Hard exoskeleton

End of Chapter Questions

1. The outer covering of ecdysozoans is
a. always hard and rigid.
b. always thin and flexible.
c. present at some stage in the life cycle but not always among adults.
d. ranges from very thin to hard and rigid.
e. prevents animals from changing shape.

2. The primary support for members of several small phyla of marine worms is
a. their exoskeletons.
b. their internal skeletons.
c. their hydrostatic skeletons.
d. the surrounding sediments.
e. the bodies of other animals within which they live.

3. Roundworms are abundant and diverse because
a. they are both parasitic and free-living and eat a wide variety of foods.
b. they are able to molt their exoskeletons.
c. their thick cuticle enables them to move in complex ways.
d. their body cavity is a pseudocoelom.
e. their segmented bodies enable them to live in many different places.

4. The arthropod exoskeleton is composed of a
a. mixture of several kinds of polysaccharides.
b. mixture of several kinds of proteins.
c. single complex polysaccharide called chitin.
d. single complex protein called arthropodin.

e. mixture of layers of proteins and a polysaccharide called chitin.

5. The arthropod phyla that have unjointed legs are
 a. Trilobita, and Onychophora.
 b. Onychophora and Tardigrada.
 c. Trilobita and Tardigrada.
 d. Onychophora and Chelicerata.
 e. Tardigrada and Chelicerata.

6. The crustacean group whose members are probably the most abundant of all animals is
 a. Decapoda.
 b. Amphipoda.
 c. Copepoda.
 d. Cirripedia.
 e. Isopoda.

7. The body plan of insects is composed of which of the three following parts?
 a. Head, abdomen, and trachea
 b. Head, abdomen, and cephalothorax
 c. Cephalothorax, abdomen, and trachea
 d. Head, thorax, and abdomen
 e. Abdomen, trachea, and mantle

8. Insects that hatch from eggs into juveniles that resemble the adults are said to have
 a. instars.
 b. neopterous development.
 c. simple development.
 d. incomplete metamorphosis.
 e. complete metamorphosis.

9. Which of the following groups of insects cannot fold their wings back against the body?
 a. Beetles
 b. True bugs
 c. Earwigs
 d. Stone flies
 e. Mayflies

10. Insects may have undergone such remarkable evolutionary diversification because
 a. the terrestrial environments penetrated by insects were ecologically empty.
 b. insects evolved the ability to fly.
 c. some lineages of insects evolved complete metamorphosis.
 d. insects evolved effective means of delivering oxygen to their internal tissues.
 e. All of the above

Student Web Site Self-Quiz Questions

1. Which of the following statements regarding cuticles in ecdysozoans is *false*?
 a. Cuticles protect the animal.
 b. Cuticles provide support for the animal.
 c. Cuticles, in conjunction with longitudinal muscles, provide a mechanism for movement in the animal.
 d. Cuticles allow the exchange of gases.
 e. Cuticles allow the exchange of water.

2. Which of the following statements regarding the kinorhynchs is *false*?
 a. They have protective cuticles.
 b. They are relatively large worms, greater than 5mm in length.
 c. They feed by ingesting materials in the surrounding substrate.
 d. They are segmented.
 e. They undergo molting.

3. Which of the following statements regarding the arrow worms is *false*?
 a. They are members of the phylum Chaetognatha.
 b. They have a body plan based on a coelom divided into a head, trunk, and tail.
 c. They have a complete digestive tract.
 d. They lack a circulatory system.
 e. Cilia line the coelomic cavity.

4. Roundworms are in the phylum __________.
 a. Annelida
 b. Platyhelminthes
 c. Nematoda
 d. Onychophora
 e. Trilobita

5. A characteristic of the arthropods is their rigid exoskeleton. Which of the following statements regarding the exoskeleton is *false*?
 a. It contains chitin.
 b. The rigidity of the exoskeleton prohibits wormlike movement.
 c. The exoskeleton works in conjunction with internal muscles to provide movement.
 d. Most arthropod appendages are jointed.
 e. Segments in the arthropods share muscles to generate movement.

6. Which of the following features does *not* apply to all arthropods and related lineages?
 a. Chitin
 b. Exoskeleton
 c. Hemocoel
 d. Segmentation
 e. Jointed appendages

7. Arachnids are a major type of arthropod known as chelicerates. Which one of the following is *not* also a chelicerate?
 a. Decapods
 b. Harvestmen (daddy longlegs)
 c. Horseshoe crabs
 d. Scorpions
 e. Ticks

8. Which of the following statements regarding arachnids is *false*?
 a. Most arachnids have a short larval stage.
 b. They are abundant in terrestrial environments.
 c. They are predators.
 d. They form elaborate webs.
 e. Some species of arachnids give birth to live young.

9. Which of the following is *not* a crustacean (phylum Crustacea)?
   a. Shrimp
   b. Barnacle
   c. Horseshoe crab
   d. Sow bug
   e. Sand flea

10. A shrimp is a typical crustacean. Which of the following statements about crustaceans is *false*?
    a. Crustaceans are the dominant arthropods of the oceans.
    b. Most crustaceans have the same body regions as the insects.
    c. There are marine, freshwater, and terrestrial crustaceans.
    d. Each of the multiple thoracic and abdominal segments contains two pairs of legs.
    e. Some crustaceans are sessile as adults.

11. Which of the following statements about crustacean larvae is *false*?
    a. Only some species release larvae.
    b. Some species release fertilized eggs into the environment from which the larvae hatch.
    c. Larvae contain a single simple eye.
    d. The crustacean larva is called a trocophore.
    e. The larva has three pairs of appendages.

12. Grasshoppers are in the order Orthoptera. If you used the following terms to classify a grasshopper beginning with the most inclusive term and ending with the least inclusive term, which term would be fourth in sequence?
    a. Animalia
    b. Arthropoda
    c. Insecta
    d. Protostomia
    e. Uniramia

13. The millipede is a member of the Myriapoda subphylum. Which of the following statements about the myriapods is *true*?
    a. Myriapods have three pairs of legs.
    b. Myriapods have three body regions: head, thorax, and abdomen.
    c. Myriapods are strictly predatorial.
    d. Myriapods are small in size, only a few centimeters in size.
    e. Myriapods have a flexible, segmented trunk.

14. Insects are the dominant uniramians. Which one of the following features is *not* a characteristic of the subphylum Insecta?
    a. Tracheae
    b. Presence of a single pair of antennae on the head and three pairs of legs attached to the thorax.
    c. Metamorphosis
    d. Body plan consisting of head, thorax, and abdomen
    e. Exoskeleton

15. The order Orthoptera, which includes grasshoppers, contains a variety of organisms. Which of the following organisms is *not* a member of the order Orthoptera?
    a. Cricket
    b. Mantid
    c. Earwig
    d. Walking stick
    e. Roach

TEST FILE

> ★ *Indicates a **difficult** question*
> ■ *Indicates a **conceptual** question*

33 Deuterostomate Animals

Fill in the Blank

1. The bony fishes are capable of regulating their buoyancy because of an organ called the **swim bladder**.
2. One group of Osteichthyes, the Crossopterygii, gave rise to the class **Amphibia**, which are capable of exchanging gases through their skin.
3. The first major animal group that could live completely out of water is the class **Reptilia**.
4. The reptiles were extraordinarily successful in part due to the evolution of the **amniotic egg**, which allowed their offspring to develop in a protected watery environment.
5. The birds (subclass **Aves**) are in the class Dinosauria.
6. The class **Mammalia** includes monotremes, marsupials, and eutherians.
7. **Arrow worms** are members of the phylum Chaetognatha that swim in the open sea, have lateral and caudal fins, possess grasping spines, prey on protists and young fish, and have a tripartite body plan.
8. The evolutionary lineage leading to the **chordates** lost the lophophore and proboscis, replacing them with large pharyngeal slits as a feeding device.
9. Members of the phylum **Chordata** are bilaterally symmetrical and possess pharyngeal slits at some stage of their life cycle. They also possess an internal skeleton, a dorsal nervous system, a ventral heart, and a tail that extends beyond the anus.
10. **Placentas** are structures that nourish developing embryos and evolved in reptilian species.
11. The evolution of primitive **lungs (or lunglike sacs)** in the early bony fishes (Osteichthyes) set the stage for the invasion of land because they provided an alternative method of gaseous exchange.
12. In the evolutionary lineage leading to dinosaurs and birds, and in the lineage leading to mammals, two factors were important in enabling those organisms to run for long periods of time at fast rates. These factors were a more vertical position assumed by the legs and special, **ventilatory** muscles that could operate independently of locomotory muscles.
13. The heavily armored fishes that evolved jaws were known as the **placoderms**.
14. The expanded mental abilities of humans are largely responsible for the development of **culture**, the process by which knowledge and traditions are passed from one generation to another.
15. The echinoderms have a network of calcified hydraulic canals that function in gas exchange, locomotion, and feeding called a **water vascular system**.
16. Vertebrates have a jointed, dorsal **vertebral column** that replaced the notochord as their primary support.
17. The cartilaginous fishes belong to the class **Chondrichthyes** and include the sharks, skates, and rays.
18. Females of the group **Marsupialia** have ventral pouches in which they carry and feed their offspring.

Multiple Choice

1. Which of the following statements conserning echinoderms is *not* true?
   a. Echinoderms are radially symmetrical as adults.
   b. Echinoderms have an external skeleton.
   c. Echinoderms lack a brain.
   d. Echinoderms are radially symmetrical as larvae.
   e. Echinoderms have an extensive fossil record.
2. Which of the following phyla has a water vascular system?
   a. Mollusca
   b. Chordata
   c. Annelida
   d. Echinodermata
   e. a and d

★3. The major change in the evolutionary lineage leading to the chordates was
   a. evolution of the water vascular system.
   b. loss of the lophophore and enlargement of the pharyngeal gill slits for feeding.

c. calcification of an internal skeleton.
d. development of the lophophore as an adaptation to predatory life.
e. the ability to "suck mud" (extract food from mud or sand).

4. Which of the following is *not* found in the phylum Chordata?
a. Bilateral symmetry
b. A dorsal hollow nerve chord
c. An external skeleton
d. Gill slits at some stage during development
e. A notochord at some stage during development

5. Which of the following is *not* a characteristic of the vertebrate body plan?
a. A vertebral column
b. Two pairs of appendages
c. A large coelom in which body organs are found
d. A well-developed circulatory system
e. A large coelom that serves as a hydrostatic skeleton

6. The important evolutionary novelty that evolved from the jawless fishes and is present in Placodermi, Chondrichthyes, and Osteichthyes is
a. heavily armored skin.
b. the jaw.
c. fins.
d. the ability to swim.
e. All of the above

7. The jaw found in sharks is most likely evolved from
a. a gill arch.
b. dermal bone.
c. the skull.
d. the backbone.
e. the mouth plates of echinoderms.

8. The cartilaginous fishes, including sharks, skates and rays, and chimaeras,
a. belong to the class Osteichthyes.
b. have heavy external armor.
c. have an open circulatory system.
d. have a few bones, but mostly cartilage, in their skeletons.
e. have less external armor and are faster swimmers than their ancestors.

▪*9. A major ecological difference between bony fishes (class Osteichthyes) and Chondrichthyes is that bony fishes
a. do not have swim bladders.
b. have no cartilage in their skeletons.
c. need more dissolved oxygen in the water than sharks.
d. evolved in fresh water.
e. All of the above

▪10. Major differences between the classes Chondrichthyes and Osteichthyes include which of the following?
a. Chondrichthyes have a lung or swim bladder, whereas Osteichthyes do not.
b. Chondrichthyes have a cartilaginous skeleton, whereas Osteichthyes have a bony skeleton (with true bone).
c. Chondrichthyes do not have paired fins, whereas Osteichthyes do.
d. Chondrichthyes evolved in fresh water, whereas Osteichthyes evolved in salt water.
e. b and d

11. The swim bladders of bony fishes
a. evolved from lunglike sacs that supplemented the gills in respiration.
b. are used for respiration in most contemporary species.
c. are organs of buoyancy that help the fish control its depth in the water column.
d. prevented bony fishes from existing in a marine environment.
e. a and c

12. Lobe-finned fishes (Crossopterygians) had several adaptations that were instrumental in the transition to a life on land. Adaptations found in lobe-finned fishes that were important in the evolution of the amphibians include
a. primitive lungs.
b. jointed fins with strong muscular support.
c. watertight skin.
d. a and b
e. All of the above

13. Amphibians breathe air by which two means?
a. Gills and swim bladders
b. Gills and thin skin
c. Lungs and gills
d. Lungs and thin skin
e. Swim bladders and lungs

14. Reptiles are the first group of animals to be completely liberated from a need to return to water for some portion of their life cycle. Adaptations first found in reptiles that were important to this shift include which of the features below?
a. A hard-shelled (amniotic) egg
b. Parental care
c. Watertight skin
d. a and c
e. All of the above

15. The single most characteristic feature of birds is
a. the ability to fly.
b. the ability to lay eggs that will not dry out.
c. feathers.
d. the enormous amount of parental care they provide their young.
e. All of the above

16. The birds, or Aves,
a. descended most recently from an amphibian ancestor.
b. should be included in the class Mammalia.
c. should include some modern reptiles.
d. should include all modern reptiles.
e. are in the class that includes dinosaurs and crocodiles.

17–19. Match the groups of organisms from the list below with their most important evolutionary novelty.
a. Reptiles
b. Birds
c. Osteichthyes
d. Chondrichthyes
e. Echinoderms

17. Swim bladders **(c)**

18. Powered flight **(b)**

19. Jaws **(d)**

▪*20. In several important ways, deuterostome and protostome evolution are similar. Which of the following is *not* true of both protostome and deuterostome evolution?
a. Both lineages exploited the abundant food supplies buried in soft marine substrates.
b. Many groups in both lineages developed elaborate structures for extracting prey from water.
c. In both lineages, a coelomic cavity evolved and became divided into compartments that allowed better control of body shape and movement.
d. Both lineages evolved locomotor abilities.
e. Both groups invaded land and evolved into very large terrestrial animals.

21. The oldest fossil remains of members of our genus, *Homo*, suggest that early relatives of humans lived
a. near the oceans, where fish are plentiful throughout the year.
b. near rivers, where fish and fresh water are plentiful.
c. in the Midwestern United States, where there is some of the most fertile soil in the world.
d. in the American tropics, where there are long growing seasons and many species of fruits and berries.
e. in dry African savannas, eating roots, bulbs, tubers, and animals.

22. To which class of echinoderms do the brittle stars belong?
a. Crinoidea
b. Ophiuroidea
c. Asteroidea
d. Echinoidea
e. Holothuroidea

23. From the following list of echinoderm classes, select the class that was much more abundant in the past than it is today.
a. Crinoidea
b. Ophiuroidea
c. Asteroidea
d. Echinoidea
e. Holothuroidea

24. Which feature does *not* characterize animals in the phylum Echinodermata?
a. Bilaterally symmetric larvae
b. Radially symmetric adults
c. Presence of a water vascular system
d. An external skeleton of calcified plates
e. A tubular digestive system

25. Which class of echinoderms has tube feet at their anterior end modified into tentacles that are used for feeding?
a. Crinoidea
b. Ophiuroidea
c. Asteroidea
d. Echinoidea
e. Holothuroidea

26. Which class of echinoderms is herbivorous and has no arms?
a. Crinoidea
b. Ophiuroidea
c. Asteroidea
d. Echinoidea
e. Holothuroidea

27. The chordates
a. all have a backbone.
b. include some animals without a nervous system.
c. all pass through a developmental stage with gill slits.
d. are poorly represented in the fossil record.
e. None of the above

28. Which of the following is *not* a characteristic unique to all members of the phylum Chordata?
a. A notochord
b. A ventral heart
c. Vertebrae
d. An endoskeleton
e. A hollow dorsal nerve cord

29. Which of the following structures is seen in both the adult and tadpole-like larvae of the tunicates?
a. Pharyngeal gill slits
b. Notochord
c. Hollow dorsal nerve cord
d. A tunic
e. An atrial siphon

*30. Evolution of jaws first occurred in the group of fish known as the
a. ostracoderms.
b. placoderms.
c. cartilaginous fishes.
d. Chondrichthyes.
e. Osteichthyes.

31. The sharks, skates, and rays are members of the vertebrate class known as
a. Osteichthyes.
b. Agnatha.
c. Placodermi.
d. Chondrichthyes.
e. Ascidiacea.

*32. Select the following adaptation that was *not* first evolved in the group of fish known as the Osteichthyes.
a. Lungs
b. A swim bladder
c. Paired lateral fins
d. Bony skeletons
e. None of the above

33. Which of the following statements about the crossopterygian fishes is *not* true?
   a. They are commonly known as the lobe-fin fishes.
   b. *Latimeria chalumnae* is a well-known fossil crossopterygian fish.
   c. Crossopterygian fishes have lungs, but no gills.
   d. The bones in their pectoral and pelvic limbs have well-formed joints.
   e. The crossopterygian fishes gave rise to the amphibians during the Devonian period.

34. Which of the following statements about the amphibians is *not* true?
   a. The class Amphibia was more abundant in the past.
   b. Waterproof coverings on the skin of amphibians permit them to be truly terrestrial.
   c. Most amphibians must reproduce in or near water.
   d. Amphibian eggs are very sensitive to drying.
   e. Living amphibians belong to three major classes.

*35. Select the following feature that was *not* a significant evolutionary advancement of the class Reptilia.
   a. Water-impermeable skin
   b. Maternal care of the young
   c. Better separation of oxygenated and deoxygenated blood in the heart
   d. More effective ventilation of the lungs
   e. None of the above

36. Which of the following subclasses of the class Reptilia has the fewest living members?
   a. Squamata
   b. Sphenodontida
   c. Chelonia
   d. Crocodylia
   e. Aves

37. Select the following group of modern reptiles that includes entirely carnivorous species.
   a. Lizards
   b. Turtles and tortoises
   c. Snakes
   d. Caecilians
   e. None of the above

38. Which of the following statements comparing *Archaeopteryx* with modern birds is *not* true?
   a. The breastbone of *Archaeopteryx* lacked a keel.
   b. Unlike modern birds, *Archaeopteryx* had a long tail.
   c. Unlike modern birds, *Archaeopteryx* had long fingers on its forearms.
   d. Although *Archaeopteryx* could fly, it lacked true feathers.
   e. Unlike *Archaeopteryx*, no modern birds have teeth.

39. Which of the following characteristics is unique to the marsupial mammals?
   a. They are the only egg-laying mammals.
   b. Their mammary glands have no nipples.
   c. They have no placentas.
   d. Gestation is short, and the young complete development outside the uterus in a special pouch.
   e. They are only found in Australia.

40. Which of the following hominids is oldest?
   a. *Homo habilis*
   b. *Homo erectus*
   c. *Homo sapiens*
   d. *Neanderthals*
   e. Cro-Magnons

*41. Which of the following was *not* a change that accompanied the transition from the australopithecines to *Homo habilis*?
   a. Increase in brain size
   b. Change in diet
   c. Increase in body size
   d. Use of tools
   e. Dispersal of populations to Europe and Asia

*42. Which of the following evolutionary themes did *not* occur in both the protostomes and deuterostomes?
   a. Evolution of structures for filtering food from water
   b. Evolution of wormlike burrowing forms
   c. Evolution of adaptations needed for invasion of the land
   d. Evolution of large terrestrial species
   e. Evolution of jointed appendages for improved locomotion

43. Which of the following is *not* characteristic of deuterostomes?
   a. Indeterminate or regulatory cleavage
   b. Formation of mesoderm from an outpocketing of the embryonic gut
   c. A blastopore that becomes the anus
   d. Greater number of species than the protostomes
   e. Triploblastic with well-developed coelomic body cavities

44. Members of the phylum __________ exhibit radial symmetry that evolved in association with a complex body plan.
   a. Porifera
   b. Chordata
   c. Annelida
   d. Cnidaria
   e. Echinodermata

▪45. The principal reason we consider tunicates similar to ancestors of all chordates is that
   a. the body plan of adult tunicates parallels that of chordates.
   b. tunicate larvae reveal close evolutionary relationships with chordates.
   c. tunicates have a lophophore-style mouth.
   d. a and c
   e. None of the above

▪46. We sometimes refer to birds as "feathered dinosaurs" because
   a. birds are descendants of a lineage of dinosaurs.
   b. dinosaurs had feathers as well as scales.
   c. birds are warm-blooded and preserve warmth via feathers, while dinosaurs were also warm-blooded.
   d. a and c
   e. None of the above

47. The calcification of an internal skeleton is characteristic of the phylum
   a. Chaetognatha.
   b. Echinodermata.
   c. Hemichordata.
   d. Brachiopoda.
   e. Phoronida.

▪*48. The now-living species *Latimeria chalumae* belongs to a subclass that is an important link in evolution because it was the first subclass of
   a. fish with lungs.
   b. fish with paired fins.
   c. fish with lobed fins.
   d. bony fish.
   e. fish with cartilage.

▪49. A difference between amphibians and reptiles is
   a. amphibian eggs can survive out of water and reptile eggs cannot.
   b. amphibians have thin skins and reptiles have thick skins.
   c. amphibians have gills and lungs and reptiles have only lungs.
   d. a and b
   e. a, b, and c

*50. Cartilaginous fishes
   a. rely on swimming to stay afloat.
   b. have swim bladders.
   c. belong to the class Osteichthyes.
   d. evolved in fresh water.
   e. are relatively poor swimmers.

51. Which of the following is *not* true concerning general characteristics of deuterostomate animal phyla?
   a. All have three body layers.
   b. All have closed circulatory systems.
   c. All have coelomic body cavities.
   d. All have complete digestive tracts.
   e. All of the above

*52. Which of the following is *not* one of the major traits that distinguish the primates from other mammals?
   a. Dexterous hands with opposable thumbs that can grasp branches and manipulate food
   b. Nails rather than claws
   c. Maternal care of the young
   d. Eyes on the front of the face that provide good depth perception
   e. b and c

53. Which of the following was *not* one of the skeletal modifications that accompanied the evolution of the small mammals from their larger reptilian ancestors?
   a. Bones from the lower jaw were incorporated into the middle ear.
   b. The number of bones in the skull was decreased.
   c. The bulk of the limbs and the bony girdles from which they are suspended were reduced.
   d. The development of a vertebral column
   e. All of the above

54. The human population has experienced __________ major phase(s) of increase.
   a. one
   b. two
   c. three
   d. four
   e. five

55. Members of which group are probably most similar to the ancestors of all chordates?
   a. Urochordata
   b. Cephalochordata
   c. Vertebrata
   d. Amphibia
   e. Reptilia

*56. Put the following human ancestors in order of their probable appearance.
   I. *Homo habilis*
   II. *Australopithecus afarensis*
   III. *Homo erectus*
   IV. *Australopithecus garhi*
   a. I, II, III, IV
   b. IV, II, I, III
   c. IV, II, I, IIII
   d. II, IV, I, III
   e. II, IV, III, I

57. Which of the following is *not* accurate in describing the Old World primates?
   a. A prehensile tail
   b. Arboreal species
   c. Terrestrial species
   d. Species that live and travel in large groups
   e. a and b

Study Guide Questions

1. Which of the following is not a characteristic of species in the phylum Echinodermata?
   a. They have a water vascular system.
   b. They have an internal skeleton.
   c. They are protostomes.
   d. They have bilateral symmetry.

2. All but one of the following classes of echinoderm belongs to the subphylum Eleutherozoa. Select the exception.
   a. Asteroidea
   b. Crinoidea
   c. Asteroidea
   d. Holothuroidea

3. Which of the following traits is *not* shared by both the hemichordates and the chordates?
   a. Pharyngeal slits
   b. Deuterostomes
   c. Notochord
   d. Triploblasts

4. The primary anatomical characteristic that preadapted pharyngeal slits for use as feeding structures was their
   a. large surface area.
   b. location anterior to the mouth.
   c. bilateral symmetry.
   d. ability to provide support for the internal skeleton.

5. Which of the following groups of chordates evolved prior to the appearance of cartilaginous fishes?
   a. Bony fishes and sea squirts
   b. Tunicates, lancelets, and agnathans
   c. Tunicates, lancelets, and bony fishes
   d. Agnathans and bony fishes

6. The vertebrate body plan includes all of the following traits except
   a. a ventral spinal cord.
   b. an internal skeleton.
   c. a well-developed circulatory system.
   d. organs suspended in the coelom.

7. The swim bladder of many fishes evolved from a lunglike sac. What important function does the swim bladder provide?
   a. Aids in prey capture
   b. Controls swimming speed
   c. Controls buoyancy
   d. Aids in reproduction

8. Which one of the following traits do the Condrichthyes and Osteichthyes share?
   a. The gills are the major site of gas exchange.
   b. The skeleton is composed of cartilage.
   c. The outer surface is covered with scales.
   d. They have a swim bladder.

9. The transition from aquatic to terrestrial lifestyles required many adaptations in the vertebrate lineage. Which of the following is *not* one of those adaptations?
   a. Switch from gill respiration to air-breathing lungs
   b. Improvements in water resistance of skin
   c. Alteration in mode of locomotion
   d. Development of feathers for insulation

10. The amniotes evolved the ability to reproduce by laying eggs. What is the major advantage to laying eggs?
    a. The embryo needs only a small amount of yolk for development.
    b. The shelled egg does not have to be laid in a moist environment.
    c. Evaporation from the egg is increased.
    d. The nitrogenous wastes can be excreted across the shell.

11. How do birds differ from reptiles?
    a. Birds have a lower metabolic rate.
    b. Reptiles lay eggs and birds give birth to live young.
    c. Birds can breathe and run at the same time, while reptiles cannot.
    d. They do not share a common ancestor.

12. Which of the following is *not* a trait that you could use to identify an animal as a mammal?
    a. Mammary glands
    b. Hair
    c. Sweat glands
    d. Kidneys

13. Dinosaurs dominated the Earth for nearly 150 million years before becoming extinct. Which of the following vertebrate groups is arguably a living representative of the dinosaur lineage?
    a. Amphibians
    b. Birds
    c. Lobe-finned fishes
    d. Snakes

14. The now extinct hominid most like ourselves is
    a. *H. habilis.*
    b. *H. erectus.*
    c. Cro-Magnon.
    d. Neanderthal.

15. Which of the following statements about human evolution is *false*?
    a. Bipedalism was a hominid adaptation for life on the ground.
    b. Increases in the size of hominid brains preceded the appearance of language and culture.
    c. The extinction of the Neanderthal people was caused by the emergence of *H. habilis.*
    d. Humans are not the direct descendants of modern-day chimpanzees.

End of Chapter Questions

1. Which of the following are deuterostomate phyla with a three-part body plan?
   a. Rotifera, Phoronida, Bryozoa, and Brachiopoda
   b. Phoronida, Bryozoa, Brachiopoda, and Hemichordata
   c. Phoronida, Bryozoa, Hemichordata, and Chordata
   d. Echinodermata, Bryozoa, Brachiopoda, and Chordata
   e. Phoronida, Bryozoa, Hemichordata, and Echinodermata

2. The structure used by brachiopods to capture food is a
   a. pharyngeal gill basket.
   b. proboscis.
   c. lophophore.
   d. mucous net.
   e. radula.

3. The water vascular system of echinoderms is a series of
   a. seawater channels derived by enlargement and extension of a coelomic cavity.
   b. seawater channels derived by enlargement and extension of the pharyngeal cavity.
   c. channels derived by enlargement and extension of a coelomic cavity and filled with coelomic fluid.
   d. channels derived by enlargement and extension of a coelomic cavity and filled with fresh water.
   e. channels that can be filled to different levels with water, enabling the animal to control its buoyancy.

4. The pharyngeal gill slits of chordates originally functioned as sites for
   a. uptake of oxygen only.
   b. release of carbon dioxide only.
   c. both uptake of oxygen and release of carbon dioxide.
   d. removal of small prey from the water.
   e. forcible expulsion of water to move the animal.

5. The key to the vertebrate body plan is a
   a. pharyngeal gill basket.
   b. vertebral column to which internal organs are attached.
   c. vertebral column to which two pairs of appendages are attached.
   d. vertebral column to which a pharyngeal gill basket is attached.
   e. pharyngeal gill basket and two pairs of appendages.

6. Which of the following fishes do *not* have a cartilaginous skeleton?
   a. Chimaeras
   b. Lungfishes
   c. Sharks
   d. Skates
   e. Rays

7. In most fishes, lunglike sacs evolved into
   a. pharyngeal gill slits.
   b. true lungs.
   c. coelomic cavities.
   d. swim bladders.
   e. None of the above

8. Most amphibians return to water to lay their eggs because
   a. water is isotonic to egg fluids.
   b. adults must be in water while they guard their eggs.
   c. there are fewer predators in water than on land.
   d. amphibians need water to produce their eggs.
   e. amphibian eggs quickly lose water and desiccate if their surroundings are dry.

9. The horny scales that cover the skin of reptiles prevent them from
   a. using their skin as an organ of gas exchange.
   b. sustaining high levels of metabolic activity.
   c. laying their eggs in water.
   d. flying.
   e. crawling into small spaces.

10. Which statement about bird feathers is *not* true?
    a. They are highly modified reptilian scales.
    b. They provide insulation for the body.
    c. They arise from well-defined tracts.
    d. They help birds fly.
    e. They are important sites of gas exchange.

11. Monotremes differ from other mammals in that they
    a. do not produce milk.
    b. lack body hairs.
    c. lay eggs.
    d. live in Australia.
    e. have a pouch in which the young are raised.

12. Bipedalism is believed to have evolved in the human lineage because bipedal locomotion is
    a. more efficient than quadrupedal locomotion.
    b. more efficient than quadrupedal locomotion, and it frees the hands to manipulate objects.
    c. less efficient than quadrupedal locomotion, but it frees the hands to manipulate objects.
    d. less efficient than quadrupedal locomotion, but bipedal animals can run faster.
    e. less efficient than quadrupedal locomotion, but natural selection does not act to improve efficiency.

Student Web Site Self-Quiz Questions

1. The phylum Echinodermata includes organisms such as the sea stars. Which of the following features is *not* characteristic of the echinoderms?
   a. Biradially symmetrical larvae
   b. Calcified internal skeleton
   c. Biradial symmetry as adults
   d. Water vascular system
   e. Marine habitats only

2. What major feature do hemichordates share with the chordate lineage?
   a. Tube feet
   b. A three-part body plan
   c. Pharyngeal slits
   d. A notochord
   e. A proboscis modified for digging

3. Adult tunicates are sessile organisms that filter seawater through their pharyngeal baskets. Their tadpolelike larvae share all but which of the following features with the lancelets?
   a. Notochord
   b. Dorsal, hollow nerve cord
   c. Pharyngeal slits
   d. Notochord-attached muscles
   e. Lack of mobility

4. Present-day agnathans include the lamprey and the hagfish. Which of the following characteristics probably does *not* apply to the first vertebrates (class Agnatha)?
   a. Jawless
   b. Development of small teeth to grind food items
   c. Were motile
   d. Filter feeders
   e. Some groups developed external body armor

5. The first important group of jawed fishes was the __________. Their jaws were derived from __________.
   a. ostracoderms, gill arches
   b. ostracoderms, skull bones
   c. cartilaginous fishes, skull bones
   d. placoderms, gill arches
   e. placoderms, skull bones

6. The bony fishes are the most diverse group of fishes alive today. Which of the following features of bony fishes is also characteristic of cartilaginous fishes?
   a. Swim bladders
   b. Paired lateral unjointed fins
   c. Flaps covering their gill slits
   d. No need to swim to move water over their gills
   e. Occupy all aquatic habitats (marine, brackish, freshwater)

7. Amphibians include the legless, burrowing caecilians, the salamanders, and the frogs and toads. Which of the following statements about amphibians is *false*?
   a. Fewer living amphibians exist than are known from fossils.
   b. Amphibians are tetrapods.
   c. All amphibians exchange gases primarily through the use of well-developed lungs.
   d. The exact causes of the ongoing decline in amphibian populations are unknown.
   e. Some amphibians are entirely aquatic; some are entirely terrestrial.

8. Lizards, along with the other reptiles, arose from early tetrapods in the Carboniferous period. Which of the following morphological changes was *not* an adaptation that enabled reptiles to live in terrestrial habitats?
   a. The amniote egg
   b. Placentas
   c. Skin covered with horny scales
   d. Separation of oxygenated and deoxygenated blood within the heart
   e. A rib cage that can actively ventilate the lungs

9. The present-day reptiles include the turtles, tortoises, tuataras, and lizards and snakes. Which of the following groups is most closely related to the turtles?
   a. Birds
   b. Dinosaurs
   c. Crocodiles
   d. Mammals
   e. All of the above

10. The birds have undergone an extensive adaptive radiation. Which of the following statements about the class Aves is *false*?
    a. All of the "opposite birds" are extinct.
    b. The earliest known avian fossil, *Archaeopteryx*, came from a relatively highly evolved bird.
    c. Unlike the dinosaurs, birds were little affected by the mass extinction at the end of the Cretaceous period.
    d. Feathers were an adaptation for insulation.
    e. The relatively large brain in birds is mostly attributable to enlargement of sight and muscular coordination centers.

11. Which of the following statements about the class Mammalia is *false*?
    a. Reduction in the bulk of pectoral and pelvic girdles occurred as part of the simplification of mammalian skeletons.
    b. Little evidence from the fossil record is available on the appearances of nonskeletal mammalian features.
    c. Mammals have fewer but more diversified teeth than reptiles.
    d. The monotremes lay eggs and lack mammary glands.
    e. Most living mammals are eutherians.

12. The primates include the lemurs, New World monkeys, Old World monkeys, and great apes. Which of the following statements about the primates is *false*?
    a. No prosimians occur in the New World.
    b. Binocular vision is an adaptation for an arboreal life style.
    c. Arboreal species of the New World and Old World monkeys have prehensile tails.
    d. Primates deliver small litters that receive extended parental care.
    e. *Australopithecus garhi* probably gave rise to the genus *Homo* about 2.5 mya.

13. Which of the following statements regarding human evolution is *false*?
    a. The brains of the earliest members of *Homo sapiens* were larger than earlier *Homo* species.
    b. Communication was an important development during human evolution.
    c. Spirituality developed during this period.
    d. Neanderthals were direct ancestors of modern humans.
    e. The manipulation and use of tools played an important role in human development.

★ *Indicates a* ***difficult*** *question*

■ *Indicates a* ***conceptual*** *question*

34 The Plant Body

Fill in the Blank

1. Unlike dicots, monocots have their vascular bundles in a scattered arrangement in their stems and possess a central region called the **pith** inside the xylem layer in their roots.
2. When a tree is cut down, the cut surface of the stump often shows **annual rings** due to variations in the size of the vessels.
3. The most common type of undifferentiated cells in plant bodies are **parenchyma** cells.
4. Just inside the epidermis of the root is a region of unspecialized cells called the **cortex**.
5. **Sclerenchyma** cells have thickened secondary cell walls and actually function when dead.
6. The **endodermis** is a layer of cells waterproofed by a layer of suberin in their cell walls.
7. Some vascular bundles contain a single layer of actively dividing cells known as the vascular **cambium.**
8. In the epidermis of a leaf are pores called **stomata** whose opening is controlled by the action of a pair of **guard cells** surrounding the pore.
9. Division of cells that will form all of the plant's organs occurs from regions called **meristems**.
10. **Suberin** is the waxy substance that waterproofs the inner core of vascular tissues in the root and stem.
11. Potatoes are a portion of the **stem** of the plant, and their eyes contain lateral **buds**.
12. Cork cambium, the bark on trees, arises from **phloem**.

Multiple Choice

1. Unlike primary cell walls, secondary cell walls have
   a. plasmodesmata.
   b. deposits of lignin or suberin.
   c. deposits of cellulose.
   d. pit pairs.
   e. permeability to small molecules.
2. In vessel elements and sieve tube elements, only sieve tube elements
   a. are dead at maturity.
   b. are stacked end-to-end.
   c. transport substances through the plant.
   d. often have companion cells.
   e. occur in all plant organs.
3. Sieve tube elements are unusual cells because they lack
   a. cell walls.
   b. cytoplasm.
   c. water.
   d. nuclei.
   e. plasma membranes.
4. _________ are generally narrow-leaved flowering plants such as grasses.
   a. Monocots
   b. Dicots
   c. Eudicots
   d. Angiosperms
   e. Gymnosperms
5. _________ are broad-leaved flowering plants such as roses.
   a. Monocots
   b. Grasses
   c. Angiosperms
   d. Eudicots
   e. Gymnosperms
6. A _________ is the point where a leaf attaches to a stem.
   a. internode
   b. bud
   c. node
   d. apical bud
   e. petiole

*7. Which of the following is characteristic of heartwood but not of sapwood?
   a. Lighter color
   b. Stores resin
   c. Conducts water and minerals
   d. Younger wood
   e. Has knots

*8. Cacti thorns result from a modification of the same plant organ that produces
a. coconut trunks.
b. maple leaves.
c. strawberry runners.
d. potato tubers.
e. corn adventitious roots.

9. Tracheids, vessel elements, and sclereids are similar in that they all
a. lack secondary cell walls.
b. conduct water and minerals.
c. harden seed coats.
d. have open ends.
e. function when dead.

10. A _________ is an embryonic shoot.
a. node
b. petiole
c. blade
d. bud
e. root

11. A root is called adventitious if it
a. forms a mycorrhizal association.
b. belongs to a fibrous root system.
c. originates from a stem or leaf.
d. is modified for storage.
e. is actively growing.

12. The widening of a tree trunk is mostly due to the activity of its
a. apical meristem.
b. secondary phloem.
c. phelloderm.
d. vascular cambium.
e. primary xylem.

13. In each vascular bundle, the tissue nearest the center of the stem is
a. collenchyma.
b. fibers.
c. phloem.
d. vascular cambium.
e. xylem.

14. Moving from the center of a tree trunk outward, the order of vascular tissues is
a. primary xylem, secondary xylem, vascular cambium, secondary phloem, and primary phloem.
b. secondary xylem, primary xylem, vascular cambium, primary phloem, and secondary phloem.
c. primary xylem, primary phloem, secondary xylem, secondary phloem, and vascular cambium.
d. primary xylem, primary phloem, vascular cambium, secondary phloem, and secondary xylem.
e. secondary xylem, secondary phloem, vascular cambium, primary xylem, and primary phloem.

15. A _________ leaf has multiple blades arranged along an axis or radiating from a central point.
a. simple
b. compound
c. star
d. common
e. complex

16. The pull of gravity is detected by a root's
a. apical meristem.
b. cap.
c. endodermis.
d. pericycle.
e. region of elongation.

17. The collective term for phelloderm, cork cambium, and cork is
a. pericycle.
b. periderm.
c. phloem.
d. procambium.
e. protoderm.

18. What are the major functions of plant roots?
a. Absorption and support
b. Absorption and reproduction
c. Anchoring and support
d. Anchoring and absorption
e. Absorption and transport

19. Which of the following describes a fibrous root system?
a. Deep-growing
b. Thick roots
c. Holds soil well
d. Typical of many dicots
e. Food storage organ

20. The meristem is
a. the tip of the stem.
b. the site on the stem where a bud forms.
c. supporting tissue.
d. growing tissue.
e. the base of the leaves.

21–25. Match the following descriptions with one of the cell types from the list below.
a. Parenchyma cells
b. Sclerenchyma cells
c. Collenchyma cells

21. Photosynthesis occurs in these cells. **(a)**

22. These cells function when dead. **(b)**

23. Stone cells of pears **(b)**

24. Cells that support growing organs **(c)**

25. Bulk of root cells **(a)**

26. The xylem tissue of advanced angiosperms typically is distinguished by its
a. elongate tracheids.
b. short stacked vessel elements.
c. sieve tube elements.
d. companion cells.
e. thick-walled fiber cells.

27. Stems function in support and transport. Which type of cell accomplishes most of this function?
a. Collenchyma
b. Companion
c. Parenchyma
d. Sclerenchyma
e. Vessel elements

28. The vascular tissue system of plants has the same function as what animal system?
a. Circulatory
b. Digestive
c. Excretory
d. Reproductive
e. Respiratory

29. A layer of cells that protects the plant is the
a. cuticle.
b. endoderm.
c. epidermis.
d. ground tissue.
e. pericycle.

30. The vascular cambium is located between which two tissues?
a. Phloem and cork cambium
b. Xylem and cork cambium
c. Phloem and bark
d. Xylem and phloem
e. Phloem and ground tissue

▪31. Which is characteristic of secondary growth but not of primary growth?
a. Growth in plant diameter
b. Growth in plant height
c. Typical of meristems
d. Growth by cell elongation
e. Occurs in all dicots

32. Which of the following is typical of cork cells?
a. Interior to cork cambium
b. Waxy suberin
c. Water storage
d. Active cell division
e. Abundant in monocots

33. Root hairs are adaptations that
a. increase surface area.
b. defend the plant.
c. reduce water loss.
d. provide active growth.
e. support the plant.

34. In the course of the development of an individual root cell, the typical sequence is
a. elongation, division, and differentiation.
b. differentiation, elongation, and division.
c. division, differentiation, and elongation.
d. elongation, differentiation, and division.
e. division, elongation, and differentiation.

*▪35. Grasses can grow back after their tips are removed by clipping or grazing because their
a. stems elongate at the tips.
b. stems elongate at the bases of internodes.
c. leaf primordia serve as a protective tip.
d. stems thicken with growth.
e. vascular bundles are scattered through the stem.

36. A root or stem increases in diameter when
a. primary xylem cells divide.
b. phloem cells divide.
c. secondary xylem is deposited external to the vascular cambium.
d. vascular cambium cells divide.
e. phloem cells elongate.

▪37. Annual rings are seen in temperate-zone trees because
a. heartwood cells alternate with sapwood cells.
b. resin is deposited in rings in the stem.
c. cork is deposited in rings in the stem.
d. xylem activity varies with season.
e. xylem cell size varies with season.

38. Lenticels are spongy regions on the surface of some woody stems that function in
a. water uptake.
b. water conservation.
c. gas exchange.
d. protection of growing layers.
e. support of the plant.

39. What is the advantage of the spongy arrangement of mesophyll cells in the lower leaf layer?
a. Maximum absorption of sunlight for photosynthesis
b. Maximum diffusion of carbon dioxide in the leaf
c. Maximum movement of water to leaf cells
d. Minimum water loss from the leaf
e. Minimum exchange of oxygen within the leaf

*40. The veins of a leaf consist of
a. mesophyll cells.
b. guard cells.
c. stomata.
d. vascular cells.
e. epidermal cells.

41. Cacti are plants with stems modified for water storage. Which type of tissue is well developed in cacti for this function?
a. Cork
b. Xylem
c. Phloem
d. Parenchyma
e. Epidermis

42. Conducting cells called ________ elements are the part of xylem where water and minerals are transported.
a. tracheary
b. vascular
c. vessel
d. xylem
e. phloem

43. Unlike fibrous root systems, taproot systems
 a. maximize surface area.
 b. are used as food storage organs.
 c. anchor the plant.
 d. are better at holding soil.
 e. transport water and minerals to the stem.

44. The region of cell division in a primary root is located
 a. in the root cap.
 b. in the apical meristem.
 c. in the region of elongation.
 d. in the area containing differentiated tissues.
 e. throughout the root.

45. A common function of stems but not roots would be
 a. anchorage.
 b. transport.
 c. storage.
 d. support.
 e. absorption.

46. Which of the following is *least* likely to be formed from a lateral bud?
 a. A flower
 b. A branch
 c. A runner
 d. A leaf
 e. A root

*47. Technically, simple tissues in plants
 a. are composed of only one type of cell.
 b. have only one function.
 c. only produce primary cell walls.
 d. can only be found in apical meristems.
 e. are only found in primitive plants.

48. "An organized group of plant cells, working together as a functional unit" best defines a(n)
 a. organism.
 b. organ.
 c. organ system.
 d. tissue.
 e. tissue system.

49. Pit pairs allow plasmodesmata to travel through
 a. the primary cell wall.
 b. the secondary cell wall.
 c. both the primary and secondary cell walls.
 d. neither the primary nor the secondary cell walls.
 e. the secondary and sometimes the primary cell walls.

50. Which of the following is *not* true of parenchyma cells?
 a. They are the most common cell in the plant.
 b. They may contain chloroplasts.
 c. They are commonly used for food storage.
 d. They help support leaves in nonwoody plants.
 e. They usually have thick cell walls.

51. Compared to sclerenchyma, collenchyma cells
 a. have more secondary cell wall materials.
 b. are variously shaped.
 c. can be found in bundles.
 d. are used to support the plant.
 e. are more flexible.

52. Unlike tracheids, vessel elements
 a. function when dead.
 b. are spindle-shaped.
 c. are found primarily in gymnosperms.
 d. lose part or all of the end walls.
 e. evolved to be progressively longer.

53. In angiosperm phloem,
 a. both the sieve tube elements and the companion cells have nuclei.
 b. the sieve tube elements have nuclei but the companion cells do not.
 c. the companion cells have nuclei, but the sieve tube elements do not.
 d. neither the companion cells nor the sieve tube elements have nuclei.
 e. the sieve tube elements have nuclei, but the companion cells may or may not have nuclei.

54. One primary function of the ground tissue in a plant is
 a. photosynthesis.
 b. to protect the plant.
 c. to anchor the plant.
 d. for water conduction.
 e. for conduction of sugars.

55. The shoot epidermis secretes a layer of wax-covered cutin, the __________, which helps retard water loss from stems and leaves.
 a. lignan
 b. suberin
 c. cuticle
 d. stomata
 e. bark

56. In the development of a root, the protoderm gives rise to the
 a. cortex.
 b. root hairs.
 c. endodermis.
 d. xylem and phloem.
 e. pith.

57. In a young root, you would observe xylem cells in the
 a. root cap.
 b. apical meristem.
 c. zone of cell division.
 d. zone of cell elongation.
 e. zone of cell differentiation.

58. Which of the following is *not* part of a monocot stele?
 a. Xylem
 b. Phloem
 c. Endodermis
 d. Pericycle
 e. Pith

59. The __________ is centermost tissue in a dicot stem.
 a. pith
 b. xylem
 c. phloem
 d. pericycle
 e. endodermis

60. Branch roots arise from the
    a. epidermis.
    b. pericycle.
    c. endodermis.
    d. cortex.
    e. pith.

★■61. In which of the following states would you be *least* likely to find annual rings in dicot tree trunks?
    a. Maine
    b. Washington
    c. Kansas
    d. Arizona
    e. Hawaii

62. The periderm of a tree functions primarily to
    a. transport sugars.
    b. form branches.
    c. absorb water.
    d. protect the inner tissues.
    e. support the leaves.

63. In a woody stem, gas exchange occurs through
    a. stomata.
    b. the waxy cuticle.
    c. lenticels.
    d. the cork cambium.
    e. bundle sheath cells.

64. In a typical dicot leaf, most of the chloroplasts would be found in the
    a. upper epidermal cells.
    b. palisade mesophyll cells.
    c. bundle sheath cells.
    d. phloem cells.
    e. guard cells.

65. The primary function of a typical leaf is
    a. photosynthesis.
    b. food storage.
    c. support.
    d. anchorage.
    e. absorption.

66. Guard cells
    a. protect the plant from herbivores.
    b. secrete a waxy cuticle to prevent evaporation.
    c. contain chemicals that poison insects.
    d. control gas exchange.
    e. inhibit germination of fungal spores.

67. The purpose of vascular rays is to transport
    a. nutrients through the phloem to storage cells.
    b. water from the roots to the xylem.
    c. water from the leaves to the phloem.
    d. nutrients from the sclerenchyma to the phloem.
    e. carbon dioxide from the leaves to the phloem.

68. Growth in the diameter of the stems and roots, produced by vascular and cork cambia, is called
    a. outgrowth.
    b. primary growth.
    c. secondary growth.
    d. tertiary growth
    e. ingrowth.

69. Which of the following is *not* a function of cork cambium?
    a. Protection from microorganisms
    b. Minimize water loss
    c. Secondary growth of stems and roots
    d. Mineral uptake
    e. To break off and allow expansion of tree trunks

Study Guide Questions

1. You are studying tropical plants in a Costa Rican cloud forest. You have identified a plant that has not been studied before. It has compound leaves with netted veins, vascular bundles arranged in a cylinder, and five petals on its very large and showy flower. This plant is most likely a(n)
    a. monocot.
    b. eudicot.
    c. gymnosperm.
    d. None of the above

2. The plant described above grows to approximately 40 m at maturity and is in a forest ecosystem. This plant most likely has which of the following root types?
    a. Fibrous
    b. Taproot
    c. Adventitious
    d. Rhizoids

3. You planted sweet peas along your back garden fence. You note that they are very effective at climbing the fence. Upon closer inspection you notice that the plants are attached to the fence by tendrils. These tendrils are modifications of
    a. stems.
    b. roots.
    c. branches.
    d. leaves.

4. Plant cells are easily distinguished from animal cells because they have
    a. rigid cell walls.
    b. plastids.
    c. large vacuoles.
    d. All of the above

5. Plant cells that are photosynthetically active are found in the ____________ layer of the leaf and are ____________ cells.
    a. mesophyll, parenchyma
    b epidermis, parenchyma
    c. mesophyll, sclerenchyma
    d. epidermis, sclerenchyma

6. Water is conducted in ____________ tissue, and carbohydrates and nutrients are transported in ____________ tissue.
    a. xylem, phloem
    b. phloem, xylem
    c. parenchyma, phloem
    d. parenchyma, xylem

7. Plants are capable of indeterminate growth because of
   a. meristems.
   b. regions of continually dividing cells.
   c. their modular nature.
   d. All of the above

8. Which of the following best describes the origin of wood?
   a. Xylem cells enlarge and deposit large amounts of lignin.
   b. Primary meristems increase the amount of xylem deposited.
   c. Secondary meristems allow for continuous increases in vascular tissue.
   d. None of the above

9. Which of the following statements concerning wood is *true*?
   a. All woody plants show annual growth rings.
   b. Patterns of both cylindrical secondary growth and lateral vascular rays are visible in wood.
   c. Sapwood and hardwood result from patterns in primary xylem.
   d. All of the above.

10. Which of the following best describes the function of the cork cambium?
    a. It lays down a protective cork covering over exposed phloem tissue.
    b. It allows for the sloughing of epidermal tissue.
    c. It allows for diameter expansion in stems and roots.
    d. All of the above

11. Sieve tube members have sieve plates where they join other sieve tube members. Which of the following best describes the sieve plates?
    a. Sieve plates are enlargements of the plasmodesmata.
    b. Sieve plates are necessary to allow conduction between sieve tube cells.
    c. Sieve plates allow joining of cytoplasm between adjacent sieve tube cells.
    d. All of the above

12. Plants regulate gas exchange and water loss via
    a. the cuticle.
    b. guarded stomata.
    c. coated pits.
    d. sieve plates.

End of Chapter Questions

1. Which of the following is *not* a difference between monocots and dicots?
   a. Dicots more frequently have broad leaves.
   b. Monocots commonly have flower parts in multiples of three.
   c. The veins in monocot leaves are arranged in a parallel fashion.
   d. The vascular bundles of monocots are commonly arranged in a ring.
   e. Dicot embryos commonly have two cotyledons.

2. Roots
   a. always form a fibrous root system that holds the soil.
   b. possess a root cap at their tip.
   c. grow in an indeterminate manner.
   d. are commonly photosynthetic.
   e. do not show secondary growth.

3. The plant cell wall
   a. lies immediately inside the plasma membrane.
   b. is an impenetrable barrier between cells.
   c. is always waterproofed with either lignin or suberin.
   d. always consists of a primary wall and a secondary wall, separated by a middle lamella.
   e. contains cellulose and other polysaccharides.

4. Which statement about parenchyma cells is *not* true?
   a. They are alive when they perform their functions.
   b. They typically lack a secondary wall.
   c. They often function as storage depots.
   d. They are the most numerous cells in young plants.
   e. They can be photosynthetic in leaves.

5. Tracheids and vessel elements
   a. die before they become functional.
   b. are important constituents of all plants.
   c. have walls consisting of middle lamella and primary wall.
   d. are always accompanied by companion cells.
   e. are found only in the secondary plant body.

6. Which statement about sieve tube members is *not* true?
   a. Their end walls are called sieve plates.
   b. They die before they become functional.
   c. They link end-to-end, forming sieve tubes.
   d. They form the system for translocation of foods.
   e. They lose the membrane that surrounds their central vacuole (tonoplast).

7. The pericycle
   a. separates the stele from the cortex.
   b. is the tissue within which lateral roots arise.
   c. consists of highly differentiated cells.
   d. forms a star-shaped structure at the very center of the root.
   e. is part of the endodermis.

8. Secondary growth of stems and roots
   a. is brought about by the apical meristems.
   b. is common in both monocots and dicots.
   c. is brought about by vascular and cork cambia.
   d. produces only xylem and phloem.
   e. is brought about by vascular rays.

9. Which of the following statements regarding the formation of annual rings is *false*?
   a. Annual rings are formed as a result of seasonal environmental conditions.
   b. Tracheids/vessel elements are larger during periods when water is abundant.
   c. Tracheids/vessel elements have thicker walls during periods of water deprivation.

d. Wood from trees in wet tropics do not display annual rings.
e. Wood formed in previous years is darker than newer wood.

10. Which statement about leaf anatomy is *not* true?
a. Stomata are controlled by paired guard cells.
b. The cuticle is secreted by the epidermis.
c. The veins contain xylem and phloem.
d. The cells of the mesophyll are packed together, minimizing air space.
e. C_3 and C_4 plants differ in leaf anatomy.

Student Web Site Self-Quiz Questions

1. Complete the following sentences from the choices given below. On a typical eudicot plant, the point at which a leaf is attached to the stem is called __________, and __________ are also usually located there.
a. an internode, lateral buds
b. an internode, terminal buds
c. an internode, lenticels
d. a node, lateral buds
e. a node, terminal buds

2. Angiosperm leaves come in a variety of forms. Which one of the following is *not* a function of the leaves of some plants?
a. Protection from herbivores
b. Protection of the inner layers from damaging UV light rays
c. Water storage
d. Support of the plant
e. Storage of energy-rich molecules

3. Select below the sequence of structures you would encounter in moving from the inside to the outside of a typical cell.
a. Plasma membrane, primary cell wall, secondary cell wall, middle lamella
b. Primary cell wall, secondary cell wall, middle lamella
c. Secondary cell wall, primary cell wall, plasma membrane, middle lamella
d. Secondary cell wall, primary cell wall, middle lamella
e. Middle lamella, secondary cell wall, primary cell wall

4. Which one of the following is *not* a feature or function of parenchyma cells?
a. Photosynthesis
b. Rigid structural support
c. Large central vacuole
d. Starch and lipid storage
e. Contain leucoplasts

5. Select the statement about the primary growth of a root that is *false.*
a. Root hairs are derived from the procambium via the pericycle.
b. The root apical meristem produces the primary root meristems.
c. The procambium produces xylem and phloem.
d. There is no clear delineation between the three growth zones of the root.
e. The procambium is interior to the ground meristem.

6. How many of the following six tissues are present within the stele of a eudicot root? cortex, endodermis, pericycle, phloem, pith, xylem
a. 2
b. 3
c. 4
d. 5
e. 6

7. Which of the following tissues/regions is *not* present in both typical monocot roots and eudicot roots?
a. Pith
b. Endodermis
c. Pericycle
d. Cortex
e. Xylem

TEST FILE

★ *Indicates a **difficult** question*

■ *Indicates a **conceptual** question*

35 Transport in Plants

Fill in the Blank

★1. The effect of increasing dissolved solutes is to make the solute potential (**more**/less) negative.

2. Minerals, taken up in their ionic form, are moved into root cells against their concentration gradient by the process of **active** transport.

3. When osmotic **potential** increases inside a cell, water moves into the cell passively by osmosis.

4. More negative water potential in the xylem of the root draws water into the xylem, generating a force called **root pressure**.

5. Cell walls and spaces between cells make up a compartment called the **apoplast**, and the continuous meshwork of living cells connected by plasmodesmata is called the **symplast**.

6. When water moves outside cells, the movement is (regulated/**unregulated**).

7. Evaporative water is lost through pores in the leaf, which are called **stomata**.

■8. Water in the xylem is pulled up to replace water lost by evaporation because of the **cohesion** of water molecules, a physical property of water due to the **hydrogen** bonding between water molecules.

■9. In CAM plants, carbon dioxide is taken in through the **stomata** during the **night** but is not immediately used in photosynthesis. Instead, it is made into **organic acids** until photosynthesis can resume in the daytime. The advantage of CAM is that these plants can **conserve water**.

10. Water moves toward the region of more **negative** water potential.

11. Parenchymal cells known as **transfer** cells help transport mineral ions from the symplast into the apoplast.

12. Water transport through the xylem results from **evaporation** in the leaves and subsequent **tension** in the xylem, which pulls water up.

13. On a hot day, water is lost via evaporation through the shoot. This is called **transpiration**.

14. Endodermal cells are lined with waxy structures, called **Casparian strips**, which prevent water and ions from moving between cells.

15. The movement between living cells of a plant occurs through **plasmodesmata**.

Multiple Choice

1. The tendency for water to move toward greater solute concentration is an example of
   a. active transport.
   b. osmolarity.
   c. diffusion.
   d. reverse osmosis.
   e. passive transport.

2. Pumping protons (H^+) out of a cell can trigger which movements of potassium ion (K^+) and chloride ion (Cl^-)?
   a. K^+ out of the cell and Cl^- into the cell
   b. K^+ into the cell and Cl^- out of the cell
   c. both K^+ and Cl^- out of the cell
   d. both K^+ and Cl^- into the cell
   e. K^+ into the cell and no movement of Cl^-

3. Water tends to move into a cell that has a(n)
   a. high turgor pressure due to cell wall rigidity.
   b. high, positive water potential.
   c. interior solute concentration like that of distilled water.
   d. more negative water potential.
   e. low turgor pressure.

4. The wilting of plant tissue occurs when
   a. water potential is high.
   b. turgor pressure is high.
   c. interior solute concentration is high.
   d. osmotic potential is high.
   e. turgor pressure is low.

5. Which of the following is the Casparian strip?
   a. The layer of endodermal cells
   b. The layer of epidermal cells
   c. The apoplast
   d. The symplast
   e. The waxy layer between endodermal cells

▪6. Endodermal cells differ from other cells in the root in that they
a. lack a symplast region.
b. are nonselective with regard to solute uptake.
c. have a high rate of water transport.
d. are completely surrounded by a waxy layer.
e. prevent water and ions from moving between them.

7. Water enters the xylem tissue from surrounding root cells via
a. active transport.
b. facilitated diffusion.
c. osmosis.
d. pressure pumping.
e. guttation.

8. Guttation is most commonly observed under conditions of
a. high atmospheric humidity and plentiful soil water.
b. low atmospheric humidity and plentiful soil water.
c. high atmospheric humidity but little soil water.
d. low atmospheric humidity and little soil water.
e. varying atmospheric humidity and plentiful soil water.

9. The phenomenon of guttation is related to
a. active transport.
b. osmosis.
c. root pressure.
d. transpiration.
e. translocation.

10. Water moves from the soil into the root by
a. active transport.
b. passive transport.
c. facilitated transport.
d. simple diffusion.
e. facilitated diffusion.

11. Pure water under no applied pressure is defined as having a water potential of
a. +10.
b. +1.
c. 0.
d. –1.
e. –10.

▪12. A plant cell placed in distilled water will
a. expand until the osmotic potential reaches that of distilled water.
b. become more turgid until the osmotic potential reaches that of distilled water.
c. become less turgid until the osmotic potential reaches that of distilled water.
d. become more turgid until the pressure potential of the cell reaches its osmotic potential.
e. become less turgid until the pressure potential of the cell reaches the outside water potential.

▪*13. Water will move from the root hairs through the cortex to the xylem if the water potentials are
a. root hairs = 0; cortex = 0; xylem = 0.
b. root hairs = –1; cortex = –1; xylem = –1.
c. root hairs = –2; cortex = –1; xylem = 0.
d. root hairs = 0; cortex = +1; xylem = +2.
e. root hairs = 0; cortex = –1; xylem = –2.

14. Which of the following is part of the apoplast?
a. Cell wall
b. Plasma membranes
c. Plasmodesmata
d. Cytoplasm
e. Vacuole

15. Cell walls impregnated with water-repellent suberin are found in the cells of the
a. root hairs.
b. cortex.
c. endodermis.
d. pericycle.
e. tracheids.

16. Plants in habitats where excess water can be lost by transpiration might show which of the following adaptations to conserve water loss during gas exchange?
a. Stomata that close at night
b. Stomata concentrated on the upper sides of leaves
c. A uniformly high density of stomata
d. Stomata that close during the day
e. Stomata that open only during photosynthesis

17. Succulent plants of the family Crassulaceae have unusual gas exchange patterns in that they
a. do not take in carbon dioxide through stomata.
b. give off oxygen at night.
c. accumulate carbon dioxide stored as organic acid.
d. retain oxygen in the leaf.
e. obtain gases mainly through their roots.

18. The value of CAM (crassulacean acid metabolism) plant reactions is that
a. photosynthesis can proceed in darkness.
b. carbon dioxide can be concentrated in the leaf for photosynthesis.
c. stomata can remain open during daylight.
d. sugar formation can occur during the night.
e. excess water can be eliminated.

19. According to the pressure flow model for translocation,
a. sugar concentration is highest near the sink area.
b. water enters the sieve tube by osmosis.
c. sugar is transported out of the sieve tubes near the source area.
d. osmosis accomplishes the bulk flow of water and nutrients.
e. little ATP expenditure is required.

20. The movement in plants due to differences in pressure potential is called
a. osmosis.
b. passive transport.
c. diffusion.
d. active transport.
e. bulk flow.

21. On moderately dry, hot days, recently watered plants have
a. high transpiration rates and low root pressures.
b. low transpiration rates and high root pressures.
c. low transpiration rates and low root pressures.
d. high transpiration rates and high root pressures.
e. moderate transpiration rates and moderate root pressures.

22. To initiate stomatal opening, potassium ions
a. passively diffuse into guard cells.
b. passively diffuse out of guard cells.
c. are driven into guard cells.
d. are actively transported out of guard cells.
e. bond to receptor sites on guard cell walls.

▪23. The advantage to CAM plants of forming organic acids from carbon dioxide rather than storing carbon dioxide directly is that
a. concentrated carbon dioxide is toxic to cells.
b. organic acids can be used directly in photosynthesis.
c. more energy is required to store carbon dioxide than organic acids.
d. the presence of organic acids keeps stomata closed.
e. storage as organic acids make it possible to keep stomata closed during daylight.

*24. The active transport of sucrose molecules from a leaf's apoplast to its phloem requires the sucrose to be cotransported with
a. protons.
b. amino acids.
c. potassium ions.
d. chloride ions.
e. water.

25. As a tree begins transpiring in the morning, tension pressure occurs first in
a. the leaves.
b. the branches.
c. the trunk.
d. the roots.
e. all regions of the tree.

▪26. How do mineral nutrients enter the plant body directly from the environment?
a. Uptake through the roots
b. Uptake by the leaves
c. Uptake from digested food molecules
d. Uptake into vascular tissue
e. Uptake through the stems

▪27. When the ion concentration inside root cells is higher than ions in the soil solution, such ions can enter root cells by what processes?
a. Active transport only
b. Simple diffusion only
c. Simple diffusion and facilitated diffusion
d. Facilitated diffusion and active transport
e. Simple diffusion and active transport

▪28. The facilitated diffusion of ions from the soil solution into root cells requires
a. that the concentration of ions at the root cells be the same outside and inside.
b. that the concentration of ions outside the root cells be lower than that inside.
c. the expenditure of ATP.
d. specific channel proteins in the membranes.
e. cellular respiration.

▪29. When a large amount of water enters a plant cell, what happens?
a. Entry of water increases as the water potential increases.
b. Entry of water is opposed by turgor pressure.
c. Water moves toward the region of more positive water potential.
d. Entry of water reduces the turgor pressure.
e. Entry of water causes increased active transport into the cell.

▪30. Which of the following is true about the apoplast (the transport route through intercellular spaces)?
a. Osmosis of water is involved.
b. Movement of materials is regulated by membranes.
c. Water and solutes can move by bulk flow.
d. Plasmodesmata are involved.
e. Water and solutes enter the stele via this channel.

*31. The uptake of ions in plant cells is influenced by
a. the electrical gradient.
b. the concentration gradient.
c. the ionic balance.
d. pumping of protons.
e. All of the above

32. In order to transport K^+ ions into their cells, plants
a. pump protons out.
b. use the sodium/potassium pump.
c. pump K^+ in.
d. pump water in.
e. More than one of the above

33. When bulk flow occurs in plants, ________ move(s).
a. water and minerals
b. water and organic molecules
c. dissolved minerals and organic molecules
d. water
e. organic molecules

34. To drive water into root cells, energy comes most directly from
a. ATP.
b. the sun.
c. $NADPH^+$.
d. bulk flow.
e. chloroplasts.

*35. Per Scholander used a __________ to measure tension in xylem sap.
a. balloon on a bell jar
b. water column
c. pressure gauge
d. barometer
e. pressure bomb

36. Which technique allowed researchers to conclude that the fibrous proteins in sieve tube elements are normally dispersed and only obstruct sieve plates in response to cell damage?
a. Analyzing phloem sap extruded from aphid stylets
b. Freezing phloem tissue before cutting and examining it
c. Watering the plant before cutting and examining it
d. Using patch clamping to examine sieve plates
e. Exposing plants to blue light before examination

*37. To study the tension in limbs high in trees, Scholander used
a. lumberjacks to cut down trees.
b. expert tree climbers.
c. a ladder.
d. aphids.
e. sharp shooters.

▪38. Which of the following causes the root pressure that moves water upward in plant xylem?
a. Negative water potential in the xylem sap
b. High pressure potential of water in the soil
c. Movement of water from root cells into the soil
d. Active transport of minerals from soil to root cells
e. High atmospheric humidity

▪39. Which force accounts for the movement of water upward through a narrow tube?
a. Cohesion of water molecules via hydrogen bonding
b. Negative water potential in the xylem
c. Active transport of water molecules
d. Passive osmosis of water following ion movement
e. Pumping of water into the phloem

40. The evaporation-tension-cohesion mechanism explains how
a. water is lost from leaf openings.
b. water is transported in the xylem.
c. water and minerals enter the root.
d. mineral ions move through the xylem.
e. leaf epidermal cells minimize water loss.

41. Bulk water flow is stopped by
a. metabolic inhibitors.
b. closed stomata.
c. accumulation of K^+.
d. blue light.
e. surface tension.

42. Which of the following increases a plant's intake of carbon dioxide for photosynthesis?
a. Thick waxy cuticle
b. Loss of water vapor
c. Darkness
d. Electric imbalance
e. Open stomata

43. Stomata begin to open when potassium
a. enters guard cells, followed by water, and they become turgid.
b. leaves guard cells and they become less turgid.
c. enters guard cells and they become less turgid.
d. leaves guard cells and they become turgid.
e. reaches equilibrium between guard cells and their surroundings.

44. What structure of the leaf minimizes water loss?
a. Stoma
b. Epidermis
c. Cuticle
d. Phloem
e. Xylem

45. Most water moving through the apoplast from the soil into the stele cells first crosses a plasma membrane in the cells of the
a. root hairs.
b. cortex.
c. endodermis.
d. pericycle.
e. tracheids.

46. A transfer cell has knobby growths extending into the cell that facilitate movement of minerals from its
a. cell wall into its cytoplasm.
b. cytoplasm into its cell wall.
c. cell wall into the next cell's wall.
d. cytoplasm into the next cell's cytoplasm.
e. cell wall into the next cell's cytoplasm.

▪47. What process makes the water potential in a leaf more negative?
a. The pressure placed on the leaf by the cuticle
b. The evaporation of water from mesophyll cells
c. The movement of water into the leaf by root pressure
d. The increased potassium pumped out of guard cells
e. The movement of water from the veins into the leaf

48. The evaporative loss of water from the shoot is called
a. translocation.
b. transformation.
c. transportation.
d. transpiration.
e. transcention.

49. Cohesion is the tendency of water molecules to attract
a. other water molecules by covalent bonds.
b. other water molecules by hydrogen bonds.
c. cellulose molecules by covalent bonds.
d. cellulose molecules by hydrogen bonds.
e. lignin molecules by covalent bonds.

▪50. If xylem sap in a stem is under tension, what will occur if you cut the stem?
a. Xylem sap will spurt out.
b. Xylem sap will stay at the cut surface.

c. Air will be pulled into the xylem.
d. The cut surface will form bubbles if placed under water.
e. Xylem sap will run out if placed under water.

▪51. What happens to potassium ions to initiate stomatal closing?
a. Ions are actively transported into the guard cells.
b. Ions are actively transported out of the guard cells.
c. Ions are actively transported from one guard cell to another.
d. Ions diffuse into the guard cells.
e. Ions diffuse out of the guard cells.

52. CAM plants have adapted to dry areas by
a. only opening their stomata on cool days.
b. having a more efficient Calvin–Benson cycle.
c. keeping stomata closed during the daylight hours.
d. evolving an active pump for carbon dioxide.
e. fixing carbon dioxide in one type of cell, then transferring the product to another cell for sugar formation.

53. A good antitranspirant
a. closes stomata.
b. reduces water evaporation, not CO_2 uptake.
c. is a leaf coating.
d. is dry soil.
e. increases CO_2 release.

54. When cells in the leaf release abscisic acid, it is in response to
a. negative water potential.
b. not enough carbon dioxide.
c. reduction of light (nighttime).
d. increase of chloride ions in the stomata.
e. increase in potassium ions in the stomata.

55. The effect of abscisic acid is
a. to store CO_2 for daytime use.
b. reduce water loss.
c. open stomata.
d. increase proton pumping in guard cells.
e. lower pH.

▪56. When sugars are actively transported into a cell, what happens to the turgor pressure inside that cell as a result?
a. No change; sugar concentration has no effect on turgor pressure
b. Increases, because sugar concentration directly affects turgor pressure
c. Increases, because water enters and affects turgor pressure
d. Decreases, because water exits and affects turgor pressure
e. Decreases, because sugar concentration directly affects turgor pressure

▪57. At the site where sugars are to be used, how do sugars move from the sieve tubes into the tissue?
a. By diffusion
b. Via the apoplast
c. By active transport
d. By osmosis
e. Via translocation

58. Transport through both the xylem and the phloem
a. stops if the tissue is killed.
b. requires ATP.
c. can occur simultaneously in both directions.
d. requires negative pressure (tension).
e. involves long, thin channels.

59. Plant physiologists can obtain pure phloem sap from individual phloem cells by
a. using tiny drills and capillary pipettes.
b. collecting material from aphid stylets.
c. obtaining liquids oozing from cut stem surfaces.
d. analyzing the contents of droplets formed by guttation.
e. gathering the mycorrhizal fungi for subsequent analysis.

60. The pressure flow model of translocation depends entirely on the existence of mechanisms for loading sugars into phloem at the ________ regions and for unloading them at the ________ regions.
a. sink, sink
b. sink, source
c. source, source
d. source, sink
e. source, source or sink

61. According to the pressure flow model, during fruit development, photosynthesizing leaves would be the ________ and the fruit would be the ________.
a. sink, sink
b. sink, source
c. source, source
d. source, sink
e. source, source or sink

62. Sugars pass from cell to cell in the leaf, starting in the ________ of the mesophyll, through the ________ of other cells, and finally into the ________ of the sieve tube element.
a. symplast, symplast, symplast
b. apoplast, apoplast, apoplast
c. symplast, apoplast, symplast
d. apoplast, symplast, symplast
e. apoplast, symplast, apoplast

*63. Normally, plasmodesmata allow molecules as large as ________ daltons to pass.
a. 10
b. 100
c. 1,000
d. 10,000
e. 100,000

Study Guide Questions

1. The primary function of the Casparian strip is to
   a. force water and minerals through the membranes of endodermal cells.
   b. prevent entry into the stele solely through the apoplast.
   c. provide regulation for water and mineral movement in the plant.
   d. All of the above

2. The primary difference between the apoplast and the symplast is that the
   a. apoplast is nonliving spaces and cell walls.
   b. apoplast relies on active transport.
   c. symplast is nonliving spaces and cell walls.
   d. apoplast prevents passive diffusion.

3. Which of the following regarding water transport is *true*?
   a. Root pressure is sufficient to drive xylem sap movement.
   b. Bulk flow is not a mechanism by which water and minerals are transported.
   c. The cohesive nature of water is central to water movement in a plant.
   d. None of the above

4. Tension is a result of which of the following?
   a. Transpiration at the leaf surface
   b. The cohesive nature of water
   c. The narrowness of the xylem tube
   d. All of the above

5. The fact that water transport continues as long as leaves are alive and active indicates that
   a. leaves pump water.
   b. leaves are necessary for transport of water.
   c. roots are active.
   d. water is not needed for leaves to remain alive.

6. Which of the following is true regarding transport in phloem?
   a. Transport in phloem is always in the direction of leaves to roots.
   b. Transport in phloem is from source tissue to sink tissue.
   c. Transport in phloem cells requires energy.
   d. None of the above

7. If the pressure potential is + 0.16 megapascals (MPa) and the osmotic potential is – 0.24 megapascals, then the water potential would be
   a. + 0.04 MPa
   b. + 0.08 MPa
   c. – 0.08 MPa
   d. – 0.24 MPa

8. If you were to order the water potential of the following root cells/regions from least to most negative, which cell/region would be third?
   a. Xylem
   b. Soil next to root
   c. Cortex apoplast
   d. Stele apoplast

9. The movement of water up the stems of tall plants is least dependent on which of the following factors?
   a. Guttation
   b. Transpiration
   c. Cohesion of water molecules
   d. Tension within water columns

10. Which of the following characteristics applies to both xylem and phloem transport?
    a. It is a passive, nonenergy requiring process.
    b. It involves only living cells.
    c. It follows a water potential gradient.
    d. It can occur in both directions.

11. Which of the following regulates stomatal opening and closing?
    a. Abscisic acid levels
    b. Light levels
    c. Carbon dioxide concentrations
    d. All of the above

12. Regulators of stomatal opening and closing work by activating the
    a. proton pump in guard cells.
    b. proton pump in stomata.
    c. sodium–potassium pump.
    d. All of the above

End of Chapter Questions

1. Osmosis
   a. requires ATP.
   b. results in the bursting of plant cells placed in pure water.
   c. can cause a cell to become turgid.
   d. is independent of solute concentrations.
   e. requires a cell wall.

2. Water potential is
   a. equal to the solute potential.
   b. analogous to the air pressure in an automobile tire.
   c. the movement of a solvent through a membrane.
   d. the tendency of a solution to take up water from another solution across a membrane.
   e. defined as 1.0 for pure water under no applied pressure.

3. Which statement about proton pumping across the plasma membranes of plants is *not* true?
   a. It requires ATP.
   b. The region inside the membrane becomes positively charged with respect to the region outside.

c. It enhances the movement of K^+ ions into the cell.
d. It pushes protons out of the cell against a proton concentration gradient.
e. It can drive the secondary active transport of negatively charged ions.

4. A pressure bomb can be used to measure the
a. opening of stomata.
b. turgor pressure inside guard cells.
c. sap tension in phloem.
d. negative pressure potential of sap present in xylem.
e. loss of water in stems present at high elevations.

5. Which statement is *not* true?
a. The symplast is a meshwork consisting of the (connected) living cells.
b. Transfer cells are specialized for transporting water across their cell walls.
c. The Casparian strips prevent water from moving between endodermal cells.
d. The endodermis is a cell layer in the cortex.
e. Water can move freely in the apoplast without entering cells.

6. Which of the following is *not* part of the transpiration–cohesion–tension model?
a. Water evaporates from the walls of mesophyll cells.
b. Removal of water from the xylem exerts a pull on the water column.
c. Water is remarkably cohesive.
d. The wider a tube, the greater the tension its water column can withstand.
e. At each step, water moves to a region with a more strongly negative water potential.

7. Stomata
a. control the opening of guard cells.
b. release less water to the environment than do other parts of the epidermis.
c. are usually most abundant on the upper epidermis of a leaf.
d. are covered by a waxy cuticle.
e. close when water is being lost at too great a rate.

8. Plants that perform crassulacean acid metabolism
a. incorporate carbon dioxide into organic acids.
b. have leaves that become more acidic during the day.
c. close their stomata at night.
d. are also called C_4 plants.
e. must live in environments where water is plentiful.

9. Which statement about phloem transport is *not* true?
a. It takes place in sieve tubes.
b. Contents in a sieve tube move can move bidirectionally.
c. It stops if the phloem is killed by heat.
d. Sucrose is actively transported into sieve tube members at sources.
e. A high turgor pressure is maintained in the sieve tubes.

10. The fibrous protein in sieve plates
a. clogs the sieve plates at all times.
b. never clogs the sieve plates.
c. serves no known function.
d. may caulk leaks when a plant is damaged.
e. provides the motive force for transport in the phloem.

Student Web Site Self-Quiz Questions

1. If you compare the cells of a wilted tomato plant with its cells when the plant is not wilted, which one of the following choices in the table below shows the expected results? Note: in selecting an answer, be sure to take the sign of the values into account.

	Osmotic Potential	Turgor Pressure
a.	**wilted < not wilted**	**wilted < not wilted**
b.	wilted > not wilted	wilted < not wilted
c.	wilted < not wilted	wilted > not wilted
d.	wilted > not wilted	wilted > not wilted
e.	wilted = not wilted	wilted < not wilted

2. Using a pressure bomb and the methods of Per Scholander, you obtain twigs from three heights (20, 40, and 60 m) from a tree at sunrise and at noon. At what height and time of day would you expect to require the greatest gas pressure in the pressure bomb to force sap out at the twig's cut surface?
a. 20 m, sunrise
b. 20 m, noon
c. 40 m, sunrise
d. 20 m, sunrise
e. 60 m, noon

3. Various factors affect movement of potassium ions into and out of guard cells, including light cues. Which of the following does *not* open stomata or keep them open?
a. Low CO_2 levels within the leaf
b. Water potential of leaf cells becoming less negative.
c. Acidification of guard cells
d. Low levels of abscisic acid
e. Exposure to light

4. Which of the following events does *not* take place as stomata close?
a. Potassium ions diffuse out of the guard cells.
b. Turgor pressure increases.
c. Water potential becomes less negative.
d. Chloride and negative organic ions diffuse out of the guard cells.
e. Guard cells go limp.

TEST FILE

★ *Indicates a* ***difficult*** *question*

■ *Indicates a* ***conceptual*** *question*

36 Plant Nutrition

Fill in the Blank

1. Decomposed plant litter produces a dark-colored organic material called **humus**.
2. The source from which plants derive their carbon is the **atmosphere**.
3. Nitrogen fixation is catalyzed by the enzyme **nitrogenase,** which cannot function in the presence of the element oxygen.
4. Plants acquire their essential mineral nutrients from the **soil**.
5. Some essential elements that occur as positive ions in soils may be traded with H^+ ions in soil solutions by the process of **ion exchange**.
6. Because of leaching and crop production, soils may become depleted of nutrients and require addition of **fertilizer**.
7. The three elements most commonly added to agricultural soils are **nitrogen, phosphorus**, and **potassium**.
8. Nitrogen fixation is the conversion of atmospheric nitrogen to **ammonia** and is catalyzed by a single enzyme called **nitrogenase**.
9. All nitrogen fixation is carried out by **prokaryotes**, and some of them live in symbiosis with other organisms.
10. The process that is the opposite of nitrogen fixation is called **denitrification**.

■11. Some plants living in boggy areas of low pH have adaptations for carnivory in order to increase their intake of **nitrogen**. These carnivorous plants are considered to be **autotrophs** because they acquire energy from photosynthesis.

12. The large *Rhizobium* cells that live within plant root nodules are called **bacteroids**.
13. The minerals required by plants in concentrations of less than 100 µg per gram of dry matter are called **micronutrients**.
14. Nutrients necessary for plant growth and reproduction, not replaceable by another nutrient, and directly required by the plant are called **essential elements**.
15. The layers in soil are called **horizons**.
16. Most soils have at least **two** horizons.
17. **Cyanobacteria** are bacteria that both photosynthesize and fix nitrogen.

★18. Soils with high clay content have the disadvantage that they lack **oxygen**.

Multiple Choice

1. The four elements found in highest concentration in plants are
   a. phosphorus, calcium, hydrogen, and carbon.
   b. magnesium, phosphorus, calcium, and potassium.
   c. magnesium, iron, phosphorus, and potassium.
   d. carbon, hydrogen, oxygen, and nitrogen.
   e. carbon, hydrogen, oxygen, and potassium.

2. Hydrogen in hydrocarbons enters the living system from
   a. the atmosphere.
   b. oil and coal.
   c. water.
   d. H_2S.
   e. All of the above

★3. Which of the following organisms are autotrophs?
   a. Carnivorous organisms
   b. Herbivorous organisms
   c. Chemosynthetic organisms
   d. Organisms that require organic nutrients
   e. All of the above

4. Which of the following serves as an energy source for chemosynthetic bacteria?
   a. Sugars
   b. Carbon dioxide
   c. Hydrogen sulfide
   d. Oxygen
   e. Protein

5. A photoautotroph acquires its carbon
   a. from the soil
   b. from the air.
   c. dissolved in water.
   d. from the sun.
   e. from carbon-fixing prokaryotes.

6. Plants acquire minerals from the soil by
   a. recycling them.
   b. growing.
   c. rain water.
   d. soil microbes.
   e. All of the above

*7. If a new nutrient is discovered, it would be
   a. a macronutrient
   b. a micronutrient.
   c. a rare gas.
   d. a small organism.
   e. It could be any of the above.

8. The process of ion exchange is the means by which
   a. carbonic acid is added to soils.
   b. positive ion nutrients are replaced by H^+.
   c. negative ion nutrients are incorporated into soils.
   d. positive ions are replaced by negative ions.
   e. neutral atoms become ions in soils.

*9. Plants in nature tend to
   a. have adequate nitrogen supplies.
   b. have inadequate nitrogen supplies.
   c. always have surplus nitrogen supplies.
   d. have inadequate nitrogen supplies if leaves are yellow, adequate if green.
   e. usually have surplus nitrogen supplies.

10. Nitrogen deficiencies cause __________ leaves to turn __________ in many crops.
   a. old, yellow
   b. young, yellow
   c. old, orange
   d. young, orange
   e. old, brown spotted

11. Iron deficiencies cause __________ leaves to turn __________.
   a. old, yellow
   b. young, yellow
   c. old, orange
   d. young, orange
   e. old, brown spotted

12. The form of nitrogen from the soil that most plants prefer is
   a. found in amino acids.
   b. ammonia.
   c. N_2.
   d. nitrate.
   e. nitrite.

13. Which of the following nitrogen compounds is used directly by plants to build proteins?
   a. Ammonia
   b. N_2
   c. Nitrate
   d. Nitrite
   e. Nitrous oxide

14. One example of a nutrient in reduced form is the
   a. carbon in carbon dioxide.
   b. hydrogen in water.
   c. nitrogen in ammonia.
   d. phosphorus in phosphate.
   e. sulfur in sulfate.

15. The most common gas in the atmosphere is
   a. carbon dioxide.
   b. nitrogen.
   c. oxygen.
   d. ozone.
   e. water vapor.

16. Some bacteria function as denitrifiers; they
   a. oxidize ammoniun ions to nitrate.
   b. oxidize nitrate to nitrite.
   c. reduce N_2 to ammonia.
   d. reduce nitrates to ammonia.
   e. oxidize nitrate to N_2.

17. Where do plant roots get oxygen?
   a. Air spaces in the soil
   b. Water in soil
   c. From the leaves by way of the phloem
   d. The leaves get oxygen from the air and then send it to the roots
   e. All of the above

18. Most plants continue to obtain new sources of mineral nutrients by
   a. breaking down organic matter.
   b. growing longer roots.
   c. shading the plants below them.
   d. evolving more elaborate photosystems.
   e. absorbing minerals through leaves.

▪19. One of the defining characteristics of an essential element is that it
   a. is only necessary for early growth of the seedling.
   b. can be replaced by another element.
   c. has a direct function in the plant.
   d. may function by relieving the toxicity of another element.
   e. is found in relatively high concentrations in the environment.

20. Plants do *not* obtain which of the following elements from soil?
   a. Carbon
   b. Nitrogen
   c. Potassium
   d. Sulfur
   e. Zinc

21. The five elements that comprise most proteins are
   a. carbon, oxygen, sulfur, phosphorus, and potassium.
   b. carbon, hydrogen, oxygen, nitrogen, and phosphorus.
   c. carbon, hydrogen, oxygen, nitrogen, and sulfur.
   d. carbon, hydrogen, nitrogen, sulfur, and potassium.
   e. carbon, hydrogen, oxygen, phosphorus, and potassium.

22. Which of the following is a micronutrient in plants?
    a. Potassium
    b. Sulfur
    c. Calcium
    d. Iron
    e. Magnesium

23. Soil scientists recognize ________ major zones: ________.
    a. two; A and B
    b. four; top, top middle, middle, and lower
    c. seven; 1, 2, 3,4, 5, 6, and 7
    d. two; upper and lower
    e. three; A, B, and C

24. The maximum diameter of a clay particle is
    a. 2,000,000 micrometers.
    b. 2,000 micrometers.
    c. 2 micrometers.
    d. 0.002 micrometer.
    e. 0.000002 micrometer.

25. Most earthworms can be found in
    a. topsoil.
    b. subsoil.
    c. topsoil and subsoil.
    d. bedrock.
    e. subsoil and bedrock.

26. Which of the following plants does *not* have nitrogen-fixing root nodules containing *Rhizobium*?
    a. Alfalfa
    b. Beans
    c. Clover
    d. Peas
    e. Rice

27. Which mineral is deficient in a plant whose growth is stunted and whose oldest leaves turn yellow and die prematurely?
    a. Calcium
    b. Iron
    c. Magnesium
    d. Nitrogen
    e. Phosphorus

28. The three elements most commonly added to agricultural soils in fertilizers are
    a. nitrogen, phosphorus, and iron.
    b. nitrogen, potassium, and iron.
    c. potassium, sulfur, and iron.
    d. nitrogen, potassium, and phosphorus.
    e. nitrogen, sulfur, and iron.

29. What is the advantage of adding organic fertilizers to soils instead of inorganic fertilizers?
    a. Improved physical properties of soil
    b. Use of specific nutrient formulas for specific problems
    c. More rapid increase in nutrients
    d. Leaching of the soils
    e. Increase of clay particles

30. Chemical weathering, an important part of soil formation, includes the
    a. splitting of clays.
    b. effects of freezing and thawing.
    c. hydrolysis of rock.
    d. crushing of rock.
    e. drying of soils.

31. Which of the following is *not* a consequence of adding lime to soil?
    a. The soil's pH is raised.
    b. Nutrient availability increases.
    c. Calcium (Ca^{2+}) is added to the soil.
    d. Clay particles release hydrogen ions (H^+).
    e. The soil's ability to retain water increases.

▪32. One advantage of organic fertilizers over inorganic fertilizers is that organic fertilizers
    a. improve the physical properties of the soil.
    b. provide an almost instantaneous supply of soil nutrients.
    c. increase the soil's pH by liming.
    d. contain higher concentrations of essential nutrients.
    e. contain chemically active clay particles.

33–36. Match the mineral nutrient in the list below with the following description. Each item may be used once, more than once, or not at all.
    a. Magnesium
    b. Nitrogen
    c. Phosphorus
    d. Potassium
    e. Iron

33. When bonded to oxygen atoms, important in many energy-storing and energy-releasing pathways **(c)**

34. Pumped into cells to create osmotic potential **(d)**

35. Constituent of both proteins and amino acids **(b)**

36. Used as a cofactor by many enzymes **(a)**

▪37. What is the effect of the ionization of carbonic acid on soil?
    a. It triggers the release of mineral ions from clay.
    b. It lowers the pH of the soil.
    c. It reduces the leaching of phosphates and nitrates.
    d. It causes laterization of the A horizon.
    e. It increases the amount of soil carbon available to plants.

38. Clay has the advantage that it
    a. provides a reservoir of cation nutrients.
    b. holds moisture
    c. resists pH changes.
    d. All of the above
    e. None of the above

39. Which of the following elements does *not* reversibly attach to the surface of clay particles?
    a. Calcium
    b. Chloride
    c. Hydrogen
    d. Magnesium
    e. Potassium

40. The best topsoils are composed of
a. clay.
b. sand.
c. organic matter.
d. All of the above
e. None of the above

41. Most of the nitrogen on Earth is in the form of
a. ammonia.
b. nitrate ions.
c. nitrogen gas (N_2).
d. amino acids.
e. proteins.

42. Nitrogen fixers convert
a. ammonia to N_2.
b. N_2 to ammonia.
c. ammonia to nitrate.
d. nitrate to ammonia.
e. N_2 to nitrate.

43. All nitrogen fixers belong to the kingdom(s)
a. Eubacteria.
b. Plantae.
c. Eubacteria and Plantae.
d. Protista.
e. Protista and Plantae.

44. Cyanobacteria are known to fix nitrogen in association with all of the following *except*
a. fungi in lichens.
b. ferns.
c. bryophytes.
d. cycads.
e. wheat.

45. Nitrogenase enzymes react with the substrate N_2 with _________ hydrogen atoms before releasing the product.
a. two
b. three
c. four
d. five
e. six

46. Nitrogenase enzymes are extremely sensitive to _________ molecules.
a. hydrogen
b. oxygen
c. water
d. carbon dioxide
e. calcium carbonate

47. Nodules that are actively fixing nitrogen are pink, demonstrating the presence of
a. iron.
b. chlorophyll.
c. leghemoglobin.
d. anthocyanin.
e. alkaloids.

▪48. Recombinant DNA technology is now being used in an attempt to "teach" new plants how to
a. produce their own nitrogenase enzymes.
b. fix nitrogen without using ATP.
c. produce ammonia under anaerobic conditions.
d. convert nitrates into ammonium ions.
e. recycle their nitrogen instead of excreting it.

49. Denitrifying bacteria are part of nature's nitrogen cycle, converting
a. ammonia to N_2.
b. N_2 to nitrate.
c. ammonia to nitrate.
d. nitrate to ammonia.
e. nitrate to N_2.

50. Plants take up sulfur in the _________ form and phosphorus in the _________ form.
a. reduced, oxidized
b. oxidized, oxidized
c. oxidized, reduced
d. oxidized, reduced or oxidized
e. reduced, reduced or oxidized

51. The process of nitrogen fixation is the
a. uptake of atmospheric nitrogen by plants.
b. conversion of atmospheric nitrogen into ammonia.
c. production of nitrogen-bearing compounds in plants.
d. release of nitrogen into the atmosphere.
e. release of ammonia into the atmosphere.

52. Root nodules on plants of the legume family contain
a. cyanobacteria.
b. *Nitrosococcus* bacteria.
c. *Rhizobium* bacteria.
d. *Pseudomonas* bacteria.
e. *Nitrobacter* bacteria.

▪53. Which of the following statements about the chemical process of nitrogen fixation in cells is *true*?
a. All three bonds between nitrogen atoms are broken simultaneously.
b. Hydrogen atoms are added in one reaction at a time.
c. Very little energy in the form of ATP is required.
d. A different enzyme catalyzes each of the many reactions.
e. It is enhanced by high oxygen concentrations.

▪54. How do nitrogen-fixing microbes first become symbiotic with their plants?
a. They are carried in the seed.
b. They are attracted by chemicals on root hairs.
c. They move in via openings in the plant cell walls.
d. They move into the vascular system and multiply.
e. They enter root nodules previously produced by the plant.

55. Legume root nodules represent a symbiosis between the legume plant, which receives fixed nitrogen, and the bacteria, which receive
   a. sugars.
   b. oxygen.
   c. carbon dioxide.
   d. nitrogenase.
   e. leghemoglobin.

56. The industrial production of nitrogen-containing fertilizer is currently limited by
   a. its high energy expense.
   b. the inability to insert nitrogenase genes into plants.
   c. the lack of nitrogenase for the industrial process.
   d. the limited supply of N_2 gas.
   e. the need to exclude free oxygen in the process.

57. The process that is the opposite of nitrogen fixation is
   a. nitrification.
   b. denitrification.
   c. aerobic breakdown of amino acids.
   d. release of ammonia.
   e. nitrate reduction.

58. The products of nitrogen-fixing organisms can be oxidized to form nitrites and nitrates by
   a. all living organisms.
   b. most living organisms that utilize oxygen.
   c. many types of microorganisms.
   d. a few specific genera of soil bacteria.
   e. only the nitrogen-fixing bacteria.

▪59. The big drawback in recombinant DNA for making plants that produce their own nitrogenase is
   a. bacterial genes cannot be inserted into plants.
   b. the DNA degrades when inserted into a plant.
   c. plants can't exclude oxygen, so the enzyme is ineffective.
   d. plants can't take up N_2, so the enzyme is ineffective.
   e. the amount of energy needed for the reaction is very great.

60. An advantage to using inorganic fertilizers is that
   a. they also improve physical properties of the soil.
   b. they can be taken up almost instantaneously.
   c. their quality is better than organic fertilizers.
   d. the minerals are in their proper ionic form.
   e. None of the above

61. The main problem with how nitrogen fertilizers are made is
   a. the large energy requirement.
   b. the pollution of lakes and streams.
   c. their availability to plants when applied to soil.
   d. the cost of shipping.
   e. they are not organic sources.

62. The capture and digestion of insects allows carnivorous plants to
   a. pollinate their flowers.
   b. absorb nitrogen compounds.
   c. disperse their fruits.
   d. overcome insect parasitism.
   e. neutralize acidic soils.

▪*63. Although mistletoes are green, they are considered to be parasites because
   a. they depend on other plants for water and minerals.
   b. they cling to woody plants for physical support.
   c. they capture and digest insects.
   d. they have root nodules containing nitrogen-fixing *Rhizobium*.
   e. their chlorophyll is not functional.

Study Guide Questions

1. Which of the following nutrients are not considered essential for plant growth?
   a. Cadmium
   b. Nitrogen
   c. Manganese
   d. Potassium

2. Macronutrients are __________ than micronutrients.
   a. larger molecules
   b. needed in greater quantities
   c. more essential
   d. more important for growth

3. Carbon and oxygen are acquired through __________ and nitrogen and potassium are acquired from __________ .
   a. consumption, soil solution
   b. photosynthesis, nitrification
   c. photosynthesis, soil solution
   d. photosynthesis, fertilizer

4. You notice that the young leaves of your tomato plants are very yellow. What type of deficiency does this suggest?
   a. Nitrogen
   b. Carbon
   c. Water
   d. Iron

5. Years of cotton farming in the South stripped away much of the A horizon of the soils. Subsequent agriculture has been difficult for farmers because the
   a. A horizon contains most of the available nutrients.
   b. B horizon contains significantly less available nutrients.
   c. A horizon is most conducive to root growth.
   d. All of the above

6. Clay particles in soils are important for
   a. holding soil together.
   b. ion exchange.
   c. holding water.
   d. All of the above

7. Most clays form from the __________ of rock.
   a. mechanical weathering
   b. chemical weathering
   c. heaving
   d. All of the above

8. You purchase a commercial fertilizer at your local garden center. The label says that it is 10-20-10. This label refers to the
   a. **percentages of nitrogen, phosphate, and potassium.**
   b. percentages of nitrogen, carbon, and oxygen.
   c. rate at which nitrogen is released from the fertilizer.
   d. ratio of organic to inorganic matter in the fertilizer.

9. The relationship between *Rhizobium* and the roots of legumes can best be described by which of the following terms?
   a. Parasitic
   b. Symbiotic
   c. **Mutualistic**
   d. Carnivorous

10. Nitrogen gas is reduced to ammonia by which of the following enzymes?
   a. Rhizobium
   b. **Nitrogenase**
   c. Nitrification
   d. Denitrification

11. Plants are unique in that they can synthesize the amino acids ____________ and ____________, while all animals must consume these amino acids.
   a. **cysteine, methionine**
   b. threonine, methionine
   c. uracil, guanine
   d. cytosine, methionine

12. Carnivorous plants are often found in acidic and nutrient-poor environments. The main selective pressure for carnivory is
   a. lack of nitrogen and phosphorous sources.
   b. **lack of iron and calcium sources.**
   c. incomplete ion exchange.
   d. All of the above

End of Chapter Questions

1. Which of the following is *not* an essential mineral element for plants?
   a. Potassium
   b. Magnesium
   c. Calcium
   d. **Lead**
   e. Phosphorus

2. Fertilizers
   a. are often characterized by their N–P–O percentages.
   b. are not required if crops are removed frequently enough.
   c. **restore needed mineral nutrients to the soil.**
   d. are needed to provide carbon, hydrogen, and oxygen to plants.
   e. are needed to destroy soil pests.

3. Which of the following is *not* an important step in soil formation?
   a. **Removal of bacteria**
   b. Mechanical weathering
   c. Chemical weathering
   d. Clay formation
   e. Hydrolysis of soil minerals

4. Which of the following statements regarding essential elements is *false*?
   a. The majority of essential elements come from the soil.
   b. Essential elements are necessary for growth of plants.
   c. Essential elements cannot be replaced by other elements.
   d. The requirement of an essential element must be direct.
   e. **Under times of nutritional stress, plants can synthesize essential elements as needed.**

5. Nitrogen fixation is
   a. performed only by plants.
   b. the oxidation of nitrogen gas.
   c. catalyzed by the enzyme nitrogenase.
   d. **a multi-step chemical reaction.**
   e. possible because N^2 is a highly reactive substance.

6. Nitrification is
   a. performed only by plants.
   b. **the reduction of ammonia to nitrate ions.**
   c. the reduction of nitrate ions to nitrogen gas.
   d. catalyzed by the enzyme nitrogenase.
   e. performed by specialized plant cells located in the root.

7. Nitrate reduction
   a. **is performed by plants.**
   b. takes place in mitochondria.
   c. is catalyzed by the enzyme nitrogenase.
   d. includes the reduction of nitrite ions to nitrate ions.
   e. is performed by specialized bacteria in the soil.

8. Which statement about sulfur is *not* true?
   a. All living things require it.
   b. **It is a component of DNA and RNA.**
   c. It is a constituent of two amino acids.
   d. Its metabolism is similar to the metabolism of nitrogen.
   e. Plants obtain it from the soil.

9. Which of the following is a parasite?
   a. Venus flytrap
   b. Pitcher plant
   c. Sundew
   d. **Dodder**
   e. Tobacco

10. All heterotrophic seed plants
   a. are parasites.
   b. are carnivores.
   c. are incapable of photosynthesis.
   d. **can derive some of their nutrition from other organisms.**
   e. do not depend on essential elements for survival.

Student Web Site Self-Quiz Questions

1. Plants need nutrients from their environment. Which of the following is a **nonmineral** nutrient **obtained from the soil**?
   a. Carbon
   b. Oxygen
   c. Nitrogen
   d. Phosphorus
   e. Chlorine

2. In the choices below, which is *not* a correct pairing of a macronutrient and the major functions it performs in the life of a plant?
   a. Potassium; enzyme activation, water balance, ion balance
   b. Calcium; activity of membranes and cytoskeleton, second messenger
   c. Sulfur; in proteins and coenzymes
   d. Iron; in active sites of many redox enzymes and electron carriers
   e. Magnesium; in chlorophyll, required for many enzymes

3. In the typical approach to the identification of essential elements in plants, a plant growing in a solution containing all known nutrients is transplanted to a solution _________. This approach is least effective at detecting nutrients needed in _________ amounts.
   a. lacking a single potential nutrient, small
   b. lacking a single potential nutrient, large
   c. lacking all known nutrients, small
   d. lacking all known nutrients, large
   e. having a single potential nutrient, small

4. Which of the following statements about soils and soil formation is *false*?
   a. A soil's profile consists of its horizons.
   b. Extreme temperatures play a part of the soil-forming process.
   c. Clay particles result from physical weathering.
   d. Chemical alterations of rock materials.
   e. Soil consists of living as well as nonliving components.

5. Ion exchange is an important process that facilitates the uptake of nutrients by plants. Which of the following statements about ion exchange is *false*?
   a. Ion exchange is most important in soils with high clay content.
   b. Clay particles have a permanent negative charge.
   c. Mineral cations bind more strongly to clay particles than do protons.
   d. Negative ions such as phosphate, nitrate, and sulfate are leached from the soil.
   e. Effective ion exchange depends on soil that is aerated.

6. If a farmer has a soil with little phosphorus and a very high pH, what should be applied?
   a. 5-10-5 fertilizer and sulfate ions.
   b. 10-5-5 fertilizer and sulfate ions.
   c. 5-10-5 fertilizer and lime.
   d. 10-5-5 fertilizer and lime.
   e. 5-5-10 fertilizer and lime.

7. Conifers produce a type of plant litter called _________ that is rich in _________.
   a. humus, nitrogen
   b. humus, carbon
   c. humus, phosphorus
   d. fertilizer, carbon
   e. fertilizer, nitrogen

8. Which of the following groups of microorganisms do *not* include representatives that can fix nitrogen?
   a. Free-living *Rhizobium*
   b. Nodule-inhabiting *Rhizobium*
   c. Filamentous bacteria called actinomycetes
   d. Cyanobacteria in the oceans
   e. Cyanobacteria in fresh water

9. Which of the following statements about nitrogen fixation is *false*?
   a. Nitrogenase is only catalytic under anaerobic conditions.
   b. The energy for nitrogen fixation can be provided by either photosynthesis or respiration.
   c. In nitrogen fixation, nitrogen is reduced by the addition of three successive pairs of hydrogen atoms.
   d. Most nitrogen fixing microbes are aerobic.
   e. Within a nodule, oxygen levels are kept at a low level.

10. In the list given below, which would be fourth in the sequence of steps in nodule formation in the mutualism between *Rhizobium* species and a legume?
    a. Bacteroids are formed.
    b. The root produces flavonoids that attract *Rhizobium* bacteria to the vicinity of the root.
    c. Flavonoids trigger transcription of the bacterial *nod* gene.
    d. The plant produces leghemoglobin.
    e. *Nod* factors secreted by the bacteria cause cell division leading to the formation of primary nodule meristem.

11. Which of the following choices is the correct name for the process leading to the form of fixed nitrogen preferred by most plants?
    a. Nitrogen fixation
    b. Nitrification
    c. Reduction
    d. Nitrate reduction
    e. Denitrification

TEST FILE

★ *Indicates a* ***difficult*** *question*

■ *Indicates a* ***conceptual*** *question*

37 Plant Growth Regulation

Fill in the Blank

1. A germinating grass seedling embryo produces **gibberellins** that mobilize stored nutrients.

2. The hormone **auxin** is responsible for the phenomenon of apical dominance in plants.

3. It is thought that auxins control growth by **cell elongation,** while cytokinins cause growth by **cell division**.

4. The hormone that generally shuts down or inhibits plant activity is **abscisic acid**.

5. Application of **cytokinins** to leaves can keep them green and delay senescence.

6. Fruit shippers apply **ethylene** to speed up ripening.

7. Growth of a shoot tip toward the light is called **phototropism** and is thought to be mediated by the actions of the hormone **auxin**.

8. Bending toward the light occurs when the auxin moves to the side **away from** the light and has the effect of **loosening** the cell walls in that region, causing asymmetric growth and thus bending.

9. The process of seed coat modification to increase ease of germination is called **scarification**.

10. Molecules active at small concentrations that regulate plant development are called **hormones**.

11. Dicot seedlings form an aprical **hook** to protect the stem while it grows through the soil.

Multiple Choice

1. The longest period seeds of any plant species can remain dormant is
   a. one year.
   b. two years.
   c. a decade.
   d. a century.
   e. a millenium.

2. To make seedless grapes grow to normal size, they are
   a. sprayed with cytokinins.
   b. treated with NO.
   c. treated with ethylene.
   d. sprayed with auxin solution.
   e. sprayed with gibberellin solution.

3. Auxin is transported __________, __________ from the apical meristem.
   a. down, asymmetrically
   b. down, symmetrically
   c. upward, symmetrically
   d. evenly, toward the roots
   e. upward, toward the tip

★4. Production of auxin in response to light most likely involves
   a. phytochrome.
   b. photoreceptors.
   c. *Arabidopsis*.
   d. phototropin.
   e. There is no data to provide a clue.

5. Phototropin is classified as a(n)
   a. phytochrome.
   b. cryptochrome.
   c. gibberellin.
   d. cytokinin.
   e. abscisic acid.

6. A protein receptor for auxin is
   a. auxin A.
   b. kinetin.
   c. ABP1.
   d. phototropin.
   e. chlorophyll.

7. Abscisic acid closes stomata by
   a. stimulating proton pumping.
   b. triggering potassium pumps to remove potassium from the cytosol.
   c. releasing calcium into the cytoplasm.
   d. pumping water from the cell.
   e. the "stress effect."

8. A steroid-like hormone that has some of the same effects as auxin is
   a. adinosteroid.
   b. estrogen.
   c. brassinosteroids.
   d. testosterone.
   e. *det2*.

9. A first response to attack by bacteria is
   a. release of oligosaccharins.
   b. release of auxin.
   c. production of abscisic acid.
   d. production of ethylene.
   e. conversion of all P_{fr} to P_r.

*10. Auxins cause elongation by
   a. triggering proton release.
   b. indirectly activating expansins.
   c. indirectly increasing plasticity.
   d. All of the above
   e. None of the above

11. Plant growth substances generally
   a. have a single specific role.
   b. affect mainly the cells that produce them.
   c. are species-specific.
   d. are produced in many parts of the plant.
   e. elicit rapid responses.

12. As a grass seed germinates, the embryonic plant secretes gibberellins that
   a. absorb light.
   b. cause elongation.
   c. mobilize stored foods.
   d. take up water.
   e. direct the shoot upward.

13. The hormone responsible for phototropism is
   a. abscisic acid.
   b. auxin.
   c. ethylene.
   d. gibberellin.
   e. phytochrome.

14. Gibberellins were discovered by studying the "foolish seedling" disease of rice, in which seedlings
   a. grew unusually slowly.
   b. grew into tall, spindly plants.
   c. died after germination.
   d. produced seeds unusually early.
   e. had a harmful mutation.

15. Which of the following suggests that plants produce gibberellin growth hormones?
   a. Genetically dwarf corn seedlings grow tall with gibberellin treatment.
   b. Genetically tall corn plants grow even taller with gibberellin treatment.
   c. Gibberellins are produced by tall and dwarf varieties of corn.
   d. Gibberellins are produced by dwarf varieties of corn.
   e. Different varieties of corn produce different gibberellins.

16. In the Darwins' experiment, which part of the seedling was sensitive to light?
   a. The entire seedling
   b. The entire shoot above the roots
   c. The entire leaf sheath
   d. The sheath just below the tip
   e. The extreme tip of the leaf sheath

17. In plant tropisms,
   a. an imbalance in ethylene concentration causes curvature.
   b. roots grow toward the light.
   c. one side of the root or shoot grows more rapidly than the other.
   d. DNA is the light receptor or gravity receptor.
   e. auxin is the light receptor or gravity receptor.

18. Leaf abscission is the
   a. separation of leaves from stems.
   b. orientation of a leaf toward the light.
   c. regeneration of a leaf after a wound.
   d. maturation of leaf tissue.
   e. initiation of growth of new tissue.

19. The phenomenon of apical dominance is strengthened most by
   a. removal of the tip.
   b. auxin production.
   c. removal of leaves.
   d. production of fruits.
   e. removal of fruits.

20. Cytokinins are formed primarily in which area of the plant?
   a. Tips of the shoot
   b. Leaves
   c. Stems
   d. Roots
   e. Lateral buds

21. In the phenomenon of gravitropism, the growth of a plant part is
   a. toward the center of Earth.
   b. in a direction determined by gravity.
   c. in a direction opposite that of the main light source.
   d. in a direction opposite that of the growth of the shoot.
   e. toward the darkest area.

▪22. Why are cell differentiation experiments often done with pith tissue cultures?
   a. Pith cells are all unspecialized.
   b. Only pith tissue grows rapidly.
   c. Pith tissue of stem does not differentiate to root cells.
   d. Pith tissue responds primarily to auxin.
   e. Pith tissue is found in all plants.

23. Which hormone stimulates lateral buds to grow into branches?
   a. Abscisic acid
   b. Auxin
   c. Cytokinin
   d. Ethylene
   e. Gibberellin

24. Which of the following hormones is a gas?
   a. Abscisic acid
   b. Auxin
   c. Cytokinin
   d. Ethylene
   e. Gibberellin

25. Which of the following is considered to be the plant's "stress hormone"?
    a. Abscisic acid
    b. Auxin
    c. Cytokinin
    d. Ethylene
    e. Gibberellin

▪26. If a shoot cutting is treated with auxin, which of the following is likely to result?
    a. Extensive root production
    b. Suppression of apical dominance
    c. Growth of lateral buds
    d. Bolting of the shoot
    e. Nothing will happen to the cutting.

27. Which pair of hormones has opposing effects on senescence, in that the first promotes it and the second inhibits it?
    a. auxin, cytokinin
    b. ethylene, cytokinin
    c. ethylene, auxin
    d. cytokinin, auxin
    e. cytokinin, ethylene

28. Bolting, or rapid stem elongation, is induced by _________ and can be inhibited by _________.
    a. auxin, cytokinin
    b. abscisic acid, ethylene
    c. gibberellin, abscisic acid
    d. ethylene, auxin
    e. cytokinin, abscisic acid

29. When it is said that the movement of auxin in plants is "polar," this means that
    a. auxin is a chemically polar molecule.
    b. auxin is produced only at one part of the plant.
    c. auxin is transported from tip to base of the plant.
    d. auxin moves away from the light.
    e. auxin cannot move through gelatin.

30. Which of the following processes is *not* increased by ethylene?
    a. Breakdown of fruit cell walls
    b. Stimulation of leaf abscission
    c. Ripening of fruit
    d. Inhibition of stem elongation
    e. Change in leaf color from green to red or yellow

31. Which hormone has its highest concentrations in dormant (inactive) buds and seeds?
    a. Abscisic acid
    b. Auxin
    c. Cytokinin
    d. Ethylene
    e. Gibberellin

32. Plants utilize which cue to detect the onset of winter?
    a. Decreasing temperature
    b. Increasing precipitation
    c. Decreasing length of daylight
    d. Increasing length of darkness
    e. Height of the midday sun

33. In plants that germinate in response to a brief pulse of light,
    a. green light is most effective in triggering germination.
    b. far-red light is most effective in triggering germination.
    c. far-red light reverses the effect of prior exposure to red light.
    d. initiation of photosynthesis is the mechanism for germination.
    e. a rise in temperature also triggers germination.

▪34. A protein pigment called phytochrome is thought to monitor photoperiod because
    a. it is converted between two forms by specific wavelengths of light.
    b. a phytochrome that absorbs red light breaks down in darkness.
    c. the photoperiod response depends on how much light it absorbs.
    d. in darkness the pigment becomes inactive.
    e. the pigment response can only be observed in intact living plants.

*35. Which of the following is most advantageous for a young plant seedling that has not yet been exposed to light?
    a. Increased production of chlorophyll
    b. Rapid elongation of the shoot
    c. Increased production of phytochrome
    d. Rapid photosynthesis
    e. Increased uptake of water

36. Gibberellins were first discovered by a biologist studying the "foolish seedling" disease of
    a. corn.
    b. wheat.
    c. rice.
    d. barley.
    e. millet.

37. Phinney reported the first evidence that gibberellins were produced by plants. In his studies with dwarf mutant strains of corn, he demonstrated that treatment with gibberellins caused the dwarf plants _________, while the normal tall plants _________.
    a. to grow taller, also grew taller
    b. to grow taller, were virtually unaffected
    c. to stop growing, also stopped growing
    d. to stop growing, were virtually unaffected
    e. to die, stopped growing

▪38. Why are there so many gibberellins?
    a. Each is needed for a different process.
    b. Each is produced by a different part of the plant.
    c. Some are no longer important to the plant's development.
    d. Most are simply intermediates.
    e. Most are produced by fungi, not the plants in which they are found.

*39. Spraying biennials like cabbage with gibberellin causes them to bolt, a process first observed as an increase in
a. leaf senescence.
b. stem elongation.
c. the number of flowers.
d. the size of fruit.
e. the number of seeds per fruit.

40. The discovery of auxin is traced back to the work of Charles and Francis Darwin. In their experiments, they studied
a. photosynthesis.
b. photorespiration.
c. photophosphorylation.
d. phototropism.
e. photoperiodism.

41. In the Darwins' experiment with grass coleoptiles, they observed that the photoreceptor was __________; the actual bending took place __________.
a. in the tip, at the tip also
b. below the tip, below the tip also
c. in the tip, below the tip
d. below the tip, throughout the coleoptile
e. at the tip, throughout the coleoptile

*42. In a classic experiment, a gelatin block containing auxin was placed on one edge of a decapitated coleoptile. The result was that the coleoptile grew
a. straight up if in the dark.
b. more on the side with the block if in the dark.
c. more on the side away from the block if in the dark.
d. more on the side with the block if light was shining on the side with the block.
e. more on the side with the block if light was shining on the side opposite the block.

43. Auxin transport is
a. from apex to base.
b. from base to apex.
c. in either direction.
d. primarily from apex to base, but a little in reverse.
e. primarily from base to apex, but a little in reverse.

44. Your cat knocks over the Coleus plant you were keeping in a dark closet. After a few days on its side, you notice that the shoots are growing upright again. You have just observed
a. positive gravitropism.
b. negative gravitropism.
c. positive phototropism.
d. negative phototropism.
e. positive thigmotropism.

45. Removal of the auxin source demonstrates that leaf abscission is __________ by auxin, and apical dominance is __________ by auxin.
a. promoted, promoted
b. inhibited, inhibited
c. promoted, inhibited
d. inhibited, promoted
e. promoted, unaffected

46. An ideal herbicide to kill weeds in a wheat field would kill __________ and break down __________ in soil.
a. monocots and dicots, slowly
b. monocots but not dicots, slowly
c. dicots but not monocots, slowly
d. monocots but not dicots, rapidly
e. dicots but not monocots, rapidly

47. A cell wall is a network of crystalline __________ molecules in a jelly-like matrix of __________ molecules.
a. cellulose, starch
b. starch, other polysaccharide
c. cellulose, other polysaccharide
d. starch, nonpolysaccharide
e. cellulose, nonpolysaccharide

48. The growth of a plant cell is driven primarily by the
a. breakdown of ATP to ADP.
b. uptake of water into the vacuole.
c. strengthening of cell wall components.
d. deposition of new cell wall materials on the outside of the cell wall.
e. forces in transpirational pull.

*49. The "wall-loosening factor" from the cytoplasm is now believed to be
a. auxins.
b. cellulose-digesting enzymes.
c. starch-digesting enzymes.
d. expansins.
e. calcium ions.

50. Cytokinins are believed to form primarily in the plant's
a. roots.
b. stems.
c. leaves.
d. vegetative apical meristem.
e. floral apical meristem.

51. Unlike other plant hormones, ethylene
a. is not produced by plants.
b. exerts a number of effects.
c. inhibits plant development.
d. acts as either an inhibitor or a promoter.
e. is a gas.

52. Leaf senescence is important for the survival of a plant. Which of the following statements about senescence is *not* true?
a. The delicate leaves could be a liability during winter.
b. Amino acids in the leaves are exported to the stems.
c. Leaf senescence is a reversible process.
d. The hormone ethylene promotes leaf senescence.
e. Controlled leaf abscission costs the plant little and benefits it greatly.

53. Abscisic acid concentrations are __________ in some dormant seeds and __________ in buds during winter dormancy.
a. high, high
b. high, low
c. low, high

d. low, nonexistent
e. low, low

54. Which of the following is true for phytochrome P_r?
a. It absorbs green light.
b. It looks red in a test tube.
c. It is the active form of phytochrome.
d. It spontaneously converts to the other form in the dark.
e. It controls a variety of plant responses.

55. When seeds germinate below the soil surface, the young etiolated seedlings
a. grow very slowly.
b. produce chlorophyll.
c. elongate so that the apical meristem is the first part of the shoot to break through the soil surface.
d. have small, unexpanded leaves.
e. keep the cotyledons enclosed in a coleoptile.

56. Which of the following does *not* serve to break dormancy and initiate seed germination?
a. Scarification by abrasion
b. Scarification by fire
c. Exposure to water
d. Abscisic acid
e. Growth promoters

57. A shift away from rapid vegetative growth often accompanies which of the following phenomena?
a. Repeated mitotic division
b. Production of new leaves
c. Production of flowers
d. Increased root development
e. Increased rate of photosynthesis

58. Which of the following is *not* a way to break seed dormancy?
a. Soaking with water
b. Singeing with flame
c. Passing through an animal's gut
d. Tumbling with stones
e. Absorption of carbon dioxide

59. Seed dormancy is usually an adaptation to ensure that
a. the embryo is mature.
b. germination occurs at a favorable time.
c. seeds germinate near the parent plant.
d. levels of abscisic acid are high enough.
e. plenty of other seeds are ready to germinate.

60. All of the following occur during the earliest stages of germination *except*
a. intake of water.
b. activation of enzymes.
c. cell division.
d. lengthening of the root.
e. mobilization of food reserves.

61. The adaptive advantage of seed dormancy includes all of the following *except* that
a. it may result in germination at a favorable time.
b. it may increase the probability of a seed germinating in the right place.
c. it may be a way to avoid competition.
d. it may result in an increase in the likelihood of dispersal.
e. it may ensure the success of a population by germinating while still attached to the parent plant.

62. During the initial stages of seed germination, all of the following increase *except*
a. cell size.
b. respiration.
c. RNA synthesis.
d. DNA synthesis.
e. protein synthesis.

63. Which of the following allows more energy to be stored in a smaller space?
a. Protein
b. Lipid
c. Starch
d. Sugar
e. Amino acid

64. As a seed germinates, DNA synthesis begins
a. when gibberellins are secreted by the embryo.
b. when the embryonic root, or radicle, begins to grow.
c. as imbibition takes place.
d. when the endosperm starts metabolizing starches, proteins, and lipids.
e. when the aleurone layer assembles enzymes, proteases, and ribonucleases.

65. Treatment of some plants with gibberellin or auxin causes parthenocarpy; this is
a. flowers with petals in multiples of four.
b. the shoot dividing into two separate shoots in one plant.
c. fruit with only one seed instead of many.
d. fruit formation without fertilization.
e. formation of an embryo without fertilization.

▪66. Two cytokinins are kinetin and zeatin. What is the difference between these two?
a. Kinetin is found only in aged DNA, and zeatin is found only in fresh DNA.
b. Kinetin is the active form of zeatin.
c. Zeatin is the active form of kinetin.
d. Zeatin is a naturally occurring plant cytokinin; kinetin is not.
e. Zeatin is a synthetic cytokinin, and kinetin is naturally occurring.

▪67. Why do florists use silver thiosulfate?
a. To delay abscission of petals caused by ethylene action
b. To delay abscission of leaves caused by abscisic acid action
c. To keep the petals from turning brown because of ethylene action
d. To keep the leaves green longer
e. To promote flower fertilization

68. Abscisic acid promotes the formation of bud scales; these are for
a. retaining more water in dry areas.
b. waterproofing the leaf primordia and stem during winter.
c. cell elongation at the apical meristem.
d. helping leaf drop during autumn.
e. trapping more carbon dioxide in the leaves.

69. Key to seed germination is that the seed
a. dehydrates.
b. is in the ground.
c. imbibes.
d. is activated by germisin.
e. pumps protons.

*70. The possibility of an ideal lawn that rarely needs mowing could become a reality applying the knowledge of the *bas-1* gene. The growth of grass would be stimulated by
a. applying auxin.
b. applying cytokinins.
d. applying bassinosteroid.
d. controlling water supply.
e. All of the above are possible ways.

71. The physiologically active form of phytochrome is
a. P_{fr}.
b. P_r.
c. G protein.
d. cryptochrome A.
e. 730 nm.

Study Guide Questions

1. Plant growth is ultimately regulated by which of the following?
a. Environmental cues
b. Hormones
c. Signal transduction pathways
d. The plant's genome

2. Which of the following schemes best represents how environmental cues are transduced to changes in a plant?
a. Receptors receive environmental cues, a signal transduction pathway is initiated, there is an alteration in the particular genes that are transcribed and translated, and a cellular response is generated.
b. Receptors are triggered, hormones are released, signal transduction pathways are initiated, there is an alteration in expression of genes, and a cellular response is generated.
c. Neither a nor b
d. a and b

3. Which of the following triggers seed germination?
a. Seeds imbibe water.
b. Seeds are released from fruit.
c. Seeds undergo chemical changes.
d. All of the above

4. Which of the following may function to break dormancy in seeds?
a. Penetration of the seed coat
b. Leaching of inhibitory compounds by water
c. Exposure to fire
d. All of the above

5. Which of the following hormones are responsible for bud break in the spring in deciduous trees?
a. Auxins
b. Cytokinens
c. Gibberellins
d. Ethylene.

6. Auxin regulates cell growth by which of the following mechanisms?
a. Altering the elasticity of cell walls
b. Altering the plasticity of cell walls
c. Synthesizing new cell walls
d. Breaking down cell walls in growing cells

7. Which of the following hormones are responsible for maintaining bud dormancy in deciduous trees?
a. Auxins
b. Cytokinens
c. Gibberellins
d. Abscisic acid

8. You have installed an outdoor gas burning grill on your back patio next to your favorite camellia bush. After the first few chilly nights of using your grill you notice that your camellia, which does not normally lose its leaves, is beginning to do so. Which of the following is the best explanation for what is happening?
a. The bush is getting too warm next to your grill.
b. Ethylene is a by-product of the gas you are burning and is causing senescence in your plant.
c. Abscisic acid is a by-product of the gas you are burning and is causing senescence in your plant.
d. The plant is a biennial and is bolting.

9. You are slicing a green pepper for the pizza you are making at home. As you slice into it you notice lots of tiny pepper plants emerging from the seeds of the pepper. This pepper is exhibiting __________ and may be lacking in __________.
a. parthenocarpy, gibberellins
b. parthenocarpy, abscisic acid
c. vivapary, gibberellins
d. vivapary, abscisic acid

10. Which of the following light receptors is responsible for absorbing blue and ultraviolet light?
a. Phytochrome P_r
b. Photochrome P_{fr}
c. Cryptochrome
d. Photropin

11. Etiolated seedlings are produced by germinating seeds and keeping them in total darkness. Under which of the following conditions will plants kept in the dark begin to synthesize chlorophyll?
    a. After being given a pulse of blue light
    b. After being given a pulse of red light
    c. After being given a pulse of red light followed by a pulse of far-red light
    d. After being given a pulse of far-red light followed by a pulse of red light

12. Match the major site of production on the left with the plant growth substance produced at that site on the right. A substance may be produced at multiple sites.

b.	Abscisic Acid	a. Embryo
c., f.	Auxin	b. Buds
e.	Cytokinin	c. Leaves
d.	Ethylene	d. Fruit
a.	Gibberellin	e. Root tips
		f. Shoot tips

End of Chapter Questions

1. Which of the following is *not* an advantage of seed dormancy?
   a. It makes the seed more likely to be digested by birds that disperse it.
   b. It counters the effects of year-to-year variations in the environment.
   c. It increases the likelihood that a seed will germinate in the right place.
   d. It ensures seed survival through unfavorable conditions.
   e. It may result in germination at a favorable time of year.

2. Which of the following are *not* involved in seed germination?
   a. Imbibition of water
   b. Metabolic changes
   c. Growth of the radicle
   d. Mobilization of food reserves
   e. Extensive mitotic divisions

3. To mobilize its food reserves, a germinating barley seed
   a. becomes dormant.
   b. undergoes senescence.
   c. secretes gibberellins into its endosperm.
   d. converts glycerol and fatty acids into lipids.
   e. embryo takes up proteins from the endosperm.

4. The gibberellins
   a. are responsible for phototropism and gravitropism.
   b. are gases at room temperature.
   c. are produced only by fungi.
   d. cause bolting in some biennial plants.
   e. inhibit the synthesis of digestive enzymes by barley seeds.

5. In coleoptile tissue, auxin
   a. is transported from base to tip.
   b. is transported from tip to base.
   c. can be transported toward either the tip or the base, depending on the orientation of the coleoptile with respect to gravity.
   d. is transported by simple diffusion, with no preferred direction.
   e. is not transported, because auxin is used where it is made.

6. Which process is *not* directly affected by auxin?
   a. Apical dominance
   b. Leaf abscission
   c. Synthesis of digestive enzymes by barley seeds
   d. Root initiation
   e. Parthenocarpic fruit development

7. Plant cell walls
   a. are strengthened primarily by proteins.
   b. often make up more than 90 percent of the total volume of an expanded cell.
   c. can be loosened by an increase in pH.
   d. become thinner and thinner as the cell grows longer and longer.
   e. are made more plastic by treatment with auxin.

8. Which statement about cytokinins is *not* true?
   a. They promote bud formation in tissue cultures.
   b. They delay the senescence of leaves.
   c. They usually promote the elongation of stems.
   d. They cause certain light-requiring seeds to germinate in the dark.
   e. They stimulate the development of branches from lateral buds.

9. Ethylene
   a. is antagonized by silver salts.
   b. is liquid at room temperature.
   c. delays the ripening of fruits.
   d. generally promotes stem elongation.
   e. inhibits the swelling of stems, in opposition to cytokinin effects.

10. Phytochrome
    a. is a nucleic acid.
    b. exists in two forms interconvertible by light.
    c. is a pigment that is colored red or far red.
    d. is sometimes called the stress hormone.
    e. is the photoreceptor for phototropism.

Student Web Site Self-Quiz Questions

1. Which of the following is *not* a factor that regulates plant development?
   a. Environment
   b. Photoreceptor
   c. Photosynthesis
   d. Hormones
   e. Enzymes

2. Which of the following chemicals is a nonhormone involved in flowering or fruit production?
   a. Auxin
   b. Cytokinins
   c. Ethylene
   d. Gibberellins
   e. Photoreceptor

3. Seed dormancy affords adaptive advantages. Which of the following features would *not* be adaptive for the seeds of a desert plant that grows in disturbed areas and requires deeply buried seeds?
   a. Large seeds
   b. Seeds requiring an extended period of imbibition
   c. Seeds requiring softening
   d. Light inhibited seeds
   e. Seeds with water-soluble chemical inhibitors that need to be leached by water

4. Which one of the following was *not* observed in studies done on dwarf strains of plants?
   a. Applications of extracts of normal strains promoted growth of dwarf strains.
   b. Dwarf strains grew normally if additional fertilizer was applied.
   c. Application of gibberellin A_1 promoted growth of dwarf strains.
   d. Gibberellins cause little additional growth of normal strains.
   e. Dwarf strains are homozygous recessive for an allele involved in gibberellin A_1 biosynthesis.

5. Gibberellins have many effects. Which one of the following is *not* an effect of gibberellins in plant development?
   a. Promotion of normal fruit growth
   b. Bolting
   c. Promotion of seed germination in some plants
   d. Inducing dormancy
   e. Promotion of fruit development in unfertilized flowers

6. Which of the following is *not* a conclusion that resulted from the Darwins' studies of phototropism in grass seedlings?
   a. Light is sensed in the tip of the coleoptile.
   b. Bending occurred in a region below the tip.
   c. A signal travels from the tip downward.
   d. Light caused movement of a growth-promoting substance from the tip to the growing region.
   e. Light did not directly affect the growing region where bending occurred.

7. In the experiments conducted by Frits W. Went, which of the following was the primary conclusion?
   a. Growth hormones are not involved in the process of coleoptile growth.
   b. Gravity plays an important role in the growth of coleoptiles.
   c. Hormones present in the original cut coleoptile diffuse into the gelatin block.
   d. The plant hormone is indoleacetic acid.
   e. Light is not required for phototropism.

8. Cytokinins and ethylene have opposite effects on
   a. leaf senescence.
   b. maintenance of the apical hook in dicot seedlings.
   c. elongation of stems.
   d. lateral swelling of stems.
   e. winter dormancy.

9. Which of the following statements regarding ethylene in plants is *false*?
   a. Ethylene promotes plant senescence.
   b. Ethylene speeds the ripening of fruit.
   c. Ethylene causes the formation of an apical hook by stimulating the elongation of cells on the inner surface of the hook.
   d. Ethylene synthesis in seedlings stops when exposed to light.
   e. Ethylene promotes lateral swelling in stems.

10. A potent inhibitor of seed germination is
    a. abscisic acid.
    b. gibberellins.
    c. brassinosteroids.
    d. auxin.
    e. ethylene.

11. Two seedlings, A and B, are etiolated because they have been kept in the dark since germination. Seedling A is placed in red light; seedling B is placed in far-red light. Which of the following statements comparing the seedlings is true?
    a. After light exposure, seedling A has more P_r than seedling B.
    b. After light exposure, seedling B has more P_{fr} than seedling A.
    c. Before light exposure, the phytochrome of both seedlings is in the P_r form.
    d. After light exposure, seedling B will begin chlorophyll synthesis.
    e. After light exposure, seedling A will develop an apical hook.

★ *Indicates a **difficult** question*

■ *Indicates a **conceptual** question*

38 Reproduction in Flowering Plants

Fill in the Blank

1. Angiosperm plants are characterized by double fertilization, in which one sperm nucleus fertilizes the **ovum** to begin the embryo and the other fertilization results in production of **endosperm** tissue.
2. The male gametophyte in seed plants is the **pollen grain**, and the mature female gametophyte is an embryo sac with **eight** haploid nuclei.
3. The process of transfer of pollen grains to the stigma is called **pollination**.
4. The major role of the fruit of a flowering plant is to facilitate **seed dispersal**.
5. The production of progeny all having identical genotypes to the parent is called, in general, **asexual** reproduction, while vegetative reproduction is the modification of a **vegetative** part of the plant to produce new individuals.
6. The asexual production of seeds is called **apomixis**.
7. One agricultural industry in which grafting is an important technique is **fruit** production.
8. Instead of germinating immediately after release, many seeds undergo an inactive period of **dormancy**.
9. Experiments on the effect of light cues on flowering have shown that the significant cue that is sensed or measured by plants is **night length**.
10. The physiological mechanism by which plants measure photoperiod involves a pigment called **phytochrome**. This pigment alternates between two forms, which absorb **red** and **far-red** light.
11. Most plants do not rely on light cues for inducing flowering and are called **day-neutral** plants.
12. An opening in the ovule called the **micropyle** is where the pollen tube grows toward the egg.
13. Pollination before the flower bud opens is an example of **self-fertilization**.
14. Angiosperms are unique in that the endosperm is **triploid**.
15. Floral meristems produce flowers; this is **determinate** growth.
16. A northern tree that is placed too far south will not flower well; this is an example of **vernalization**.

Multiple Choice

1. Lifestyle categories for flowering plants include all the following *except*
   a. annual.
   b. biennial.
   c. perennial.
   d. axial.
   e. All of the above

2. An apomictic seed contains an embryo that is
   a. produced when two sperm fertilize one egg.
   b. developed from one egg alone.
   c. the result of parental self-fertilization.
   d. genetically identical to its parent.
   e. homozygous for most genetic traits.

3. A plant's transition to a flowering state is often marked by
   a. an increased rate of photosynthesis.
   b. a decrease in vegetative growth.
   c. an increase in root development.
   d. a decreased rate of respiration.
   e. an increase in lateral bud growth.

4. After pollination, which of the following events is crucial for fertilization?
   a. Sperm swim to the egg and the polar nuclei.
   b. Petals close around the reproductive parts.
   c. Meiosis occurs within the pollen grain.
   d. A pollen tube grows from the stigma to the ovule.
   e. An insect delivers pollen to the stigma.

5. Two modifications of vegetative parts used for asexual reproduction are short, vertical stems called
   a. rhizomes and tubers.
   b. tubers and corms.
   c. bulbs and corms.
   d. bulbs and rhizomes.
   e. corms and rhizomes.

6. Flowering is triggered by (yet to be discovered)
   a. florigen.
   b. NO.
   c. ethylene.
   d. abscisin.
   e. phytochrome.

7. After fertilization of the egg, the flower's integument develops into the
   a. cotyledons.
   b. embryo.
   c. endosperm.
   d. fruit.
   e. seed coat.

▪8. In some dicots, no distinct endosperm can be seen. Why?
   a. The embryo has digested the endosperm.
   b. The cotyledons have absorbed the endosperm.
   c. The seeds never produced endosperm.
   d. The endosperm has become the seed coat.
   e. The fruit has incorporated the endosperm.

9. The loss of water from a developing seed causes it to
   a. produce a root.
   b. die.
   c. be released from the fruit.
   d. become dormant.
   e. be protected from animal predators.

*10. The following experiment supports the theory that a specific flower-initiating hormone is produced in plants: When a short-day plant (SDP) and a long-day plant (LDP) are grafted together and the SDP is exposed to a photoperiod that causes it to flower, the LDP flowers as well. The proper control for this experiment would be to repeat the same design and
   a. omit the grafting.
   b. omit the LDP.
   c. omit the SDP.
   d. omit the SDP photoperiod.
   e. include an LDP photoperiod.

11. Which of the following is a gametophyte of a flowering plant?
   a. Flower
   b. Egg
   c. Pollen grain
   d. Anther
   e. Entire plant

12. The megagametophyte of flowering plants consists of the
   a. pollen grain.
   b. pollen tube.
   c. eight-nucleate embryo sac.
   d. ovule.
   e. megasporangium and cells within it.

*13. From megasporocyte to egg cell, what processes are required?
   a. Meiosis followed by mitosis
   b. Mitosis followed by meiosis
   c. Several meiotic divisions only
   d. Several mitotic divisions only
   e. Several nuclear fusion events

14. Within which of the following structures does meiosis occur?
   a. Petal
   b. Ovule
   c. Stigma
   d. Sepal
   e. Pollen grain

*15. The advantage of self-fertilization in plants is
   a. increased genetic recombination.
   b. that meiosis can occur.
   c. greater efficiency of pollination.
   d. that no flowering is needed.
   e. that only asexual reproduction is necessary.

16. The advantage of cross-fertilization in plants is
   a. increased genetic recombination.
   b. that meiosis can occur.
   c. greater efficiency of pollination.
   d. that no flowering is needed.
   e. that only asexual reproduction is necessary.

17. Plants with wind-pollinated flowers tend to have
   a. colorful petals.
   b. smooth stigmas.
   c. large quantities of pollen.
   d. large quantities of nectar.
   e. pollination before the bud opens.

18. Where does fertilization occur in flowering plants?
   a. Where pollen lands on the stigma
   b. Where the pollen tube germinates
   c. Inside the pollen tube
   d. At the base of the embryo sac
   e. Inside the seed

19. Winter wheat can be planted in the spring for fall harvest if it is
   a. sprayed with gibberellins.
   b. soaked in water.
   d. planted on a full moon.
   d. stored in the dark for 50 days.
   e. vernalized.

20. The egg can be fertilized by
   a. one tube nucleus.
   b. one sperm nucleus.
   c. two sperm nuclei.
   d. one generative nucleus.
   e. two synergid nuclei.

*21. What is the fate of the seven cells of the embryo sac?
   a. All but one disintegrate upon fertilization.
   b. Two become fertilized; the others disintegrate.
   c. Two become fertilized; the others fuse to form endosperm.
   d. All are involved in nuclear fusion events.
   e. They all become part of the seed tissue.

22. What nuclei fuse to form the endosperm?
    a. Egg and sperm nucleus
    b. Egg and two sperm nuclei
    c. Synergid nuclei and sperm nucleus
    d. Polar nuclei and generative nucleus
    e. Polar nuclei and sperm nucleus

23. Which of the following describes the ploidy of the components of a seed?
    a. Diploid embryo, triploid endosperm, diploid seed coats
    b. Haploid embryo, triploid endosperm, diploid seed coats
    c. Diploid embryo, triploid endosperm, haploid seed coats
    d. Diploid embryo, diploid endosperm, diploid seed coats
    e. Haploid embryo, diploid endosperm, haploid seed coats

24. The term "suspensor" applies to which of the following?
    a. The pollen tube
    b. The organ supporting the ovule
    c. The base of the flower
    d. The narrow part of the embryo
    e. The stalk of the stamen

25. The fruit generally develops from which part of the flower?
    a. Petals
    b. Sepals
    c. Ovary
    d. Stamens
    e. Pedicel

26. What process early in development initiates cell specialization in the embryo?
    a. Mitotic division of the zygote
    b. Uneven distribution of cell contents
    c. Absorption of food from endosperm
    d. Seed germination
    e. Maturation of the fruit

27. Which of the following is a distinguishing characteristic of all angiosperms?
    a. Double cotyledons
    b. Fleshy cotyledons
    c. Seed with nutrients
    d. Double fertilization
    e. Pollen production

28. What is the function of the nutritious flesh of many fruits?
    a. To nourish the embryo
    b. To attract seed eaters
    c. To attract pollinators
    d. To attract seed dispersers
    e. To ensure that the seeds fall close to the parent plant

29. Asexual reproduction is the best strategy for plants
    a. that are well adapted to their stable environment.
    b. as winter approaches.
    c. when new genes must be introduced.
    d. that have underground stems.
    e. that have low seed production that season.

30. What is necessary for successful grafting to occur?
    a. Each section must be able to form roots.
    b. The grafted section must be able to form seeds.
    c. Fusion of the two vascular tissues must occur.
    d. Fusion of the two cambial tissues must occur.
    e. Each section must be from the same species.

▪*31. Fruit-eating bats tend to feed extremely rapidly on fruits and have relatively inefficient digestion (sometimes defecating seeds as early as an hour after feeding on them). Why are they good seed dispersal agents?
    a. Seed survival in bat guts is low.
    b. Undigested seeds are deposited in a heap at the bats' roost site.
    c. Undigested seeds are deposited at various bat feeding sites.
    d. Undigested seeds are deposited near the same plant that produced them.
    e. Digested seeds are dispersed in bat waste products.

32. Which of the following is a characteristic of wind-dispersed seeds?
    a. Abundant pollen
    b. Hooked extensions
    c. Fleshy fruit
    d. Air chambers
    e. Flat, winged extensions

33. Bamboo reproduces by
    a. rhizomes.
    b. tubers.
    c. corms.
    d. stolons.
    e. wind-dispersed seed germination.

34. Navel oranges are reproduced
    a. by cross-hybridization.
    b. forced seed germination.
    c. apomixis.
    d. cloning.
    e. vernalization.

▪35. Which of the following probably involves asexual reproduction?
    a. A water hyacinth population fills a pond in midsummer.
    b. A potato plant grows from the "eye" of a potato.
    c. Genetically identical, interconnected aspen trees produce flowers.
    d. Apomictic seed production in dandelions
    e. A bamboo plant gives rise to a forest of plants.

36. The process of grafting involves which of the following?
    a. Allowing a piece of one plant to grow onto the root of another
    b. Allowing cross-fertilization between two plants
    c. Preparing several cuttings from a plant to grow into individual plants

d. The production of xylem and phloem from the same cambium layer
e. Interbreeding of two species of plants

37. In a flower, the microsporangia are found in the
a. **anther.**
b. filament.
c. stigma.
d. ovule.
e. ovary.

38. Which is the correct order of events for female gametophytes?
a. Megagametophyte; megasporocyte; megaspore
b. Megagametophyte; megaspore; megasporocyte
c. Megasporocyte; megaspore; megagametophyte
d. Megaspore; megasporocyte; megagametophyte
e. Megaspore; megagametophyte; megasporocyte

39. In the mature embryo sac, the cells closest to the micropyle are the
a. polar nuclei.
b. synergids.
c. eggs.
d. egg and polar nuclei.
e. egg and synergids.

40. In flowering plants, the pollen is transferred to the
a. stigma.
b. style.
c. ovary.
d. ovule.
e. micropyle.

41. A flower that is wind-pollinated would be least likely to
a. have numerous anthers.
b. have sticky or feather-like stigmas.
c. produce large numbers of pollen grains.
d. have a colorful corolla.
e. have smooth wall sculpturing on its pollen.

42. Pollination is the
a. fusion of the egg and sperm nuclei.
b. transfer of pollen from the anther to the stigma.
c. development of the two-celled pollen grain.
d. growth of the pollen tube after pollen germination.
e. division of the generative nucleus to produce two sperm nuclei.

43. The three nuclei in a mature pollen grain are formed by
a. one meiotic division and one mitotic division.
b. two meiotic divisions and one mitotic division.
c. one meiotic division and two mitotic divisions.
d. two meiotic divisions and two mitotic divisions.
e. one meiotic division in which one of the four cells degenerates.

44. The "embryo sac" is also called the
a. megaspore.
b. megasporangium.
c. megasporocyte.
d. megasporophyll.
e. megagametophyte.

45. Double fertilization results in the formation of
a. two diploid embryos.
b. one diploid embryo and a diploid endosperm.
c. two diploid embryos and a haploid endosperm.
d. one diploid embryo and a triploid endosperm.
e. two diploid embryos and a diploid seed coat.

46. The integuments of the ovule develop into the _________, while the carpels ultimately become the wall of the _________.
a. cotyledons, endosperm
b. seed coats, fruit
c. cotyledons, seed coats
d. endosperm, seed coats
e. cotyledons, fruit

47. The function of a fleshy fruit is to
a. feed the new embryo.
b. attract pollinators.
c. disperse the seeds.
d. protect the immature embryo.
e. keep the rest of the plant from being eaten.

48. Coconut fruits are dispersed by
a. monkeys.
b. wind.
c. fruit bats.
d. water.
e. birds.

49. Self-pollination results in progeny that
a. are identical to the parent.
b. are somewhat different because mutations are common.
c. may express a recessive gene if the parent is heterozygous.
d. may be heterozygous in a locus where the parent is homozygous.
e. may be as varied as from cross-pollination.

50. A clone of white potatoes may be derived from underground
a. stolons.
b. tubers.
c. rhizomes.
d. bulbs.
e. root suckers.

51. The major advantage of asexual reproduction is
a. that it results in no genetic variation.
b. that it results in increased dispersal.
c. production of more progeny.
d. the ability to invade new environments.
e. only observed when the habitat is unstable.

52. The ability of plants to measure night length was determined in what sort of experiment?
a. Growing plants in 12 hours of darkness alternating with 12 hours of light
b. Growing plants in continuous light
c. Interrupting the dark period with a brief pulse of light

d. Measuring flowering in plants of different ages placed in the same light–dark schedule
e. Keeping a 24-hour cycle but increasing the relative length of light

53. Which of the following photoperiods would induce flowering in a short-day plant with a critical day length of 15 hours?
a. Twelve hours of light alternating with 12 hours of darkness
b. Sixteen hours of light alternating with 8 hours of darkness
c. Fourteen hours of light alternating with 8 hours of darkness
d. Eight hours of light alternating with 8 hours of darkness
e. Fifteen hours of light alternating with 9 hours of darkness, interrupted by one short burst of white light

54. Vernalization refers to the requirement of which of the following before flowering can occur?
a. Exposure to cold
b. Availability of soil calcium
c. Minimal day length
d. Sufficient moisture for a minimum period
e. One full year of growth

55. Short-day annuals usually flower
a. in the spring.
b. in midsummer.
c. in late summer.
d. in midsummer and late summer.
e. throughout the summer.

56. Technically, short-day plants flower when the
a. light period exceeds a critical period.
b. light period is less than a critical period.
c. light period equals the critical period.
d. dark period exceeds a critical period.
e. dark period is less than a critical period.

57. Long-day plants will not flower if exposed to
a. a long day.
b. a long day interrupted by a dark period.
c. a long day interrupted by a prolonged period of red light.
d. a long night interrupted by a light period of far-red light.
e. a long day interrupted by a brief period of red light.

58. Circadian rhythms
a. are always exactly 24-hour cycles in nature.
b. are only found in multicellular organisms.
c. have a period that is remarkably sensitive to temperature.
d. do not continue when placed in complete darkness.
e. can be made to coincide with the light–dark regime.

59. Although it has never been discovered, there is evidence that the mysterious "flowering hormone" is synthesized in the
a. floral meristem.
b. vegetative apical meristem.
c. stems.
d. leaves.
e. roots.

60. A benefit of sexual reproduction in plants is
a. greater number of progeny.
b. that pollination is easier.
c. better adaptation to new environments.
d. that the haploid plant becomes diploid.
e. that progeny are dispersed farther.

61. Long–short-day plants bloom
a. in the spring.
b. in the fall.
c. in the summer.
d. only after a cold winter.
e. only after the second year of the plant's life.

62. Initial resistance of the French wine grape *Vitis vinifera* to plant lice was accomplished by
a. self-fertilization of resistant plants.
b. grafting the scion onto resistant plants' roots.
c. recombinant DNA techniques using the resistance gene.
d. planting seeds in California where there are no plant lice.
e. taking cuttings, or slips, of the plants and inserting them into California soil, without lice, where they formed roots and grew.

63. Nutrients for the seedling are generally stored in what form in a seed?
a. As monomers in solution
b. As monomers in fat storage
c. As macromolecules
d. As cellular enzymes
e. As cellular organelles

64. As a pollen tube grows into the female organ, the nucleus that enters the synergid first is called the
a. sperm nucleus.
b. generative nucleus.
c. tube nucleus.
d. pollen nucleus.
e. microspore.

65. The most critical determinant in light–dark cycle length for regulation of flowering is
a. temperature.
b. dark length for short and light for long.
c. light length for short and dark for long.
d. length of the uninterrupted dark cycle.
e. length of the uninterrupted light cycle.

Study Guide Questions

1. You manage a greenhouse that produces roses for Valentine's Day. Roses normally bloom in June. Which of the following will most likely be the best lighting schedule for your roses?
   a. 16 hours of light, 8 hours of interrupted dark
   b. 16 hours of light, 8 hours of uninterrupted dark
   c. 10 hours of light, 14 hours of dark
   d. None of the above

2. After setting the correct photoperiod, you still don't have roses. Which of the following has most likely contributed to the problem?
   a. The heating system is allowing fluctuations in temperature of between 20° and 25° C.
   b. The furnace mechanic accidentally turned off the lights for an hour two days in a row.
   c. The cleaning crew turned the lights on for an hour three nights in a row.
   d. None of the above

3. You have moved into a new house. During the first summer you notice lots of plants that do not bloom. During the second summer your yard is a sea of blooms. It is now spring of the third year, and there are no plants. This can best be explained by which of the following?
   a. Your plants were annuals.
   b. Your plants were biennials.
   c. Your plants were perennials.
   d. Your plants were affected by drought.

4. You mother gives you a houseplant. You notice it sending out long stems with what looks like "little plants" attached. You allow one of these to rest in a cup of water and note that roots form. This is an example of
   a. asexual reproduction.
   b. apomixis.
   c. heterospory.
   d. parthenogenesis.

5. If plants produce pollen and mature eggs at the same time, how are self-pollination and fertilization prevented?
   a. Some plants have self-incompatibility genes.
   b. Some plants rely on physical barriers.
   c. a and b
   d. None of the above

6. Most wine grape vines are grafted onto rootstock of another species. How does this help grape yields?
   a. A hardy rootstock can replace a weak rootstock.
   b. A high-producing vine stock can replace a low-producing vine stock.
   c. The process allows vintners to select for pest resistance without losing grape quality.
   d. All of the above

7. The induction of flowering by means of exposure to low temperature is called
   a. vernalization.
   b. frigidation.
   c. apomixis.
   d. None of the above

8. The cyclic variation in an organism that corresponds to a set light–dark cycle is referred to as an organism's
   a. biorhythm.
   b. circadian rhythm.
   c. period.
   d. photoperiod.

9. The production of seeds without fertilization is called
   a. apomixis.
   b. parthenogenesis.
   c. conception.
   d. circadian rhythm.

10. Identify the following structures as sporophyte or gametophyte.

a. Embryo sac	**gametophyte**
b. Antipodal cells	**gametophyte**
c. Polar nuclei	**gametophyte**
d. Integument	**sporophyte**
e. Receptacle	**sporophyte**
f. Ovary	**sporophyte**
g. Anther	**sporophyte**
h. Pollen grain	**gametophyte**

End of Chapter Questions

1. Sexual reproduction in angiosperms
   a. is by way of apomixis.
   b. requires the presence of petals.
   c. can be accomplished by grafting.
   d. gives rise to genetically diverse offspring.
   e. cannot result from self-pollination.

2. The typical angiosperm female gametophyte
   a. is called a megaspore.
   b. has eight nuclei.
   c. has eight cells.
   d. is called a pollen grain.
   e. is carried to the male gametophyte by wind or animals.

3. Pollination in angiosperms
   a. never requires water.
   b. never occurs within a single flower.
   c. always requires help by animal pollinators.
   d. is also called fertilization.
   e. makes most angiosperms independent of water for reproduction.

4. Which statement about double fertilization is *not* true?
   a. It is found in most plants, including angiosperms.
   b. It occurs only by accident.
   c. One of its products is a triploid nucleus.
   d. One sperm nucleus fuses with the egg nucleus.
   e. One sperm nucleus fuses with two polar nuclei.

5. The suspensor
   a. gives rise to the embryo.
   b. is heart-shaped in dicots.
   c. separates the two cotyledons of dicots.
   d. ceases to elongate early in embryo development.
   e. is larger than the embryo.

6. Which statement about photoperiodism is *not* true?
   a. It is related to the biological clock.
   b. Phytochrome plays a role in the timing process.
   c. It is based on measurement of the length of the night.
   d. Most angiosperm species are day-neutral.
   e. It is limited to the plant kingdom.

7. Although florigen has never been isolated, we think it exists because
   a. night length is measured in the leaves, but flowering occurs elsewhere.
   b. it is produced in the roots and transported to the shoot system.
   c. it is produced in the coleoptile tip and transported to the base.
   d. we think that gibberellin and florigen are the same compound.
   e. it may be activated by prolonged (more than a month) chilling.

8. Which statement about vernalization is *not* true?
   a. It may require more than a month of low temperature.
   b. The vernalized state generally lasts for about a week.
   c. Vernalization makes it possible to grow wheat in previously hostile regions.
   d. It is accomplished by subjecting moistened seeds to chilling.
   e. It was of interest to Russian scientists because of their native climate.

9. Which of the following does *not* participate in asexual reproduction?
   a. Stolon
   b. Rhizome
   c. Fertilization
   d. Tuber
   e. Apomixis

10. Apomixis includes
   a. sexual reproduction.
   b. meiosis.
   c. fertilization.
   d. a diploid embryo.
   e. no production of a seed.

Student Web Site Self-Quiz Questions

1. If you arranged the following structures of a typical flower from most inclusive to least inclusive, which structure would be third in sequence?
   a. Carpel
   b. Egg
   c. Megasporangium
   d. Pistil
   e. Ovary

2. Complete the following sentence about sexual reproduction in flowering plants: Pollination occurs on the surface of the __________; fertilization occurs within the __________.
   a. ovule, female gametophyte
   b. ovule, male gametophyte
   c. stigma, female gametophyte
   d. stigma, male gametophyte
   e. stigma, receptacle

3. Which of the following statements about pollen and pollination is *false*?
   a. Evolution of the pollen grain negated the need for swimming sperm in flowering plants.
   b. At maturity, the pollen grain consists of two sperm nuclei and a tube nucleus.
   c. The pollen tube enters the female gametophyte through the style.
   d. The pollen grain makes twice the genetic contribution to endosperm cells than it does to the cells of the embryo.
   e. Growth of the pollen tube is not simply the result of gravitropism.

4. Which of the following statements about fruits and fruit-bearing plants is *false*?
   a. Minimally, a fruit consists of an ovary and its seeds.
   b. Fruits help disperse seeds.
   c. Just as pollinators can benefit from their association with the plant being pollinated, seed dispersers can also benefit.
   d. Seeds remain viable in fruits only if uneaten by animals.
   e. Some plants produce a dry, winged fruit that may be dispersed by wind.

5. *Chrysanthemum* sp. forms flower buds more readily when the day is shorter than a critical maximum. This species can best be described as a
   a. short-day plant.
   b. long-day plant.
   c. day-neutral plant.
   d. short-long-day plant.
   e. long-short-day-plant.

6. Some of the results of studies done on flowering in cocklebur indicate that cocklebur is an __________, and that the length of the __________ triggers flowering.
   a. SDP, day
   b. SDP, night
   c. LDP, day
   d. LDP, night
   e. long–short-day plant, night

7. Select below the unexpected response to one of the five light treatments (1–5) for an LDP.
   a. 1 = flowering
   b. 2 = no flowering
   c. 3 = no flowering
   d. 4 = no flowering
   e. 5 = no flowering

8. Studies of circadian rhythms have shown that the period of the rhythm can be __________ to cycles that differ from 24 hours, and the amplitude of the rhythm may be changed by __________.
   a. entrained, brief light exposure
   b. entrained, temperature
   c. phase-shifted, brief light exposure
   d. phase-shifted, temperature
   e. phase-shifted, total darkness

9. Grafting is an example of asexual reproduction. Which of the following choices is an example of asexual reproduction involving nonvegetative parts of a plant?
   a. Apomixis
   b. Production of corms
   c. Production of bulbs
   d. Production of rhizomes
   e. Production of stolons

TEST FILE

★ *Indicates a **difficult** question*

■ *Indicates a **conceptual** question*

39 Plant Responses to Environmental Challenges

Fill in the Blank

★1. Synthesis of additional **polysaccharides** acts in part by blocking plasmodesmata to prevent the spreading of viral pathogens.

2. Plants adapted to dry environments are called **xerophytes**.

3. Plants with long dormant periods interrupted by short periods of rapid growth and reproduction typically live in **dry** environments.

4. When cells accumulate the amino acid **proline,** their water potential becomes more/**less** negative.

5. Some swamp plants have root extensions called **pneumatophores** that grow into the air and deliver **oxygen** to the rest of the root system.

6. The major stress encountered by plants living in waterlogged soil is lack of soil **oxygen**.

7. In cacti, leaves are modified to form **spines**, and photosynthesis is carried out in the **stem** region.

8. Globally, the toxic substance that most restricts plant growth is **sodium chloride**.

9. Halophytes accumulate chloride and sodium ions and transport these substances to their **leaves**.

10. Plants with fleshy, water-storing leaves are called **succulents**.

11. Plants produce small molecules called **phytoalexins** within hours of infection by fungi or bacteria.

12. **Secondary** products produced by plants are useful to humans as fungicides and pharmaceuticals.

13. Large molecules, called **pathogenesis-related** proteins, are produced as a defense against infection and in cleanup operations.

14. One problem with cold is its effect on **membrane** fluidity.

Multiple Choice

1. Phytoalexin production is triggered by
   a. high temperature.
   b. high salt concentration.
   c. oligosaccharins.
   d. suberin.
   e. heat shock protein.

2. The secondary plant products that prevent the normal development of insect herbivores are
   a. alkaloids.
   b. flavonoids.
   c. phenolics.
   d. quinones.
   e. steroids.

■3. Some plants produce the amino acid canavanine, which is toxic to many insects because
   a. insects lack a tRNA specific for canavanine.
   b. canavanine is a component of alkaloids.
   c. canavanine prevents cells from synthesizing the amino acid arginine.
   d. insects lack large vacuoles for the storage of secondary products.
   e. insect proteins that incorporate canavanine function poorly.

4. In defense against tissue damage caused by pathogens,
   a. animals repair damaged tissues and plants seal off damaged tissues.
   b. animals seal off damaged tissues and plants repair damaged tissues.
   c. both plants and animals sometimes seal off and sometimes repair damaged tissues.
   d. both plants and animals repair damaged tissues.
   e. both plants and animals seal off damaged tissues.

5. Phytoalexins
   a. are always present in plants.
   b. occur in uniform concentration throughout a plant.
   c. are toxic to many fungi and bacteria.
   d. have no effect on viral infections.
   e. cause the plants to seal off areas of damaged tissue.

6. A plant's hypersensitive reaction to infection may include all of the following, except
   a. The production of phyoalexins by cells around the infection
   b. Long-term resistance to the infective agent
   c. Synthesis of pathogenesis-related proteins
   d. Death of both infected cells and cells near the infection
   e. Transport of phytoalexins to all parts of the plant.

7. Salicylic acid in plants does not
   a. increase resistance to pathogens.
   b. trigger the production of pathogenesis-related proteins.
   c. poison fungi and bacteria.
   d. protect against tobacco mosaic virus.
   e. play a role in the hypersensitivity response.

8. Non-water-soluble or hydrophobic poisons are stored in a plant's
   a. chloroplasts.
   b. epidermal waxes.
   c. Golgi bodies.
   d. mitochondria.
   e. vacuoles.

▪*9. Plants can produce the respiratory poison cyanide without poisoning themselves because plants
   a. do not respire.
   b. store a cyanide precursor in one compartment and activating enzymes in another.
   c. store water-soluble cyanide in laticifers.
   d. possess enzymes that are unaffected by cyanide.
   e. also produce proteins that bind and inhibit cyanide.

▪10. Which evidence best supports the hypothesis that the presence of toxic latex in leaves deters insects from feeding on a plant?
   a. Many insects do not feed on latex-producing plants.
   b. Latex-producingplants release milky latex when their leaves are damaged.
   c. Beetles that drain latex out of part of a leaf can then feed on that part.
   d. Beetles that cut veins in the leaves can then feed on the latex released.
   e. Latex-producing plants have high survival rates.

11. Plants can be treated with __________ to stimulate the production of pathogenesis-related proteins.
   a. salicylic acid
   b. PR inducer
   c. phytoalexin
   d. cellulose
   e. willowgen

12. Once secondary compounds are formed by plants, where are these defensive compounds usually stored?
   a. Vacuoles
   b. Nuclei
   c. Cell walls
   d. Cytoplasm
   e. Bound to membrane proteins

*13. Oil of wintergreen *might* stimulate production of PR protein in
   a. leaves.
   b. roots.
   c. flowering regions.
   d. the plant that harboring an infection.
   e. plants neighboring the the one that is infected.

14. In order for an *R* gene to confer resistance, the predator must have a corresponding
   a. *R* gene.
   b. tRNA.
   c. virus.
   d. *Avr* gene.
   e. bacteria.

15. Which of the following are "secondary products"?
   a. Proteins
   b. Lipids
   c. Alkaloids
   d. Carbohydrates
   e. Nucleic acids

16. Steroids produced by plants may
   a. attract pollinators and animals that disperse seeds.
   b. affect nervous systems of animals.
   c. inhibit fungal action.
   d. prevent normal development of insects.
   e. impair growth of competing plants.

17. Polysaccharides serve to
   a. store water in plants.
   b. defend against pathogens.
   c. repel predators because they are toxic to animals.
   d. act as salt glands.
   e. strengthen cell walls to form a barrier against invasion of a pathogen.

18. Laticifers are
   a. specialized cells for containing sodium ions.
   b. latex-containing tubes for storing hydrophobic products.
   c. waxy cells in the epidermis.
   d. cells that produce poisons such as alkaloids.
   e. cells in roots that take up water in dry environments.

19. In the gene-for-gene resistance mechanism, if a plant has a dominant resistance gene and a pathogen has a dominant avirulence gene
   a. the pathogen's gene overrides the plant's gene and infects the plant.
   b. epistasis occurs and the plant is infected.
   c. epistasis occurs and the plant is resistant to the pathogen.
   d. the plant is resistant to all pathogens whether or not they have the dominant avirulence gene.
   e. None of the above

▪20. Why aren't plants affected by their production of canavanine?
   a. Their tRNA molecules don't bind it.
   b. They keep it sequestered in vacuoles.

c. It isn't activated until it leaves the plant.
d. Plants don't use amino acids like animals do.
e. They produce it in such small amounts that it isn't detrimental to the plant.

▪*21. Scientists discovered a protein called arcelin in wild bean plants. This protein is useful to the plant because it
a. is toxic to herbivore predators.
b. helps the plants in dry environments.
c. is a secondary product that is beneficial to humans.
d. is responsible for making plant seeds resistant to predation by weevils.
e. makes the plant unpalatable to weevils and other herbivores.

22. Grazing can increase photosynthesis production by
a. increasing branching.
b. removing old or dead leaves.
c. providing increased root nutrients to the remaining leaves.
d. All of the above
e. None of the above

23. The difference between systemin and jasmonates is that systemin is a ________, and jasmonates is a ________.
a. nucleic acid, steroid
b. steroid, nucleic acid
c. fatty acid derivative, peptide
d. nucleic acid, steroid
e. peptide, fatty acid derivative

24. Plants protect themselves from toxins they produce by all of the following *except*
a. compartmentalization of the toxin.
b. building up a tolerance to them.
c. timing of toxin production.
d. storage in waxes.
e. storage in vacuoles.

25. To convert *Arabidopsis* from nonhalophyte to halophyte, plants were altered to
a. have salt glands.
b. not express Na^+/H^+ symport.
c. not express Na^+/H^+ antiport
d. overexpress Na^+/H^+ antiport.
e. overexpress Na^+/H^+ symport.

26. Which of the following is *not* an adaptation to a dry environment?
a. Thick cuticle
b. Vertically hanging leaves
c. Stomata in sunken cavities
d. Epidermal hairs
e. Salt glands in leaves

27. The typical environment for annual plants with a brief growing period and seeds capable of long dormant periods is a
a. desert.
b. salt marsh.
c. freshwater marsh.
d. environment contaminated with heavy metals.
e. grazed field.

28. Corn and related grasses roll up their leaves in response to
a. excess water.
b. lack of water.
c. excess salt.
d. heavy metals.
e. herbivores.

▪29. Compared to plants in moderate environments, xerophytes carry out photosynthesis more
a. slowly due to interference from the amino acid proline.
b. slowly because transpiration occurs more quickly.
c. slowly because their adaptations minimize carbon dioxide uptake.
d. quickly because their stems are also photosynthetic.
e. quickly because they have short periods of intense growth.

30. Swamp plants typically have root systems that
a. grow quickly.
b. penetrate deeply into the soil.
c. can carry out alcoholic fermentation.
d. alternate periods of growth and die-back.
e. accumulate the amino acid proline.

31. The pneumatophores of swamp plants are modified
a. flowers.
b. leaves.
c. roots.
d. spines.
e. stems.

32. Leaf parenchyma tissue with large spaces between cells are possessed by
a. aquatic plants to provide buoyancy.
b. aquatic plants to decrease transpiration rates.
c. desert plants to store water.
d. desert plants to form succulent leaves.
e. halophytes to excrete salt.

33. By accumulating the amino acid proline, plants
a. become toxic to most herbivores.
b. can carry out alcoholic fermentation.
c. can extract more water from the soil.
d. can prevent toxic effects from sodium.
e. attract animals that disperse seeds.

▪34. If a nonhalophyte and a halophyte are both placed in a salty environment, which will accumulate more sodium internally, and why?
a. The halophyte, because the nonhalophyte will not absorb much sodium.
b. The halophyte, because it requires sodium as a nutrient.
c. The halophyte, because as a succulent, it has more internal storage.
d. The nonhalophyte, because it cannot excrete sodium after absorption.
e. The nonhalophyte, because it requires sodium to create a negative water potential.

35. Which of the following adaptations is *not* found in both xerophytes and halophytes?
   a. High root-to-shoot ratios
   b. Sunken stomata
   c. Large air spaces in the leaf parenchyma
   d. Reduced leaf area
   e. Thick cuticles

▪36. The reason that certain plants can grow in soils contaminated with high levels of heavy metals is because they
   a. do not take up the heavy metals.
   b. excrete the heavy metals.
   c. have a genetic tolerance to the heavy metals.
   d. use the heavy metals for normal biochemical functions.
   e. are toxic to herbivores due to the heavy metals.

37. Which of the following is *not* true of plants that tolerate heavy metals?
   a. They take up the heavy metals.
   b. Different populations of plants will all tolerate heavy metals equally.
   c. Tolerant populations can evolve rapidly.
   d. A plant's tolerance is determined by its genotype.
   e. They usually experience little competition.

▪38. Grazing increases photosynthetic rates in certain plant species because
   a. more light reaches younger, more active leaves.
   b. the remaining leaves can transport sugars more slowly to the roots.
   c. the roots will be able to take up more nitrogen.
   d. older, dying leaves are sugar sinks.
   e. competition for atmospheric carbon dioxide is reduced.

39. The ability of grasses to grow from the base of the shoot and leaf is an adaptation to
   a. dry environments.
   b. soil fungi.
   c. heavy metals.
   d. grazing.
   e. saline environments.

40. Grazed plants may exhibit increased productivity in all of the following ways, except
   a. faster photosynthetic rates.
   b. growth of more stems.
   c. greater seed distribution.
   d. growth later into the season.
   e. growth of more roots.

41. Secondary products
   a. are essential for basic biological reactions.
   b. are similar in all plants.
   c. occur more often in animals than in plants.
   d. attract or inhibit other organisms.
   e. are usually of high molecular weight.

42. Which of the following is not a special adaptation of leaves to dry environments?
   a. Modification into spines
   b. Stomata in sunken cavities
   c. Dense epidermal hairs
   d. Fleshy leaves
   e. Horizontal leaves

43. The presence of pneumatophores in plants is an adaptation for success in which type of habitat?
   a. Desert
   b. Mountain
   c. Grassland
   d. Seashore
   e. Swamp

44. Some halophytic plants have salt glands that
   a. accumulate salt in their roots.
   b. serve as barriers to salt intake.
   c. maintain high salt concentrations in the plant.
   d. secrete salt onto the leaf surface.
   e. increase water loss from the plant.

45. What combination of adaptations is often observed in plants that occur in saline environments?
   a. Salt glands and succulence
   b. Salt glands and broad leaves
   c. CAM metabolism and succulence
   d. Spines and thin cuticles
   e. Dense stomata and thick cuticle

46. Certain plants concentrate the harmless amino acid proline in their cells. What effect does this have on the plant?
   a. Increased negative water potential
   b. Increased rate of transpiration
   c. Increased positive water potential in leaves
   d. Decreased salt loss
   e. Decreased water uptake

47. When some leaves are removed from a plant, what typically happens?
   a. Less light is available to the leaves.
   b. Less nitrogen is obtained from the soil.
   c. The plant dies.
   d. The rate of photosynthesis in remaining leaves increases.
   e. The transport of sugar from the remaining leaves decreases.

48. In conditions where water is plentiful but oxygen scarce, plants respond by
   a. rapidly growing roots that penetrate deeply into the soil.
   b. fermenting sugars to lactic acid.
   c. inhibiting the production of ATP.
   d. forming aerenchyma tissue.
   e. producing oxygen from water.

49. All of the following are adaptations to saline environments, except
   a. accumulation and transport of sodium ions.
   b. sequestering of sodium ions in the roots.
   c. salt glands in leaves.
   d. fleshy, gummy leaves.
   e. smaller leaves with smaller cells.

50. One way that grazing increases the productivity of a plant is by
 a. supporting food chains in nature.
 b. reducing the rate of photosynthesis in remaining leaves.
 c. increasing the number of sinks for absorbed nitrogen.
 d. shading younger leaves.
 e. causing the production of more replacement stems.

51. Many halophytes accumulate the amino acid
 a. proline.
 b. arginine.
 c. glycine.
 d. methionine.
 e. chlorine.

52. High temperatures trigger heat shock proteins, which include
 a. nucleases.
 b. kinases.
 c. chaperonins.
 d. antifreeze proteins.
 e. proline.

Study Guide Questions

1. Primary nonspecific defensive strategies in plants include which of the following?
 a. PR proteins
 b. Suberin
 c. Cutin
 d. b and c

2. Plants exhibit systemic acquired resistance much the same way people acquire resistance to pathogens; however, the mechanism is quite different. Which of the following have a role in acquired resistance in plants?
 a. Salicylic Acid
 b. PR proteins
 c. Methyl salicylate
 d. All of the above

3. Gene-for-gene resistances depends on which of the following?
 a. Compatible alleles in the plant and the pathogen
 b. Incompatible alleles in the plant and the pathogen
 c. Recessive *Avr* genes
 d. Recessive *R* genes

4. Upon infection by a pathogen, plant cells increase synthesis of polysaccharides. The function of these is to
 a. reinforce cell walls and seal off plasmodesmata.
 b. synthesize antibodies to the pathogen.
 c. isolate the pathogen in the invaded tissue.
 d. a and c

5. Canavanine is toxic to many herbivores but not to plants. This differential toxicity can best be explained by which of the following?
 a. Canavanine is confused with arginine in the predators.
 b. Canavanine is incorporated into proteins and causes them to fold improperly.
 c. Plants are able to differentiate between arginine and canavanine.
 d. All of the above

6. Which of the following best describes how plants produce their own insecticide?
 a. Wounded cells release systemin, systemin directly induces synthesis of protease inhibitors, protease inhibitors act as insecticides.
 b. Wounded cells release jasmonates, jasmonates stimulate systemin synthesis, systemin causes the production of protease inhibitors, protease inhibitors act as insecticides.
 c. Wounded cells release systemin, systemin causes membrane breakdown, membrane breakdown releases jasmonates, jasmonates induce synthesis of protease inhibitors, protease inhibitors act as insecticides.
 d. Wounded cells release systemin, systemin causes membrane breakdown, membrane breakdown releases jasmonates, jasmonates act as insecticides.

7. Which of the following are strategies for coping with drought conditions?
 a. Water storing tissues
 b. Leaf loss during drought
 c. Sunken stomata
 d. All of the above

8. Halophytes are different from all other types of plants in that they
 a. can accumulate sodium and chloride ions.
 b. have a positive water potential.
 c. are unaffected by saline conditions.
 d. All of the above

9. Which of the following conditions stimulates the production of heat shock proteins?
 a. Abnormally high temperatures only
 b. Abnormally low temperatures only
 c. Both abnormally high or abnormally low temperatures
 d. They are continually available in the plant.

10. The main function of heat shock protein is to
 a. stabilize proteins necessary to a cell's survival.
 b. reinforce membranes that lose fluidity.
 c. cause the plant to enter dormancy.
 d. act as an antifreeze compound.

End of Chapter Questions

1. Which of the following is *not* a common defense against bacteria, fungi, and viruses?
 a. Deposition of additional polysaccharides in the cell wall.
 b. Phytoalexins
 c. A waxy covering
 d. The hypersensitive response
 e. The conversion of recessive *R* genes to dominant *R* genes.

2. Plants sometimes protect themselves from their own toxic secondary products by
   a. producing special enzymes that destroy the toxic substances.
   b. storing precursors of the toxic substances in one compartment and the enzymes that convert precursors to toxic products in another compartment.
   c. storing the toxic substances in mitochondria or chloroplasts.
   d. distributing the toxic substances to all cells of the plant.
   e. performing crassulacean acid metabolism.

3. Grazing
   a. is predation by plants on animals.
   b. always reduces plant growth.
   c. usually increases the rate of photosynthesis in the remaining leaves.
   d. reduces the rate of transport of photosynthetic products from the remaining leaves.
   e. always is lethal to the affected plant.

4. Which statement about secondary plant products is *not* true?
   a. Some attract pollinators.
   b. Some are poisonous to herbivores.
   c. Most are proteins or nucleic acids.
   d. They help plants compensate for being unable to move.
   e. Some mimic the hormones of animals.

5. Which statement about latex is *not* true?
   a. It is sometimes contained in laticifers.
   b. It is typically white.
   c. It is often toxic to insects.
   d. It is a solid material found on the surface of leaves.
   e. Milkweeds produce it.

6. Which of the following is *not* an adaptation to dry environments?
   a. A less negative osmotic potential in the vacuoles
   b. Hairy leaves
   c. A heavier cuticle over the leaf epidermis
   d. Sunken stomata
   e. A root system that grows each rainy season and dies back when it is dry

7. Some plants adapted to swampy environments meet the oxygen needs of their roots by means of a specialized tissue called
   a. parenchyma.
   b. aerenchyma.
   c. collenchyma.
   d. sclerenchyma.
   e. chlorenchyma.

8. Halophytes
   a. all accumulate proline in their vacuoles.
   b. have water potentials that are more negative than those of other plants.
   c. have thinner cuticles to facilitate water transport.
   d. have low root-to-shoot ratios.
   e. rarely accumulate sodium.

9. Which of the following is *not* a commonly toxic heavy metal?
   a. Copper
   b. Lead
   c. Cadmium
   d. Potassium
   e. Aluminum

10. Plants that tolerate heavy metals commonly
   a. grow faster than non-tolerant plants under all conditions.
   b. do not take up the heavy metal ions.
   c. are tolerant to all heavy metals.
   d. are slow to colonize an area rich in heavy metals.
   e. have adapted mechanisms to deal with specific heavy metals.

Student Web Site Self-Quiz Questions

1. Which one of the following categories of substances produced by plants is toxic to many bacteria and fungi and is usually produced within hours of infection?
   a. Phytoalexins
   b. PR proteins
   c. Alkaloids
   d. Enzyme
   e. Salicylic acid

2. Some plants are a rich source of salicylic acid. Which of the following statements about the hypersensitive response or salicylic acid is *false*?
   a. Necrotic lesions or "dead spots" are often associated with the hypersensitive response.
   b. The acquired resistance that sometimes follows the hypersensitive response is limited to the pathogen that triggered the response.
   c. Salicylic acid is found in most plants, not just willows.
   d. Treatment of plants with aspirin can lead to the production of PR proteins.
   e. The hypersensitive response is a nonspecific defense system.

3. Which of the following is *not* a response of some plants to grazing by herbivores?
   a. Increased rate of photosynthesis of the leaves
   b. Production of a bushier growth form
   c. Reduction in the export of sugars from leaves
   d. Evolution of growth from base of shoots and leaf types
   e. Lengthening of the growing season

4. Plants produce a variety of secondary products. For example, many plants produce flowers with odorous or colored chemicals that attract pollinators. Which of the following choices does *not* correctly pair a class of secondary plant product with a role those products typically play?
   a. Alkaloids—affect herbivore nervous systems
   b. Phenolics—attract pollinators
   c. Quinones—inhibit growth of competing plants

d. Terpenes—act as fungicides and insecticides
e. Steroids—mimic animal hormones

5. Which of the following is *not* a typical use of secondary products by plants?
a. Reduce competition with neighboring plants
b. Inhibit fungal growth
c. Attract pollinators
d. Disrupt the normal development of insects
e. Produce and store arginine within seeds

6. Which of the following is *not* a typical mechanism used by plants to protect themselves from their own toxic secondary products?
a. Storage of hydrophobic secondary products in laticifers
b. Production of the secondary product in an inactive form
c. Activation of the secondary product only in damaged tissue
d. Modified enzymes or receptors that only recognize the secondary product
e. Isolation of the secondary product in a special compartment

7. Which one of the following is *not* an adaptation seen in xerophytes?
a. Succulence
b. Leaves modified into spines
c. Increased rate of photosynthesis
d. Leaves that can roll up
e. Leaves with a heavy cuticle layer

8. Mesquite trees are common in some very dry deserts. Which of the following is *not* a typical xerophytic adaptation?
a. Rapid growth and a short life cycle from seed to seed
b. Shallow but extensive fibrous root system
c. Deep taproot systems
d. Keep the water potential of cells less negative
e. Vertically oriented leaves

9. Aerenchyma is commonly seen in plants __________, and the spaces seen in the tissue are normally filled with __________.
a. that are fully submerged, water
b. that are fully submerged, air
c. with submerged root systems, air
d. in dry environments, water
e. in dry environments, proline

10. Black mangrove trees grow with their roots submerged in salt water. Which of the following adaptations would *not* be useful for these plants, given their environment?
a. Salt glands
b. Ability to develop large negative water potential within root cells
c. Reduced activity of antiport proteins in tonoplast
d. Accumulation of salt in central vacuole
e. Thickened cuticle

11. Which of the following is *not* a feature that many halophytes and xerophytes share?
a. Accumulation of proline
b. Succulence
c. Thick cuticle
d. Large root-to-shoot ratio
e. Open stomata during the daytime

12. Soils contaminated with heavy metals are *not* suitable for most plants, although some plants are tolerant. Which of the following statements about heavy metal tolerance in plants is *false*?
a. Differences between tolerant and nontolerant plants usually have a genetic basis.
b. Most tolerant plants are able to avoid being poisoned by excluding the heavy metal.
c. Tolerant plants may be useful agents for bioremediation.
d. Tolerance is usually specific to a particular type of heavy metal.
e. Evolution of heavy metal tolerance can be fairly rapid.

13. While some plants may be adapted to survival at high temperatures, others may be adapted for survival at cold temperatures. Which of the following is *not* an adaptation for survival in cold environments?
a. Cold-hardening
b. Heat shock proteins
c. Cross-protection.
d. Antifreeze proteins
e. Increased production of protective saturated fats

TEST FILE

★ *Indicates a* ***difficult*** *question*

■ *Indicates a* ***conceptual*** *question*

40 Physiology, Homeostasis, and Temperature Regulation

Fill in the Blank

1. The four types of tissues are epithelial, muscle, nervous, and **connective**.

★2. In certain "hot" fish such as bluefin tuna, heat is exchanged between vessels carrying blood in opposite directions. This adaptation is called **countercurrent heat exchange**.

3. The metabolic rate of a resting animal at a temperature within the thermoneutral zone is called the **basal metabolic rate**.

4. Small endotherms can extend the period over which they can survive without food by dropping body temperature. This adaptive hypothermia is called **daily torpor**.

5. The maintenance of more or less constant physiological conditions within an organism is called **homeostasis**.

6. **Positive** feedback amplifies a response.

7. Animals whose temperature fluctuates to match that of the environment are termed **poikilotherms**.

8. Animals that can affect their body temperature by generating metabolic heat are called **endotherms**.

9. In the vertebrate animal, the thermostat is located in the **hypothalamus**.

10. During the condition called **hypothermia**, when an endotherm's body temperature is far below normal, metabolic rates become **lower**. Although this condition can be harmful to endotherms, it becomes a useful strategy for some overwintering mammals called **hibernators**.

11. In mammals, most nonshivering heat production occurs in the adipose tissue called **brown fat**.

12. The linings of many organs consist of **epithelial** tissue.

13. The most abundant protein in our bodies is **collagen**, found in connective tissue.

14. Cells with the same characteristics form **tissues**, and different tissues form discrete structures with specific functions called **organs**.

★15. **Pyrogens**, derived from bacteria or viruses that invade the body, cause fever.

Multiple Choice

1. Which of the following statements about tissues is *false*?
   a. The protein fibrinogen functions in clotting.
   b. An organ is usually comprised of a single type of tissue.
   c. Stratified epithelium consists of several cell layers.
   d. In many vertebrates, the early skeleton of cartilage is gradually replaced by bone.
   e. None of the above

2. Which of the following statements about heat exchange is *false*?
   a. Conduction is the direct transfer of heat when two objects of different temperatures come in contact.
   b. Evaporation of water from the surface of the body heats the body.
   c. Ectotherms and endotherms can change the rate of heat exchange between their bodies and the external environment by changing blood flow to the skin.
   d. Animals may lose heat by convection when exposed to a wind with a temperature below that of their body surface.
   e. None of the above

■3. Homeostasis
   a. favors a constant internal physiological environment regardless of the changes in the external environment.
   b. keeps vital organs working at their maximum potential.
   c. keeps all cells working at the same metabolic rate.
   d. keeps the body's metabolic rate constant in varying environmental temperatures.
   e. keeps the body's temperature constant in varying environmental temperatures.

4. Which term refers to organisms that largely depend on external sources of heat to maintain body temperature?
   a. Homeothermic
   b. Endothermic
   c. Heterothermic
   d. Ectothermic
   e. None of the above

5. Compared to an ectothermic organism's body temperature response to a 10°C rise in environmental temperature, an endotherm's body temperature will
   a. rise at a constant rate.
   b. fall at a constant rate.
   c. fall to a point, then become stable.
   d. rise to a point, then become stable.
   e. remain relatively constant.

▪6. Which of the following adaptations would *not* favor an increase in an ectotherm's body temperature?
   a. Muscle contractions
   b. Cluster or huddling behavior
   c. Decreased surface-to-surface contact with the cold environment
   d. Circulatory changes to maintain core or internal temperatures greater than the animal's peripheral temperatures
   e. Metabolic compensation

7. Readjustment of an organism's metabolic rate to compensate for seasonal thermal change is termed
   a. homeostasis.
   b. negative feedback.
   c. metabolic compensation.
   d. acclimatization.
   e. regulation.

8. Which term best describes an organism that maintains a constant body temperature?
   a. Ectotherm
   b. Homeotherm
   c. Heterotherm
   d. Poikilotherm
   e. All of the above

▪9. A desert lizard slowly crawls from its burrow after cool nighttime temperatures and postures on a rock warmed by the mid-morning sun. Which one of the following is most likely *not* part of the lizard's thermoregulatory responses?
   a. Orientation to the sun to maximize exposure to solar radiation
   b. Increased peripheral circulation via dilation of surface blood vessels
   c. Increased metabolic heat production
   d. Increased physical contact between the warm rock and the relatively cool lizard
   e. Orientation of the lizard on the rock into more suitable microenvironments to avoid cold air currents

10–13. The endocrine and nervous systems are both responsible for integration of information and control of organs and organ systems. This control differs with each system. Fill in the blanks using the following answer choices.
   a. Hormones
   b. Neurons
   c. Glial cells
   d. Ductless glands

10. The principal organ(s) of the endocrine system are __________. **(d)**

11. The nervous system routes control to specific targets by way of __________. **(b)**

12. The chemical messages of the endocrine system are called __________. **(a)**

13. __________ are cells of the nervous system that do not conduct signals. **(c)**

14. What are the normal value(s) for most biological Q_{10}'s?
   a. 1
   b. 1–2
   c. 2–3
   d. 2–10
   e. 0–45

15–21. Use the following answer choices to correctly indicate the probable metabolic response of a mammal exposed to a 3–5°C environmental temperature change.
   a. Increased metabolic rate
   b. Decreased metabolic rate
   c. No change in the metabolic rate
   d. Death

▪15. The environmental temperature increases above the upper critical temperature. **(a)**

▪16. The environmental temperature increases above the lower critical temperature. **(c)**

▪17. The environmental temperature decreases far below zero. **(d)**

▪18. The environmental temperature decreases below the lower critical temperature. **(a)**

▪19. The environmental temperature decreases below the upper critical temperature. **(c)**

▪20. The environmental temperature fluctuates between the upper and the lower critical temperature. **(c)**

▪21. The environmental temperature increases above zero. **(b)**

*22. An endotherm's thermoneutral zone can be variously described. Which of the following does *not* describe the limits of the thermoneutral zone?
   a. A range of environmental temperatures between the upper critical and lower critical temperature.
   b. A range of environmental temperatures over which the organism exhibits a basal metabolic rate.
   c. A range of body temperatures at which the metabolic rate is maximum.
   d. A range of environmental temperature over which the organism's metabolic rate neither increases nor decreases for thermoregulation.
   e. A range of environmental temperatures over which the organism produces minimal metabolic activity to support itself.

23. Increased heat for thermoregulation or thermogenesis is produced either by shivering or by nonshivering mechanisms. Nonshivering thermogenesis is dependent upon all of the following, *except*
    a. brown fat.
    b. thermogenin.
    c. production of ATP.
    d. rich blood supply.
    e. increased mitochondria.

24. The hypothalamus in part serves as an integrated thermoregulatory center defining an organism's response to changes in its thermal environment. Since the hypothalamus normally serves to produce metabolic changes to reverse the direction of environmental temperature change, the control is termed
    a. positive feedback.
    b. metabolic compensation.
    c. negative feedback.
    d. feedforward.
    e. metabolic torpor.

▪25. Which of the following is *not* true of hibernation?
    a. It may be interrupted by brief returns to normal body temperature.
    b. It is a form of regulated hypothermia.
    c. Body temperature is turned down to a low level.
    d. Metabolic rate is reduced to only a fraction of the basal metabolic rate.
    e. It is found in more species of birds than species of mammals.

26. Which of the following is an organ system that processes information?
    a. Urinary system
    b. Nervous system
    c. Endocrine system
    d. **b and c**
    e. a, b, and c

▪27. Most physiological processes
    a. occur more rapidly at higher temperatures.
    b. occur less rapidly at higher temperatures.
    c. are not temperature-sensitive.
    d. first increase in rate and then decrease in rate at high temperatures.
    e. None of the above

28. _________ is the process of physiological and biochemical change that an animal undergoes in response to seasonal changes in climate.
    a. Acclimatization
    b. Homeostasis
    c. Sublimity
    d. Metabolic compensation
    e. Hybridization

29. Adaptations used by endotherms to reduce heat loss include which of the following?
    a. Rounder body shapes
    b. Shorter appendages
    c. Increased thermal insulation
    d. a, b, and c
    e. **b and c**

30. The vertebrate thermoregulatory center ("thermostat") is located within the central nervous system in the
    a. pons.
    b. cerebellum.
    c. hypophysis.
    d. medulla.
    e. hypothalamus.

31. Homeostasis refers to the tendency to keep the body systems
    a. matched to the external environment.
    b. the same relative to one another.
    c. at a steady state over time.
    d. under the control of the brain.
    e. at the same specific temperature.

32. Which type of tissue is responsible for secreting digestive enzymes?
    a. Connective
    b. Epithelial
    c. Matrix
    d. Muscle
    e. Nervous

33. Bone is which type of tissue?
    a. Connective
    b. Epithelial
    c. Matrix
    d. Muscle
    e. Nervous

34. Blood is which type of tissue?
    a. Connective
    b. Epithelial
    c. Matrix
    d. Muscle
    e. Nervous

35. Connective tissues differ from each other mostly in their
    a. cellular structure.
    b. function in support.
    c. matrix properties.
    d. location in the body.
    e. cell packing.

*36. In regulatory systems, the phenomenon of negative feedback
    a. is the least common type of feedback mechanism.
    b. stimulates a return to set point.
    c. amplifies a response.
    d. disrupts homeostasis.
    e. None of the above

*▪37. Which of the following would serve to increase the set point of a regulatory system?
    a. Negative feedback
    b. Positive feedback
    c. Feedforward
    d. Insensitivity to information
    e. None of the above

38. Cellular functions are generally limited to what temperature range (in °C)?
a. 20–100
b. 20–45
c. 0–20
d. 0–45
e. 0–100

39. The upper temperature limit at which cells can function is determined by the
a. boiling point of water.
b. melting point of water.
c. melting point of fats.
d. denaturation point of nucleic acids.
e. denaturation point of proteins.

40. The Q_{10}, which describes the sensitivity of a reaction to temperature, is calculated as the
a. rate of a process at a certain temperature divided by its rate at 10°C.
b. rate of a process at a certain temperature divided by its rate at a temperature 10° lower.
c. rate of a process at a certain temperature divided by its rate at a temperature 10° higher.
d. temperature at which the rate of a certain process doubles.
e. temperature at which the rate of a certain process becomes insignificant.

▪41. The rate of a particular biological function is X at 15°C. Which of these statements is true about the rate of that function at 25°C?
a. If the Q_{10} were 1, the rate would be X.
b. If the Q_{10} were 1, the rate would be 25.
c. If the Q_{10} were 2, the rate would be X.
d. If the Q_{10} were 3, the rate would be 2X.
e. If the Q_{10} were 3, the rate would be 30.

▪42. If an animal exhibits metabolic compensation, this means that
a. it adapts its physiology to local environmental conditions.
b. its metabolic rate is always slightly higher in summer.
c. its Q_{10} stays constant despite seasonal temperature change.
d. its body temperature and its metabolic rate are independent of each other.
e. it uses the same reaction pathways at all temperatures.

43. Poikilothermic animals are those whose body temperature
a. is maintained at a constant level some of the time.
b. is maintained at a constant level all of the time.
c. fluctuates with environmental temperature.
d. is kept at a higher temperature than the environment.
e. is kept at a lower temperature than the environment.

44. Which of the following animals are endotherms?
a. Fish
b. Amphibians
c. Birds
d. Mammals
e. c and d

45. Which of the following animals would be most successful with the part-time temperature regulation strategy of heterothermy?
a. A desert lizard that basks in the sun
b. A deep ocean fish in an environment with little temperature change
c. An insect that is warm-blooded in flight and cold-blooded at rest
d. An amphibian that is sometimes in water and sometimes on land
e. A hummingbird that drops its body temperature at night

46. Which of the following animals is behaving as an endotherm in order to warm its body?
a. A moth that shivers its wings before flight
b. A black beetle that absorbs solar radiation.
c. A snake that lies on a warm blacktop road
d. A fish that moves to a warm, shallow part of a pond
e. An insect that positions its body for maximum exposure to sunlight

▪47. A certain desert lizard thermoregulates as an ectotherm. Which of the following phenomena should *not* be a part of its thermoregulatory behavior?
a. Staying in a burrow when the surface temperature is below 10°C
b. Basking in the sun during the early morning hours
c. Moving into the shade during the midday hours
d. Pressing its body to the ground during the midday hours
e. Consuming prey that is ectothermic

▪48. In which case would a poikilothermic animal have a higher body temperature than a homeotherm?
a. When resting in an underground burrow
b. At midday on the desert floor
c. When swimming in the cold ocean
d. When basking in the early morning sun
e. When the poikilotherm is much larger

49. Which of the following exemplifies adaptive thermoregulatory behavior in humans?
a. A cowboy wearing a broad-brimmed hat
b. An Eskimo wearing lightweight white clothing
c. A desert home with large windows facing the sun
d. Swimming in winter weather
e. Humans are endothermic and don't exhibit thermoregulatory behavior

50. Which physiological control mechanism is a response to a rise in body temperature?
a. Slower heart rate
b. Increased blood flow to the skin
c. Constriction of blood vessels in the skin
d. Contraction of muscles
e. Retention of water

51. What is the adaptive advantage of honeybee workers clustering to maintain warm temperatures in the hive?
   a. Protection of the queen bee
   b. Protection of the comb structure
   c. Optimal pollen collection
   d. Optimal digestion of honey
   e. Optimal brood development

*52. Within a range of environmental temperatures called the thermoneutral zone, the metabolic rate of an endotherm is
   a. constant.
   b. low and independent of temperature.
   c. high and independent of temperature.
   d. near the basal metabolic rate.
   e. dependent upon the temperature.

*▪53. In a mammal, metabolic rate is highest in which of the following situations?
   a. The animal is at rest; the temperature is within the thermoneutral zone.
   b. The animal is active; the temperature is below the thermoneutral zone.
   c. The animal is active; the temperature is within the thermoneutral zone.
   d. The animal is at rest; the temperature is above the thermoneutral zone.
   e. The animal is at rest; the temperature is below the thermoneutral zone.

54. Which of the following is *not* true of brown fat?
   a. It contains abundant mitochondria.
   b. It provides most energy for shivering.
   c. It produces less ATP in metabolism.
   d. It converts more fuel to heat.
   e. It is more abundant in hibernating animals.

*55. The mechanism of heat production in brown fat depends on
   a. inefficient use of ATP in metabolism.
   b. more rapid breakdown of protein.
   c. more rapid breakdown of fatty acids.
   d. additional shivering of skeletal muscles.
   e. uncoupling proton movement from ATP production.

56. Which of the following is an important adaptation of animals to cold climates?
   a. Increased tendency to shiver
   b. Thinner layers of body fat
   c. Reduced density of fur or feathers
   d. Reduced surface area–to–volume ratio
   e. Increased flow of blood to surface

57. The elephant is better adapted to tropical habitats than to cold climates because of its
   a. sparse hair.
   b. large size.
   c. stocky appendages.
   d. vegetarian diet.
   e. thick skin.

58. Why is evaporative cooling used only as a last resort by animals in dry environments?
   a. It is ineffective at dissipating heat.
   b. Water may be a limiting resource.
   c. Sweating requires little energy expenditure.
   d. It requires an insulating layer in the skin.
   e. Resetting the thermostat is required.

59. When the temperature of the hypothalamus rises, which thermoregulatory responses result?
   a. Increased metabolic heat production.
   b. Resetting of the thermostat higher.
   c. Dilation of blood vessels in the skin.
   d. Overall increase in body temperature.
   e. Initiation of shivering movements.

*▪60. Which of the following accurately describes the thermoregulatory set point?
   a. The set point for shivering is the same as the set point for panting.
   b. In a given individual, the set points are relatively constant.
   c. The set points are the same for all members of the same species.
   d. Information on skin temperature can change the metabolism set point.
   e. Temperature information serves as a positive feedback signal.

61. Which of the following statements about thermoregulation is *false*?
   a. Thermoregulatory set points are higher during sleep.
   b. Entrance into hibernation begins with a decrease in metabolic rate.
   c. During a fever, aspirin can lower the hypothalamic set point.
   d. Hypothermia may result from starvation.
   e. None of the above

▪62. Single-celled and simple multicellular animals meet their needs by having every cell exposed to the environment; more complex animals have an internal environment. Which of the following is *not* an advantage of an internal environment?
   a. Each cell is capable of performing every function.
   b. Cells can be specialized for one function.
   c. Cells can be more efficient.
   d. It can be maintained independently of the external environment.
   e. Complex animals can occupy otherwise inhospitable habitats.

63. Which is true of smooth muscle?
   a. It controls the pumping of blood through the heart.
   b. It is responsible for our behavior.
   c. It is not under conscious control.
   d. It is what connects bones to bones.
   e. It makes up the ciliated lining of the gut.

64. When scientists measured the Q_{10} of the fish in the lab, their prediction of the metabolic rate differed from the observed rate because
    a. the fish acclimated to the temperature change in the lake.
    b. metabolic rate varied in different parts of the fish.
    c. the fish do not produce metabolic compensation.
    d. the internal temperature of the fish is higher than the surrounding environment.
    e. it is impossible to figure the Q_{10} of a poikilotherm.

65. "Hot" fish, such as bluefin tuna, keep a higher temperature difference between their body and the surrounding water than cold fish because
    a. these fish are actually endotherms.
    b. they use a countercurrent heat exchange system of veins and arteries.
    c. they use shivering to create more heat.
    d. they have many brown fat tissues.
    e. they have a large dorsal aorta that keeps them warmer.

66. Your body responds to an infection by producing __________, which cause many of the symptoms of sickness such as fever and feeling crummy.
    a. pyrogens
    b. interleukins
    c. prostaglandins
    d. a and b
    e. b and c

67. As the environmental temperature increases, up to 25°C, the metabolic rate of an ectotherm __________ and an endotherm __________.
    a. increases, increases
    b. increases, decreases
    c. decreases, increases
    d. decreases, decreases
    e. stays the same, decreases

Study Guide Questions

1. Which of the following contributes to the proper maintenance of homeostasis in the bodies of animals?
    a. pH buffering of the blood
    b. Control of blood flow to the skin
    c. Production of hormones
    d. All of the above

2. In an environment with an ambient temperature lower than an animal's core body temperature, which of the following would be an inappropriate physiological or behavioral response if the animal needed to eliminate excess body heat?
    a. Wallowing in a pool of water
    b. Preventing blood flow to peripheral vessels
    c. Sweating
    d. All of the above

3. A lizard lives in a desert environment where the temperature at night is low and during the day is high. What might this animal do to maintain the highest stable body temperature?
    a. Stay in a burrow during the night and shuttle between the sun and shade on the surface during the day
    b. Stay on the surface during the night and move to a burrow during the day
    c. Increase metabolism and heat production during the night and decrease it during the day
    d. None of the above

4. On a warm day your dog has been running around fetching his frisbee and returning it to you. You notice that he is panting heavily. Which of the following does not describe the condition of your dog?
    a. Your dog is using evaporative cooling to lower core body temperature.
    b. Your dog is currently operating at a basal metabolic rate.
    c. Your dog is at the upper end of its thermoneutral zone.
    d. All of the above

5. Evaporative cooling is an effective way to increase heat loss, but carries with it the physiological drawback of
    a. increased use of ATP and substantial water loss.
    b. lowering the hypothalamic thermal set point.
    c. decreased use of ATP and substantial water gain.
    d. exhaustion of the supply of brown fat.

6. Fever
    a. is a higher than normal body temperature that is always dangerous.
    b. decreases the metabolic rate of the body to conserve energy.
    c. is regulated by chemicals that reset the thermostat of the body to a higher set point.
    d. causes the liver to release large amounts of calcium, which seems to inhibit bacterial growth.

7. Which of the following is a characteristic unique to connective tissue?
    a. Highly modified cells that show the special property of contractility
    b. Diverse anatomy with distinct specializations for information transfer
    c. Lining of the inner surfaces of the intestines and lungs in mammals
    d. Loose array of cells embedded in some kind of extracellular matrix

8. Elephants use their ears to dump heat to the environment. What mechanisms might they employ to increase heat loss from the ears?
    a. Increased convection due to flapping of the ears
    b. Moving into the sun
    c. Increased blood flow to the ears
    d. a and c

9. Which organ system is not involved in the maintenance of homeostasis in humans?
    a. Endocrine system
    b. Digestive system
    c. Muscle system
    d. All are involved in maintenance of homeostasis

10. What is the difference between a negative feedback mechanism and a positive feedback mechanism?
    a. Negative feedback mechanisms only exist in the circulatory system.
    b. Negative feedback mechanisms stabilize toward a set point and positive feedbacks move away from a set point.
    c. Negative feedback mechanisms move away from a set point and positive feedbacks stabilize toward a set point.
    d. Negative feedback mechanisms stabilize toward a set point and positive feedbacks reset the set point.

11. A number of physiological processes can undergo acclimatization. Which of the following statements about acclimatization is false?
    a. Acclimatization occurs in response to seasonal temperature changes.
    b. Acclimatization of metabolic rate occurs because enzyme expression changes.
    c. Acclimatization involves the changing of a set point.
    d. All are correct.

12. In fast-swimming cool-water fishes such as sharks and tunas, which of the following contribute to the generation and maintenance of core body temperatures in excess of ambient water temperatures?
    a. Specializations in the size and location of their blood vessels
    b. Low rates of activity in the swimming muscles
    c. An ability to acclimatize rapidly to the surrounding water
    d. Metabolic rates that are insensitive to temperature change

13. Which of the following statements about hypothalamic function is false?
    a. Circulating blood temperature is monitored by the hypothalamus.
    b. Hypothalamic thermal set points never change.
    c. The hypothalamus is a part of the central nervous system.
    d. Different thermoregulatory responses have different hypothalamic set points.

14. Some animals use brown fat as a source of heat generation. Which of the following statements about brown fat is *true*?
    a. Brown fat is involved in nonshivering thermogenesis.
    b. The protein thermogenin is involved in uncoupling ATP production and proton movement.
    c. Brown fat is found in endotherms.
    d. All of the above

15. During childbirth, the pressure exerted on the mother's cervix by the emerging infant leads to increased contraction of the uterus. This interaction could be explained by
    a. negative feedback.
    b. feedforward feedback.
    c. positive feedback.
    d. metabolic compensation.

End of Chapter Questions

1. If the Q_{10} of the metabolic rate of an animal is 2, then
   a. the animal is better acclimatized to a cold environment than if its Q_{10} is 3.
   b. the animal is an ectotherm.
   c. the animal consumes half as much oxygen per hour at 20°C as it does at 30°C.
   d. the animal's metabolic rate is not at basal levels.
   e. the animal produces twice as much heat at 20°C as it does at 30°C.

2. Which statement about brown fat is *true*?
   a. It produces heat without producing ATP.
   b. It insulates animals acclimatized to cold.
   c. It is a major source of heat production for birds.
   d. It is found only in hibernators.
   e. It provides fuel for muscle cells responsible for shivering.

3. What is the most important and most general difference between endotherms (mammals and birds) adapted to cold climates in comparison to endotherms adapted to warm climates?
   a. Higher basal metabolic rates
   b. Higher Q_{10} values
   c. Brown fat
   d. Greater insulation
   e. Ability to hibernate

4. Which of the following would cause a decrease in the hypothalamic temperature set point for metabolic heat production?
   a. Entering a cold environment
   b. Taking an aspirin when you have a fever
   c. Arousing from hibernation
   d. Getting an infection that causes a fever
   e. Cooling the hypothalamus

5. Mammalian hibernation
   a. occurs when animals run out of metabolic fuel.
   b. is a regulated decrease in body temperature.
   c. is less common than hibernation in birds.
   d. can occur at any time of year.
   e. lasts for several months, during which body temperature remains close to environmental temperature.

6. Which of the following is an important difference between an ectotherm and an endotherm of similar body size?
   a. The ectotherm has higher Q_{10} values.
   b. Only the ectotherm uses behavioral thermoregulation.
   c. Only the endotherm can constrict and dilate the blood vessels to the skin to alter heat flow.
   d. Only the endotherm can have a fever.
   e. At a body temperature of 37°C, the ectotherm has a lower metabolic rate than the endotherm.

7. The function of the countercurrent heat exchanger in "hot" fish is to
   a. **trap heat in the muscles.**
   b. produce heat.
   c. heat the blood returning to the heart.
   d. dissipate excess heat generated by powerful swimming muscles.
   e. cool the skin.

8. What is the difference between a winter- and a summer-acclimatized fish that is termed metabolic compensation?
   a. The winter-acclimatized fish has a higher Q^{10}.
   b. The winter-acclimatized fish develops greater insulation.
   c. The winter-acclimatized fish hibernates.
   d. The summer-acclimatized fish has a countercurrent heat exchanger.
   e. **The summer-acclimatized fish has a lower metabolic rate at any given water temperature.**

9. Which of the following is an important characteristic of epithelial cells?
   a. They generate electric signals.
   b. They contract.
   c. They have an extensive extracellular matrix.
   d. They have secretory functions.
   e. **They cover the surface of the body and line the body cavities.**

10. Negative feedback
   a. works in opposition to positive feedback to achieve homeostasis of a physiological variable.
   b. always turns off a process.
   c. **reduces an error signal in a regulatory system.**
   d. is responsible for metabolic compensation.
   e. is a feature of the thermoregulatory systems of endotherms but not of ectotherms.

Student Web Site Self-Quiz Questions

1. Select below a function that is *not* performed by some types of epithelial tissue.
   a. Secretion
   b. Absorption
   c. Translocation of materials
   d. **Movement of cells**
   e. Protection

2. Select below a choice that is *not* a component or type of connective tissue.
   a. Collagen fiber
   b. **Glial cells**
   c. Adipose tissue
   d. Red blood cells
   e. Elastin

3. Which of the following basic vertebrate tissue types is *not* found in the stomach?
   a. Epithelial tissue
   b. Connective tissue
   c. Muscle tissue
   d. Nervous tissue
   e. **All of the above are found in the stomach**

4. Which of the following is *not* a correct association between the operation of a car and homeostasis concepts?
   a. Controlled system—brakes/accelerator
   b. Regulatory system—speedometer/driver
   c. Set point—posted speed limit
   d. Error signal—difference between speed limit and current rate of speed
   e. **Negative feedback—cruise control**

5. Many lizards show behavioral thermoregulation based on how they orient themselves relative to the sun. Which one of the following is a *not* statement about adaptations in animals for temperature regulation?
   a. Some endotherms use behavior in thermoregulation.
   b. Some ectotherms use physiological thermoregulation.
   c. Some ectotherms produce heat.
   d. In certain "hot" fish, anatomical adaptations in the circulatory system conserve metabolic heat.
   e. **In the countercurrent heat exchanger seen in some fish, the temperature of the blood in the arteries and veins is the same.**

6. Many cold-adapted endotherms share common morphological features. Which of the following is *not* an adaptation for life in the cold?
   a. Decreased surface-to-volume ratio of the body
   b. Decreased appendage length
   c. **Use of water as an insulator**
   d. Thicker insulation
   e. Shunting of blood away from skin.

7. Experiments have been done by artificially cooling and warming the hypothalamus and observing the effects on a ground squirrel's metabolic rate. Which of the following statements is *not* supported by these sorts of experiments?
   a. Temperature of the hypothalamus affects the thermoregulatory behavior of ectotherms.
   b. The hypothalamic temperature is a negative feedback signal.
   c. Different thermoregulatory responses have different set points.
   d. **Like the thermostat-heater analogy, once the set point initiating metabolic heat production is reached, the response is all-or-none.**
   e. Skin temperature provides feedforward information to the hypothalamus.

8. Experiments involving lizards gave us insight into how the body reacts when exposed to bacterial pyrogens. Which of the following statements regarding fever is not *true*?
   a. Fevers help to fight infection.
   b. Pyrogens increase the hypothalamic heat set point.
   c. **Taking aspirin reduces fevers by raising the hypothalamic set point.**
   d. Pyrogens stimulate the body's macrophages to produce interleukins.
   e. Extreme fevers can be fatal.

TEST FILE

★ *Indicates a **difficult** question*

■ *Indicates a **conceptual** question*

41 Animal Hormones

Fill in the Blank

1. In an emergency, the adrenal glands produce **epinephrine** immediately and then **cortisol** for prolonged response to stress.

2. A prominent second messenger that responds to hormones is **cyclic AMP, or cAMP**.

*3. **Histamine** mediates inflammation and is an example of a paracrine hormone.

4. Hormonal and other systems in which production is shut off when the product becomes abundant are known as **negative** feedback systems.

5. **Prolactin** is a pituitary hormone that stimulates the secretion of milk.

6. A lobed gland located around the windpipe is called the **thyroid**.

7. Sweat glands and other glands with ducts are called **exocrine** glands.

8. Overproduction or underproduction of **growth hormone** in children can cause gigantism or dwarfism.

9. Pituitary peptide hormones that affect other target glands are called **tropic** hormones.

10. Local or circulating hormones are capable of acting on target cells only if target cells are equipped with response components termed **receptors**.

11. The hormone in *Rhodnius* produced by the corpora allata that prevents metamorphosis into an adult is called **juvenile hormone**.

12. The **pituitary gland** is an important endocrine gland derived embryonically from an outpocketing of the mouth region of the digestive tract and a downgrowth of the floor of the brain.

13. The three classes of steroid hormones produced by the adrenal cortex are glucocorticoids, **mineralocorticoids**, and **sex steroids**.

14. Protein hormones do not cross the cell membrane, but **steroid** hormones do.

*15. When a hormone's receptor is on the secreting cell, it is called a(n) **autocrine** message.

16. When people fail to get adequate amounts of iodine in their diet, they develop goiter. The hormone that is nonfunctional is called **thyroxine**.

17. If the circulating levels of calcium are high, **calcitonin** stimulates the osteoblasts to resorb it.

18. The steroid hormones are synthesized from **cholesterol**.

Multiple Choice

1. Which of the following is a local hormone?
   a. Adrenaline
   b. Estrogen
   c. Histamine
   d. Insulin
   e. Thyroxine

■2. The hormones of invertebrates
   a. differ in structure from vertebrate hormones.
   b. are mostly involved in controlling growth.
   c. play no role in molting and metamorphosis.
   d. have different functions from those in vertebrates.
   e. require large quantities to have an effect.

■3. Which of Wigglesworth's observations of *Rhodnius* bugs helped describe the role of insect hormones?
   a. A blood meal triggers molting in these bugs.
   b. If decapitated immediately following a blood meal, the bug will molt.
   c. When two bugs are connected, they molt simultaneously.
   d. When two bugs are connected, the feeding status of one can trigger molting in the other.
   e. When two bugs are decapitated and connected, they will never molt into adults.

4. Insect brain hormone serves to
   a. stimulate the prothoracic gland to release molting hormone.
   b. stimulate the corpora cardiaca to release molting hormone.
   c. directly stimulate molting if food reserves are adequate.

d. inhibit molting until the insect is a certain size.
e. have a general inhibitory effect on insect growth.

5. Which of the following hormones directly stimulates an insect larva to molt?
a. Brain hormone
b. Ecdysone
c. Juvenile hormone
d. Moltin
e. Prolactin

▪6. Why can juvenile hormone be used by humans as an effective control of insect populations?
a. It causes juvenile insects to die.
b. It causes the insects to fail to molt.
c. It causes the insects to fail to develop into adults.
d. It causes even tiny insects to pupate.
e. It causes insects to molt too quickly.

7. The neurohormones vasopressin (antidiuretic hormone) and oxytocin are produced by the
a. anterior pituitary and released by the posterior pituitary.
b. hypothalamus and released by the pituitary.
c. pituitary and signal to the hypothalamus.
d. hypothalamus and signal to the brain.
e. pituitary and signal to the reproductive organs.

8. Which of the following is an effect of oxytocin?
a. Stimulation of uterine contractions at birth
b. Increased reabsorption of water in the kidney
c. Stimulation of tropic hormone release
d. Increased productivity of the hypothalamus
e. Increased rate of ovulation in the ovary

*9. Which of these pairs is a correct match of a pituitary hormone and its target organ?
a. Oxytocin; breast tissue
b. Melanocyte-stimulating hormone; kidney
c. Endorphins; brain
d. Growth hormone; skin and hair
e. Luteinizing hormone; thyroid

10. The best source of pituitary hormones for medical uses today is
a. from the slaughter of sheep or cattle.
b. by extraction from human cadavers.
c. by synthesis from amino acids in the laboratory.
d. from genetically engineered bacteria.
e. by extraction from human blood samples.

▪11. Under which of these conditions would a mammal need to increase thyroxine levels?
a. In a female following childbirth
b. During illness and fever
c. When blood glucose levels are high
d. During sleep and rest
e. When exposed to cold

12. Which of the following would signal a reduction in thyrotropin release?
a. Increased levels of thyrotropin
b. Decreased levels of thyrotropin
c. Increased levels of thyroxine
d. Decreased levels of thyroxine
e. Decreased activity of the thyroid

13. Which type of goiter (enlarged thyroid) can be reduced by addition of iodine to the diet?
a. Hyperthyroid goiter, in which the thyroxine does not turn off the pituitary
b. Hyperthyroid goiter, in which the thyroid is activated
c. Hypothyroid goiter, involving low functional thyroxine and high thyrotropin
d. Hypothyroid goiter, involving high functional thyroxine and low thyrotropin
e. All types of goiter

14. In addition to thyroxine, the mammalian thyroid gland also produces
a. adrenaline.
b. alcitonin.
c. iodine.
d. prolactin.
e. thyrotropin.

15. The parathyroid glands are involved in regulation of blood levels of
a. alcium.
b. glucose.
c. iodine.
d. sodium.
e. thyroxine.

16. Why does a lack of insulin cause diabetes mellitus?
a. Insulin is required for excretion of glucose.
b. Insulin is required for glucose breakdown.
c. Insulin is required for glucose uptake.
d. Insulin is required for converting glucose to glycogen.
e. Insulin is required for synthesizing glucose.

17. Which of the following is *not* produced by the pancreas?
a. Cortisol
b. Glucagon
c. Insulin
d. Somatostatin
e. Digestive enzymes

18. The adrenal medulla develops from nervous tissue and produces the hormone
a. epinephrine.
b. adrenocorticotropin.
c. aldosterone.
d. cholesterol.
e. cortisol.

19. Cortisol, the stress hormone, has all of the following effects, *except*
a. metabolizing fats for energy.
b. increasing blood pressure.
c. slowing down digestion.
d. stimulating the immune response.
e. slowing down protein synthesis.

20. Muscle-building anabolic steroids are known to cause all of the following effects, *except*
   a. irregular menstrual periods in women.
   b. increase in body and facial hair in women.
   c. increase in body and facial hair in men.
   d. breast enlargement in men.
   e. kidney disease.

21. The hormones secreted by the gonads are synthesized from
   a. complex carbohydrates.
   b. amino acids.
   c. cholesterol.
   d. hemoglobin.
   e. nucleic acids.

22. What determines whether a developing mammalian gonad will become an ovary or a testis?
   a. Any gonad with cells containing Y chromosomes will become a testis.
   b. A steady high level of estrogens causes a gonad to become an ovary.
   c. The absence of estrogens causes a gonad to become a testis.
   d. The absence of androgens causes a gonad to become an ovary.
   e. Androgen release at a critical fetal stage causes a testis to develop.

▪23. Which of the following is the earliest event in puberty?
   a. The pituitary produces more gonadotropins
   b. The level of circulating androgens rises in males
   c. Initiation of the menstrual cycle in females
   d. Differentiation of the gonads into testes or ovaries
   e. Increased subcutaneous fat in males

▪24. Why do some hormones (first messengers) need to trigger a "second messenger" to activate a target cell?
   a. The first messenger requires activation by ATP.
   b. The first messenger is not a water-soluble molecule.
   c. The first messenger binds to too many types of cells.
   d. The first messenger cannot cross a plasma membrane.
   e. There are no specific cell surface receptors for the first messenger.

*25. Which of the following is *not* true of cyclic AMP (cAMP)?
   a. It regulates formation of the slug in slime molds.
   b. It is a second messenger.
   c. It mediates responses within cells via phosphorylation of enzymes.
   d. It is found only within cells of vertebrates.
   e. None of the above

▪26. Which of the following statements about hormones is *false*?
   a. The same hormone can cause different responses in different types of cells.
   b. Hormone structure evolves more rapidly than does hormone function.
   c. The receptor for a hormone may be the secretory cell itself.
   d. Pheromones are chemical messages that influence other individuals.
   e. None of the above

*▪27. Which of the following statements about hormones is *false*?
   a. Growth factors are paracrine hormones.
   b. Neurons are an example of paracrine cells.
   c. Hormones often control long-term physiological processes.
   d. All hormones travel in the blood to target cells.
   e. None of the above

*28. Steroid hormones initiate the production of target cell substances in which manner?
   a. They initiate second messenger activity.
   b. They bind with membrane proteins.
   c. They initiate DNA transcription.
   d. They activate enzyme pathways.
   e. They bind with membrane phospholipids.

29. A ________ stimulus releases hormones that produce and coordinate major developmental, physiological, and behavioral changes in the male cichlid fish of Lake Tanganyika. The hormone transforms "wimpy" males into big, brightly colored, aggressive, sexually attractive, "macho" males.
   a. chemical
   b. physical
   c. behavioral
   d. a, b, and c
   e. a and b

30. Hormones are secreted by
   a. endocrine glands like the thyroid gland.
   b. individual cells like those lining portions of the digestive tract.
   c. cells in the nervous system (neurohormones).
   d. a and c
   e. a, b, and c

▪31. Which of the following is the chronological order of events in the molting of *Rhodnius*, as determined by Sir Vincent Wigglesworth?
   a. Blood meal, brain hormone release, and ecdysone release
   b. Blood meal, ecdysone release, and brain hormone release
   c. Ecdysone release, blood meal, and brain hormone release
   d. Brain hormone release, blood meal, and ecdysone release
   e. None of the above

32. Hormones belong to a number of distinct chemical groups. Which of the following is *not* a chemical group to which hormones belong?
   a. Steroids
   b. Proteins
   c. Amino acids and peptides
   d. Carbohydrates
   e. Modified fatty acids

33. Which of the following hormones is *not* produced within the islets of Langerhans of the pancreas?
a. Somatostatin
b. Insulin
c. Glucagon
d. Calcitonin
e. None of the above

34. Sweat and saliva are secretions from
a. the prothoracic gland.
b. target glands.
c. endocrine glands.
d. ductless glands.
e. exocrine glands.

*35. What was the target for the brain hormone demonstrated by Sir Wigglesworth's experiments with Rhodnius?
a. Prothoracic gland
b. Pupal stage
c. Larval stage
d. Corpora cardiaca
e. Corpora allata

36. Hormones released by the posterior pituitary are produced in the
a. anterior pituitary.
b. hypothalamus.
c. pineal.
d. thymus.
e. thyroid.

37. Which of the following hormones is produced by the posterior pituitary gland?
a. Prolactin
b. Oxytocin
c. Endorphins
d. Growth hormone
e. Luteinizing hormone

38. Tropic hormones control other endocrine glands. Which of the following is *not* a tropic hormone?
a. Thyrotropin
b. Adrenocorticotropin
c. Luteinizing hormone
d. Enkephalins
e. Follicle-stimulating hormone

39. Which of the following is an effect of increased levels of vasopressin in the blood?
a. It stimulates the letdown of milk from mammary tissues.
b. It stimulates uterine contractions during birth.
c. It stimulates water conservation by the kidney.
d. It stimulates the liver to produce somatomedins.
e. It stimulates the hypothalamus to produce enkephalins.

40. Which of the following hormones are *not* correctly paired with their target organ?
a. Oxytocin; endocrine adrenal cortex
b. Prolactin; endocrine mammary tissue
c. Endorphin; endocrine spinal cord neurons
d. Vasopressin; endocrine kidneys
e. Luteinizing hormone; endocrine gonads

*▪41. Which of the following statements about growth hormone is *false*?
a. Growth hormone stimulates cells to take up amino acids.
b. Growth hormone is not a tropic hormone.
c. Pituitary dwarfism is now preventable.
d. Neurohormones of the hypothalamus influence release of growth hormone.
e. None of the above

*42. The 1972 Nobel prize in medicine was awarded to Roger Guillemin and Andrew Schally for discoveries relating to
a. the cause of hypopituitary dwarfism.
b. hypothalamic releasing factors.
c. production of growth hormone using genetically engineered bacteria.
d. the cause of acromegaly.
e. the discovery of a cure for hypothyroidism.

43. Malfunctioning of the thyroid gland in the human adult can result in
a. cretinism.
b. diabetes.
c. dwarfism.
d. goiter.
e. hermaphroditism.

44. Which of the following hormones is produced by the parathyroid glands?
a. Parathormone
b. Thyroxine
c. Calcitonin
d. Oxytocin
e. Pitressin

45–55. Match the hormones from the list below with the correct function or action.
a. Insulin
b. Glucagon
c. Epinephrine
d. Somatostatin
e. Cortisol

45. First extracted by Frederick Banting and Charles Best **(a)**

46. Produced by the adrenal medulla **(c)**

47. Released in response to rapid rises of glucose and amino acids in the blood **(d)**

48. Responsible for stimulation of cells to use glucose as a metabolic fuel and to convert excess glucose into fat and glycogen storage **(a)**

49. Responsible for the conversion of glycogen into glucose when serum glucose levels fall **(b)**

50. Inhibits the release of insulin and glucagon **(d)**

51. Slowly released in response to short-term stress **(e)**

52. Rapidly released in response to immediate stress **(c)**

53. Insufficient levels result in diabetes mellitus **(a)**

54. Inhibits the release of growth hormone **(d)**

55. Steroidal hormone synthesized from cholesterol **(e)**

*▪56. Which of the following best describes the function of the corticosteroid hormones?
a. Depresses the immune response
b. Stimulates sexual and reproductive activity
c. Influences blood glucose concentrations
d. Influences ionic and osmotic concentration of the blood
e. All of the above

▪57. Which of the following does *not* describe a major difference between water-soluble and lipid-soluble hormones?
a. Lipid-soluble hormones pass readily through the plasma membrane.
b. Lipid-soluble hormones act by stimulating the synthesis of new kinds of proteins through gene activation.
c. Many water-soluble hormones normally require compounds such as cAMP and cGMP to function properly.
d. Water-soluble hormones exert their effect by altering the activity of enzymes normally present in the target cell.
e. Lipid-soluble hormone action is characterized by a cascade of regulatory steps resulting in an amplification of the effect of a single hormone molecule.

58. Hormones produced by the adrenal cortex are
a. water-soluble.
b. proteins.
c. steroids.
d. carbohydrates.
e. tropic hormones.

59–62. Match the appropriate hormone from the list below with the correct gland, target organ, function, or action.
a. Cholecystokinin
b. Melatonin
c. Aldosterone
d. Testosterone
e. Atrial natriuretic hormone

59. Increases sodium ion excretion **(e)**

60. Increases reabsorption of sodium ions **(c)**

61. An androgen **(d)**

62. Secreted by the pineal **(b)**

▪63. Which of the following is *not* true of a hormone?
a. One hormone can have different effects on different cells.
b. Its actions are as fast as a neural impulse.
c. It can act on the same cell that secretes it.
d. It can enter a cell's nucleus.
e. It is usually present in very small amounts.

*64. Portal blood vessels connect the __________ to the __________.
a. hypothalamus; brain
b. hypothalamus; posterior pituitary
c. hypothalamus; anterior pituitary
d. anterior pituitary; posterior pituitary
e. pancreas; liver

65. The hormone responsible for releasing calcium from bone into the bloodstream is
a. insulin.
b. calcitonin.
c. parathormone.
d. thyroxine.
e. somatostatin.

66. Which of the following statements about sex steroids is *false*?
a. Presence of the Y chromosome causes the embryonic gonads of mammals to produce androgens.
b. Gonads produce hormones and gametes.
c. Androgens are required to trigger male development in mammals and birds.
d. The human embryo has the potential to develop into either a male or a female until about the seventh week of development.
e. None of the above

67. Which of the following statements about biological rhythms is *false*?
a. Melatonin influences biological rhythms.
b. Melatonin is released by the pineal gland during the day.
c. Photoperiodicity is the phenomenon whereby seasonal changes in day length cause physiological changes in animals.
d. Melatonin is inhibited by exposure to light.
e. None of the above

*68. Which of the following statements about regulation of hormone receptors is *false*?
a. Upregulation of receptors is a negative feedback mechanism.
b. Type II diabetes is characterized by a downregulation of insulin receptors.
c. Downregulation is more common than upregulation.
d. The upregulation of FSH receptors accelerates maturation of the ovum.
e. None of the above

69. In order for a cell to be responsive to a lipid-soluble hormone, it must have
a. a specific cell surface receptor.
b. a specific receptor in the cytoplasm or nucleus.
c. cAMP.
d. G protein.
e. a specific DNA sequence that the hormone binds to.

*70. Which of the following statements about responses to hormones is *false*?
   a. A dose-response curve quantifies response to a hormone.
   b. The half-life of epinephrine in the blood is a few minutes.
   c. Hormones are degraded by enzymes in the liver.
   d. Hormones that bind to carrier proteins have shorter half-lives than do those that circulate as free molecules.
   e. None of the above

Study Guide Questions

1. A paracrine hormone is
   a. a local hormone that acts on the cell that released it.
   b. always acting on a wide variety of target tissues.
   c. a local hormone produced at one site but active at a different site in the body.
   d. None of the above

2. A target cell's response to a hormone depends on
   a. the amount of hormone reaching that target cell.
   b. the number of receptors on that target cell.
   c. the target cell's receptor-affinity for hormone.
   d. All of the above

3. "Upregulation" of hormone receptors refers to
   a. increase in hormone receptor numbers with low hormone levels.
   b. increase in hormone receptor numbers with high neurotransmitter levels.
   c. increase in hormone levels produced by increase in hormone receptor numbers.
   d. decrease in hormone levels produced by decrease in hormone receptor numbers.

4. Hormones have various regulatory functions. Which of the following statements does not describe how a hormone functions?
   a. Hormones act in very low concentration.
   b. Hormones act at sites distant from where they are produced.
   c. Hormones are transported in the blood.
   d. None of the above

5. Which of the following is a common "second messenger" substance in hormone action?
   a. Thyroid hormone
   b. ADH
   c. Cyclic AMP
   d. Epinephrine

6. A hormone circulating in the blood with target tissues 1, 2, and 3 could act selectively on only target tissue 1 if
   a. target tissue 1 had fewer hormone receptors than 2 and 3.
   b. target tissues 2 and 3 had hormone receptors with a higher hormone-affinity than 1.
   c. target tissue 1 had hormone receptors with a higher hormone-affinity than 2 and 3.
   d. the hormone was released only in the vicinity of target tissue 1.

7. Which of the following glands is considered the "master" endocrine gland in vertebrates?
   a. Adrenal glands
   b. Thyroid gland
   c. Hypothalamus
   d. Pituitary gland

8. Which gland is involved in the function of our "biological clock"?
   a. Pineal gland
   b. Thyroid gland
   c. Adrenal glands
   d. Ovaries

9. The half-life of hormones in the plasma ranges from
   a. minutes to hours.
   b. minutes to days.
   c. days to weeks.
   d. hours to months.

10. The target tissues of hormones are those tissues that
   a. the particular hormone can actually penetrate.
   b. have specific enzymes with which hormones directly interact.
   c. have high concentrations of the "second messenger."
   d. have receptors for the particular hormone.

11. Which of the following sets of vertebrate hormones are all produced in the anterior pituitary gland?
   a. Somatostatin, vasopressin, insulin
   b. Prolactin, growth hormone, enkephalins
   c. Oxytocin, prolactin, adrenocorticotropin
   d. Estrogen, progesterone, testosterone

12. Hormones that are secreted by one endocrine gland and control the activities of another endocrine gland are called
   a. growth hormones.
   b. obstructive.
   c. tropic.
   d. selective.

13. Which of the following secretes hormones but is not considered a traditional glandular tissue?
   a. Pancreas
   b. Heart
   c. Testes
   d. Adrenal glands

End of Chapter Questions

1. Which statement is true for all hormones?
   a. They are secreted by glands.
   b. They have receptors on cell surfaces.
   c. They may stimulate different responses in different cells.
   d. They target cells distant from their site of release.
   e. When the same hormone occurs in different species, it has the same action.

2. The hormone ecdysone
   a. is released from the posterior pituitary.
   b. stimulates molt and metamorphosis in insects.
   c. maintains an insect in larval stages unless brain hormone is present.
   d. stimulates the secretion of juvenile hormone from the prothoracic glands.
   e. keeps the insect exoskeleton flexible to permit growth.

3. The posterior pituitary
   a. produces oxytocin.
   b. is under the control of hypothalamic releasing neurohormones.
   c. secretes tropic hormones.
   d. secretes neurohormones.
   e. is under feedback control by thyroxine.

4. Growth hormone
   a. can cause adults to grow taller.
   b. stimulates protein synthesis.
   c. is released by the hypothalamus.
   d. can be obtained only from cadavers.
   e. is a steroid.

5. Both epinephrine and cortisol are secreted in response to stress. Which of the following statements is also true for both of these hormones?
   a. They act to increase blood glucose.
   b. Their receptors are on the surfaces of target cells.
   c. They are secreted by the adrenal cortex.
   d. Their secretion is stimulated by adrenocorticotropin.
   e. They are secreted into the blood within seconds of the onset of stress.

6. Before puberty
   a. the pituitary secretes luteinizing hormone and follicle-stimulating hormone, but the gonads are unresponsive.
   b. the hypothalamus does not secrete much gonadotropin-releasing hormone.
   c. males can stimulate massive muscle development through a vigorous training program.
   d. testosterone plays no role in development of the male sex organs.
   e. genetic females will develop male genitals unless estrogen is present.

7. Which of the following contributes to a long half-life for a circulating hormone?
   a. The number of receptors on the target cells.
   b. The fact that it is lipophobic.
   c. The sensitivity to the hormone as expressed by its dose-response curve.
   d. Its binding to carrier proteins in the blood.
   e. A rapid rate of uptake by liver cells.

8. Steroid hormones
   a. are all produced by the adrenal cortex.
   b. have only cell surface receptors.
   c. are lipophobic.
   d. act through altering the activity of proteins in the target cell.
   e. act through stimulating the production of new proteins in the target cell.

9. Which of the following is a likely cause of goiter?
   a. The thyroid gland is producing too much parathormone.
   b. Circulating levels of thyrotropin are too low.
   c. There is an inadequate supply of functional thyroxine.
   d. There is an oversupply of functional thyroxine.
   e. The diet contains too much iodine.

10. Parathormone
    a. stimulates osteoblasts to lay down new bone.
    b. reduces blood calcium levels.
    c. stimulates calcitonin release.
    d. is produced by the thyroid gland.
    e. is released when blood calcium levels fall.

Student Web Site Self-Quiz Questions

1. Which of the following statements about chemical messages that are classified as circulating hormones is *false*?
   a. Hormones can have more than one type of target cell.
   b. Different cells can respond differently to the same hormone.
   c. Hormones must be present in high concentrations to cause their intended effect.
   d. The action of circulating hormones is rapid.
   e. A given hormone may have very different functions in different tissues.

2. The endocrine system of humans is specialized at secreting substances. Which of the following is a key distinguishing feature of endocrine glands?
   a. They release non-hormone secretions through ducts that lead outside of the body.
   b. They are present in the brain.
   c. They secrete their products directly into the blood.
   d. They secrete hormones and lack ducts.
   e. Each person has a total of 10 discrete endocrine glands in the body.

3. Experiments by Wigglesworth with the bug *Rhodnius* have led to our current understanding of the hormonal control of insect growth and development. Select one of the following choices that is *not* the expected outcome of the experimental treatment.
   a. A third instar bug that is decapitated one week after a blood meal—molts into a miniature adult.
   b. A third instar bug that is decapitated one hour after a blood meal—does not molt.
   c. A third instar bug that is decapitated one hour after a blood meal and injected with ecdysone—molts into a miniature adult.
   d. An intact fourth instar bug is given a blood meal and injected with the ground corpora allata of a first instar bug—molts into an adult.
   e. A third instar bug decapitated one hour after a blood meal is joined to a fourth instar bug which had been decapitated one week after a blood meal—both molt into adults.

4. Which of the following statements regarding complete metamorphosis is *false*?
   a. If juvenile hormone concentrations are high, the larvae develop and molt into pupae.
   b. When juvenile hormone concentration drops, larvae develop into pupae.
   c. In pupae, when juvenile hormone is absent, they molt into adults.
   d. Juvenile hormone can be used as a form of pest control.
   e. The structure of juvenile hormone is highly conserved.

5. Regarding the thyroid and parathyroid glands, a decrease in blood calcium would stimulate the __________ to release its hormone, which would activate __________.
   a. thyroid, osteoblasts
   b. thyroid, osteoclasts
   c. parathyroid, osteoblasts
   d. parathyroid, osteoclasts
   e. thyroid, triiodothyronine (T_3)

6. After a vigorous hike during which your appetite increases, you eat a large meal and then relax. If you were to arrange the following events into a chronological order, which event would be fourth? Note: begin immediately after the hike, when your blood glucose levels are below normal.
   a. High blood glucose levels
   b. Conversion of glycogen to glucose
   c. Pancreas secretes insulin
   d. Pancreas secretes glucagons
   e. Conversion of glucose to glycogen

7. The adrenal gland would produce which of the following pairs of hormones in response to being chased by a grizzly bear?
   a. Epinephrine, aldosterone
   b. Epinephrine, cortisol
   c. Epinephrine, adrenocorticotropin
   d. Adrenocorticotropin releasing hormone, cortisol
   e. Progesterone, aldosterone

8. The hormone melatonin is involved in biological rhythms and photoperiodicity. Melatonin is produced in the __________.
   a. Hypothalamus
   b. Thyroid
   c. Pancreas
   d. Adrenal cortex
   e. Pineal gland

TEST FILE

★ *Indicates a **difficult** question*

■ *Indicates a **conceptual** question*

42 Animal Reproduction

Fill in the Blank

1. A behavior that transfers sperm into the female's reproductive tract is **copulation**.

2. The uterus opens into the vagina at a muscular neck region called the **cervix**.

3. A **follicle** consists of one egg cell and the surrounding ovarian cells.

4. In humans, fertilization typically occurs when the egg is located in the **oviduct**.

5. Females of most mammals have a period of sexual receptivity called **estrus**.

6. The corpus luteum secretes the hormones **progesterone** and **estrogen**.

7. Following fertilization, the human chorionic gonadotropin secreted by the **blastocyst or embryo** keeps the corpus luteum functional.

8. Birth control pills contain synthetic hormones that prevent the ovarian cycle through negative feedback to the **pituitary**.

■9. A major evolutionary step allowing the first reptiles to succeed on land was the **shelled egg**.

10. **Artificial insemination** is a reproductive technology that involves placing sperm in the correct location in the female reproductive tract for fertilization to occur. This technique is commonly used in the production of livestock.

11. As opposed to viviparous animals, **oviparous** animals lay eggs in the environment, and their offspring go through embryonic stages outside the body of the mother.

12. Very early in the life of a new mammalian embryo, a population of cells arises during the first few cell divisions. These special cells, called **germ cells**, ultimately end up in the region of the developing gonads where they populate and become gametes.

13. The **ejaculate or semen** of the human male contains sperm and secretions from accessory glands, seminal vesicles, and the prostate gland.

14. **Parthenogenesis** is a specialized type of asexual reproduction in which offspring develop from unfertilized eggs.

15. Some vertebrates have a **cloaca**, a common opening for products of the digestive, urinary, and reproductive systems.

16. The male sexual response includes a **refractory** period following orgasm, in which he cannot achieve a full erection.

Multiple Choice

1. What is a disadvantage of asexual reproduction?
   a. Only mitosis is necessary for cell division.
   b. Populations can grow until limited by resources.
   c. Single individuals can produce offspring.
   d. All the offspring are identical in a changing environment.
   e. No energy expenditure is required for mating and fertilization.

■2. In order for an organism to reproduce by regeneration, what must be true?
   a. Each cell must be able to give rise to a new organism.
   b. A fragment must be broken off at a particular place.
   c. A fragment must contain all essential tissues.
   d. Gamete formation must occur.
   e. From one individual, two are developed.

■3. Parthenogenetic reproduction involves
   a. meiosis but not fertilization.
   b. neither meiosis nor fertilization.
   c. fertilization but not meiosis.
   d. both meiosis and fertilization.
   e. identical copies of the same fertilized egg.

■4. In populations that alternate between periods of asexual and periods of sexual reproduction, when would asexual reproduction be most advantageous?
   a. When adults of both sexes are numerous
   b. During the most stressful season
   c. When there is a threat of hybridization
   d. When conditions are favorable and stable
   e. When the population is most dense

*5. In male gametogenesis, the second meiotic division produces four haploid
a. germ cells.
b. primary spermatocytes.
c. secondary spermatocytes.
d. spermatids.
e. spermatogonia.

*6. Unlike spermatogenesis, oogenesis in humans
a. is continuous over the life of the woman.
b. begins when the woman is a fetus.
c. produces four haploid gametes.
d. occurs at a more rapid rate.
e. results in swimming cells.

7. Hermaphrodites are organisms that
a. possess both male and female reproductive systems.
b. breed for a time as one sex, then change to the other.
c. develop offspring from unfertilized eggs.
d. usually fertilize themselves.
e. have abnormal reproductive organs.

▪8. Reproductive systems with external fertilization are most common in
a. terrestrial animals.
b. populations with many more males.
c. animals that are sessile.
d. animals producing few gametes.
e. animals that are widely dispersed.

9. For what organisms is a penis necessary?
a. For all male animals
b. For all but hermaphrodites
c. For species with external fertilization
d. For species with internal fertilization
e. For species whose courtship precedes mating

10. What is the advantage of mammalian testes being located in a sac outside the body cavity?
a. A shorter distance for semen to be ejaculated
b. A shorter distance for sperm to swim
c. The testes can be held at a constant temperature
d. The body temperature is too high for sperm production
e. It allows testis enlargement with sexual maturity

11. How does the penis become erect during arousal in human males?
a. A bone is moved into place.
b. Spongy tissue is engorged.
c. Smooth muscles contract.
d. By emission of glandular fluid.
e. Intercellular fluid accumulates.

*12. Which part of the testis produces the male sex hormones?
a. Epididymis
b. Leydig cells
c. Scrotum
d. Seminiferous tubules
e. Vas deferens

*13. Which of the following is *not* true of mammals?
a. All mammals have internal fertilization.
b. Some species of mammals are oviparous.
c. Most mammals are eutherians.
d. Some species of mammals are ovoviviparous.
e. None of the above

14. A human female has the maximum number of primary oocytes in her ovaries
a. at birth.
b. just prior to puberty.
c. early in her fertile years.
d. midway through her fertile years.
e. at menopause.

15. Following ovulation, if the egg is not fertilized, what becomes of the follicle cells?
a. They move with the egg in the oviduct.
b. They degenerate.
c. They grow into a corpus luteum.
d. They stop secreting hormones.
e. They begin to develop a new egg.

16. By what means does the egg move through the oviduct?
a. It is propelled by cilia on its surface.
b. It is propelled by its flagellum.
c. It is propelled by oviduct contractions.
d. It is propelled by cilia lining the oviduct.
e. It is propelled by its amoeboid motion.

▪17. In which way is the human female different from females of other mammalian species?
a. She cycles at the slowest rate per year.
b. She reabsorbs the uterine lining.
c. She must copulate in order to ovulate.
d. She can be continuously sexually receptive.
e. Her cycles are under hormonal control.

18. Female reproductive events are coordinated by two pituitary gonadotropins,
a. follicle-stimulating hormone and estrogen.
b. estrogen and progesterone.
c. follicle-stimulating hormone and progesterone.
d. luteinizing hormone and progesterone.
e. luteinizing hormone and follicle-stimulating hormone.

19. In the early half of the menstrual cycle, before the events of ovulation, which hormone is at its highest level in the blood?
a. Estrogen
b. Follicle-stimulating hormone
c. Luteinizing hormone
d. Progesterone
e. Testosterone

20. What triggers ovulation and corpus luteum formation?
a. A peak in estrogen
b. A peak in luteinizing hormone
c. A peak in progesterone
d. Presence of sperm in the reproductive tract
e. Readiness of the endometrium lining the uterus

*21. Why do new follicles not mature as long as the corpus luteum is maintained?
   a. Corpus luteum hormones inhibit gonadotropin release.
   b. Corpus luteum hormones inhibit ovarian secretion.
   c. The ovary receives negative feedback from the uterine endometrium.
   d. The ovary receives negative feedback as long as the egg is viable.
   e. New follicle development depends upon corpus luteum hormones.

22. In the female sexual response, the structure that is most analogous to the male's penis is the
   a. breast.
   b. clitoris.
   c. labium.
   d. uterus.
   e. vagina.

23. With a few exceptions, mammals are
   a. oviparous.
   b. ovoviviparous.
   c. viviparous.
   d. multiparous.
   e. marsupial.

24. Which of the following statements about the prostate gland is *false*?
   a. Semen contains secretions from the prostate gland.
   b. Benign prostate hyperplasia (BHP) results from enlargement of the prostate gland.
   c. Prostate cancer is the second most common cancer in men.
   d. The prostate gland surrounds the urethra as it leaves the bladder.
   e. None of the above

▪25. Which sexually transmitted disease can be cured/treated with antibiotics?
   a. AIDS
   b. Hepatitis B
   c. Genital warts
   d. Genital herpes
   e. Chlamydia

26. The intrauterine device is a means of birth control that prevents
   a. ovulation.
   b. fertilization.
   c. implantation.
   d. ejaculation.
   e. sperm reaching the egg.

▪27. Which of the following is *not* a part of asexual reproduction?
   a. Identical Genetics
   b. Fertilization
   c. Parthenogenesis
   d. Regeneration
   e. Budding

28. Sexual reproduction has an evolutionary advantage since it
   a. promotes both males and females of a species.
   b. does not allow for rapid reproduction.
   c. promotes genetic variability to cope with changes in the environment.
   d. is controlled by many hormonal mechanisms.
   e. protects and nurtures the embryo.

*29. Which of the following is the correct order of structures that sperm passes through during copulation?
   a. Vas deferens, seminiferous tubules, epididymis, urethra
   b. Epididymis, seminiferous tubules, vas deferens, urethra
   c. Seminiferous tubules, vas deferens, epididymis, urethra
   d. Seminiferous tubules, epididymis, vas deferens, urethra
   e. Vas deferens, urethra, epididymis, seminiferous tubules

*30. Semen, which is the fluid matrix for sperm during emission and ejaculation, contains all of the following *except*
   a. seminal fluid from the seminal vesicles.
   b. glucose to serve as an energy source for the sperm.
   c. alkaline secretions from the prostate gland.
   d. hormone-like substances termed prostaglandins.
   e. clotting enzymes.

31. Which of the following is a probable advantage of hermaphroditism?
   a. For a given species, every organism is a potential mate.
   b. The reduction of cultural strife resulting from societies that contain individuals of only one sex.
   c. Simpler behavior patterns are required for mate selection.
   d. More rapid reproduction of genetically successful individuals is possible through asexual reproduction.
   e. Complex hormonal control and feedback mechanisms are not required.

32–37. Choose the appropriate cell stage from the list below. There may or may not be more than one correct answer.
   a. Oogonia
   b. Spermatogonia
   c. Spermatid
   d. Primary spermatocyte
   e. Secondary spermatocyte

*32. Germ cell from the female gonad **(a)**

*33. Diploid germ cell of the male gonad **(b)**

*34. Cell resulting from the initial mitotic division of a germ cell **(d)**

*35. First haploid germ cell produced in spermatogenesis **(e)**

*36. Cell resulting from the second meiotic division **(c)**

*37. The terminal cell type that, at the end of spermatogenesis, differentiates into a sperm cell **(c)**

38–42. Choose the appropriate cell stage from the list below. There may or may not be more than one correct answer.
a. Second polar body
b. Primary oocyte
c. Secondary oocyte
d. First polar body
e. Ootid

*38. Stage during which the energy, raw materials, and RNA needed for the first cell divisions after fertilization is acquired **(b)**

*39. First essential haploid germ cell produced during oogenesis **(c)**

*40. Daughter cell to the secondary oocyte **(a and e)**

*41. Largest haploid cell resulting from second meiotic division of oogenesis **(e)**

*42. The terminal cell type that, at the end of oogenesis, differentiates into the mature ovum **(e)**

43. After spermatogenesis, sperm cells are generally stored in the
a. spermatophore.
b. prostate gland.
c. vas deferens.
d. seminiferous tubule.
e. epididymis.

▪44. Progeny inherit all the characteristics of a single parent through
a. copulation.
b. asexual reproduction.
c. gametogenesis.
d. sexual reproduction.
e. fertilization.

45. Generally, among vertebrates
a. cells are haploid during embryological development, becoming diploid at birth.
b. cells are diploid during embryological development, becoming haploid at birth.
c. cells are diploid during adult life, haploid as a blastocyst.
d. only gametic cells are haploid.
e. only gametic cells are diploid.

46. Progesterone and estrogen are hormones produced by the
a. anterior pituitary.
b. posterior pituitary.
c. Sertoli cells.
d. hypothalamus.
e. corpus luteum.

47. The blastocyst normally implants within the
a. vas deferens.
b. ovarian follicle tissue.
c. endometrium of the uterus.
d. myometrium of the vagina.
e. chorion.

48. Which of the following describes the principal difference between menstrual and estrous cycles?
a. The estrous cycle occurs if the female is fertilized, and the menstrual cycle occurs if the female isn't fertilized.
b. The menstrual cycle occurs if the female is fertilized, and the estrous cycle occurs if the female isn't fertilized.
c. The estrous cycle lacks menstruation.
d. The menstrual cycle occurs in mammals; the estrous cycle occurs in birds.
e. Only the menstrual cycle is controlled by estrogen.

49. Which of the following is the most effective form of birth control?
a. Vasectomy
b. Condom
c. Intrauterine device
d. Rhythm method
e. "The pill"

50. A vasectomy is a minor operation that involves removing a small section of the ________ and tying off the loose ends.
a. epididymis
b. urethra
c. oviduct
d. vas deferens
e. seminiferous tubules

51. In most male mammals, what anatomical feature is shared by both the reproductive and excretory systems?
a. Prostate gland
b. Seminiferous tubules
c. Vas deferens
d. Urethra
e. Epididymis

52. In mammals, which is the correct sequence of structures that sperm passes through before fertilizing an egg?
a. Uterus, vagina, cervix, oviduct, ovary
b. Vagina, cervix, uterus, oviduct
c. Oviduct, uterus, vagina, ovary, body cavity
d. Vagina, cervix, oviduct, uterus
e. Vagina, uterus, oviduct, body cavity

53. The hormone that normally triggers mammalian ovulation is
a. progesterone.
b. follicle-stimulating hormone.
c. luteinizing hormone.
d. estrogen.
e. prolactin.

*54. Which of the following statements about human reproduction is *false*?
a. Women are most likely to be fertile from day 10 to day 20 of the ovarian cycle.
b. Ovulation can be detected by tracking body temperature.
c. An ovum remains viable for 2 to 3 days after ovulation.
d. Sperm may survive for up to 6 days in the female reproductive tract.
e. None of the above

55. The birth control method that prevents ovulation is
a. a vasectomy.
b. an intrauterine device.
c. a condom.
d. the rhythm method.
e. "the pill."

56. Most methods of birth control focus on
a. preventing oogenesis.
b. preventing spermatogenesis.
c. preventing fertilization.
d. decreasing libido.
e. decreasing testosterone levels.

57. The end result of spermatogenesis is the production of four haploid spermatids. The end result of oogenesis is the production of
a. four haploid ootids.
b. two haploid ootids.
c. one diploid ootid.
d. two diploid ootids.
e. one haploid ootid with two polar bodies.

58. The terms "monoecious" and "dioecious" refer to
a. patterns of inheritance.
b. circulatory physiology.
c. patterns of cleavage in developing embryos.
d. the evolutionary origin of species.
e. male and female reproductive systems.

*59. At birth, a female has about one million primary oocytes in each ovary. During a woman's fertile years, about ________ of these oocytes will mature completely into eggs and be released at ovulation.
a. 100
b. 200
c. 300
d. 450
e. None of the above

60. Which of the following is *not* true of asexual reproduction?
a. It occurs mostly in invertebrates.
b. Asexually reproducing species tend to live in variable environments.
c. Asexually reproducing species tend to be sessile.
d. Asexually reproducing species live in sparse populations.
e. One method is parthenogenesis.

61. The contraceptive method that also helps prevent the spread of sexually transmitted disease is
a. the diaphragm.
b. the cervical cap.
c. the condom.
d. the IUD.
e. "the pill."

62. The contragestational pill, RU-486, blocks ________ receptors, preventing implantation of the embryo in the uterine lining.
a. luteinizing hormone
b. progesterone
c. estrogen
d. follicle-stimulating hormone
e. gonadotropin-releasing hormone

63. Pregnancy tests detect the presence of which hormone?
a. Luteinizing hormone
b. Follicle-stimulating hormone
c. Human chorionic gonadotropin
d. Estrogen
e. Progesterone

*64. Which of the following methods of contraception prevents the egg from entering the uterus?
a. Coitus interruptus
b. Condom
c. RU-486
d. Diaphragm
e. Tubal ligation

*65. In which of the following procedures does fertilization occur within the female reproductive tract?
a. In vitro fertilization
b. Gamete intrafallopian transfer
c. Intracytoplasmic sperm injection
d. b and c
e. None of the above

66. Some species of mites and scorpions use indirect fertilization, excreting a gelatinous container of sperm called a(n)
a. spermatophore.
b. spermatogonium.
c. spermatid.
d. cloaca.
e. amplexus.

67. The human sexual response consists of four phases in the order
a. excitement, refractory, plateau, and orgasm.
b. plateau, excitement, orgasm, and resolution.
c. plateau, excitement, refractory, and orgasm.
d. excitement, plateau, orgasm, and resolution.
e. excitement, plateau, refractory, and orgasm.

68. Gamete intrafallopian transfer is
a. removing eggs from a woman's oviduct and fertilizing them in a culture medium.
b. moving eggs from the ovaries to the oviduct when the ovaries are blocked.

c. injecting eggs and sperm into the oviduct when the oviduct entrance is blocked.
d. injecting an embryo into the oviduct.
e. removing an embryo form the oviduct and injecting it into the uterus when the oviducts are blocked.

Study Guide Questions

1. Asexual reproduction is an effective strategy in stable environments because
   a. gametogenesis is most efficient under these conditions.
   b. the offspring, genetically identical to their parents, are preadapted to the environment.
   c. asexual parthenogenesis produces a large amount of genetic diversity.
   d. animal cells tend to be more totipotent under stable conditions.

2. An important difference between a sperm and an egg is
   a. their size.
   b. the amount of cytoplasm they contain.
   c. whether or not they are motile.
   d. All of the above

3. Which of the following would *not* be present in the body or lifestyle of a terrestrial male animal that reproduces sexually?
   a. Internal fertilization
   b. A secondary sexual organ as part of the genitalia
   c. Synchrony of reproductive physiology with the female
   d. Oogenesis

4. Which of the following best represents the normal path of a sperm cell as it makes its way from the point of entry into the female's reproductive tract to the place where fertilization typically occurs?
   a. Cervix, vagina, ovary, oviduct
   b. Vagina, cervix, uterus, oviduct
   c. Uterus, cervix, vagina, oviduct
   d. Vagina, uterus, cervix, oviduct

5. The function of the seminal vesicle is to
   a. produce a solution of fructose to provide energy for the mitochondria of the sperm.
   b. secrete alkaline fluids that neutralize the acidity of the female's reproductive tract.
   c. initiate the muscular contractions leading to emission.
   d. produce prostaglandins that stimulate contractions of the male reproductive organs.

6. Which of the following is an example of positive feedback control in the reproductive cycle of males or females?
   a. The increased response of the hypothalamus and anterior pituitary gland in response to estrogen
   b. The decreased response of the hypothalamus and anterior pituitary gland in response to estrogen
   c. The inhibition of luteinizing hormone by high levels of testosterone
   d. The stimulation of luteinizing hormone by low levels of testosterone

7. Which of the following statements about oogenesis is *false*?
   a. The polar bodies degenerate after the second meiotic division.
   b. The ovum produced is haploid.
   c. The major growth phase of the primary oocyte occurs in prophase I.
   d. The primary oocyte is haploid.

8. Which of the following is not an accessory sex organ?
   a. Penis
   b. Uterus
   c. Gonads
   d. Cloaca

9. For approximately how long during the human female's menstrual cycle are progesterone concentrations high enough to maintain the uterus in a proper condition for pregnancy?
   a. All of the cycle
   b. None of the cycle
   c. During the first half of the cycle
   d. During the second half of the cycle

10. Which of the following is *not* true about the birth control pill?
    a. The pill works by preventing ovulation.
    b. The pill works by preventing implantation.
    c. The ovarian cycle is suspended by the birth control pill.
    d. The birth control pill contains low doses of estrogen and progesterone.

11. Which one of the following is not a biological or medical reality for human reproduction?
    a. Negative feedback control of the hypothalamus by sex hormones
    b. Contraception through complete chemical blockage of sperm production
    c. Surgical transfer of gametes from one area of the reproductive tract to another
    d. Susceptibility of reproductive organs to cancer

12. If you compared the genetic makeup of an animal produced by parthenogenesis with that of its mother, which of the following would you expect?
    a. About 100 percent genetic similarity
    b. About 50 percent genetic similarity
    c. No genetic similarity
    d. Parthenogenetic animals have no mother.

13. Which of the following definitions about oviparity is *false*?
    a. Only birds and reptiles are oviparous.
    b. The large amount of yolk provides the nutrients for the developing embryo.
    c. The shell protects the egg from dehydrating.
    d. Both oxygen and carbon dioxide can diffuse through the shell.

14. During spermatogenesis a single male germ cell produces _________ sperm cell(s).
    a. 1
    b. 2
    c. 3
    d. 4

End of Chapter Questions

1. Match each of the following modes of asexual reproduction with the statement or description that characterizes it. (Each letter may be used more than once, and more than one letter may apply to each statement.)
    a. Budding
    b. Regeneration
    c. Parthenogenesis
    (i) This form of asexual reproduction usually follows an animal's being broken by an external force, but it can also be initiated by the animal itself. **(b)**
    (ii) Many freshwater sponges produce clusters of undifferentiated cells that eventually "escape" the parent and become free-living organisms genetically identical to the parent. **(a)**
    (iii) Offspring develop from unfertilized eggs. **(c)**
    (iv) This process requires totipotent cells. **(a, b, c)**
    (v) Species that reproduce in this way may also engage in sexual reproduction. **(a, b, c)**

2. A species in which the individual possesses both male and female reproductive systems is termed (choose all that apply)
    a. dioecious.
    b. parthenogenetic.
    c. hermaphroditic.
    d. diploid.
    e. monoecious.

3. The major advantage of internal fertilization is that
    a. it ensures paternity.
    b. it permits the fertilization of many gametes.
    c. it reduces the incidence of destructive competitive interactions between the members of a group.
    d. it results in the formation of a stable pair bond between mates.
    e. it gives the developing organism a greater degree of protection during the early phases of development.

4. Which statement about oocytes is true?
    a. At birth, the human female has produced all the oocytes she will ever produce.
    b. At the onset of puberty, ovarian follicles produce new ones in response to hormonal stimulation.
    c. At the onset of menopause, the human female stops producing them.
    d. They are produced by the human female throughout adolescence.
    e. Those produced by the female are stored in the seminiferous tubules.

5. Spermatogenesis and oogenesis differ in that
    a. spermatogenesis produces gametes with greater energy stores than those produced by oogenesis.
    b. spermatogenesis produces four equally functional diploid cells per meiotic event and oogenesis does not.
    c. oogenesis produces four equally functional haploid cells per meiotic event and spermatogenesis does not.
    d. spermatogenesis produces many gametes with meager energy reserves, whereas oogenesis produces relatively few, well-provisioned gametes.
    e. in humans, spermatogenesis begins before birth, whereas oogenesis does not start until the onset of puberty.

6. Semen contains all of the following except
    a. fructose.
    b. mucus.
    c. clotting enzymes.
    d. substances to reduce the pH of the uterine environment.
    e. substances to increase the motility of the uterine muscles.

7. During oogenesis in mammals, the second meiotic division occurs
    a. in the formation of the primary oocyte.
    b. in the formation of the secondary oocyte.
    c. before ovulation.
    d. after fertilization.
    e. after implantation.

8. One of the major differences between the sexual response cycles in human males and females is
    a. the increase in blood pressure in males.
    b. the increase in heart rate in females.
    c. the presence of a refractory period in females after orgasm.
    d. the presence of a refractory period in males after orgasm.
    e. the increase in muscle tension in males.

9. Which of the following is true of sexually transmitted diseases?
    a. They are always caused by viruses or bacteria.
    b. Using any form of contraception will prevent them.
    c. The organisms that cause them have evolved to depend on intimate physical contact as their means of transmission.
    d. Their mode of transmission has a high probability of failure.
    e. You cannot catch one from someone you love.

10. Contractions of muscles in the uterine wall and in the breasts are stimulated by
    a. progesterone.
    b. estrogen.
    c. prolactin.
    d. oxytocin.
    e. human chorionic gonadotropin.

Student Web Site Self-Quiz Questions

1. Budding in hydra is an example of asexual reproduction in animals. Which of the following statements about asexual reproduction in animals is *false*?
   a. Asexual reproduction is common among sessile organisms.
   b. Asexually reproducing species are more likely to be found in relatively constant environments.
   c. New individuals formed through budding and regeneration arise from undifferentiated cells that have not lost totipotency.
   d. Parthenogenesis is a type of asexual reproduction involving development of offspring from unfertilized eggs.
   e. In some parthogenetic species, the female can reproduce sexually or asexually.

2. In humans, what stage of female reproductive cell is fertilized by a sperm?
   a. Oogonium
   b. Primary oocyte
   c. Secondary oocyte
   d. Ootid
   e. Ovum

3. Earthworms have both testes and ovaries but mate to exchange sperm. Select the most specific term to describe this condition.
   a. Dioecious
   b. Monoecious
   c. Sequential hermaphroditism
   d. Simultaneous hermaphroditism
   e. Amplexus

4. The oviduct (fallopian tube) in the female reproductive system carries eggs to the uterus. What analogous male reproductive structure transports sperm from the testes?
   a. Epididymis
   b. Seminal vesicle
   c. Ureter
   d. Urethra
   e. Vas deferens

5. Reproductive organs of the male and female respond to hormones produced by the anterior pituitary, but also may have their own endocrine functions. Which of the following is *not* a correct pairing of a hormone and its source in a reproductive structure?
   a. Estrogen—follicle cells
   b. Inhibin—Sertoli cells
   c. Luteinizing hormone (LH)—endometrium
   d. Progesterone—corpus luteum
   e. Testosterone—Leydig cells

6. Which feature of the male sexual response is *not* present in the female?
   a. Excitement
   b. Orgasm
   c. Plateau
   d. Refractory period
   e. Resolution

7. Most birth control pills consist of synthetic forms of estrogen and progesterone. Most of these oral contraceptives work by
   a. maintaining a corpus luteum in a mature condition and suspending the ovarian cycle.
   b. keeping levels of gonadotrophins low.
   c. acting on the ovary to block formation of primary oocytes.
   d. inducing menopause.
   e. suspending the uterine cycle.

8. If we define contraception as the prevention of fertilization, then which of the following birth control strategies is *not* a method of contraception?
   a. Vasectomy (right)
   b. Rhythm method
   c. Intrauterine device
   d. Coitus interruptus
   e. Birth control pills

9. A variety of reproductive technologies have been used to assist couples who have had trouble conceiving a child. Which of the following is *not* a correct pairing of reproductive technologies and descriptions?
   a. In vitro fertilization—Eggs harvested from female, fertilized in lab, replaced into female.
   b. Gamete intrafallopian transfer (GIFT)—harvested eggs and sperm co-injected into oviduct.
   c. Artificial insemination—insertion of sperm into female reproductive tract.
   d. Assisted reproductive technologies (ART)—injection of sperm directly into the oviduct for enhanced fertilization
   e. Intracytoplasmic sperm injection (ICSI) – harvested eggs are injected directly with a sperm cell

TEST FILE

★ *Indicates a **difficult** question*

■ *Indicates a **conceptual** question*

43 Animal Development

Fill in the Blank

1. The time from conception to birth, or the period of pregnancy, is called **gestation**.

★2. The primitive streak of bird and mammalian embryos is analogous to the **blastopore** of amphibians and sea urchins.

★3. The cells that surround unfertilized mammalian eggs when they first arrive in the oviducts are **cumulus** cells.

4. The glycoprotein shell that surrounds the eggs of many mammalian species is called **zona pellucida**.

5. The membrane of an egg is sometimes called the **vitelline** envelope.

6. The **acrosome** contains enzymes that assist the sperm in penetrating the zona pellucida.

★7. The region of the frog egg that is darkly pigmented is known as the **animal** pole.

8. **Spina bifida** is a birth defect caused by the failure of the closure of the posterior region of the neural tube.

9. The process in development by which germ layers form and take specific positions relative to each other is **gastrulation**.

10. The rod that forms from the chordomesoderm and gives structural support to the developing embryo is called the **notochord**.

★11. In chicks, the first extraembryonic membrane to form is the **yolk sac**.

★12. In mammals, the first extraembryonic membrane to form is the **trophoblast**.

Multiple Choice

■1. Development occurs
   a. only during growth of the organism.
   b. throughout the life of the organism.
   c. only in nondividing cells.
   d. in ectoderm and endoderm, but not in mesoderm.
   e. only in animals.

2. Bird eggs have large amounts of yolk. This results in
   a. gradual metamorphosis.
   b. complete metamorphosis.
   c. incomplete cleavage.
   d. incomplete mitosis during cleavage.
   e. bicoid larvae.

3. If one of the blastomeres is removed from a developing mouse embryo, the remaining cells will go on to develop into a normal mouse. This is an example of
   a. regulated development.
   b. the zone of polarizing activity.
   c. cleavage.
   d. gastrulation.
   e. ingression.

4. During cleavage, the number of cells in a developing frog embryo increases. The cytoplasm in these new cells
   a. comes from the egg cytoplasm.
   b. is synthesized by the blastomeres.
   c. does not contain any yolk.
   d. is the vegetal pole.
   e. undergoes mitosis.

5. The mesoderm
   a. is located on the outside of the embryo.
   b. lies between the endoderm and the ectoderm.
   c. is found in blastula-stage embryos.
   d. gives rise to the linings of the gut.
   e. is formed during cleavage.

6. The formation of the endoderm during gastrulation in frogs results from
   a. movement of cells from the surface layer to the interior.
   b. migration of cells within the blastocoel.
   c. formation of columnar cells at the vegetal pole.
   d. rapid cell division.
   e. migration of secondary mesenchyme cells.

7. During the second trimester of pregnancy,
   a. the blastocyst implants in the uterine lining.
   b. the mother goes through noticeable hormonal responses.
   c. the fetal digestive system starts to function.
   d. the fetal brain undergoes cycles of sleep and waking.
   e. limbs of the fetus elongate, and fingers and toes become well formed.

8. In bird eggs, cells migrate into the interior of the embryo through the primitive streak. In nonmammalian embryos with smaller amounts of yolk, this involution occurs at the
   a. archenteron.
   b. blastopore.
   c. notochord.
   d. mesenchyme.
   e. endoderm.

9. Which of the following occur(s) during fertilization in mammals?
   a. Final meiotic division of egg nucleus
   b. Fusion of egg and sperm nuclei
   c. Acrosomal reaction
   d. Blocks to polyspermy
   e. All of the above

*▪10. If cells from the neural tube of a frog embryo are transplanted onto the ventral surface of a second embryo, the transplanted tissue will still go on to develop into tissues of the nervous system. The transplanted cells are
   a. differentiated.
   b. totipotent.
   c. discontinuous.
   d. determined.
   e. endodermal.

▪11. When the blastopore dorsal lip is grafted from one frog embryo onto a second embryo, the second dorsal lip will
   a. change the polarity of the adjacent segments.
   b. block gastrulation.
   c. change the developmental fate of the surrounding cells.
   d. change the prospective potency of the surrounding cells.
   e. cause rapid cell division.

▪12. Because of the differences between regulative and mosaic development, which of the following animals would not develop as twins?
   a. Humans
   b. Frogs
   c. Sea urchins
   d. Insects
   e. Birds

13. Involution
   a. requires unique cell movement.
   b. creates unique positions for new cell interactions.
   c. occurs in sea urchins.
   d. forms the archenteron in echinoderms.
   e. All of the above

14. Gastrulation is the stage of development
   a. when neural tube formation begins.
   b. when sea urchins begin to form primary mesenchyme.
   c. when new embryonic tissue begins to form in the frog embryo.
   d. that precedes cleavage.
   e. b and c

15. The gray crescent
   a. is observable in the zygote and two-celled frog embryo.
   b. is a homeobox gene.
   c. controls cellular affinities.
   d. induces the optic cup to form.
   e. can be mimicked by retinoic acid.

16. An organism (such as the chick) with extensive yolk
   a. has complete cleavage.
   b. has incomplete cleavage.
   c. forms a blastoderm but no blastocoel.
   d. has yolk evenly distributed in the egg.
   e. fails to synthesize DNA during cleavage.

▪17. Because the human embryo is able to split at the 64-cell level of organization to produce two viable progeny, it is said to exhibit _________ development.
   a. mosaic
   b. determinative
   c. definitive
   d. classical vertebrate
   e. regulated

18. The gray crescent experiments of Hans Spemann provided experimental evidence for the
   a. cell theory of life.
   b. concept of determination.
   c. unequal distribution of cytoplasmic determinants.
   d. basis of genetic engineering.
   e. None of the above

19. Place the following developmental events in their proper chronological sequence: (1) formation of the neural tube, (2) movement of neural folds, and (3) thickening of neural ectoderm.
   a. 1, 2, and 3
   b. 2, 1, and 3
   c. 3, 1, and 2
   d. 3, 2, and 1
   e. None of the above

20. The gray crescent is the region of the egg
   a. that is opposite the site of sperm penetration.
   b. that later is where the gastrulation begins.
   c. that was pigmented before the cytoplasm rearranged.
   d. where the dorsal lip of the blastopore will be.
   e. All of the above

21. The sea urchin differs from insects in early cleavage. The sea urchin undergoes _________, whereas the insects undergo _________.
   a. complete cleavage, superficial cleavage
   b. complete cleavage, incomplete cleavage
   c. superficial cleavage, incomplete cleavage
   d. incomplete cleavage, complete cleavage
   e. superficial cleavage, complete cleavage

22. When frog eggs are placed in water, which end is up?
    a. It would be random.
    b. The animal pole would face up.
    c. The vegetal pole would face up.
    d. The eggs would lay on their side.
    e. None of the above

23. The embryos of complex multicellular animals must establish spatial coordinates in order for development to progress. An important reference coordinate for developing frogs is the
    a. location of the vegetal pole.
    b. location of the animal pole.
    c. point of sperm penetration.
    d. location where gastrulation begins.
    e. direction of the sun.

*24. Which of the cells below are the ones that actually develop into the embryo?
    a. Trophoblast
    b. Extraembryonic
    c. Inner cell mass
    d. Cumulus
    e. All of the above

25. The initial formation of the nervous system occurs during neurulation. The important inducing structure is (are)
    a. the somites.
    b. imaginal discs.
    c. Hensen's node.
    d. the notochord.
    e. the spinal column.

26. The structure in birds and mammals that is most analogous to the dorsal lip of the blastopore found in frogs is
    a. primitive streak.
    b. the archenteron.
    c. the yolk plug.
    d. Hensen's node.
    e. the notochord.

27. The trophoblast cells in frogs are important
    a. to form the placenta.
    b. to protect the egg from sunlight.
    c. for gas exchange.
    d. to help with implantation.
    e. Frogs do not have trophoblast cells.

28. Gastrulation in birds and frogs differs quite a lot. The one main feature that both have in common is that
    a. three germ layers are established.
    b. the gut is formed from the migration of cells from the surface of the blastocyst.
    c. the neural tube forms.
    d. somites are created.
    e. All of the above

29. The fate of the mesodermal cells after gastrulation is to contribute predominantly to the developing
    a. brain and nervous system.
    b. skeletal system and muscles.
    c. inner lining of the gut and respiratory tract.
    d. sweat glands.
    e. liver and pancreas.

30. The fate of the ectodermal cells after gastrulation is to contribute predominantly to the developing
    a. brain, nervous system, and sweat glands.
    b. skeletal system and muscles.
    c. inner lining of the gut and respiratory tract.
    d. liver and pancreas.
    e. None of the above

31. The fate of the endodermal cells after gastrulation is to contribute predominantly to the developing
    a. brain, nervous system, and nails.
    b. skeletal system and muscles.
    c. inner lining of the gut and respiratory tract.
    d. sweat glands and milk secretory glands.
    e. None of the above

*32. An interesting difference between identical twins formed prior to the formation of the trophoblast versus after is
    a. if prior to trophoblast formation, each developing embryo has its own chorion.
    b. if after trophoblast formation, each developing embryo has its own chorion.
    c. if prior to trophoblast formation, each developing embryo has its own amnion.
    d. if after trophoblast formation, each developing embryo has its own amnion.
    e. a and c

33. Which of the following is *not* true of the umbilical cord?
    a. It contains blood vessels from the embryo.
    b. It contains blood vessels from the mother.
    c. It is derived from the allantois.
    d. It joins the placenta and the embryo.
    e. It carries nutrients and wastes.

34. What is the source of high levels of estrogen and progesterone during the latter half of pregnancy?
    a. Corpus luteum
    b. Chorion
    c. Follicle cells
    d. Placenta
    e. Uterus

35. The formation of the zygote is synonymous with
    a. copulation.
    b. fertilization.
    c. intercourse.
    d. recombination.
    e. capacitation.

36. Which of the following is *not* true of the first trimester of pregnancy?
    a. There is rapid cell division.
    b. It is when the embryo is the most sensitive to radiation and drugs.
    c. The fetus grows rapidly in size.
    d. The mother has symptoms of nausea and mood swings.
    e. There are high levels of estrogen and progesterone circulating in the mother's blood.

37. During labor, uterine contractions are stimulated by
    a. estrogen.
    b. oxytocin.
    c. progesterone.
    d. a and b
    e. a, b, and c

38. Bottle cells are
    a. cells of the trophoblast.
    b. cells of the inner cell mass.
    c. fluid-filled cells used for storage.
    d. cells with the shape of a bottle found around the blastopore.
    e. another name for granulosa cells.

39. The movement of cells toward the blastopore is called
    a. mass cellular migration.
    b. embryonic cellular integration migration.
    c. fulfilling cellular destiny.
    d. furrowing.
    e. epiboly.

40. One of Hans Spemann's important experiments involved
    a. dividing human embryos into equal halves.
    b. killing a single cell in a sea urchin embryo to study the effects.
    c. dividing a frog embryo in two using a human hair.
    d. removing cytoplasm from muscle cells and transplanting it into a fertilized egg.
    e. All of the above

41. The epiblast and the hypoblast are structures found during
    a. human development.
    b. frog development.
    c. human and frog development.
    d. chicken development.
    e. human and chicken development.

42. In humans, the amnion forms from the
    a. hypoblast.
    b. epiblast.
    c. chorion.
    d. trophoblast.
    e. yolk sac.

43. The _________ is the structure from which the ectoderm, endoderm, and mesoderm cells are derived in mammals and birds.
    a. trophoblast
    b. cumulus cells
    c. blastocyst
    d. epiblast
    e. hypoblast

44. The cells that form the neural tube come from the
    a. notochord.
    b. mesoderm.
    c. endoderm.
    d. ectoderm.
    e. neuroderm.

45. The _________ is an important structure for waste storage in birds and some mammals including pigs, but not in humans.
    a. allantois
    b. yolk sac
    c. placenta
    d. umbilical cord
    e. amnion

46. Eggs are triggered into activity by
    a. sperm fusion.
    b. sodium influx.
    c. ovulation.
    d. ionic disequilibrium.
    e. actozymes.

*47. The fast block to polyspermy includes the rise in
    a. intracellular potassium ions.
    b. extracellular potassium ions.
    c. intracellular sodium ions.
    d. release of cortical vesicles.
    e. All the above

48. In many mammalian species, before the sperm can penetrate the zona pellucida the sperm must
    a. capacitate.
    b. activate.
    c. decapitate.
    d. swim for hours.
    e. penetrate the oviducts.

49. In sea urchins, slow block takes
    a. a day.
    b. a few seconds.
    c. 1 hour or more.
    d. 20 seconds or more.
    e. None of the above

50. _________ contain(s) enzymes that are released during the slow block to polyspermy.
    a. Cortical granules
    b. Exocytotic granules
    c. Acrosome
    d. Endoplasmic reticulum
    e. Mitochondria

51. The sperm contributes _________ to the zygote.
    a. a nucleus
    b. half of the mitochondria and a nucleus
    c. a centriole and a nucleus
    d. cilium and a nucleus
    e. All of the above

52. In frogs, the sperm penetrates
    a. in the region of the vegetal pole.
    b. in the region of the animal pole.
    c. in the region of the gray crescent.
    d. anywhere.
    e. somewhere on the border of the animal and vegetal pole.

53. If a cell is removed from an 8-cell embryo and a part ends up missing in the organism, the development is termed
   a. regulative.
   b. controlled.
   c. banished.
   d. irreversible.
   e. mosaic.

54. ________ are used by mesenchyme cells to move along extracellular matrix molecules.
   a. Amoeboids
   b. "Walking feet"
   c. Filopodia
   d. Sliding cell receptors
   e. None of the above

55. The segmented characteristic of human embryonic development is evident from these bricklike structures that form along the notochord.
   a. Neural tubes
   b. Mesoderm
   c. Ectoderm
   d. Somites
   e. Somatomeres

56. Which of the following ensure that eggs are fertilized by sperm from the correct species?
   a. Mating behavior
   b. Acrosomal reaction
   c. Bindin
   d. Zona pellucida
   e. All of the above

57. Which of the following does *not* occur during cleavage?
   a. Rapid DNA synthesis
   b. Cell growth
   c. Differential distribution of nutrients and information molecules
   d. Rapid series of cell divisions
   e. Formation of blastula

58. Which of the following statements about neurulation is *false*?
   a. Neurulation initiates the nervous system.
   b. The notochord gives structural support to the embryo.
   c. Bulges at the posterior end of the neural tube become the brain.
   d. The incidence of neural tube defects can be lowered if pregnant women take folic acid.
   e. None of the above

59. Which of the following statements about gestation is *false*?
   a. Small mammals have shorter gestation periods than do large mammals.
   b. Amniocentesis is performed during the first trimester of pregnancy.
   c. Chorionic villus sampling can be used to detect genetic diseases at as early as 8 weeks of gestation.
   d. Human pregnancy has a duration of about 9 months.
   e. None of the above

60. Which of the following statements about neurulation is *false*?
   a. Neural crest cells give rise to peripheral nerves.
   b. Body segmentation develops during neurulation.
   c. Hox genes control differentiation along the anterior–posterior body axis.
   d. Neural crest cells are migratory.
   e. None of the above

Study Guide Questions

1. Which of the following is *not* a mechanism used to block polyspermy?
   a. The formation of a fertilization envelope
   b. Changes in membrane potential upon fertilization
   c. The phosphatidylinositol signaling system
   d. The process of genomic imprinting

2. The sperm and the egg make different contributions to the zygote. Which statement about their contributions is *false*?
   a. The sperm contributes most of the mitochondria.
   b. The egg contributes most of the cytoplasm.
   c. Both the sperm and the egg contribute a haploid nucleus.
   d. All of the above

3. What must a mammalian sperm penetrate before it can fertilize the egg?
   a. The cumulus
   b. The cumulus and the zona pellucida
   c. The jelly coat and the zona pellucida
   d. The cumulus and the vitelline envelope

4. Why do some species have complete cleavage and others incomplete cleavage?
   a. Incomplete cleavage occurs in species with small volumes of cytoplasm.
   b. Complete cleavage, found in mammals, is a more evolved characteristic.
   c. Incomplete cleavage occurs in species with large amounts of yolk.
   d. Complete cleavage occurs only in eggs that have been fertilized by two sperm.

5. A fate map can be used to map out the tissues and organs that the germ layers will become in a frog blastula. What is the fate of the dorsal lip of the blastopore?
   a. The lining of the gut
   b. The notochord
   c. The epidermal layer of skin
   d. The heart

6. Whereas the location of the ________ determines the anterior–posterior axis of the embryo, the ________ determines the embryo's bilateral symmetry.
   a. primitive streak, epiblast
   b. blastopore, neural plate
   c. vegetal hemisphere, notochord
   d. hypoblast, blastopore

7. What is the order of the germ layers from the inside to the outside?
   a. Mesoderm, ectoderm, endoderm
   b. Endoderm, ectoderm, mesoderm
   c. Ectoderm, mesoderm, endoderm
   d. Endoderm, mesoderm, ectoderm

8. Birds develop from a blastodisc while frogs develop from a blastula. During gastrulation in frogs, a dorsal lip forms that will eventually form the notochord. What is the equivalent in the chicken blastodisc?
   a. Epiblast
   b. Hypoblast
   c. Hensen's node
   d. Bottle cells

9. Birds develop extraembryonic membranes during development. Which statement about avian extraembryonic membranes is *false*?
   a. The yolk sac surrounds the yolk and provides nutrients.
   b. The amnion and chorion are derived from ectoderm and mesoderm.
   c. The allantois stores nutrients.
   d. The chorion exchanges gases and water between the embryo and the environment.

10. Which of the following statements about the mammalian blastocyst is *false*?
    a. The trophoblast gives rise to the embryo proper.
    b. Maternal genes are expressed during cleavage.
    c. The blastocyst implants in the mother's uterus.
    d. Early mammalian development is slow.

11. The role of the corpus luteum and the hormones it produces ends
    a. at the time of embryonic implantation in the uterus.
    b. with the formation of the placenta.
    c. during the second trimester of pregnancy.
    d. at birth.

12. In which of the following animal eggs would you expect to find a zona pellucida covered by a cumulus?
    a. Human
    b. Frog
    c. Sea urchin
    d. Chicken

13. Why did Hans Spemann call the dorsal lip of the blastopore the embryonic organizer?
    a. It is the point where gastrulation begins.
    b. It becomes part of the nervous system.
    c. It becomes part of the notochord.
    d. It leads to the establishment of the embryonic axes.

14. During its development, the human embryo is contained within a fluid-filled chamber bounded by the membranous
    a. yolk sac.
    b. amnion.
    c. chorion.
    d. allantois.

End of Chapter Questions

1. Fertilization involves all of the following except
   a. second meiotic division of the sperm nucleus.
   b. metabolic activation of the egg.
   c. breakdown of the acrosomal membrane.
   d. change in electric charge across the cell membrane of the egg.
   e. binding of the sperm cell to coatings surrounding the egg.

2. Which of the following does *not* occur during cleavage in frogs?
   a. High rate of mitosis
   b. Reduction in the size of cells
   c. Expression of genes critical for blastula formation
   d. Orientation of cleavage planes at right angles
   e. Unequal division of cytoplasmic determinants

3. How does cleavage in mammals differ from cleavage in frogs?
   a. Slower rate of cell division
   b. Formation of tight junctions
   c. Control involving the embryo's genome
   d. Early separation of cells that will not contribute to the embryo
   e. All of the above

4. Which statement about gastrulation is *true*?
   a. In frogs, it begins in the animal hemisphere.
   b. In sea urchins, it produces the notochord.
   c. In birds, it produces an epiblast and hypoblast.
   d. In mammals, the blastopore is a long groove in the epiblast.
   e. In sea urchins, it produces only two germ layers.

5. Which of the following was a conclusion from the experiments of Spemann and Mangold?
   a. Cytoplasmic determinants of development are homogeneously distributed in the amphibian zygote.
   b. In the late blastula, regions of cells are determined to form skin or nervous tissue.
   c. The dorsal lip of the blastopore can be isolated and will form a complete embryo.
   d. The dorsal lip of the blastopore can initiate gastrulation.
   e. The dorsal lip of the blastopore gives rise to the neural tube.

6. The acrosome of the sperm
   a. carries genetic information.
   b. provides energy for movement.
   c. carries the enzymes that facilitate fertilization.
   d. induces ovulation.
   e. prevents polyspermy.

7. Which of the following characterizes neurulation?
   a. Chordomesoderm forms a neural tube.
   b. Ectoderm forms a neural tube.
   c. A neural tube forms around the notochord.
   d. The neural tube forms somites.
   e. In chicks, the neural tube forms from the primitive groove.

8. Which statement about trophoblast cells is *true*?
   a. They are capable of producing monozygotic twins.
   b. They are derived from the hypoblast of the blastocyst.
   c. They are endodermal cells.
   d. They secrete proteolytic enzymes.
   e. They prevent the zona pellucida from attaching to the oviduct.

9. Which membrane is part of the embryonic contribution to placenta formation?
   a. Amnion
   b. Chorion
   c. Yolk sac
   d. Allantois
   e. Zona pellucida

10. A major factor in the determination and differentiation of tissues along the anterior-posterior axis of the mouse is
    a. The differential expression of Hox genes.
    b. The concentration gradient of β-catenin.
    c. The differential expression of the *sonic hedgehog* gene.
    d. The distance of the tissue from the gray crescent.
    e. The distribution of GSK-3 which degrades β-catenin.

Student Web Site Self-Quiz Questions

1. If you arrange the following events of sperm-egg interactions in the sea urchin in chronological order, which event would be third in sequence?
   a. Substances in jellycoat trigger acrosomal reaction.
   b. Acrosomal process forms.
   c. Fertilization cone forms.
   d. Enzymes digest jelly coat.
   e. Bindin and bindin receptors interact.

2. Many organisms have mechanisms that prevent multiple fertilization events. Which of the following events is *not* part of the slow block to polyspermy in the sea urchin?
   a. Change in electrical potential of membrane
   b. Release of Ca^{2+} from the endoplasmic reticulum (ER) into the cytoplasm
   c. Breaking of bonds attaching the vitelline envelope to the plasma membrane
   d. Osmotic influx of water into space between vitelline envelope and plasma membrane
   e. Exocytosis of cortical vesicles

3. Regarding the rearrangement of cytoplasm of frog eggs after fertilization, which of the following statements about this process is *false*?
   a. Fertilization establishes bilateral symmetry in the zygote.
   b. The cortical cytoplasm of the egg rotates.
   c. The gray crescent results when heavily pigmented cortical cytoplasm overlies the vegetal hemisphere.
   d. The site of sperm entry will become the ventral region of the embryo.
   e. The gray crescent will become the tail end of the embryo.

4. Early cleavage in mammals is unique. Which of the following choices is *not* a feature unique to mammals?
   a. During early cleavage, the embryo's genes are inactive and development is directed almost entirely by the maternal genome.
   b. At the 8-cell stage, a very tight mass of cells forms.
   c. At the 32-cell stage, the embryo has an outer cell layer called the trophoblast.
   d. The blastocyst contains the hollow blastocoel.
   e. The zona pellucida—functionally similar to the vitelline membrane of sea urchins—prevents premature implantation of the blastocyst.

5. Which of the following structures present in the gastrula of a frog is *not* also found in a sea urchin gastrula?
   a. Archenteron
   b. Blastocoel
   c. Blastopore
   d. Mesenchyme
   e. Plug of yolk-rich cells

6. Which of the following terms does *not* apply to development in birds?
   a. Blastodisc
   b. Complete cleavage
   c. Epiblast
   d. Henson's node
   e. Primitive streak

7. The process of childbirth (parturition) involves an elaborate sequence of events. Which of the following statements regarding parturition is *false*?
   a. Progesterone inhibits the contraction of uterine muscle.
   b. Estrogen stimulates the contraction of uterine muscle.
   c. Oxytocin stimulates the contraction of uterine muscle.
   d. Mechanical stimulation during childbirth is generated by pressure of the fetal head on the cervix.
   e. Mechanical stimuli increase the release of progesterone from the pituitary.

TEST FILE

★ *Indicates a **difficult** question*
■ *Indicates a **conceptual** question*

44 Neurons and Nervous Systems

Fill in the Blank

1. The part of the neuron specialized for receiving impulses is the **dendrite**.

2. The initial membrane event of an action potential is the flow of **sodium** ions across the membrane.

3. Following an action potential, a neuron has a **refractory period** during which it cannot be stimulated.

4. In myelinated axons of vertebrate neurons, breaks in the insulation occur at points called the **nodes of Ranvier**.

5. Deadly nerve gases are examples of inhibitors of the enzyme **acetylcholinesterase**.

★6. In vertebrates, the two most common inhibitory neurotransmitters are **GABA** and **glycine**.

7. The information that flows through the nervous system consists of chemical and **electrical** messages.

8. When a neuron contacts another neuron or a muscle or gland, special junctions called **synapses** transmit the message carried by the incoming neuron.

9. Special glial cells, called **astrocytes**, surround the smallest, most permeable blood vessels in the brain, thereby participating in the formation of the blood–brain barrier.

★10. The nicotinic receptors of acetylcholine are found in **skeletal** muscles.

11. The depolarization of a neuron must rise above the **threshold** before an action potential is achieved.

12. In simple organisms, such as the sea anemone, the nervous system consists of a **nerve net**.

★13. Neurotransmitters that depolarize the postsynaptic membrane bring about an **excitatory** postsynaptic potential.

Multiple Choice

1. A human being has the maximum number of functioning neurons
   a. as a fetus.
   b. just after birth.
   c. at about college age.
   d. at middle age.
   e. just before death.

2. The functions of glial cells include all of the following *except*
   a. supporting developing neurons.
   b. supplying nutrients.
   c. conducting nerve impulses.
   d. consuming foreign particles.
   e. insulating nerve tissue.

3. About how many neurons are in the human brain?
   a. 100 thousand
   b. 1 million
   c. 100 million
   d. 1 billion
   e. 100 billion

■4. Electrical synapses
   a. are fast.
   b. are also called gap junctions.
   c. cannot be inhibitory.
   d. do not integrate information well.
   e. All of the above

5. Which of the following describes the resting potential of the neuronal cell membrane?
   a. The inside is 60 millivolts more positive than the outside.
   b. The outside is 60 millivolts more positive than the inside.
   c. The inside is 30 millivolts more positive than the outside.
   d. The outside is 30 millivolts more positive than the inside.
   e. The inside has about the same charge as the outside.

6. Which of the following can carry electric charges across the cell membrane?
   a. Electrons
   b. Protons
   c. Water
   d. Ions
   e. Proteins

7. What generally maintains the electric charge across the neuronal membrane?
   a. Sodium–potassium pump
   b. Action potential
   c. Resting potential
   d. Voltage-gated channels
   e. Negative ion pump

8. Which of the following describes the mechanism of voltage-gated channel proteins?
   a. Depending upon the membrane voltage, ions are pumped through.
   b. Depending upon the membrane voltage, ions can diffuse through.
   c. Ions can move through only if the overall membrane voltage stays the same.
   d. Ions are pumped through in order to maintain same membrane voltage.
   e. When the gates close, membrane voltage changes.

9. What happens if Na^+ channels open and sodium ions diffuse into the cell?
   a. The cell becomes hyperpolarized.
   b. Other sodium ions will move out of the cell.
   c. Voltage-gated channels will remain closed.
   d. The charge across the nearby membrane will change.
   e. Action potentials will be triggered.

10. Following depolarization, the neural membrane potential is restored when
    a. Na^+ ions rush outward through the membrane.
    b. K^+ ions rush outward through the membrane.
    c. Cl^- ions rush inward through the membrane.
    d. a pump moves ions to their original concentrations.
    e. the membrane becomes freely permeable to many ions.

▪11. Which of the following would an electrode record as an action potential moves along an axon membrane?
    a. The inside membrane voltage becomes negative.
    b. The membrane voltage permanently changes.
    c. The action potential may reverse its direction.
    d. The action potential moves at a constant speed.
    e. The height of the action potential does not change.

12. Which of the following nerves would have the most rapid action potentials?
    a. Those with the thinnest axon diameters
    b. Those with sheaths of myelin
    c. Those in invertebrate animals
    d. Those with a greater membrane potential
    e. Those with the most ion channels

13. Saltatory conduction results when
    a. continuous propagation of the nerve impulse speeds up.
    b. a nerve impulse jumps from one neuron to another.
    c. the threshold for an action potential is suddenly increased.
    d. action potentials spread from node to node down the axon.
    e. the direction of an action potential suddenly changes.

14. When an action potential arrives at the axon terminal, the voltage-gated calcium channels there
    a. release calcium into the synaptic cleft.
    b. actively transport neurotransmitter into the synaptic cleft.
    c. cause vesicles to release neurotransmitter into the synaptic cleft.
    d. depolarize the membrane at the axon terminal.
    e. cause the membrane receptors to bind neurotransmitter.

15. When the neurotransmitter diffuses across the synaptic cleft,
    a. it automatically causes depolarization of the postsynaptic membrane.
    b. it can be excitatory or inhibitory depending on the type of postsynaptic membrane.
    c. a single molecule is sufficient to trigger activation of the postsynaptic membrane.
    d. only a few molecules make it to the postsynaptic membrane.
    e. it must move through nodes in the myelin sheath.

16. Neurotransmitters
    a. have multiple types of receptors.
    b. may be excitatory or inhibitory.
    c. may have different effects in different tissues.
    d. include dopamine and serotonin, which are monoamines.
    e. All of the above

17. What is the effect of inhibiting acetylcholinesterase?
    a. Neurotransmitter release from the presynaptic membrane is inhibited.
    b. Synthesis of neurotransmitter in cells is inhibited.
    c. Breakdown of neurotransmitter in the synapse is inhibited.
    d. Stimulation of the postsynaptic membrane is inhibited.
    e. Cholinergic receptors are inhibited.

18. Some organisms, such as flatworms and earthworms, have clusters of neurons called
    a. the spinal cord.
    b. the central nervous system.
    c. the peripheral nervous system.
    d. ganglia.
    e. None of the above

*19. Which of the following is a neurotransmitter that is *not* released synaptically?
    a. Glycine
    b. Norepinephrine
    c. Nitric oxide
    d. Adenosine
    e. None of the above

20. The two primary cell types of the nervous system are
    a. fibroblasts and chondrocytes.
    b. neurons and glial cells.

c. epithelial cells and glandular cells.
d. neurons and epithelial cells.
e. neuromuscular cells and epithelial cells.

21. Synaptic clefts can be cleansed of neurotransmitters in which of the following ways?
a. Enzymatic degradation
b. Simple diffusion
c. Active transport
d. a and c
e. a, b, and c

22. Which of the following ions is most responsible for generating an action potential?
a. Na^+
b. K^+
c. Cl^-
d. H^+
e. OH^-

23. Most nerve cells communicate with others through
a. electric signals that pass across at synapses.
b. chemical signals that pass across at synapses.
c. bursts of pressure that "bump" the postsynaptic cell membrane.
d. Na^+ ions, released from one cell and entering the next.
e. None of the above

24. The resting potential of a neuron is produced by
a. voltage-gated channels in the membrane.
b. chemically gated channels in the membrane.
c. permanently open potassium channels in the membrane.
d. the concentration difference in Na^+ across the membrane.
e. blocking the sodium–potassium pump.

25. The myelin sheath that surrounds some axons in the peripheral nervous system is formed by
a. neurons.
b. Schwann cells.
c. bacteria that have invaded the nervous system.
d. synapses.
e. None of the above

26. Which of the following limits the frequency at which a single neuron can "fire" action potentials?
a. The number of synapses that the neuron forms
b. The number of other cells that the neuron contacts
c. The refractory period for the neuron's Na^+ channel
d. The length of the axon of the neuron
e. The number of dendrites on the neuron

27. Which of the following can be said to be involved in triggering synaptic transmission?
a. The action potential
b. The opening of Ca^{2+} channels at the synaptic terminal
c. The entry of Ca^{2+} into the presynaptic terminal
d. Fusion of synaptic vesicles with the presynaptic membrane
e. All of the above

28. Neurons
a. have a uniform shape throughout the nervous system.
b. are more numerous than glial cells in the nervous system.
c. are found only in mammals and birds.
d. communicate with other cells at synapses.
e. All of the above

29. Dopamine is
a. involved in schizophrenia.
b. involved in Parkinson's disease.
c. a neurotransmitter of the central nervous system.
d. a monoamine.
e. All of the above

30. Narcotic drugs, such as opium and heroin, activate the receptors of
a. substance P.
b. endorphins.
c. enkephalins.
d. b and c
e. a, b, and c

*31. Long-term potentiation
a. was discovered by neurobiologists working with brain slice preparations.
b. involves an enhanced postsynaptic response.
c. results from repeated stimulation of a presynaptic cell.
d. involves activation of the NMDA receptor.
e. All of the above

32. The action potential
a. begins with an increased permeability to potassium.
b. returns to resting when the sodium channels open.
c. can be triggered in very rapid succession, with no delay.
d. involves voltage-gated channels in the membrane.
e. propagates only because chloride ions move through the membrane.

33. The action potential
a. travels along all axons at the same speed.
b. is slowed down if a nerve cell has myelin around it.
c. is blocked at the nodes of Ranvier.
d. causes a brief depolarization of the membrane potential.
e. triggers a change in potential simultaneously along the entire axon.

*34. Mice with modified NMDA receptors
a. ran through mazes more slowly than did normal mice.
b. remembered mazes for longer periods of time than did normal mice.
c. failed to learn tasks.
d. remembered mazes for shorter periods of time than did normal mice.
e. a, c, and d

▪35. Choose the correct statement about vertebrate nervous systems.
a. The nervous system consists only of brain and spinal cord.
b. They can carry out many tasks at the same time.
c. All nervous system functions are voluntary.
d. Animals of similar body mass have brains of similar size.
e. Ions are equally distributed across nerve cell membranes.

36. Choose the statement about acetylcholine that is *false*.
a. Acetylcholine is a neurotransmitter.
b. Acetylcholine is found at mammalian neuromuscular junctions.
c. Both smooth muscles and skeletal muscles respond to acetylcholine.
d. Acetylcholine is degraded by acetylcholinesterase.
e. The acetylcholine receptor of the motor end plate of muscle cells is a metabotropic receptor.

37. Which of the following statements about the nervous system is *false*?
a. The nervous system is the most complex system of the human body.
b. Oligodendrocytes cover the axons of neurons in the peripheral nervous system.
c. Effectors are muscles or glands.
d. Sensory cells transduce information into the electric signals that can be transmitted by neurons.
e. Brain stem structures serve basic physiological functions.

▪38. Neurons
a. fire action potentials only on the basis of the number of excitatory inputs they receive.
b. sum excitatory and inhibitory postsynaptic potentials.
c. make the "decision" to fire in the dendrites of the neuron.
d. make a spatial summation, but not a temporal one.
e. never form synapses on synapses.

39. Patch clamping is used to
a. fix a break in a cell membrane.
b. record electrical activity inside a cell.
c. record ion movements through a single channel.
d. record ion movement through the entire neuron.
e. record an action potential through a single channel.

40. When an action potential arrives at an axon terminal, it causes the opening of _________ channels, which triggers fusion of a neurotransmitter vesicle with the cell membrane.
a. calcium
b. sodium
c. potassium
d. chloride
e. acetylcholine

*41. Muscarinic receptors of acetylcholine are
a. found in heart muscle.
b. tend to be inhibitory.
c. an example of a metabotropic receptor.
d. a and b
e. a, b, and c

*42. Another name for long-term potentiation is
a. synaptic memory.
b. inhibitory postsynaptic potential.
c. temporal summation.
d. spatial summation.
e. None of the above

43. Which of the following neurotransmitters has been shown to be involved with sleep/wake cycles?
a. Acetylcholine
b. Norepinephrine
c. GABA
d. Serotonin
e. Dopamine

*44. Which of the following neurotransmitters is a peptide?
a. Acetylcholine
b. Norepinephrine
c. Serotonin
d. Glycine
e. Endorphin

45. When a postsynaptic cell adds together information from synapses at different sites, this is called a(n)
a. excitatory postsynaptic potential.
b. inhibitory postsynaptic potential.
c. spatial summation.
d. temporal summation.
e. action potential.

46. The most critical area in a neuron for "decision making" is the
a. axon hillock.
b. presynaptic terminal.
c. postsynaptic terminal.
d. cell body.
e. synapse.

*47. Some glial cells communicate with each other through
a. a myelin sheath.
b. axons.
c. dendrites.
d. gap junctions.
e. tight junctions.

48. Anesthetics and alcohol can permeate the blood–brain barrier because
a. they are small molecules.
b. they are water-soluble.
c. they are fat-soluble.
d. they pass through gated channels.
e. there are receptors for them on blood vessels.

49. The sodium–potassium pump
a. needs energy to work.
b. expels three sodium ions for every two potassium ions imported.
c. works against a concentration gradient.

d. a and b
e. a, b, and c

50. Prozac, a commonly prescribed drug for depression,
a. enhances the activity of serotonin at the synapse.
b. slows the reuptake of serotonin.
c. increases endorphins.
d. a and b
e. a, b, and c

Study Guide Questions

1. The extensions of postsynaptic neurons that provide the main receptive surface for presynaptic neurons are
a. nuclei.
b. somas.
c. axons.
d. dendrites.

2. The substance that wraps the axon of many neurons and provides for increased conduction speed is
a. dendrase.
b. histamine.
c. acetylcholine.
d. myelin.

3. The long extensions from the neurons that provide the pathway for action potentials to synapse are
a. dendrites.
b. cell bodies.
c. axons.
d. presynaptic membranes.

4. The threshold of a neuron is
a. the amount of inhibitory neurotransmitter required to inhibit an action potential.
b. the membrane voltage at which an axon potential will be suppressed.
c. the amount of excitatory neurotransmitter required to elicit an action potential.
d. the membrane voltage at which the membrane potential develops into an action potential.

5. When a membrane is at the resting potential, the concentration of
a. sodium and potassium ions is higher on the inside of its membrane.
b. sodium and potassium ions is higher on the outside of its membrane.
c. sodium ions is higher on the inside of its membrane and of potassium ions is higher on the outside.
d. sodium ions is higher on the outside of its membrane and of potassium ions is higher on the inside.

6. Glial cells are specialized to do all of the following *except*
a. receive neural impulses.
b. insulate the axons.
c. supply neurons with nutrients.
d. help maintain proper ionic environment for the neuron.

7. The cells that create the blood–brain barrier keeping toxic substances from entering the brain are ________ and belong to a type of neural tissue called ________.
a. endothelial cells, Schwann cells
b. astrocytes, glial cells
c. glial fibers, axons
d. dendrites, synapses

8. A particular disease of the nervous system specifically involves the Ca^{2+} ion channels at the chemical synapses of motor neurons where a neurotransmitter is stored and released. In other words, this disease affects the
a. axon terminals of the presynaptic cell and the release of acetylcholine.
b. axon terminals of the postsynaptic cell and the release of K^+ ions.
c. electrical synapses.
d. axon terminals of the presynaptic cell and the release of K^+ ions.

9. Which of the following statements about gap junctions or electrical synapses is *false*?
a. Connexons form molecular tunnels between two cells.
b. Electrical synapses cannot be inhibitory.
c. Electrical synapses do not allow for temporal summation.
d. Electrical transmission is very slow and is bidirectional.

10. One group of neurotransmitter receptors is the metabotropic receptors. Which one of the following statements about metabotropic receptors is *false*?
a. Both muscanaric and β-adrenergic receptors are metabotropic receptors.
b. Metabotropic receptors are coupled with A proteins.
c. Metabotropic receptors can stimulate gene expression.
d. Muscanaric receptors bind with acetylcholine.

11. The rapid depolarization of a neuron during the first half of an action potential is due to the
a. exit of K^+ ions from the cell through gated potassium channels.
b. rapid reversal of ion concentration caused by the action of the sodium–potassium pump.
c. entry of Na^+ ions into the cell through gated sodium channels.
d. movement of both Na^+ and K^+ ions through appropriate open channels.

12. The refractory period of a neuron results from
a. the period when the sodium–potassium pump is nonfunctional.
b. activation of voltage-gated chloride channels.
c. closing of inactivated voltage-gated sodium channels.
d. the action potential reaching the synapse.

13. Which of the following statements about the process of summation in a neuron is *false*?
   a. Slight perturbations of the membrane potential spread across the postsynaptic cell body.
   b. Axons that terminate closer to the axon hillock have more influence on the summation process.
   c. Summation consists essentially of comparing the total of excitatory and inhibitory postsynaptic inputs.
   d. The concentration of voltage-gated sodium channels is highest in the dendrites of the postsynaptic cell.

14. The amount of neurotransmitter released from a presynaptic cell is a function of
   a. the voltage of the individual action potential in the presynaptic neuron.
   b. the frequency of arriving action potentials in the presynaptic neuron.
   c. whether the action potential is excitatory or inhibitory.
   d. the receptors in the postsynaptic cell.

15. The electrical events labeled as EPSPs are the result of
   a. hyperpolarization of the postsynaptic membrane.
   b. depolarization of the postsynaptic membrane.
   c. hyperpolarization of the presynaptic membrane.
   d. depolarization of the presynaptic membrane.

End of Chapter Questions

1. In the nervous system, the most abundant cell type is the
   a. motor neuron.
   b. sensory neuron.
   c. preganglionic parasympathetic neuron.
   d. glial cell.
   e. preganglionic sympathetic neuron.

2. Within the nerve cell, information moves from
   a. dendrite to cell body to axon.
   b. axon to cell body to dendrite.
   c. cell body to axon to dendrite.
   d. axon to dendrite to cell body.
   e. dendrite to axon to cell body.

3. The resting potential of a neuron is due mostly to
   a. local current spread.
   b. open Na^+ channels.
   c. synaptic summation.
   d. open K^+ channels.
   e. unequal pumping of K^+ and Na^+ ions across the membrane.

4. Which statement about synaptic transmission is *not* true?
   a. The synapses between neurons and muscles use acetylcholine as their neurotransmitter.
   b. A single vesicle of neurotransmitter cannot cause a muscle fiber to contract.
   c. The release of neurotransmitter at the neuromuscular junction causes the motor end plate to fire action potentials.
   d. In vertebrates, the synapses between motor neurons and muscle fibers are always excitatory.
   e. Inhibitory synapses cause the resting potential of the postsynaptic membrane to become more negative.

5. Which statement accurately describes an action potential?
   a. Its magnitude increases along the axon.
   b. Its magnitude decreases along the axon.
   c. All action potentials in a single neuron are of the same magnitude.
   d. During an action potential the transmembrane potential of a neuron remains constant.
   e. It permanently shifts a neuron's transmembrane potential away from its resting value.

6. A neuron that has just fired an action potential cannot be immediately restimulated to fire a second action potential. The short interval of time during which restimulation is not possible is called
   a. hyperpolarization.
   b. the resting potential.
   c. depolarization.
   d. repolarization.
   e. the refractory period.

7. The rate of propagation of an action potential depends on
   a. whether or not the axon is myelinated.
   b. the axon's diameter.
   c. whether or not the axon is insulated by glial cells.
   d. the cross-sectional area of the axon.
   e. All of the above

8. The binding of neurotransmitter to the postsynaptic receptors in an inhibitory synapse results in
   a. depolarization of the transmembrane potential.
   b. generation of an action potential.
   c. hyperpolarization of the transmembrane potential.
   d. increased permeability of the membrane to sodium ions.
   e. increased permeability of the membrane to calcium ions.

9. Whether a synapse is excitatory or inhibitory depends on the
   a. type of neurotransmitter.
   b. presynaptic terminal.
   c. size of the synapse.
   d. nature of the postsynaptic neurotransmitter receptors.
   e. concentration of neurotransmitter in the synaptic space.

10. Which of the following is a likely mechanism for long-term potentiation?
   a. When the neurotransmitter glutamate reaches post synaptic AMPA receptors, it activates G-proteins that trigger intracellular changes.
   b. When glutamate binds to NMDA receptors, it allows Mg^{2+} ions to enter the cell which initiate intracellular changes.

c. When sufficient glutamate is released by the presynaptic cell, it causes an increase in the number of AMPA receptors on the postsynaptic cells.
d. When sufficient glutamate is released, both AMPA and NMDA receptors are activated, and NMDA receptors allow Ca^{2+} as well as Na^+ to enter the cell thus initiating intracellular changes.
e. When both glutamate and acetylcholine are released together, they create a long lasting depolarization of the postsynaptic cell.

Student Web Site Self-Quiz Questions

1. Which of the following terms does not apply to the nervous systems of either the sea anemone or the squid?
a. Brain
b. CNS
c. Ganglia
d. Nerve net
e. Neurons

2. One of the functions of glial cells is to wrap axons with myelin. Which of the following is *not* a function of any type of glial cells?
a. Create the blood-brain barrier
b. Generate nerve impulses
c. Insulate neurons
d. Increased speed of nerve signal
e. Nourish neurons

3. Ion pumps and channels generate resting and action potentials. If some of the gated K^+ channels in the cell membrane of a resting neuron were to open, the immediate effect would be to
a. depolarize the membrane.
b. make the membrane potential equal to 0.0 mV.
c. make the membrane potential equal to 60 mV.
d. make the membrane potential equal to -60 mV.
e. hyperpolarize the membrane.

4. What property of the neural membrane prevents an action potential from reversing itself?
a. Its refractory period
b. Change in the ratio of Na^+ and K^+ ions
c. Voltage-gated K^+ channels
d. Shortage of neurotransmitter
e. Shutdown of the Na^+-K^+ pump

5. Presynaptic and postsynaptic membranes of the neuromuscular junction differ in a number of important ways. Which of the following is *not* a way in which they differ?
a. Presence or absence of voltage-gated Ca^{2+} channels
b. Presence or absence of voltage-gated Na^+ channels
c. Ability to become depolarized
d. Presence or absence of acetylcholine receptors
e. Ability to propagate an action potential

6. Nicotine, the active ingredient in tobacco, binds to acetylcholine (ACh) receptors in skeletal muscle. Which of the following statements about neurotransmitters and drugs affecting neurotransmitter action is *false*?
a. Neurotransmitter actions depend on their receptors.
b. A drug that inhibits acetylcholinesterase would cause extreme weakness of skeletal muscles.
c. Some neurotransmitters in the central nervous system are amino acids.
d. Muscarine is a toxin produced by mushrooms.
e. Natural products are an important source of drugs that interact with neurotransmitters or their receptors.

7. Which of the following statements about synaptic transmission is *true*?
a. The postsynaptic membranes of the neuromuscular junctions are identical to presynaptic membranes.
b. Both motor end plate and presynaptic membranes fire action potentials
c. Synapses between motor neurons and skeletal muscles are always inhibitory.
d. A single molecule of acetylcholine is enough to trigger a response from the muscle cell.
e. Voltage-gated ion channels are found only at the presynaptic membrane.

8. Which of the following is *not* a key difference between the two receptor types AMPA and NMDA?
a. The molecules to which they bind.
b. The timing of their responses to stimulation
c. Whether the receptor channels are blocked at resting potential
d. Whether the receptor channels allows the influx of Na^+.
e. The level of stimulus required to activate the receptors are activated.

★ *Indicates a **difficult** question*

■ *Indicates a **conceptual** question*

45 Sensory Systems

Fill in the Blank

1. Equilibrium organs called **statocysts** use hair cells to respond to gravity.

2. Fish can detect pressure waves in water through their **lateral line** sensory system.

3. The three tiny bones of the middle ear transmit vibrations from the **tympanic membrane** to the oval window.

4. The inner ear is a long, coiled structure called the **cochlea**.

5. The molecule **rhodopsin** is the basis for photosensitivity.

6. A dense layer of photoreceptor cells at the back of the vertebrate eye forms the **retina**.

7. Besides vertebrates, the other group of animals that has eyes that form images like cameras is the **cephalopod mollusk** group.

★8. Mammals can alter the shape of the lens by contracting the **ciliary muscles**.

★9. Just prior to transmission to the brain via the optic nerve, visual signals are processed by **ganglion cells**.

10. The phenomenon of emitting sounds and creating images from reflections of those sounds is **echolocation**.

■11. **Action potentials** arriving in the visual cortex are interpreted as light; in the auditory cortex as sound; and in the olfactory cortex as smell.

12. An important characteristic of many sensory cells is that they can stop being excited by a stimulus that initially caused them to be active. This ability to ignore background or unchanging conditions while remaining sensitive to changes or new information is called **adaptation**.

13. The snake's forked tongue fits into cavities in the roof of its mouth that are richly endowed with olfactory sensors. Thus the snake is really using its tongue to **smell** its environment.

14. A common cause of **nerve deafness** (or **deafness)** is cumulative and permanent damage to the hair cells of the organ of Corti.

15. A hawk's vision is about eight times more sensitive than that of humans because it has about six times more photoreceptors per square millimeter of **fovea** than humans.

16. A receptor potential that causes action potentials by causing voltage-gated sodium channels to open is called a **generator** potential.

17. Hair cells have a set of projections called **stereocilia** that look like organ pipes.

18. Many animals can focus sounds by moving their ear **pinnae** toward the sound.

Multiple Choice

1. A sensory cell converts a stimulus
   a. from one form of physical stimulus to another.
   b. into some type of membrane potential.
   c. from an action potential to a synaptic signal.
   d. by summing incoming action potentials.
   e. into different forms to be sent to the brain.

■2. The magnitude of a receptor potential
   a. depends on the strength of the incoming action potential.
   b. remains high even after a long period of stimulation.
   c. is the same no matter what the type of stimulus.
   d. depends on the amount of neurotransmitter released.
   e. affects the frequency of resulting action potentials.

3. Which of the following is an example of adaptation of sensory cells?
   a. Going into deep sleep
   b. Discriminating different colors
   c. Ignoring your shoes as you walk
   d. Ignoring a boring lecture
   e. Detecting sound and light simultaneously

4. Pheromones are chemical signals that can signal, for example, from
   a. one neuron to another.
   b. the peripheral nervous to the central nervous system.
   c. prey to predator.
   d. parasite to host.
   e. female to male.

5. The chemosensory hairs that flies use to detect and taste food are located
   a. near the mouthparts.
   b. at the base of the wings.
   c. on the tip of the proboscis.
   d. on the feet.
   e. on the antennae.

6. How does a male silkworm moth locate a female at a distance?
   a. He flies toward a chemical signal.
   b. He flies toward a sound signal.
   c. He flies toward a shape like a female moth.
   d. He emits a sound and she approaches.
   e. He emits a chemical and she approaches.

7. Sensitivity of the sense of smell is proportional to the
   a. amount of mucus in the nose.
   b. surface area of nasal epithelium.
   c. density of olfactory nerve endings.
   d. number of capillaries in the nose.
   e. typical body temperature.

*8. The greatest intensity of perceived smell comes from the
   a. enzyme that binds with the most odorant molecules.
   b. odorant that binds to the most receptors.
   c. most different types of odorant molecules.
   d. greatest threshold of depolarization.
   e. greatest number of odorant molecules entering the cell.

9. Which of the following are *not* mechanoreceptors?
   a. Stretch receptors
   b. Hair cells
   c. Pressure receptors
   d. Olfactory receptors
   e. Airflow receptors

10. Action potentials are generated in a mechanoreceptor when
    a. ion channels close in response to membrane distortion.
    b. ion channels open in response to membrane distortion.
    c. receptors bind chemicals in response to pressure.
    d. sensitivity of the membrane to neurotransmitters increases.
    e. signals from other mechanoreceptors are summated.

11. Which of the following is most descriptive of the Pacinian corpuscle?
    a. It has high two-point discrimination ability.
    b. It responds to steady pressure for a long time.
    c. It has concentric layers of connective tissue.
    d. It has extensive, long dendritic processes.
    e. It is extremely sensitive to light touch.

12. When the stereocilia of hair cells are bent, those hair cells then
    a. trigger muscle contraction.
    b. release neurotransmitter.
    c. undergo action potentials.
    d. contract their stereocilia.
    e. become less sensitive.

*13. Stereocilia in hair cells in the canals of the fish lateral line
    a. are embedded in gelatinous material.
    b. are moved individually by pressure.
    c. are immobilized within a cupula.
    d. bend only after electrical stimulation.
    e. bend under the influence of gravity.

14. In invertebrates, the functional equivalent of the vertebrate vestibular apparatus is the
    a. hair cell.
    b. lateral line.
    c. Meissner's corpuscle.
    d. statocyst.
    e. stretch receptor.

15. Vertebrates rely on information from which sensory structure to keep their balance?
    a. Eustachian tube
    b. Otoliths
    c. Semicircular canal
    d. Statocyst
    e. Tympanic membrane

16. The middle ear serves which auditory function?
    a. It converts air pressure waves into fluid pressure waves.
    b. It converts fluid pressure waves into air pressure waves.
    c. It converts air pressure waves into nerve impulses.
    d. It converts fluid pressure waves into nerve impulses.
    e. It converts pressure waves into hair cell movements.

17. Hair cells in the ear that give auditory information are concentrated in the
    a. oval window.
    b. tympanic membrane.
    c. organ of Corti.
    d. semicircular canals.
    e. pinnae.

*18. What is the basis for auditory response distinguishing different sound frequencies?
    a. The three bones of the middle ear respond differentially.
    b. The loops of the semicircular canals respond differentially.
    c. The oval window and round window respond differentially.
    d. Different sections of the basilar membrane respond differentially.
    e. Individual hair cells have different peak frequency responses.

*19. The molecular mechanism for absorbing light into visual systems is
    a. shape change in the protein opsin.
    b. depolarization of the rhodopsin molecule.

c. isomerization of the molecule retinal.
d. a combination of opsin and retinal.
e. oxidation of the rhodopsin molecule.

*20. When an individual rod cell is stimulated with light, its membrane potential
a. becomes more negative than before stimulation.
b. becomes more positive than before stimulation.
c. becomes more positive than other neurons.
d. begins to generate action potentials.
e. begins to reduce membrane polarization.

*21. The activation of a rhodopsin molecule sets off a chain reaction that leads to the
a. opening of a sodium channel.
b. closing of a million sodium channels.
c. formation of an activated phosphodiesterase molecule.
d. activation of a million transducin molecules.
e. activation of other rhodopsin molecules.

22. Arthropods have evolved compound eyes; these consist of large numbers of
a. retinas.
b. cones.
c. eye cups.
d. ommatidia.
e. pupils.

23. The amount of light entering the eye can be decreased when
a. an optician puts in atropine drops.
b. the shape of the lens changes.
c. the autonomic nervous system opens the pupil.
d. the light is focused.
e. the iris constricts.

▪24. Which of the following does *not* affect the focus of an image on the retina?
a. The shape of the lens
b. The shape of the retina
c. The ligaments suspending the lens
d. The ciliary muscles
e. The elasticity of the lens

25. What causes the blind spot in the eye?
a. An unusually high density of rod cells
b. An unusually high density of cone cells
c. The location where incoming light is focused
d. The location where neural axons exit
e. The location where photoreceptors are saturated

26. Which of the following is a difference between rods and cones?
a. Cones are more sensitive at low light intensity.
b. Rods are responsible for color vision.
c. There are more cones than rods in the human retina.
d. Strictly nocturnal animals have more cones than rods.
e. Cones provide greatest acuity of vision.

27. Which of the following best describes the visual processing that occurs within the retina?
a. After processing within individual photoreceptors, impulses travel to the brain.
b. Light signals are processed in retinal cells just in front of the photoreceptors.
c. Signals pass from photoreceptor cells directly to ganglion cells.
d. Action potentials in photoreceptor cells affect bipolar cells before reaching ganglion cells.
e. Membrane potential changes in a network of retinal cells activate ganglion cells.

28. Which of the following statements about echolocation is *false*?
a. Porpoises and dolphins use echolocation.
b. Humans cannot hear the pulses of sound emitted by echolocating bats.
c. Bats have muscles in their middle ears that decrease auditory sensitivity during emission of echolocation cries.
d. Bats use echolocation in foraging and navigation.
e. None of the above

*▪29. Which of the following describes the receptive field of a retinal ganglion cell?
a. There is no overlap with the receptive fields of neighboring ganglion cells.
b. The ganglion cell receives input from one rod or cone cell.
c. The receptive field depends on connections of horizontal and bipolar cells.
d. A signal at the edge of the receptive field is more significant than one from the center.
e. The receptive field is defined by signals from neighboring ganglion cells.

30. Which of these statements about animal sensitivity is *true*?
a. Humans and snakes can see ultraviolet rays, but insects cannot.
b. Humans and snakes can see infrared rays, but insects cannot.
c. Humans and insects can see infrared rays, but snakes cannot.
d. Humans cannot detect ultraviolet rays, but insects can.
e. Humans cannot detect ultraviolet rays, but snakes can.

31. In general, __________ are cells of the nervous system that transduce physical or chemical stimuli into signals that are transmitted to other parts of the nervous system for processing and interpretation.
a. sensory cells
b. effectors
c. glial cells of the blood–brain barrier
d. "nuclei" within the midbrain
e. None of the above

*32. Which of the following statements regarding the stimulation of receptor proteins within plasma membranes is *true*?
a. The receptor proteins of mechanosensors, thermosensors, and electrosensors are themselves the ion channels.

b. The receptor proteins of chemosensors and photosensors initiate biochemical cascades that eventually open and close ion channels.
c. Receptor proteins are integral to the sensory process.
d. a, b, and c
e. a and b

33. Receptor potentials produce action potentials in two ways: by generating action potentials within the sensory cells, or by causing the release of _________ that induces an associated neuron to generate action potentials.
a. hormone
b. ATP
c. interleukin
d. neurotransmitter
e. glucagon

34. That human beings tend to join bird-watching societies more often than mammal-smelling societies suggests
a. humans are descended from the same reptilian ancestors that gave rise to birds.
b. humans have a limited sense of smell.
c. humans are unusual among mammals in that we depend more on vision than on olfaction.
d. a and b
e. None of the above

35. Electroreceptors may be associated with the _______ system.
a. lateral line
b. visual
c. auditory
d. gustatory
e. None of the above

36. Which of the following statements about sensory cells is *false*?
a. Most sensory cells are modified neurons.
b. Specific sensory cells can respond to all types of stimuli.
c. Changes in the stimulus strength lead to changes in the sensory cell's receptor potential.
d. Sensory cells show the phenomenon called adaptation.
e. How information from sensory cells is interpreted depends on where the information is sent.

37. Which of the following is *not* a necessary part of the definition of a sensory cell?
a. It is specialized for detecting specific kinds of stimuli.
b. It transduces energy into action potentials.
c. It causes the opening of sodium channels in the membrane.
d. It has a receptor potential.
e. It can become insensitive to a source of continuous stimulation.

38. As a male silkworm moth nears a female that is releasing bombykol,
a. the female starts to release more bombykol.
b. the action potentials in the antennal nerve increase.
c. 200 bombykol-sensitive hairs are being stimulated per second.
d. a larger number of the bombykol-sensitive hairs undergo adaptation.
e. the receptor potential in bombykol-sensitive hairs is reduced.

39. Which of the following events is triggered by the binding of an odorant molecule to a receptor protein?
a. Opening of sodium channels
b. Increase of second messenger in cytoplasm of sensory cell
c. Activation of G protein
d. Depolarization of sensory cell
e. All of the above

40. Which of the following statements about gustation is *false*?
a. All taste buds are specialized for one of four distinct tastes: sweet, sour, salty, and bitter.
b. Stimulated taste sensory cells respond by releasing neurotransmitter.
c. Microvilli increase the surface area of taste sensory cells.
d. Individual taste buds are replaced every few days.
e. Taste sensory cells form synapses with dendrites of neurons.

41. Which of the following statements about mechanoreceptors is *false*?
a. Mechanoreceptors transduce mechanical forces into changes in receptor potential.
b. If a mechanoreceptor is subject to increased distortion, more ion channels within its membrane open.
c. Usually, sensory nerves are not involved in circuits returning from mechanoreceptors.
d. If the receptor potential of a mechanoreceptor rises above a threshold, an action potential is propagated.
e. Stimulus strength determines the rate of generated action potentials.

42. Which of the following receptors is located deep in the skin and is specifically adapted for sensing pressure?
a. Meissner's corpuscle
b. Pacinian corpuscle
c. Expanded-tip tactile receptors
d. Neuron-wrapped hair follicles
e. Bare nerve endings

43. Hair cells are associated with all but one of the following sensory systems. Select the exception.
a. Golgi tendon organ
b. Lateral line sensory systems
c. Statocysts
d. Vestibular apparatus
e. Semicircular canals

*44. Which of the following statements about the lateral line system is *false*?
a. The cupulae of the lateral line system each contain several hair cells.
b. Movement of mucus within the canal of the lateral line system displaces the cupulae.

c. The lateral line system allows the fish to sense the presence of other fish even in total darkness.
d. The lateral line system responds to pressure waves in the surrounding water.
e. The canal of the lateral system has numerous openings to the external environment.

45. Which of the following structures of the mammalian auditory system is responsible for signal amplification?
a. The tympanic membrane
b. The ear ossicles
c. The oval window
d. The cochlea
e. The round window

★46. Which of the following statements about the auditory functioning of the cochlea is *false*?
a. The flexing of the round window follows the flexing of the oval window in a delayed fashion.
b. The hair cells on the organ of Corti are moved against the rigid tectorial membrane.
c. The intensity of the sound determines how many hair cells will be stimulated.
d. The frequency of the sound determines which hair cells will be stimulated.
e. Lower frequency sounds result in the stimulation of hair cells closer to the round window.

47. Which of the following structures of the mammalian auditory system is involved in transduction of pressure changes into changes in receptor potential?
a. The tympanic membrane
b. The ear ossicles
c. The oval window
d. The organ of Corti
e. The basilar membrane

★48. Which of the following statements about rhodopsin is *false*?
a. Rhodopsin consists of a protein and a light-absorbing molecule.
b. When 11-*cis*-retinal absorbs light, it becomes all-*trans*-retinal.
c. Energy is required to convert all-*trans*-retinal into 11-*cis*-retinal.
d. Rhodopsin is a transmembrane protein.
e. All animals that can sense light do so using rhodopsin.

49. Which of the following statements about the functioning of a rod cell is *false*?
a. Rhodopsin is located in the stack of disks in the end of the rod cell that is farthest from the light source.
b. A rod cell is a modified neuron.
c. The resting potential of a rod cell in the dark is less negative than a typical neuron.
d. A membrane potential of a rod cell exposed to light becomes more positive.
e. The plasma membrane of a rod cell is fairly permeable to Na^+ ions.

★50. Which of the following statements about the molecular events of photoreception is *false*?
a. A single photon of light can excite a rhodopsin molecule.
b. A rod cell in the light would have most of its sodium channels open.
c. cGMP keeps the sodium channels open.
d. Activated phosphodiesterase (PDE) catalyzes the reaction hydrolyzing cGMP into GMP.
e. Activated transducin activates PDE.

▪51. If a planarian is positioned relative to a stationary light source so that more light-sensitive cells in its right eye cup are stimulated than those in its left eye cup, the planarian will
a. turn to the left.
b. turn to the right.
c. make a complete clockwise circle.
d. make a complete counterclockwise circle.
e. stop moving.

52. Which of the following statements about the compound eyes of arthropods is *false*?
a. Ommatidia are the optical units of compound eyes.
b. The number of ommatidia per eye can vary greatly in different species.
c. The light-sensitive cell of the ommatidium is called a rhabdom.
d. A single sensory neuron leaves each ommatidium.
e. Ommatidia have a lens system that focuses the light on the rhabdom.

53. Which of the following statements about the functioning of the vertebrate eye is *false*?
a. The cornea is transparent so it can transmit light.
b. The size of the pupil varies with light levels.
c. The iris is under control of the autonomic nervous system.
d. The lens focuses the image on the retina.
e. The fovea is the part of the retina with the lowest density of photoreceptors.

★▪54. Which of the following statements about accommodation is *false*?
a. Accommodation is the process whereby objects from different portions of the visual field are focused on the retina.
b. Vertebrates like fish and reptiles accommodate by moving the lens relative to the retina.
c. The suspensory ligaments keep the lens flattened.
d. The ciliary muscles contract when you attempt to focus on a distant object.
e. The ciliary muscles change the shape of the eye by counteracting the action of the suspensory ligaments.

55. Which of the following statements about vertebrate vision is *false*?
a. Visual acuity varies with the density of photoreceptors in the retina.
b. Some vertebrates have two foveas per eye.
c. A blind spot is always located where the optic nerve leaves the eye.

d. Unlike a camera, the vertebrate eye does not project inverted images on the retina.
e. Two major types of photoreceptor cells are found in the retina.

56. Which of the following statements about the vertebrate retina is *false*?
a. There are more rods than cones in the retinas of humans.
b. There is a higher proportion of cones than rods in the foveas of humans.
c. Cones give us our highest visual acuity.
d. Our peripheral vision involves more rod than cone cells.
e. Some animals that are entirely nocturnal have only cones in their retinas.

*57. Which of the following statements about color vision is *false*?
a. There are three different types of cone cells in the human retina.
b. The absorption spectra of cone cells differ because of molecular differences in the retinal molecules.
c. Rods do not contribute to color vision.
d. There are several different genes that code for opsin molecules in humans.
e. The proportion of the several different cone cells that respond to light determines the color that is perceived.

*58. Which of the following cell layers of the retina is responsible for the prevention of backscattering of light?
a. Horizontal
b. Pigmented
c. Bipolar
d. Amacrine
e. Ganglion

59. Which of the following cell layers of the retina is responsible for producing action potentials that are sent to the brain?
a. Horizontal
b. Pigmented
c. Bipolar
d. Amacrine
e. Ganglion

▪60. Which of these is the correct order of flow of information?
a. Stimulus, ion channel, action potential, receptor protein, neurotransmitter release
b. Stimulus, neurotransmitter release, action potential, ion channel, receptor protein
c. Stimulus, receptor protein, ion channel, neurotransmitter release, action potential
d. Stimulus, action potential, neurotransmitter release, receptor protein, ion channel
e. Stimulus, receptor protein, action potential, neurotransmitter release, ion channel

61. Chemosensors
a. are universal among animals.
b. can cause strong behavioral responses.
c. do not undergo adaptation.
d. a and b
e. a, b, and c

62. Meissner's corpuscles
a. adapt very slowly.
b. are present uniformly on skin surfaces.
c. sense pressure.
d. have concentric layers of connective tissue.
e. sense light touch.

63. The Golgi tendon organ
a. causes muscles to relax and protects against tearing.
b. senses light touch.
c. increases muscle contraction.
d. is found in high densities on lips and fingertips.
e. provides steady-state information about pressure.

▪64. Which of the following statements is *not* true of receptor potentials?
a. They can cause the release of a neurotransmitter.
b. They can cause an action potential.
c. They are a change in membrane potential of the sensory cell.
d. They can spread over long distances.
e. They can be amplified.

*65. Conduction deafness is caused by loss of function of
a. the inner ear.
b. the eustachian tube.
c. the tympanic membrane.
d. hair cells in the organ of Corti.
e. the cochlea.

Study Guide Questions

1. An electrode is inserted into a chemosensory nerve leading away from a taste bud in the mouth of a dog. A mild acid solution is then flushed continuously over the sensors associated with this nerve. Initially the nerve responds to this stimulation but over time ceases to carry action potentials. This observation would best be explained by
a. translocation.
b. adaptation of the sensory cells.
c. depletion of neurotransmitter in the sensory nerve.
d. second messenger influences that increase cell membrane potentials.

2. There are five different types of sensory cells that contribute to the sensory system. Which of the following statements about sensory cells is *false*?
a. Mechanosensors open ion channels in response to a pressure change.
b. Electrosensors control ion channels in response to changes in electric charge.

c. Thermosensors involve an intracellular messenger cascade.
d. Photoreceptors exhibit a conformational change when stimulated by light.

3. Silkworms use chemosensory molecules for mate attraction that are know as
a. rhodopsin.
b. hormones.
c. pheromones.
d. G proteins.

4. Stretch receptors in the aorta and carotid artery sense changes in arterial pressure. These sensors are
a. chemosensors.
b. thermosensors.
c. electrosensors.
d. mechanosensors.

5. All but which one of the following mechanosensor systems employs hair cells as its transducer?
a. Meissner's corpuscle
b. Statocyst
c. Organ of Corti
d. Vestibular apparatus

6. Which of the following statements about human gustation is *false*?
a. Taste sensors are relatively short-lived cells because of the high degree of abrasion they encounter.
b. Lemon juice is tasted most clearly on the sides of the tongue.
c. Receptors on the microvilli of taste sensors bind to food molecules.
d. Humans perceive only three categories of tastes: sweet, sour, and bitter.

7. Which of the following statements about the photosensitive molecule rhodopsin is *false*?
a. Opsin is converted from the 11-*cis* to the all-*trans* form upon light-induction.
b. The retinal is the light-absorbing group.
c. Photoexcited rhodopsin binds with transducin.
d. None of the above.

8. Which of the following cells in the human visual system send information directly to the brain?
a. Amacrine cells
b. Bipolar cells
c. Ganglion cells
d. Rods and cones

9. Through which of the following cell layers must a photon of light pass before striking a cone cell in the eye of a hawk?
a. Amacrine
b. Bipolar
c. Ganglion
d. All of the above

10. Which of the following statements about animal vision is *false*?
a. Photoreceptors in simple invertebrates produce image vision.
b. There are several types of cones in vertebrate retinas, each of which detects different wavelengths.
c. An ommatidium is only one visual element within the compound eye of an arthropod.
d. Although retinal actually absorbs the photon, the wavelength it absorbs best is determined by opsin.

11. Which of the following is true about stimulation of an off-center receptive field?
a. Stimulating the surround inhibits it.
b. Stimulating the surround excites it.
c. Stimulating the center excites it.
d. a and c

12. Which of the following is *not* a membrane found in the cochlea?
a. Reissner's membrane
b. Tectorial membrane
c. Tympanic membrane
d. Basilar membrane

13. Which of the following about generator potentials is *false*?
a. They create an action potential by opening sodium voltage-gated channels.
b. The receptor potential of a sensory cell becomes a generator potential at the axon hillock.
c. Generator potentials help transmit sensory stimuli.
d. A generator potential is created by neurotransmitter release at the axon hillock.

End of Chapter Questions

1. Which statement about sensory systems is *not* true?
a. Sensory transduction in vertebrate sensory systems involves the conversion (direct or indirect) of a physical or chemical stimulus into action potentials.
b. In general, a stimulus causes a change in the flow of ions across the plasma membrane of a sensory cell.
c. The term "adaptation" is given to the process by which a sensory system becomes insensitive to a continuing source of stimulation.
d. The more intense a stimulus, the greater the magnitude of each action potential fired by a sensory neuron.
e. Sensory adaptation plays a role in the ability of organisms to discriminate between important and unimportant information.

2. The female silkworm moth releases a chemical called bombykol from a gland at the tip of her abdomen. Bombykol is
a. a sex hormone.
b. detected by the male only when present in large quantities.
c. not species-specific.

d. **detected by hairs on the antennae of male silkworm moths.**
e. a chemical basic to the taste process in arthropods.

3. Which statement about olfaction is *not* true?
   a. **Dogs are unusual among mammals in that they depend more on olfaction than on vision as their dominant sensory modality.**
   b. Olfactory stimuli are recognized by the interaction between the stimulus and a specific macromolecule on olfactory hairs.
   c. The more odorant molecules binding to receptors, the more action potentials generated.
   d. The greater the number of action potentials generated by an olfactory receptor, the greater the intensity of the perceived smell.
   e. The perception of different smells results from the activation of different combinations of olfactory receptors.

4. The touch receptors that are located very close to the skin surface
   a. are relatively insensitive to light touch.
   b. **adapt very quickly to stimuli.**
   c. are uniformly distributed throughout the surface of the body.
   d. are called Pacinian corpuscles.
   e. adapt slowly and only partially to stimuli.

5. The membrane that gives us the ability to discriminate different pitches of sound is the
   a. round window.
   b. oval window.
   c. tympanic membrane.
   d. tectorial membrane.
   e. **basilar membrane.**

6. Which statement is *not* true?
   a. The transmembrane potential of a rod cell becomes more negative when the rod cell is exposed to light after a period of darkness.
   b. A photoreceptor releases the most neurotransmitter (per unit time) when in total darkness.
   c. Whereas in vision the intensity of a stimulus is encoded by the degree of hyperpolarization of photoreceptors, in hearing the intensity of a stimulus is encoded by changes in firing rates of sensory cells.
   d. Stiffening of the ossicles in the middle ear can lead to deafness.
   e. **The interaction among hammer (malleus), anvil (incus), and stirrup (stapes) conducts sound waves across the fluid-filled middle ear.**

7. In primates, the region of the retina where the central part of the visual field falls is called the
   a. central ganglion cell.
   b. **fovea.**
   c. optic nerve.
   d. cornea.
   e. pupil.

8. The region of the vertebrate eye where the optic nerve passes out of the retina is called the
   a. fovea.
   b. iris.
   c. **blind spot.**
   d. pupil.
   e. optic chiasm.

9. Which statement about the cones in a human eye is *not* true?
   a. They are responsible for high-acuity vision.
   b. They encode color vision.
   c. **They are more sensitive to light than rods are.**
   d. They are fewer in number than rods.
   e. They exist in high numbers at the fovea.

10. The color in vision results from the
    a. ability of each cone to absorb all wavelengths of light equally.
    b. lens of the eye acting like a prism and separating the different wavelengths by light.
    c. different absorption of wavelengths of light by different classes of rods.
    d. **three different isomers of opsin in different classes of cone cells.**
    e. absorption of different wavelengths of light by amacrine and horizontal cells.

Student Web Site Self-Quiz Questions

1. The membranes of sensory cells are specialized to respond to a variety of stimuli. Which of the following features do all sensory cells have in common?
   a. They are modified neurons.
   b. **They detect the stimulus by modifying the flow of ions across the cell membrane.**
   c. They produce action potentials.
   d. They amplify the signal via intracellular messenger cascades.
   e. They show adaptation.

2. Olfactory sensor cells and gustatory sensors have several features in common. Which of the following is a *false* statement about human olfactory and gustatory sensors?
   a. They are both chemoreceptors.
   b. In both, stimulus strength is transduced into the rate of action potential firing.
   c. Both systems depend on numerous specific receptor proteins.
   d. Both sensors have adaptations for increasing the membrane surface for receiving stimuli.
   e. **Both sensors are modified neurons.**

3. The skin is endowed with a variety of mechanosensors. Which of the following statements about mechanosensors is *false*?
   a. The body monitors blood pressure using mechanosensors.
   b. Meissner's corpuscles are very sensitive to light touch and adapt very rapidly.

c. Mechanosensors are incorporated into the mechanism for hearing in vertebrates.
d. Pacinian corpuscles sense heat and cold.
e. Mechanosensors transduce stimuli by acting directly on ion channels.

4. Stretch receptors help animals receive information about the stress and position of its limbs. Which of the following statements about stretch receptors is *false*?
a. Stretch receptors send information to the CNS only when stimulated.
b. Muscle spindles are modified muscle fibers innervated with a sensory neurons.
c. Stretch receptors are involved in coordination of limb movement.
d. Golgi tendon organs are found in ligaments.
e. Golgi tendon organs inhibit muscle contraction.

5. Which of the following structures is *not* involved in gathering sound, directing it to the sensors, or amplifying it?
a. Ear ossicles
b. Ear pinnae
c. Eustachian tube
d. Organ of Corti
e. Cochlear canal

6. Many mechanosensors are functionally dependent on hair cells and involved in equilibrium. Which of the following organs does *not* depend on hair cells for providing information on equilibrium and orientation?
a. Ampulla
b. Cochlea
c. Lateral line
d. Statocyst
e. Vestibular apparatus

7. Which of the following statements about the rhodopsin-initiated cascade is *false*?
a. Photons are absorbed by rhodopsin.
b. Na^+ channels are fully open in a stimulated photosensor.
c. An excited rhodopsin can activate many transducin molecules.
d. An activated PDE can hydrolyze many cyclic GMP molecules.
e. cGMP is not a second messenger in this system.

8. Which of the following structures of the vertebrate eye determines the ability of mammals and birds to accommodate?
a. Cornea
b. Fovea
c. Lens
d. Pupil
e. Sclera

TEST FILE

★ *Indicates a **difficult** question*

■ *Indicates a **conceptual** question*

46 The Mammalian Nervous System: Structure and Higher Functions

Fill in the Blank

1. The basic functional unit of the brain is the **neuron**.
2. The brain and spinal cord together are called the **central nervous system**.
3. The **peripheral nervous system** is made up of a network of nerves throughout the body.
4. Vision, hearing, touch, and balance make up **afferent** information.
5. **Efferent** information is sent from the brain to the muscles and glands.
6. Efferent pathways that are involuntary are also called the **autonomic** division.
7. ■ The nervous system can engage in many tasks at the same time. This is called **parallel** processing.
8. A spinal **reflex** occurs when afferent information is converted to efferent information without involvement of the brain.
9. A group of neurons that is anatomically or neurochemically distinct is called a(n) **nucleus**.
10. The transfer of short-term memory to long-term memory is the function of the **hippocampus**.
11. ★ The cortex is folded into ridges called **gyri** and valleys called **sulci**.
12. ★■ If a person is blind in one eye, they have difficulty discriminating **distances**.
13. Sleep researchers use an **electroencephalograph** to measure electric potential differences between neurons.
14. When experiences modify behavior, it is called **learning;** the ability of the brain to retain this is **memory**.
15. When two unrelated stimuli are linked to the same response, it is called a **conditioned** reflex.
16. ■ **Procedural** memory cannot be consciously recalled. It is remembering how to perform a motor skill, such as using a computer keyboard.
17. ★ A deficit in the ability to use or understand words is called a(n) **aphasia**.
18. The autonomic nervous system is crucial to the maintenance of **homeostasis** in the body.
19. The **neocortex** is an evolutionarily more recent part of the telencephalon found in birds and mammals.
20. In the visual cortex, cells that receive information from both eyes are called **binocular** cells.

Multiple Choice

1. What is the brain mostly made of?
   a. Neurons
   b. Axons
   c. Water
   d. Gray matter
   e. White matter

2. Afferent information flows __________ the CNS, and efferent information flows __________ the CNS.
   a. to, to
   b. to, from
   c. from, to
   d. from, from
   e. from and to, to

3. The hindbrain develops into which structure?
   a. Medulla
   b. Pons
   c. Cerebellum
   d. All of the above
   e. None of the above

4. The forebrain develops into which structure?
   a. Telencephalon
   b. Diencephalon
   c. Cerebellum
   d. a and b
   e. All of the above

5. The thalamus and hypothalamus develop from the
   a. telencephalon.
   b. diencephalon.
   c. cerebrum.
   d. cerebellum.
   e. hindbrain.

*6. Select the correct direction of flow of afferent information.
a. **Medulla, pons, midbrain, thalamus**
b. Medulla, pons, thalamus, midbrain
c. Telencephalon, thalamus, pons, medulla
d. Telencephalon, midbrain, pons, medulla
e. Pons, medulla, thalamus, midbrain

7. In general, the more autonomic functions are found in the _________, and the more complex functions are found in the _________.
a. forebrain, hindbrain
b. telencephalon, diencephalon
c. thalamus, hypothalamus
d. midbrain, hindbrain
e. **hindbrain, forebrain**

8. The biggest difference between the brains of humans and fish is the size of the
a. medulla.
b. cerebellum.
c. **cerebrum.**
d. diencephalon.
e. thalamus.

9. In the spinal cord, the gray matter contains the _________, and the white matter contains the _________.
a. axons, cell bodies
b. **cell bodies, axons**
c. dorsal horn, ventral horn
d. ventral horn, dorsal horn
e. afferent information, efferent information

*10. Efferent nerves leave the spinal cord through the
a. **ventral roots.**
b. dorsal roots.
c. gray matter.
d. interneurons.
e. ventral and dorsal horns.

*11. Interneurons are found in the
a. midbrain.
b. thalamus.
c. white matter of the spinal cord.
d. **gray matter of the spinal cord.**
e. telencephalon.

12. The function of the reticular system is to
a. conduct impulses through the spinal cord.
b. distribute information to its proper location in the forebrain.
c. **regulate the level of arousal of the nervous system.**
d. regulate physiological drives and emotion.
e. transfer short-term memory to long-term memory.

13. The reticular system is located in the
a. spinal cord.
b. midbrain.
c. hindbrain.
d. **b and c**
e. a, b, and c

14. The function of the limbic system is
a. regulation of instincts, emotions, and physiological drives.
b. to transfer short-term memory to long-term memory.
c. to regulate levels of arousal.
d. **a and b**
e. a, b, and c

15. What structure constitutes the largest part of the human brain?
a. **Telencephalon**
b. Diencephalon
c. Medulla
d. Pons
e. Cerebellum

16. The telencephalon is divided into two hemispheres covered by a sheet of gray matter called the
a. hippocampus.
b. reticular system.
c. sulci.
d. gyri.
e. **cerebral cortex.**

*17. The part of the brain that is involved in high-order information processing is called the
a. **association cortex.**
b. thalamus.
c. limbic system.
d. central sulcus.
e. hippocampus.

18. A patient can see and hear, but cannot recognize a familiar face. Which lobe of the cerebrum is most likely damaged?
a. **Temporal**
b. Occipital
c. Parietal
d. Frontal
e. Cerebellum

19. A patient cannot feel pressure applied to her hand even though nothing is wrong with her hand. Which lobe of the cerebrum is most likely damaged?
a. Temporal
b. Occipital
c. **Parietal**
d. Frontal
e. Cerebellum

20. A patient cannot see motion, even though nothing is wrong with his eyes. Which lobe of the cerebrum is most likely damaged?
a. Temporal
b. **Occipital**
c. Parietal
d. Frontal
e. Cerebellum

21. A patient suffers from a personality defect and cannot plan for future events. Which lobe of the cerebrum is most likely damaged?
a. Temporal
b. Occipital
c. Parietal
d. Frontal
e. Cerebellum

*22. A patient suffers from contralateral neglect syndrome, ignoring stimuli from the left side of the body. Which part of the brain has most likely been damaged?
a. The left parietal lobe
b. The right parietal lobe
c. The left frontal lobe
d. The right frontal lobe
e. The left temporal lobe

*23. The primary motor cortex is found in the _________ lobe and controls _________.
a. parietal, feeling sensation
b. parietal, movement
c. temporal, movement
d. frontal, movement
e. frontal, feeling sensation

*24. The primary somatosensory cortex is located in the _________ lobe and controls _________
a. parietal, feeling sensation
b. parietal, movement
c. temporal, feeling sensation
d. temporal, movement
e. frontal, feeling sensation

25. _________ have the largest brains in the animal kingdom, but _________ have the largest brain-to-body size ratio.
a. Dolphins, humans
b. Elephants, whales
c. Elephants, humans
d. Dolphins, humans
e. Humans, rodents

26. The fight-or-flight mechanisms are the function of the _________ branch of the autonomic nervous system.
a. sympathetic
b. parasympathetic
c. contralateral
d. efferent
e. afferent

27. The parasympathetic division controls
a. fight-or-flight response.
b. increased heart rate and blood pressure.
c. increased digestion and decreased heart rate.
d. increased release of epinephrine and glucose production.
e. memory.

*28. Preganglionic neurons use _________ as the neurotransmitter, while postganglionic neurons use _________ as the neurotransmitter.
a. norepinephrine, acetylcholine
b. acetylcholine, norepinephrine
c. norepinephrine or acetylcholine, norepinephrine
d. acetylcholine, norepinephrine or acetylcholine
e. norepinephrine or acetylcholine, norepinephrine or acetylcholine

29. The anatomy of the sympathetic and parasympathetic divisions differs. For instance, the preganglionic neurons of the sympathetic division meet their postganglionic connections
a. mostly in the brain stem.
b. in the upper regions of the spinal cord.
c. in ganglia arranged like chains along the spinal cord.
d. near the target organs.
e. in the midbrain.

30. Visual information follows the pathway:
a. eye, thalamus, occipital lobe, optic chiasm.
b. eye, thalamus, optic chiasm, occipital lobe.
c. eye, optic chiasm, occipital lobe, thalamus.
d. eye, thalamus, occipital lobe, optic chiasm.
e. eye, optic chiasm, thalamus, occipital lobe.

31. Complex cells in the visual cortex are stimulated by
a. specific colors.
b. bars of light with specific orientations and locations on the retina.
c. bars of light with any orientation at a specific location on the retina.
d. bars of light with specific orientations at any location on the retina.
e. any type of light flashed on the retina.

*32. Each retina sends _________ axons to the brain, received by about _________ neurons in the visual cortex.
a. 1 million, 200 million
b. 1 million, 2 million
c. 100 million, 2 million
d. 100 million, 200 million
e. 200 million, 1 million

▪33. David Hubel and Torsten Wiesel, in their studies on vision, found
a. many areas of the retina can stimulate a single cell in the visual cortex.
b. cells in the visual cortex respond to a receptive field on the retina.
c. cats can see bars of light at specific orientations only.
d. simple cells make connections to complex cells in the visual cortex.
e. visual information crosses over the optic chiasm.

34. During REM sleep, _________ occurs.
a. dreaming
b. loss of motor output by the brain
c. sleepwalking
d. a and b
e. a, b, and c

*35. On the cellular level, sleep occurs because of
a. hyperpolarization of the cells of the thalamus and cortex.
b. depolarization of the cells of the thalamus and cortex.
c. increased synaptic input between axons and neurons in the thalamus and cortex.
d. desynchronization of electrical impulses in the cortex.
e. decreased opening of potassium and calcium channels in the membrane of cortical cells.

36. Increased levels of adenosine in the brain probably contribute to
a. REM sleep.
b. muscle twitching during sleep.
c. deep, slow-wave sleep.
d. depolarization of cells in the thalamus and cortex.
e. waking after sleep.

*37. Long-term potentiation is
a. increased sensitivity to an electrical stimulation.
b. decreased sensitivity to an electrical stimulation.
c. habituation to a stimulus.
d. the application of a high-frequency electrical stimulation.
e. a decreased entry of calcium ions into the postsynaptic cell.

38. Long-term depression is
a. increased sensitivity to an electrical stimulation.
b. decreased sensitivity to an electrical stimulation.
c. the application of a continuous low-level stimulus.
d. the inability of neurons to fire an electrical impulse.
e. the inability to retain long-term memory.

▪39. The Russian physiologist Ivan Pavlov and his dog became famous for demonstrating
a. how short-term memory converts to long-term memory.
b. associative learning.
c. long-term potentiation.
d. eye blink reflex.
e. muscle twitches during REM sleep.

*40. In eye blink reflex studies, the conditioned reflex was localized to a region in the
a. medulla.
b. spinal cord.
c. cerebellum.
d. thalamus.
e. frontal lobe.

41. In order to try to cure a patient's severe epilepsy, both sides of his hippocampus were removed. An unfortunate side effect was
a. he could no longer feel emotions.
b. he lost his short-term memory.
c. he couldn't convert short-term memory into long-term memory.
d. he lost his immediate memory.
e. he could no longer recognize faces.

42. ________ memory is almost perfectly photographic.
a. Immediate
b. Short-term
c. Long-term
d. Declarative
e. Procedural

43. Short-term memory lasts about
a. a few seconds.
b. 5 or 10 minutes.
c. 20 or 30 minutes.
d. about one hour.
e. a few days.

44. Roger Sperry won the Nobel prize for his work on
a. the hippocampus and memory loss.
b. the conditioned reflex.
c. eye puffs on rabbits.
d. lateralization of language to the left hemisphere.
e. REM sleep.

45. When the corpus callosum is cut in a young child's brain, a likely result would be
a. language functions are learned by the right hemisphere.
b. each hemisphere would take on a separate personality.
c. loss of long-term memory.
d. a and b
e. a, b, and c

46. Damage to Broca's area results in loss of
a. understanding language.
b. speech or poor speech.
c. ability to recognize faces.
d. ability to read.
e. hearing.

47. Damage to Wernicke's area results in loss of
a. understanding language.
b. speech or poor speech.
c. ability to recognize faces.
d. hearing.
e. long-term memory.

48–51. From the following list, match the structure with its function in the following questions.
a. Medulla
b. Cerebellum
c. Diencephalon
d. Telencephalon

48. Contains the final relay station for sensory information going to the telencephalon **(c)**

49. Controls physiological functions such as breathing **(a)**

50. Orchestrates and refines motor commands **(b)**

51. Plays major roles in conscious behavior, learning, and memory **(d)**

52. Which of the following is *not* a function of the spinal cord?
   a. Generation of repetitive motor patterns
   b. Reflexes
   c. Conduction of motor impulses from the brain
   d. Refinement of motor and behavioral processes
   e. Conversion of afferent to efferent information

53. The knee jerk reflex
   a. involves a sensory neuron that synapses with a motor neuron in the ventral horn of the spinal cord.
   b. can be readily checked by a physician.
   c. is a monosynaptic reflex.
   d. involves the leg extensor muscle.
   e. All of the above

54. Norepinephrine __________ the heart rate, and acetylcholine __________ the heart rate.
   a. decreases, increases
   b. increases, decreases
   c. increases, increases
   d. decreases, decreases
   e. increases, neither increases nor decreases

*55. The human brain has about __________ neurons.
   a. 1 million
   b. 10 million
   c. 1 billion
   d. 10 billion
   e. 100 billion

▪56. Which of the following statements about nerves is *false*?
   a. A nerve is a bundle of axons.
   b. Some axons in a nerve may be carrying information to the central nervous system, while other axons in the same nerve are carrying information from the central nervous system to organs.
   c. A nerve is the axon of a single neuron.
   d. a and b
   e. None of the above

*57. Paresthesia
   a. is an enhancement of sensation.
   b. is a loss of motor function.
   c. may result from damage to the brain stem below the reticular system.
   d. results when the hippocampus is damaged.
   e. None of the above

▪58. Ivan Pavlov
   a. discovered that if he rang a bell after presenting food to a dog, the dog would eventually salivate at the sound of the bell alone.
   b. discovered operant conditioning.
   c. discovered that if he rang a bell before presenting food to a dog, the dog would eventually salivate at the sound of the bell alone.
   d. discovered a form of learning in which two very similar stimuli become linked to the same response.
   e. None of the above

Study Guide Questions

1. A man has damage to his brain that affects his ability to recognize the faces of people he knows. Where did the damage occur?
   a. Hypothalamus
   b. Temporal lobe
   c. Parietal lobe
   d. Frontal lobe

2. The secretion of hormones from an endocrine gland is most likely under the control of which of the following components of the nervous system?
   a. Autonomic
   b. Voluntary
   c. Dendritic
   d. Limbic

3. The motor cortex and somatosensory cortex of the cerebrum are
   a. mapped from the head region on the upper side of the cortex to the lower part of the body on the lower side of the cortex.
   b. areas of the brain where the most sensitive body areas are represented by the largest areas of the cortex.
   c. located in the parietal lobe.
   d. None of the above.

4. Which of the following statements about the sympathetic division of the autonomic nervous system is *false*?
   a. It increases heart rate.
   b. It relaxes the urinary bladder.
   c. It dilates blood vessels.
   d. It stimulates secretion by the sweat glands.

5. Which of the following is *not* part of the central nervous system?
   a. Brain stem
   b. Spinal gray matter
   c. Cerebellum
   d. Neuronal cell body of a sensory afferent

6. Observations of patients with aphasias indicate that
   a. only Broca's area is essential for normal language skills.
   b. language skills depend on proper neural flow between the temporal lobes and motor cortex.
   c. the right hemisphere dominates the production and use of language in humans.
   d. most people can control language functions from either cerebral hemisphere.

7. A friend wakes you from sleep, and you have the sensation of having just experienced a vivid dream. Which of the following would be *false*?
   a. Your hands and feet have been twitching slightly.
   b. Your eyes have been twitching.
   c. Most of your voluntary body muscles have been inactive.
   d. Your cerebral cortex was not as active as when you were awake.

8. If you want to find a complex cell in a cat's visual system, you need to look in
   a. Broca's area.
   b. the spinal cord.
   c. the occipital cortex.
   d. the reticular activating system.

9. Which one of the following statements about the peripheral nervous system is *false*?
   a. Afferent portions of the peripheral nervous system carry information to the CNS.
   b. Efferent portions of the peripheral nervous system carry information from the CNS.
   c. The peripheral nervous system communicates only with the circulatory and digestive systems.
   d. The efferent portion is broken down into voluntary and involuntary divisions.

10. Development of the vertebrate brain begins early in development. Which one of the following statements about the developing CNS is *false*?
    a. The CNS develops from a solid neural cylinder.
    b. The midbrain becomes part of the brain stem.
    c. The forebrain develops into both the diencephalon and the telencephalon.
    d. The hindbrain develops into the medulla, the pons, and the cerebellum.

11. In which of the following cortical lobes can association areas and long-term memory be found?
    a. Temporal
    b. Parietal
    c. Occipital
    d. All of the above

12. Which of the following animals would likely have the smallest ratio of telencephalon size to body size?
    a. Fishes
    b. Amphibians
    c. Mammals
    d. Reptiles

13. Which of the following statement about the knee-jerk reflex is *false*?
    a. It is a monosynaptic reflex.
    b. The leg extends in response to a hammer tap.
    c. Chemoreceptors sense the hammer tap.
    d. The afferent nerve travels from the receptor to the spinal cord.

14. The ridges of the cerebral cortex are called __________ and the valleys on the cerebral cortex are the __________.
    a. gyri, interneuron
    b. gyri, sulci
    c. sulci, gyri
    d. interneuron, gyri

End of Chapter Questions

1. Which of the following describes the route of sensory information from the foot to the brain?
   a. ventral horn, spinal cord, medulla, cerebellum, midbrain, thalamus, parietal cortex
   b. dorsal horn, spinal cord, medulla, pons, midbrain, hypothalamus, frontal cortex
   c. dorsal horn, spinal cord, medulla, pons, midbrain, thalamus, parietal cortex
   d. ventral horn, spinal cord, pons, cerebellum, midbrain, thalamus, parietal cortex
   e. dorsal horn, spinal cord, medulla, pons, midbrain, thalamus, frontal cortex

2. Which statement about the reticular system is *not* true?
   a. Increased activity in the reticular system induces sleep.
   b. It is located in the brain stem.
   c. Lesions of the reticular system in the midbrain can result in coma.
   d. Information from the spinal cord is routed to different brain stem nuclei and to the forebrain in the reticular system.
   e. There are groups of neurons called nuclei in the reticular system.

3. Which statement is *not* true?
   a. Sensory afferents carry information of which we are consciously aware.
   b. Visceral afferents carry information about physiological functions of which we are not consciously aware.
   c. The voluntary motor division of the efferent side of the peripheral nervous system executes conscious movements.
   d. The cranial nerves and spinal nerves are part of the peripheral nervous system.
   e. Afferent and efferent axons never travel in the same nerve.

4. Which statement is *not* true?
   a. In the spinal cord, the white matter contains the axons conducting information up and down the spinal cord.
   b. The limbic system is involved in basic physiological drives, instincts, and emotions.
   c. The limbic system consists of primitive forebrain structures.
   d. The vast majority of the nerve cell bodies in the human nervous system are contained within the limbic system.
   e. In humans, a part of the limbic system is necessary for the transfer of short-term memory to long-term memory.

5. Which of the following describes the largest amount of the human cerebral cortex?
   a. the frontal lobes
   b. the primary somatosensory cortex
   c. the temporal cortex
   d. the association cortex
   e. the occipital cortex

6. Which statement about the autonomic nervous system is *true*?
   a. The sympathetic division is afferent, and the parasympathetic division is efferent.
   b. The transmitter norepinephrine is always excitatory and acetylcholine is always inhibitory.
   c. Each pathway in the autonomic nervous system includes two neurons, and the neurotransmitter of the first neuron is acetylcholine.
   d. The cell bodies of many sympathetic preganglionic neurons are in the brain stem.
   e. The cell bodies of most parasympathetic postganglionic neurons are in or near the thoracic and lumbar spinal cord.

7. Which statement about cells in the visual cortex is *not* true?
   a. Many cortical cells receive inputs directly from single retinal ganglion cells.
   b. Many cortical cells respond most strongly to bars of light falling at a specific location on the retina.
   c. Some cortical cells respond most strongly to bars of light falling anywhere over large areas of the retina.
   d. Some cortical cells receive inputs from both eyes.
   e. Some cortical cells respond most strongly to an object when it is a certain distance from the eyes.

8. Which of the following characterizes non-REM sleep?
   a. dreaming
   c. EEG slow waves
   d. rapid and jerky eye movements
   e. it makes up about 20 percent of total sleep time

9. Which conclusion was supported by experiments on split-brain patients?
   a. Language abilities are localized mostly in the left cerebral hemisphere.
   b. Language abilities require both Wernicke's area and Broca's area.
   c. The ability to speak depends on Broca's area.
   d. The ability to read depends on Wernicke's area.
   e. The left hand is served by the left cerebral hemisphere.

10. In the knee jerk reflex
   a. spinal interneurons inhibit the motor neuron of the antagonistic muscle.
   b. the cell body of the muscle stretch receptor is in the dorsal horn of the spinal cord.
   c. the cell body of the motor neuron is in the dorsal horn of the spinal cord.
   d. action potentials in the sensory neuron release inhibitory neurotransmitter onto the motor neurons.
   e. the sensory neuron forms a monosynaptic loop with the motor neuron to the antagonistic muscle.

Student Web Site Self-Quiz Questions

1. Which of the following statements about the human nervous system is *false*?
   a. The central nervous system (CNS) includes the brain, spinal cord, cranial nerves, and spinal nerves.
   b. The human brain contains approximately 10 billion nerve cells or neurons.
   c. A nerve can contain axons from many different neurons.
   d. A nerve can contain neurons conveying information from the CNS to the body and from the body to the CNS.
   e. The peripheral nervous system (PNS) reaches every tissue of the body.

2. Which of the following structures is *not* a part of the brain stem?
   a. Pons
   b. Telencephalon
   c. Medulla
   d. Midbrain
   e. Reticular system

3. Which of the following statements about the reticular system is *false*?
   a. The reticular system consists of a number of distinct nuclei distributed throughout the brain stem.
   b. The reticular system receives sensory input from many areas of the body.
   c. High levels of activity in the reticular system would adversely affect your ability to sleep.
   d. Damage to the reticular activating system may cause a person to become comatose.
   e. If a person with damage to the brainstem suffers from paralysis but patterns of sleep and wakefulness are normal, you can conclude that the damage is likely to be above the reticular system.

4. Which one of the major regions of the cerebral cortex has association areas involved with recognition, identification, and naming of objects?
   a. Central sulcus
   b. Frontal lobe
   c. Occipital lobe
   d. Parietal lobe
   e. Temporal lobe

5. Complete the following sentence about the autonomic innervation of the heart: The preganglionic neuron of the sympathetic nervous system releases __________, the postganglionic neuron releases __________, and the effect is to __________ the heart rate.
   a. acetylcholine, norepinephrine, increase
   b. acetylcholine, norepinephrine, decrease
   c. acetylcholine, acetylcholine, increase
   d. acetylcholine, acetylcholine, decrease
   e. norepinephrine, acetylcholine, increase

6. Which of the following statements about the integration of visual information in the visual cortex is *false*?
   a. Both simple cells and complex cells are located in the visual cortex.
   b. Simple cells receive input from several ganglion cells in the retina whose receptive fields are lined up in a row.
   c. Simple cells respond best when stimulated by a spot of light moving in a particular direction.
   d. Complex cells can receive input from ganglion cells with receptive fields in widely separated parts of the retina.
   e. Some complex cells respond best when a bar of light moves in a particular direction across their receptive fields.

7. Which of the following statements about the anatomy of binocular vision is *false*?
   a. Visual information from the right eye is projected onto the visual cortex of the left hemisphere and vice versa.
   b. Axons from ganglion cells in the retinas of both eyes cross to the opposite side of the brain at the optic chiasm.
   c. Binocular cells are located on the borders between the columns of cortical cells.
   d. Binocular cells interpret distance by measuring the disparity between where the same stimulus falls on the two retinas.
   e. Some binocular cells respond best to a stimulus falling on both retinas.

8. Which of the following statements about sleep in humans is *false*?
   a. Stages 3 and 4 of non-REM sleep are characterized by a deep, restorative slow-wave sleep pattern.
   b. REM sleep is associated with almost complete paralysis of skeletal muscle.
   c. High levels of adenosine in the extracellular fluid of the brain increases the depth and duration of non-REM sleep.
   d. During sleep, neural activity of certain brainstem nuclei decreases, resulting in depolarization of the thalamus and cortex.
   e. Sleepwalking usually occurs during non-REM sleep.

9. Complete the following sentence. The part of the limbic system called the __________ is involved with __________.
   a. thalamus, relaying sensory information going to the telencephalon
   b. reticular system, activating higher brain regions
   c. neocortex, associative learning
   d. hippocampus, transfer from short-term to long-term memory
   e. hippocampus, procedural memory

10. If a split-brain patient with a severed corpus callosum is blindfolded and a pencil is placed in his left hand, the patient would
    a. be able to use and name the pencil.
    b. be able to use the pencil, but not to name it.
    c. be able to name the pencil, but not to use it.
    d. not be able to use or name the pencil.
    e. be able to use and name the pencil but not remember having done so

★ *Indicates a **difficult** question*

■ *Indicates a **conceptual** question*

47 Effectors: Making Animals Move

Fill in the Blank

1. Microtubules are comprised of subunits of **tubulin** protein.

2. Intercalated discs join together cells in **cardiac** muscle tissue.

3. Huxley and Huxley proposed the **sliding filament** theory of muscle contraction.

4. An action potential spreading through the muscle fiber causes a minimal contraction known as a **twitch**.

5. **Tetanus** is the maximum tension of a muscle contraction.

6. Simple **hydrostatic** skeletons consist of a fluid-filled cavity controlled by muscle movement.

7. Cartilage tissue consists of **collagen** fibers in a rubbery matrix of proteins and polysaccharides.

8. Cells that break down or reabsorb bone are the **osteoclasts**.

9. Muscles at joints are often arranged in antagonistic pairs of extensors and **flexors**.

10. A type of cellular effector, **nematocysts** are fired like miniature missiles by jellyfish to capture prey.

11. **Effectors** are adaptations that animals use to respond to information that is sensed, integrated, and transmitted by their neural and endocrine systems.

12. **Smooth** muscle provides the contractile forces for most of our internal organs, which are under control of the autonomic nervous system.

★13. The **sarcoplasmic reticulum** is a highly specialized network of intracellular membranes of striated muscle cells that enables them to sequester and release calcium for muscle contraction.

14. The living cells of bone that are responsible for the remodeling of bone structure are called **osteoblasts**.

15. The heartbeat is **myogenic**, generated by the heart muscle itself.

16. **Chitin**, a nitrogen-containing polysaccharide, is found in the endocuticle of arthropods.

17. In humans, the **axial** skeleton includes the skull, vertebral column, and ribs.

Multiple Choice

1. What mechanism is responsible for movements of cilia and flagella?
   a. Muscle contraction
   b. The microfilament system
   c. The microtubule system
   d. Crystal formation
   e. The skeletal system

2. Which of the following does *not* rely on ciliary movement?
   a. Movement of an egg through an oviduct
   b. Movement of particles through the airways of lungs
   c. Feeding movements of a paramecium
   d. Movement of particles over the gills of a clam
   e. Amoeboid movement

3. The movement of eukaryote flagella resembles
   a. turning a corkscrew.
   b. a swimmer's stroke.
   c. cracking a whip.
   d. an airplane propeller.
   e. rowing a boat.

★4. The beating of cilia and flagella occurs by
   a. contraction of individual tubules.
   b. contraction of individual tubulin molecules.
   c. attraction of sliding filaments to each other.
   d. moving dynein attachments between filaments.
   e. contraction of dynein in individual tubules.

5. Which of the following substances is required for any movement of cilia or flagella?
   a. Actin
   b. ATP
   c. Auxin
   d. Proteolytic enzymes
   e. a and b

6. Which of the following statements about fast-twitch and slow-twitch fibers is *false*?
   a. The most important factor determining the proportion of fast- and slow-twitch fibers in skeletal muscles is genetic heritage.
   b. Aerobic training increases the oxidative capacity of fast-twitch fibers.
   c. A single muscle may contain fast- and slow-twitch fibers.
   d. Fast-twitch fibers fatigue slowly.
   e. None of the above

7. The extension of a pseudopod in amoeboid movement occurs by
   a. extension of actin microfilaments next to the membrane.
   b. contraction of the plasmagel next to the membrane.
   c. the changing of plasmagel to plasmasol, which expands and bulges outward.
   d. microfilament-generated cytoplasmic streaming of the plasmasol.
   e. rapid expansion of cell membrane and filling with plasmagel.

*8. The resting potential of smooth muscle cells is
   a. not subject to forming action potentials.
   b. affected by stretching the cells.
   c. more negative than in most cells.
   d. unaffected by nearby potential change.
   e. nearly zero.

9. The striated appearance of skeletal muscle is due to the
   a. dark color of myosin.
   b. multiple nuclei per fiber.
   c. regular arrangement of filaments.
   d. dense array of microtubules.
   e. dense packing of ATP molecules.

10. An individual sarcomere unit consists of
   a. a stack of actin fibers.
   b. a stack of myosin units.
   c. overlapping actin and membrane.
   d. overlapping myosin and membrane.
   e. overlapping actin and myosin.

▪11. How do muscle fibers shorten during contraction?
   a. Individual protein filaments contract.
   b. More cross-bridges between filaments are formed.
   c. Arrays of filaments overlap each other.
   d. Protein filaments coil up more tightly.
   e. Subunits of protein polymers detach.

12. How do actin and myosin molecules interact?
   a. Globular myosin heads bind to actin filaments.
   b. Globular actin heads bind to myosin filaments.
   c. Other proteins connect between the two.
   d. Myosin filaments bend to connect to actin.
   e. Actin filaments bend to connect to myosin.

13. Why do muscles stiffen in rigor mortis when animals die?
   a. Without ATP, muscles cannot contract.
   b. Without ATP, actin and myosin cannot bind.
   c. Without ATP, actin and myosin cannot separate.
   d. ATP is required for synthesis of protein filaments.
   e. ATP forms cross-bridges between filaments.

14. Vertebrate skeletal muscles are excitable cells because they
   a. can be stimulated by ATP.
   b. can be stimulated by an electric charge.
   c. can secrete neurotransmitter.
   d. possess voltage-gated sodium channels.
   e. can attain a high level of activity.

*15. What is the role of the sarcoplasmic reticulum in muscle contraction?
   a. It stores calcium ions for release during contraction.
   b. It surrounds and protects the muscle filaments.
   c. It provides sites of ATP synthesis.
   d. It depolarizes when stimulated by an impulse.
   e. It synthesizes actin and myosin filaments.

16. How does tropomyosin control muscle contraction?
   a. It provides a bridge between actin and myosin.
   b. It provides a site where ATP can be utilized.
   c. Changing its position exposes myosin binding sites.
   d. It transmits electric charge to the filaments.
   e. Changing its shape opens membrane channels.

▪17. How can muscle fibers show a range of responses to different levels of stimulation?
   a. Each muscle fiber contraction is all or none.
   b. Calcium ion availability sets an upper limit.
   c. A new contraction can occur only after resting condition is reached.
   d. Following a stimulation, the fiber stays contracted.
   e. Individual twitches in the same fiber can summate.

18. The legs of cross-country skiers and long-distance runners are likely to have
   a. almost all slow-twitch fibers.
   b. almost all fast-twitch fibers.
   c. about the same number of slow-twitch as fast-twitch fibers.
   d. more slow-twitch fibers.
   e. more fast-twitch fibers.

19. Fast-twitch skeletal muscle fibers, called white muscle, are characterized by
   a. high concentration of myoglobin.
   b. abundant mitochondria.
   c. rapid development of high tension.
   d. sustaining activity for a long time.
   e. higher oxygen requirements.

20. What is a feature of cardiac muscle that helps the heart withstand high pressures without tearing?
   a. The fibers branch and intertwine.
   b. The fibers are arranged in parallel.
   c. Each fiber has a single nucleus.
   d. The fibers have gap junctions between them.
   e. Some fibers have pacemaking functions.

21. The exoskeleton of clams contains
    a. calcium carbonate.
    b. cartilage.
    c. chitin.
    d. collagen.
    e. cuticle.

22. A disadvantage of an arthropod's exoskeleton is that it
    a. protects against abrasion.
    b. prevents water loss.
    c. provides attachment sites for muscles.
    d. bends at the animal's joints.
    e. is soft just after molting.

23. An advantage of an endoskeleton over an exoskeleton is that it
    a. provides muscle attachment sites.
    b. provides protection.
    c. grows as the animal grows.
    d. supports the animal's weight.
    e. gives structure to the animal.

24. What is the role of cartilage in the skeleton?
    a. To bear a heavy load
    b. To add flexibility
    c. To be lightweight
    d. To sustain vibrations
    e. To grow rapidly

25. Bone tissue consists of __________ cells in a matrix of __________.
    a. osteocyte, polysaccharide
    b. osteocyte, collagen and calcium phosphate
    c. osteocyte, myosin and actin
    d. collagen, calcium phosphate
    e. collagen, polysaccharide

26. How do osteoblasts cause bone to grow?
    a. They lay down new matrix until surrounded.
    b. They increase the number of cells within the matrix.
    c. They lay down new matrix on bone surface.
    d. They tear down old bone and deposit new bone.
    e. They tear down old cartilage and deposit bone.

27. Haversian systems consist of
    a. osteocytes that connect different cavities.
    b. osteoblasts that will give rise to osteoclasts.
    c. a meshwork of cancellous bone tissue.
    d. cylindrical units surrounding a canal of blood vessels.
    e. a reinforced system of bone that resists fracturing.

28. The role of tendons is to join __________ together.
    a. two bones
    b. two ligaments
    c. bone and ligament
    d. muscle and ligament
    e. muscle and bone

▪29. Which of the following does *not* act as an effector?
    a. The poisonous secretion of pufferfish
    b. The mercaptan spray of skunks
    c. Pheromone secretion by butterflies
    d. Sensory neuron stimulation
    e. The secretion of saliva in the mouth

30. Which of the following functions is not governed by microfilaments?
    a. Formation of daughter cells following mitosis
    b. Support of intestinal cell microvilli
    c. Skeletal muscle contraction
    d. Changes in cell shape
    e. Phagocytosis

31. Whether a muscle contraction is strong or weak depends both on how many motor neurons to that muscle are firing, and the rate at which those neurons are firing. These two factors can be thought of as spatial __________ and temporal __________, respectively.
    a. transmission, transduction
    b. transduction, transmission
    c. organization, summation
    d. summation, summation
    e. coordination, summation

▪32. What function does the hydrostatic skeleton of invertebrates perform?
    a. Skeletal muscle contraction
    b. Skeletal muscle expansion
    c. Locomotion
    d. Propulsion
    e. b, c, and d

33. Which of the following joints is *not* found in humans?
    a. Ball-and-socket joint
    b. Hinge joint
    c. Pivot joint
    d. Saddle and ellipsoid joints
    e. All of the above are found in humans.

34. One group of invertebrate effectors that enables animals to change color is called
    a. nematocysts.
    b. chromatophores.
    c. electroplates.
    d. poison glands.
    e. retractable stingers.

▪35. Which of the following general statements about effectors is *not* true?
    a. Any mechanism that an animal uses to respond is an effector.
    b. All effector action that involves movement is dependent on the interaction of microfilaments or microtubules.
    c. Muscle action depends on the movement of microtubules.
    d. Movement due to microfilaments and microtubules depends on the sliding action of long protein molecules.
    e. Effectors rely on energy made available by ATP.

▪36. Which of the following statements about cilia and flagella is *false*?
    a. The cilium encounters the greatest resistance to movement during the power stroke.
    b. Flagella are longer and less numerous than cilia.

c. The same molecular mechanism powers both the cilium and the flagellum.
d. The pattern of movement of cilia and flagella is the same.
e. Flagella are used to create feeding currents in sponges and swimming in sperm of most animals.

▪37. In which of the following locations would you *not* expect to find cilia?
a. Within the air passages of the lungs in humans
b. Within the female reproductive tract in humans
c. On the surface of intestinal cells in the human digestive tract
d. On the surface of a clam gill
e. On a filter-feeding animal

*38. Which of the following statements about the molecular mechanism powering both the cilium and the flagellum is *not* true?
a. Nine pairs of microtubules are found in the axonemes of both cilia and flagella.
b. The globular protein tubulin is polymerized to form the hollow, tubular microtubules.
c. Radial cross-arms of microtubules consist of a molecule of dynein.
d. A mechanoenzyme converts chemical energy of ATP into movement.
e. The microtubules in a cilium cannot slide past each other because at any given time some cross-arms link the microtubules together.

*▪39. What happens to axonemes that are isolated from the cell when all components are removed except for microtubules and the dynein cross-arms?
a. The microtubules will flex normally if ATP and Ca^{2+} are available.
b. The microtubules will telescope past each other if ATP and Ca^{2+} are available.
c. The microtubules lose all motility even if ATP and Ca^{2+} are available.
d. The microtubules will flex normally even without ATP and Ca^{2+}.
e. The microtubules will telescope past each other even without ATP and Ca^{2+}.

40. The polymerization and depolymerization of tubulin is involved in helping to cause all of the following phenomena except one. Select the exception.
a. Movement of chromosomes during mitosis and meiosis
b. Growth of neurons during development
c. Changing cell shape
d. Amoeboid movement
e. Movement of organelles

41. The dominant microfilament component in cells is
a. actin.
b. myosin.
c. tubulin.
d. dynein.
e. troponin.

42. Microfilaments are involved in helping to cause all of the following phenomena except one. Select the exception.
a. Movement of chromosomes during mitosis and meiosis
b. The division of cytoplasm during animal cell mitosis
c. Changing cell shape
d. Amoeboid movement
e. Phagocytosis and pinocytosis

▪43. Which of the following statements about amoeboid movement is *false*?
a. Plasmasol is located in the interior of the cell.
b. Plasmagel contains a network of actin microfilaments that interact with myosin.
c. The cell moves by flowing into its pseudopodia.
d. During amoeboid movement, the plasmasol contracts.
e. The pseudopod stops forming when the cytoplasm at the leading edge converts to gel.

*44. Which of the following statements about bone is *false*?
a. When astronauts spend long periods of time in zero gravity, they experience decalcification of their bones.
b. Osteoclasts are derived from the lineage of cells that produce red blood cells.
c. Most bones have compact and cancellous regions.
d. Bones may thicken or thin depending on the amount of stress placed on them.
e. None of the above

45. Which of the following statements about smooth muscle is *false*?
a. Smooth muscle is under the control of the autonomic nervous system.
b. Smooth muscle cells are multinucleate.
c. Smooth muscle does not have a striated appearance when viewed with the microscope.
d. Gap junctions are common in smooth muscle.
e. Stretched smooth muscle will contract.

*▪46. Which of the following events does *not* occur during muscle contraction?
a. The distance between Z lines increases.
b. The sarcomere shortens.
c. The H zone is reduced.
d. The I band is reduced.
e. The area with both actin and myosin increases.

47. Which of the following statements about the molecular arrangement of actin and myosin in myofibrils is *false*?
a. A thin filament consists of actin and tropomyosin.
b. Two chains of actin monomers are twisted into a helix.
c. Two strands of tropomyosin lie in the grooves of the actin.
d. Troponin forms the head of the myosin molecule.
e. The myosin heads have ATPase activity and interact with the actin.

■48. Starting with the arrival of an action potential at the neuromuscular junction, which of the following is the correct order of events?
a. Calcium is released from the sarcoplasmic reticulum, an action potential travels down the T tubules, depolarization spreads through the T tubule, and myosin binds actin.
b. An action potential travels down the T tubules, depolarization spreads through the T tubule, calcium is released from the sarcoplasmic reticulum, and myosin binds actin.
c. An action potential travels down the T tubules, depolarization spreads through the T tubule, calcium is taken up by the sarcoplasmic reticulum, and myosin binds actin.
d. An action potential travels down the T tubules, depolarization spreads through the T tubule, ATP binds to myosin, and myosin binds actin.
e. A T tubule is depolarized, calcium is released from the sarcoplasmic reticulum, an action potential is created in the muscle cell, and myosin binds actin.

49. Which of the following statements about a muscle twitch is *false*?
a. A twitch is usually caused by a single action potential.
b. A twitch is an all-or-none response.
c. During tetanus, the twitching of many muscle cells keeps the whole muscle in a state of tension.
d. During a twitch, the actin and myosin filaments return to their resting positions before the next stimulus.
e. During a twitch, the Ca^{2+} are pumped back into the sarcoplasmic reticulum before the next stimulus.

50. During tonus
a. different muscles are taking responsibility for maintaining posture.
b. the muscle has generated maximum tension.
c. the muscle will eventually exhaust its supply of ATP.
d. a small but changing number of motor units are active.
e. the maximum number of action potentials is being received by the muscle.

51. Fast- and slow-twitch fibers differ in all of the following ways except one. Select the exception.
a. Number of mitochondria
b. Amount of myoglobin
c. Amount of glycogen and fat
d. Number of neuromuscular junctions
e. Strength of twitch

52. Which of the following phenomena does *not* involve a hydrostatic skeleton?
a. Burrowing in earthworms
b. Swimming in scallops
c. Retraction of tentacles and body in the sea anemone
d. Jet propulsion in squid
e. b and d

53. Which of the following statements about exoskeletons is *false*?
a. During molting, the old exoskeleton is reabsorbed and a new exoskeleton forms.
b. A clam shell is a exoskeleton.
c. The outer layer of the arthropod cuticle prevents water loss in terrestrial species.
d. The inner layer of the arthropod exoskeleton is used for muscle attachment.
e. The arthropod exoskeleton is a continuous covering that is thinned at the joints.

54. Which of the following statements about the endoskeletons of vertebrates is *false*?
a. The muscles are attached to a living support structure.
b. The rib bones are part of the appendicular skeleton.
c. Molting is unnecessary in animals with endoskeletons.
d. Endoskeletons consist of bone and cartilage.
e. At least two bones are required to form a joint.

55. Which of the following statements about cartilage and bone is *false*?
a. Some vertebrates are composed entirely of cartilage and no bone.
b. Some vertebrates are composed entirely of bone and no cartilage.
c. Bone has crystals of calcium phosphate.
d. The endoskeleton in many vertebrates serves as a reservoir of calcium.
e. The principal protein in cartilage is collagen.

56. Long bones begin as __________ bones. At maturity the shaft of a long bone consists of a __________ of __________ bone.
a. dermal, cylinder, compact
b. dermal, solid rod, cancellous
c. cartilage, cylinder, compact
d. cartilage, cylinder, cancellous
e. cartilage, solid rod, compact

57. Which of the following statements about skeletal systems as a series of joints and levers is *false*?
a. An extensor muscle moves a bone closer to the body.
b. Antagonistic pairs include an extensor muscle and a flexor muscle associated with the same joint.
c. Ligaments hold joints together.
d. Tendons attach muscles to bones.
e. Ligaments sometimes hold tendons in position.

★■58. If you were designing a skeletal lever system for maximum power, you would
a. insert both the muscle and the load force close to the joint.
b. insert both the muscle and the load force far from the joint.
c. insert the muscle close to the joint and place the load far from the joint.
d. insert the muscle far from the joint and place the load close to the joint.
e. use a large muscle.

*59. Which of the following mechanisms is *not* known to occur in chromatophore-mediated color adaptation?
   a. Rapid synthesis or destruction of pigment within chromatophores
   b. Dispersal or aggregation of pigment granules within chromatophores
   c. Shape change of chromatophores due to amoeboid movement
   d. Shape change of chromatophores due to interaction with muscle tissue
   e. All of the above

▪60. Glands are involved in all of the following, *except*
   a. intracellular communication.
   b. intercellular communication.
   c. communication between different individuals.
   d. defense.
   e. temperature regulation.

61. Which of the following statements about electric organs is *false*?
   a. In some species production of electric fields is used for defense or prey capture.
   b. In some species production of electric fields is used for orientation.
   c. Electric organs are derived from nervous tissue.
   d. Nerves, muscles, and electric organs use the same mechanisms to produce electric potentials.
   e. Large electric currents are achieved by coordinating the discharge of the cells in the electric organ.

62. Which of these muscle types has gap junctions?
   a. Smooth
   b. Cardiac
   c. Skeletal
   d. a and b
   e. a, b, and c

63. Which of these muscle types is multinucleated?
   a. Smooth
   b. Cardiac
   c. Skeletal
   d. b and c
   e. a, b, and c

64. Which of these muscle types has the least regular arrangements of actin and myosin?
   a. Smooth
   b. Cardiac
   c. Skeletal
   d. a and b
   e. a, b, and c

65. Which of these muscle types uses calcium to trigger actin–myosin interactions for movement?
   a. Smooth
   b. Cardiac
   c. Skeletal
   d. a and b
   e. a, b, and c

66. Which of these muscle types has pacemaking function?
   a. Smooth
   b. Cardiac
   c. Skeletal
   d. a and b
   e. a, b, and c

▪67. __________ complexes with calcium and acts as a second messenger in muscle cells.
   a. Myosin
   b. Actin
   c. Troponin
   d. cAMP
   e. Calmodulin

68. Which type of joint is found in the elbow?
   a. Ball and socket
   b. Pivotal
   c. Hinge
   d. Saddle
   e. Plane

69. Which type of joint allows almost complete rotational movement?
   a. Ball-and-socket
   b. Pivotal
   c. Hinge
   d. Saddle
   e. Plane

70. An annelid moves by
   a. stretching its longitudinal muscles and pushing the fluid-filled body cavity forward.
   b. forming pseudopods.
   c. ciliary movement.
   d. alternating contractions of longitudinal and circular muscles.
   e. contraction of circular muscles, which puts pressure on the hydrostatic skeleton causing it to extend.

71. When troponin binds calcium,
   a. it allows tropomyosin to bind actin.
   b. ion channels open and sodium rushes into the muscle cells.
   c. it changes conformation, twisting tropomyosin and exposing the actin–myosin binding site.
   d. it changes conformation, exposing the ATP binding site and allowing the actin–myosin bond to break.
   e. calmodulin binds to calcium and starts a cascade by activating myosin kinase.

Study Guide Questions

1. Both cilia and flagella are used for movement in many organisms. What protein uses the energy from ATP hydrolysis to cause the cilia or flagella to move?
   a. Calmodulin
   b. Dynein
   c. Axoneme
   d. None of the above

2. A motor unit is best described as:
   a. all the nerve fibers and muscle fibers in a single muscle bundle.
   b. one muscle fiber and its single nerve fiber.
   c. a single motor neuron and all the muscle fibers that it innervates.
   d. the neuron that provides the CNS with information on the state of contraction of the muscle.

3. Which of the following statements is *false*?
   a. Cardiac muscle is striated.
   b. Smooth muscle does not contain actin.
   c. Skeletal muscle is considered voluntary.
   d. Smooth muscle is found in the iris of the eye and the walls of the bladder.

4. The compound acting as an oxygen store in skeletal muscle is
   a. myoglobin.
   b. hemoglobin.
   c. ATP.
   d. myokinase.

5. The action potential that triggers a muscle contraction travels deep within the muscle cell by means of
   a. sarcoplasmic reticulum.
   b. transverse tubules.
   c. synapses.
   d. motor end plates.

6. A sarcomere is best described as a
   a. moveable structural unit within a myofibril bounded by H zones.
   b. fixed structural unit within a myofibril bounded by Z lines.
   c. fixed structural unit within a myofibril bounded by A bands.
   d. moveable structural unit within a myofibril bounded by Z lines.

7. ATP provides the energy for muscle contraction by allowing for
   a. an action potential formation in the muscle cell.
   b. cross-bridge detachment of myosin from actin.
   c. cross-bridge attachment of myosin to actin.
   d. release of calcium by sarcoplasmic reticulum.

8. Ca^{2+} binds to __________ in skeletal muscle and leads to exposure of the binding site for __________ on the __________ filament.
   a. troponin, myosin, actin
   b. troponin, actin, myosin
   c. actin, myosin, troponin
   d. tropomyosin, myosin, actin

9. Tropmyosin is moved by which of the following proteins?
   a. Calmodulin
   b. Acetylcholine
   c. Actin
   d. Troponin

10. Summation of frequent muscle twitches to give a smooth sustained contraction is called
    a. motor unit summation.
    b. treppe.
    c. facilitation.
    d. tetanus.

11. __________ are responsible for regulating extracellular calcium balance by releasing or sequestering Ca^{2+} ions in the bone matrix.
    a. Osteoblasts
    b. Osteoblasts and osteoclasts
    c. Osteoclasts and osteocytes
    d. Osteoblasts, osteoclasts, and osteocytes

12. The primary difference between an endoskeleton and an exoskeleton has to do with
    a. the presence of both circular and longitudinal muscles.
    b. whether or not the skeleton is on the inside of the body.
    c. the presence or absence of joints.
    d. the amount of fluid in the body.

13. A soccer player suffers a knee injury that damages the tissue holding his upper and lower leg bones together. The damaged tissue is probably a kind of
    a. muscle.
    b. tendon.
    c. ligament.
    d. cartilage.

End of Chapter Questions

1. The movement of cilia and flagella is due to the
   a. polymerization and depolymerization of tubulin.
   b. making and breaking of cross-bridges between actin and myosin.
   c. contractions of microtubules.
   d. changes in conformations of dynein molecules.
   e. use of ATP by spokes of the axoneme to contract.

2. Smooth muscle differs from both cardiac and skeletal muscle in that
   a. it can act as a pacemaker for rhythmic contractions.
   b. contractions of smooth muscle are not due to interactions between neighboring microfilaments.
   c. neighboring cells can be in electrical continuity through gap junctions.
   d. neighboring cells are tightly coupled by intercalated discs.
   e. the membranes of smooth muscle cells are depolarized by stretching.

3. Fast-twitch fibers differ from slow-twitch fibers in that
   a. they are more common in the leg muscles of champion sprinters.
   b. they have more mitochondria.
   c. they fatigue less rapidly.
   d. they abundance is more a product of genetics than of training.
   e. they are more common in the leg muscles of champion cross-country skiers.

4. The role of Ca^{2+} in the control of muscle contraction is to
   a. cause depolarization of the T tubule system.
   b. change the conformation of troponin, thus exposing myosin-binding sites.
   c. change the conformation of myosin heads, thus causing microfilaments to slide past each other.
   d. bind to tropomyosin and break actin–myosin cross-bridges.
   e. block the ATP-binding site on myosin heads, enabling muscle to relax.

5. Which statement about muscle contractions is *not* true?
   a. A single action potential at the neuromuscular junction is sufficient to cause a muscle to twitch.
   b. Once maximum muscle tension is achieved, no ATP is required to maintain that level of tension.
   c. An action potential in the muscle cell activates contraction by releasing Ca^{2+} into the sarcoplasm.
   d. Summation of twitches leads to a graded increase in the tension that can be generated by a single muscle fiber.
   e. The tension generated by a muscle can be varied by controlling how many of its motor units are active.

6. Which statement about the structure of skeletal muscle is *true*?
   a. The bright bands of the sarcomere are the regions where actin and myosin filaments overlap.
   b. When a muscle contracts, the A bands of the sarcomere (dark regions) lengthen.
   c. The myosin filaments are anchored in the Z lines.
   d. When a muscle contracts, the H bands of the sarcomere (light regions) shorten.
   e. The sarcoplasm of the muscle cell is contained within the sarcoplasmic reticulum.

7. The long bones of our arms and legs are strong and can resist both compressional and bending forces because
   a. they are solid rods of compact bone.
   b. their extracellular matrix contains crystals of calcium carbonate.
   c. their extracellular matrix consists mostly of collagen and polysaccharides.
   d. they have a very high density of osteoclasts.
   e. they consist of lightweight cancellous bone with an internal meshwork of supporting elements.

8. If we compare the jaw joint with the knee joint as lever systems,
   a. the jaw joint can apply greater compressional forces.
   b. their ratios of power arm to load arm are about the same.
   c. the knee joint has greater rotational abilities.
   d. the knee joint has a greater ratio of power arm to load arm.
   e. only the jaw is a hinged joint.

9. Which statement about skeletons is *true*?
   a. They can consist only of cartilage.
   b. Hydrostatic skeletons can be used only for amoeboid locomotion.
   c. An advantage of exoskeletons is that they can continue to grow throughout the life of the animal.
   d. External skeletons must remain flexible, so they never include calcium carbonate crystals as do bones.
   e. Internal skeletons consist of four different types of bone: compact, cancellous, dermal, and cartilage.

10. Which of the following effectors can be used for avoiding predators and for communication?
    a. chromatophores.
    b. nematocysts.
    c. electric organs.
    d. light emitting organs.
    e. poison glands.

Student Web Site Self-Quizzes

1. The movement of cilia and flagella are both based on the subcellular structure called a __________. The central structure of a cilium and flagellum is __________.
   a. microfilament, an axoneme
   b. microfilament, a sarcomere
   c. microtubule, an axoneme
   d. microtubule, a sarcomere
   e. microvillus, an axoneme

2. Which of the following components is *not* necessary for microtubule movement?
   a. ATP
   b. Ca^{2+} ions
   c. Central pair of microtubules
   d. Dynein
   e. Tubulin

3. The microfilaments actin and myosin are the major proteins in muscle. Microfilaments also underlie all of the following movement-related phenomena *except*
   a. amoeboid movement.
   b. the contractile ring.
   c. support of or structure of microvilli.
   d. movement of chromosomes during mitosis.
   e. phagocytosis and pinocytosis.

4. Which one of the following is *not* a feature of smooth muscle?
   a. Smooth muscle is under involuntary control.
   b. Smooth muscle is organized into regular structures.
   c. Contraction of smooth muscle is based on interacting actin and myosin filaments.
   d. Smooth muscle cells have gap junctions.
   e. Smooth muscles depolarize when stretched.

5. If you arranged the following components of skeletal muscle tissue from most to least inclusive, which component would be fourth in order?
   a. A band
   b. Muscle fiber

c. Myosin filament
d. Myofibril
e. Sarcomere

6. Which one of the following components of actin and myosin filaments binds Ca^{2+} ions?
   a. Actin monomer
   b. Calmodulin
   c. Myosin head
   d. Tropomyosin
   e. Troponin

7. If you arranged the following steps of the process of muscle contraction in chronological order, which step would be fourth?
   a. Ca^{2+} binds troponin.
   b. Ca^{2+} leaves sarcoplasmic reticulum.
   c. Myosin binding site is exposed.
   d. Myosin head delivers power stroke.
   e. ATP is hydrolyzed.

8. Which of the following statements about different types of muscle fibers is *false*?
   a. Slow-twitch fibers have more myoglobin than fast-twitch fibers.
   b. Slow-twitch fibers have more mitochondria than fast-twitch fibers.
   c. The leg muscle of a sprinter would have a higher proportion of slow-twitch fibers than the same muscle of a long-distance runner.
   d. The force output of fast-twitch fibers is greater than slow-twitch fibers.
   e. Fast-twitch fibers fatigue more rapidly than slow-twitch fibers.

9. The earthworm has a __________ skeleton, and contraction of its __________ muscles would shorten the segments and make them thicker.
   a. axial, circular
   b. axial, longitudinal
   c. hydrostatic, circular
   d. hydrostatic, longitudinal
   e. endoskeleton, circular

10. Which of the following statements about exoskeletons is *false*?
    a. Some mollusks have exoskeletons consisting of only two parts.
    b. In the arthropod exoskeleton, cuticle covers the outer surface of the body.
    c. Like endoskeletons, exoskeletons consist of a series of pliable joints.
    d. The endocuticle provides a waxy layer that protects the body from drying out.
    e. Arthropods must molt their exoskeletons in order to grow.

11. Adaptations that allow an organism to respond to information about the environment or its internal state are called effectors. Some common effectors like cilia cause movement, like the unidirectional flow of water through a clam. From the following list of adaptations, select the one that is *not* an effector.
    a. Chromatophore
    b. Electric organ
    c. Nematocyct
    d. Mercaptan
    e. Poison gland of a pit viper

★ *Indicates a **difficult** question*

■ *Indicates a **conceptual** question*

48 Gas Exchange in Animals

Fill in the Blank

1. External **gills** are highly branched folds of the body surface for gas exchange.
2. The surface openings of an insect's breathing tubes are on its **abdomen**.
3. The lungs of **birds** are the most efficient lungs of vertebrates.
4. In fish gas exchange systems, blood flows through gill lamellae in a direction opposite to water flow, a phenomenon known as **countercurrent exchange**.
5. In the mammalian lung, the amount of air being moved in normal breathing is called the **tidal volume**.
6. After extreme exhalation, the lungs and airways still contain a **residual volume** of air.
7. The oxygen-binding molecule abundant in muscle is **myoglobin**.
8. Hemoglobin consists of **four** protein chains linked together.
9. The notation P_{O_2} stands for the **partial pressure** of oxygen.

★10. The shape of the oxygen dissociation curve of hemoglobin is said to be **sigmoid**.

11. An unusual feature of the lungs of **birds** is that they expand and contract relatively little during a breathing cycle. They also contract during inhalation and expand during exhalation.
12. Energy to carry out essential functions comes in the form of ATP, which can be sustained only where **oxygen** gas is present.

■13. There are no active transport mechanisms for respiratory gases. Therefore, **diffusion** is the only means by which respiratory gases are exchanged.

★14. In Fick's law of diffusion, the term $(C_1 - C_2)/L$ is a **concentration gradient**.

15. Respiration is unique in birds. In addition to lungs, birds have **air sacs** at several locations in their bodies.
16. Mammalian lungs have two adaptive features: the production of mucus, and **surfactant**, a substance that reduces the surface tension of the liquid lining the insides of the alveoli.
17. Breathing mechanisms **ventilate** the environmental side of the gas exchange surface, while circulatory systems **perfuse** the internal side of the surface.

Multiple Choice

1. Breathing provides the body with oxygen required to support the energy metabolism of all cells. Breathing also eliminates __________, one of the waste products of cell metabolism.
   a. carbon monoxide
   b. carbon dioxide
   c. carbon tetrachloride
   d. calcium carbonate
   e. carbonic acid

2. Which of the following explains why oxygen can be exchanged more easily in air than in water?
   a. The oxygen content of air is higher than that of water.
   b. Oxygen diffuses more slowly in water than in air.
   c. More energy is required to move water than air because water is denser.
   d. a and b
   e. a, b, and c

■3. Aquatic __________ are in a double bind, because as the temperature of their environment increases, so does their demand for oxygen, but the oxygen content of water declines with increasing water temperatures.
   a. insects
   b. ectotherms
   c. endotherms
   d. plants
   e. None of the above

4. Which factor accounts for the efficiency of gas exchange in fish gills?
   a. Maximizing surface area
   b. Minimizing path length for diffusion
   c. Countercurrent flow of blood and water over opposite sides of the gas exchange surfaces
   d. a and b
   e. a, b, and c

5. Gas exchange in animals always involves
   a. cellular respiration.
   b. breathing movements.
   c. neural control of exchange.
   d. diffusion across membranes.
   e. active transport of gases.

6. The amount of gas exchange required to support an animal's metabolism increases with decreased
   a. oxygen in the air.
   b. temperature.
   c. body size.
   d. water breathing.
   e. body movement.

7. Rapid gas exchange is easier in air than in water because
   a. the oxygen content of water is higher than that of air.
   b. the carbon dioxide content of water is higher than that of air.
   c. oxygen diffuses more rapidly in water.
   d. water is more dense and viscous than air.
   e. more energy is required to move air than water.

▪8. Why do humans have difficulty breathing at high elevations?
   a. Oxygen makes up a lower percentage of the air there.
   b. The temperature is lower there.
   c. The barometric pressure is higher there.
   d. The partial pressure of oxygen is lower there.
   e. The air is drier there.

9. Small insects may take a bubble of air underwater when they dive. The bubble can serve as an air tank for some time because
   a. as carbon dioxide increases, the bubble inflates.
   b. as oxygen is consumed, air pressure in the bubble decreases.
   c. as oxygen is consumed, oxygen diffuses into the bubble.
   d. the partial pressure of nitrogen remains constant.
   e. for each oxygen molecule used, a carbon dioxide molecule replaces it.

10. The respiratory system of insects consists of
   a. branched air tubes called spiracles that supply capillaries.
   b. branched air tubes called tracheae that supply capillaries.
   c. branching gill systems that end in openings called tracheae.
   d. branching gill systems that end in openings called spiracles.
   e. extensive layers of gas exchange tissue just under the exoskeleton.

*11. The delicate gills of fish are supported by
   a. opercular flaps and gill arches.
   b. opercular flaps and gill filaments.
   c. gill filaments and gill arches.
   d. opercular flaps and a diaphragm.
   e. gill arches and a diaphragm.

12. Which of the following is *not* true about the air sacs of birds?
   a. They connect with each other.
   b. They occur in anterior and posterior pairs.
   c. They make the bird's respiratory system more efficient than a mammal's.
   d. They allow for one-way air flow through the lungs.
   e. They provide extra gas exchange surface.

13. Which of the following characterizes the lungs of mammals?
   a. Joined with air sacs
   b. Tidal ventilation system
   c. Complete emptying in exhalation
   d. Crosscurrent air flow
   e. Countercurrent air flow

14. Which of the following structures is the site of gas exchange in the lungs?
   a. Alveoli
   b. Bronchi
   c. Bronchioles
   d. Trachea
   e. a and b

15. Which factor maximizes the rate of gas exchange in mammals?
   a. Exceedingly high partial pressures of oxygen in the blood
   b. The larynx (voice box) opening into the trachea
   c. Enormous surface area for gas exchange
   d. The small increase in size during inhalation
   e. Production of mucus and surfactants

*16. Surfactant produced by cells lining the alveoli serves to
   a. increase surface tension.
   b. increase stretching of the alveolar walls.
   c. assist muscular movements of breathing.
   d. reduce cohesion of surface molecules.
   e. reduce ciliary movement.

17. One effect of cigarette smoking on the lungs is to
   a. remove mucus from the bronchi.
   b. immobilize cilia lining the bronchi.
   c. speed air flow through the bronchi.
   d. increase surfactant action.
   e. increase surface area in alveoli.

18. The lower side of the lung cavity is formed by the
   a. diaphragm.
   b. esophagus.

c. stomach.
d. ribs.
e. intercostal muscles.

▪19. The lungs expand in inhalation because
a. the diaphragm contracts upward.
b. the shoulder girdle moves upward.
c. the volume of the thoracic cavity increases.
d. lung tissue actively stretches.
e. the lung tissue rebounds from exhalation.

20. The process of exhalation is begun mainly due to
a. the contraction of intercostal muscles.
b. the relaxation of muscles.
c. the contraction of the diaphragm.
d. complete collapse of the lung tissue.
e. low pressure in the thoracic cavity.

*21. If the thoracic wall is punctured, air leaking in will cause the lung to collapse. Thus in the normal, intact thoracic cavity
a. punctured lungs will collapse.
b. air pressure is the same as outside.
c. a slight suction keeps lungs inflated.
d. breathing movements keep lungs inflated.
e. lung expansion is passive.

22. What part of the blood is most efficient at carrying oxygen?
a. Blood plasma solution
b. Blood plasma proteins
c. Blood platelets
d. Membrane molecules of red blood cells
e. Hemoglobin molecules of red blood cells

23. Which of the following would increase the amount of oxygen diffusing from the lungs into the blood?
a. Increasing the binding rate of oxygen to hemoglobin
b. Decreasing the partial pressure of oxygen in the lung
c. Increasing the partial pressure of oxygen in the blood
d. Decreasing the red blood cell count of the blood
e. Increasing the water vapor of air in the lungs

24. Each molecule of hemoglobin when fully saturated carries how many molecules of oxygen?
a. 1
b. 2
c. 4
d. 20
e. Nearly 100

25. Hemoglobin delivers oxygen to body cells from the red blood cells
a. until the hemoglobin is depleted of oxygen.
b. until the partial pressures of oxygen in the cells are equivalent.
c. although the hemoglobin is never completely saturated.
d. by releasing oxygen to cells with higher partial pressure of oxygen.
e. until the fluid pressure in the red blood cells is lower.

26. In which of the following P_{O_2} environments will hemoglobin release its oxygen most easily?
a. 30 mm Hg
b. 50 mm Hg
c. 70 mm Hg
d. 90 mm Hg
e. 110 mm Hg

*27. What is the advantage of having myoglobin in muscle cells?
a. It binds with oxygen just as easily as hemoglobin does.
b. It contributes to the dark color of flight muscle in birds.
c. It increases the effectiveness of muscles used in short bursts.
d. It releases bound oxygen at lower P_{O_2} conditions than hemoglobin does.
e. It uses hemoglobin as a reserve source of oxygen.

28. What is the advantage of having fetal hemoglobin?
a. Fetal hemoglobin pumps oxygen from adult hemoglobin to the fetus.
b. Fetal hemoglobin releases oxygen at a lower partial pressure than adult hemoglobin does.
c. In the placenta, oxygen diffuses from adult hemoglobin to fetal hemoglobin.
d. At low oxygen pressures, adult hemoglobin is more likely to pick up oxygen.
e. Fetal hemoglobin occurs in higher density in red blood cells than adult hemoglobin does.

29. In rapidly metabolizing tissues where conditions are more acidic, how does hemoglobin respond in comparison to its action in less acidic environments?
a. It releases more oxygen.
b. It releases less oxygen.
c. It releases more carbon dioxide.
d. It releases less carbon dioxide.
e. There is no difference in response.

30. After humans acclimate to high altitudes, their hemoglobin
a. is more often saturated with oxygen.
b. delivers more oxygen to the tissues.
c. is more concentrated in red blood cells.
d. shifts its oxygen-binding curve to the left.
e. resembles fetal hemoglobin in oxygen binding.

31. Most of the carbon dioxide in the blood is transported
a. bound with hemoglobin.
b. as dissolved gas.
c. as bicarbonate ion.
d. as carbonic acid.
e. as calcium carbonate.

32. Carbonic anhydrase is an enzyme in red blood cells that catalyzes a reaction between carbon dioxide and
a. bicarbonate.
b. carbonic acid.
c. hemoglobin.
d. oxygen.
e. water.

33. Neural control of breathing is in the
   a. cerebrum.
   b. diaphragm.
   c. medulla.
   d. olfactory lobe.
   e. spinal cord.

34. The breathing center initiates ventilation in response to
   a. a decrease in air pressure.
   b. a decrease in oxygen.
   c. an increase in carbon dioxide.
   d. the time since the last breath.
   e. the rate of gas exchange in the alveoli.

35. Nodes that are sensitive to oxygen and blood pressure levels are located in the
   a. aorta and carotid arteries.
   b. brain stem capillaries.
   c. hypothalamus.
   d. pulmonary artery.
   e. pulmonary vein.

36. Aquatic animals that lack an internal system for transporting oxygen
   a. often have flat, leaflike bodies.
   b. are usually small.
   c. may have a very thin body built around a central cavity through which water circulates.
   d. have cells all of which are close to the respiratory medium.
   e. All of the above

37. Which of the following is *not* an expected response from a fish to a drop in water temperature?
   a. Metabolism increases
   b. Amount of oxygen available increases
   c. Blood flow decreases
   d. Oxygen consumption decreases
   e. All of the above

*▪38. A planet is discovered where the barometric pressure is 2,000 mm of mercury and the air is 15% oxygen. What is the partial pressure of oxygen at sea level on this planet?
   a. 15%
   b. 15 mm Hg
   c. 133 mm Hg
   d. 300 mm Hg
   e. 30%

*39. Which of the following statements does *not* explain why diffusion of CO_2 from an animal is not as great a problem as the diffusion of O_2 into the animal?
   a. CO_2 is more soluble in water than O_2.
   b. Cellular respiration produces less CO_2 than the O_2 that is consumed.
   c. The CO_2 content of air is less than the O_2 content.
   d. The atmospheric partial pressure of O_2 is greater than the atmospheric partial pressure of CO_2.
   e. There is a greater CO_2 concentration gradient from the cell to the atmosphere than is true for oxygen.

40. Which one of the following organisms does *not* require both external ventilation of its respiratory surfaces with the medium containing O_2 and internal ventilation of its respiratory surfaces with blood?
   a. A crayfish
   b. A rabbit
   c. An insect
   d. A squid
   e. A bird

41. Which of the following statements about respiratory adaptations is *false*?
   a. Internalization of respiratory surfaces leads to the need for ventilation.
   b. External gills are found only in invertebrates.
   c. Some fish ventilate their gills by constantly swimming with their mouth open.
   d. Most fish ventilate the external surface of their gills using a two-pump mechanism.
   e. Desiccation of the respiratory surface is more likely to occur in lungs than gills.

42. Which of the following statements about insect respiration is *false*?
   a. Spiracles are the air tubes that carry the respiratory gases.
   b. Because oxygen diffuses much faster in air than water, the insect respiratory system is highly efficient.
   c. Aquatic insects that carry air bubbles with them underwater are using the air bubble as a source of oxygen.
   d. Some insects have external gills.
   e. None of the above

43. Which of the following statements about the structure of the fish gill is *false*?
   a. Afferent vessels bring blood to the gills, while efferent vessels take blood away from the gills.
   b. Exchange of respiratory gases occurs within the lamellae of the gill filaments.
   c. The efferent and afferent vessels are the countercurrent flow system of the gill.
   d. The lamellae greatly increase the surface area for gas exchange.
   e. The opercular flaps enclose the gill chambers.

44. Which of the following adaptations is *not* seen in fish gills?
   a. A countercurrent exchange system
   b. Bidirectional ventilation of the gills
   c. Morphological features to increase the surface area available for gas exchange
   d. Morphological features to decrease the path length for diffusion of the respiratory gases
   e. Morphological features to maximize the efficiency of oxygen extraction

45. In the respiratory cycle of birds, air flows through the trachea, then the bronchi, then
   a. the parabronchi.
   b. back through the trachea.
   c. the air capillaries.

d. the posterior air sacs.
e. the anterior air sacs.

46. Bird and fish respiratory systems are similar because
a. both employ a countercurrent exchanger.
b. both have air sacs.
c. both have unidirectional flow of the environmental medium over the gas exchange membranes.
d. both are infoldings of the body.
e. All of the above

47. In the following equation, which quantity represents the volume of your normal breath? Total lung capacity = residual volume + expiratory reserve volume + inspiratory reserve volume + tidal volume.
a. Total lung capacity
b. Residual volume
c. Expiratory reserve volume
d. Inspiratory reserve volume
e. Tidal volume

48. Which of the following is *not* an adverse consequence of tidal breathing as seen in mammals?
a. Precludes countercurrent gas exchange
b. Dead space
c. Residual volume
d. Short diffusion path length
e. Limits the O_2 concentration gradient between air and blood

49. Carbon monoxide (CO)
a. binds to hemoglobin with a higher affinity than does oxygen.
b. may be released by faulty furnaces or the burning of charcoal or kerosene in unventilated areas.
c. exposure may cause death from lack of oxygen.
d. All of the above
e. None of the above

50. Which of the following statements about surfactants in the lungs is *false*?
a. Water coating the respiratory surfaces of the lungs creates surface tension.
b. Surfactants increase the cohesive forces between water molecules.
c. Certain cells in the alveoli produce surfactants.
d. Surface tension influences the amount of effort required to inflate the lungs.
e. Premature infants suffer respiratory distress syndrome if natural surfactants are not present within the lungs.

▪51. Which of the following statements about the mechanics of ventilation is *false*?
a. If the pleural cavity is punctured, the lung may collapse.
b. As the diaphragm relaxes, air is expelled from the respiratory system.
c. As the diaphragm relaxes, the pressure within the pleural cavities increases.
d. Less energy is expended during the exhalation phase.
e. There is a slight positive pressure within the pleural cavities between breaths.

52. Which of the following statements about the transport of respiratory gases by the blood is *false*?
a. The amount of oxygen that can dissolve directly in plasma is sufficient to support the resting metabolism.
b. By binding oxygen, hemoglobin helps to maintain a steeper oxygen concentration gradient.
c. Internal ventilation of the gas exchange surfaces is dependent on blood flow.
d. Each hemoglobin molecule can carry a maximum of four O_2 molecules.
e. Molecules of O_2 are associated with the heme groups of hemoglobin.

53. Which of the following statements about the binding of oxygen by hemoglobin is *false*?
a. The percent of completely oxygen-saturated hemoglobin increases as P_{O_2} increases.
b. The oxygen-binding curve is S-shaped.
c. There is a narrow range of the oxygen-binding curve where oxygen dissociation-association is rapid.
d. It requires a very low P_{O_2} to cause all four hemoglobin subunits to bind to O_2.
e. If one subunit of a completely oxygen-saturated hemoglobin loses its O_2, the affinity of the remaining subunits for their O_2 increases.

*54. Which of the following statements about oxygen-binding curves is *false*?
a. It requires a smaller decrease in P_{O_2} to go from 100% to 75% saturated hemoglobin than it does to go from 75% to 50%.
b. During normal metabolism, the body is only using the upper one-fourth of its oxygen-binding curve.
c. The blood normally carries an enormous reserve of O_2.
d. The steepest part of the oxygen-binding curve results in a great release of O_2 for only a modest drop in P_{O_2}.
e. The P_{O_2} of blood only goes below about 40 mm Hg during extreme exertion.

55. Which statement about myoglobin is *false*?
a. Iron in the myoglobin molecule can bind to oxygen.
b. Myoglobin is found in muscle cells.
c. Myoglobin has a lesser affinity for oxygen than hemoglobin does.
d. Diving mammals have high concentrations of myoglobin in their muscles.
e. Fast- and slow-twitch muscle differ in their myoglobin content.

56. If you plotted the oxygen-binding curves of the following molecules, which curve would be to the left of all the others?
a. Fetal human hemoglobin
b. Adult llama hemoglobin
c. Adult human hemoglobin
d. Myoglobin
e. Fetal llama hemoglobin

*57. If you plotted the oxygen-binding curves of the following hemoglobins subject to the pH levels shown, which curve would be to the right of all the others?
a. Fetal human hemoglobin at pH 7.6
b. Adult human hemoglobin at pH 7.6
c. Adult human hemoglobin at pH 7.2
d. Fetal human hemoglobin at pH 7.2
e. Adult llama hemoglobin at pH 7.2

*58 If you plotted the oxygen-binding curves of adult hemoglobins subject to the following conditions, which curve would be to the right of all the others?
a. pH 7.6, little diphosphoglyceric acid
b. pH 7.6, much diphosphoglyceric acid
c. pH 7.2, little diphosphoglyceric acid
d. pH 7.2, much diphosphoglyceric acid
e. pH 7.4, little diphosphoglyceric acid

59. Most of the CO_2 is transported in the blood as __________, and it is located mainly in the __________.
a. CO_2, plasma
b. H_2CO_3, plasma
c. HCO_3^-, plasma
d. carboxyhemoglobin, erythrocytes
e. HCO_3^-, erythrocytes

*60. Due to the activity of carbonic anhydrase, within the lung there is a concentration gradient of CO^2 from the erythrocyte to the __________, and a concentration gradient of __________ from the plasma to the __________.
a. plasma, HCO_3^-, plasma
b. plasma, HCO_3^-, lung
c. lung, CO_2, lung
d. lung, HCO_3^-, erythrocyte
e. lung, HCO_3^-, lung

61. The breathing rhythm is generated in the __________ and is influenced by variation in levels of __________ in the blood.
a. medulla, CO_2 and O_2
b. medulla, CO_2
c. medulla, O_2
d. frontal lobe, CO_2 and O_2
e. frontal lobe, CO_2

62. The Hering–Breuer reflex
a. begins with stretch receptors in the trachea.
b. involves the forebrain.
c. occurs when mucus blocks the airways.
d. is an override reflex that prevents the breathing muscles from overdistending and damaging the lungs.
e. None of the above

Study Guide Questions

1. Which of the following is *not* a reason that breathing air is easier than breathing water?
a. A given volume of air is easier to move across the respiratory organs than the same volume of water.
b. Air holds more oxygen per unit volume than water.
c. Water breathers have a difficult time ridding themselves of CO_2 because CO_2 does not dissolve well in water.
d. Temperature increases affect the O_2 content of water more than it does that of air.

2. External gills, tracheae, and lungs all share which of the following sets of characteristics?
a. Part of gas-exchange system; exchange both CO_2 and O_2; increase surface area for diffusion
b. Used by water breathers; based on countercurrent exchange; use negative pressure breathing
c. Exchange only O_2; are associated with a circulatory system; found in vertebrates
d. Found in insects; employ positive-pressure pumping; based on crosscurrent flow

3. Which of the following represents a larger volume of air than is normally found in the resting tidal volume of a human lung?
a. Residual volume
b. Inspiratory reserve volume
c. Expiratory reserve volume
d. All of the above

4. Which of the following is shared by bird and mammal lungs?
a. Both need two cycles of inhalation and exhalation for air to move through the lungs.
b. Both contain alveoli as the terminal end of the lung.
c. Both have an anatomical dead space.
d. Both exchange O_2 and CO_2 with blood in capillaries.

5. According to Fick's law of diffusion, what does not play a role in the diffusion of oxygen across a membrane?
a. Surface area
b. Volume
c. Difference in concentration or partial pressure
d. Diffusion distance

6. Because of the relatively high altitude of Antonito, Colorado, the town has a normal barometric pressure of about 600 mm Hg rather than 760 mm Hg as at sea level. The partial pressure of oxygen in Antonito's air is approximately
a. 75 mm Hg.
b. 126 mm Hg.
c. 160 mm Hg.
d. 760 mm Hg.

7. Gas flows into the lungs of mammals during inspiration because
a. the pressure in the lungs falls below atmospheric pressure.
b. the volume of the lungs decreases.
c. the pressure in the lungs rises above atmospheric pressure.
d. the diaphragm moves upward toward the lungs.

8. The movement of O_2 and CO_2 between the blood in the tissue capillaries and the cells in tissues depends most directly upon
a. active transport of O_2 and CO_2.

b. total atmospheric (barometric) pressure differences across the cell membranes.
c. diffusion of O_2 and CO_2 down a concentration gradient.
d. diffusion of O_2 and CO_2 down a partial pressure gradient.

9. The alveoli of the lungs do not contain "air" because
a. we normally do not ventilate our lungs at a high enough rate.
b. the lungs have too many alveoli to ventilate.
c. there is "dead space" in the trachea and bronchi.
d. the trachea and bronchi are too small in volume.

10. Which of the statements about hemoglobin is *false*?
a. Hemoglobin allows the blood to carry a large amount of O_2.
b. Hemoglobin contains a single polypeptide chain with very high affinity for O_2.
c. Hemoglobin is packaged inside red blood cells
d. Fetal hemoglobin is structurally different from adult hemoglobin.

11. As blood becomes fully O_2-saturated, hemoglobin is combining with __________ molecule(s) of oxygen.
a. 1
b. 2
c. 4
d. 8

12. The Bohr shift describes
a. the outward movement of Cl^- from the blood cell in exchange for HCO_3^- moving into the cell.
b. the leftward shift of the entire oxygen equilibrium curve when temperature rises.
c. the rightward shift of the entire oxygen equilibrium curve when pH rises.
d. the rightward shift of the entire oxygen equilibrium curve when pH falls.

13. The presence of CO_2 in blood will lower pH because CO_2 combines with __________ with the rate of reaction increased by __________.
a. H_2O to form H^+ and HCO_3^-, carbonic anhydrase
b. H_2O to form only HCO_3^-, carbonic anhydrase
c. H_2O to form only H^+, carbonic ions
d. H^+ to form HCO_3^-, oxyhemoglobin

14. The largest proportion of CO_2 carried by the blood is in the form of
a. bicarbonate ions (HCO_3^-) carried in the plasma.
b. molecular CO_2 dissolved in the plasma.
c. bicarbonate ions (HCO_3^-) carried within the red blood cells.
d. molecular CO_2 chemically bound to hemoglobin.

15. Which of the following is not involved in the neural control of ventilation?
a. Neurons in the medulla
b. The vagus nerve
c. The contraction state of the diaphragm
d. Chemosensors on the surface of the medulla

End of Chapter Questions

1. Which statement is *not* true?
a. Respiratory gases are exchanged by diffusion only.
b. Oxygen has a lower rate of diffusion in water than in air.
c. The oxygen content of water falls as the temperature of water rises, all other things being equal.
d. The amount of oxygen in the atmosphere decreases with increasing altitude.
e. Birds have evolved active transport mechanisms to augment their respiratory gas exchange.

2. Which statement about the gas exchange system of birds is *not* true?
a. Respiratory gases are not exchanged in the air sacs.
b. It can achieve more complete exchange of O_2 from air to blood than the human gas exchange system can.
c. Air passes through birds' lungs in only one direction.
d. The gas exchange surfaces in bird lungs are the alveoli.
e. A breath of air remains in the system for two breathing cycles.

3. In fish
a. blood flows over the gas exchange surfaces, opposite of the flow of water.
b. gases are exchanged across the gill filaments.
c. ventilation of the gills is tidal.
d. less work is needed to ventilate gills in warm water than in cold water.
e. the path length for diffusion of respiratory gases is determined by the length of the gill filaments.

4. In the human gas exchange system
a. the lungs and airways are completely collapsed after a forceful exhalation.
b. the average O_2 concentration of air inside the lungs is always lower than that in the air outside the lungs.
c. the P_{O_2} of the blood leaving the lungs is greater than the P_{O_2} of the exhaled air.
d. the amount of air that is moved per breath during normal, at-rest breathing is termed the total lung capacity.
e. oxygen and carbon dioxide are actively transported across the alveolar and capillary membranes.

5. Which statement about the human gas exchange system is *not* true?
a. During inhalation a negative pressure exists in the space between the lung and the thoracic wall.
b. Smoking one cigarette can immobilize the cilia lining the airways for hours.
c. The respiratory control center in the medulla responds more strongly to changes in arterial O_2 concentration than to changes in arterial CO_2 concentrations.
d. Without surfactant, the work of breathing is greatly increased.
e. The diaphragm contracts during inhalation and relaxes during exhalation.

6. The hemoglobin of a human fetus
   a. is the same as that of an adult.
   b. has a higher affinity for O_2 than adult hemoglobin has.
   c. has only two protein subunits instead of four.
   d. is supplied by the mother's red blood cells.
   e. has a lower affinity for O_2 than adult hemoglobin has.

7. The amount of oxygen carried by hemoglobin depends on the partial pressure of oxygen in the blood. Hemoglobin in active muscles
   a. becomes saturated with oxygen.
   b. takes up only a small amount of oxygen.
   c. readily unloads oxygen.
   d. tends to decrease the partial pressure of oxygen in the muscle tissues.
   e. is denatured.

8. Most carbon dioxide is carried in the blood
   a. in the cytoplasm of red blood cells.
   b. dissolved in the plasma.
   c. in the plasma as bicarbonate ions.
   d. bound to plasma proteins.
   e. in red blood cells bound to hemoglobin.

9. Myoglobin
   a. binds O_2 at P_{O_2} values at which hemoglobin is releasing its bound O_2.
   b. has a lower affinity for O_2 than hemoglobin does.
   c. consists of four polypeptide chains, just as hemoglobin does.
   d. provides an immediate source of O_2 for muscle cells at the onset of activity.
   e. can bind four O_2 molecules at once.

10. Carbon dioxide, a product of cellular respiration, is carried in the bloodstream. When the level of CO_2 in the blood becomes greater than the set operating range, the
   a. rate of respiration decreases.
   b. pH of the blood rises.
   c. respiratory centers become dormant.
   d. rate of respiration increases.
   e. blood becomes more alkaline.

Student Web Site Self-Quiz Questions

1. A simple solution to gas exchange in some aquatic animals is to keep the body very thin. Which of the statements about respiratory gas exchange is *false*?
   a. No active transport mechanisms move gases across membranes.
   b. Fish spend a larger percentage of their energy budget on ventilation than do terrestrial vertebrates.
   c. Limits imposed by the slow diffusion of O_2 molecules in water are completely avoided in animals that breathe air.
   d. A given volume of water holds much less oxygen than the same volume of air.
   e. A fish breathes water.

2. Which of the following is a *false* statement about water temperature or altitude?
   a. Warm water holds less oxygen than cold water.
   b. The O_2 consumption of fish increases as water temperature increases.
   c. Barometric pressure declines with increasing altitude.
   d. The percentage of oxygen in air at 5,000 meters is about one-half of what it is at sea level.
   e. The rate of diffusion of O_2 across a gas exchange surface depends on the difference between the P_{O_2} of the air and the P_{O_2} of the body.

3. Differences in ventilation and perfusion of the respiratory systems of a rabbit and an insect would most directly affect which of the following components of Fick's law describing the rate, Q, at which a substance diffuses between two locations?
   $Q = DA(C_1\text{-}C_2)/L$
   Where:
   D = diffusion coefficient of medium
   A = cross-sectional area of diffusion
   C_1 = concentration of substance at location 1
   C_2 = concentration of substance at location 2
   L = distance between locations
   a. D
   b. A
   c. $(C_1\text{-}C_2)/L$
   d. D and A
   e. L

4. Insects have a tracheal gas exchange system. Which of the following would *not* be an important component of the gas exchange system of all insects?
   a. Air bubble
   b. Air capillaries
   c. Tracheoles
   d. Spiracles
   e. Tracheae

5. Complete the following sentence about water flow over the gill surfaces in fish: The negative-pressure pump for _________ of the gills in fish involves the _________.
   a. perfusion, mouth cavity
   b. perfusion, opercular flaps
   c. ventilation, gill arches
   d. ventilation, mouth cavity
   e. ventilation, opercular flaps

6. If a bird is made to inhale pure O_2, when the bird inhales the second time, the previous breath with elevated O_2 will be in the _________ and during inhalation the bird's lungs are _________.
   a. anterior air sacs, contracting
   b. anterior air sacs, expanding
   c. lungs, expanding
   d. posterior air sacs, contracting
   e. posterior air sacs, expanding

7. Which component of the human respiratory system is associated with respiratory distress syndrome in premature babies?
   a. Alveoli
   b. Bronchi
   c. Diaphragm
   d. Pleural membranes
   e. Capillaries

8. Which of the following statements about ventilation of human lungs is *false*?
   a. A collapsed lung can result from the puncture of a pleural membrane.
   b. Inhalation depends on positive pressure, exhalation depends on negative pressure.
   c. Surfactant makes ventilation of the lungs easier.
   d. Ventilation of human lungs is tidal in nature.
   e. Exhalation is usually a passive process.

TEST FILE

★ *Indicates a* ***difficult*** *question*

■ *Indicates a* ***conceptual*** *question*

49 Circulatory Systems

Fill in the Blank

■1. From fishes to amphibians to reptiles to mammals and birds, the complexity and number of chambers in the heart increases. An important consequence of this increased complexity is the gradual separation of the circulatory system into two independent circuits—**pulmonary** and **systemic**.

2. The "lub-dub" sounds of the cardiac cycle are caused by the **shutting of the heart valves.**

3. A blood clot that becomes established within the lumen of an artery is termed a thrombus. A piece of a thrombus breaking loose can travel through the arteries and eventually become lodged in a vessel of small diameter. If this occurs in the brain, the resulting condition is called a **stroke**.

4. The interstitial fluid that accumulates outside the capillaries is returned to the heart by a separate system of vessels called the **lymphatic** system.

5. In a human being under normal conditions, the bone marrow can produce about 2 million **red blood cells, or erythrocytes,** every second.

6. In open circulatory systems, fluid enters the **heart** through holes called ostia.

7. In humans, blood is pumped from heart to lungs and back through the **pulmonary** circuit.

8. The walls of large **arteries** have elastic fibers and can withstand high pressures.

9. The condition of **edema** results from accumulation of interstitial fluids and tissue swelling.

10. If too little blood is pumped to the brain, **fainting** will result.

11. Blood cells form from **stem cells** in the bone marrow.

12. **Platelets** are cell fragments responsible for blood clotting.

13. The brain's cardiovascular control center is in the **medulla**.

■14. The **diving reflex** slows the heart when the individual is plunged into cold water.

15. **Arteries** are vessels that carry blood away from the heart, and **veins** are vessels that carry blood back to the heart.

16. The lymphatic system empties intercellular fluid into the **thoracic duct**, which is emptied into the superior vena cava.

17. In humans, contraction of the right ventricle pumps blood into the **pulmonary artery**.

18. Snakes, lizards, and turtles can bypass the lung circuit and pump all blood around the body because of the incomplete division of their **ventricles**.

Multiple Choice

1. A cardiovascular system is *not* necessary in the hydra because it
   a. is an aquatic animal.
   b. has no skeleton.
   c. is only two cells thick.
   d. has tentacles to move water.
   e. does not move rapidly.

2. In an open circulatory system,
   a. there is no heart.
   b. there is a gastrovascular cavity.
   c. there are no blood vessels.
   d. blood flows out of the body.
   e. there is no distinction between blood and tissue fluid.

3. An example of an animal with an open circulatory system is the
   a. bat.
   b. earthworm.
   c. hydra.
   d. snail.
   e. lizard.

4. In humans, which vessel(s) empty into the right atrium?
   a. Pulmonary veins
   b. Inferior vena cava

c. Superior vena cava
d. Pulmonary artery
e. b and c

5. In vertebrates, exchange of substances between the blood and the interstitial fluids occurs in the
a. arteries.
b. arterioles.
c. capillaries.
d. veins.
e. venules.

6. In the fish circulatory system, blood
a. moves from the muscular ventricle to the gills.
b. entering the aorta is under high pressure.
c. leaving the heart moves to body tissues.
d. leaving the gills is under high pressure.
e. is received from the body into a muscular atrium.

▪7. Circulatory systems of adult amphibians demonstrate which of the following advantages over those of fishes?
a. A four-chambered heart
b. Heart metamorphosis
c. Partial separation of pulmonary and systemic circulation
d. A pocket of the gut that serves as an air bladder
e. Separation of oxygenated from deoxygenated blood

▪8. The advantage of having a heart with two atria is that
a. oxygenated and deoxygenated blood are separated.
b. blood is pumped directly from heart to tissues.
c. blood can be slowed down before going to tissues.
d. the body can support a higher blood pressure.
e. there are two muscular regions for pumping.

9. In which way are crocodilians different from other reptiles?
a. They have a separate pulmonary circulation.
b. They have lungs and no gills.
c. They have an open circulatory system.
d. They have more muscular atria.
e. They have a four-chambered heart.

10. In the human heart, blood is pumped from the left ventricle into the
a. right ventricle.
b. left atrium.
c. right atrium.
d. pulmonary circuit.
e. systemic circuit.

11. What vessel transports oxygenated blood from the lung into the heart?
a. Pulmonary artery
b. Pulmonary vein
c. Superior vena cava
d. Inferior vena cava
e. Coronary artery

*12. In the cardiac cycle, blood pressure is at a maximum when the
a. atria are contracting during systole.
b. atria are contracting during diastole.
c. ventricles are contracting during systole.
d. ventricles are relaxing during systole.
e. ventricles are relaxing during diastole.

13. The heart beats because of cardiac muscle's unique properties, such as
a. high resistance to outside stimulation.
b. cell communication via gap junctions.
c. low level of electric activity.
d. external pacemaking system.
e. rapid chemical communication.

14. The specific location of the heart pacemaker is the
a. sinoatrial node.
b. atrioventricular node.
c. Purkinje fibers.
d. bundle of His.
e. ventricular mass.

*15. The timing of the spread of the action potential from atrium to ventricle is controlled by the
a. sinoatrial node.
b. atrioventricular node.
c. Purkinje fibers.
d. bundle of His.
e. ventricular mass.

16. The diameter of a capillary is about the same as that of a(n)
a. arteriole.
b. nerve.
c. red blood cell.
d. valve.
e. venule.

*17. As blood enters the capillaries from the arterioles, the blood pressure
a. decreases and the osmotic potential decreases.
b. decreases and the osmotic potential increases.
c. increases and the osmotic potential decreases.
d. decreases and the osmotic potential remains steady.
e. increases and the osmotic potential remains steady.

*18. Histamine causes swelling by
a. making blood vessels expand and decreasing the permeability of capillaries.
b. making blood vessels contract and decreasing the permeability of capillaries.
c. making blood vessels expand and increasing the pressure in capillaries.
d. decreasing the permeability of and increasing the pressure in capillaries.
e. increasing the permeability of and increasing the pressure in capillaries.

19. What is the function of lymph nodes?
a. They are sites where bacteria can grow.
b. They are sites where inflammation occurs.
c. They are sites regulating blood fluid.
d. They are sites where new cells enter.
e. They are sites of mechanical filtering.

20. What causes blood to move in the veins toward the heart?
    a. Gravity
    b. The contraction of venous walls
    c. Pulsing movement from the heart
    d. The contraction of nearby muscles
    e. Venous capacitance

*21. The Frank–Starling law suggests that during exercise the
    a. flow of blood through the heart speeds up.
    b. heart is stretched and contracts more forcefully.
    c. veins expand to handle increased blood flow.
    d. heart fills to capacity with blood at each beat.
    e. valves in veins can no longer prevent backflow.

22. The hematocrit, or percentage of the blood made up of cells, in normal humans is about
    a. 10%.
    b. 25%.
    c. 40%.
    d. 70%.
    e. 90%.

23. The most abundant cells in the blood are
    a. erythrocytes.
    b. leukocytes.
    c. phagocytes.
    d. platelets.
    c. None of the above

24. The hormone erythropoietin is released by the kidney
    a. to remove old red blood cells from circulation.
    b. in response to high levels of oxygen in circulation.
    c. in response to low levels of hemoglobin.
    d. to stimulate production of red blood cells.
    e. to stimulate platelet formation.

25. The lifetime of an individual red blood cell is approximately
    a. 4 hours.
    b. 4 days.
    c. 4 weeks.
    d. 40 days.
    e. 4 months.

26. The mature red blood cells of humans
    a. consist of over 75% hemoglobin.
    b. contain no nuclei or endoplasmic reticulum.
    c. have a nearly spherical shape.
    d. have a relatively low surface area–to–volume ratio.
    e. have a rigid shape and low flexibility.

27. When is a platelet activated to initiate clotting?
    a. During a histamine reaction
    b. When leukocytes are activated
    c. When collagen fibers are encountered
    d. When blood flow rates drop
    e. In the presence of prothrombin

28. How does the circulating protein fibrinogen contribute to blood clotting?
    a. It polymerizes to form fibrin threads.
    b. It acts as a catalyst for thrombin activation.
    c. It binds to red blood cells.
    d. It activates the platelets.
    e. It triggers phagocytosis by the leukocytes.

*29. Transferrin molecules in blood plasma function in
    a. carrying oxygen.
    b. carrying carbon dioxide.
    c. acting as a buffer.
    d. energy storage.
    e. carrying iron.

30. Which of the following will lead to autoregulatory increase of blood flow to a tissue?
    a. High blood pressure
    b. High carbon dioxide concentration
    c. High oxygen concentration
    d. High ATP concentration
    e. High glucose concentration

31. Which of the following statements about gastrovascular cavities is *not* true?
    a. Organisms with gastrovascular cavities tend to be only a few cells thick.
    b. Gastrovascular cavities are always highly branched.
    c. Gastrovascular cavities are filled with interstitial fluid.
    d. The diffusion path length in organisms with gastrovascular cavities is usually short.
    e. Large organisms do not usually have gastrovascular cavities.

32. Which of the following statements about open circulatory systems is *not* true?
    a. A pump is usually present.
    b. Blood is pumped out of openings in the heart called ostia.
    c. The blood is the interstitial fluid.
    d. Tissue fluid bathes the tissues directly.
    e. Many mollusks and all arthropods have open circulatory systems.

33. Which of the following structures is *not* part of the circulatory system of an earthworm?
    a. Contractile vessels
    b. A major dorsal vessel
    c. Capillary beds in the lungs
    d. Capillary beds in each segment
    e. A major ventral vessel

34. The left ventricle in humans is more muscular than the right ventricle because
    a. resistance is higher in the systemic circuit.
    b. resistance is higher in the pulmonary circuit.
    c. it pumps more blood.
    d. it pumps more viscous blood.
    e. None of the above

35. Which of the following vertebrates have a three-chambered heart with both pulmonary and systemic circuits?
    a. Fish
    b. Lungfish
    c. Amphibians
    d. Crocodiles
    e. b and c

36. Which of the following structures in a vertebrate with a four-chambered heart would have blood with the lowest oxygen concentration?
    a. Pulmonary vein
    b. Pulmonary artery
    c. Left atrium
    d. Aorta
    e. Arteriole end of a capillary

*37. When submerged, frogs receive most of their O_2 from capillaries within the skin. Which of the following structures would contain blood with the highest O_2 concentration in a submerged frog that is not able to breath using its lungs?
    a. Right atrium
    b. Left atrium
    c. Aorta
    d. Ventricle
    e. Pulmonary vein

38. Which of the following carries oxygenated blood?
    a. Aorta
    b. Pulmonary vein
    c. Pulmonary artery
    d. a and b
    e. a and c

39. A blood pressure represented as 140/100 means that the
    a. pressure during ventricular contraction is 140, while the pressure during ventricular relaxation is 100.
    b. pressure during ventricular contraction is 100, while the pressure during ventricular relaxation is 140.
    c. pressure during ventricular contraction is 140, while the pressure during atrial contraction is 100.
    d. pressure during atrial contraction is 140, while the pressure during ventricular contraction is 100.
    e. diastolic pressure is 140 and the systolic pressure is 100.

*40. During the cardiac cycle, the blood pressure in the aorta is at a minimum
    a. at the end of systole.
    b. at the beginning of systole.
    c. at the beginning of diastole.
    d. when the physician hears the "dub" sound with the stethoscope.
    e. None of the above

*41. Which of the following heart regions is third in sequence of action potential propagation during a normal heartbeat?
    a. Bundle of His
    b. Atrioventricular node
    c. Purkinje fibers
    d. Sinoatrial node
    e. Ventricular muscle cells

42. Which of the following statements concerning the heartbeat is *true*?
    a. Only the ventricles contract together.
    b. Only the atria contract together.
    c. The atria contract together, and the ventricles contract together.
    d. Only the atrium and ventricle on the right side of the heart contract together.
    e. Only the atrium and ventricle on the left side of the heart contract together.

43. Gap junctions
    a. electrically isolate cardiac muscle cells.
    b. allow large numbers of cardiac muscle cells to contract in unison.
    c. provide enough strength to withstand the large pressure generated by the ventricles.
    d. are only found in the cardiac muscle cells of the ventricles.
    e. are especially abundant in the muscles between the atria and ventricles.

44. The blood vessels with the greatest total cross-sectional area are the
    a. arteries.
    b. arterioles.
    c. capillaries.
    d. venules.
    e. veins.

45. The __________ have blood with the lowest pressure.
    a. arteries
    b. arterioles
    c. capillaries
    d. venules
    e. veins

*46. In the electrocardiogram (EKG) of a normal cardiac cycle, the wave called T corresponds to
    a. depolarization of the atria.
    b. depolarization of the ventricles.
    c. repolarization of the atria.
    d. repolarization of the ventricles.
    e. More than one of the above

*47. Blood pressure is __________ than osmotic pressure at the arterial end of a capillary bed, and the process of __________ occurs there.
    a. greater, reabsorption
    b. greater, filtration
    c. lesser, reabsorption
    d. lesser, filtration
    e. lesser, edema

*48. Liver failure results in a
    a. lower concentration of protein in the plasma.
    b. lower osmotic potential in the plasma.
    c. buildup of tissue fluid in the abdomen and extremities.
    d. loss of water from capillaries.
    e. All of the above

★49. Which of the following statements about the lymphatic system is *not* true?
a. The lymphatic system conducts lymph from the lymph capillaries to the thoracic duct, where it enters the circulatory system.
b. Lymph is identical to blood except it does not contain erythrocytes.
c. Lymphoid capillaries are distributed throughout the body.
d. Contraction of skeletal muscles propels the lymph.
e. Lymph nodes are an important part of the defense system that combats infections.

50. All of the following except one are mechanisms that facilitate return of venous blood to the heart. Select the exception.
a. Fainting
b. Contraction of skeletal muscles
c. Breathing
d. Contraction of smooth muscle in most of the veins of the body
e. Valves in veins

51. The blood vessels with the greatest capacity to store blood are the
a. arteries.
b. arterioles.
c. capillaries.
d. venules.
e. veins.

52. Which of the following statements about blood cells is *not* true?
a. There are at least 500 times as many red as white blood cells.
b. Erythrocytes are anucleate in mammals.
c. The biconcave shape of the erythrocyte creates a large surface area for gas exchange.
d. Red blood cells are generated by stem cells in the spleen.
e. Leukocytes destroy foreign cells.

★53. Which of the following circulatory components is *not* normally found circulating in the plasma?
a. Platelets
b. Thrombin
c. Albumin
d. Fibrinogen
e. Transferrin

54. The purpose of precapillary sphincters is to
a. increase blood pressure to the arteries.
b. decrease blood pressure to the veins.
c. shut off the supply of blood to the capillary bed.
d. increase blood pressure to the capillaries.
e. provide blood to smooth muscles surrounding capillary beds.

55. Which of the following statements about veins is *true*?
a. As much as 80% of total blood volume may be in the veins at any one time.
b. Veins are capacitance vessels.
c. The walls of veins are more expandable than the walls of arteries.
d. Veins are resistance vessels.
e. a, b, and c

56. Which one of the following responses characterizes the diving reflex?
a. Reduction in the heartbeat rate
b. Increase in systolic pressure
c. Increase in diastolic pressure
d. Tachycardia
e. Dilation of arteries to most organs

▪57. Advantages of closed circulatory systems over open circulatory systems include which of the following?
a. Exchange occurs more rapidly.
b. Closed systems can direct blood to specific tissues.
c. Cells and large molecules can be kept separate from the animal's intercellular material.
d. Closed circulatory systems can support higher levels of metabolic activity.
e. All of the above

58. In the evolution of the vertebrate circulatory system, the ________ is the organism that reveals the transition step leading to separate pulmonary and systemic circuits.
a. lungfish
b. ancient ostracoderm
c. bird
d. amphibian
e. reptile

59. Systole and diastole describe the ________ and ________, respectively, of the ________ in mammals.
a. relaxation, contraction, ventricles
b. contraction, relaxation, ventricles
c. relaxation, contraction, atria
d. contraction, relaxation, atria
e. None of the above

60. Which of the following is *not* a risk factor for developing atherosclerosis?
a. A high fat and high cholesterol diet
b. Smoking
c. High altitude
d. Sedentary living
e. Hypertension

61. Which of the following statements about valves is *false*?
a. Defective heart valves produce sounds called heart murmurs.
b. Veins and lymphatic vessels have two-way valves.
c. Varicose veins result when valves within veins can no longer prevent backflow of blood.
d. The pulmonary valve is located between the right ventricle and the pulmonary artery.
e. Atrioventricular valves prevent backflow of blood into the atria when the ventricles contract.

62. When a piece of thrombus breaks loose and lodges in a smaller vessel, blocking the flow of blood, it is called a(n)
   a. thrombosis.
   b. infarction.
   c. embolism.
   d. atherosclerosis.
   e. hematocrit.

63. When whole blood is centrifuged, it is separated into a liquid component called __________ and a bottom layer of __________.
   a. hematocrit, cells
   b. plasma, cells
   c. plaque, cells
   d. hematocrit, plasma
   e. packed cell volume, plasma

64. The function of leukocytes is
   a. defense against infection.
   b. transport of respiratory gases.
   c. blood clotting.
   d. to distribute nutrients to tissues.
   e. to produce all the different types of blood cells.

65. Which of these hormones causes constriction of the blood vessels?
   a. Epinephrine
   b. Vasopressin
   c. Angiotensin
   d. b and c
   e. a, b, and c

66. The purpose of stretch receptors in the aorta and carotid arteries is to
   a. detect atherosclerosis.
   b. detect changes in blood pressure and relay to the brain.
   c. constrict arteries when available oxygen is low.
   d. detect when a flexor muscle is stretched too far and pump more blood to the muscle.
   e. None of the above

67. Which of these factors contributes to increased blood flow to tissues?
   a. Low oxygen concentration
   b. Increased concentration of lactate
   c. High concentration of carbon dioxide
   d. a and b
   e. a, b, and c

Study Guide Questions

1. Which of the following is not one of the reasons that closed circulatory systems are more efficient than open circulatory systems?
   a. Closed systems rely exclusively on simple diffusion for transport, whereas open systems rely on pumping mechanisms.
   b. Transport within closed systems is more rapid than in open systems.
   c. Blood can easily be directed to specific areas in closed systems, but not in open systems.
   d. Closed systems operate better under higher pressure than do open systems.

2. How does the pattern of blood flow change when reptiles undergo periods of intermittent breathing?
   a. Blood is shunted from the systemic circuit to the pulmonary circuit.
   b. Blood is shunted specifically to the brain.
   c. Blood is shunted from the pulmonary circuit to the systemic circuit.
   d. The pattern does not change during intermittent breathing.

3. In which of the following would you record the highest blood pressure?
   a. The ventricle supplying blood to the gills of a fish
   b. The anterior dorsal artery of an ant
   c. The pulmonary vein of a frog
   d. The ventricle supplying blood to the systemic circuit of a bird

4. Blood consists of a fluid fraction consisting of __________ and a solid fraction consisting of __________.
   a. plasma; water, erythrocytes, and platelets
   b. erythrocytes; leukocytes, macrophages, and platelets
   c. plasma; erythrocytes, platelets, and leukocytes
   d. leukocytes; erythrocytes and platelets

5. Red blood cells are
   a. biconcave cells containing hemoglobin.
   b. spherical cells containing hemoglobin.
   c. spherical cells capable of ameboid motion and containing hemoglobin.
   d. biconcave cells that contain platelets.

6. Blood clotting pathways cause
   a. conversion of vitamin K to prothrombin.
   b. conversion of fibrin to fibrinogen.
   c. conversion of thrombin to prothrombin.
   d. None of the above

7. Which of the following regions of the vascular bed is the actual site of gas exchange with surrounding tissue?
   a. Arteries
   b. Capillaries
   c. Lymphatic vessels
   d. Veins

8. Which is the correct sequence of parts through which cardiac action potentials pass?
   a. Purkinje fibers, AV node, SA node, bundle of His, atrial fibers
   b. AV node, atrial fibers, SA node, bundle of His, Purkinje fibers
   c. SA node, bundle of His, atrial fibers, AV node, Purkinje fibers
   d. SA node, atrial fibers, AV node, bundle of His, Purkinje fibers

DHS

9. The purpose of the AV node is to _________ and the purpose of the Purkinje fibers is to _________.
   a. create simultaneous atrial and ventricular depolarization; speed up transmission of the cardiac impulse into the ventricle
   b. delay ventricular depolarization relative to atrial depolarization; insulate the cardiac impulse from the general ventricular fibers
   c. delay ventricular depolarization relative to atrial depolarization; transmit the cardiac impulse to very small localized groups of ventricular fibers
   d. delay atrial depolarization relative to ventricular depolarization; transmit the cardiac impulse to very small localized groups of ventricular fibers

10. The atrial walls are _________ than the ventricular wall, and pressure generated in the atrial chambers is _________ than in the ventricles.
    a. thinner, higher
    b. thinner, lower
    c. thicker, higher
    d. thicker, lower

11. The left ventricle exceeds the right ventricle in
    a. the amount of blood that enters during heart contraction.
    b. the volume expelled during contraction.
    c. the pressure developed during contraction.
    d. All of the above

12. Which of the following structures of the lymphatic system acts primarily as a filter for detecting and destroying microorganisms in lymph traveling through major lymph vessels?
    a. Lymph nodes
    b. Thymus
    c. Lymph capillaries
    d. Tonsils, but not the appendix

13. The net loss of fluid from blood capillaries will increase if
    a. plasma filtration decreases.
    b. the osmotic pressure of plasma increases.
    c. blood (hydrostatic) pressure increases in the capillaries.
    d. the colloid osmotic pressure of interstitial fluid decreases.

14. Which of the following statements about the control of circulation in humans is *false*?
    a. Sympathetic nerve input to skeletal muscle causes the blood vessels in the muscle to dilate.
    b. Blood flow can be regulated by autonomic nerve signals emanating from cardiovascular control centers in the medulla of the brain.
    c. Carotid artery chemosensors detect low O_2 levels in the blood and promote increased blood pressure.
    d. Hormones such as angiotensin and vasopressin cause venules to constrict.

End of Chapter Questions

1. An open circulatory system is characterized by
   a. the absence of a heart.
   b. the absence of blood vessels.
   c. blood with a composition different from that of intercellular fluid.
   d. the absence of capillaries.
   e. a higher-pressure circuit through gills than to other organs.

2. Which statement about vertebrate circulatory systems is *not* true?
   a. In fish, oxygenated blood from the gills returns to the heart through the left atrium.
   b. In mammals, deoxygenated blood leaves the heart through the pulmonary artery.
   c. In amphibians, deoxygenated blood enters the heart through the right atrium.
   d. In reptiles, the blood in the pulmonary artery has a lower oxygen content than the blood in the aorta.
   e. In birds, the pressure in the aorta is higher than the pressure in the pulmonary artery.

3. Which statement about the human heart is *true*?
   a. The walls of the right ventricle are thicker than the walls of the left ventricle.
   b. Blood flowing through atrioventricular valves is always deoxygenated blood.
   c. The second heart sound is due to the closing of the aortic valve.
   d. Blood returns to the heart from the lungs in the vena cava.
   e. During systole the aortic valve is open and the pulmonary valve is closed.

4. Pacemaker actions of cardiac muscle
   a. are due to opposing actions of norepinephrine and acetylcholine.
   b. are localized in the bundle of His.
   c. depend on the gap junctions between cells that make up the atria and those that make up the ventricles.
   d. are due to spontaneous depolarization of the plasma membranes of some cardiac-muscle cells.
   e. result from hyperpolarization of cells in the sinoatrial node.

5. Blood flow through capillaries is slow because
   a. lots of blood volume is lost from the capillaries.
   b. the pressure in venules is high.
   c. the total cross-sectional area of capillaries is larger than that of arterioles.
   d. the osmotic pressure in capillaries is very high.
   e. red blood cells are bigger than capillaries and must squeeze through.

6. How are lymphatic vessels like veins?
   a. Both have nodes where they join together into larger common vessels.
   b. Both carry blood under low pressure.
   c. Both are capacitance vessels.
   d. Both have valves.
   e. Both carry fluids rich in plasma proteins.

7. The production of red blood cells
   a. ceases if the hematocrit falls below normal.
   b. is stimulated by erythropoietin.
   c. is about equal to the production of white blood cells.
   d. is inhibited by prothrombin.
   e. occurs in bone marrow before birth and in lymph nodes after birth.

8. Which of the following does *not* increase blood flow through a capillary bed?
   a. High concentration of CO_2
   b. High concentration of lactate and hydrogen ions
   c. Histamine
   d. Vasopressin
   e. Increase in arterial pressure

9. Blood clotting
   a. is impaired in patients with hemophilia because they don't produce platelets.
   b. is initiated when platelets release fibrinogen.
   c. involves a cascade of factors produced in the liver.
   d. is initiated by leukocytes forming a meshwork.
   e. requires conversion of angiotensinogen to angiotensin.

10. Autoregulation of blood flow to a tissue is due to
    a. sympathetic innervation.
    b. the release of vasopressin by the hypothalamus.
    c. increased activity of baroreceptors.
    d. chemoreceptors in carotid and aortic bodies.
    e. the effect of local environment on arterioles.

Student Web Site Self-Quiz Questions

1. A mollusk has a(n) _________ circulatory system. In this type of system, blood and tissue fluid are _________.
   a. closed, continuous
   b. closed, separate
   c. open, hemolymph
   d. open, continuous
   e. open, separate

2. Which of the following statements about the circulatory system of the earthworm is *false*?
   a. The earthworm has a closed circulatory system.
   b. The dorsal and ventral vessels are connected by a single pair of connecting vessels.
   c. The direction of blood flow in the earthworm's circulatory system is determined by the orientation of one-way valves.
   d. Small vessels connect the dorsal and ventral vessels.
   e. The ventral vessel carries blood away from the hearts.

3. Which of the following statements about the circulatory systems of crocodilians is *false*?
   a. Crocodilians have a four-chambered heart.
   b. Unlike reptiles, crocodilians cannot shunt blood away from the pulmonary circuit when they are not breathing.
   c. The crocodilian's right aorta carries mostly mixed blood.
   d. Only deoxygenated blood is sent to the lungs.
   e. Structurally and functionally, the crocodilian heart is most like the reptilian heart.

4. Beginning in a vein in the kidney and ending in the aorta, if you order the following structures of the human heart in the sequence in which they would be encountered, which structure would be **fourth**?
   a. Inferior vena cava
   b. Left ventricle
   c. Pulmonary artery
   d. Pulmonary vein
   e. Right atrium

5. Heart block is the failure of stimulation to the ventricles following atrial contraction. Which one of the following heart structures could *not* be involved in heart block?
   a. Atrioventricular node
   b. Bundle of His fibers
   c. Bundle branches
   d. Purkinje fibers
   e. Sinoatrial node

6. Which one of the following could *not* lead to edema?
   a. Blockage of a lymph vessel
   b. Decrease in filtration
   c. Decrease in blood proteins
   d. Presence of histamine
   e. Increased blood pressure

7. Which of the following statements about blood is *false*?
   a. An individual with anemia (fewer erythrocytes than normal) would have a low hematocrit.
   b. One effect of living at high altitude is decreased erythropoietin production.
   c. Plasma is blood minus its cellular components.
   d. Stem cells produce erythrocytes.
   e. Plasma is very similar to intercellular fluid except that the protein concentration of plasma is higher.

8. Which of the following conditions in the capillary bed would *not* be expected to cause the smooth muscles of precapillary sphincters to relax in a capillary bed serving a skeletal muscle?
   a. High CO_2 concentration
   b. High lactate concentration
   c. Low O_2 concentration
   d. Sympathetic neuron stimulation in skeletal muscle
   e. Angiotensin

9. Which of the following would *not* be expected to increase heart rate and blood flow?
   a. Decreased activity in the stretch receptors in the walls of the aorta
   b. Diving reflex
   c. Information from the carotid bodies that blood O_2 levels are falling
   d. Vasopressin
   e. Input from higher brain center

TEST FILE

★ *Indicates a **difficult** question*
■ *Indicates a **conceptual** question*

50 Animal Nutrition

Fill in the Blank

1. The crowns of the teeth of mammals are covered with a hard material called **enamel**.

2. **Hydrolytic** enzymes break down macromolecules into their monomeric units by adding water.

★3. Many digestive enzymes are produced in an inactive **zymogen** form.

4. A wave of smooth muscle contraction called **peristalsis** pushes food along the digestive tract.

5. In addition to serving as an endocrine gland, the pancreas secretes the digestive enzyme precursor **trypsinogen**.

6. Many species of **bacteria** live in the human large intestine.

7. The initial digestion of protein in the vertebrate digestive tract begins in the **stomach**.

8. The **liver** is the organ that manages most of the balance between circulating and storing nutrients.

9. The liver converts pyruvate and other molecules to glucose by the process of **gluconeogenesis**.

10. The regulation of blood glucose is accomplished mainly by hormones produced by the **pancreas**.

■11. Most plants, some monerans, and some protists are autotrophs. Animals, however, are **heterotrophs** because they derive both their energy and their structural molecules from their food.

★12. Most animals process their food through **extracellular** digestion in a digestive cavity called a gut.

13. If 1 kcal = 4.184 joules, then the number of kcal found in a diet consisting of 13,398 joules is **3,200** kcal.

14. From acetyl units obtained from carbohydrates or fats, we can synthesize almost all the lipids required by the body, but we must have a dietary source of essential **fatty acids**, notably linoleic acid.

15. **Bile** prevents fat droplets from aggregating, or clumping together, so that the maximum surface area is exposed to lipase action.

16. Crabs and earthworms that actively feed on dead organic matter are called **detritivores**.

17. An unusual feature of the vertebrate digestive tract is that it has its own intrinsic **nervous system**.

18. **Saprobes** are organisms like protists and fungi that absorb nutrients from dead organic matter.

19. The three sections of the small intestine are the **duodenum, jejunum,** and **ileum.**

Multiple Choice

1. The order in which stored fuels are utilized during starvation is
   a. fats, then glycogen, then proteins.
   b. glycogen, then proteins, then fats.
   c. proteins, then fats, then glycogen.
   d. fats, then proteins, then glycogen.
   e. glycogen, then fats, then proteins.

2. Why are certain amino acids called essential amino acids?
   a. They are required for making protein.
   b. They cannot be made from other amino acids.
   c. They are universally needed by all animals.
   d. They are essential as an energy source.
   e. They are required for making nucleic acids.

3. Why are vitamins essential nutrients for cells?
   a. Vitamins are used as an energy source.
   b. Vitamins are used to digest foods.
   c. Animals cannot synthesize any vitamins.
   d. Some vitamins are fat soluble.
   e. Vitamins function as coenzymes.

■4. Organisms that derive both their energy and molecular nutrients from other organisms are called
   a. autotrophs.
   b. herbivores.
   c. heterotrophs.
   d. photosynthetic.
   e. protists.

5. The energy content of food is described in terms of calories because
   a. the amount of energy in food depends on the temperature.

b. food heats up as it is being digested.
c. the energy in food ultimately becomes heat.
d. heat is the main product of digestion.
e. heat is the main product of respiration.

6. The major form of stored energy in animal bodies is
a. protein, because it is a long-term energy storage form.
b. glycogen, because it breaks down into readily usable carbohydrates.
c. glycogen, because it is lightweight.
d. fat, because it has the highest energy content per gram.
e. fat, because it is readily stored with water.

7. Vitamin C has all *except* which of the following properties?
a. Functions in collagen production
b. Important for healthy skin
c. An essential vitamin for humans
d. A fat-soluble vitamin
e. Excess will be excreted

8. Vitamin D is obtained by different people in various ways, including
a. from high-latitude sunlight by dark-skinned people.
b. from tropical sunlight by dark-skinned people.
c. from sunlight by Eskimos.
d. by absorption of sunlight by protein molecules.
e. from water-soluble components of foods.

9. The nutritional disease kwashiorkor results from
a. protein deficiency.
b. vitamin deficiency.
c. calorie deficiency.
d. overdose of fat-soluble vitamins.
e. overdose of thyroxine.

10. Which of the following diseases is not due to a vitamin deficiency?
a. Pellagra
b. Rickets
c. Goiter
d. Pernicious anemia
e. Beriberi

11. All of the following represent feeding adaptations of carnivores *except*
a. spider webs.
b. rattlesnake venom.
c. bat echolocation.
d. elephant trunks.
e. jellyfish tentacles.

12. How does a filter feeder acquire food items?
a. By using poison to restrain prey
b. By using claws and jaws to restrain prey
c. By ingesting mud and extracting particles
d. By filtering food substances from blood
e. By extracting particles suspended in water

▪13. A mammal with a diet of grain and leaves would be expected to have what kind of teeth?
a. Prominent canine teeth and small molars
b. Prominent molars and small incisors
c. Prominent molars and canine teeth
d. A balanced set of incisors, molars, and canines
e. Prominent canine teeth and small incisors

14. For most animals, digestion of food occurs
a. intracellularly.
b. in the gastrovascular cavity.
c. in the coelomic body cavity.
d. in the midgut.
e. in the crop.

*15. Most animals digest dietary proteins to their constituent amino acids and then synthesize new proteins because
a. macromolecules like proteins cannot be readily transported through plasma membranes.
b. protein function often varies with species.
c. foreign proteins are considered invaders and attacked by the immune system.
d. All of the above
e. None of the above

▪16. Which of the following is *true* about digestive systems in all animals?
a. Food in the gut is digested extracellularly.
b. There are two openings to the tract.
c. Animals break food into smaller pieces before ingestion.
d. All digestion occurs inside the gut.
e. The digestive tract has specialized segments.

*17. Which of the following structures does not serve to increase surface area for nutrient absorption?
a. Intestinal villi
b. Earthworm typhlosole
c. Shark spiral valve
d. Bird gizzard
e. Cell microvilli

18. An endolipase is an enzyme that breaks down
a. carbohydrates.
b. nucleic acids.
c. proteins.
d. fat molecules by cutting in the middle.
e. fat molecules by snipping at the ends.

19. Most ulcers are caused by
a. stress.
b. an infectious bacterium.
c. oversecretion of digestive juices
d. old age.
e. an acidic diet.

*20. In the small intestine, the blood and lymph vessels that carry away absorbed nutrients lie in which layer?
a. Microvilli
b. Mucosa
c. Submucosa
d. Circular muscle layer
e. Longitudinal muscle layer

*21. Movement of food from the stomach into the esophagus is normally prevented by
a. peristalsis.
b. reverse peristalsis.
c. the pyloric sphincter.
d. a sphincter.
e. the pharynx.

22. The major enzyme produced by the stomach is
a. amylase.
b. chyme.
c. mucus.
d. pepsin.
e. trypsin.

23. What activates the inactive form of stomach enzymes?
a. Activating enzymes
b. ATP
c. Low pH
d. The appropriate substrate molecule
e. The presence of water

24. Most absorption of nutrients in the digestive tract takes place in the
a. stomach.
b. small intestine.
c. large intestine.
d. liver.
e. pancreas.

*25. Bile aids in the breakdown of lipids by
a. hydrolyzing lipids.
b. activating hydrolytic enzymes.
c. aggregating droplets of lipids.
d. emulsifying lipids.
e. making lipids water-soluble.

*26. What neutralizes the acidic chyme in the small intestine?
a. Bicarbonate from the pancreas
b. Buffers from the jejunum
c. Bile from the liver
d. Trypsin activation
e. A variety of zymogens

27. How does sodium cotransport facilitate glucose absorption?
a. Active transport of sodium aids in salt uptake.
b. When sodium diffuses into cells, the carrier also binds a glucose.
c. When sodium is pumped into cells, glucose moves out.
d. Sodium and glucose both diffuse into cells from the gut.
e. Glucose is actively transported into the gut.

28. The major function of the colon or large intestine is
a. digestive breakdown of foods.
b. nutrient absorption of foods.
c. housing parasitic bacteria.
d. secretion of bile and enzymes.
e. reabsorption of water.

29. The hormone secretin is a chemical message
a. secreted by the pancreas.
b. secreted by the stomach.
c. whose release is stimulated by the nervous system.
d. that triggers the intestine to release enzymes.
e. that triggers pancreatic secretion.

30. Insulin is released by the pancreas when
a. blood glucose falls.
b. blood glucagon falls.
c. blood glucose rises.
d. blood glucagon rises.
e. blood insulin falls.

31. The effects of glucagon include
a. stimulating glucose uptake into cells.
b. stimulating liver cells to break down glycogen.
c. controlling blood glucose in the absence of insulin.
d. stimulating cells to store energy as fat.
e. None of the above

32. Which of the following statements about energy storage is *true*?
a. Carbohydrates are stored in the liver and in muscle as glycogen. The total glycogen stores are usually not more than the equivalent of a day's energy requirements.
b. Fat is an important form of stored energy.
c. Fat has the highest energy content per gram.
d. Protein is the most important energy storage component.
e. a, b, and c

33. Digestive enzymes are classified according to the substances they hydrolyze and where the enzyme cleaves the given molecule. An endopeptidase hydrolyzes a __________ at __________ site along the length of the molecule.
a. carbohydrate, an external
b. protein, an internal
c. fat, an internal
d. carbohydrate, an internal
e. peptide, an external

34. The gut of an animal is often described as an elongated tube consisting of four layers of different cell types. These layers are
a. submucosa, cartilage, mucosa, and endoplasmic reticulum.
b. smooth muscle layers, submucosa, mucosa, and cartilage.
c. cartilage, smooth muscle layers, mucosa, and circular muscle.
d. mucosa, submucosa, circular muscle, and longitudinal muscle.
e. cartilage, mucosa, endoplasmic reticulum, and submucosa.

35. __________ ingest both plant and animal food and process their food through __________ digestion.
a. Predators, intracellular
b. Omnivores, intracellular

c. Carnivores, intracellular
d. Herbivores, extracellular
e. Omnivores, extracellular

36. Which of the following statements about undernourishment is *not* true?
a. One-fifth of the world's population is undernourished.
b. Several weeks of fasting are required to deplete glycogen reserves.
c. Self-imposed starvation is called anorexia nervosa.
d. The loss of blood proteins during starvation leads to edema.
e. During starvation, fat reserves are metabolized before body protein.

*37. Herbivores have __________ essential amino acids than carnivores and consequently are less likely to suffer from __________ than carnivores.
a. fewer, undernourishment
b. fewer, malnourishment
c. more, undernourishment
d. more, malnourishment
e. more, overnourishment

*38. Which of the following statements about the essential amino acids is *not* true?
a. Most animals have some essential amino acids.
b. Ingesting a surplus of one essential amino acid cannot compensate for a shortage of another.
c. Some plant foods such as legumes supply all eight of the essential amino acids in humans.
d. Acetyl groups can be combined with amino groups to produce many of the nonessential amino acids.
e. As with amino acids, some fatty acids are also essential.

39. The enzyme lactase
a. cleaves lactose into glucose and galactose.
b. is produced by the small intestine.
c. is often not produced by humans after four years of age.
d. cleaves milk sugar.
e. All of the above

*40. A strictly vegetarian diet with no vitamin B_{12} supplements can lead to
a. beriberi.
b. pellagra.
c. pernicious anemia.
d. scurvy.
e. night blindness.

41. The nutritional deficiency disease beriberi is caused by an inadequate supply of the water-soluble vitamin
a. B_1, thiamin.
b. B_2, riboflavin.
c. B_{12}, cobalamin.
d. folic acid.
e. C, ascorbic acid.

42. The nutritional deficiency disease rickets is caused by an inadequate supply of the fat-soluble vitamin
a. A, retinol.
b. D, calciferol.
c. E, tocopherol.
d. K, menadione.
e. biotin.

43. The nutritional deficiency disease simple goiter is caused by inadequate supply of the micronutrient
a. fluoride.
b. iodine.
c. chromium.
d. zinc.
e. copper.

44. Which of the following statements about vitamins is *not* true?
a. Vitamins, like essential amino acids and fatty acids, are organic molecules, but they are needed in micro-quantities.
b. Most vertebrates require the very same vitamins.
c. Vitamins function mostly as, or as parts of, coenzymes.
d. Vitamins are required only in very small amounts.
e. Humans require more water-soluble vitamins than fat-soluble vitamins.

*45. Which of the following is the true stomach of a ruminant?
a. Cecum
b. Reticulum
c. Rumen
d. Omasum
e. Abomasum

46. The root of a typical tooth contains __________ and __________ but *not* __________.
a. enamel, dentine, a pulp cavity
b. enamel, a pulp cavity, dentine
c. dentine, a pulp cavity, enamel
d. dentine, enamel, a pulp cavity
e. cement, dentine, a pulp cavity

47. Which of the following teeth are *least* likely to be found in animals with a diet consisting mainly of plants?
a. Incisors
b. Canines
c. Premolars
d. Molars
e. Cheek teeth

48. Which one of the following structures of a tubular digestive tract is mainly involved with water and ion recovery?
a. Gizzard
b. Buccal cavity
c. Stomach
d. Hindgut
e. Midgut

49. All of the following structures found in a gut tube except one facilitate the absorption of nutrients. Select the exception.
    a. Villi
    b. Spiral valve
    c. Crop
    d. Microvilli
    e. Typhlosole

50. Which layer of the vertebrate gut shows adaptations for increasing absorptive surface area?
    a. Lumen
    b. Mucosa
    c. Submucosa
    d. Serosa
    e. Peritoneum

51. Which layer of the vertebrate gut is responsible for peristalsis?
    a. Smooth muscle layer
    b. Mucosa
    c. Submucosa
    d. Serosa
    e. Peritoneum

52. Which structure is *not* encountered by a food bolus when being swallowed?
    a. Epiglottis
    b. Larynx
    c. Soft palate
    d. Pharynx
    e. Esophagus

*53. Which of the following statements about movement of food in the gut is *not* true?
    a. Three sphincters are found in the vertebrate gut.
    b. Peristalsis can move food in both directions.
    c. Stretched smooth muscle contracts.
    d. Peristalsis begins when food enters the glottis.
    e. The muscle in a sphincter is normally contracted.

54. The sweetness that develops if you continually chew a piece of bread is due to the enzyme
    a. maltase.
    b. sucrase.
    c. lactase.
    d. amylase.
    e. pepsin.

*55. Pepsinogen is converted into pepsin by
    a. low pH.
    b. chyme.
    c. enterokinase.
    d. trypsinogen.
    e. amylase from the salivary glands.

56. Which of the following is *not* brought about by the HCl secreted in the stomach?
    a. Activation of the principal zymogen of the stomach
    b. Proper pH for the digestive enzyme of the stomach
    c. Breakdown of ingested tissues
    d. Formation of chylomicrons
    e. Death of ingested bacteria

*57. Which of the following statements about digestion in the small intestine is *not* true?
    a. Most digestion occurs in the duodenum of the small intestine.
    b. Bile is produced by the liver, stored in the gallbladder, and emulsifies fat in the small intestine.
    c. Bile molecules have one end that is lipophobic and one end that is hydrophilic.
    d. The pancreas secretes zymogens and bicarbonate ions into the duodenum.
    e. Enterokinase is secreted by intestinal mucosal cells.

*58. Which one of the following proteases is produced by the small intestine?
    a. Dipeptidase
    b. Chymotrypsin
    c. Trypsin
    d. Carboxypeptidase
    e. Pepsin

59. Which of the following statements about the role of the large intestine in digestion is *not* true?
    a. If too little water is reabsorbed from the feces, diarrhea results.
    b. The last section of the large intestine is called the colon.
    c. *Escherichia coli* is a normal inhabitant of the large intestine.
    d. Flatulence results from the metabolism of intestinal bacteria.
    e. The appendix is a vestigial cecum.

*60. Which of the following activities is *not* carried out by the liver in its involvement in the regulation of fuel metabolism?
    a. Conversion of nutrients into glycogen and fat
    b. Gluconeogenesis
    c. Synthesis of plasma proteins from amino acids
    d. Production of high-density lipoproteins (HDL) for deposition in adipose tissue
    e. Processing of chylomicrons from the small intestine

*61. Which of the following statements about the hormonal control of fuel metabolism is *not* true?
    a. During the absorptive period, the liver converts glycogen into glucose.
    b. During the postabsorptive period, the body cells preferentially use fatty acids for metabolic fuel.
    c. Cells of the nervous system depend almost exclusively on glucose for metabolic fuel.
    d. Glucagon plays a major role during the postabsorptive period by causing the liver to convert glycogen into glucose and stimulating gluconeogenesis.
    e. Adrenaline and glucocorticoid cortisol have effects similar to glucagon.

62. People living in impoverished regions and chronic alcoholics frequently have a niacin deficiency called
    a. beriberi.
    b. kwashiorkor.
    c. pellagra.
    d. scurvy.
    e. rickets.

63. Primates should eat citrus fruit to prevent scurvy, a deficiency of
    a. niacin.
    b. thiamin.
    c. calciferol.
    d. ascorbic acid.
    e. biotin.

*64. Ruminant animals, such as goats and cows,
    a. produce enzymes in their guts that break down cellulose.
    b. practice coprophagy.
    c. have a four-chambered stomach for increased surface area for better absorption of nutrients.
    d. produce hydrogen sulfide, an important greenhouse gas.
    e. get much of their protein from the digestion of microorganisms.

65. These molecules accept cholesterol and probably remove it from the tissues to the liver. Smoking lowers their levels.
    a. Chylomicrons
    b. Very-low-density lipoproteins
    c. Low-density lipoproteins
    d. High-density lipoproteins
    e. Cholecystokinin

66. Which of the following molecules contain mostly triglycerides and transport them to fat cells in tissues.
    a. Chylomicrons
    b. Very-low-density lipoproteins
    c. Low-density lipoproteins
    d. High-density lipoproteins
    e. Cholecystokinin

67. Which of the following is true of lipid-soluble compounds?
    a. They dissolve in water.
    b. They are often stored for a long time.
    c. They metabolize quickly.
    d. They are easily filtered by the kidney.
    e. They are hydrophilic.

▪68. Which of the following is an example of bioaccumulation?
    a. A bird that eats fish, which eat invertebrates, which eat algae, which pick up a pesticide in a stream
    b. Lead building up in a child's liver
    c. A mouse eating poison in a mousetrap
    d. A factory constantly dumping toxins in a river
    e. A bird accidentally getting sprayed with pesticide and infecting its offspring

69. The purpose of cytochrome P450 is to
    a. make energy in the electron transport chain.
    b. break down cytochrome into simple diglycerides.
    c. detoxify synthetic chemicals by adding an —OH or an —SO^3 group.
    d. break down toxins into smaller, less harmful molecules.
    e. break down glycogen into glucose in the liver.

70. Polychlorinated biphenyls are dangerous because they
    a. cause hormones to shut off.
    b. mimic hormones, causing loss of control of hormonal functioning.
    c. bind vitamin B_1 and cause a deficiency.
    d. cause a thinning of bird eggshells.
    e. replace iron in blood and calcium in bones.

71. The protein leptin
    a. is produced by cells of the hypothalamus.
    b. signals the brain about the status of body fat reserves.
    c. is found in lower than normal levels in most obese humans.
    d. serves as a positive feedback signal to the brain to limit food intake.
    e. is an environmental toxin that accumulates in body fat.

Study Guide Questions

1. Certain amino acids are essential to the diet of animals because
    a. they prevent overnourishment.
    b. they are cofactors and coenzymes that are required for normal physiological function.
    c. an animal cannot directly synthesize them through the transfer of an amino group to an appropriate carbon skeleton.
    d. animals need these substances in order to make the stored fats that are used during hibernation and migration.

2. Which of the following descriptions describes digestive characteristics of a sheep?
    a. It is a saprobe: It engulfs food and performs intracellular digestion.
    b. It is an autotroph: It synthesizes organic nutrients and performs extracellular digestion.
    c. It is an herbivore: It ingests food and performs extracellular digestion
    d. It is a detritivore: It ingests food and performs intracellular digestion.

3. Protection of the walls of the stomach against the action of its own digestive juices
    a. results from the presence of an anti-enzyme chemical formed by the gastric glands.
    b. results from the nervous reactions of the lining of the stomach.
    c. is controlled by a center in the medulla of the brain.
    d. results from the neutralizing, buffering, and a coating mucus covering its inner surface.

4. Chylomicrons are produced in the
    a. mouth.
    b. stomach.
    c. lumen of the small intestines.
    d. epithelial cells of the small intestines.

5. The gallbladder
    a. produces bile.
    b. is part of the liver.

c. stores bile which is produced in the liver.
d. produces cholecystokinin.

6. The pancreas
   a. produces exocrine products involved in chyme digestion.
   b. is exclusively an endocrine gland which produces salivary amylase.
   c. contains villi to increase surface area.
   d. produces urobiligen (a bile pigment).

7. Hydrochloric acid
   a. is secreted by the gastric glands of the liver.
   b. is secreted by the gastric glands of the stomach.
   c. produces a low pH in the small intestine.
   d. a and c

8. Bile produced in the liver is associated with which of the following?
   a. Emulsification of fats into tiny globules in the small intestine
   b. Digestive action of pancreatic amylase
   c. Emulsification of fats into tiny globules in the stomach
   d. Digestion of proteins into amino acids

9. Most of the chemical digestion of food in humans is completed in the
   a. small intestine.
   b. appendix.
   c. ascending colon.
   d. stomach.

10. Which one of the following does *not* contribute to the large surface area available for nutrient absorption in the small intestines?
    a. Villi
    b. Intestinal length
    c. Microvilli
    d. Bile duct

11. Waves of muscle contractions that move the intestinal contents are
    a. caused by contraction of skeletal muscle.
    b. regulated by liver secretions.
    c. called peristalsis.
    d. voluntary.

12. What is the function of enterokinase?
    a. It converts pepsinogen to pepsin.
    b. It converts trypsinogen to trypsin.
    c. It digests proteins.
    d. It activates HCL.

13. Digestive enzymes responsible for breaking down disaccharides include
    a. pepsin, trypsin, and trypsinogen.
    b. amylase, pepsin, and lipase.
    c. sucrase, lactase, and maltase.
    d. pepsin, trypsin, and chymotrypsin.

14. Which of the following is characteristic of the large intestine?
    a. It has almost no bacterial populations.
    b. It contains chyme.
    c. It absorbs much of the water remaining in waste materials.
    d. It is the site of most of digestion.

15. The innermost layer of the digestive tract is the
    a. serosa membrane.
    b. mucosa membrane.
    c. submucosa membrane.
    d. lumen.

16. Which of the following hormones affects the rate of gluconeogenesis?
    a. Glucagon
    b. Insulin
    c. Epinephrine
    d. All of the above

End of Chapter Questions

1. Most of the metabolic energy that a bird requires for a long-distance migratory flight is stored as
   a. glycogen.
   b. fat.
   c. protein.
   d. carbohydrate.
   e. ATP.

2. Which statement about essential amino acids is true?
   a. They are not found in vegetarian diets.
   b. They are stored by the body for the times when they are needed.
   c. Without them, one is undernourished.
   d. All animals require the same ones.
   e. Humans can acquire all of theirs by eating milk, eggs, and meat.

3. Which statement about vitamins is true?
   a. They are essential inorganic nutrients.
   b. They are required in larger amounts than are essential amino acids.
   c. Many serve as coenzymes.
   d. Vitamin D can be acquired only by eating meat or dairy products.
   e. When vitamin C is eaten in large quantities, the excess is stored in fat for later use.

4. The digestive enzymes of the small intestine
   a. do not function best at a low pH.
   b. are produced and released in response to circulating secretin.
   c. are produced and released under neural control.
   d. are all secreted by the pancreas.
   e. are all activated by an acidic environment.

5. Which statement about nutrient absorption across the gut epithelium is *true*?
   a. Carbohydrates are absorbed as disaccharides.
   b. Fats are absorbed as fatty acids and monoglycerides.
   c. Amino acids move across only by diffusion.
   d. Bile salts transport fats across.
   e. Most nutrients are absorbed in the duodenum.

6. Chylomicrons are like the tiny particles of dietary fat in the lumen of the small intestine in that both
   a. are coated with bile salts.
   b. are lipid-soluble.
   c. travel in lacteals.
   d. contain triglyceride.
   e. are coated with lipoproteins.

7. Microbial fermentation in the gut of a cow
   a. produces fatty acids as a major nutrient for the cow.
   b. occurs in specialized regions of the small intestine.
   c. occurs in the cecum, from which food is regurgitated, chewed again, and swallowed into the true stomach.
   d. produces methane as a major nutrient.
   e. is possible because the stomach wall does not secrete hydrochloric acid.

8. Which of the following is stimulated by cholecystokinin?
   a. Stomach motility
   b. Release of bile
   c. Secretion of hydrochloric acid
   d. Secretion of bicarbonate ions
   e. Secretion of mucus

9. During the absorptive period
   a. breakdown of glycogen supplies glucose to the blood.
   b. glucagon secretion is high.
   c. the number of circulating lipoproteins is low.
   d. glucose is the major metabolic fuel.
   e. the synthesis of fats and glycogen in muscle is inhibited.

10. During the postabsorptive period
    a. glucose is the major metabolic fuel.
    b. glucagon stimulates the liver to produce glycogen.
    c. insulin facilitates the uptake of glucose by brain cells.
    d. fatty acids constitute the major metabolic fuel.
    e. liver functions slow down because of low insulin levels.

Student Web Site Self-Quiz Questions

1. In animal nutrition, the acetyl group is a(n)
   a. carbon skeleton.
   b. essential amino acid.
   c. fatty acid.
   d. macronutrient.
   e. vitamin.

2. Vitamins are required for normal growth and metabolism. Vitamin C is a ________ vitamin. Its deficiency causes ________.
   a. fat-soluble, anemia
   b. fat-soluble, night blindness
   c. water-soluble, beriberi
   d. water-soluble, pellagra
   e. water-soluble, scurvy

3. Which of the following terms best describes organisms that actively feed on dead matter, such as the earthworm?
   a. Saprobe
   b. Detritivore
   c. Herbivore
   d. Filter feeder
   e. Fluid feeder

4. Which of the following structures is the functional equivalent of the shark's spiral valve?
   a. Crop of a cockroach
   b. Radula of a snail
   c. Gizzard of a bird
   d. Rectum of a rabbit
   e. Villi of a human

5. If the action of the enzyme trypsin is to hydrolyze peptide linkages within a protein, then we can describe trypsinogen as ________ and trypsin as ________.
   a. chyme, a hydrolytic enzyme
   b. chyme, an exoprotease
   c. a zymogen, a hydrolytic enzyme
   d. a zymogen, an endoprotease
   e. a zymogen, an exoprotease

6. Which layer in the vertebrate gut can constrict the gut lumen?
   a. The mucosa
   b. The submucosa
   c. The circular muscle layer
   d. The longitudinal muscle layer
   e. The serosa

7. Regarding ingestion in humans, the tube leading directly to the stomach is the ________ and movement of food in this tube is caused by ________.
   a. esophagus, peristalsis
   b. esophagus, a sphincter
   c. esophagus, swallowing
   d. pharynx, peristalsis
   e. pharynx, a sphincter

8. Which of the following is *not* an activity of the stomach?
   a. Absorption
   b. Mechanical breakdown of food
   c. Secretion of enzymes
   d. Secretion of HCl
   e. Secretion of mucus

9. Which of the following is *not* a function of the pancreas?
   a. Production of carbohydrate-digesting enzymes
   b. Production of fat-digesting enzymes
   c. Production of protein-digesting enzymes
   d. Neutralization of acidic chyme
   e. Production of fat-dissolving bile

10. Which of the following would *not* be an expected event resulting from chyme delivery to the small intestine?
    a. Release of secretin by the intestinal mucosa
    b. Release of cholecystokinin by the intestinal mucosa.
    c. Stimulation of gastrin release caused by acidity
    d. Stimulation of bile release by cholecystokinin
    e. Stimulation of bicarbonate release from the pancreas

11. One result of a diet with increased calories may be increased adipose tissue (human adipose tissue). Which of the following statements about fat metabolism is *false*?
    a. The liver can convert glucose into either glycogen or fat.
    b. Very low density lipoproteins (VLDL) have less fat than low-density lipoproteins (LDL).
    c. Cigarette smoking lowers HDL levels; exercise increases HDL levels.
    d. Insulin increases fat synthesis.
    e. Lipoproteins in the blood are the equivalent of micelles in the gut cavity.

12. Which of the following would *not* be expected to occur during the absorptive period of fuel metabolism?
    a. Increase insulin production
    b. Use of glucose for metabolic fuel
    c. Liver cells synthesize glycogen
    d. Increased release of glucagon by the pancreas
    e. Increased fat synthesis in fat cells

13. Obesity is a major health issue facing the United States. Much of this problem relates to social and lifestyle factors; however, some of this problem may be genetic in nature. One gene that has been identified as playing a role in obesity is the *ob* gene, first identified in mice, which codes for the protein leptin. Which of the following statements regarding the ob gene, leptin, and its effects is *false*?
    a. Mutations in the *ob* gene result in overweight mice.
    b. Leptin is produced in fat cells.
    c. Leptin is a signal that travels to the hypothalamus.
    d. Excessive levels of leptin in the blood causes obesity.
    e. High insulin levels cause increased fat synthesis in fat cells.

14. Which of the following is *not* considered a danger produced by the presence of toxins in food supplies?
    a. Bioaccumulation
    b. Thinning eggshells
    c. PCB-induced immune system depression
    d. The rapid accumulation and bioaccumulation of water-soluble toxins
    e. Increased production of synthetic toxins

TEST FILE

★ *Indicates a* ***difficult*** *question*
■ *Indicates a* ***conceptual*** *question*

51 Salt and Water Balance and Nitrogen Excretion

Fill in the Blank

1. Organisms that allow their tissue fluids to have the same osmolarity as the environment are called **osmoconformers.**

2. A cell is **hypertonic** to its environment if its fluids are more concentrated than the environment.

★3. Marine bony fishes actively excrete salt from their kidneys and gills, and nitrogenous wastes are lost as ammonia from the **gills**.

4. Tadpoles living in fresh water excrete nitrogenous wastes in the form of **ammonia.**

5. The Malpighian tubules of insects conserve water and excrete **uric acid.**

6. In the nephron of the vertebrate kidney, the glomerulus is a dense knot of thin-walled **capillaries**.

7. In humans, urine from the collecting ducts passes through the **ureter** to the urinary bladder.

8. Resorption in the renal tubules is facilitated by **microvilli**, providing extra surface area in the cells lining the tubule.

★■9. A normal kidney has several **autoregulatory** mechanisms that are adaptations to monitor and maintain kidney functions.

10. High levels of antidiuretic hormone cause the collecting ducts to become (**more**/less) permeable to water, and highly concentrated urine is produced.

11. The human kidneys filter about 180 liters of blood per day, but produce about 2 to 3 liters of urine per day. Therefore, the percentage of fluid volume that ends up in urine is about **1 to 2% (2 to 3 L/180 L)**.

★12. Sharks and rays secrete salt from the **rectal gland**.

13. Renin is a regulatory enzyme released by the kidney. It acts on a circulating protein to begin converting that protein into an active hormone called **angiotensin**.

14. In addition to water and carbon dioxide, the metabolism of proteins and nucleic acids also produces nitrogenous waste found most frequently in the form of **ammonia (or NH_3).**

15. Birds that eat marine animals ingest an excess of salts, which they excrete through their **nasal salt** glands.

16. The hormone produced by the heart that influences kidney function is **atrial natriuretic hormone**.

Multiple Choice

1. Which of the following is *true* of the extracellular fluids in our bodies?
   a. They have the same composition as seawater.
   b. They have much higher osmotic concentration than fluids in cells.
   c. They contain no proteins.
   d. They have a fixed water concentration.
   e. **They can supply water and nutrients to the cells.**

2. Water moves into tissues as a result of all of the following *except*
   a. osmotic gradient.
   b. fluid pressure.
   c. **active transport.**
   d. principles of diffusion.
   e. salinity gradient.

3. Ammonia and urea are waste products derived from the metabolic breakdown of
   a. carbohydrates.
   b. lipids.
   c. sugars
   d. **proteins.**
   e. salts.

■4. Marine invertebrates in which the salinity of body fluids changes with the osmotic potential of their environments are known as
   a. **osmoconformers.**
   b. osmoregulators.
   c. osmoexcretors.
   d. hypotonic.
   e. hypertonic.

5. Which of the following molecules is most toxic to cells?
   a. Water
   b. Sodium chloride
   c. **Ammonia**
   d. Urea
   e. Uric acid

6. Which of the following animals is most likely to excrete mostly urea or uric acid instead of ammonia?
   a. Freshwater fishes
   b. Saltwater fishes
   c. Tadpoles
   d. Shrimp
   e. **Seagulls**

*7. Which of the following substances is the *least* soluble in water?
   a. Ammonia
   b. **Uric acid**
   c. Urea
   d. Sodium chloride
   e. Amino acids

▪8. Organisms that are ionic regulators maintain an
   a. osmotic concentration that may differ from the environment.
   b. osmotic concentration that is the same as the environment.
   c. **ion concentration that may differ from the environment.**
   d. ion concentration that is the same as the environment.
   e. ion concentration by diffusion only.

9. Which of the following operates by filtering the body fluids into a tube, then secreting or resorbing specific substances?
   a. Flame cells of flatworms
   b. Metanephridia of annelid worms
   c. Malpighian tubules of insects
   d. Vertebrate nephrons
   e. **All of the above**

*10. In which of the following are nitrogenous waste solutes eliminated via the gut?
   a. Flame cells of flatworms
   b. Metanephridia of annelid worms
   c. **Malpighian tubules of insects**
   d. Vertebrate nephrons
   e. All of the above

11. The functional unit of the kidney is the
   a. Bowman's capsule.
   b. capillary.
   c. glomerulus.
   d. **nephron.**
   e. renal tubule.

12. What drives the process of filtration from the capillaries into the glomerulus?
   a. Active transport
   b. **Arterial blood pressure**
   c. Venous blood pressure
   d. Osmotic pressure
   e. Secretion

13. During filtration, which does *not* enter the Bowman's capsule from the bloodstream?
   a. Water
   b. Glucose
   c. Ions
   d. Amino acids
   e. **Plasma proteins**

▪14. Analysis of kidney function in simple vertebrates suggests that the earliest vertebrate nephron functioned mainly to
   a. **remove excess fluid.**
   b. remove excess salts.
   c. remove excess nutrients.
   d. filter nitrogenous wastes.
   e. conserve water.

15. Marine bony fishes acquire excess salt when they drink seawater. This salt load is handled in all of the following ways *except*
   a. **producing very dilute urine.**
   b. producing very little urine.
   c. excreting ions from gills.
   d. secreting salt from the renal tubules.
   e. a reduced number of glomeruli.

16. Cartilaginous fishes maintain an osmotic concentration in their tissue fluids that matches that of seawater by
   a. excreting large amounts of water.
   b. drinking large amounts of seawater.
   c. concentrating seawater salts in their blood.
   d. **concentrating urea in their tissue fluids.**
   e. excreting salts at their gills.

17. Water conservation adaptations of reptiles include all of the following *except*
   a. scaly skin.
   b. excretion of uric acid.
   c. shelled eggs.
   d. internal fertilization of gametes.
   e. **excretion of salts.**

18. Water conservation adaptations of amphibians living in dry environments include all of the following *except*
   a. waxy secretions on the skin.
   b. estivation during dry periods.
   c. **production of concentrated urine.**
   d. burrowing into the ground.
   e. large urinary bladders.

19. Which part of nephrons is found in the inner medulla of the kidney?
   a. Bowman's capsule
   b. Convoluted tubule
   c. Glomerulus
   d. **Loop of Henle**
   e. a and d

20. Of the fluid that is filtered into the Bowman's capsule, approximately how much of it is resorbed back into the blood within the kidney?
   a. Less than 5%
   b. About 25%
   c. About 50%
   d. About 75%
   e. **Over 95%**

21. The cells lining the proximal convoluted tubule have numerous microvilli and mitochondria. Such structures indicate that which of the following activities are conducted in these cells?
a. Rapid diffusion of water
b. **Active transport**
c. Conservation of water
d. Storage of salts
e. Production of urea

22. Valuable molecules like glucose, amino acids, and vitamins are resorbed into the blood at which location in the nephron?
a. Bowman's capsule
b. Collecting duct
c. Glomerulus
d. Loop of Henle
e. **Convoluted tubule**

*▪23. The loops of Henle are considered to function as a countercurrent multiplier because
a. ascending and descending limbs have different permeabilities.
b. the amount of water conserved is related to the length of the loop.
c. **as urine flows in the loop, a concentration gradient is established in the medulla.**
d. as urine flows in the loop, it becomes more concentrated in the nephron.
e. active transport in the ascending and descending limbs multiplies the concentration gradient.

*24. Which of the following responses would *not* correct for a drop in glomerular blood pressure?
a. Angiotensin elevates the body's blood pressure.
b. Angiotensin stimulates thirst in the brain.
c. **Angiotensin constricts arterioles entering the kidney.**
d. Stretch receptors trigger antidiuretic hormone release.
e. Aldosterone stimulates sodium resorption.

25. The effect of antidiuretic hormone is to
a. reduce permeability in the loop of Henle.
b. reduce permeability of the collecting ducts.
c. reduce blood volume.
d. increase the volume of urine.
e. **increase the concentration of urine.**

26. Antidiuretic hormone secretion increases when the hypothalamus is stimulated by
a. angiotensin receptors.
b. glucose receptors.
c. **osmoreceptors.**
d. renin receptors.
e. stretch receptors.

27. Urine that is more concentrated than their body fluids is produced by
a. **birds.**
b. catfish.
c. crocodiles.
d. frogs.
e. sharks.

28. In adult humans, urination is controlled by
a. a smooth muscle sphincter at the base of the urethra.
b. the autonomic nervous system.
c. a skeletal muscle sphincter at the base of the urethra.
d. the voluntary nervous system.
e. **All of the above**

29. Which of the following is a function of excretory systems?
a. Help regulate osmotic potential and the volume of extracellular fluids
b. Excrete molecules that are present in excess
c. Conserve molecules that are valuable or in short supply
d. Eliminate toxic waste products of phosphorus metabolism
e. **a, b, and c**

30. Excretory systems control
a. filtration.
b. excretion.
c. resorption.
d. homeostasis.
e. **All of the above**

31. Which of the following is true of the human kidney?
a. Functional units are called hepatocytes.
b. Nephrons consist of proximal and distal convoluted tubules as well as collecting ducts and loops of Henle.
c. Most of the substances and water filtered from the blood in the glomerulus return to the venous blood draining the kidney.
d. Anything coming into the kidney has to come from the renal artery.
e. **b, c, and d**

*32. The kidneys help regulate acid-base balance by controlling the level of __________ in the blood.
a. CO_2
b. H^+
c. HCO_3^-
d. **b and c**
e. All of the above

33. The terms "ammonotelic," "ureotelic," and "urecotelic" are used to describe
a. the actions of hormones on the excretory system.
b. **the types of nitrogenous waste produced by various classes of vertebrates.**
c. pathways of kidney evolution.
d. modifications of kidney tubules to enhance excretion.
e. modes of excretory system development.

34. Organisms living in a freshwater environment normally
    a. excrete copious dilute urine and retain salts.
    b. excrete a small volume of dilute urine and retain salts.
    c. excrete copious concentrated urine.
    d. excrete small amounts of concentrated urine.
    e. conserve both water and salts.

35. A ureotelic organism excretes most of its nitrogenous waste as
    a. ammonia.
    b. water.
    c. carbon dioxide.
    d. uric acid.
    e. urea.

36. Terrestrial organisms must conserve water. The least amount of water is lost with the excretion of which nitrogenous waste product?
    a. Carbon dioxide
    b. Uric acid
    c. Ammonia
    d. Salt
    e. Urea

37. Gout results from an accumulation of __________ in the joints.
    a. carbon dioxide
    b. water
    c. ammonia
    d. uric acid
    e. urea

38–42. Choose the appropriate excretory system or component from the list below.
    a. Protonephridia
    b. Nephron
    c. Malpighian tubule
    d. Green gland
    e. Metanephridia

38. Annelid worms are associated with which type of excretory system? **(e)**

39. The vertebrate system is associated with which type of functional unit? **(b)**

40. Flatworms are distinguished by which type of excretory unit? **(a)**

41. Insects and most terrestrial arthropods have an excretory system composed of which type of functional unit? **(c)**

42. The primary excretory system of humans utilizes what type of functional unit? **(b)**

43. Which of the following is *not* part of a nephron?
    a. Podocyte
    b. Flame cell
    c. Renal corpuscle
    d. Glomerulus
    e. Bowman's capsule

44. Blood enters a nephron's vascular component by way of the
    a. peritubular capillaries.
    b. glomerulus.
    c. efferent arteriole.
    d. afferent arteriole.
    e. renal vein.

45–54. Choose the appropriate anatomical section from the list below. There may or may not be more than one correct answer.
    a. Renal pyramids
    b. Proximal convoluted tubule
    c. Distal convoluted tubule
    d. Loop of Henle
    e. Collecting duct

45. Reside(s) anatomically closest to Bowman's capsule **(b)**

46. Function(s) as a countercurrent multiplier system **(d)**

47. Comprise(s) the internal core of the medulla **(a)**

48. Site of glucose and amino acid resorption **(b)**

49. Site of major water resorption **(b)**

50. Permeability is under the control of antidiuretic hormone **(e)**

51. Length of this particular section relates to the potential maximum urine concentration possible **(d)**

52. Connect(s) the proximal and distal convoluted tubules **(d)**

53. Segment where microvilli are located **(b)**

54. Located principally within the medulla **(a, d, and e)**

55. The proper sequence of urine passage from the kidney is
    a. renal pyramid, ureter, bladder, rectal gland, and urethra.
    b. ureter, renal pyramid, rectal gland, and bladder urethra.
    c. renal pyramid, urethra, bladder, and ureter.
    d. renal pyramid, ureter, bladder, and urethra.
    e. rectal gland, renal pyramid, ureter, bladder, and urethra.

56. A person with diabetes insipidus fails to respond to ADH. What is a symptom of a person with this condition?
    a. Urine contains glucose
    b. Copious hyperosmotic urine
    c. Copious dilute urine
    d. Small volume of concentrated urine
    e. Failure to urinate

57. Beer contains ethyl alcohol, which inhibits ADH production. After consuming beer, the kidney
    a. produces copious concentrated urine.
    b. produces a small volume of concentrated urine.
    c. produces a small volume of dilute urine.

d. produces copious dilute urine.
e. does nothing.

58. A freshwater fish is continuously faced with regulating its internal environment. The fish tends to __________ water because it is __________ to fresh water.
a. gain, hypoosmotic
b. gain, hyperosmotic
c. lose, hypoosmotic
d. lose, hyperosmotic
e. lose, isoosmotic

59. To osmoregulate in freshwater, a fish must __________ salts and produce __________ urine.
a. conserve, copious dilute
b. excrete, copious dilute
c. conserve, small volumes of dilute
d. excrete, small volumes of dilute
e. absorb, large volumes of concentrated

*60. The brine shrimp, *Artemia*, can live in extremely high environmental salinities because it
a. has active transport of salt across its gill membranes.
b. has active transport of water across its gill membranes.
c. excretes salts through its nasal salt glands.
d. osmoconforms with its surroundings.
e. excretes concentrated uric acid, so it conserves water.

61. Why don't terrestrial animals excrete nitrogen as ammonia?
a. It takes more energy to convert nitrogen to ammonia than to urea.
b. They lose more water by excreting ammonia.
c. Ammonia is very toxic to terrestrial animals.
d. It is not stable and quickly converts to urea.
e. It is insoluble in water and so cannot be excreted as urine.

*62. Which molecule contains the most nitrogen?
a. Ammonia
b. Uric acid
c. Urea
d. b and c
e. All have the same number of nitrogen atoms.

63. This excretory system contains beating cilia and is found in flatworms.
a. Protonephridia
b. Metanephridia
c. Malpighian tubules
d. Nephron
e. Aquaporin

64. This excretory system actively transports uric acid and ions into blind tubules and is found in insects.
a. Protonephridia
b. Metanephridia
c. Malpighian tubules
d. Nephron
e. Aquaporin

65. This excretory system has capillary beds that come into close contact with podocytes, where there is a transfer of water and small molecules. It is found in amphibians.
a. Protonephridia
b. Metanephridia
c. Malpighian tubules
d. Nephron
e. Aquaporin

66. This excretory system has ciliated openings in the coelom, called nephrostomes, which lead to a tubule that opens to the outside of the animal. It is found in annelids.
a. Protonephridia
b. Metanephridia
c. Malpighian tubules
d. Nephron
e. Aquaporin

Study Guide Questions

1. A marine fish lives in a hypertonic environment. What mechanisms does a marine fish use to maintain homeostasis?
a. It drinks copious amounts of water, excretes only small amounts of water, and pumps sodium out of the body.
b. It drinks little water, excretes large amounts of water, and pumps sodium into the body.
c. It drinks little water, excretes only small amounts of water, and pumps sodium out of the body.
d. None of the above

2. Which of the following statements about the excretory system of insects is *not* true?
a. Active transport moves materials from the coelomic fluid into the Malpighian tubules.
b. The Malpighian tubules can produce a highly concentrated urea solution, allowing insects to inhabit some of Earth's driest habitats.
c. Resorption of salts takes place mostly in the gut.
d. Water resorption is by osmotic movement only.

3. What pathway is taken by water and solutes as they travel through a nephron?
a. Glomerulus, to Bowman's capsule, to proximal tubule, to loop of Henle, to distal tubule, to collecting ducts
b. Bowman's capsule, to glomerulus, to distal tubule, to loop of Henle, to proximal tubule, to collecting ducts
c. Glomerulus, to Bowman's capsule, to distal tubule, to loop of Henle, to proximal tubule, to collecting ducts
d. Glomerulus, to Bowman's capsule, to proximal tubule, to collecting ducts, to distal tubule, to loop of Henle

4. Na^+ and Cl^- are actively transported out of the tubules to help set up the countercurrent multiplier. Which of the following are sites of active Na^+ and Cl^- transport in the nepheron?
   a. **Proximal tubule, ascending limb of the loop of Henle**
   b. Descending limb of the loop of Henle, ascending limb of the loop of Henle
   c. Ascending limb of the loop of Henle, proximal tubule
   d. Collecting duct, descending limb of the loop of Henle

5–9. Use the following diagram to complete the statements about the mammalian nephron.

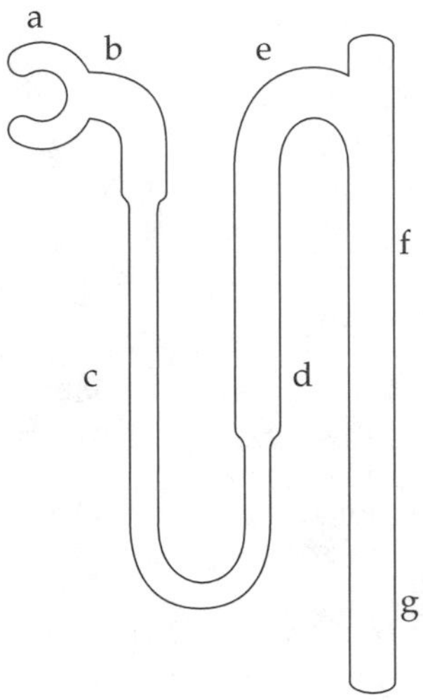

5. The composition of the filtrate would be most like plasma in the tubule next to letter **(a)**.

6. The NaCl concentration in the extracellular fluid would be greatest in the area of letter **(g)**.

7. The osmolarity of the filtrate next to letters **(a, b, e)** is similar to the osmolarity of blood plasma.

8. The urine would be most concentrated in the collecting duct next to letter **(g)**.

9. Most of the glomerular filtrate is resorbed into peritubular capillary blood next to letter **(b)**.

10. The sole mechanism for water resorption by the renal tubules is
    a. active transport.
    b. **osmosis.**
    c. cotransport with sodium ions.
    d. cotransport with bicarbonate ions.

11. A number of hormones help to regulate water and solute uptake and release in the nephron. Antidiuretic hormone (ADH) promotes __________ in response to __________.
    a. active transport of Cl^-, increased solute concentration
    b. active transport of Na^+, increased blood pressure
    c. **increased permeability of the collecting duct to water, increased blood pressure**
    d. decreased permeability of the collecting duct to water, increased solute concentration

12. Which of the following is not a normal constituent of the glomerular filtrate?
    a. **Red blood cells**
    b. Urea
    c. Sodium ion
    d. Glucose

13. If the afferent arteriole that supplies blood to the glomerulus becomes dilated,
    a. the protein concentration of the filtrate decreases.
    b. hydrostatic pressure in the glomerulus decreases.
    c. **the glomerular filtration rate increases.**
    d. All of the above

14. Osmoreceptors in the brain detect changes in blood ion concentration which can reflexively result in
    a. water retention by the kidneys.
    b. water loss by the kidneys.
    c. ion retention by the kidneys.
    d. **All of the above**

15. If the human kidneys filter 150 liters of plasma in a 24-hour period, what is the typical amount of urine produced and eliminated in that time period?
    a. 0.15 liters
    b. **1.5 liters**
    c. 15 liters
    d. 30 liters

End of Chapter Questions

1. Which statement about osmoregulators is *true*?
   a. Most marine invertebrates are osmoregulators.
   b. **All freshwater invertebrates are hypertonic osmoregulators.**
   c. Cartilaginous fishes are hypotonic osmoregulators.
   d. Bony marine fishes are hypertonic osmoregulators.
   e. Mammals are hypotonic osmoregulators.

2. The excretion of nitrogenous wastes
   a. **by humans can be in the form of urea and uric acid.**
   b. by mammals is never in the form of uric acid.
   c. by marine fishes is in the form of urea.
   d. does not contribute to the osmotic potential of the urine.
   e. requires more water if the waste product is the rather insoluble uric acid.

3. How are earthworm metanephridia like mammalian nephrons?
   a. Both process coelomic fluid.
   b. Both take in fluid through a ciliated opening.
   c. Both produce hypertonic urine.
   d. **Both employ tubular secretion and reabsorption to control urine composition.**
   e. Both deliver urine to a urinary bladder.

4. What is the role of renal podocytes?
   a. They control the glomerular filtration rate by changing the resistance of renal arterioles.
   b. They reabsorb most of the glucose that is filtered from the plasma.
   c. They prevent red blood cells and large molecules from entering the renal tubules.
   d. They provide a large surface area for tubular secretion and reabsorption.
   e. They release renin when the glomerular filtration rate falls.

5. Which of the following are *not* found in a renal pyramid?
   a. Collecting ducts
   b. Vasa recta
   c. Peritubular capillaries
   d. Convoluted tubules
   e. Loops of Henle

6. Which part of the nephron is responsible for most of the difference in mammals between the glomerular filtration rate and the urine production rate?
   a. The glomerulus
   b. The proximal convoluted tubule
   c. The loop of Henle
   d. The distal convoluted tubule
   e. The collecting duct

7. For mammals of the same size, what feature of their excretory systems would give them the greatest ability to produce a hypertonic urine?
   a. Higher glomerular filtration rate
   b. Longer convoluted tubules
   c. Increased number of nephrons
   d. More-permeable collecting ducts
   e. Longer loops of Henle

8. Which of the following would *not* be a response stimulated by a large drop in blood pressure?
   a. Constriction of afferent renal arterioles
   b. Increased release of renin
   c. Increased release of antidiuretic hormone
   d. Increased thirst
   e. Constriction of efferent renal arterioles

9. Which statement about angiotensin is *true*?
   a. It is secreted by the kidney when the glomerular filtration rate falls.
   b. It is released by the posterior pituitary when blood pressure falls.
   c. It stimulates thirst.
   d. It increases permeability of the collecting ducts to water.
   e. It decreases glomerular filtration rate when blood pressure rises.

10. Birds that feed on marine animals ingest a lot of salt, but they excrete most of it by means of
    a. Malpighian tubules.
    b. rectal salt glands.
    c. gill membranes.
    d. hypertonic urine.
    e. nasal salt glands.

Student Web Site Self-Quiz Questions

1. Some organisms, like the brine shrimp (*Artemia* sp.), are able to adapt to a wide range of salinities. All organisms need to be able to maintain a proper water balance. Which of the following statements about salt and water balance is *false*?
   a. If the solute potential of the extracellular fluid is less negative than a cell, then that cell has a greater solute concentration than the fluid.
   b. Active transport of water out of cells is an important mechanism for many animals living in fresh water.
   c. Most excretory organs consist of a series of tubules.
   d. Extracellular fluid acted upon by secretion and reabsorption becomes urine.
   e. The mechanism's filtration, secretion, and reabsorption are used in excretory systems that excrete water and conserve salts, and in systems that conserve water and excrete salts.

2. In some species, different developmental stages may have different forms of nitrogen excretion. Frogs and toads are __________ as tadpoles, but __________ as adults.
   a. ammonotelic, ureotelic
   b. ammonotelic, uricotelic
   c. ureotelic, ammonotelic
   d. ureotelic, uricotelic
   e. uricotelic, ureotelic

3. The flame cell excretory system in Planaria flatworms is a __________ and the fluid leaving the system via the excretory pore is __________ to the animal's body fluid.
   a. metanephridium, hypertonic
   b. metanephridium, hypotonic
   c. nephron, hypertonic
   d. protonephridium, hypertonic
   e. protonephridium, hypotonic

4. In the metanephridia of earthworms, filtration takes place from the __________ while secretion and reabsorption occur in the __________.
   a. capillaries, collecting tubules
   b. capillaries, bladder
   c. nephrostome, collecting tubules
   d. nephrostome, bladder
   e. nephridiopore, collecting tubules

5. In the insect excretory system, uptake of uric acid takes place in the __________ and reabsorption of water occurs in the __________.
   a. Malpighian tubules, hindgut
   b. Malpighian tubules, rectum
   c. midgut, Malpighian tubules
   d. midgut, hindgut
   e. midgut, rectum

6. Of the following components of the vertebrate nephron, which one.is *not* involved in filtration?
   a. Bowman's capsule
   b. Glomerulus
   c. Peritubular capillaries
   d. Podocytes
   e. Renal corpuscle

7. Which of the following statements about excretory systems of marine vertebrates is *false*?
   a. Amphibians drink little water and produce copious dilute urine.
   b. Marine bony fish meet their water needs by drinking seawater.
   c. Marine bony fish produce concentrated urine.
   d. Cartilaginous fishes are isotonic to seawater.
   e. Cartilaginous fish actively excrete salt.

8. Which of the following statements about the excretory systems of tetrapod vertebrates is *false*?
   a. Most amphibians produce large amounts of dilute urine and conserve salts.
   b. Most reptiles lose less water through their skin than do terrestrial amphibians.
   c. Reptiles are uricotelic.
   d. Only birds can produce urine that is hypertonic to their body fluids.
   e. Birds and reptiles share many similar water conservation adaptations.

9. A molecule of urea enters the glomerulus. If you were to follow the path of this molecule to its excretion from the body, what would be the correct order for the structures numbered below?
   1. Ascending limb of loop of Henle
   2. Collecting duct
   3. Proximal convoluted tubule
   4. Ureter
   5. Urethra

   a. 23145
   b. 13245
   c. 31245
   d. 31254
   e. 31452

10. Of the following components of the kidney, which is entirely within the medulla region?
    a. Bowman's capsule
    b. Distal convoluted tubule
    c. Collecting duct
    d. Peritubular capillaries
    e. Vasa recta

11. Which of the following descriptions of mammalian nephron components is incorrect?
    a. Cells of the proximal convoluted tubules have many microvilli and mitochondria.
    b. The osmolarity of the urine is controlled by changing the rate of NaCl transport in the thick ascending limb of Henle.
    c. Cells of the thick segment of the ascending limb of the loop of Henle resorb NaCl out of the tubular fluid.
    d. Urine is less concentrated when it leaves the loop of Henle than when it entered it.
    e. Cells of the collecting duct conduct little active transport.

12. Which of the following statements about the production of hypertonic urine by the mammalian nephron is *false*?
    a. Glomeruli filter large volumes of blood plasma.
    b. Blood pressure in the glomerulus determines how concentrated the urine can become.
    c. The proximal convoluted tubule reabsorbs most of the volume of fluid as well as needed solutes.
    d. The loop of Henle is a countercurrent multiplier system.
    e. As urine moves from the cortex to the tip of the medulla, it encounters an increasingly hypertonic extracellular fluid.

13. Regarding the relationship between glomerular filtration rate and arterial pressure, which of the following events would *not* be expected to occur when arterial pressure is less than 85 mm Hg?
    a. Dilation of afferent renal arterioles
    b. The kidney releases the enzyme renin
    c. The efferent renal arterioles dilate
    d. Angiotensin stimulates thirst
    e. Enhanced secretion of antidiuretic hormone (ADH) by the hypothalamus.

14. In addition to regulating water and salt balance, kidneys also play a role in regulating the pH of blood. Which of the following statements about pH balance in the blood is *false*?
    a. The kidneys control the level of CO_2 in the blood.
    b. HCO_3^- is the major buffer present in the blood.
    c. Buffers serve to supply or absorb excess H^+, depending upon conditions.
    d. Maintaining a constant pH is important since pH can alter protein structure.
    e. Carbonic acid is a precursor to HCO_3^-

TEST FILE

★ *Indicates a **difficult** question*

■ *Indicates a **conceptual** question*

52 Animal Behavior

Fill in the Blank

1. **Releasers** are simple stimuli that elicit highly stereotyped, species-specific patterns of behavior.

2. Scientists in the field of **ethology** usually study animals under natural conditions and focus on behavior that is highly stereotyped and species-specific.

3. Male songbirds learn, and later express, their songs under the influence of the hormone **testosterone**.

4. The **round** dance is performed by honeybees when food is within 80 meters of the hive.

5. Learning a particular releaser during a critical period is the phenomenon of **imprinting**.

6. A bird's song is said to be **crystallized** when it has been learned in the form in which it will be sung thereafter.

7. Konrad Lorenz investigated **courtship** behavior in hybrid ducks.

8. If behavioral patterns can be artificially selected for in laboratory populations, then behavior in that species must be at least partially under **genetic** control.

9. Many observations indicate that organisms have an internal or **endogenous** clock.

10. A circadian rhythm can be reset with an environmental cue by a process called **entrainment**.

11. The simplest form of navigation, using landmarks, is called **piloting**.

★12. If the **pineal** gland of a bird is removed, the bird will no longer have circadian rhythms.

13. The regular departure and return of organisms seasonally is called **migration**.

■14. Morphological structures, physiological apparatus, and behavior of organisms are all shaped by **natural selection (or evolution)**.

15. Molecules used for chemical communication between animals are called **pheromones**.

■16. **Culture** is the transmission of learned behavior through generations.

Multiple Choice

■1. Imprinting
a. must occur during a critical period in development.
b. always occurs between parents and offspring.
c. is genetically determined without a component of learning.
d. is an encoding of simple information.
e. occurs equally well with a variety of stimuli.

★2. What outcome is expected when adult rats have their ovaries or testes removed and are subsequently treated with female sex steroids?
a. Both sexes will exhibit lordosis (female mating behavior).
b. Neither sex will exhibit lordosis.
c. Females will exhibit lordosis, and males will exhibit no response.
d. Females will exhibit lordosis, and males will exhibit mounting behavior (male mating behavior).
e. Females will exhibit no response, and males will exhibit lordosis.

★3. What outcome, in terms of lordosis (female mating behavior) or mounting (male mating behavior), is expected when newborn rats have their ovaries or testes removed, are treated with testosterone as newborns, and are treated again with testosterone as adults?
a. Both exhibit mounting behaviors.
b. Neither sex exhibits mounting or lordosis behaviors.
c. Males exhibit mounting, and females have no response.
d. Males exhibit mounting, and females exhibit lordosis.
e. Both sexes exhibit lordosis.

4. Experiments on songbirds determined that the reason female birds normally do *not* sing is due to
a. lack of the proper muscles.
b. lack of specific parts of the nervous system.
c. the absence of testosterone.

d. the inability to learn the song.
e. learning sex-specific behavior from a nonsinging mother.

5. Anatomical comparisons of male and female songbirds reveal that __________ of male birds increase(s) in volume during periods of active singing.
a. the throat muscles
b. specific regions of the brain
c. the pineal gland
d. the suprachiasmatic nuclei
e. the auditory nerves

*6. When purebred nonhygienic bees are crossed with purebred hygienic bees, they produce dihybrid nonhygienic offspring. If these offspring are backcrossed with the purebred hygienic bees, which type of behavior is *not* exhibited in the next set of offspring?
a. Nonhygienic
b. Nonhygienic, but they will remove pupae from uncapped cells
c. Nonhygienic, but they will uncap cells of dead pupae
d. Hygienic, but only when neighboring bees are hygienic
e. Hygienic

7. Male gypsy moths travel thousands of meters to reach female gypsy moths. Their behavior is a response to
a. visual signals.
b. auditory signals.
c. pheromones.
d. random search patterns.
e. electric signals.

8. The observation that potato washing spread through a population of Japanese macaques was important because it demonstrated that
a. animals are capable of associative learning.
b. animals other than humans are capable of transmitting learned behavior through generations.
c. animals can optimally forage.
d. behavior can be inherited.
e. None of the above

9. Chemical territory marking can convey all of the following information *except*
a. height.
b. direction of travel.
c. individual identity.
d. reproductive status.
e. elapsed time.

10. Which of the following is *not* a feature of visual communication?
a. Easy to produce
b. Comes in an endless variety
c. Effective over long distances
d. Can be changed rapidly
e. Indicates the position of the signaler

11. When a food source is located toward the sun, the direction of a honeybee's waggle dance on a vertical surface will be
a. straight up.
b. straight down.
c. 90° to the right.
d. 90° to the left.
e. 45° to the right.

12. The feature of a honeybee's waggle dance that indicates the distance to a food source is the __________ of the waggles.
a. frequency
b. speed
c. direction
d. size
e. shape

13. Insects use their antennae in which two forms of communication?
a. Tactile and auditory
b. Tactile and chemical
c. Chemical and auditory
d. Visual and tactile
e. Visual and chemical

14. Which of the following are most likely to communicate by sound?
a. Fish
b. Amphibians
c. Reptiles
d. Insects
e. Mammals

15. Visual communication is better than auditory communication
a. in dark environments.
b. at conveying complex information.
c. at going around objects.
d. at getting the attention of the receiver.
e. at communicating over long distances.

▪16. Which of the following would suggest that learning plays a role in a spider's spinning a web?
a. A newly hatched spider spins a perfect web.
b. The web structure is the same every time.
c. Web structure varies from species to species.
d. The web size changes as the individual grows.
e. The web is changed after contact with particular prey.

▪17. If behavior is under partial genetic control,
a. no variability in the behavior will be seen.
b. it is not subject to natural selection.
c. it may be modifiable by experience.
d. it is acquired only through learning.
e. it will be species-specific.

18. What is the objective of a deprivation experiment?
a. To determine the effect of isolation on an animal
b. To determine if a behavior is inherited
c. To determine how an animal learns
d. To determine the influence of parents on behavior
e. To determine the influence of mates on behavior

▪19. Which of the following is an example of a deprivation experiment?
a. Removing a bird's mate in the middle of courtship
b. Removing half of a litter of puppies
c. Transferring a newborn rat to a different mother
d. Hand-rearing a bird from the egg
e. Taking all nest material away from nesting birds

20. A releaser
a. is required in order for learning to occur.
b. triggers stereotypic, species-specific behavior.
c. triggers the onset of aggressive behavior.
d. facilitates recognition of appropriate mates.
e. causes the animal to remember the full stimulus.

21. Experiments on interactions between herring gulls and their chicks showed that
a. simple models of releasers were effective.
b. the more lifelike the releaser, the more likely the behavior.
c. chicks preferred parents with contrasting marks on their bills.
d. appropriate sounds as well as visual stimuli were required.
e. parents gave more food to visually discriminating chicks.

22. Questions about the selective pressures that shaped a particular behavior focus on the
a. stimuli that elicit the behavior.
b. underlying neural and hormonal mechanisms.
c. proximate causes of the behavior.
d. ultimate causes of the behavior.
e. All of the above

23. Which of the following is critical to the production of normal song by adult male white-crowned sparrows?
a. Auditory feedback after the song has crystallized
b. Auditory feedback during the time the bird is matching its vocal output to memorized song
c. Exposure to the song of his mother
d. Exposure to the song of an adult male during a critical period
e. b and d

▪24. Which of the following behaviors would be expected to be genetically programmed?
a. Behavior that can be learned by trial and error
b. Behavior that leads to successful mating
c. Behavior that does not endanger the animal
d. Behavior that involves social cues
e. Behavior that is important in variable circumstances

25. Which of the following statements about the singing behavior of birds is *true*?
a. Male and female birds have song control regions in their brains.
b. Rising levels of testosterone cause song control regions to decrease in size in the spring.
c. Males isolated after their songs have crystallized produce abnormal song.
d. Males deafened after their songs have crystallized produce abnormal song.
e. None of the above

26. What can behaviorists learn from hybridization experiments?
a. A behavior is often coded by a single gene.
b. Behavioral traits are usually all inherited from a single parent.
c. A hybrid may show behavior intermediate between that of its parents.
d. Hybrids show behavior patterns completely different than either parent.
e. Only simple animals have genetically programmed behavior.

27. One behavior that is inherited in a simple Mendelian manner is
a. fruit fly courtship.
b. lovebird nest building.
c. honeybee brood cleaning.
d. dog retrieving behavior.
e. rattlesnake prey capture.

28. Circadian rhythms are characterized by
a. seasonal changes.
b. lack of synchrony with the environment.
c. tendency to lengthen with time.
d. persistence in the absence of cues.
e. changes with the animal's age.

29. When a person changes time zones suddenly, as in jet travel, the circadian rhythm
a. is destroyed temporarily.
b. needs entrainment.
c. adjusts within a few hours.
d. becomes random.
e. takes on a different periodicity.

30. The circadian clock in mammals appears to be located in the
a. hypothalamus.
b. suprachiasmatic nucleus (SCN).
c. pituitary.
d. reticular activating system.
e. visual cortex.

31. A bird will be unable to entrain its circadian rhythms if light is obscured from its
a. eyes.
b. optic nerve.
c. pineal gland.
d. pituitary.
e. visual cortex.

32. The adaptive advantage of circannual rhythms is
a. in timing breeding to coincide with peak resources.
b. maximum in tropical species.
c. that it entrains circadian rhythms.
d. minimal in migratory birds.
e. maximal in photoperiodic species.

33. Homing pigeons released at a remote, unfamiliar site find their way home by
   a. following visual landmarks.
   b. retracing the route along which they were carried.
   c. searching randomly until they locate home.
   d. following other birds.
   e. navigating without visual images.

34. Which of the following is an experimental method to test how migrating animals navigate?
   a. Deprive them of foodstores
   b. Blindfold them
   c. Capture and displace them
   d. Keep them in captivity
   e. Get them to migrate at the wrong season

*35. Biological rhythms play a role in migratory behavior, as displayed by the
   a. tendency to migrate over great distances.
   b. presence of migratory restlessness.
   c. tendency to migrate at night.
   d. presence of orientation toward the sun.
   e. tendency to form flocks.

*36. Clock-shifting experiments revealed that birds tell direction from
   a. the sun.
   b. the stars.
   c. the wind direction.
   d. sound cues.
   e. air pressure cues.

37. Experiments with young songbirds in a planetarium suggested that birds can use star clues if
   a. they are tested in a rotating star pattern.
   b. they learned in a stationary star pattern.
   c. they learned in a rotating star pattern.
   d. they learned in the Northern Hemisphere.
   e. they do not see the stars until adulthood.

38. What would be the advantage of orienting with respect to a fixed point of light in the night sky?
   a. It is easier to fly to a steady coordinate.
   b. In the Northern Hemisphere it would always be in the north.
   c. It would give information about the sun compass.
   d. It would give information if the night were overcast.
   e. It would be the brightest spot in the sky.

39. Which of the following patterns is seen in animal orientation systems?
   a. Each species has one sensory mechanism for orientation.
   b. All seem to rely on light systems.
   c. They seem to rely on bicoordinate navigation.
   d. They are based on redundant cues.
   e. All seem to need an internal clock.

40. What information suggests that pigeons use magnetic clues for orientation?
   a. A neurophysiological magnetic transducer has been described.
   b. They can orient even on cloudy days.
   c. Attachment of small magnets disrupts their homing.
   d. Magnetic field lines could give latitude information.
   e. Magnetite particles have been found in their cells.

41. Which of the following suggests that humans have some innate, unlearned motor patterns?
   a. All humans feel anger and fear.
   b. All cultures teach their young.
   c. Humans have a large capacity to learn.
   d. Blind infants do not smile normally.
   e. All cultures have similar facial expressions.

▪42. Genetically determined behaviors are expected to be common in species in all of the following circumstances *except*
   a. when there is no opportunity to learn.
   b. when it is possible to learn the wrong behavior.
   c. when environmental conditions are complex and changing.
   d. when generations do not overlap.
   e. when dealing with dangerous prey.

43. Human behavior is largely influenced by
   a. releasers.
   b. genetic traits.
   c. instinct.
   d. learning.
   e. orienting.

44. Which of the following statements about spiders' web building is *not* true?
   a. Webs are built perfectly the first time.
   b. Instinctive behavior, such as web building, is not usually included within the province of the field of ethology.
   c. The building of webs in spiders is genetically determined.
   d. Young spiders build their first webs without prior experience.
   e. Web design and construction is species-specific.

45. Which of the following is *not* a part of the definition of instinctive behavior?
   a. Stereotypic
   b. Genetically determined
   c. Not modified by learning
   d. Inherited in a Mendelian fashion
   e. Species-specific

46. Which of the following is *not* an example of a stereotyped, species-specific behavior?
   a. Web construction in orb-weaving spiders
   b. Food burying in squirrels
   c. Behaviors that are expressed during a deprivation experiment
   d. Song acquisition in the white-crowned sparrow
   e. Bill pecking in herring gull chicks

47. If an animal shows a particular behavior during a deprivation experiment, then the
   a. behavior may be instinctive.
   b. behavior has a strong learning component.

c. correct releaser was present.
d. behavior has substantial genetic determinants.
e. a, c, and d

48. Which of the following is *not* a releaser?
a. Shape of the adult head in herring gull chicks
b. The bill spot of the adult in herring gull chicks
c. A tuft of red feathers for a male European robin
d. Image of a duckling's parent
e. The presence of a nut for a tree squirrel

49. Which of the following is true of cricket song?
a. Cricket songs are species-specific.
b. Only male crickets "sing."
c. Song pattern is genetically determined.
d. Female preference for song is genetically determined.
e. All of the above

50. Which of the following statements about imprinting is *not* true?
a. Imprinting is rapid learning that occurs during a critical period.
b. Imprinting depends on either visual or auditory cues in mammals.
c. Imprinting makes it possible to encode complex information in the nervous system.
d. Imprinting occurs during song acquisition of some species of birds.
e. Parent–offspring recognition is often dependent on imprinting.

51. Which of the following statements about hygienic behavior in honeybees is *not* true?
a. Hygienic behavior in honeybees can be explained in a simple Mendelian manner.
b. The strain showing hygienic behavior is resistant to a disease that kills larvae.
c. In a cross of the hygienic and nonhygienic strains, all of the F_1 offspring are of one phenotype.
d. The offspring from a backcross show a 3:1 ratio.
e. Behavior of the hybrids suggested that there are not separate genes for uncapping cells and for removing infected larvae.

52. The *per* or "period" gene in fruit flies
a. alters the frequency of wing vibrations in male courtship displays.
b. is homologous to clock genes found in many other organisms.
c. shows circadian rhythms during transcription.
d. may form part of the molecular basis for circadian rhythms in this species.
e. All of the above

53. Select the ethologist from the list below.
a. Konrad Lorenz
b. Niko Tinbergen
c. Karl von Frisch
d. All of the above
e. None of the above

54. Which of the following statements about circadian rhythms is *not* true?
a. Circadian rhythms are seldom exactly 24 hours.
b. Entrainment is the process of bringing two rhythms into phase.
c. A rhythm that is less than 24 hours must be phase-advanced to remain in phase with the 24-hour cycle of light and dark.
d. Animals in constant darkness exhibit circadian rhythms.
e. The period of a rhythm is the length of one cycle.

55. Which of the following statements about endogenous clocks is *not* true?
a. There are limits to the entrainment ability of endogenous clocks.
b. The endogenous clock of mammals is located in the suprachiasmatic nuclei.
c. Endogenous clocks are found in virtually every animal group.
d. Cells in the two eyes of birds function as endogenous clocks.
e. Many plants have endogenous clocks.

*56. A blind bird with a circadian rhythm of 22 hours is fitted with a black cap that covers the top of its head. It is placed into an environmental chamber with a light–dark period of exactly 12 hours. After several days, the bird will
a. become phase-advanced.
b. become phase-delayed.
c. lose all circadian rhythms.
d. develop a free-running rhythm.
e. be unaffected.

57. Which of the following statements about piloting is *not* true?
a. Piloting is used by some wasps to find their nests.
b. Piloting is only useful for short-distance migrations.
c. Gray whales use piloting while moving from their wintering grounds to their summering grounds.
d. Water currents can be used as piloting cues.
e. Wind patterns can be used as piloting cues.

58. Which of the following statements about homing is *not* true?
a. Simple piloting can explain many cases of homing.
b. Homing pigeons fitted with frosted contact lenses were still able to home.
c. Piloting is not sufficient to explain homing in the homing pigeon.
d. Marine birds show many examples of homing over great distances.
e. In most migratory bird species, the young learn to home by following the adults.

59. Studies done on European starlings transported and released from different areas during their migratory period suggest that the starlings use the type of navigation called
a. piloting.
b. distance-and-direction navigation.

c. bicoordinate navigation.
d. random search navigation.
e. homing.

*60. Research has shown that there is a correlation between the duration of __________ and the required __________ to the goal.
a. migratory restlessness, distance
b. migratory restlessness, direction
c. homing, distance
d. homing, direction
e. circadian rhythms, distance

61. In bicoordinate navigation, the animal knows all of the following *except*
a. its initial latitude and longitude.
b. the latitude of its goal.
c. the distance to the goal.
d. the sun's position at its current location and at its goal.
e. its initial longitude.

62. A sun compass requires that the animal know
a. its current location.
b. its latitude and longitude.
c. the current time.
d. the sun's direction at the goal.
e. the latitude and longitude of its goal.

*63. A bird is trained to feed from the southeast end of a circular cage. At sunrise, a mirror is used to shift the sun 45° to the right of it real position. Where will the bird attempt to feed?
a. North
b. East
c. South
d. West
e. Southwest

*64. A bird is trained to feed from the west end of a circular cage. At sunrise, a fixed light is set up at the south end of the cage. If the bird treats the light as the sun, where will the bird attempt to feed at noon?
a. North
b. East
c. South
d. West
e. Northwest

*65. A bird is trained to feed from the south end of a circular cage. The bird is then placed into a light-controlled room and phase-advanced by 6 hours. If the bird is returned to the circular cage at noon, where will it search for food?
a. North
b. East
c. South
d. West
e. Northwest

*66. An animal is displaced such that the sun rises earlier and is lower in the sky than it would be at its home position. What direction should the animal move in order to home?
a. Southwest
b. Southeast
c. Northwest
d. Northeast
e. South

67. A king snake must clamp down on the mouth of its prey, the rattlesnake, to avoid being bitten and killed by the rattlesnake. This type of behavior is
a. learned.
b. genetically programmed.
c. imprinting.
d. dependent upon hormone release.
e. entrainment.

68. The fruit fly, *Drosophila*, makes a good subject for hybridization studies because
a. it has a short life span.
b. it produces large numbers of offspring.
c. its behavior traits can be traced to simple Mendelian segregation.
d. a and b
e. a, b, and c

69. Which of the following conveys information about direction of a food source in honeybees?
a. Round dance
b. Odor on the body of a returning forager
c. Angle of the straight run of the waggle dance
d. Speed of the waggle dance
e. Speed of the round dance

70. Glass knife fish communicate by __________ since they live in murky waters and cannot see well.
a. sound waves
b. chemical signals
c. tactile signals
d. electric signals
e. water vibrations

71. Which of the following is true of animal communication?
a. It is shaped by natural selection.
b. It may benefit the sender and the receiver.
c. The particular channel of communication is shaped by the environment.
d. Some species exploit the communication systems of other species.
e. All of the above

72. In most cases, the behaviors of animals
a. are instinctive.
b. are learned.
c. have both genetic and learned components.
d. are maladaptive.
e. are not shaped by natural selection.

Study Guide Questions

1. A newborn male rat is castrated and then allowed to mature. When he reaches adulthood, he is given an injection of male sex steroids and exposed to a normal adult female rat exhibiting lordosis. Which of the following will occur?
   a. The male will mount the female and attempt to copulate.
   b. The male will show lordosis.
   c. The male will attempt to copulate and show lordosis.
   d. The male will show no sexual behavior at all toward the female.

2. Select the following example that is least likely to be a fixed action pattern.
   a. Egg laying in a turtle
   b. Courtship display
   c. Predator's prey-search behavior
   d. Web spinning in a spider

3. Which of the following statements about hormonal influences on bird song is *false*?
   a. Male songbirds castrated as nestlings do not sing proper songs as adults.
   b. Female songbirds can be induced to sing by treatment with testosterone.
   c. Females learn their species song, but do not normally express it under natural conditions.
   d. Testosterone is not necessary for singing in males once they have learned their song.

4. Studies of which of the following organism types have led most directly toward connecting a specific gene with a specific behavior?
   a. Honeybees
   b. Dabbling ducks
   c. *Drosophila* fruit flies
   d. Domestic animals

5. Which of the following is not a valid conclusion resulting from studies done on hygienic behavior in honeybees?
   a. The inheritance of this behavior can be explained in a simple Mendelian manner.
   b. The uncapping and larval-removal components of this behavior seemed to be determined by a single gene.
   c. Since all the F_1 bees resulting from a cross between hygienic and nonhygienic strains were nonhygienic, the trait must be determined by recessive alleles.
   d. A backcross of the F_1 to the nonhygienic strain should yield one-quarter hygienic to three-quarters nonhygienic individuals.

6. Which of the following statements about the action of releasers in herring gull chicks is *false*?
   a. The red dot on the mother's beak acts as a releaser.
   b. The shape of the head is not important in eliciting a response from the chick.
   c. This behavior is learned rather than inherited genetically.
   d. Begging is the behavior triggered by the releaser.

7. Which of the following statements about genetically determined behaviors is *false*?
   a. Inherited behaviors that involve mating and courtship are highly adaptive.
   b. Hybridization experiments mate closely related species to determine the behavior of the offspring.
   c. The web spinning in spiders is a natural deprivation experiment.
   d. Most behaviors have only a genetic component.

8. Animals exhibit daily rhythms in their behavior and physiology. Which of the following is *false* about circadian rhythms?
   a. Animals that are active at night do not display circadian rhythms.
   b. The circadian clock can be entrained by environmental cues.
   c. The circadian clock is located in the suprachiasmatic nuclei in mammals.
   d. In fruit flies the loss of the *tim* gene will result in the loss of circadian rhythms.

9. Which of the following is an example of piloting?
   a. Silkworms following a trail of hormones
   b. Marine migration over the featureless ocean
   c. Bees learning the directions to a food source from a dance
   d. Gray whale migration along the pacific coast of America

10. There are a number of different navigational methods used by animals. Which definition of a navigational method listed below is correct?
    a. Distance-and-direction navigation requires knowledge of latitude and longitude.
    b. Bicoordinate navigation may involve using the circadian clock to determine direction in relation to the sun.
    c. The stars offer two sources of information about direction, fixed points and moving constellations.
    d. Piloting involves the magnetic mineral magnetite.

11. Which one of the following animals would not be expected to have an endogenous circannual rhythm?
    a. A ground squirrel that hibernates
    b. An Arctic hare
    c. A cave salamander
    d. A bird that summers in North American and migrates to South America for the winter

12. In studies using fruit flies it has been found that the __________ gene is involved in the development of male courtship behavior, and the __________ gene is involved in anatomical differentiation of males.
    a. *sex-lethal, transformer*
    b. *sex-lethal, doublesex*
    c. *fruitless, doublesex*
    d. *fruitless, transformer*

End of Chapter Questions

1. Which of the following is *true* about the building of a web by a spider?
   a. Spiders use a different design depending on the environment.
   b. A young spider learns to build a web by copying the web of its mother.
   c. A young spider imprints on its mother's web and when it is sexually mature it replicates that design.
   d. The motor patterns for web building are largely inherited.
   e. Female spiders select mates on the basis of the quality of their webs.

2. If you do not see courtship behavior in a deprivation experiment investigating the proximate cause of sexual behavior, you can conclude that
   a. the animal is not sexually mature.
   b. the animal has low sexual drive.
   c. it is the wrong time of year.
   d. the appropriate releaser is not present.
   e. none of the above

3. Which statement about releasers is *true*?
   a. The appropriate releaser always triggers a response.
   b. They are simple subsets of sensory cues available to the animal.
   c. They are learned through imprinting.
   d. They trigger learned behavior patterns.
   e. An animal responds to a releaser only when it is sexually mature.

4. Which statement about the genetics of behavior is *true*?
   a. Approximately 20 genes control the courtship displays of male dabbling ducks.
   b. The loss of a single gene can eliminate male sexual behavior in fruitflies.
   c. Genes for retrieving, pointing, and herding have been described in dogs.
   d. Inherited behaviors are highly modifiable because learning can influence gene expression.
   e. Hygienic behavior in bees has been shown to be controlled by two dominant genes.

5. Which of the following statements about communication is *true*?
   a. Complex information can be conveyed most rapidly by pheromones.
   b. Visual signalling is advantageous in complex environments.
   c. A disadvantage of auditory communication is that it always reveals the location of the signaller.
   d. An advantage of pheromones is that the message can persist through time.
   e. The dance of bees is an example of using visual signalling to communicate.

6. If the sun were to come up earlier than expected on the basis of a circadian rhythm,
   a. it could cause symptoms of jet lag.
   b. it could phase-advance the circadian rhythm.
   c. the animal could be east of home.
   d. it could entrain the circadian rhythm.
   e. all of the above

7. To be able to pilot, an animal must
   a. have a time-compensated solar compass.
   b. orient to a fixed point in the night sky.
   c. know the distance between two points.
   d. know landmarks.
   e. know its longitude and latitude.

8. Birds that migrate at night
   a. inherit a star map.
   b. determine direction by knowing the time and the position in the sky of a star constellation.
   c. orient to the fixed point in the sky.
   d. imprint on one or more key constellations.
   e. determine distance, but not direction, from the stars.

9. The most likely explanation for the observation that humans from entirely different societies smile when they greet a friend is that
   a. they share a common culture.
   b. they have imprinted on smiling faces when they were infants.
   c. they have learned that smiling does not stimulate aggression.
   d. smiling is an inherited behavior pattern.
   e. smiling is a learned behavior.

10. If (1) a bird is trained to seek food on the western side of a cage open to the sky, (2) the bird's circadian rhythm is then phase-delayed by 6 hours, and (3) after phase-shifting the bird is returned to the open cage at noon real time, it will seek food in the
    a. north.
    b. south.
    c. east.
    d. west.

Student Web Site Self-Quiz Questions

1. Web spinning in spiders is a complex behavior. Which of the following statements about web-spinning patterns is *false*?
   a. Web-spinning patterns are stereotypic.
   b. If a spider is interrupted during the web-spinning process, the spider can modify the design to overcome the obstacle.
   c. Web patterns are species-specific.
   d. Web patterns require no learning.
   e. Web patterns are not modified by experience.

2. Some behaviors, such as the escape jump of the kangaroo rat in response to the sound of a striking rattlesnake, cannot be learned. Which of the following types of behavior could be learned or modified by learning?
   a. Species-specific bird song
   b. Behaviors in species with nonoverlapping generations
   c. Behavior in situations where mistakes are costly or dangerous

d. Behavior in an environment in which both correct and incorrect behavioral models exist
e. Behavior used by a predator to capture deadly prey

3. Under which of the following conditions would you expect a white-crowned sparrow to correctly learn its song?
a. Reared in acoustic isolation until after it matures
b. Being allowed to only hear songs of other species
c. Deafened before maturity
d. Deafened after hearing its species-specific song, but before expressing it as a juvenile
e. Deafened after expressing its species-specific song as a juvenile

4. Which of the following statements about the control of learning and expression of song in birds is *false*?
a. Both male and female songbirds imprint on their species-typical song as nestlings.
b. Testosterone stimulates the male to begin to sing and gradually improve his species-typical song.
c. Females lack the muscular and nervous capabilities to sing.
d. Testosterone causes seasonal changes in the brains of male birds.
e. During the nonbreeding season, the male brain is similar to the female brain.

5. Hybridization studies by Konrad Lorenz on courtship displays of different duck species have shown the influence of genes on behavior. Which of the following statements about the genetics of behavior is *false*?
a. The effect of a gene on a specific behavior is usually complex.
b. In crickets, only male hybrids show new combinations of behaviors.
c. In Lorenz's studies of duck courtship behavior, the hybrid always showed new combinations of behaviors.
d. Studies of *Drosophila* courtship behavior have shown the importance of multiple genes that influence behavior indirectly.
e. Two genes determining hygienic behavior in honeybees are inherited in a simple Mendelian fashion.

6. The red throat pouch of the male frigate bird probably evolved from throat-fluttering behavior during thermoregulation. Which of the following statements about displays or signals is *false*?
a. Displays or signals can consist of behaviors, anatomical features, or physiological responses.
b. Intention movements are a source of raw material for evolution of displays.
c. Autonomic responses are a source of raw material for evolution of displays.
d. Displacement behaviors are are a source of raw material for evolution of displays.
e. Many displays represent novel behaviors that evolved specifically for the communication function that they perform.

7. Which of the following features is *not* characteristic of olfactory communication?
a. Oldest form of animal communication
b. Ability to communicate specific messages
c. Ease of orientation
d. Fast communication
e. Longevity

8. Animals have evolved a variety of specialized structures for display purposes. Which of the following signaling modes is particularly good for communicating well over a distance?
a. Chemical signals
b. Visual signals
c. Auditory signals
d. Tactile signals
e. Electric signals

TEST FILE

★ *Indicates a **difficult** question*

■ *Indicates a **conceptual** question*

53 Behavioral Ecology

Fill in the Blank

1. Many **fish** protect their young, but do not feed them.
2. Species with social systems having sterile classes are termed **eusocial**.
3. Members of social groups often have **increased** competition for food compared to solitary animals.
4. Members of social groups usually have **increased** protection from predators compared to solitary animals.
5. **Sexual selection** is the spread of traits that confer advantages to their bearers in competition for mates or for resources needed for courtship.
6. **Territorial behavior** involves the defense of an area that contains food, nesting sites, or other resources.
7. In white-fronted bee-eaters, about half the nonbreeding individuals help at the nests of other individuals. Helping relatives **increases** fitness, and thus white-fronted bee-eaters usually help relatives rather than nonrelatives.
8. Ecologists often analyze their observations of animal behavior in terms of the costs and **benefits** to the performer.
9. An important variable in determining the costs and benefits of a behavior is the **relatedness** between the performer and recipient.
10. The **energetic** cost of a behavior is the difference between the energy the animal would have expended had it rested, and the energy expended in performing the behavior.
11. The **risk** cost of a behavior is the increased chance of being injured or killed as a result of performing it, compared with resting.
12. The **opportunity** cost of a behavior is the sum of the benefits the animal forfeits by not being able to perform other behaviors during the same time interval. For example, an animal forfeits the time to feed when it defends its territory.
13. **Reciprocal altruism** evolves if the performer is in turn the recipient of beneficial acts from the individuals it has helped.
14. In order for behaviors to evolve, they must have a **genetic** basis.
15. The environment in which an organism normally lives is called its **habitat**.
16. In several species of weaverbirds, the males are **polygynous**, having more than one mate.

Multiple Choice

1. An elephant seal's environment is defined as
   a. all other elephant seals in the same area.
   b. all other organisms that influence it.
   c. the physical factors of its area.
   d. all other organisms of its food web.
   e. all the physical and biological factors that influence it.

2. An appropriate study in the field of ecology would be
   a. a comparison of vertebrate skeletal structures.
   b. the effect of adrenaline on human heart rate.
   c. an investigation of the roles of hormones in plant growth.
   d. the effect of fire on forest animals' populations.
   e. a classification of fungi into various groups.

3. __________ ecology is the study of how animals make decisions that influence their survival and reproductive success.
   a. Natural
   b. Social
   c. Behavioral
   d. Physical
   e. Environmental

4. The term for the environment in which an organism lives is its
   a. community.
   b. ecology.
   c. habitat.
   d. niche.
   e. population.

5. Red abalone larvae will settle on coralline algae substrates only after recognizing a peptide molecule that the algae produce. As an example of a habitat selection cue, one would expect that this peptide is
   a. required for abalone reproductive success.
   b. produced by the algae in order to attract the abalone.
   c. a good predictor of conditions suitable for abalone survival.
   d. a food source for the abalone.
   e. a waste by-product of the algae.

6. Bluegill sunfish are put in two tanks with their prey, water fleas. Each tank contains an equal proportion of small, medium, and large water fleas. One tank has a low density of water fleas, and the other has a high flea density. According to foraging theory, the bluegill sunfish should eat
   a. an equal proportion of the three flea sizes at both densities.
   b. a greater proportion of large fleas at both densities.
   c. equal proportions of the three flea sizes at low density and mostly large fleas at high density.
   d. mostly large fleas at low density and equal proportions of the three flea sizes at high density.
   e. mostly medium and large fleas at low density and mostly large fleas at high density.

7. Which of the following is *not* a necessary assumption of foraging theory?
   a. Efficient predators spend less time on predation than nonefficient predators.
   b. Superior predators can produce more offspring.
   c. Superior foraging ability can be genetically based.
   d. Predators choose prey in ways that maximize their energy intake rate.
   e. Efficient predators always choose the most abundant prey.

8. A hyena that takes food away from another hyena is exhibiting _________ behavior.
   a. altruistic
   b. cooperative
   c. selfish
   d. spiteful
   e. territorial

9. An example of reciprocal altruism is
   a. two chimpanzees picking parasitic fleas off one another.
   b. a parent bird feeding its young.
   c. one penguin guarding its territory against other penguins.
   d. wild dogs hunting together to capture prey.
   e. two male deer competing for the same female.

10. When wood pigeons draw together in a tight flock in response to the presence of a hawk, they are exhibiting _________ behavior.
    a. altruistic
    b. cooperative
    c. selfish
    d. spiteful
    e. territorial

11. When one dog defends its food by attacking a second dog, and meanwhile a third dog takes the food, the first dog has exhibited _________ behavior.
    a. altruistic
    b. cooperative
    c. selfish
    d. spiteful
    e. territorial

12. Kin selection is
    a. the mating of relatives.
    b. the recognition of relatives among societal animals.
    c. the adoption of young by generally unrelated adults.
    d. a process of forcing young males out of a society while keeping the females.
    e. a behavior that increases the survivorship of an individual's relatives.

13. Altruistic acts are most likely to evolve into behavior patterns when the participants
    a. are capable of learning.
    b. have individual fitness.
    c. are genetically related.
    d. are largely nonsocial.
    e. are mating partners.

▪14. What is the most likely reason a female animal who is reproductively mature would reject a courting male of the same species?
    a. She determined the male was monogamous.
    b. She determined the male was not from her social group.
    c. She determined the male was polygynous.
    d. She determined the male was not healthy.
    e. None of the above

15. Although exceptions exist, in general a female animal would _________ her reproductive success by mating with many males.
    a. reduce
    b. increase
    c. double
    d. not change
    e. None of the above

16. Together, individual fitness and kin selection determine the _________ fitness of an individual.
    a. social
    b. selective
    c. exclusive
    d. inclusive
    e. reciprocal

17. Male courtship behavior accomplishes all of the following *except* to
    a. induce a female to mate with the male.
    b. display that the male is in good health.
    c. show that the male is a good provider.
    d. convey that the male has successfully mated in the past.
    e. signal that the male has a good genotype.

18. After copulation, a male may remain with a female for a time and prevent her from copulating with other males. Although this behavior has its benefits, it also carries a high __________ cost.
a. energetic
b. opportunity
c. risk
d. competitive
e. altruistic

▪19. After copulation, many male insects are genetically programmed to stay near the female for a time. This behavior is most adaptive because it
a. guards the female against injury.
b. prevents the female from copulating again until after fertilization.
c. allows the male to recover from the energetic cost of mating.
d. defends a nesting territory.
e. creates a social bond between the male and female.

20. The most widespread social system among animals is called a
a. family.
b. pack.
c. clan.
d. flock.
e. group.

21. Females may choose a mate based on all of the following *except*
a. the quality of the resources in his territory.
b. his display of food items.
c. his apparent good health.
d. his physical features.
e. the increased probability that his sperm will fertilize the eggs.

22. Female reproductive success is improved by choosing a male with all of the following features *except* his
a. high genetic quality.
b. good health.
c. interaction with other female mates.
d. high parental care.
e. control of abundant resources.

23. Parental investment
a. increases the chances of survival of the parent.
b. increases the parent's ability to produce additional offspring.
c. increases the chances of survival of the offspring.
d. decreases as the number of offspring increases.
e. decreases when costs to parents are lowered.

▪24. Which of the following characteristics is most likely to occur in a bird species having brightly colored females and dull-colored males?
a. The females pursue the males during courtship.
b. The species has a monogamous mating system.
c. Offspring are raised by relatives of the parents.
d. The species forms large nesting colonies.
e. The relative size of the females is small compared to the males.

▪25. Which of the following statements does *not* support the theory that the large size of males in many vertebrate species is a result of sexual selection?
a. Larger males tend to defend larger territories.
b. Larger males win fights with smaller males.
c. Larger males attract more females than smaller males.
d. Polygynous species have larger males than similar monogamous species.
e. Parents feed larger male offspring more food than smaller offspring.

26. The most widespread social system among animals is
a. parents and offspring.
b. peer grouping.
c. several females and their offspring.
d. several pairs and their offspring.
e. a dominant pair, their offspring, and unrelated subordinates.

27. Among mammals, which offspring, if any, tend to leave their parents' group?
a. Both male and female offspring leave.
b. Females leave, but males leave only if a new territory becomes available.
c. Males leave but females do not leave.
d. Neither males nor females leave.
e. Males leave, but females leave only if a new set of offspring are born.

28. Large numbers of sterile individuals occur in all of the following *except*
a. termites.
b. ants.
c. bees.
d. beetles.
e. wasps.

29. __________ theory tries to answer questions related to how an animal looks for and acquires food.
a. Storage
b. Foraging
c. Social
d. Animal
e. None of the above

30. In species with diploid females and haploid males, which pair of relatives is most similar genetically?
a. Mother and daughter
b. Mother and son
c. Father and son
d. Full sisters
e. Full brothers

31. Which of the following groups exhibits eusocial behavior and has diploid females and haploid males?
a. Penguins
b. Bees
c. Jackals
d. Termites
e. Naked mole rats

32. W. D. Hamilton hypothesized that eusociality evolved because sterile worker females are genetically more related to their sisters than to any other relatives. According to this hypothesis, which one of the following would *not* be expected?
    a. Queens should produce equal numbers of sons and daughters.
    b. Workers should feed their sisters better than their brothers.
    c. If the original queen is lost, the workers would not favor the new queen's daughters.
    d. Worker females would have the same father and mother.
    e. Both males and females are diploid.

33. Which of the following is *not* true of eusociality?
    a. Eusocial species form elaborate nests or burrows.
    b. Helping behavior may be a necessary prerequisite for the evolution of eusociality.
    c. Some eusocial species contain only diploid members.
    d. Eusociality may occur in one species and be lacking in a closely related species.
    e. Eusocial individuals typically live part of their lives away from the society.

34. Large, hoofed, plant-eating mammals tend to
    a. feed preferentially on high-protein foods.
    b. live in herds.
    c. feed in forests.
    d. have monogamous mating patterns.
    e. have higher metabolic rates than smaller mammals.

35. Which of the following features is *not* common among forest-dwelling birds?
    a. Feed alone
    b. Build well-hidden nests
    c. Monogamy
    d. Sexes have identical plumage
    e. Nest in colonies

36. Which is *not* true of primate social systems?
    a. Nocturnal species tend to feed alone.
    b. Day-active species tend to eat animal food when available.
    c. Arboreal forest-dwelling species live in large groups.
    d. Females usually remain with their natal group.
    e. Both males and females may develop dominance hierarchies.

37. Which is *not* true of social systems?
    a. Social systems can be studied by asking how individuals in a system benefit by it.
    b. Social systems are dynamic; individuals' relationships with one another constantly change.
    c. Relationships within a social system are partly determined by genetic relatedness.
    d. Social systems have evolved primarily because societies have an increased success at obtaining food.
    e. Social systems continue to evolve in relation to animals' sizes, diets, and habitats.

38. The benefits lost by an animal that is not able to feed while defending its breeding territory is an example of ________ cost.
    a. opportunity
    b. energetic
    c. risk
    d. feeding
    e. breeding

39–41. Match the social acts in the list below with the correct descriptions.
    a. Altruistic act
    b. Selfish act
    c. Cooperative act
    d. Political act
    e. Spiteful act

39. Willow leaf beetles from the same clutch feed together. Beetles in groups can initiate feeding sites on tough leaves more easily than lone beetles can, and there are group defenses against predators. **(c)**

40. A tadpole eats a conspecific egg and becomes diseased as a result of doing so. **(e)**

41. A bird sees a predator and gives out a warning call to other birds in the area. The caller may be at increased risk of being eaten by the predator. **(a)**

42. A willow leaf beetle larvae eats a conspecific egg and gains a nutritional benefit **(b)**

43. Which of the following is *not* a consequence of parental investment?
    a. Increased chance of survival of the offspring
    b. Increased care of each offspring
    c. Increased chance of survival of the parent
    d. Increased fitness of the parent
    e. Decreased chance of production of additional offspring

▪44. Paternal investment by mammals is often less than that of birds. This may be because
    a. young mammals are less likely to be eaten by predators and thus need less care.
    b. paternity is more certain in birds than in mammals.
    c. there is less cost to birds for paternal care.
    d. male mammals do not have functional mammary glands and thus cannot feed the young until they are weaned.
    e. Newborn mammals are more self-sufficient than are newly-hatched birds.

45. Fitness for each sex is usually maximized in a mating cycle when males copulate ________ and females copulate ________.
    a. many times, once
    b. once, once
    c. many times, many times
    d. a few times, many times
    e. a few times, once

46. Group living
    a. can result in foraging benefits.
    b. reduces exposure to disease.
    c. is usually advantageous when food is scarce.
    d. has no effect on reproductive success.
    e. usually decreases competition between individuals.

47. A cooperative act benefits
    a. the performer.
    b. spiteful individuals.
    c. the performer and another individual.
    d. selfish individuals and their mates.
    e. nearest kin only.

48. What is the purpose a Thomson's gazelle performing "stotting" behavior?
    a. It is signaling to a predator that it has been seen.
    b. It has seen a predator and is now alerting other gazelles.
    c. Attraction of mates
    d. Territory defense
    e. None of the above

49. To construct a hypothesis about how a foraging animal should behave, a scientist first specifies the objective of the behavior and then attempts to determine the behavioral choices that would best achieve that objective. This process is called __________ modeling.
    a. minimal
    b. optimality
    c. behavioral
    d. forager
    e. natural

50. Parental investment is an example of __________ behavior.
    a. selfish
    b. spiteful
    c. territorial
    d. altruistic
    e. reciprocal

▪51. Females are usually choosier than males with respect to mates because
    a. they typically invest more energy per offspring than do males.
    b. they typically have more total offspring than do males.
    c. males are more variable, so females have more to choose from.
    d. only females can nurse young in most animal species.
    e. they pass on more genes to their offspring than do males.

52. Males are most likely to provide care for young when
    a. fertilization is internal.
    b. the males are mammals.
    c. offspring survival is unaffected by the care.
    d. the males mate with many females.
    e. certainty of paternity is high.

53. In some birds, individuals other than the parents help in rearing offspring. These helpers are usually
    a. unrelated to the parents.
    b. close relatives of the parents.
    c. unrelated females that have lost their own offspring.
    d. members of a different altruistic species.
    e. cuckolding males.

54. When helpers participate in rearing young, the fitness of parental Florida scrub jays
    a. decreases due to competition with the helpers.
    b. decreases due to reduced interaction with their offspring.
    c. increases because the helpers often feed the parents.
    d. increases due to increased offspring survival.
    e. is usually unchanged.

55. Groups of grooved-bill anis (ground-dwelling birds) defend communal nests. When one female begins to lay eggs, she throws all other eggs in the nest out before laying her own. This is an example of __________ behavior.
    a. displacement
    b. altruistic
    c. intention
    d. selfish
    e. helping

56. In many mammals, including primates,
    a. females are more likely than males to remain in natal groups.
    b. males are more likely than females to remain in natal groups.
    c. both males and females usually leave natal groups.
    d. both males and females typically remain in natal groups.
    e. natal groups rarely persist once the young can be independent.

57. Baboons travel in groups that are defended from predators by males. This behavior of the males probably
    a. increases their fitness because it increases their survival.
    b. increases their fitness because it increases survival of their offspring.
    c. decreases their fitness because it decreases their survival.
    d. decreases their fitness because it decreases survival of their offspring.
    e. is unrelated to fitness.

58. An example of reciprocal altruism is
    a. female anis throwing each other's eggs out of the communal nest.
    b. primates grooming one another.
    c. male bees stealing one another's mates.
    d. male elephant seals defending harems against one another.
    e. parent birds feeding their offspring.

▪59. It was shown that the female African long-tailed widowbird was more attracted to males with elongated tails. Why don't the male widowbirds have even longer tails?
a. It was genetically impossible to create longer tails in males.
b. The cost of producing the longer tails was more than the benefit.
c. The shorter-tailed males were jealous and fought off the longer-tailed males.
d. The longer-tailed males were sterile.
e. None of the above

60. Behavioral ecology deals with
a. how animals make decisions about their survival.
b. the choice of mates.
c. how animals select food and shelter.
d. group interactions.
e. All of the above

61. Which of the following is *not* a criterion a predator uses to choose its prey?
a. How long it takes to capture the prey
b. The energy it receives from eating the prey
c. The abundance of the prey
d. Whether another predator is after the same prey
e. Whether the prey contains the proper nutrients for the predator

Study Guide Questions

1. Which of the following is *not* usually included within the domain of ecology?
a. Interactions between conspecifics
b. Modifications of the environment by organisms
c. Evolution of different social organizations
d. Modification of the environment by physical processes

2. Which of the concepts listed below best describes the following situation: Birds spend some of their time scanning the horizon for predators. While scanning, they cannot be foraging for food.
a. Cooperative hunting
b. Energetic cost
c. Risk cost
d. Opportunity cost

3. Two species of mice live in the same geographical region, but species 1 prefers open fields, whereas species 2 lives in forests. When presented in an experiment with simulated "fields" and "forests," individuals of each species born in a laboratory preferred the environment in which they normally lived. This experiment illustrates the concept of
a. habitat selection.
b. optimal foraging strategy.
c. territoriality.
d. a and b

4. Which of the following animals would likely have a territory that is used exclusively for a nest site?
a. The song sparrow
b. The Florida scrub jay
c. The red-winged blackbird
d. A colonial marine bird

5. You perform a series of experiments in which the density of prey species 1 is varied while the density of prey species 2 is kept constant. At each prey species 1 density, you determine how many prey of species 1 and 2 the predator takes. The two curves (a and b) plotted in the following graph show the outcome of this study.

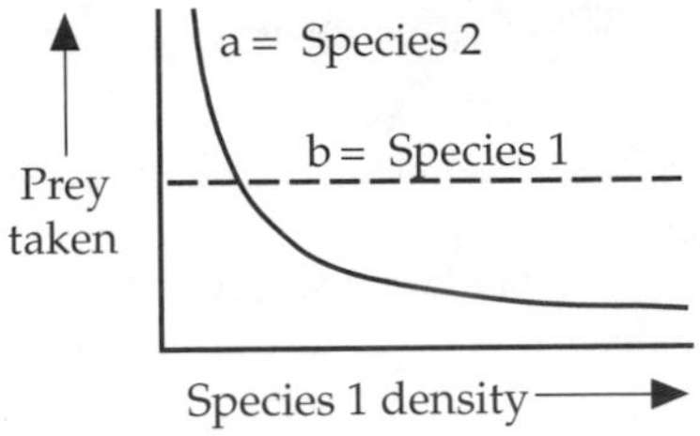

Select the statement that correctly interprets these curves.
a. Curve a indicates that species 2 is the preferred prey.
b. Curve b indicates that species 1 is the preferred prey.
c. Curve b indicates that species 1 is not the preferred prey.
d. Neither of these curves provides insight into the prey preference of the predator.

6. You observe an example of an apparently altruistic act by an animal that seems to reduce its near-term likelihood of reproductive success. What would be the *least* plausible explanation for its behavior?
a. The act aids the reproductive success of individuals sharing a high proportion of genes with the altruistic individual.
b. The act is only apparently altruistic; over the long term, the behavior actually contributes to individual fitness.
c. The behavior is based on the expectation that performing the act increases the probability of being repaid in the future by other members of the social group.
d. The act is advantageous because it helps the species to survive and reproduce, even if the altruist himself does not.

7. Some birds give a species-specific vocalization called an "alarm call" when they see a predator, although this call may direct the predator toward them. Other members of their species respond to these calls by taking cover. This would be an example of altruistic behavior that is beneficial to the calling bird if
a. the bird giving the vocalization survives the attack.
b. the inclusive fitness of the bird giving the vocalization is increased.
c. all birds survive the attack.
d. the birds benefiting are offspring of the bird giving the vocalization.

8. A dominant individual displaces a subordinate individual from food that the subordinate has discovered. This is an example of a(n)
   a. selfish act.
   b. cooperative act.
   c. altruistic act.
   d. spiteful act.

9. For a behavioral act to be considered an example of reciprocal altruism,
   a. both individuals must potentially benefit from the act at some time.
   b. the act must have a genetic basis.
   c. the individuals involved must be closely related.
   d. a and b

10. Which of the characteristics listed below is (are) shared by all eusocial species?
    a. A sex determination system in which males are haploid and females are diploid
    b. Queens that mate with a single male
    c. Presence of sterile classes
    d. All of the above

11. Which of the following statements describing the social system of the Florida scrub jay is *false?*
    a. It is a eusocial organization.
    b. Most helpers at the nest are offspring from the previous breeding season that remained with their parents.
    c. The social system probably evolved because offspring have little chance of establishing new territories on their own.
    d. Only the parent birds breed.

12. Of the following factors, select the one that was *not* mentioned as being important in the evolution of social organization in African hoofed animals.
    a. Metabolic needs
    b. Predation
    c. Dispersal of breeding territories
    d. Dispersal of food resources

13. Which of the following choices would *not* be a factor that can influence the social organization typical of a species?
    a. Presence of a dominance hierarchy
    b. Availability of nest sites
    c. Sex determination mechanism
    d. Asymmetry in reproductive investment

End of Chapter Questions

1. Which of the following is not a component of the cost of performing a behavioral act?
   a. Its energetic cost
   b. The risk of being injured
   c. Its opportunity cost
   d. The risk of being attacked by a predator
   e. Its information cost

2. An almost universal cost associated with group living is
   a. increased risk of predation.
   b. interference with foraging.
   c. higher exposure to diseases and parasites.
   d. poorer access to mates.
   e. poorer access to sleeping sites.

3. An act is said to be altruistic if it
   a. confers a benefit on the performer by inflicting some cost on some other individual.
   b. confers a benefit both on the performer and on some other individual.
   c. inflicts a cost both on the performer and on some other individual.
   d. confers a benefit on another individual at some cost to the performer.
   e. imposes a cost on the performer without benefiting any other individual

4. Which of the following statements about male and female roles in social systems is *false*?
   a. Females invest more in gamete production, but they may invest more or less than males in care of offspring.
   b. Biparental care is prevalent among birds.
   c. Males of most mammal species help feed offspring.
   d. Males with a high probability of parentage invest more in parental care than males that are less certainly related to the offspring of their mates.
   e. Among fishes, if there is unequal parental care by individuals of the two sexes, it is nearly always the male that does more.

5. The basic components of an optimality model of behavior are the
   a. type of behavior and its neural control mechanisms.
   b. objective of the behavior and the choices that would best achieve it.
   c. objective of the behavior and its neural control mechanisms.
   d. goal of the behavior and the constraints imposed by the animal's structure.
   e. objective to be maximized and the currency used to measure it.

6. Choice of mating partner may be based on
   a. the inherent qualities of a potential mate.
   b. the resources held by a potential mate.
   c. both the inherent qualities of a potential mate and the resources it holds.
   d. the success of individuals of the opposite sex in courtship.
   e. all of the above

7. A lek is a
   a. territory held by a single male.
   b. territory held by two or more males.
   c. display ground at which a single male displays.
   d. display ground at which two or more males display.
   e. territory held by two or more females.

8. Among social birds, there are usually more male than female helpers in those species with helpers because
   a. males are better helpers than females.
   b. males typically receive greater benefits from helping.
   c. males survive better than females.
   d. mothers do not allow their daughters to help.
   e. males often need to wait for an unoccupied territory elsewhere.

9. Small African hoofed mammals are usually solitary because
   a. they feed on scattered high-quality foods in forested environments.
   b. the low quality of their food does not permit them to assemble in groups.
   c. they are too small to defend themselves against predators.
   d. they are too small to follow the rains to areas where grass growth is best.
   e. they are usually driven from their natal groups.

10. Species whose social groups include sterile individuals are said to be:
   a. eusocial
   b. semisocial
   c. oligosocial
   d. sterisocial
   e. supersocial

Student Web Site Self-Quiz Questions

1. The mating system of the elephant seal is based on polygyny—one male mating with more than one female. Which of the concepts described below best describes the following situation: A male elephant seal cannot search for food when he is defending his section of the beach.
   a. Benefit
   b. Energetic cost
   c. Opportunity cost
   d. Risk cost
   e. Foraging theory

2. The male Yarrow's spring lizard displays differential territorial behavior based on the time of year. Which of the following is a *false* statement regarding this observation?
   a. Testosterone induces the territorial behavior
   b. Territorial behavior is associated with an increased risk cost.
   c. Territorial behavior is associated with a decreased energy cost.
   d. Territorial behavior is normally displayed during mating season.
   e. The benefits of reproductive success outweigh the costs associated with territorial behavior in males during mating season.

3. Studies of prey selection by predators have been important in testing foraging theories. Which of the following statements about the construction of foraging theories is *false*?
   a. Cost-benefit analysis is an important part of developing a foraging theory.
   b. There are many possible foraging theories.
   c. Because animals almost always forage in order to obtain the greatest energy input per unit of effort, this is a key feature of all theories.
   d. Optimality modeling is an important part of developing a foraging theory.
   e. The role of natural selection in modeling the behavior of animals is an underlying assumption in foraging theories.

4. Tail length in the African long-tailed widowbird results from __________ and the benefit of a longer tail is __________ than a shorter tail.
   a. optimality modeling, greater
   b. reciprocal altruism, less
   c. reciprocal altruism, greater
   d. sexual selection, less
   e. sexual selection, greater

5. Many bird species are territorial. Which one of the following types of bird would have a territory whose size would be expected to vary *most* with habitat quality?
   a. A bird that uses its territory mostly for nesting and feeding
   b. A bird species that uses its territory only for display and mating
   c. The song sparrow
   d. A bird that uses its territory only for nesting

6. Which of the following statements about the behavior of nesting red-winged blackbirds is *false*?
   a. There are fewer infertile eggs in nests with multiple fathers.
   b. About one-third of the nestlings had a father different than the owner of the territory in which the nest resides.
   c. The best strategy for males is to spend time seeking copulation with additional females.
   d. Females who copulated with more than one male were reproductively more successful than those that remained faithful to their mates.
   e. Nests of females who have multiple mates suffer less predation than females faithful to their mates.

7. A white fronted bee-eater forgoes its own breeding to help another pair breed. To maximize her inclusive fitness, should she choose to help a sister or her mother. What term best describes this type of behavior?
   a. Mother, altruistic
   b. Sister, altruistic
   c. Doesn't matter, altruistic
   d. Mother, cooperative
   e. Sister, cooperative

8. The honeybees have a society that is said to be based on __________. Female sibling honeybees share __________ of their genes on average.
   a. eusociality, 75%
   b. eusociality, 50%
   c. eusociality, 25%

d. reciprocal altruism, 25%
e. reciprocal altruism, 75%

9. The origins of all animal societies lie in the association of parents and offspring. Which of the following statements about parental care in vertebrates is *false*?
   a. Males of most mammals contribute little directly to offspring nutrition.
   b. Both males and females feed young in most bird species.
   c. The Florida scrub jay has a social system based on an extended family.
   d. In fish, females provide more care for offspring than males.
   e. Most mammals evolved socially via the extended family route.

10. Which of the following is *not* a characteristic associated with a savanna-dwelling African weaverbird species?
   a. Males and females with different appearance
   b. Monogamy
   c. Colonial nesting
   d. Seed eating
   e. Feeding in large flocks

11. The wildebeest is a large, African, hoofed animal. Of the following factors, select the one that was *not* important in the evolution of social organization in African hoofed animals.
   a. Diet
   b. Dispersal of food resources
   c. Evolutionary history
   d. Metabolic needs
   e. Predation

TEST FILE

★ *Indicates a **difficult** question*

■ *Indicates a **conceptual** question*

54 Population Ecology

Fill in the Blank

1. The proportions of individuals in each age group in a population make up its **age distribution**.
2. The stages an individual goes through during its life constitute its **life history**.
3. A group of individuals born at the same time is known as a(n) **cohort**.
4. The number of individuals of any particular species that a habitat or environment can support is that habitat's **carrying capacity**.
5. The measure of the contribution an individual of a particular age is likely to make to its population's growth rate is called **reproductive capactiy**.
6. When comparing two populations whose members differ in size, the most useful measure of their densities is to compare **biomass**.
7. The equation describes **exponential** growth.
8. The total number of individuals of a species per unit of area (or volume) is called the **population density**.
9. Plants, corals, and other organisms that are composed of repeated sets of similar units are called **modular organisms**.
10. When a single bacterium is placed in a culture vessel, the resulting population at first grows very quickly; however, as the population expands and uses up its resources, the growth rate eventually slows down and stops. This S-shaped growth pattern is best modeled mathematically as **logistic** growth.
11. Population regulation that works to decrease numbers when its members are common and increase numbers when they are rare is **density dependent**.
12. If the death rate in a population is unrelated to the number of individuals in an area, the resulting change in population size is said to be **density independent**.
13. A short-term event that disrupts populations, communities, or ecosystems by changing their environment is called a(n) **disturbance**.
14. The study of changes in the size and structure of populations is known as **demography**.
15. A(n) **population** consists of all the individuals of a species within a given area.

Multiple Choice

1. Which one of the following characteristics of whales is not a factor in the whale population's slow response to management practices?
   a. Whales have long prereproductive periods.
   b. Whales produce one offspring at a time.
   c. Whales have long life spans.
   d. Whales have long intervals between births.
   e. Whales are hunted by several nations.
2. Of the following factors that regulate population size, the most density-independent factor is
   a. food supply.
   b. predators.
   c. disease.
   d. available nesting sites.
   e. sudden temperature changes.
3. If a population's birth rate is density-__________, it will __________ as the population increases.
   a. dependent, increase
   b. dependent, decrease
   c. independent, increase
   d. independent, decrease
   e. dependent or independent, increase
4. Which of the following undergoes the most radical metamorphosis in moving from an immature to adult stage?
   a. Bee
   b. Caribou
   c. Locust
   d. Turtle
   e. Salmon

*5. Agave and yucca plants appear similar and grow in the same environments, but agaves reproduce once and die, while yuccas reproduce several times before dying. Compared to agave, each yucca seed crop should use _________ energy and produce _________ seeds.
a. more, more
b. more, fewer
c. equal amounts of, equal amounts of
d. less, more
e. less, fewer

6. Undesirable alleles are least affected by natural selection if they are expressed in
a. embryos.
b. the young.
c. adolescents.
d. young adults.
e. the old.

7. For the last 30 years, the average human age of death in the United States has _________ due to _________.
a. fallen, new strains of infectious disease
b. fallen, environmental pollution
c. remained stable, deaths resulting from genetic diseases of old age
d. risen, control of infectious diseases
e. risen, elimination of most factors that cause death

▪8. In which of these populations would the individuals be evenly spaced?
a. Penguins incubating their eggs just out of pecking range of neighboring penguins
b. Maple seedlings sprouting around the base of the parent tree
c. Coconut seeds germinating on beaches after being carried by tides and water currents
d. Beaver offspring living with their parents for several years
e. Giraffes coming to a water hole to drink

▪9. If a certain age interval in a population contains ten individuals and has a death rate of 1.000, how many individuals will be in the next older age interval?
a. None
b. One
c. Five
d. Nine
e. Ten

10. The number of individuals in a population is least affected by the rate of
a. births.
b. deaths.
c. mating.
d. immigration.
e. emigration.

11. If a population of organisms consisted of 25 individuals and over the span of a year 5 die, what can be inferred about the population?
a. The organisms that died were weak.
b. Survivorship of the population was 80%.
c. Mortality in the population was high.
d. Natural selection is at work.
e. None of the above

12. Which of the following is not true of ecological disturbance?
a. Adaptations to disturbance occur among most organisms.
b. Adaptations are most common if the disturbance is frequent.
c. Large organisms usually are affected more than small organisms.
d. The frequency of some disturbances can be influenced by organisms.
e. Examples of biological disturbance are tree falls and diseases.

13. In northern Scandinavia, rodents called lemmings have irruptions, or periodic buildups of large populations. The most likely cause of these irruptions is a(n)
a. increase in food supply.
b. decrease in predator population.
c. increase in favorable nesting sites.
d. increase in birth rate.
e. decrease in disease rate.

*14. A population, such as Sweden's, with a low birth rate and a low death rate
a. would have a relatively even distribution of individuals of different ages.
b. would have a population dominated by young individuals.
c. would have a population dominated by old individuals.
d. would have a population dominated by individuals of intermediate age, with relatively few young or old individuals.
e. could have almost any age distribution. Birth and death rates do not affect the age distribution.

*15. Which of the following is *not* an important part of the life history of all organisms?
a. Birth: When an organism is born.
b. Energy-gathering stage: When individuals gather energy and nutrients for growth and reproduction. This stage may encompass the entire life history.
c. Parental care stage: When parents provide additional care and protection during early growth.
d. Dispersal stage: When individuals move or are moved from their birthplaces to the places where they will live as adults.
e. Reproductive stage: When offspring are produced.

16. Which of the following is *not* a demographic process that determines the number of individuals in a population?
a. Death
b. Birth
c. Migration
d. Immigration
e. Emigration

▪17. For modular organisms, it is often preferable to measure the biomass of a population rather than the population density because
a. modular organisms are easy to distinguish, and most adult members of a population are similar to one another in size and shape.
b. modular organisms often reproduce quickly and die, making it difficult to obtain an accurate count of individuals.
c. modular organisms (with very few exceptions) have very high densities, making it impractical to count individuals. Biomass is measured as a means to avoid counting all of the individuals.
d. modular organisms often vary greatly in size and shape (within a species) and individuals are not easy to distinguish.
e. For modular organisms, it is almost never preferable to measure biomass.

▪18. In a life table, the number of individuals alive at the beginning of the 1-year to 2-year age interval is 800. During this interval 200 individuals die. The death rate for this interval is
a. 0.25.
b. 200.
c. 800.
d. 0.2.
e. 0.8.

19. A single bacteria put in an environment with unlimited resources and no competition would
a. reproduce logistically.
b. reproduce exponentially.
c. reproduce linearly.
d. not reproduce.
e. b and c

20. _________ capacity is the maximum number of individuals of a species an environment can support.
a. Habitat
b. Growth
c. Population
d. Carrying
e. None of the above

21. Which of the following is *true* concerning exponential growth?
a. No population can grow exponentially for long.
b. Exponential growth slows down as the population nears its maximal size.
c. Bacterial colonies have been observed to maintain exponential growth for over a month.
d. Exponential growth is commonly observed in large, slow-growing species such as humans and elephants.
e. Exponential growth includes a component of environmental resistance.

22. The intrinsic rate of increase, r_{max}, is
a. the number of individuals added to each generation in a growing population under optimal conditions.
b. the difference between the average per capita birth rate, b, and the average per capita death rate, d, under optimal conditions.
c. the number of individuals added to each generation in a growing population under the conditions that are actually occurring.
d. the average per capita birth rate, *b*, under optimal conditions.
e. the difference between the average per capita birth rate, *b*, and the average per capita death rate, *d*, under the conditions that are actually occurring.

23. The population growth equation describes a population that
a. grows without limits.
b. grows rapidly at small population sizes, but whose growth rate slows and eventually stops as the population reaches the number the environment can support.
c. rapidly overshoots the number the environment can support and then fluctuates around this number.
d. grows very rapidly and then crashes when the environmental resources are used up.
e. is declining in size.

24. Which of the following is not an assumption of the logistic growth equation?
a. An individual exerts its effects immediately at birth.
b. All individuals produce equal effects.
c. Resources in the environment are limited.
d. Each individual depresses population growth equally.
e. There is a time lag between gathering resources and reproduction.

*25. Under which circumstances might natural selection favor delaying reproduction until later in life?
a. Juvenile survival is very poor, and parents can produce large numbers of very small offspring with little cost to themselves.
b. Juvenile survival is high, and parents always die following reproduction.
c. Juvenile survival increases with the experience of the parent, and reproduction greatly reduces the chances for parental survival.
d. Juvenile survival is high, and reproduction has little effect on parental survival.
e. Juvenile survival is high, and adult survival independent of reproduction is very poor.

26. Large numbers of small offspring are favored by natural selection
a. if large offspring survive so much better than small offspring that more of them survive to reproductive maturity.
b. in environments where competition among members of the same species is intense.
c. in fluctuating environments where disturbances create unoccupied sites.
d. in populations that are always near the maximum number the habitat can support.
e. in mature communities where all of the available sites are occupied.

27. The reproductive value of an individual is greatest
  a. at birth.
  b. just before first reproduction.
  c. halfway through the reproductive stage.
  d. just after it is finished reproducing.
  e. just before death.

28. Genetic traits that are highly detrimental often do not appear until late in life. A possible explanation for this is that
  a. young individuals are physiologically strong and able to mask the effects of detrimental genetic traits. As an individual ages and becomes physiologically weaker, these traits start to be expressed.
  b. mutations are occurring throughout an individual's lifetime. The probability that an individual will have a highly detrimental mutation increases with age.
  c. traits that enhance the fitness of young individuals become detrimental to older individuals that have different ecological and physiological needs.
  d. reproduction decreases the energy available for the maintenance and growth of the parent. Because older individuals are putting so much energy into reproduction, they are unable to repair the damage caused by genetically deleterious traits.
  e. natural selection tends to favor the delay of the phenotypic expression of detrimental traits until individuals have completed most or all of their reproduction.

29. If individuals frequently move between subpopulations, immigrants may prevent declining subpopulations from becoming extinct. This process is known as the __________ effect.
  a. emigration
  b. migration
  c. rescue
  d. dilution
  e. population

*30. Which of the following is an example of density-independent population regulation?
  a. A contagious disease sweeps through a dense population of lemmings.
  b. Jaegers (predators of lemmings) search widely for places where lemmings are abundant and concentrate their hunting in those areas.
  c. An outbreak of lemmings leads to a depletion of the food supply. As a result lemmings are underfed and produce few offspring.
  d. Arctic foxes switch from primarily eating mice to primarily eating lemmings when lemmings are abundant.
  e. An early cold spell kills 80 percent of the lemmings in a particular area.

*31. A major difference between populations with density-dependent birth or death rates, or both, and populations with density-independent birth and death rates is that density-dependent populations
  a. have an equilibrium population size; density-independent populations do not have an equilibrium population size.
  b. grow more rapidly than density-independent populations.
  c. never actually occur in nature, whereas density-independent populations are frequently found in nature.
  d. are more likely to go extinct than density-independent populations.
  e. are more likely to live in disturbed habitats than density-independent populations.

32. Organisms are likely to have adaptations for tolerating a particular kind of disturbance if the disturbance is
  a. unusually severe.
  b. rare and occurs at highly unpredictable intervals.
  c. frequent relative to the organisms' life span.
  d. of very recent human origin.
  e. unusual in its timing or location.

33. Which of the following statements is *true* concerning fires as a disturbance?
  a. No plants have successfully adapted to fire as a form of disturbance.
  b. Fires normally destroy all of the seeds of the plants growing in the area.
  c. Fires are extremely unpredictable.
  d. Fires are a recent form of disturbance because they did not occur before humans evolved.
  e. The frequency of fires may be proportional to the rate at which vegetation accumulates as fuel.

*34. If we wish to maximize the number of individuals in a deer population so that many deer can be harvested, we should manage the populations so that they
  a. are at the number the environment can support to maximize the number of individuals available.
  b. are far enough below the number the environment can support to have high birth and growth rates.
  c. are very rare and don't come into contact with each other.
  d. slightly exceed the number the environment can support so that the excess can be harvested.
  e. greatly exceed the number the environment can support to maximize the number of excess individuals.

35. A species of plants has many genetically identical parts that are grouped in one location and are derived from one fertilized egg. This type organization is called a
  a. cohort.
  b. group.
  c. module.
  d. population.
  e. None of the above

▪36. Which would be the most effective way to minimize a rat population in an alley?
  a. Kill as many rats as possible by poisoning and trapping.
  b. Clean up the alley so that the rats have no garbage to feed on.
  c. Lure the rats away to another site where they will be less harmful.

d. Search out and kill very young prereproductive rats.
e. Release cats into the alley.

37. Preserving a rare, endangered species will usually be expensive unless we
a. understand all aspects of its biology and reproduction.
b. bring the species into zoos and manage it in a controlled setting.
c. continually provide additional resources such as food.
d. can make it economically profitable by harvesting the species as a food source.
e. provide it with sufficient suitable habitat.

▪38. The efts, or young, of the red-spotted newt (a salamander that breeds in small ponds in the Appalachian Mountains) sometimes emigrate from their pond of birth to find a new pond. Efts apparently travel long distances because they have colonized most Appalachian ponds. When studying the population dynamics of the red-spotted newt, the appropriate scale of study is
a. a portion of a single pond where a few of the adults habitually lay eggs.
b. a single pond.
c. a few widely spaced ponds.
d. all (or most) of the ponds within a large area.
e. all (or most) of the ponds in the United States.

39. Desert plants often compete for water. Young plants germinating near a mature plant normally die because they cannot compete successfully for water. Plants that germinate far away from a mature plant are much more successful and often grow to maturity. Based on this, you would expect the mature plants to be
a. randomly scattered.
b. aggregated into clumps.
c. evenly spaced.
d. aggregated into clumps that are evenly spaced.
e. aggregated into clumps that are randomly scattered.

40. When a time delay is added to the logistic population growth equation, the population
a. growth rate will gradually slow down until it reaches the number the environment can support, at which time the population will stop growing.
b. will grow rapidly until it reaches the number the environment can support, then abruptly stop growing.
c. will slowly decline and eventually go extinct.
d. will grow without limits.
e. may shoot past the number the environment can support, and eventually fluctuate around that number.

41. Which of the following would *not* contribute to a clumped pattern of distribution?
a. Offspring settling close to their place of birth
b. Competition among individuals for resources
c. Chance
d. The distribution of suitable food resources
e. The location of favorable habitat

▪42. The analysis of population dynamics is simpler for unitary organisms than for modular organisms because
a. unitary organisms are constructed of repeated units.
b. unitary organisms are more likely to be sessile.
c. birth rates are higher for modular organisms.
d. individuals of unitary organisms are more easily distinguished.
e. the genetic structure of populations of modular organisms is more complex.

*43. A straight line relationship with a negative slope between age and percent survival describes a survivorship pattern in which
a. most individuals die at about the same age.
b. young individuals have the highest death rates.
c. the probability of death is about the same for all ages.
d. the majority of individuals survive to old age.
e. death rates increase following reproduction.

*44. What pattern of survivorship is most likely for an organism such as the sea urchin, which produces large numbers of very small offspring and provides no parental care?
a. Most individuals survive for most of their potential life span and then die simultaneously.
b. The probability of death is low throughout the life span.
c. Survivorship is high early in life and declines slowly.
d. Young individuals have high probability of dying, but older individuals have relatively low probability of dying.
e. Probability of death is equally high throughout the potential life span.

45. The growth rate of a population in an unlimited environment
a. is described by an S-shaped curve.
b. is greater than the intrinsic rate of increase.
c. is equal to $(b - d)N$.
d. depends on competition for resources.
e. is infinite.

46. Which of the following is *not* an assumption of the logistic model for population growth?
a. Members of the population compete for resources at high density.
b. All individuals have the same effect on population growth.
c. Population growth rate increases as the size of the population approaches the carrying capacity.
d. There is no time lag between birth and the time an organism begins to affect population growth.
e. The environment imposes a carrying capacity on the population.

47. Which of the following is *not* a likely result of devoting more resources to reproduction?
   a. Increased number of offspring produced
   b. Increased probability of death for the parent
   c. Increased growth rate for the parent
   d. Increased probability of survival for the offspring
   e. Decreased probability of survival for the parent

48–51. Use the graph below to answer the following questions.

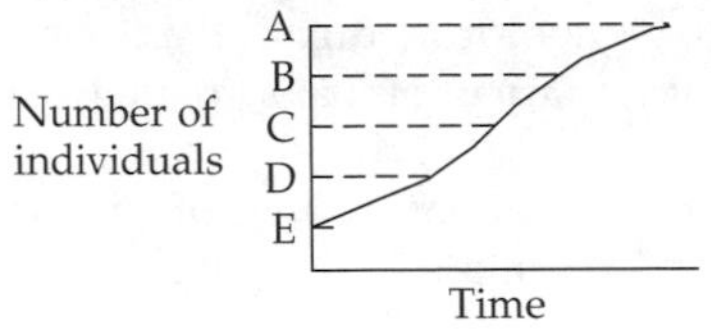

48. At what population size is the growth rate of this population greatest?
   a. A
   b. B
   c. C
   d. D
   e. E

49. At what population size is the term $(K - N)/K$ largest?
   a. A
   b. B
   c. C
   d. D
   e. E

50. At what population size is the rate of population growth beginning to slow down?
   a. A
   b. B
   c. C
   d. D
   e. E

51. What would be the most effective practical means of managing this population to maximize the number of individuals harvested over a long period of time?
   a. Increase the carrying capacity
   b. Decrease the carrying capacity
   c. Harvest until the population size is equal to B
   d. Harvest until the population size is equal to C
   e. Increase r

52. Which of the following is *true* of the carrying capacity (K)?
   a. When $N = K$, the birth rate in a population is zero.
   b. The rate of population growth in an unlimited environment is proportional to K.
   c. K is always determined by the amount of food in an environment.
   d. In a population at its K, the birth rate equals the death rate.
   e. K changes over time for each population.

53. Which of the following would be *unlikely* to contribute to large fluctuations in population size?
   a. Frequent disturbance
   b. A high death rate
   c. A density-dependent birth rate
   d. A high birth rate
   e. Seasonal variation in temperature

54. A population of trees consists of large numbers of seedlings and saplings and a few reproducing adults. Which of the following is likely to be true for this population?
   a. The birth rate is density-dependent.
   b. The population is close to K.
   c. The death rate is low.
   d. Birth and death rates of young are high.
   e. Its members are evenly spaced.

55. Which of the following is *not* true of dispersal?
   a. Individuals may disperse as either juveniles or adults.
   b. All organisms have a dispersal stage in their life history.
   c. Dispersal can be an adaptation to disturbance.
   d. Organisms disperse only once in their life cycle.
   e. Organisms can disperse in time as well as in space.

56. A graph showing the proportion of organisms of an initial group alive at different times throughout the life span is a(n)
   a. cohort.
   b. life table.
   c. survivorship curve.
   d. age structure diagram.
   e. population growth curve.

57–59. Use the following life table for a hypothetical cohort of organisms to answer the questions below.

Age interval (yr)	**Number alive at beginning**	**Number dying**	**Death rate**	**Number of offspring**
0–1	200	150	0.75	0
1–2	50	10	0.20	0
2–3	40	20	0.50	3
3–4	20	(?)	0.50	9
4–5	10	10	1.00	4
5–6	0	—	—	—

57. During which of the following age intervals does this organism have the highest risk of mortality?
   a. 0–1
   b. 1–2
   c. 2–3
   d. 3–4
   e. Cannot tell from the information in the table

58. How many individuals died during the interval 3–4?
   a. 5
   b. 10
   c. 20
   d. 40
   e. 50

59. From the life table, you can conclude that
 a. individuals begin reproducing in their first year of life.
 b. the oldest individuals have the largest clutch size.
 c. on average, clutch size is larger for age interval 3–4 than for the interval 4–5.
 d. clutch size increases continuously with age.
 e. individuals of all ages produce about the same number of offspring.

60. The intrinsic rate of increase is
 a. the rate of population growth in a limited environment.
 b. the maximum possible birth rate for a population.
 c. average birth rate minus the average death rate under optimal conditions.
 d. the average clutch size for individuals in a population in an unlimited environment.
 e. the rate at which a population colonizes a favorable new habitat.

61. A delay between the time an organism is born and the time it begins to affect other members of the population causes
 a. population size to oscillate around the carrying capacity.
 b. correlation between birth rate and death rate in the population.
 c. a decrease in the carrying capacity of the population.
 d. fluctuation in the intrinsic rate of increase.
 e. low mortality rates of the young.

62. A population whose dynamics are dominated by periodic recovery from crashes will
 a. have a lower carrying capacity than a population not subject to crashes.
 b. be well below carrying capacity most of the time.
 c. live in a very stable environment.
 d. never reach exponential growth rates.

▪63. Many fish populations can be heavily harvested on a sustained basis because
 a. there are many "excess" individuals since populations frequently overshoot their carrying capacity.
 b. they are not limited by the availability of suitable habitat.
 c. only a small number of females are needed to produce enough eggs to maintain population size.
 d. natural sources of mortality are low because their natural predators have gone extinct.
 e. they each reproduce only once and would have died soon if not harvested.

▪64. A population of seed-eating rodents infests your barn and consumes the corn stored there. Which of the following would likely be the most successful way to control the population of these pests?
 a. Leave corn out in the surrounding fields to lure the rodents away from the barn.
 b. Set large numbers of traps to kill as many rodents as possible.
 c. Introduce a voracious predator.
 d. Introduce a corn-eating competitor.
 e. Store grain in rodent-proof bins to reduce their access to corn and reduce their carrying capacity.

65. Growth of the human population from prehistoric times to the present
 a. roughly fits an exponential curve.
 b. exhibits large oscillations around *K*.
 c. follows an S-shaped curve.
 d. has slowed considerably over the last 50 years.
 e. has been strongly influenced by density-dependent regulation.

66. An example of a trade-off would be
 a. Pacific salmon, they reproduce only once and then die.
 b. the larvae of red-spotted newts, they transform into efts and live on land before returning to the water to reproduce.
 c. Dall sheep, they have low survivorship of young individuals, then survivorship increases in the middle of the life span and decreases again at the end of the life span.
 d. humans and whales, they both take many years to come to reproductive age but survival rates for the adults are higher than animals that reproduce at a younger age.
 e. the introduction of predators to reduce a population to increase its carrying capacity.

67. Currently the most important carrying capacity of Earth for humans is set by
 a. limitations on food resources.
 b. the ability to absorb CO_2 from fossil fuel consumption.
 c. disease.
 d. predation.
 e. space limitations.

Study Guide Questions

1. From an ecologist's point of view, population structure does *not* include the
 a. distribution of genotypes within a population.
 b. population density.
 c. spacing of population members.
 d. biomass of the population.

2. A random distribution of population members is usually caused by
 a. settling near one's birthplace.
 b. territoriality.
 c. competition for a uniformly distributed resource.
 d. the interaction of several factors that affect survival.

3. In the following graph, which of the expressions (*a* through *d*) of the exponential growth equation should be *increased* in order for curve 1 to become more like curve 2?

$$\frac{\Delta N}{\Delta t} = (b - d)\, N$$

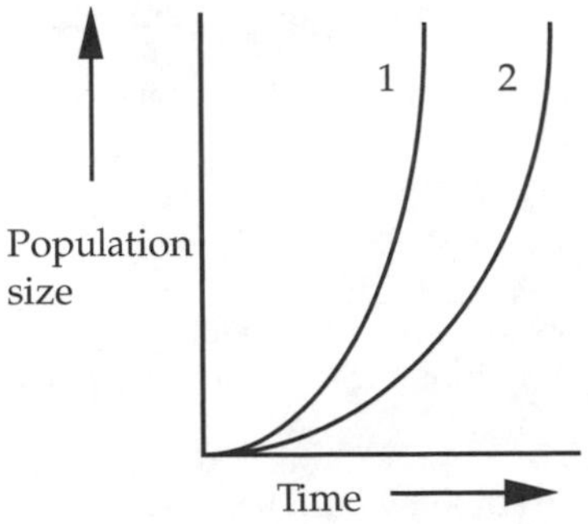

a. N
b. d
c. b
d. $(b - d)$

4. Species A and B have intrinsic rates of increase (r_{max}) of A = 0.25 and B = 0.50. In reference to the graph shown in question 3, the population growth curve for A should be more like curve **(2)**.

5. In the logistic population growth curve shown below, the rate of growth is greatest at which point? **(c)**

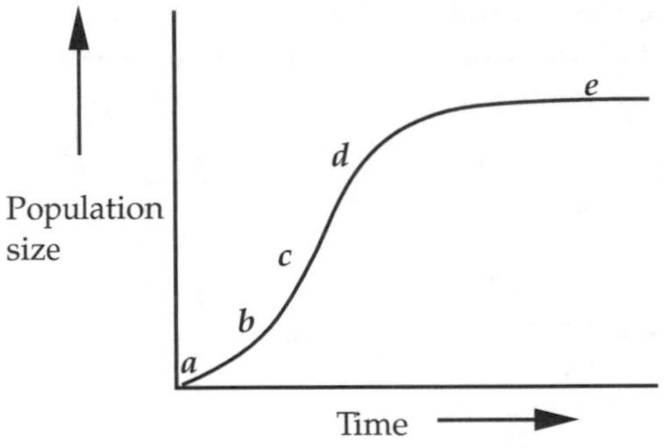

6. At which point in the graph shown above would there be zero population growth ($\Delta N/\Delta t = 0$)? **(e)**

7. Based on the following life table, during what time interval is survivorship greatest?

Age (years)	Number Alive
0	800
1	770
2	550
3	125
4	75
5	0

a. 0–1 years
b. 1–2 years
c. 2–3 years
d. 3–4 years
e. 4–5 years

8. Based on the graph shown below, which of the following events is *least likely* to be true of this population?

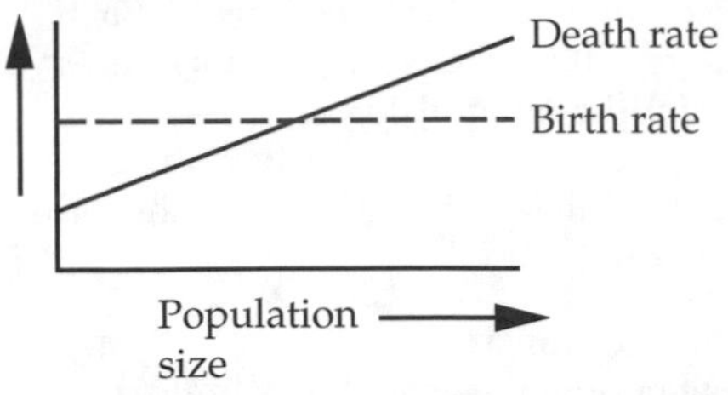

a. Parasites spread between population members.
b. Increased competition for food causes some individuals to delay reproduction.
c. The number of predators in the area varies with population density.
d. Territorial disputes led to injury and death of some males.

9. In the graph shown in Question 8, the population size where the curves for the birth and death rates intersect is
a. an estimate of the carrying capacity.
b. an estimate of the intrinsic rate of increase.
c. the point at which density-dependent regulation begins.
d. the point at which density-independent regulation begins.

10. Which of the following would probably *not* be true of a population whose dynamics are primarily influenced by density-independent factors?
a. The population's growth pattern is similar to the logistic growth curve.
b. The birth rate of the population is dependent on the nutritional status of its adult females.
c. The most important source of mortality in the population is unfavorable weather conditions.
d. a and b

11. For a population that is stable in size, the following age distribution indicates that

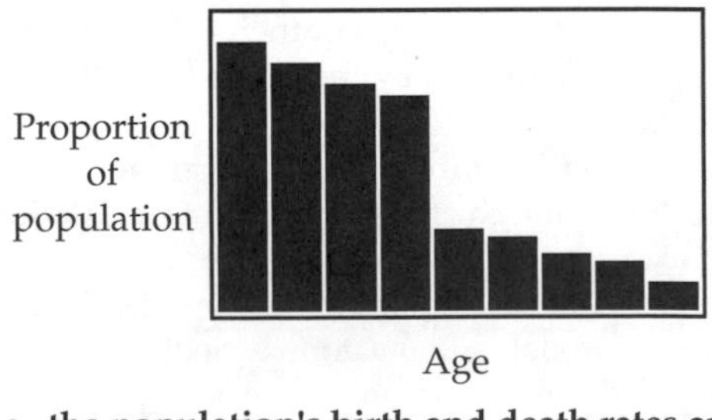

a. the population's birth and death rates are both high.
b. the population's birth and death rates are both low.
c. the population's birth rate is low but its death rate is high.
d. the population's birth rate is high but its death rate is low.

12. Select all of the following that are *not* adaptations to seasonal changes in habitat quality.
a. Migration
b. Expansion of the species range

c. Production of seeds
d. Hibernation

13. The *best* way to decrease the population size of a pest species is to
a. poison it.
b. introduce additional predators.
c. decrease the carrying capacity of the habitat for the species.
d. add competitors.

14. The number of seeds that a tree produces increases, but the resources available to the tree remain the same. Put a "+", "–", or "0" before each of the following variables to express the expected effect of this change on the variable.
(–) Weight of each seed
(–) Width of the tree's growth rings

15. In which of the following life-history stages are annual plants and Pacific salmon most similar?
a. Growth
b. Dispersal
c. Reproduction
d. Senescence

16. For a species that breeds only during one season and lives in a disturbed environment where new habitat is constantly being created, it would be best to produce
a. many small offspring only once.
b. a few small offspring only once.
c. many small offspring more than once.
d. many large offspring more than once.

End of Chapter Questions

1. The number of individuals of a species per unit of area is known as its
a. population size.
b. population density.
c. population structure.
d. subpopulation.
e. biomass.

2. The age distribution of a population is determined by
a. the timing of births.
b. the timing of deaths.
c. the timing of both births and deaths.
d. the rate at which the population is growing.
e. all of the above

3. Which of the following is not a demographic event?
a. Growth
b. Birth
c. Death
d. Immigration
e. Emigration

4. A group of individuals born at the same time is known as a
a. deme.
b. subpopulation.
c. Mendelian population.
d. cohort.
e. taxon.

5. A population grows at a rate closest to its intrinsic rate of increase when
a. its birth rates are the highest.
b. its death rates are the lowest.
c. environmental conditions are optimal.
d. it is close to the environmental carrying capacity.
e. it is well below the environmental carrying capacity.

6. Immigrants that prevent a local population from becoming extinct engage and produce a result called
a. the colonization effect.
b. the rescue effect.
c. the metapopulation effect.
d. the genetic drift effect.
e. the salvage effect.

7. Some organisms reproduce only once in their lifetimes because they
a. invest so much in reproduction that they have insufficient reserves for survival.
b. produce so many offspring at one time that they do not need to survive longer.
c. don't have enough eggs to reproduce again.
d. don't have enough sperm to reproduce again.
e. have stopped growing.

8. Which of the following is *not* true of reproductive value?
a. Reproductive value is the average number of offspring that remain to be born to an individual of a particular age.
b. The reproductive value of an individual increases until it begins to reproduce.
c. Reproductive value reaches its maximum when an individual completes reproduction.
d. Reproductive value usually reaches its maximum when an individual begins to reproduce.
e. Reproductive value usually declines during the reproductive life of an individual.

9. Density-dependent population regulation results when
a. only birth rates change in response to density.
b. only death rates change in response to density.
c. diseases spread in populations at all densities.
d. both birth and death rates change in response to density.
e. population densities fluctuate very little.

10. The best way to reduce the population of an undesirable species in the long term is to
a. reduce the carrying capacity of the environment for the species.
b. selectively kill reproducing adults.
c. selectively kill prereproductive individuals.
d. attempt to kill individuals of all ages.
e. sterilize individuals.

Student Web Site Self-Quiz Questions

1. Corals are examples of__________ organisms. The effect of such organisms on the environment is best expressed as __________.
   a. modular, shape and area of individuals
   b. modular, number of individuals per unit area
   c. modular, number of individuals per unit volume
   d. clumped, shape and area of individuals
   e. clumped, number of individuals per unit area

2. Spacing patterns tell much about populations. A random distribution of population members is usually caused by
   a. competition for a uniformly distributed resource.
   b. patchy habitat.
   c. settling near one's birthplace.
   d. territoriality.
   e. the interaction of several factors that affect survival.

3. Populations are often subject to density-dependent factors. Which of the following is *not* a density-dependent factor?
   a. Limiting food supply
   b. Harsh winters
   c. Predation
   d. Disease
   e. Territoriality

4. Which of the following statements regarding disturbances is *false*?
   a. Populations can influence the frequency of disturbances.
   b. A disturbance generally increases the carrying capacity of an environment.
   c. Tree falls are considered a type of biological disturbance.
   d. Hurricanes are considered a type of physical disturbance.
   e. Disturbances are short-term events.

5. Most life histories include stages for growth, reproduction, and dispersal. Which of the following is *not* a life history trait?
   a. Age of death
   b. The number of reproductive events
   c. Rate of locomotion
   d. Rate of growth
   e. Energy supply

6. In which one of the following life history stages are a Pacific salmon (*Onchorhyncus* sp.) and agaves of the American Southwest most similar?
   a. Growth
   b. Dispersal
   c. Metamorphosis
   d. Reproduction
   e. Senescence

7. The *least* effective way to reduce the population size of a pest species is to
   a. add competitors.
   b. decrease the carrying capacity of the habitat for it.
   c. introduce additional predators.
   d. poison it.
   e. sterilize females.

8. The growth of the human population showed steady, moderate increase until 200 years ago when the population "exploded." Which of the following has *not* contributed to the increase in the human population?
   a. Ability to artificially recycle CO_2
   b. Agricultural technology
   c. Modern medicine
   d. Plant and animal domestication
   e. Industrial revolution

★ *Indicates a **difficult** question*

■ *Indicates a **conceptual** question*

55 Community Ecology

Fill in the Blank

1. If one participant in an interaction is harmed but the other is unaffected, the interaction is **amensalism**.
2. If both partners in an interaction benefit, the interaction is **mutualism**.
3. If two organisms use the same resource, and the resource is insufficient to supply their combined needs, then the interaction is **competition**.
4. The organisms that live together in a particular area constitute a(n) **ecological community**.
5. An animal that is much smaller than its host, and may or may not kill it, is a(n) **parasite**.
6. A palatable species takes on the appearance of an inedible or noxious species. This is known as **Batesian mimicry**.
7. Two or more unpalatable species take on a similar (often bright) appearance. This is known as **Müllerian mimicry**.
8. When two species of *Paramecium* are placed together in a test tube, and one species drives the other to extinction, this is an example of **competitive exclusion**.
9. Growth on bare rock that at the beginning supports no organisms and ends up supporting a mature forest is **succession**.
10. When the dead body of a freshly killed mouse decomposes, the change in the community of decomposers in the body is called **secondary succession**.
11. Animals such as cows that eat only plant tissues are called **herbivores**.
12. Species that have mutually influenced one another's evolution are said to have **coevolved**.
13. A(n) **ecological niche** is defined as the set of environmental conditions under which a species can live.
14. When a species has a larger influence on the community than is expected by its abundance, it is called a **keystone** species.
15. Animals that eat other animals are called **predators**.
16. **Forbs** are small broad-leaved plants.

Multiple Choice

1. The relationship between a human and the fungus that causes athlete's foot is __________ when the fungus feeds only on dead skin cells, but becomes (a) __________ if the fungus penetrates the skin and feeds on living cells.
   a. amensalistic, commensalistic
   b. commensalistic, amensalistic
   c. a host–parasite interaction, commensalistic
   d. amensalistic, a host–parasite interaction
   e. commensalistic, a host–parasite interaction

2. Intraspecific competition may affect individuals in all of the following ways *except* via
   a. a reduced growth rate.
   b. a lower reproductive rate.
   c. reduced requirement for resources.
   d. exclusion from better habitats.
   e. death.

■3. Two mouse species are observed to inhabit the same forested area. Which of the following is *not* a possible reason why the two species are able to coexist?
   a. One species eats mostly berries, and the other eats mostly nuts.
   b. Predation by hawks prevents both mouse population from becoming large.
   c. The forest environment is very heterogeneous.
   d. The preferred habitat for both species is fields, but a third mouse species has excluded them from the fields.
   e. The two species nest at different times of the year.

4. Which of the following is *not* true of the compound organisms called lichens?
   a. One member provides nutrients and support.
   b. One member is photosynthetic.
   c. They are partly composed of cyanobacteria or green algae.
   d. The relationship of the two members is mutualistic.
   e. They are restricted to favorable habitats like tropical rainforests.

5. Which characteristic of moose is most significant in determining that moose are a "keystone species"?
   a. A few moose have a large effect on forest succession.
   b. Moose are the largest animals in the ecosystem.
   c. Moose browse on a large variety of tree species.
   d. Moose use both terrestrial and aquatic habitats.
   e. Moose care for their young longer than any other species in the ecosystem.

6. If the removal of a species from a community has a greater effect on the structure and functioning of the community than one would have predicted from the species' abundance, then that species is most likely a(n)
   a. keystone species.
   b. carnivore.
   c. herbivore.
   d. early successional species.
   e. mimic of another species.

7. The fact that many tree species and mycorrhizal fungi cannot survive unless they are associated with one another indicates that their relationship is
   a. amensalistic.
   b. commensalistic.
   c. competitive.
   d. mutualistic.
   e. a parasite–host interaction.

8. A female fig wasp enters the syconium of a fig, pollinates the flowers, and lays eggs in the ovaries of some of the flowers. The young larvae grow up, eat (and kill) some, but not all, of the seeds, and complete their life cycle. The fig is completely dependent on fig wasps to pollinate its flowers, and the fig wasp requires figs to complete its life cycle. The interaction between figs and fig wasps has aspects of
   a. mutualism.
   b. commensalism and mutualism.
   c. amensalism and host–parasite interaction.
   d. predator–prey and host–parasite interaction.
   e. a and d

*9. A species of wasp (not the fig wasp) lays its eggs in an already fertilized syconium of a fig. The wasp does this by inserting her long ovipositor through the wall of the syconium and laying her eggs in the developing seeds. The larvae of this wasp eat (and kill) the seeds of the fig. This wasp does not pollinate the fig or in any other manner benefit the fig. The interaction between figs and this wasp has aspects of
   a. mutualism.
   b. commensalism.
   c. amensalism.
   d. predator–prey or host–parasite interaction.
   e. competition.

10. Which of the following would *not* be considered a resource for a terrestrial animal?
    a. Food
    b. Humidity
    c. Hiding places
    d. Nest sites
    e. Water

*11. The barnacle *Chthamalus stellatus* is capable of growing at much lower depths than it is actually found in nature. Experimental studies have determined that the lower limit of *Chthamalus* is determined by another barnacle, *Balanus balanoides*. The rapidly growing *Balanus* is able to smother, crush, or undercut the slower-growing *Chthamalus*. Because of this, *Chthamalus* cannot survive in the lower depths where *Balanus* is found. The interaction between these two barnacles is an example of
    a. predator–prey interaction.
    b. mutualism.
    c. commensalism.
    d. amensalism.
    e. competition.

12. A resource required by all animals is oxygen. Which of the following statements is *true* concerning the effects of this factor on the distribution and abundance of animals?
    a. Because oxygen is such a fundamental requirement of all animals, it is critical in determining the distribution and abundance of terrestrial animals.
    b. Oxygen is unlikely to be an important factor in the structure of terrestrial communities because it is nearly always available at levels greater than the minimum required for survival.
    c. Oxygen is unlikely to be an important factor in the structure of aquatic communities because it is nearly always available at levels greater than the minimum required for survival.
    d. In terrestrial communities, many factors, including oxygen levels, affect the distribution and abundance of animals. The structure of terrestrial communities is in part, but not completely, determined by oxygen levels.
    e. In terrestrial communities, oxygen levels become an important factor in determining community structure only when animals are near the edges of their range. When animals are stressed by other factors, variation in oxygen levels becomes a critical factor.

13. In a homogeneous environment, when prey are above a certain threshold, predators can find enough to eat. As a result of this, they will produce more offspring, which will consume more prey. Eventually this will drive the prey numbers below a threshold value, and the predators will begin to starve. Nevertheless, the predators will continue to eat prey even when they are scarce, reducing the prey to an even lower level. Eventually some predators will starve, and their numbers will be reduced, allowing the numbers of prey to increase above the threshold where the predators can again reproduce. This process results in
    a. extinction of the prey, but not the predator.
    b. oscillations in the density of the prey only.

c. oscillations in the density of the predator only.
d. oscillations in the densities of both the predator and the prey.
e. stabilization of the predator and prey densities at an intermediate value.

14. Which of the following is *not* an example of an adaptation of prey to avoid predation?
a. Many tropical katydids look like dead leaves.
b. Male stickleback fish develop very bright colors during the breeding season.
c. Several moth caterpillars have sharp bristles that contain toxic chemicals.
d. Many moths have large eyespots that they flash when they are disturbed.
e. Gazelles (African antelopes) are very fast runners.

▪15. The edible viceroy butterfly looks like the inedible monarch butterfly. This similarity most likely evolved because
a. the coloration used by the viceroy butterfly and the monarch butterfly is ideal for absorbing heat on cold mornings.
b. these two butterflies are closely related and their similar coloration reflects a common ancestry.
c. there are only a limited number of color patterns a butterfly can have. It is highly likely that there will be two butterflies with similar color patterns simply by random chance.
d. the larvae of these two butterflies live on the same plant. The color pattern is a consequence of the food they eat.
e. predators frequently mistake viceroy butterflies for monarch butterflies, which they have learned to avoid.

16. Which of the following is *not* true of parasites?
a. The more slowly they kill their host, the better their chances to be transferred to another host.
b. They can only be transferred from one living host to another living host.
c. If a parasite grows too slowly in its host, it could be outcompeted by another parasite.
d. Parasites can have more than one species of host.
e. Parasites do not always kill their host.

17. Which of the following is *not* a chemical defense used by plants as a defense against herbivory?
a. Chemicals that interfere with the transmission of nerve impulses to muscle
b. Chemicals that imitate insect hormones and thereby block insect metamorphosis
c. Unusual amino acids that become incorporated into proteins and interfere with their functioning
d. Chemicals that are extremely difficult to digest
e. Odorous chemicals that attract pollinating insects

▪18. Why would it be necessary to examine more than just direct effects of two organisms on one another?
a. The organisms that facilitate the indirect effect are often unknown.
b. Indirect effects are not important when examining species–species interactions.
c. Indirect effects can alter species–species interactions.
d. Indirect effects are always positive for predators.
e. None of the above

19. An example of species-specific coevolution is
a. yucca plants and the single species of moth that pollinates them.
b. egrets feeding near water buffalo.
c. ants "milking" aphids for honeydew.
d. tropical butterflies which look more similar to each other after each generation.
e. b. and d

▪20. The sea star *Pisaster ochraceous* is an abundant predator on the rocky intertidal communities on the Pacific coast of North America. The sea star feeds preferentially on the mussel *Mytilus californianus*. In the absence of the sea star, the mussel is a dominant competitor. Predation by the sea star, however, creates bare areas. For this reason,
a. the number of species present will be changed by the presence of the sea star, but the direction of change cannot be predicted.
b. the variety of species should be unaffected by the presence or absence of the sea star.
c. the variety of species present should be greater when the sea star is present than when it is absent.
d. the variety of species present should be less when the sea star is present than when it is absent.
e. *Mytilus* should go extinct when the sea star is present.

21. Cattle egrets follow cattle around. By doing so, they are able to capture insects more efficiently, because the cattle disturb the insects as they walk, making them easier to catch. There is no cost or benefit to the cattle from this interaction. This interaction is an example of
a. commensalism.
b. amensalism.
c. mutualism.
d. parasitism.
e. competition.

22. Legumes, such as soybeans, form root nodules that become infected by *Rhizobium* bacteria. These bacteria convert nitrogen into a form that is usable by plants. The plants benefit because they have a source of nitrates, which are frequently limited in terrestrial environments. The bacteria benefit because they receive nutrients and energy from the plants. This interaction is an example of
a. commensalism.
b. amensalism.
c. mutualism.
d. parasitism.
e. predation.

*23. Which of the following is *not* an example of mutualism?
a. Ants protect aphids from predators and "milk" the plant-sucking insects for honeydew (partly digested plant sap).
b. Cleaner wrasse (a fish) enter the mouths of barracuda (a predatory fish) to remove parasites, which the wrasse eat.
c. Staphilinid beetles live inside of ant colonies and eat ant eggs, larvae, and pupae. They avoid being killed by the ants by feeling and smelling like ants.
d. Honeyguides (birds that eat beeswax) guide humans to bees' nests. The humans chop open the nests, take the honey, and leave the wax for the honeyguide.
e. Termites are able to eat wood because they have protists living in the protected environment of their guts that help them digest cellulose.

24. Which of the following is an example of diffuse coevolution?
a. Figs are pollinated only by fig wasps, which lay their eggs only in fig syconia.
b. *Pseudomyrmex* ants live in the thorns of *Acacia cornigera* and protect the acacias from herbivorous insects, mammals, and vines.
c. The edible viceroy butterfly looks like the inedible monarch butterfly.
d. Most bird-pollinated flowers are red, a color that attracts many birds.
e. *Paramecium bursaria* and *P. aurelia* are able to coexist in the same test tube if there is deoxygenated water at the bottom of the container.

25. Which of the following could *not* be used as an example of succession?
a. A landslide exposes bare rock. Eventually a forest is reestablished on the site.
b. An alien weed is introduced into the Hawaiian Islands. It eventually displaces a native plant, driving the native plant to extinction.
c. A freshly killed mouse decomposes.
d. A tree falls in the forest. Eventually a new tree replaces it.
e. The leaf litter under a pine tree is in layers, with freshly fallen litter at the top and more decomposed layers deeper down.

26. In __________ evolution, species traits are influenced by interactions with a wide variety of predators, parasites, prey, or mutualists.
a. dilute
b. general
c. regional
d. diffuse
e. species-specific

27. Keystone species may influence
a. species richness of an ecosystem.
b. the flow of energy and materials in an ecosystem.
c. the environment of an ecosystem.
d. a and b
e. None of the above

28. In the tropics there are several species of inedible butterflies that all have virtually identical bright coloration. The most likely explanation for the formation of these groups of similar-looking species is that
a. all of these similar-looking species are closely related. They are all inedible, and all look similar because they are closely related.
b. the coloration pattern is cryptic and allows the butterflies to hide. They all live in similar habitat so they look similar.
c. inedible species benefit from looking like other inedible species because predators rapidly learn to avoid all similarly colored species when they eat any member of the group.
d. all of these species have similar means of locating mates, and the coloration is an important aspect of this process.
e. all of these species live on the same host plant, and the coloration pattern results from the plant that they use as a food source.

29. In the ecological relationship between yucca plants and *Tegeticula* moths, the moth lays eggs in the yucca flowers
a. but does not associate with the yucca plant directly.
b. and kills the yucca plant.
c. and pollinates the yucca plant.
d. or in the flowers of many other species.
e. and defends the yucca against herbivores.

30. A community ecologist would most likely be concerned with
a. energy flow through an ecosystem.
b. population growth of a single species.
c. interactions between individuals of the same species living together in a small area.
d. interactions between individuals of different species living together in a small area.
e. the cycling of matter through biotic and abiotic components of an area.

31–32. Certain woodpecker-like African birds have become specialized for removing and eating ticks and their parasitic insects from the bodies of large herbivores.

31. The relationship between the birds and the ticks is an example of
a. mutualism.
b. parasitism.
c. commensalism.
d. predator–prey interaction.
e. amensalism.

32. The relationship between the birds and the herbivores is an example of
a. mutualism.
b. parasitism.
c. commensalism.
d. predator–prey interaction.
e. amensalism.

33. Elephants and other large herbivores trample many species of plants that are different from the plant species they use as food. The relationship between the elephants and the trampled plant species is an example of
   a. mutualism.
   b. parasitism.
   c. commensalism.
   d. predation.
   e. amensalism.

34. When few prey are available, predator populations generally decrease. If both predator and prey populations increase to previous levels, what is the order of events?
   a. Prey numbers increase, followed by predator numbers.
   b. Predator numbers increase, followed by prey numbers.
   c. Predator and prey numbers increase simultaneously.
   d. Prey numbers increase, predator numbers do not increase, and the predators store energy.
   e. None of the above

35. __________ compounds are chemicals contained in plants that aid in their defense from herbivores.
   a. Irritant
   b. Primary
   c. Secondary
   d. Tertiary
   e. Defense

36. Which of the following factors would be *least* important in disrupting predator–prey oscillations?
   a. Density-dependent interactions between prey and predator
   b. Spatial patchiness of the habitat
   c. Temporal patchiness of the habitat
   d. A highly motile prey
   e. The introduction of an additional prey species

37. In a typical Batesian mimicry system, the system is stable only if
   a. the mimic and model have a dissimilar appearance.
   b. the predator learns to distinguish mimic from model.
   c. the mimic converges on the model faster than the model diverges due to directional selection.
   d. the mimic converges on the model more slowly than the model diverges due to directional selection.
   e. both mimic and model become distasteful.

38. Müllerian mimicry systems make it __________ difficult for the predator to learn which prey species are distasteful or dangerous, and the relative abundance of the various prey species __________.
   a. more, does not matter
   b. less, does not matter
   c. more, matters
   d. less, matters
   e. no more or less, does not matter

39. Which of the following is *not* an example of a defense used by plants against grazers or browsers?
   a. Production of tannins
   b. Increased production of nonwoody tissue
   c. Production of nicotine
   d. Production of hairs and spines
   e. Production of hormone-like chemicals that interfere with insect metamorphosis

40. What is the type of ecological relationship that can involve either members of the same or different species and in which both participants are harmed?
   a. Mutualism
   b. Parasitism
   c. Competition
   d. Predation
   e. Amensalism

41. Which of the following was *not* usually observed by G. F. Gause in his studies on interspecific competition?
   a. Competitive exclusion occurred unless environmental heterogeneity was available.
   b. Species usually had greater growth when alone.
   c. Some species that grew very rapidly in a particular environment would be completely eliminated when grown with another species.
   d. The outcome of competition was independent of the environmental conditions in the culture.
   e. Populations grew exponentially when alone.

42. Many tropical orchid species grow on the surface of tree branches without harming or benefiting the tree. This orchid–tree relationship is an example of
   a. mutualism.
   b. parasitism.
   c. commensalism.
   d. predation.
   e. amensalism.

43. The endosymbiosis hypothesis for the evolution of eukaryotic cells, which states that modern mitochondria and chloroplasts descended from once free-living bacterial ancestors that infected larger cells, may describe a very early example of
   a. mutualism.
   b. parasitism.
   c. interspecific competition.
   d. predation.
   e. amensalism.

*44. Which of the following relationships is *not* an example of mutualism?
   a. *Rhizobium* bacteria live in the protected environment of legume roots and provide the legumes with nitrogen.
   b. A prawn cleans a coral reef fish.
   c. A plant produces "hitchhiker" fruit that are dispersed by a passing animal.
   d. *Pseudomyrmex* ants live in the hollow spines of an *Acacia* tree.
   e. A plant produces fleshy fruits with digestion-resistant seeds that are eaten by an animal.

45. _________ competition occurs when a species reduces a shared resource.
a. **Exploitative**
b. Interference
c. Diffuse
d. Random
e. Exclusion

46. Which characteristic is *not* true of ecological succession?
a. The species composition of the community changes through time.
b. The physical conditions of the community change through time.
c. **The rate of change is constant through time.**
d. Some species disappear quickly; others are more long-lived.
e. The community becomes more complex through time.

47. Acute toxins are chemicals made by plants that
a. can cause hallucinations.
b. are also known as secondary compounds.
c. disrupt herbivore metabolism.
d. can be toxic to bacteria.
e. **All of the above**

▪48. When a cigarette is smoked, the user obtains nicotine from the tobacco. Why does the tobacco plant produce nicotine?
a. For the benefit of humans who smoke
b. **To reduce grazing by herbivores**
c. To make the tobacco leaf more digestible
d. To combat tobacco mosaic virus
e. b and d

49. Which of the following is *not* true of an ecological niche?
a. Predators nearby can narrow a particular species' niche.
b. A species' niche can grow or shrink according to conditions.
c. **The area of a species' niche is not determined by the presence of another species.**
d. Physical and biotic factors can determine the area of a species' niche.
e. None of the above

▪50. Laboratory competition experiments often cannot fully imitate what happens in nature because
a. **interactions with other species not present can affect the results.**
b. resources found in nature cannot be controlled in the lab.
c. only interspecific, not intraspecific, competition happens in the lab.
d. a and b
e. All of the above

51. Which of the following is an example of an indirect effect of species interactions?
a. Lynx drive snowshoe hares to extinction and then die from starvation themselves.
b. Some birds cannot survive because all available nest sites are taken.
c. The barnacle *Chthamalus* has a larger niche when the barnacle *Balanus* is absent.
d. Fig wasps can only reproduce in one species of fig.
e. **The number of cases of hantavirus increase after a heavy winter snow, making pinecones, which are food for the virus-carrying deer mice, more abundant.**

52. In terrestrial communities, the major modifiers of the physical environment are
a. **plants.**
b. vertebrate animals.
c. invertebrate animals.
d. protists.
e. fungi.

53. Populations of a predator and its prey are found on two islands. On the first island, suitable habitat for the prey is uniformly distributed throughout the island, and the prey is found as a single uniform population. On the second island, suitable habitat for the prey is patchily distributed, and the prey is found in isolated patches. All else being equal, you would expect
a. the densities of the prey and predator populations to be more stable through time on the first (uniform) island than on the second (patchy) island.
b. **the densities of the prey and predator populations to be more stable on the second (patchy) island than on the first (uniform) island.**
c. the densities of the prey populations to be more stable on the second (patchy) island, and the densities of the predator populations to be more stable on the first (uniform) island.
d. both islands to exhibit approximately equally strong oscillations in the predator and prey populations.
e. the first (uniform) island to have very stable prey and predator populations, and the second (patchy) island to have strong fluctuations.

54. Resources whose supply is less than the demand made upon them are called _________ resources.
a. niche
b. **limiting**
c. primary
d. competitive
e. intraspecific

55. In experiments with two species of the plant *Desmodium*, when both were grown together, *D. glutinosum* grew better than *D. nudiforum*. When they were grown separately, however, *D. nudiforum* grew better than *D. glutinosum*. This shows that
a. *D. glutinosum* was more depressed by interspecific competition than was *D. nudiforum*.
b. *D. glutinosum* was unaffected by intraspecific competition.
c. ***D. nudiforum* was more depressed by interspecific competition than was *D. glutinosum*.**

d. *D. nudiforum* was more depressed by intraspecific competition than was *D. glutinosum*.
e. Both species were equally depressed by intraspecific competition.

56. In the experiment on predation of chorus frog tadpoles in ponds on islands in Lake Superior, which of the following was a conclusion?
a. Tadpoles are the only prey of dragonfly nymphs.
b. Dragonfly nymphs eat both tadpoles and dystiscid beetles.
c. Beetles and dragonfly nymphs eat different sizes of tadpoles.
d. Dragonfly nymphs can completely eliminate tadpoles from many ponds.
e. Researchers were able to determine which ponds dragonfly nymphs select to prey upon tadpoles.

Study Guide Questions

1. A bird eats the fruit of a plant species. The seeds are not digested and germinate in the bird's excrement at some distance from the parent plant. This is an example of
a. predation.
b. competition.
c. commensalism.
d. mutualism.

2. Certain birds follow swarms of foraging army ants and prey upon the insects that the ants flush. The relationship between these birds and the ants is an example of
a. competition.
b. commensalism.
c. mutualism.
d. a and b

3. According to the definition of resource developed in the text, which of the following would *not* be a resource?
a. Oxygen for a terrestrial animal
b. Sunlight for a plant species
c. Oxygen for an aquatic insect
d. Nesting boxes for a bluebird

4. As shown below, species A and B originally had nonoverlapping ranges. Species A expanded its range and now overlaps with species B in area 2. The two species still live separately in areas 1 and 3.

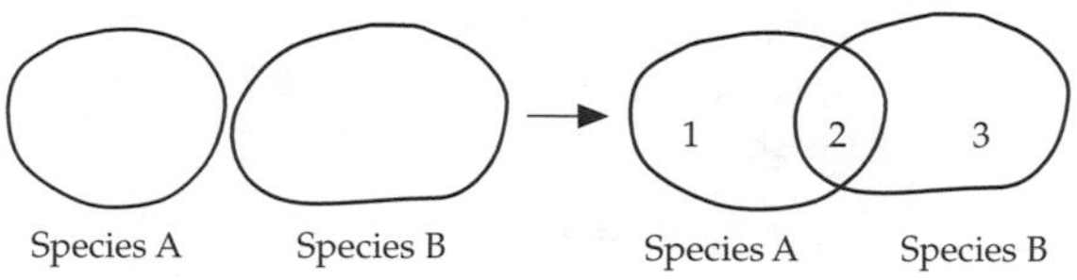

If the two species compete for common, limiting resources, which of the following statements is not a likely outcome of the range expansion of species A?
a. The range of conditions under which species B can exist will be greater in area 3 than in area 2.
b. The ecological niche of species B in area 2 would be the same as it had been before the range expansion.
c. The ecological niche of species A in area 1 would be the same as it had been before the range expansion.
d. The range of conditions under which species A can exist will be greater in area 1 than in area 2.

5. In laboratory experiments, two species of the protist *Paramecium* were grown alone and in the presence of the other species. The following graphs show growth of species 1 (left) and species 2 (right), both alone and when in mixed culture.

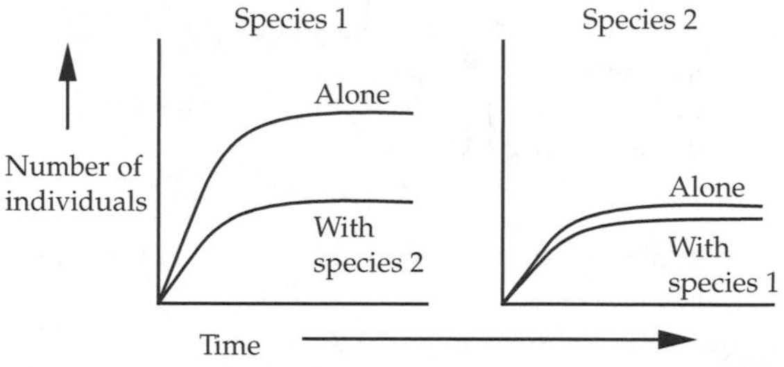

Interpretation of these graphs shows that
a. competitive exclusion occurred in these experiments.
b. both species are affected by interspecific competition but species 1 is affected less.
c. both species are affected by interspecific competition but species 2 is affected less.
d. both species are affected equally by interspecific competition.

6. Which vertical line in the following graph represents the time at which the predator population (gray) is *increasing* and the prey population (black) is *decreasing*? **(c)**

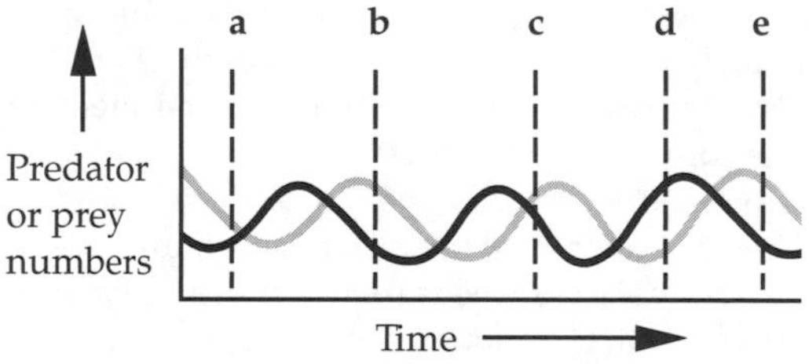

7. You are studying three butterfly species (1, 2, 3) that have very similar, bright coloration patterns. You collect the following data on the abundance of these species in two different regions, A and B. Based on your data shown below, which of the following conclusions is *most probable*?

Region	**Numbers collected for species:**		
	1	**2**	**3**
A	90	20	120
B	190	35	155

a. Species 1 is the model, species 2 and 3 are Batesian mimics.
b. Species 1 is a Batesian mimic, species 2 and 3 are Müllerian mimics.
c. Species 1 and 3 are Müllerian mimics, species 2 is a Batesian mimic.
d. All three species are Müllerian mimics.

8. If a species of plant that is trampled by an animal eventually evolves sharp spines that prevent trampling, we can say that its association with the animal has changed from
   a. amensalism to competition.
   b. amensalism to commensalism.
   c. commensalism to competition.
   d. commensalism to mutualism.

9. Which of the following is *not* an influence that plants normally have on biological communities?
   a. Moderating the climate
   b. Determining which animal species will be present
   c. Increasing CO_2 levels
   d. Reducing moisture reaching the soil

10. Which of the following is most characteristic of a keystone species?
    a. Very abundant
    b. Removal having a great effect on community structure
    c. Herbivore
    d. Uniform distribution

11. If the following list of stages of ecological succession occurring on glacial moraines in Alaska were put in correct chronological sequence, which would be *third*?
    a. Arrival of willows and alders
    b. Arrival of lichens
    c. Arrival of conifers
    d. Increase in soil nitrogen content

12. Which of the following would represent primary succession?
    a. The development of an aquatic community in a newly excavated farm pond, eventually followed by the filling in of the pond with sediment and its conversion to a forest
    b. The gradual establishment of a forest following a fire that destroyed the previous community
    c. The invasion of weeds followed by shrubs and trees on an abandoned farm field
    d. a and b

End of Chapter Questions

1. Two organisms that use the same resources when those resources are in short supply are said to be
   a. predators.
   b. competitors.
   c. mutualists.
   d. commensalists.
   e. amensalists.

2. Which of the following is not a resource?
   a. Food
   b. Space
   c. Hiding places
   d. Nest sites
   e. Temperature

3. A species' potential distribution is the range of conditions under which it could survive if
   a. there were no predators or competitors.
   b. there were no predators or other negative influences.
   c. there were no competitors, predators, or disease-causing organisms.
   d. environmental conditions were ideal.
   e. the environment were fundamentally different.

4. An animal that is much smaller than its prey and which attacks it from the inside is called a
   a. predator.
   b. parasite.
   c. commensalist.
   d. competitor.
   e. parasitoid.

5. Which of the following factors tends to stabilize populations of predators and their prey?
   a. A high birth rate of the prey
   b. A high birth rate of the predator
   c. The ability of predators to further reduce prey when they are scarce
   d. The ability of predators to search widely for prey
   e. Environmental heterogeneity

6. The convergence over evolutionary time in the appearance of two or more unpalatable species is called
   a. cladism.
   b. mutual adaptation.
   c. Müllerian mimicry.
   d. Batesian mimicry.
   e. convergent mimicry.

7. Plants are good subjects for experiments to study competition because
   a. plants don't move around.
   b. the resources for which plants compete are easily measured.
   c. the resources for which plants compete are easily manipulated.
   d. plants often compete both in nature and in the laboratory.
   e. All of the above

8. Damage caused to shrubs by branches falling from overhead trees is an example of
   a. interference competition.
   b. partial predation.
   c. amensalism.
   d. commensalism.
   e. diffuse coevolution.

9. Ecological succession is
   a. the changes in species over time.
   b. the gradual process by which the species composition of a community changes.
   c. the changes in a forest as the trees grow larger.
   d. the process by which a species becomes abundant.
   e. the buildup of soil nutrients.

10. Keystone species
    a. influence the structure of the communities in which they live more than expected on the basis of their abundance.
    b. strongly influence the species composition of communities.
    c. may speed up the rate of vegetation succession.
    d. may be herbivores or carnivores.
    e. All of the above

Student Web Site Self-Quiz Questions

1. Consider a situation where a wasp species feeds on the seeds of a particular tree. During feeding, the wasp helps the tree by transferring pollen from tree to tree (pollination). The relationship between the fig tree and wasps described best encompasses which of the two ecological interactions shown below?

		Effects on Organism 2		
		Benefit	Harm	No Effect
Effects on Organism 1	Benefit	Mutualism	Predation	Commensalism
	Harm	Predation or parasitism	Competition	Amensalism
	No Effect	Commensalism	Amensalism	—

   a. Amensalism and predation
   b. Mutualism and predation
   c. Mutualism and commensalisms
   d. Mutualism and competition
   e. Commensalism and parasitism

2. In addition to food, resources can include such things as nest holes. According to the definition of resource presented in the textbook, which of the following is *not* considered a resource?
   a. Establishment sites for barnacles
   b. Nesting boxes for bluebirds
   c. Oxygen for an aquatic insect
   d. Salinity for a marine animal
   e. Sunlight for plant species

3. Based on the discussion in the textbook about the distributions of the barnacle species, *Balanus balanoides* and *Chthamalus stallatus* in the intertidal zone of rocky North Atlantic shores, which of the following statements is *false*?
   a. The niches of the two barnacles are not the same.
   b. The presence or absence of each species affects the niche of the other.
   c. Removal of each species has an equal effect on the niche expansion of the other species.
   d. Factors other than tidal range also define the niches of the two barnacles.
   e. Larval *Chthamalus* settle in the *Balanus* zone.

4. The table below shows data from a field study to examine how Sonoran Desert rodents and ants compete and influence their food supply of seeds. Based on these results and the discussion in the textbook, which group was most affected by interspecific competition, and which group had the greatest effect on their food supply?

	Rodents Removed	**Ants Removed**	**Rodents and Ants Removed**	**Control Plots**
Number of ant colonies	543	0	0	318
Number of rodents	0	144	0	122
Density of seeds relative to control plots	1.0	1.0	5.5	1.0

   a. Ants, ants
   b. Ants, rodents
   c. Ants, can't tell
   d. Rodents, ants
   e. Rodents, rodents

5. Experiments conducted with dragonfly nymphs and frog tadpoles have shown that predators may be able to eliminate prey from certain environments, but not others. Which of the following statements regarding these experiments is *false*?
   a. Tadpoles have more than one natural predator.
   b. The presence of dragonfly nymphs in ponds corresponded with complete elimination of tadpoles.
   c. Tadpoles were more abundant in ponds with only beetles as predators.
   d. Tadpoles were less abundant in ponds containing salamander larvae.
   e. Dragonfly nymphs are able to eat tadpoles of all sizes.

6. The bumble bee moth (*Hermaris diffinis*) is an example of __________ mimicry, and for this type of mimicry to evolve, the model species must be __________ the mimic species.
   a. Batesian, less abundant than
   b. Batesian, more abundant than
   c. Müllerian, less abundant than
   d. Müllerian, more abundant than
   e. Müllerian, as abundant as

7. Butterflies that are unpalatable and make predators sick represent examples of __________ mimicry and __________ benefit from this system.
   a. Batesian, some mimic species
   b. Batesian, all mimic species
   c. Müllerian, some mimic species
   d. Müllerian, all mimic species
   e. Müllerian, all mimic and predator species

8. The best description of the relationship between species of yucca plants such as *Yucca brevifolia* and the yucca moth, *Tegeticula* sp., is
   a. commensalism evolved through species-specific coevolution.
   b. commensalism evolved through diffuse coevolution.
   c. mutualism evolved through species-specific coevolution.
   d. mutualism evolved through diffuse coevolution.
   e. predator-prey interaction evolved through species-specific coevolution.

9. In many ecosystems, beaver are keystone species through their tree-cutting, dam-building activities. Which of the following statements about keystone species is *true*?
   a. Plants that modify the physical environment are keystone species.
   b. Keystone species are usually the larger animals in the community.
   c. All members of a community are keystone species.
   d. A keystone species may influence the structure of the community.
   e. A keystone species influences only the species richness of a community.

TEST FILE

★ *Indicates a* ***difficult*** *question*

■ *Indicates a* ***conceptual*** *question*

56 Ecosystems

Fill in the Blank

1. In temperate lakes, the depth at which the temperature abruptly changes from warm surface temperatures to cold deeper temperatures is the **thermocline**.
2. The cycling of water between the oceans, atmosphere, and land is known as the **hydrological cycle**.
3. The layer of the atmosphere that humans live in (the one closest to the ground) is the **troposphere**.
4. The movement of an element through a living organism and the physical environment is called its **biochemical cycle**.
5. The addition of nutrients, especially phosphorus, to a lake will result in blooms of algae that die and decompose, using all of the oxygen in the lake. This process is called **eutrophication**.
6. With a very few exceptions, all of the energy available to living organisms ultimately comes from the **sun**.
7. The total amount of energy assimilated by photosynthesis is called **gross primary production**.
8. The amount of energy assimilated by photosynthesis after the energy used by plants for maintenance and biosynthesis is subtracted is called **net primary production**.
9. All organisms that get their energy from a common source (e.g., all herbivores) constitute a **trophic level**.
10. A set of linkages through which a plant is eaten by an herbivore, which in turn is eaten by a carnivore, and so on, is called a **food chain**.
11. The leeward side of a mountain where winds descend, warm, and absorb, rather than release moisture is called a **rain shadow**.
12. The area on Earth where the solar energy flux is greatest, and shifts with the season, is the **intertropical convergence zone**.
13. Organisms that reduce the remains of other organisms to mineral nutrients that can be taken up by plants are **detritivores**.
14. Agricultural practices that combine cultivation methods such as crop rotation, mixed plantings, and mechanical tillage of soil with biological controls such as pest-resistant strains of crops and use of predators and parasites to reduce the application of chemical pesticides is called **integrated pest management**.
15. A network of linkages connecting a set of plants with a set of herbivores and carnivores is called a **food web**.
16. Human activity has most seriously disturbed the **carbon** biogeochemical cycle.
17. The troposphere is transparent to visible light and traps most outgoing infrared light. This property results in the **warming** of the Earth.
18. The amount of heat that an area of Earth receives depends on the **angle** of the sun.

Multiple Choice

■1. The poles are cooler than the equator because
   a. the equator receives more hours of sunlight than the poles do.
   b. the equator is closer to the sun than the poles are.
   c. the angle at which the sun strikes Earth is shorter at the poles (the sun is lower on the horizon).
   d. the long polar nights cool Earth much more than the short equatorial nights.
   e. winds tend to move heat from the poles toward the equator.

■2. It tends to rain more in the mountains than in adjacent lowlands because
   a. as air travels up mountains, it cools. Cool air holds less moisture than warm air, and the excess moisture is dropped as rain.
   b. as air travels up the mountains, it is compressed. The mountains squeeze the water out of the air, which falls as rain.
   c. of their great height. Mountains are often in the clouds, and precipitation is much greater in these clouds.
   d. as air moves down mountains, it expands and does not form clouds, so no rainfall occurs.
   e. the rainfall is equal in both mountains and lowlands; however, because the rain has farther to travel from clouds to the lowlands, more evaporates before reaching the ground in the lowlands.

3. A common ecosystem on the leeward (away from the wind) side of mountains and also at 308° north or south latitude is
   a. coniferous forest.
   b. desert.
   c. rainforest.
   d. peatlands.
   e. tundra.

4. The prevailing winds in the continental United States are from the west. This is a consequence of the
   a. position of the sun.
   b. location of the major mountain ranges in the United States.
   c. Pacific Ocean being warmer than the Atlantic Ocean.
   d. currents in the Atlantic and Pacific Oceans.
   e. rotation of Earth.

*5. When the intertropical convergence zone is over an area,
   a. the trade winds prevail.
   b. the trade winds reverse direction.
   c. air sinks and the area becomes cool and dry.
   d. air rises and heavy rains fall.
   e. air stagnates and hot, dry conditions prevail.

6. In general, ocean currents circulate
   a. clockwise in both the Northern and Southern Hemispheres.
   b. counterclockwise in both the Northern and Southern Hemispheres.
   c. clockwise in the Atlantic Ocean and counterclockwise in the Pacific Ocean.
   d. counterclockwise in the Northern Hemisphere and clockwise in the Southern Hemisphere.
   e. clockwise in the Northern Hemisphere and counterclockwise in the Southern Hemisphere.

*7. North Dakota is known for its hot summers and cold winters. At the same latitude, western Washington has cooler summers and warmer winters. What is the main cause of this difference?
   a. Washington receives a steady westerly wind that leads to a more nearly constant temperature. Because the westerly winds are blocked by mountains, North Dakota receives hot southern winds in the summer and cold northern winds in the winter.
   b. Western Washington is nearly always overcast. This cloud cover prevents the sun from heating the land in the summer and prevents excess heat loss during the winter. North Dakota is very dry, and is heated and cooled more drastically.
   c. Western Washington is near the Pacific Ocean, which moderates its climate. North Dakota is in the middle of a large continent and does not receive this buffering.
   d. Western Washington is at lower elevation than North Dakota. This difference in altitude accounts for the more extreme weather in North Dakota.

*8. Which of the following elements is not a major component of living systems?
   a. Carbon
   b. Nitrogen
   c. Oxygen
   d. Hydrogen
   e. Potassium

*9. Erosion from poor farming practices leads to large losses of soil to the sea. What is the likely fate of the elements in the soil?
   a. Most of the elements will be rapidly recirculated by ocean currents and soon reintroduced into terrestrial ecosystems.
   b. Most elements will dissolve and eventually cycle into the atmosphere.
   c. Most elements will dissolve in the soil and move into freshwater ecosystems, where they will eventually be reintroduced into terrestrial ecosystems.
   d. Tidal forces will quickly return most of the soil to the shore, where the elements will collect.
   e. Most elements settle to the bottom of the oceans, and remain there for millions of years until bottom sediments are elevated by movements of Earth's crust.

10. Excluding human-induced effects, most elements enter freshwater systems from
   a. rainfall.
   b. oceans (through tidal movements).
   c. actions of animals.
   d. weathering of rocks (via groundwater).
   e. plants and plant parts decaying in streams, lakes, and rivers.

11. In a temperate zone lake in the middle of summer,
   a. most nutrients are near the surface of the lake, and oxygen is evenly distributed throughout all depths of the lake.
   b. most nutrients are near the bottom of the lake, and most oxygen is near the surface of the lake.
   c. nutrients and oxygen are evenly distributed throughout all depths of the lake.
   d. nutrients are evenly distributed throughout all depths of the lake, and oxygen is near the surface of the lake.
   e. most nutrients and oxygen are found near the surface of the lake.

12. At which of the following temperatures is water most dense?
   a. 16°C
   b. 8°C
   c. 4°C
   d. 2°C
   e. 0°C

13. Most of Earth's nitrogen is in
   a. the atmosphere.
   b. the oceans.
   c. fresh water.
   d. soil.
   e. organisms.

14. The gas that is removed from the atmosphere by plants and algae is
   a. nitrogen.
   b. oxygen.
   c. carbon dioxide.
   d. methane.
   e. water vapor.

▪15. One difference between gases such as oxygen and carbon dioxide and minerals that do not have a gas phase such as phosphorus and calcium is that
   a. there is relatively little variation from site to site in the concentration of gases, whereas solids tend to remain where they are, creating local abundances and shortages of these minerals.
   b. gases such as oxygen and carbon dioxide are never found in a rock (solid) phase or dissolved in water, whereas minerals without a gaseous phase are frequently found as rocks or dissolved in water.
   c. gaseous elements (and compounds) are essential to living organisms, whereas minerals without a gaseous phase are not.
   d. gases are modified by living organisms, whereas minerals without a gaseous phase are not.
   e. the relative proportions of different gases in the atmosphere change radically over short periods of time, whereas the relative proportions of different solid minerals remain fairly constant.

★16. Most of the world's carbon is found
   a. as carbon dioxide in the atmosphere.
   b. in living organisms.
   c. as bicarbonate and carbonate ions dissolved in the oceans.
   d. as carbonate minerals in sedimentary rock.
   e. in the decaying remains of dead organisms.

17. Photosynthesis and respiration are central to which cycle?
   a. The nitrogen cycle
   b. The carbon cycle
   c. The phosphorus cycle
   d. The sulfur cycle
   e. The hydrological cycle

18. In the past 150 years there has been a major new input to the carbon cycle. What is it?
   a. There are more humans releasing large quantities of carbon dioxide as respiration.
   b. Increased animal farming has resulted in greater carbon dioxide releases.
   c. Industrialization has resulted in the burning of fossil fuels such as oil and coal, which releases carbon dioxide into the atmosphere.
   d. Changes in ocean currents have lead to the release of large quantities of carbon dioxide.
   e. Global warming has increased the release of carbon dioxide from carbonate minerals.

▪19. Nitrogen is often in short supply in terrestrial ecosystems. Why?
   a. There is very little free nitrogen in the air.
   b. Atmospheric nitrogen is primarily in the stratosphere and does not come into contact with terrestrial ecosystems.
   c. Atmospheric nitrogen cannot be used by most organisms. It needs to be converted to useful forms by bacteria and cyanobacteria.
   d. Nitrogen solubility in water is very low and therefore atmospheric nitrogen enters cells very slowly.
   e. Atmospheric nitrogen varies widely from location to location. As a result there are frequently local shortages of nitrogen.

▪20. The phosphorus cycle differs from the carbon cycle in that
   a. phosphorus does not enter living organisms, whereas carbon does.
   b. the phosphorus cycle does not include a gaseous phase, whereas the carbon cycle does.
   c. the phosphorus cycle includes a solid phase, whereas the carbon cycle does not.
   d. the primary reservoir of the phosphorus cycle is the atmosphere, whereas the primary reservoir for the carbon cycle is in rock.
   e. phosphorus passes through living organisms many times, whereas carbon flows through living organisms only once.

21. Acid precipitation
   a. naturally occurs when clouds are formed by evaporation over land.
   b. naturally occurs when clouds are formed by evaporation over oceans.
   c. occurs as a result of the burning of fossil fuels that contain sulfur and nitrogen compounds.
   d. occurs because of carbon monoxide emissions by automobiles.
   e. occurs when large forest tracts are logged or otherwise deforested.

22. The two main human-induced causes of increased carbon dioxide levels are
   a. the burning of fossil fuels and the burning of forests.
   b. the burning of fossil fuels and the raising of cattle.
   c. modern agricultural fertilizers and the raising of cattle.
   d. pollution-induced mortality of oceanic algae and the burning of forests.
   e. modern agricultural fertilizers and pollution-induced mortality of oceanic algae.

23. If carbon dioxide levels were to double, as they may by the middle of the next century, all of the following are expected to occur except which of the following?
   a. World mean temperature will increase 3-5°C.
   b. Droughts in the central regions of continents will become more common and more severe.
   c. Precipitation in coastal areas will increase.
   d. Polar ice caps will melt, raising sea levels and flooding coastal cities and agricultural areas.
   e. The destruction of the ozone layer by carbon dioxide will allow more ultraviolet light to reach Earth.

24. Which of the following statements regarding the movement of energy and nutrients through ecosystems is *true*?
    a. Energy flows and nutrients flow.
    b. Energy flows and nutrients cycle.
    c. Energy cycles and nutrients cycle.
    d. Energy cycles and nutrients flow.
    e. Energy flows and nutrients either cycle or flow.

25. Which of the following changes would *not* result in an increase in net primary production?
    a. Increased precipitation in an arid area
    b. Increased soil fertility
    c. Increased latitude (moving from the equator toward the poles)
    d. Moving down a mountain to warmer temperatures
    e. Increasing the light in aquatic ecosystems

26. On average, how much of the energy assimilated at one trophic level is converted to production at the next trophic level (excluding the conversion of sunlight into chemical energy by plants)?
    a. 5–20%
    b. Less than 1%
    c. 20–30%
    d. 30–50%
    e. Greater than 50%

*27. Which of the following always has a "pyramidal" shape, that is, decreasing values at higher trophic levels?
    a. Pyramids of numbers only
    b. Pyramids of biomass only
    c. Pyramids of energy only
    d. Both pyramids of biomass and pyramids of energy
    e. All three pyramids

28. Which of the following is *true* concerning nutrients in modern agricultural systems?
    a. Nutrient cycles are very tight, and there is little loss of nutrients through runoff.
    b. Nutrient cycles are frequently broken because food is consumed and wastes are disposed of far from where the food was produced.
    c. Modern agricultural systems are normally carefully designed to promote efficient nutrient cycling.
    d. The addition of fertilizers, together with the natural cycling of nutrients, has resulted in a buildup in agricultural soils so that they are now far deeper and richer than they were prior to the development of modern agriculture.
    e. There is normally less loss of nutrients from agricultural fields than from natural terrestrial ecosystems.

29. A terrestrial ecosystem is directly influenced by all but which of the following?
    a. Solar energy fluxes
    b. Earlier successional stages
    c. Local species richness
    d. Global atmospheric circulation
    e. Eutrophication

30. Rain shadows form on the _________ of mountains and result in _________ rainfall there.
    a. windward side, more
    b. windward side, less
    c. leeward side, more
    d. leeward side, less
    e. peaks, more

*31. As a sea breeze is forced to rise by coastal mountains, a process occurs that is the same as that occurring in the area of the _________, except that in the latter the energy to rise is provided by the _________.
    a. trade winds, Earth's spin
    b. westerlies, sun
    c. westerlies, Earth's spin
    d. intertropical convergence zone, sun
    e. intertropical convergence zone, Earth's spin

32. The direction of the westerlies and the trade winds is directly caused by
    a. location of the intertropical convergence zone.
    b. tilt of Earth's axis.
    c. variation in the rotational speed of Earth at different latitudes.
    d. differences in heat balance between the equator and the poles.
    e. the global oceanic circulation.

33. Oceans warm and cool more _________ than land because water has a _________ specific heat.
    a. slowly, greater
    b. slowly, lesser
    c. rapidly, greater
    d. rapidly, lesser
    e. slowly, equal (but the oceans are larger)

*34. On the west coasts of continents, ocean waters are _________ and precipitation falls mostly in the _________.
    a. cool and nutrient-rich, summer
    b. cool and nutrient-rich, winter
    c. cool and nutrient-poor, winter
    d. warm and nutrient-rich, summer
    e. warm and nutrient-rich, winter

▪35. Fruit trees and other frost-sensitive plants are frequently grown on the downwind side of large lakes. This has the most to do with
    a. the fact that air that has blown over water will usually release its moisture when it encounters colder land.
    b. the high specific heat of land.
    c. wind shadows.
    d. the moderating effect that the lake has on the temperature.
    e. the moderating effect that land has on the lake.

▪36. The major differences between ocean and freshwater compartments of the global ecosystem result because
    a. of the much greater depth of the oceans.
    b. the oceans only turn over once a year.
    c. freshwater systems only receive input from groundwater.

d. temperature only varies with depth in freshwater systems.
e. nutrient recycling is less rapid in freshwater systems.

37. Which global ecosystem compartment would be characterized by very slow movement of materials and exchange of gases mostly via organisms?
a. Oceans
b. Fresh water
c. Atmospheric
d. Terrestrial
e. More than one

38. Which global ecosystem compartment is affected by Earth's rotation on its axis circulating materials and has most organisms in the uppermost layer?
a. Oceans
b. Fresh water
c. Atmospheric
d. Terrestrial
e. More than one

*39. Which of the following comparisons of the troposphere and stratosphere compartments of the global ecosystem is *true*?
a. Most water vapor resides in the troposphere.
b. Vertical circulation of air occurs mostly in the stratosphere.
c. Circulation of the stratosphere most directly influences ocean currents.
d. The stratosphere represents most of the mass of the atmosphere.
e. Most weather events occur in the stratosphere.

*40. Which of the following statements regarding the hydrological cycle is *false*?
a. Most input to the oceans occurs via runoff from rivers.
b. More water evaporates from the surface of the oceans than falls as rain over the oceans.
c. Less water evaporates from the surface of the land than falls as rain over the land.
d. Water found in sedimentary rock is constantly exchanged with the ocean.
e. The energy driving the hydrological cycle is heat from the sun.

41. In biogeochemical cycles, elements that cycle fastest
a. are found in organisms.
b. are scarce.
c. have a gaseous phase.
d. do not become fixed into sediment.
e. do not accumulate in the higher trophic levels.

42. Which of the following biogeochemical cycles is characterized by a major reservoir that is gaseous, and a major inorganic form that can only be utilized by a small group of bacteria and cyanobacteria?
a. Carbon cycle
b. Nitrogen cycle
c. Phosphorus cycle
d. Sulfur cycle
e. Oxygen cycle

43. Which of the following biogeochemical cycles is characterized by a form which is a major greenhouse gas?
a. Carbon cycle
b. Nitrogen cycle
c. Phosphorus cycle
d. Sulfur cycle
e. More than one cycle

44. Which of the following biogeochemical cycles lacks a gaseous phase?
a. Carbon cycle
b. Nitrogen cycle
c. Phosphorus cycle
d. Sulfur cycle
e. Oxygen cycle

45. Which of the following biogeochemical cycles has a gaseous phase released by volcanoes and fumaroles?
a. Carbon cycle
b. Nitrogen cycle
c. Phosphorus cycle
d. Sulfur cycle
e. Oxygen cycle

46. Which of the following biogeochemical cycles has a major reservoir in sedimentary rock?
a. Carbon cycle
b. Nitrogen cycle
c. Phosphorus cycle
d. Sulfur cycle
e. More than one cycle

47. In the human-induced condition called eutrophication, the main biogeochemical cycle that is altered is the __________ cycle, and the effect is to create __________ conditions and decrease species diversity.
a. hydrological, aerobic
b. phosphorus, anaerobic
c. hydrological, aerobic
d. phosphorus, aerobic
e. nitrogen, anaerobic

*48. After nutrient input into a lake is reduced, the time required for the lake to return to pre-eutrophication conditions depends on
a. the rate of turnover of its waters.
b. the presence of the appropriate algae-eating fish.
c. whether there is a thermocline in the lake.
d. the amount of groundwater reaching the lake.
e. the presence of an inverted energy pyramid.

*49. In the human-induced condition called acid precipitation, the main biogeochemical cycles that are altered are the __________ cycles, and one effect in lakes is the __________ of populations of nitrifying bacteria.
a. phosphorus and nitrogen, increase
b. phosphorus and nitrogen, decrease
c. nitrogen and sulfur, decrease
d. nitrogen and sulfur, increase
e. phosphorus and sulfur, decrease

50. Which of the following occurrences is *not* considered to be a likely consequence of global warming?
   a. Melting of the polar ice caps
   b. Flooding and increased precipitation in coastal areas
   c. Increased ozone depletion
   d. Average worldwide temperature increase of 3–58°
   e. Droughts in central areas of continents

51. Which of the following is not a part of either gross or net primary production in plants?
   a. Light reflected from the leaf
   b. Energy fixed into glucose
   c. Energy expended in moving material through membranes
   d. Energy used to excite a chlorophyll electron
   e. Dry weight of a plant

52. Which of the following statements about food chains and energy flow through ecosystems is *false*?
   a. A single organism can feed at several trophic levels.
   b. The lower the trophic level at which an organism feeds, the more energy is available.
   c. Detritivores feed at all trophic levels except the producer level.
   d. Food webs include two or more food chains.
   e. All organisms that are not producers are consumers.

▪53. Assuming that the energy transfer efficiency between trophic levels is 10%, how much grain would be required to produce 70 kg of human biomass if the grain is eaten by cows and the cows are eaten by humans?
   a. 210 kg
   b. 700 kg
   c. 2,100 kg
   d. 7,000 kg
   e. 70,000 kg

▪54. Grasslands can support greater grazing rates by herbivores than forests because
   a. grasslands receive more sunlight.
   b. the net production of grasslands is greater.
   c. grasslands produce less woody plant tissue.
   d. more of the grassland production is above ground.
   e. manure from large herds of herbivores increases production rate of grasslands.

55. An inverted pyramid of __________ may occasionally be observed in __________ communities.
   a. energy, grassland
   b. energy, forest
   c. biomass, marine
   d. biomass, grassland
   e. biomass, forest

56. Which of the following is *not* an objective of integrated pest management?
   a. To eliminate the use of chemicals in agriculture
   b. To develop pest-resistant strains
   c. To use natural biological control methods
   d. To reduce agriculturally-caused pollution
   e. To reduce the dependence of agriculture on fossil fuels

57. Rain and snow become acidified when the burning of fossil fuels releases the acid forms of the elements
   a. phosphorus and sulfur.
   b. sulfur and nitrogen.
   c. nitrogen and chlorine.
   d. phosphorus and chlorine.
   e. nitrogen and phosphorus.

▪58. Oceans heat and cool more slowly than air because
   a. water is thicker than air.
   b. water has a higher specific heat than air.
   c. wind circulation makes the oceans move in a westward direction, away from the sun.
   d. water is closer to the center of Earth, which is warmer.
   e. it is easier for sunlight to travel through air than through water.

▪59. In light, a plant fixes 0.12 ml of CO_2 per hour, however, in the dark the same plant releases 0.04 ml of CO_2 per hour. What is the estimated net primary production of this plant?
   a. 0.04 ml/hour
   b. 0.08 ml/hour
   c. 0.12 ml/hour
   d. 0.16 ml/hour
   e. 0.0048 ml/hour

Study Guide Questions

1. In comparing an acre of land in Colombia to an acre of land in Michigan, which of the following would *not* differ?
   a. The angle of the sun reaching the ground in the month of July
   b. The solar energy flux in the month of July
   c. The annual solar energy flux
   d. The total hours of daylight per year

2. The following diagram shows a mountain with a sea breeze blowing as indicated by the arrow. Circle the letter for the area with air that would be *both* relatively warm and dry.

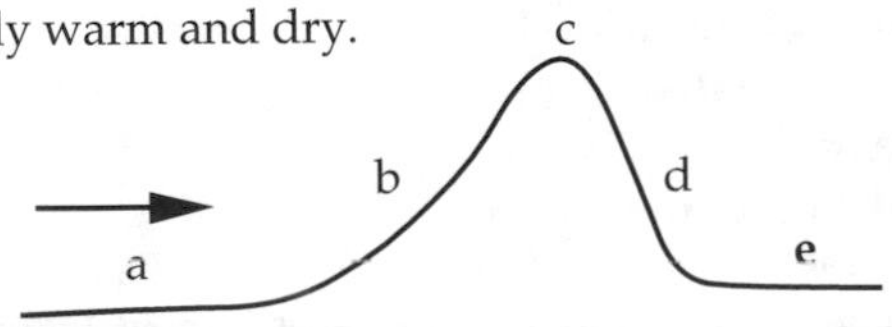

3. In what area of the above diagram in question 2 is a process occurring that is similar to the process that occurs in the intertropical convergence zone? **(b)**

4. If Earth did not spin on its axis, from what direction would the northeast trade winds blow?
   a. Northeast
   b. South
   c. North
   d. East

5. Ocean circulation patterns are influenced by all of the following *except*
   a. circulation of Earth's atmosphere.
   b. deflection by land masses.
   c. upwelling of deeper water.
   d. rotation of Earth on its axis.

6. Differences in the specific heat of water and land have most to do with creating
   a. the intertropical convergence zone.
   b. maritime climates.
   c. continental climates.
   d. oceanic upwelling.

7. Which of the following could *not* be considered an ecosystem?
   a. A small pond
   b. All the fish in a coral reef
   c. Earth
   d. A pile of dung in a pasture

8. A plant in the dark uses 0.02 ml of O_2 per minute. The same plant in sunlight releases 0.14 ml of O_2 per minute. A correct estimate of its rate of *gross* primary production is
   a. 0.02 ml of O_2 per minute.
   b. 0.12 ml of O_2 per minute.
   c. 0.14 ml of O_2 per minute.
   d. 0.16 ml of O_2 per minute.

9. Examine the following food web. Organism 9 is a(n)

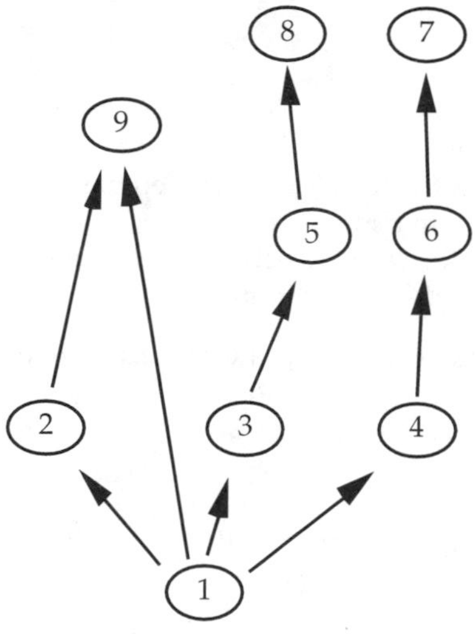

   a. herbivore.
   b. primary carnivore.
   c. secondary carnivore.
   d. omnivore.

10. How many trophic levels does the food web shown in Question 9 have?
    a. 1
    b. 2
    c. 3
    d. 4

11. Which of the following organism types would probably rank the *lowest* in terms of their expected efficiency of energy transfer between trophic levels?
    a. Herbivorous mammal
    b. Carnivorous mammal
    c. Invertebrate herbivore
    d. Invertebrate carnivore

12. Which of the following would be *least* likely to be a method employed in integrated pest management?
    a. Mixed plantings of crop plants
    b. Development of pest-resistant strains of crop plants
    c. Aerial spraying of insecticides
    d. Use of chemical attractants for controlling pest species

13. Which of the following features would be characteristic of oceans but *not* of freshwater ecosystems?
    a. Receives material from land mostly via groundwater
    b. Seasonal mixing of materials
    c. Elements buried in bottom sediments for long periods of time
    d. Bottom waters frequently lacking oxygen

14. Which of the following features would be characteristic of the stratosphere but *not* of the troposphere?
    a. Most ozone resides here.
    b. Most water vapor resides here.
    c. Most of the mass of the atmosphere lies here.
    d. Circulation of this layer influences ocean currents.

15. Which of the following compartments of the global ecosystem would be characterized by very slow movement of materials within the compartment?
    a. Oceans
    b. Fresh waters
    c. Atmosphere
    d. Land

16. In which of the following compartments of the global ecosystem would circulation of materials be affected by Earth's revòlution around the sun?
    a. Oceans
    b. Fresh waters
    c. Atmosphere
    d. All of the above

17. In the following diagram, the temperature of a freshwater lake is plotted against depth.

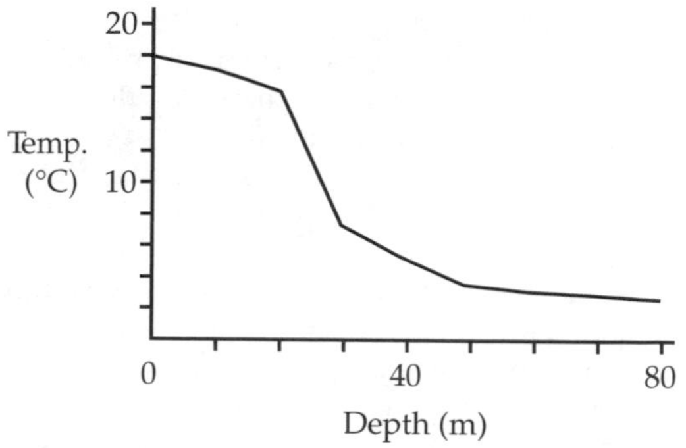

    This temperature profile would be most characteristic of the lake during
    a. winter.

b. spring.
c. summer.
d. fall.

18. Based on data in the graph from Question 17, the thermocline for the lake is located between
a. 0 and 20 meters.
b. 20 and 30 meters.
c. 30 and 50 meters.
d. 50 and 80 meters.

19. Which of the following statements about biogeochemical cycles is *not* true?
a. Most elements remain longest in the living portion of their cycle.
b. Gaseous elements cycle more quickly than elements without a gaseous phase.
c. You may have some atoms in your body that were once part of a dinosaur.
d. Biogeochemical cycles all include both organismal and nonliving components.

20. Next to each of the following features, place a *c*, *n*, *p*, or *s* if it is characteristic of the biogeochemical cycles of *carbon*, *nitrogen*, *phosphorus*, or *sulfur*. (Some features may apply to more than one cycle.)
(n) Major reservoir is atmospheric
(c, p) Major reservoir is in sedimentary rocks
(n, p) Often in short supply in ecosystems
(c) Fossil fuel reserve is part of this cycle
(s) Involved in cloud formation
(p) Lacks a gaseous phase
(c) Includes a form that is a greenhouse gas
(n) Most fluxes involve organisms

21. Which of the following statements is *not* a major concern about our altering of the carbon cycle?
a. The Greenland and Antarctic ice caps are expected to melt if global warming continues.
b. High-sulfur fuels are used by power plants because they are less expensive than low-sulfur fuels.
c. The increase in atmospheric CO_2 exceeds the ability of the oceans to absorb the increase.
d. CO_2 is a gas that traps infrared radiation.

22. Which of the following statements concerning acid precipitation is *false*?
a. Acids that enter the atmosphere primarily affect ecosystems located less than 100 kilometers from the source of the pollution.
b. The effects of acidification of lakes appear to be quickly reversible if their pH can be returned to normal.
c. While regulation of emission sources has raised the pH of precipitation in parts of the eastern United States, precipitation in many parts of the West is becoming more acidic.
d. Sulfuric acid and nitric acid from smokestack and automobile emissions are the major causes of acid precipitation.

23. The process by which a lake ecosystem is altered by eutrophication involves several stages, each causing the next. Which of the following stages would occur *second* in the chain of causation?
a. Algal blooms occur.
b. Oxygen levels drop in deeper water.
c. Phosphorus input from sewage and agricultural runoff increases.
d. Respiratory demand from decomposers increases.

End of Chapter Questions

1. Which of the following is true about the amount of sunlight and heat arriving on Earth?
a. Every place on Earth receives the same annual number of hours of sunlight and the same amount of heat.
b. Every place on Earth receives the same annual number of hours of sunlight, but not the same amount of heat.
c. Every place on Earth receives the same annual amount of heat, but not the same number of hours of sunlight.
d. Both the annual amount of sunlight and the amount of heat received vary over the surface of Earth.
e. None of the above

2. A rain shadow forms on the leeward sides of mountain ranges because
a. cool air can hold more moisture than warm air
b. cool air cannot hold as much moisture as warm air
c. temperatures fluctuate more widely on the leeward sides of mountains
d. temperatures are more constant on the leeward sides of mountains
e. actually rain shadows occur on both the windward and leeward sides of mountains

3. When an area is within the intertropical convergence zone,
a. the northeast trade winds blow steadily.
b. the southeast trade winds blow steadily.
c. air is descending and it seldom rains.
d. air is rising and heavy rains fall frequently.
e. westerly winds blow steadily.

4. Zones of marine upwelling are important because
a. they help scientists measure the chemistry of deep ocean water.
b. they bring to the surface organisms that are difficult to observe elsewhere.
c. ships can sail faster in these zones.
d. they increase marine productivity by bringing nutrients back to surface ocean waters.
e. they bring oxygenated water to the surface.

5. Which of the following is not true of the troposphere?
a. It contains nearly all atmospheric water vapor.
b. Materials enter it primarily at the intertropical convergence zone.
c. It is about 17 km deep in the tropics.

d. Most global atmospheric circulation takes place there.
e. It contains about 80 percent of the mass of the atmosphere.

6. The total amount of energy that plants assimilate by photosynthesis is called
   a. gross primary production.
   b. net primary production.
   c. biomass.
   d. a pyramid of energy.
   e. eutrophication.

7. The amount of energy reaching an upper trophic level is determined by
   a. net primary production.
   b. net primary production and the efficiencies with which food energy is converted to biomass.
   c. gross primary production.
   d. gross primary production and the efficiencies with which food energy is converted to biomass.
   e. gross primary production and net primary production.

8. Carbon dioxide is called a greenhouse gas because
   a. it is used in greenhouses to increase plant growth.
   b. it is transparent to heat radiation but opaque to sunlight.
   c. it is transparent to sunlight but opaque to heat radiation.
   d. it is transparent to both sunlight and heat radiation.
   e. it is opaque to both sunlight and heat radiation.

9. The phosphorus cycle differs from those of carbon and nitrogen in that
   a. it lacks a gaseous phase.
   b. it lacks a liquid phase.
   c. only phosphorus is cycled through marine organisms.
   d. living organisms do not need phosphorus.
   e. The phosphorus cycle does not differ importantly from the carbon and nitrogen cycles.

10. Acid precipitation results from human modifications of
    a. the carbon and nitrogen cycles.
    b. the carbon and sulfur cycles.
    c. the carbon and phosphorus cycles.
    d. the nitrogen and sulfur cycles.
    e. the nitrogen and phosphorus cycles.

11. Which of the following is not a component of integrated pest management?
    a. Use of cultural strategies such as crop rotation and mixed plantings
    b. Use of pest-resistant strains of crops
    c. Use of predators and parasites of crop pests
    d. Use of chemical attractants
    e. Use of chemical pesticides whenever pests are discovered

Student Web Site Self-Quiz Questions

1. The orbit of Earth's revolution around the sun is inclined at an angle of 23.5°, and this causes seasonal variation in the amount of solar energy reaching a given region. Which of the following does *not* differ for the tropics and polar regions?
   a. Total number of hours of sunlight per year
   b. Average angle of sunlight
   c. Mean annual air temperature
   d. Solar energy input
   e. Annual variation in day length

2. What would the southeast trade winds be called if Earth rotated on its axis from east to west?
   a. Northeast trades
   b. Northwest trade winds
   c. Southeast trade winds
   d. Southwest trade winds
   e. Westerlies

3. Ocean circulation patterns are influenced by all of the following except
   a. circulation of Earth's atmosphere.
   b. deflection by land masses.
   c. prevailing winds.
   d. rotation of Earth on its axis.
   e. Continental climate.

4. Which of the following statements about primary production in different ecosystems is *false*?
   a. Production in aquatic systems is less limited by light than in terrestrial systems.
   b. Nutrients can limit primary production in aquatic systems.
   c. Primary production in the open ocean is low.
   d. Water availability and temperature are most limiting for primary production in terrestrial systems.
   e. Tropical rainforests have the highest production of any terrestrial ecosystem.

5. The efficiency of energy transfer of different species varies widely. Which of the following statements regarding efficiency of energy transfer through food webs is *false*?
   a. A trophic level can support more herbivores than carnivores.
   b. Birds and mammals have low production efficiency.
   c. The efficiency of energy transfer depends on the total production at a particular trophic level.
   d. The efficiency of energy transfer depends on how organisms divide ingested energy.
   e. Herbivores are less efficient than carnivores.

6. Modern agricultural practices include aerial spraying of pesticides. Which of the following would *not* likely be a method employed in integrated pest management (IPM)?
   a. Crop rotation
   b. Development of pest-resistant strains of crop plants
   c. Introduction of natural predators for important pest species

d. Use of a broad-spectrum herbicide
e. Use of chemical attractants

7. Which of the following statements comparing the ocean and freshwater compartments of the global ecosystem is *false*?
   a. Upwelling in oceans and turnover in lakes are involved in the same process.
   b. Oxygen is seldom limiting in both compartments.
   c. Turnover of nutrients is more rapid in lakes than in oceans.
   d. Both oceans and lakes receive materials from land.
   e. Oceans respond much more slowly to outside disturbances than lakes.

8. Which of the following is *not* a correct pairing of an atmospheric layer and a characteristic of that layer?
   a. Troposphere, circulation of gases is mostly horizontal
   b. Stratosphere, location of the ozone layer
   c. Troposphere, most water resides here
   d. Troposphere, represents the greatest mass of the total atmosphere
   e. Troposphere, 10–17 km thick

9. Which of the following statements about biogeochemical cycles is *false*?
   a. Carbon and nitrogen cycle faster than phosphorus.
   b. All biogeochemical cycles include both organisms and nonliving components.
   c. Most elements remain longest in the living portion of their cycle.
   d. You may have atoms in your body that were once part of a dinosaur.
   e. The chemical elements used by organisms in large quantities cycle back and forth between organism and environment.

10. Burning fossil fuels returns carbon to the global carbon cycle that had been part of that cycle's nonliving reservoir. Which of the following pairings of an element and a characteristic of that element in a biogeochemical cycle is *not* true?
    a. Nitrogen, major reservoir is the atmosphere
    b. Sulfur, involved in cloud formation
    c. Phosphorus, lacks a gaseous phase
    d. Carbon, includes a form that is a greenhouse gas
    e. Nitrogen, converted to useful forms by plants

11. Which of the following statements about acid precipitation is *false*?
    a. Acid precipitation involves two major biogeochemical cycles.
    b. Because acid precipitation is restricted to a region, remediation is easier than a more widespread problem.
    c. Acid precipitation can damage plants directly, reducing their rates of photosynthesis.
    d. Some lakes are very sensitive to acid, but recover quickly.
    e. Acid precipitation is restricted mostly to the major developed countries.

TEST FILE

★ *Indicates a **difficult** question*
■ *Indicates a **conceptual** question*

57 Biogeography

Fill in the Blank

1. **Biogeography** is the study of the distribution of organisms over space and time.
2. **Historical biogeographers** study the evolutionary histories of groups of organisms, where they originated, and if and how they spread.
3. **Ecological biogeographers** study present-day interactions among organisms to determine how those relationships influence where species are found.
4. A **vicariant** distribution is one determined by the separation of formerly contiguous areas by a barrier.
5. **Australia** and **South America** are the most distinct biogeographic regions because they have been isolated from other continents for the longest time.
6. One theory of biogeography suggests that the number of species in an area is the result of a dynamic balance between **extinctions** and **colonizations**.
7. The major biome types are identified by their characteristic **vegetation**.
8. A particular community type, such as a hot desert, occurs where the **climate** is suitable for that type of vegetation.
9. Although there are **different** species of organisms in a particular biome on different continents, the species are ecologically **similar**.
10. The **tundra** biome is found in the Arctic and high in the mountains at all latitudes.
11. The most striking difference between Arctic tundra and alpine tundra in the tropics is **day length**.
12. Arctic tundra is underlain by permanently frozen ground called **permafrost**.
13. A modern biogeographic technique that shows the distribution patterns of a taxonomic group is the **area phylogeny**.
14. Major ecosystems that vary by nature of their dominant vegetation are called **biomes**.
15. The ocean's **temperature** at different levels is the biggest barrier to colonization by a species.
16. The **abyssal** zone is ocean floor below the level of sunlight penetration.

Multiple Choice

1. An endemic species is one that
   a. is found in only one region.
   b. evolved in the region where it currently lives.
   c. occurs in two distinct areas separated by a barrier.
   d. has a good fossil record showing its evolution.
   e. disperses readily from one area to another.

2. An endemic taxon may occur for all of the following reasons, except that the taxon
   a. is old and nearly extinct.
   b. has evolved only recently.
   c. lives in an isolated island habitat.
   d. has poor dispersal mechanisms.
   e. has little genetic variability.

■3. Australia has a greater percentage of endemic species than other continents because it
   a. lies near several large chains of islands.
   b. has more varied habitats than other continents.
   c. has been isolated for the longest time due to continental drift.
   d. lacks a region near the equator.
   e. has received many immigrant species due to prevailing ocean currents.

★4. A study of small flies called midges shows that one genus occurs only in New Zealand, two genera are found in Australia, South America, and New Zealand, and two other genera occur in Australia and South America only. This evidence supports the theory that the land masses drifted apart from one another in which order?
   a. New Zealand first, then Australia and South America
   b. Australia first, then New Zealand and South America
   c. South America first, then New Zealand and Australia

d. All three separated at the same time.
e. South America and Australia separated; New Zealand was never joined to either.

5. The island biogeographic model describes the relationship of the number of species present on an island with its
a. climate and vicariant events.
b. size and distance from the mainland.
c. species pool.
d. rate of extinction.
e. elevation and shape.

6. Which of the following is common to both ecological and historical biogeographers?
a. Using experiments to test hypotheses
b. Concentrating on current interactions among organisms
c. Studying the effects of time and space on species' distributions
d. Explaining the distribution of organisms in terms of continental drift
e. Constructing lineages in terms of a "molecular clock"

7. Which of the following is *not* a modern barrier between major biogeographic regions?
a. Desert
b. Mountains
c. Waterways
d. High plateau–lowland boundary
e. Tundra

8. In an experiment near the southern tip of Florida, all the arthropods were destroyed on six small islands. Which of the following statements about the recolonization of these islands by arthropods is *true*?
a. Each island regained approximately its original number of species.
b. Recolonization rates were very slow on the smaller of the six islands compared to the larger.
c. The recolonizing species were identical to the original species.
d. Equilibrium numbers of species had not been reached by the close of the experiment.
e. Rates of immigration were constant for the first year of recolonization.

9. The geographic distribution of a biome may be affected by the seasonal variation of all of the following, *except*
a. temperature.
b. latitude.
c. precipitation.
d. fire.
e. grazing.

10. If one started near Washington, D.C., on the East Coast of the United States and traveled northward toward the North Pole, one would encounter which of the following biomes, in which order?
a. Temperate deciduous forest, boreal forest, tundra
b. Chaparral, temperate deciduous forest, boreal forest
c. Savanna, boreal forest, chaparral
d. Temperate grassland, boreal forest, tundra
e. Boreal forest, tundra, chaparral

11. Forests in both the Northern and Southern Hemispheres are termed "boreal" forests when they are dominated by
a. evergreen trees with needles or small leaves.
b. cone-bearing gymnosperm trees.
c. deciduous beeches or related angiosperm trees.
d. a diverse assemblage of insect-pollinated trees.
e. evergreen trees adapted to frequent fires.

12. Which of the following biomes may lack seasonal temperature variations?
a. Tundra
b. Tropical evergreen forest
c. Boreal forest
d. Cold desert
e. Chaparral

13. Which of the following biomes lacks a pronounced seasonal variation in amounts of precipitation?
a. Tropical deciduous forest
b. Thorn forest
c. Boreal forest
d. Chaparral
e. Cold desert

14. Historical biogeographers use several types of data. Which of the following types of information would *not* be used by a historical biogeographer?
a. Phylogenies
b. Fossils
c. Distributions of living species
d. Places of origin of a group of organisms
e. Modern ecological relationships

▪*15. Biogeographers often apply the rule of parsimony when interpreting distribution patterns. Assume there are four organisms with the following degrees of similarity: Organisms A and B are very similar, and organisms C and D are very similar. Organisms A and B are similar to fossil E; organisms C and D are similar to a fossil F. Which of the following would be a parsimonious explanation for their evolutionary relationship?
a. B is more closely related to C than to A.
b. B is more closely related to C than to D.
c. B and A evolved from F; C and D evolved from E.
d. B and A evolved from E; C and D evolved from F.
e. A, B, C, and D evolved from F.

16. If a species lives in two distinct areas now separated by a barrier because it was in those areas before the barrier was imposed, it is said to have a _______ distribution.
a. dispersal
b. vicariant
c. natural
d. migration
e. static

17. A vicariant distribution is one determined by
   a. separation of formerly contiguous land masses.
   b. dispersal.
   c. extinction and recolonization.
   d. adaptive radiation.
   e. interbreeding of two similar species.

18–22. Refer to the following figure for the questions below.

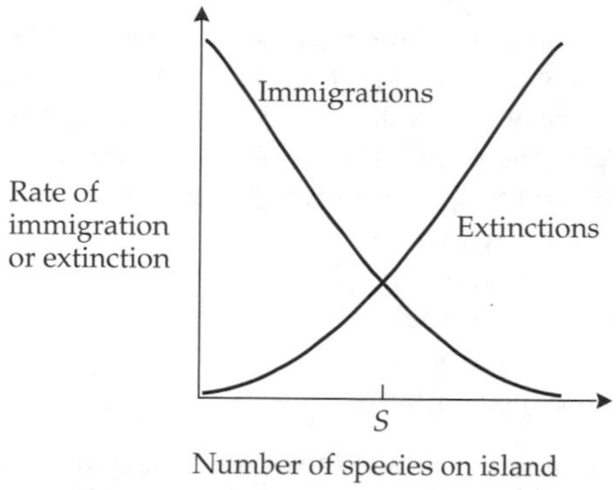

18. The above figure suggests that the rate of extinction of species
   a. decreases as the number of species increases.
   b. increases as the number of species increases.
   c. is independent of the number of species.
   d. varies for different islands.
   e. is large when the number of species is small.

19. The rate of immigration is high when the number of species is low because
   a. there is no competition.
   b. the first colonists to arrive are all "new" species.
   c. there are no predators.
   d. the island is newly formed and attracts species.
   e. there has not been time for extinctions to occur.

20. *S* represents
   a. the point at which species are numerous enough to compete for resources.
   b. the equilibrium composition of species: no new species can invade and none of the existing species will go extinct.
   c. the minimum number of species that ever existed on the island.
   d. the maximum number of species that ever existed on the island.
   e. the equilibrium number of species: extinctions will equal immigrations.

21. If the number of species on the island is $S + 3$, we would expect to see
   a. on average, three more species go extinct than invade.
   b. three species go extinct and then no changes in the biota of the island.
   c. on average, three more species invade than go extinct.
   d. three species invade and then no changes in the biota of the island.
   e. Not enough information to tell

22. If the island suddenly doubled in size, we would expect to see
   a. the extinction curve shift to the left and a decrease in *S*.
   b. the immigration curve shift to the right and an increase in *S*.
   c. the extinction curve shift to the right and a decrease in *S*.
   d. the extinction curve shift to the right and an increase in *S*.
   e. the immigration curve shift to the left and an increase in *S*.

23–27. Refer to following list of biomes for the questions below.
   a. Tundra
   b. Boreal forest
   c. Tropical evergreen forest
   d. Grassland
   e. Chaparral

23. This biome is found in many climates but all of them are relatively dry much of the year. **(d)**

24. This biome is characterized by cold temperatures, permafrost, and little rainfall. It is found in the Arctic and in high mountains at all latitudes. **(a)**

25. This is the richest of all biomes in species of both plants and animals, and it has the highest overall energy flow of all ecological communities. The soils in this biome are usually poor, however, as most of the nutrients are tied up in the vegetation. **(c)**

26. This biome is dominated by coniferous trees. It has Earth's tallest trees, and supports the highest standing biomass of wood of all ecological communities. **(b)**

27. This biome is found on the western sides of continents at moderate latitudes. The dominant vegetation is low-growing shrubs with evergreen leaves. **(e)**

*28. Which of the following statements about aquatic and terrestrial communities is *not* true?
   a. Terrestrial photosynthesizers are mostly large and provide a lot of physical structure to communities.
   b. In many aquatic communities, animals rather than plants are the largest and longest-lived organisms.
   c. The most productive aquatic communities, intertidal communities exposed to high wave action, are more productive than tropical rainforests.
   d. There is little competition for space in aquatic shoreline communities.
   e. Most aquatic producers live near the shore or in upper layers of water.

29–32. Match the aquatic habitats from the list below with the following descriptions.
   a. Marine benthic zone
   b. Marine pelagic zone
   c. Marine abyssal zone
   d. Marine littoral zone
   e. Rivers

29. Communities in this habitat may exist on and in mud and sand; others exist on rocky substrates that are sites of intense competition for space. On rocky substrates, algae, barnacles, sponges, and echinoderms are likely to be common. **(d)**

30. Zooplankton and phytoplankton are common in this region; planktonic larvae of organisms from other regions are seasonal. Most fish in this region live in large schools and are streamlined for rapid movement. **(b)**

31. This area is dynamic, and its properties change dramatically with seasons and other factors. Penetration of sunlight may be restricted by suspended solids. Most plants are firmly attached to the bottom to avoid being uprooted by currents. **(e)**

32. This region is below the level of sunlight penetration. **(c)**

33. Different regions on Earth with similar climates have
   a. similar ecological communities with similar species.
   b. similar ecological communities with identical species.
   c. identical ecological communities with identical species.
   d. very different ecological communities with very different species.
   e. similar ecological communities with very different species.

34. Which of the following is *not* a difference between aquatic and terrestrial ecosystems that has been important in the evolution of ecological communities?
   a. Compared with air, water is a less dense medium that provides little support for organisms living in it.
   b. Moving water creates greater force than moving air.
   c. Movement through water is more difficult than movement through air.
   d. Sunlight penetrates only short distances through water; thus communities at moderate depths receive little light.
   e. Oxygen is often in short supply in aquatic environments.

*35. Australia has been separated from other continents for about 65 million years. South America has been isolated from other continents for about 60 million years. North America and Eurasia have been joined for much of Earth's history. Which continents would you expect to have the most similar biota?
   a. Australia and South America
   b. North and South America
   c. North America and Eurasia
   d. Australia and both North and South America
   e. Australia and both North America and Eurasia

36. Which of the following statements is *not* an idea that dominated the early history of biogeography?
   a. The continents are fixed in position.
   b. It is better to focus on a single taxon than to study unrelated groups.
   c. Information on the ecological relationships of living species can be useful for interpreting biogeographic patterns.
   d. Dispersal of organisms explains most biogeographic patterns.
   e. Taxa originated in defined local areas.

*37. Two adjacent continents are separated by 150 km of ocean. On continent A there are 10 species of organisms closely related to 5 species on continent B. By applying the principle of __________ you would conclude that the species moved from __________.
   a. parsimony, A to B
   b. parsimony, B to A
   c. evolution, A to B
   d. evolution, B to A
   e. island biogeography, A to B

*38. Based on your knowledge of the history of continental drift on Earth, which of the following pairs of biogeographic regions should have the most similar fauna?
   a. Palearctic, Ethiopian
   b. Ethiopian, Neotropical
   c. Nearctic, Neotropical
   d. Australian, Neotropical
   e. Oriental, Ethiopian

39. Which of the following biogeographic regions has the greatest number of endemic taxa?
   a. Nearctic
   b. Neotropical
   c. Ethiopian
   d. Australian
   e. Oriental

40. If species arrive on an island at a constant rate, and there is no extinction, the rate of arrival of new species on the island should __________ as the total number of species on the island increases.
   a. increase
   b. decrease
   c. remain constant
   d. increase, then decrease
   e. decrease, then increase

41. The number of species on an island at which the arrival rate of new species equals the extinction rate of species already present is called the equilibrium species number. At this number, the species richness __________ and the species composition __________.
   a. is constant, is constant
   b. is constant, can be variable
   c. can be variable, is constant
   d. can be variable, can be variable
   e. depends on the size of the island, is constant

42. Using the model of island biogeography developed by Robert MacArthur and E. O. Wilson, which of the following can be predicted?
   a. Species composition on an island
   b. Which species will become extinct

c. Which species will arrive
d. Species richness, given island size
e. Rate of evolution of endemic species

*■43. If ISN and ISF are two small islands of the same size that are near and far from a species pool, and ILN is a large island that is near the same species pool, select the sequence in which these three islands are arranged from the smallest to the largest expected equilibrium species number.
a. ISF, ISN, ILN
b. ISN, ILN, ISF
c. ILN, ISF, ISN
d. ISF, ILN, ISN
e. ILN, ISN, ISF

44. Which of the following factors would be *least* important in determining the equilibrium species number of an island?
a. Island size
b. Distance to the mainland
c. Distance to other islands
d. Size of the mainland
e. Direction of prevailing wind and currents

45. Which of the following statements about habitat islands is *not* true?
a. A habitat island is surrounded by unsuitable habitat.
b. The arrival rates are usually lower for habitat islands than for oceanic islands.
c. The model of island biogeography developed by Robert MacArthur and E. O. Wilson also explains the effect of habitat island size and distance from a species pool on the species richness of the habitat island.
d. Adjacent lakes are habitat islands separated by intervening land.
e. Intervening areas between habitat islands may be suitable for a brief stopover.

46. Which of the following is *not* true regarding biogeographic regions and biomes?
a. Generally, similar biomes are restricted to a single biogeographic region.
b. Biogeographic regions are based on similarities in the biota.
c. Biomes are based on climate differences.
d. Many biomes are distributed latitudinally.
e. Both biomes and biogeographic regions exclude the oceans.

47. Select the biome to which the following characteristics apply: found in dry regions of different temperature; much belowground biomass.
a. Tundra
b. Boreal forest
c. Temperate deciduous forest
d. Grassland
e. Cold desert

48. Select the biome to which the following characteristics apply: dominated by coniferous, evergreen trees; short growing seasons.
a. Tundra
b. Boreal forest
c. Temperate deciduous forest
d. Grassland
e. Cold desert

49. Select the biome to which the following characteristics apply: presence of permafrost; high latitude or altitude distribution.
a. Tundra
b. Boreal forest
c. Temperate deciduous forest
d. Grassland
e. Cold desert

50. Select the biome to which the following characteristics apply: wide range of temperatures and ample precipitation year-round; animal-dispersed pollen and fruit.
a. Tundra
b. Boreal forest
c. Temperate deciduous forest
d. Grassland
e. Hot desert

51. Select the biome to which the following characteristics apply: often located in rain shadows; most precipitation falls in winter.
a. Tundra
b. Boreal forest
c. Temperate deciduous forest
d. Grassland
e. Cold desert

52. Select the biome to which the following characteristics apply: found in areas with descending warm, dry air; most precipitation falls in summer.
a. Chaparral
b. Thorn forest
c. Tropical savanna
d. Tropical deciduous forest
e. Hot desert

53. Select the biome to which the following characteristics apply: greatest species richness; highest energy flow.
a. Chaparral
b. Thorn forest
c. Tropical savanna
d. Tropical deciduous forest
e. Tropical evergreen forest

54. Select the biome to which the following characteristics apply: found on the equatorial side of hot deserts; short, intense summer rainy season.
a. Chaparral
b. Thorn forest
c. Tropical montane forest
d. Tropical deciduous forest
e. Tropical evergreen forest

55. Select the biome to which the following characteristics apply: cool, wet winters and hot, dry summers; dominated by low-growing shrubs with tough evergreen leaves.
    a. Chaparral
    b. Tropical savanna
    c. Tropical montane forest
    d. Tropical deciduous forest
    e. Tropical evergreen forest

56. Select the biome to which the following characteristics apply: longer rainy season and taller trees than neighboring thorn forest biome; many trees flower in the dry season, when they are leafless.
    a. Boreal forest
    b. Temperate deciduous forest
    c. Tropical montane forest
    d. Tropical deciduous forest
    e. Tropical evergreen forest

57. Select the biome to which the following characteristics apply: mostly wind-dispersed pollen and fruit; low species diversity.
    a. Boreal forest
    b. Temperate deciduous forest
    c. Tropical montane forest
    d. Tropical deciduous forest
    e. Tropical evergreen forest

58. Which of the following pairs of biomes is maintained by grazing?
    a. Cold desert and hot desert
    b. Chaparral and thorn forest
    c. Grassland and tropical savanna
    d. Temperate and tropical deciduous forests
    e. Cold desert and tundra

59. Which of the following is *not* true of aquatic ecosystems?
    a. Aquatic producers have extensive support tissue.
    b. Aquatic producers are very small and provide little structure to the ecosystem.
    c. Aquatic producers have adaptations for staying in the uppermost layers.
    d. Animals are usually larger and live longer than plants.
    e. Most open-water producers are cyanobacteria and algae.

60. Which of these physical events is *not* a major influence in the distribution of organisms?
    a. Continental drift
    b. Earthquakes
    c. Sea level changes
    d. Glacial retreat
    e. Formation of mountains

*61. __________ and __________ regions have more species richness than other regions.
    a. Tropical, flatland
    b. Tropical, mountainous
    c. Tropical, peninsular
    d. High latitude, peninsular
    e. High latitude, mountainous

62. The coastal zone from the uppermost limits of tidal action down to the depth where the water is thoroughly stirred by wave action is called the _______ zone.
    a. surf
    b. tidal
    c. benthic
    d. littoral
    e. reef

*▪63. Poisonous sea snakes are found in the Pacific Ocean but not in the Caribbean Sea because
    a. the water temperature differs between them too much for the snakes to live in both.
    b. the ocean's currents are counterclockwise so the eggs and larvae are carried the other direction.
    c. the Panamanian Isthmus rose and blocked the dispersal of the snakes.
    d. the Caribbean coastal zones don't get proper nutrients for the snakes.
    e. sea snakes cannot traverse deep ocean waters to get to the Caribbean.

Study Guide Questions

1. Which of the following types of information would be *least* used by a historical biogeographer?
    a. Experiments on species richness
    b. Fossils showing past distributions of organisms
    c. Application of the principle of parsimony
    d. Present distributions of organisms

2. The following diagram shows the present locations of four continents (W, X, Y, Z) and the number of endemic beetle families found on each. Three were part of a supercontinent in the past, the other was always separate.

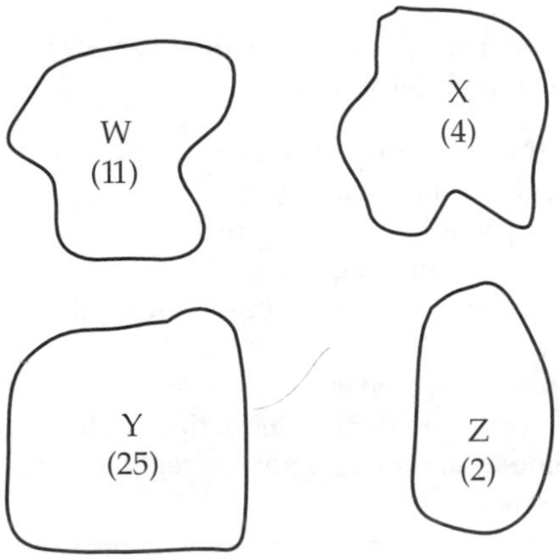

Based on this information, choose the correct conclusions.
    a. W was always separate, Y separated from the supercontinent first.
    b. X was always separate, Y separated from the supercontinent first.
    c. Y was always separate, W separated from the supercontinent first.
    d. Z was always separate, X separated from the supercontinent first.

3. The following diagram shows three islands. Islands A and B were connected in the past, C was always separate. A species of land snail is found on all three islands.

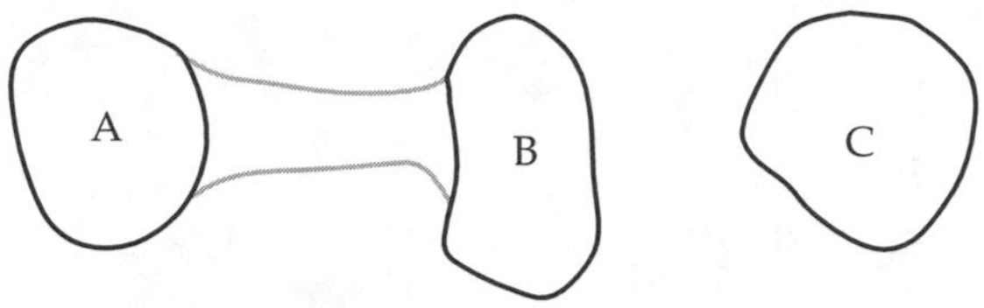

Which of the following does not correctly describe the distribution of this snail relative to the three islands?
a. Vicariant distribution relative to A and B
b. Dispersal distribution relative to A and C
c. Dispersal distribution relative to B and C
d. Vicariant distribution relative to A and C

4. If the species of snail mentioned in Question 3 is found *only* on the three islands shown in the above diagram, which statement would be correct concerning the endemism of this species?
a. Endemic relative to A and B
b. Endemic relative to C
c. Endemic relative to A, B, and C
d. All of the above

5. Which of the following biogeographical regions represents the largest area?
a. Nearctic
b. Palearctic
c. Neotropical
d. Oriental

6. Circle an area on the following graph that would most apply to the spatial and temporal scale of processes studied by ecological biogeographers.

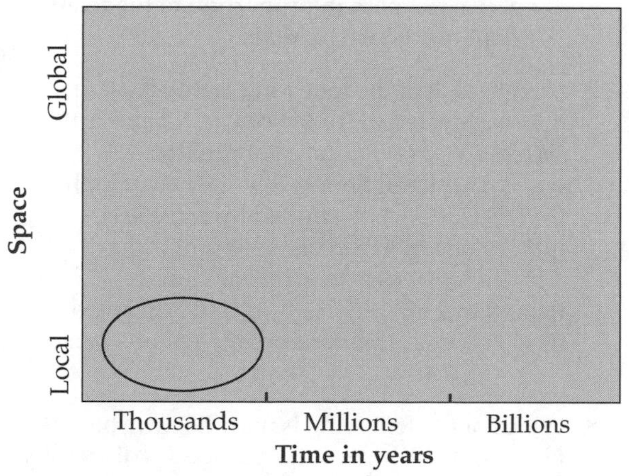

7. According to the island biogeographic model, which of the statements about the following graph showing the effect of species number on arrival and extinction rates is *not* true?

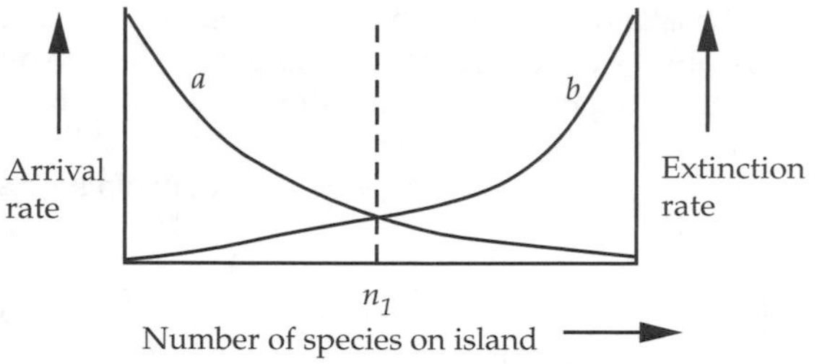

a. Curve *a* is the arrival rate curve.
b. Curve *b* is the extinction rate curve.
c. The arrival rate equals the extinction rate at n_1.
d. The extinction rate is zero at n_1.

8. The species numbers of many islands of different sizes are plotted against their distance from the mainland. Data points for islands of similar size were connected to form the four curves shown in this figure.

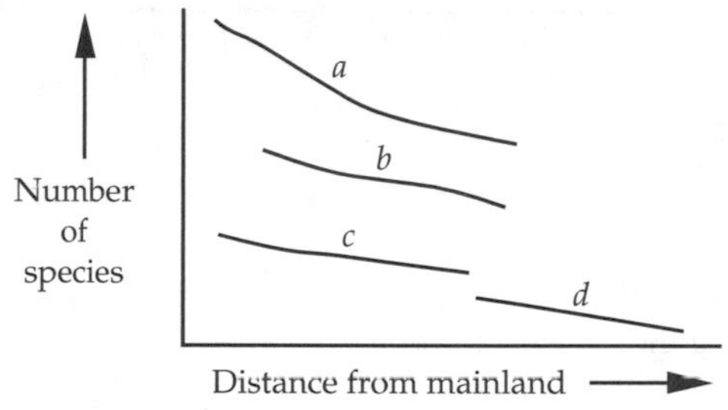

Circle the letter of the curve corresponding to the group of islands with the smallest size. **(d)**

9. A small volcanic island was destroyed by an eruption. The following graph shows recolonization by a number of different organism types.

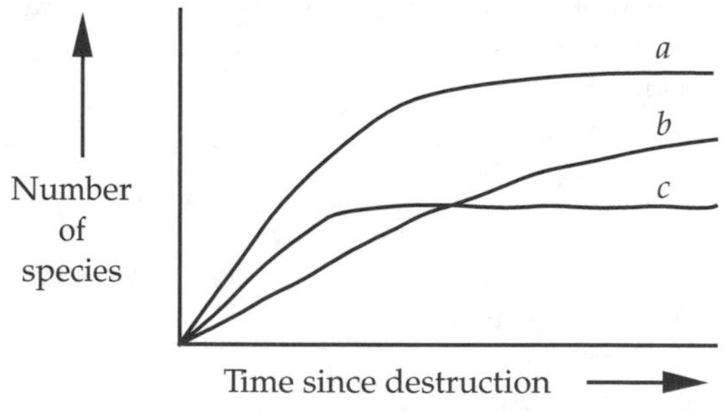

Identify the correct curve for each organism type.
(c) Birds
(b) Plants
(a) Insects

10. The island biogeographic model makes predictions about the effects of species number on the rate of extinction. In the following figure, the solid curve shows this relationship for a large island. Which of the curves shows the expected relationship for a small island?

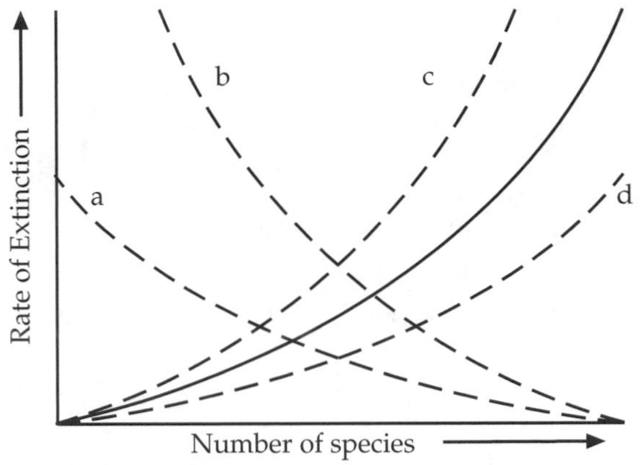

a. Curve *a*
b. Curve *b*
c. Curve *c*
d. Curve *d*

11. Which of the following locations would you expect to have the greatest number of species?
a. Mountainous forest area on a mainland in the temperate zone
b. Mountainous forest area on a mainland in the tropics
c. Lowland forest area on a mainland in the tropics
d. Mountainous forest area on a peninsula in the tropics

12. The biome that is maintained by browsers, grazers, or fire is the
a. tundra.
b. savanna.
c. cold desert.
d. tropical deciduous forest.

13. Match the letters of the following biomes with the descriptions that follow.
a. Tundra
b. Boreal forest
c. Hot desert
d. Chaparral
e. Tropical deciduous forest
(b) Mostly coniferous, wind-pollinated and wind-dispersed tree species
(e) Leaves lost during dry season; agriculturally desirable land
(d) Cool winters, hot dry summers; maritime climate
(c) Succulent plants prominent; found at 30°N and 30°S latitudes
(a) Distribution altitudinally or latitudinally determined; permafrost present

14. The following climograph shows yearly variation in rainfall and temperature for four biomes. Select the correct curve for each of the following biomes.

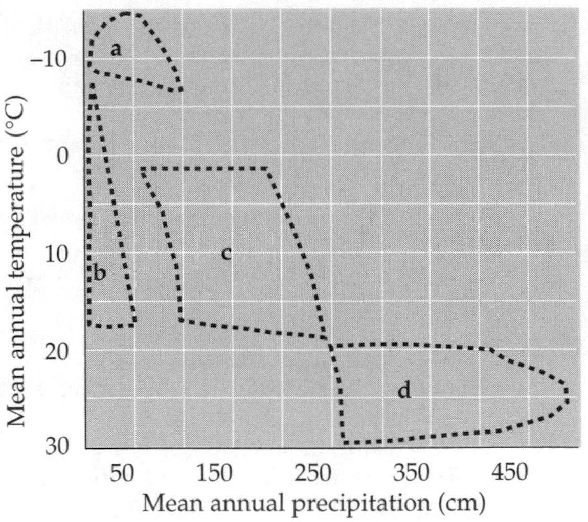

(a) Tundra
(d) Tropical evergreen forest
(c) Temperate deciduous forest
(b) Cold desert

15. The tropical evergreen forest biome has relatively constant **temperature**, but **rainfall** varies seasonally; the temperate deciduous forest biome has relatively constant **rainfall**, but **temperature** varies seasonally.

16. Which of the following comparisons of freshwater and marine habitats is *not* true?
a. About 10 percent of all aquatic species live in fresh water.
b. There are many more insect species in freshwater habitats than in marine habitats.
c. The global volume of marine habitats is much greater than that of freshwater habitats.
d. Freshwater species richness is less than marine species richness in proportion to the relative extent of the two habitats.

17. Next to each of the following features, place *p*, *l*, *b*, or *a* if it is characteristic of the *pelagic*, *littoral*, *benthic*, or *abyssal* zones of marine communities.
(a) The ocean floor below the penetration of light
(l) Coastal zone affected by wave action
(p) Zone in which organisms must have adaptations to avoid sinking
(p, b) Divided into a lighted and dark region
(l) Globally, the zone occupying the smallest total area

18. Which of the following is most likely to be true of a family of freshwater fish with species distributed on several continents?
a. They are a recently derived group.
b. They have an ancient lineage that was widely distributed in Laurasia or Gondwana.
c. Their distribution was not affected by continental drift.
d. Their distribution can be explained by movements of glaciers.

19. Which of the following statements about the biogeography of marine biomes is *false*?
    a. Vertical gradients of light and temperature divide the ocean into different zones.
    b. In interpreting the biogeography of the marine biome, oceanic currents are more important than physical barriers.
    c. Organisms adapted for life in nutrient-poor waters usually flourish in nutrient-rich areas.
    d. Continental drift does not explain the distribution of many marine organisms.

End of Chapter Questions

1. Biogeography as a science began when
    a. eighteenth-century travelers first noted intercontinental differences in the distributions of organisms.
    b. Europeans went to the Middle East during the Crusades.
    c. phylogenetic methods were developed.
    d. the fact of continental drift was accepted.
    e. Charles Darwin proposed the theory of natural selection.

2. Historical and ecological biogeography differ in that
    a. only historical biogeography is concerned with history.
    b. historical biogeography is concerned with longer time periods and larger spatial scales.
    c. both are concerned with the same time scales, but historical biogeography deals with larger spatial scales.
    d. both are concerned with the same spatial scales, but historical biogeography deals with longer time scales.
    e. historical biogeography is not concerned with the current distributions of organisms.

3. Marine biogeographic regions are less distinct than terrestrial ones because
    a. the ocean biota is more poorly known than the terrestrial biota.
    b. there are currently fewer barriers to dispersal of marine organisms.
    c. most marine families and higher taxa evolved before the oceans were separated by continental drift.
    d. we know less about the distributions of marine organisms.
    e. oceanic circulation is faster than atmospheric circulation.

4. A parsimonious interpretation of a distribution pattern is one that
    a. requires the smallest number of undocumented vicariant events.
    b. requires the smallest number of undocumented dispersal events.
    c. requires the smallest total number of undocumented vicariant plus dispersal events.
    d. accords with the cladogram of a group.
    e. accounts for centers of endemism.

5. The only major biogeographic region that today is isolated by water from other regions is
    a. Greenland.
    b. Africa.
    c. South America.
    d. Australasia.
    e. North America.

6. Equilibrium species richness is reached in the island biogeographic model when
    a. immigration rates of new species and extinction rates of species are equal.
    b. immigration rates of all species and extinction rates of species are equal.
    c. the rate of vicariant events equals rates of dispersal.
    d. the rate of island formation equals the rate of island loss.
    e. No equilibrium number of species exists in that model.

7. Chaparral vegetation is dominated by
    a. deciduous trees.
    b. evergreen trees.
    c. deciduous shrubs.
    d. evergreen shrubs.
    e. grasses.

8. Which of the following is not true of tropical evergreen forests?
    a. They have large numbers of species of trees.
    b. Most plant species are animal-pollinated.
    c. Most plant species have animal-dispersed fruits.
    d. Biological energy flow is very high.
    e. Productivity depends on a rich supply of soil nutrients.

9. At all depths, the bottom of the ocean is known as the
    a. benthic zone.
    b. abyssal zone.
    c. pelagic zone.
    d. interoceanic convergence zone.
    e. subtidal zone.

10. Biogeography exerted a strong influence on human history because
    a. humans first evolved in Africa.
    b. Eurasia had more species of plants and animals that were easily domesticated.
    c. old World mountain ranges are oriented in an east-west direction whereas New World mountain ranges are oriented in a north-south direction.
    d. horses were found only in Eurasia.
    e. All of the above.

Student Web Site Self-Quiz Questions

1. Which of the following statements about continental drift and biogeography is *false*?
   a. **Based on plant and animal distributions, early biogeographers knew that the continents had moved, although the mechanism for drift was not discovered until recently.**
   b. In addition to continental drift, other important global effects on species' distributions include glaciation, sea level changes, and mountain building.
   c. Continental drift helps to explain the distributions of many African and South American plants and animals.
   d. Continents continue to drift today.
   e. Historical biogeographers are more interested in continental drift than the ecological relationships of contemporary species.

2. Which of the following statements regarding biogeographic regions is *false*?
   a. The boundaries separating the biogeographic regions correspond to areas where species compositions change over short distances.
   b. The boundaries shown are based on taxonomic similarities among organisms living there.
   c. The regions shown are not separated based upon their general appearances.
   d. Most biogeography regions are not separated by water.
   e. **Plant and zoogeographers agree on the boundaries shown.**

3. Which of the following statements about biomes is *false*?
   a. Biomes are named for their characteristic vegetation.
   b. **Although other groups within a biome differ geographically, the plant species present are the same wherever the biome is found.**
   c. The distribution of biomes is strongly influenced by annual temperature and rainfall.
   d. The occurrence of some biomes is determined by the seasonal distribution of rainfall.
   e. Movement of the intertropical convergence zone gives some biomes their characteristic rainy and dry seasons.

4. Which of the following statements about the biogeography of marine biomes is *false*?
   a. Vertical gradients of light and temperature divide the ocean into different zones.
   b. Ocean currents are a more important influence on the biogeography of marine biomes than physical barriers.
   c. **Organisms adapted for life in nutrient-poor waters usually flourish in nutrient-rich areas.**
   d. Continental drift has been a minor influence on the distribution of marine organisms.
   e. Recent vicariant events, such as changes in sea level, are important in determining the distribution of marine organisms.

TEST FILE

★ *Indicates a **difficult** question*
■ *Indicates a **conceptual** question*

58 Conservation Biology

Fill in the Blank

1. When ecosystems, such as agricultural lands, are managed so as to divert most of their primary production to certain species intended for human use, we say that their production is **co-opted**.
2. In the 1950s and '60s peregrine falcons were extirpated from much of the eastern United States. The main cause of this was **DDT**.
3. Nearly all of the mammals and birds native to Madagascar are found only on that island. Species with distributions like this are said to be **endemic**.
4. **Edge effects** are the phenomena that are influenced by adjacent habitats, and increase as patch size decreases.
5. An **ecoregion** is an area that has relatively uniform climate and a biota dominated by a group of widely distributed species.
6. The subdiscipline of conservation biology that is concerned with converting disturbed areas back into natural areas (e.g., converting cropland into a native prairie) is called **restoration ecology**.
7. In the past (over a thousand years ago) the primary means by which humans caused the extinction of animals was **overhunting**.
8. The most important cause of extinction in the next century is almost certain to be **habitat loss**.
9. The 13 species of flightless moas became extinct in New Zealand due to **overhunting** by humans.

Multiple Choice

1. Many scientists believe that during the next two decades,
   a. we will have stabilized the Earth's environments, and no species that are currently living will go extinct.
   b. we will have developed technology that will allow us to re-create all of the species that have gone extinct.
   c. all species except *Homo sapiens* will be go extinct.
   d. less than 1 percent of the currently living species will go extinct.
   e. from 10 to 50 percent of the currently living species will go extinct.
2. In recent years the number of species driven to extinction has increased dramatically. Which of the following is *not* a reason for this?
   a. Overexploitation
   b. Habitat destruction
   c. Introduction of predators
   d. Natural predation
   e. Introduction of diseases
3. When early Polynesian people first settled in Hawaii, they drove at least 39 species of endemic land birds to extinction. The primary reason for this was
   a. introduction of predators.
   b. overhunting.
   c. habitat destruction.
   d. introduction of diseases.
   e. competition for food.
4. A species-area relationship is used by ecologists to
   a. Determine the population density of a species in a certain habitat.
   b. Examine how human populations are growing.
   c. Estimate the numbers of species extinctions resulting from habitat destruction.
   d. Produce a population model.
   e. None of the above
5. Human agricultural ecosystems (including forestry) account for________ percent of Earth's land surface.
   a. 5
   b. 10
   c. 30
   d. 60
   e. 80
6. The Furbish's Lousewort depends on constant__________to maintain its populations.
   a. water temperature
   b. river flow
   c. rainfall
   d. disturbance
   e. wind

7. Which parameter of a population do ecologists measure to assess extinction risk for a population?
   a. genetic variation
   b. behavior
   c. physiology
   d. a and b
   e. a, b, and c

8. When a species goes extinct in one area, it is often desirable to reintroduce the species from other populations. A major problem with this approach is that
   a. genetic diseases can easily be introduced when the species is reintroduced.
   b. populations are often adapted to local conditions and may not survive when moved to a different location.
   c. the community will have adapted to the extinct species' absence. Reintroduction may seriously disrupt the community.
   d. it is difficult to get an adequate sample of individuals to properly reestablish the population.
   e. animals frequently attempt to return to their native habitat.

9. Captive propagation in zoos and botanical gardens has been quite successful for several endangered species. Nevertheless, captive propagation is considered to be only a partial or temporary solution to the biodiversity crisis. Which of the following is *not* a reason for this?
   a. There is not enough space in existing zoos and botanical gardens to maintain populations of more than a small fraction of rare and endangered species
   b. Captive propagation projects in zoos have been influential in raising public awareness of the biodiversity crisis.
   c. A species in captivity can no longer evolve together with the other species in its ecological community.
   d. Extensive research on nutrition, reproduction, etc., is required for each species.
   e. Small captive populations tend to have low genetic diversity.

▪10. In 1942 there were 350 pairs of peregrine falcons breeding in the United States east of the Mississippi River. By 1960, the breeding population had entirely disappeared. Today peregrines are once again found in this area. Their reestablishment occurred because
   a. peregrines migrated in from areas west of the Mississippi.
   b. nonbreeding birds began to pair when nesting conditions again improved.
   c. the eastern United States went through a very warm period. Recently the climate has begun to cool, and peregrines are migrating down from Canada.
   d. extensive natural forest areas have been restored on the eastern seaboard, allowing peregrines to once again nest successfully.
   e. captive-reared birds have been systematically released in the eastern United States, reestablishing populations.

▪11. Why do species extinctions matter?
   a. They don't matter unless the species is a human food source.
   b. Many pharmaceutical products are derived from natural products, loss of species could mean loss of therapeutic drugs.
   c. Soil erosion may increase if certain plants go extinct.
   d. Animals and plants are nice to look at.
   e. b and c

12. __________ are undisturbed portions of forest that allow animals to pass from patch to patch.
   a. Windows
   b. Transects
   c. Corridors
   d. Pathways
   e. Portals

*13. What effect, if any, would the extinction of a mutualistic pollinator have on the plant it pollinates?
   a. No effect.
   b. Decreased pollination of the plant.
   c. The plant would die immediately.
   d. No effect because another pollinator could take the place of the extinct one.
   e. The plant would not be pollinated.

▪14. Species native to islands such as Madagascar are particularly susceptible to the effects of habitat destruction because
   a. island populations are more susceptible to disease.
   b. many species found on islands are found nowhere else.
   c. habitats are more easily destroyed on islands than on continents.
   d. habitat fragmentation is more serious on islands because islands normally have a single continuous habitat.
   e. island populations tend to have more mutualistic relationships than mainland populations.

15. A(n) __________ is an area that has a relatively uniform climate and a biota dominated by a group of widely distributed species.
   a. ecoregion
   b. habitat
   c. reserve
   d. megareserve
   e. sanctuary

*16. A lumber company proposes to clear-cut a large area of forest, but they aim to leave small patches of forest to provide habitat for forest animals. Which of the following is *not* a problem with this method of conserving natural habitat?
   a. Small patches cannot support populations of species that require large areas.
   b. Small patches can harbor only small populations of the species that can survive there.
   c. Many small patches will disperse the populations of the affected species over a larger geographical area.

d. Species that live in the clear-cut areas can invade the edges of the patches to compete with or prey on the species living there.
e. In small patches temperatures will be higher, winds stronger, and the environment will generally be more extreme than it was in the original forest.

17. __________ ecology is a sub-discipline of conservation biology the goal of which is to mend damaged and degraded ecosystems.
a. Conservation
b. Historical
c. **Restoration**
d. Renovation
e. Redistribution

18. A major difficulty faced by countries in the tropics, such as Costa Rica, that are trying to maintain a national park system is that
a. most of the land in these countries has already been destroyed and there is little to preserve.
b. national park land is so remote that the parks are frequently little more than marks on a map.
c. **the high rates of population growth put heavy pressures on parks from agricultural settlers and poachers.**
d. the high rainfall in these areas makes it difficult to maintain an active conservation program.
e. tropical ecosystems are not in any current danger and there is little incentive to establish parks.

19. Which of the following methods could be used to restore a population of animals from a few male and female individuals.
a. Cross breeding
b. Interbreeding
c. **Captive breeding**
d. Selective breeding
e. The population could not be restored from a few individuals.

▪*20. If a demographic model of an at risk population of animals indicates that they would be in danger of extinction if their numbers dropped below 1,500 individuals and another model suggested that a very cold winter (the main factor that causes decline in the animal's population) could kill 400 members of the population, what is the minimal population necessary to keep the animals from becoming endangered.
a. **1,900**
b. 1,500
c. 2,000
d. 3,000
e. 5,000

21. Which of the following is not an essential service for humans that is provided by natural ecosystems, such as wetlands?
a. **Absorption of oxygen and release of carbon dioxide**
b. Reduction of erosion and water runoff
c. Treatment and purification of wastewater
d. Production of fish, waterfowl, and other wildlife
e. Absorption of pollutants such as sulfates

*22. Of the following organisms which has the highest conservation priority?
a. A plant found on North America and in Europe.
b. **A plant endemic to Australia**
c. A plant found on the Galapagos islands and in Brazil.
d. A plant found world wide.
e. A plant found in North and South America.

23. Which of the following statements is *true* concerning the relationship between humans and the rest of the living world?
a. Modern technology has made it so that we no longer depend on other living organisms.
b. We are dependent on artificial ecosystems, such as agroecosystems, but gain no benefit from natural ecosystems.
c. We are dependent on natural ecosystems at present, but the technology exists to completely replace natural ecosystems so that we will no longer depend on them.
d. **Our survival is tightly linked to the survival of natural ecosystems throughout the world.**
e. Nonindustrial human societies are dependent on natural ecosystems, but industrial societies are not.

24. Traditional agroecosystems differ from modern agroecosystems in that
a. traditional agroecosystems had a greater fossil fuel usage rate, while achieving a lower overall yield.
b. traditional agroecosystems are far more destructive to the soil than modern agroecosystems.
c. traditional agroecosystems make greater use of fertilizers and insecticides.
d. **traditional agroecosystems are more diverse, because many more economically valuable species are grown together and because these species support many other species incidentally.**
e. traditional agroecosystems achieve a lower yield because the land is not carefully tuned to the crops being grown and because pesticides are not used. Because of these poor agricultural techniques, more land is necessary, leading to more species being driven to extinction.

▪25. Today, populations of some Galápagos tortoises are maintained only by humans who remove eggs and rear the young tortoises in captivity. Why?
a. These populations are harvested by humans for meat and eggs.
b. These primitive animals have such a slow reproductive rate that they have not been able to adapt to global warming.
c. There is such a high demand for these tortoises in zoos that they are raised this way to maximize their reproductive rate.
d. This is the best way to assure that these endangered animals are properly counted and marked for ongoing scientific studies.
e. **Introduced rats and pigs excavate all of the nests and devour all of the eggs that are laid.**

26. The Yellowstone to Yukon initiative is an
   a. ecoregional conservation project.
   b. example of small scale conservation.
   c. untested theory.
   d. impossible task.
   e. example of physical ecology.

27. As of the mid-1990s, the number of humans on Earth is approximately
   a. 0.5 billion.
   b. 1.0 billion.
   c. 2.5 billion.
   d. 6.0 billion.
   e. 10.5 billion.

■*28. Hawaii has many species of long-lived *Lobelia* plants and many species of honeycreeper birds that pollinate the plants. Assume that *Lobelia* species A is pollinated exclusively by honeycreeper species A. If honeycreeper species A becomes extinct, what is the most likely fate of *Lobelia* species A?
   a. The *Lobelia* species will become extinct for lack of a pollinator.
   b. An empty niche will be created by the bird extinction, and another honeycreeper species will fill it by pollinating the plants.
   c. The plant flower shape will evolve toward a more generalized wind-pollinating form.
   d. The plants will evolve asexual methods of reproducing.
   e. The species will switch to self-crossing (pollen landing on the female structures in the same flower) in order to survive.

29. Mitigation projects that attempt to create replacement wetlands
   a. are usually carried out by experienced wetland ecologists.
   b. are becoming less common due to lack of earlier success.
   c. must be located at the site of the previous wetland.
   d. currently show high success rates due to recent advances in modeling wetlands.
   e. provide developers a means to get building permits that otherwise would be denied for sites that contain wetlands.

30. If global warming causes average North American temperatures to rise 2–5°C in the next century, the most likely effect on the American beech will be that
   a. the species will not be able to extend its range northward fast enough to remain in a suitable climate.
   b. the trees will produce more seeds due to increased metabolic rates and become more common in forest communities.
   c. adult trees will die first, while shaded seedlings in the understory will survive longer.
   d. as animal seed predators migrate northward, the beech population will increase due to greater germination rates.
   e. the trees will grow faster, begin reproduction sooner, and die at a younger age.

31. The number of species that become extinct due to habitat destruction is greatest in _________ ecosystems with many _________ species.
   a. temperate, migratory
   b. tropical, endemic
   c. temperate, keystone
   d. tropical, migratory
   e. temperate, endemic

32. Major causes of human-induced extinctions of species include all of the following, except
   a. climate modification.
   b. overexploitation.
   c. habitat destruction.
   d. captive propagation.
   e. introduction of predators and diseases.

33. Which of the following is not true of the textbook's examples of extinctions that occurred at least 1000 years ago?
   a. Many exterminated species were large mammals.
   b. Humans had recently arrived in the area.
   c. Habitat destruction by humans was the cause.
   d. Many flightless birds were exterminated.
   e. The exterminated species were initially numerous.

34. Which of the following is likely to be the most important cause of species extinctions in the next century?
   a. Habitat destruction
   b. Overexploitation
   c. Introduced predators
   d. Introduced disease
   e. Climate modifications

35. Which of the following statements about captive propagation is *false*?
   a. Separating the organism from its ecological community can lead to difficulties.
   b. Captive propagation may cause inbreeding depression.
   c. Successful captive propagation programs may make it possible to reduce the habitat needed for an endangered species.
   d. Artificial insemination has been successfully used in some captive propagation programs.
   e. Captive propagation can be used until the external threat to the species is corrected.

36. Which of the following statements about the extirpation and subsequent reintroduction of the peregrine falcon into the eastern United States is *false*?
   a. Extirpation was caused by high adult mortality due to DDT accumulation.
   b. The habitat is now safe for reintroduction of the peregrine falcon.

c. Released birds usually remain in the same area to breed in subsequent years.
d. Thinned eggshells caused the eggs to break easily.
e. Captive propagation efforts are expensive.

37. All of the following characteristics except one are usually associated with rare species. Select the *exception*.
a. A recent arrival in a new area
b. An endemic species on an island
c. A keystone species (not in this chapter)
d. A top predator
e. A species requiring a large foraging area

38. In forest ecosystems, edge effects within a patch should __________ and environmental variation should __________ with a decrease in habitat patch size.
a. increase, decrease
b. increase, increase
c. increase, remain constant
d. decrease, increase
e. decrease, decrease

39. Which of the following statements about the size effects of habitat fragmentation is *false*?
a. Species with large home ranges and poor dispersal rates disappear as patch size decreases.
b. Edge effects increase uniformly as patches become smaller.
c. Death related to dispersal between suitable habitat patches increases as patches become smaller.
d. The adverse effects of smaller habitat patches decrease if the patches are connected to larger habitat areas.
e. Ten one-hectare habitat patches are not the same as one ten-hectare patch in terms of the biological community that can be supported.

40. Which of the following is *not* a reason to protect biodiversity?
a. The aesthetic value
b. Because of mutualistic relationships, whole communities could be endangered by the extinction of one species.
c. Important medicinal compounds can be found only in certain species.
d. Extinction is irreversible.
e. None of the above

41. Fynbos, the endemic shrubs of the hills of Western Cape Province in South Africa, and the primary vegetation of the watershed that provides the region's water supply, have been invaded by alien plants that grow taller and faster. A movement to remove these aliens by cutting and digging them up has begun. Which of the following is *not* a reason for removing them?
a. It is cheaper than obtaining water by another method.
b. Tourism money comes into the country from people wanting to see these plants.
c. It is the most technologically advanced method of supplying water to the region.
d. The fynbos can be sold as cut flowers.
e. The intensity and severity of fires is lessened when the fynbos are present.

42. An example in the textbook described a study in Belize that demonstrated that both preservation of forest-dwelling species and economic development could be supported by harvesting
a. agricultural plants, such as corn and squash.
b. lumber from logging of the forests.
c. medicinal plants.
d. cattle.
e. parrots that are sold as pets.

43. Which of the following organismal groups would be *least* likely to exhibit extensive endemism?
a. Plants in a small, isolated continental region
b. Snails in a series of isolated lakes
c. Birds in a small, isolated continental region
d. Reptiles in a small, isolated continental region
e. Invertebrates in tropical montane forests

Study Guide Questions

1. The concepts of conservation biology come mainly from all of the following fields *except*
a. ecology.
b. evolutionary biology.
c. population genetics.
d. immunology.

2. The fynbos shrub community in South Africa
a. is threatened by introduced species of taller, faster-growing plants.
b. is a fire-adapted community.
c. helps to maintain a regional supply of high-quality water.
d. All of the above

3. Which of the following is *not* currently a major cause of the global reduction in biodiversity?
a. Overexploitation
b. Global warming
c. Habitat destruction
d. Introduction of foreign predators and disease

4. Based on species-area relationships, ecologists predict that
a. 50 percent of Earth's species may become extinct in the next 50 years.
b. a 90 percent loss of habitat will result in loss of about nine percent of the species living there.
c. the area required by most species will need to be reduced.
d. extinction is inevitable.

5. In the United States, the groups of organisms with the highest proportion of endangered or extinct species live in
a. grasslands.
b. the deciduous forest biome.
c. freshwater habitats.
d. deserts.

6. In the following graph, select the curve (a, b, or c) that *correctly* shows the expected relationship between habitat patch area and the proportion influenced by edge effects. **(a)**

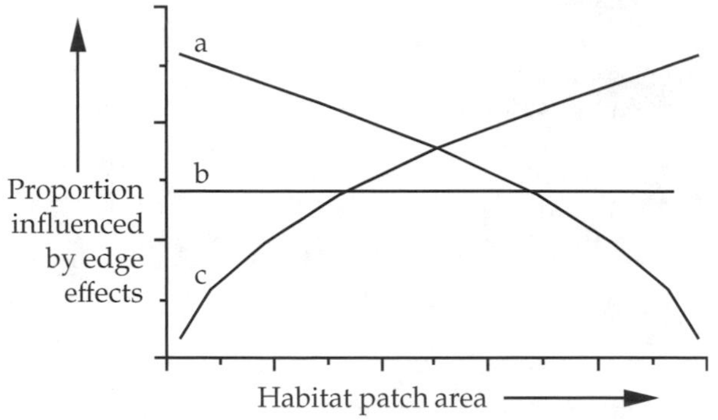

7. Why do conservation biologists believe that global warming may lead to extensive decimation of species?
   a. Since little change in plant community composition has occurred in the past, we cannot expect present communities to adapt to climate change.
   b. The magnitude of climate change will be much greater than past periods of climatic change.
   c. Many plant species may not be able to shift their ranges at the same pace as the northern movement of temperature zones.
   d. All of the above

8. Which of the following is endangered because of the loss of pollinating agents?
   a. Furbish's lousewort
   b. Madagascar rosy periwinkle
   c. American chestnut
   d. Hawaiian *Lobelia*

9. Two endangered species are Kirtland's warbler and the California sea otter. What do the recovery plans for these species have in common?
   a. The populations of both species were studied carefully to determine the reasons for their endangerment.
   b. The recovery plans for both species entail manipulation of their habitats.
   c. Captive propagation is an important part of the recovery plans of both species.
   d. All of the above

10. Which of the following statements about the successful reintroduction of the peregrine falcon is *not* true?
    a. The peregrine's habitat became more suitable after use of DDT was restricted.
    b. DDT adversely affected the peregrine's reproductive physiology.
    c. Captive propagation was an important part of the peregrine reintroduction effort.
    d. Reintroducing peregrines into habitats where they were not at the top of the food chain was critical to the success of the program.

11. Which of the following characteristics would cause a country to be considered a high-priority region for conservation efforts?
    a. A high degree of endemism
    b. Having low species richness
    c Having little natural habitat remaining
    d. All of the above

End of Chapter Questions

1. Which of the following is not currently a major cause of species extinctions?
   a. Habitat destruction
   b. Climate change
   c. Overexploitation
   d. Introduction of predators
   e. Introduction of diseases

2. The most important cause of endangerment of species in the United States currently is
   a. pollution.
   b. exotic species.
   c. overexploitation.
   d. habitat loss.
   e. loss of mutualists.

3. When ecosystems are managed to favor strongly those species intended for human use, we say that their production is
   a. modified.
   b. diverted.
   c. co-opted.
   d. channeled.
   e. managed.

4. People care about species extinctions because
   a. more than half of the medical prescriptions written in the United States contain a natural plant or animal product.
   b. people derive esthetic pleasure from interacting with other organisms.
   c. causing species extinctions raises serious ethical issues.
   d. biodiversity helps maintain ecosystem services.
   e. All of the above

5. Which of the following is not an ecosystem service?
   a. Production of carbon dioxide
   b. Flood control
   c. Water purification
   d. Air purification
   e. Preservation of biological diversity

6. Conservation biologists are concerned about global warming because
   a. the rate of change in climate is projected to be faster than the rate at which many species can shift their ranges.
   b. it is already too hot in the tropics.
   c. climates have been so stable for thousands of years that many species lack the ability to tolerate variable temperatures.

d. climate change will be especially harmful to rare species.
e. None of the above

7. A species that is found only in a particular region is said to be
a. an indicator species for that region.
b. a restricted species.
c. a vulnerable species.
d. endemic to that region.
e. demographically constrained.

8. Captive propagation is a useful conservation tool provided that
a. there is space in zoos, aquaria, and botanical gardens for breeding a few individuals.
b. if the genetic pedigree of all individuals is known.
c. if the threats that endangered the species are being alleviated so that captively reared individuals can later be released back into the wild.
d. there are sufficient caretakers.
e. captive propagation should not be used because it directs attention away from the need to protect the species in their natural habitats.

9. As a habitat patch gets smaller it
a. cannot support populations of species that require large areas.
b. supports only small populations of many species.
c. is influenced to an increasing degree by edge effects.
d. is invaded by species from surrounding habitats.
e. All of the above.

10. Restoration ecology is an important discipline because
a. many areas being incorporated into reserves have been highly degraded.
b. many areas being incorporated into reserves are vulnerable to global climate change.
c. many species suffer from demographic stochasticity.
d. many species are genetically impoverished.
e. fire is a threat to many reserves.

Student Web Site Self-Quiz Questions

1. Conservation biology is concerned with biodiversity and how to preserve it. Which of the following concepts would be least useful to a conservation biologist?
a. Evolution
b. Biogeography
c. Phylogenetics
d. Population genetics
e. Endemism

2. Which of the following statements about species–area relationships and their consequences for extinction rates discussed in the text is *false*?
a. Species–area relationships can be used to predict future extinction rates.
b. The number of species present increases with the size of an area.
c. 50% of Earth's species may become extinct in the next 50 years.
d. Extinction rates are generally higher on mainlands than on islands.
e. The relationship between area size and number of species can be used to estimate the number of extinctions that will occur in an area.

3. Of the studies to predict the survival of populations of Furbishe's lousewort, which of the following concepts was *not* mentioned as important?
a. Rescue effect
b. Sub-population structure
c. Bank slumping
d. Ice scour
e. Disturbance events

4. Populations of many species, including woodland birds, have been steadily decreasing. Reductions in the populations of some woodland birds, such as the wood thrush, have been linked to increased nest parasitism by brown-headed cowbirds, a species that lives in more open habitats. If true, this may be due to increased __________, due to __________ habitat patch size.
a. corridors, increased
b. corridors, decreased
c. edge effects, increased
d. edge effects, decreased
e. phylogenetics, decreased

5. Which of the following pairings of causes of endangerment or extinction and an organism is *not* correct?
a. Habitat destruction: Kirtland's warbler
b. Introduced predators: Galápagos tortoise (right)
c. Loss of mutualists: The American chestnut tree
d. Overexploitation: passenger pigeon
e. Environmental degradation through pollution: peregrine falcon

6. Overexploitation has driven many species to extinction. Which of the following is *not* a major cause of extinction of our current biota?
a. Intensive modern agriculture
b. Global warming
c. Habitat destruction
d. Overhunting
e. Loss of mutualists

7. Which of the following predictions about the effects of global climate change is unlikely?
a. A 1°C increase in average annual temperature will shift climate zones north by 150 km.
b. Since trees are more stable, they will be the first to adapt to the climate zone shifts.
c. Large-seeded trees disperse slowly.
d. Global warming may result in the development of new climates.
e. Global warming will not simply result in a northward shift of climate zones.

8. The remaining forests of Madagascar are home to many endemic species. Which of the following statements regarding endemism and endemic species is *false*?
   a. Endemic species are common on islands.
   b. Centers for endemism are the same for all groups of organisms.
   c. The United States has many endemic regions.
   d. Endemic specie often originate in the region where they are currently located.
   e. Regions of high endemism are prime targets for conservation efforts.

9. Which of the following statements regarding ecoregional conservation is *false*?
   a. An ecoregion is an area that has a relatively uniform climate.
   b. An ecoregion is an area that has a rich biota.
   c. An ecoregion is an area containing a concentrated, small number of endangered species.
   d. Information on species distribution is used to establish sites of highest conservation priority within an ecoregion.
   e. Ecoregional conservation involves the establishment of suitable corridors for movement of large mammals.

10. Which of the following statements about attempts to restore the tropical deciduous forest in Guanacaste National Park, in northwestern Costa Rica, is *false*?
   a. The tropical deciduous forest is the most threatened ecosystem in Central America.
   b. Creation of pasture land has been the biggest cause of the disappearance of the tropical deciduous forest biome in Guanacaste.
   c. Seeds of trees disperse slowly into the pastures that separate the forest patches with the help of cattle grazing.
   d. Fire is a major threat to the restoration effort.
   e. Prohibiting any cattle grazing is a key feature of the restoration program in Guanacaste.

11. Which of the following statements regarding habitat restoration is *false*?
   a. Restoring damaged habitats is an important conservation effort.
   b. Most wetland restoration programs have had good success.
   c. Building permits are easy to obtain as long as a simple plan is provided to "create" replacement habitat elsewhere.
   d. The National Research Council has advised that wetland restoration not be considered adequate compensation for destruction of other wetlands.
   e. The preservation of habitats is the only reliable method for maintaining biological diversity.